Y0-BRD-742

PRE-VESALIAN ANATOMY

MEMOIRS OF THE

AMERICAN PHILOSOPHICAL SOCIETY

Held at Philadelphia

For Promoting Useful Knowledge

Volume 104

FIG. 1. Alessandro Achillini. Courtesy of the University Library, Bologna.

STUDIES IN
PRE-VESALIAN ANATOMY
BIOGRAPHY, TRANSLATIONS, DOCUMENTS

L. R. LIND

Department of Classics
University of Kansas

THE AMERICAN PHILOSOPHICAL SOCIETY
Independence Square
Philadelphia

1975

Library of Congress Catalog
Card Number 74-78093

International Standard Book Number 0–87169–104–3
US ISSN 0065–9738

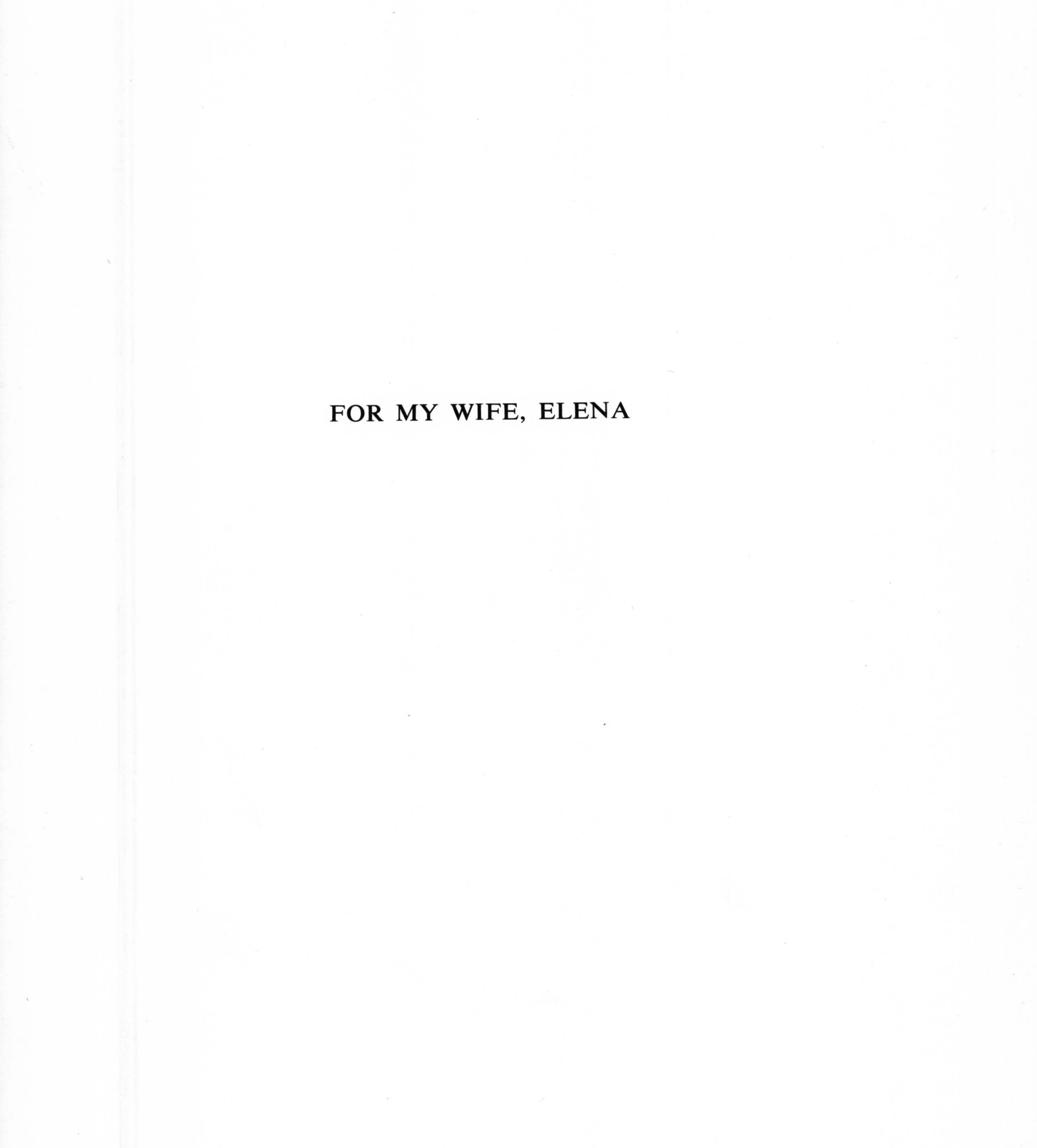

FOR MY WIFE, ELENA

PREFACE AND ACKNOWLEDGMENTS

The plan for this book evolved during the summer of 1959 while I was serving as research professor in the history of medicine in the Medical Center, University of California, Los Angeles, at the kind invitation of Dr. H. W. Magoun and the Department of Anatomy. There with the help of my research assistant, Miss Judy Hoffberg, and the excellent staff of the UCLA School of Medicine Library, headed by Miss Louise Darling, I gathered the bibliography and much of the material from which I have worked. In October, 1960, I received a U.S. Public Health Service grant (GM–07202–1–3) from the National Institutes of Health, Department of Health, Education, and Welfare through the National Advisory Council which enabled me to spend the year 1961–1962 on leave of absence from my regular duties as professor and chairman of the Department of Classics and Classical Archaeology at the University of Kansas. This year was spent half in Arlington, Virginia, and half in Bologna, Italy, where I gathered much unpublished material in Latin.

During my stay at Arlington I used the Library of Congress and the National Library of Medicine, still in its old quarters on Independence Avenue, Washington, D.C. I wish to acknowledge the expert assistance of its staff as well as that of Miss Gertrude Annan and her staff at the New York Academy of Medicine; and of Mr. C. S. T. Cavanagh, formerly director of the University of Kansas Medical Library and now of Duke University.

During my stay in Italy (March-August, 1962) I worked chiefly in the three major libraries of Bologna, the Archiginnasio, Universitaria, and Archivio di Stato. I am deeply grateful to those in charge there. My special gratitude must be expressed to Signor Luigi D' Aurizio, vice-direttore of the Biblioteca Universitaria, Bologna, for his many kindnesses and sympathetic understanding of my needs in providing bibliographical assistance as well as photostats and microfilm; to Signor Mario Fanti, of the manuscript division, Biblioteca Comunale dell' Archiginnasio; to Signor Benedetto Nicolini, direttore, Archivio di Stato, Bologna, and professor of the history of religion and of the Church at the University of Bologna; to Professors Gianfranco Orlandelli, professor of palaeography, Istituto di Paleografia, University of Bologna, and Giovanni Bronzino, of the same Institute; to Signori Alfredo Bernardini, photographer, of the Archivio di Stato, and Anteo Badiale, for his patience in bringing me mountains of manuscripts and documents, and to the Signorine Zanni and Continelli for their assistance; to Dottore Lucia Rossetti, conservatrice of the Archivio antico dell' Università di Padova; to Monsignor Giuseppe Turrini, librarian of the Biblioteca Capitolare at Verona, with whom I spent a delightful afternoon in pursuit of rare anatomical plates; to Signorina Rosetta Catenacci, librarian, Biblioteca Universitaria, Pavia; Signor Antonio Lodi, librarian, Archivio di Stato, Modena; Signor Paolo Samben, director, Archivio di Stato, Padua; Signor Renato Turci, librarian, Biblioteca Malatestiana, Cesena; Signorina Lia Sbriziolo, in charge of the manuscript room, Biblioteca Nazionale Marciana, Venice; Dottore Mario Carrara, direttore, Biblioteca Civica, Verona; to the staffs of the Biblioteche Comunali at Imola, Forlì, Modena, and Mantova; of the Biblioteca Malatestiana at Cesena; of the Archivio di Stato, Archivio del Comune, Biblioteca Capitolare, and Biblioteca Estense at Modena; and of the Archivio di Stato at Mantova.

To the late Professor Ladislao Münster and to Professor Alessandro Simili, practicing physicians and professors of the history of medicine at Ferrara and Bologna, I am particularly grateful for their friendship and kindness shown me on various occasions and for long conversations in which both drew deeply from their wide knowledge for my benefit as well as to Professor Vincenzo Busacchi, professor of the history of medicine, University of Bologna, and to Doctor Uberto Cuzzelli, director of the Accademia Virgiliana at Mantova, for special kindness.

I wish to acknowledge the extraordinary and expert assistance given me at the National Library of Medicine in Bethesda, Maryland, during my stay as resident scholar there in the summers of 1963–1965 by John B. Blake, chief of the History of Medicine Division, and particularly by Richard J. Durling, of the Division's staff, who shared generously with me his specialized knowledge of bibliography and other aspects of medical history and made corrections and additions of great value to my work. Charles M. Goss, M.D., managing editor of *The Anatomical Record,* has read the manuscript and suggested changes in accordance with the terms recognized in *Nomina Anatomica* (1956). Dr. Nikolaus Mani, of the Department of the History of Medicine, University of Wisconsin, has read part of the manuscript and made useful suggestions. The *Nomina Anatomica* have been inserted wherever the structure was identifiable by Dr. Goss for Achillini, by Howard A. Matzke, professor and chairman of the Department of Anatomy, University of Kansas Medical School, for Massa and Benedetti, and for Canano by C. W. Asling, professor and chairman of the Department of Anatomy, University of California Medical School, Berkeley, California. I am deeply grateful to all these men for their valuable assistance.

In connection with the new documentary material in this book I am encouraged by the wise words of my late friend, Dr. Ladislao Münster, published in his article on Achillini (*Rivista della storia di medicina,* etc., etc., **24** [1933]: p. 8): "Let us not forget something important: true historical research is achieved in the archives, the rest is compilation. The publication of a single important document hitherto unknown and unpublished has a much greater value than the compilation of a large historical text on the basis of documents furnished by others. . . ."

The material thus presented should, together with my published translations of Vesalius, *Epitome* (New York, Macmillan, 1949) and of Berengario da Carpi, *A Short Introduction to Anatomy* (*Isagogae Breves*), (University of Chicago Press, 1959), provide in the form of a sizable and representative corpus of annotated material the opportunity for a wider view and understanding of pre-Vesalian anatomy than was heretofore possible to anyone except a few scholars familiar with the Latin texts. So rare is an understanding of scientific Latin of the Renaissance even among serious students of the history of medicine that I decided when I began this series of studies in biographies, texts, and documents that several translations of the key works were necessary for the general reader as well as for his more experienced colleagues.

I am indebted, finally, to the expert staff of the University of Kansas Library for much kind assistance during the twelve years I have spent on this labor, which has been carried on without financial subsidy from any source since October 31, 1963; it grew too large to be completed during the term of the NIH grant described above. I am extremely grateful to two men who have shown a kindly interest in this book: to James H. Cassedy, executive secretary, History of the Life Sciences Study Section, Division of Research Grants, National Institutes of Health, and to my friend, Professor C. W. Asling, Department of Anatomy, University of California, Berkeley, for reading the manuscript and making certain suggestions; he will be recalled for the excellent notes and chapter introductions which he contributed to my translation of Vesalius's *Epitome*. To my wife, who has borne the brunt of my vigils and peregrinations and who has striven to provide me with the time for writing this book, I am wholly indebted in a very special way.

L. R. Lind

University of Kansas, Lawrence, Kansas
January, 1973

Contents

Illustrations

Introduction

I. THE HISTORICAL BACKGROUND OF PRE-VESALIAN ANATOMY*

PRE-VESALIAN anatomy was written during one of the most exciting periods of European history. A number of the influential ideas, movements, and cultural developments of the Renaissance were coming into existence or were being rapidly expanded in the years between 1490, when Girolamo Manfredi published his *Anothomia*, and 1543, the date of publication of Vesalius's *De Humani Corporis Fabrica*. It was a half-century of amazing achievement not only in science but in art, literature, philosophy, exploration, political organization, imperialism, nationalism, and religion. The invention of printing less than fifty years before this period had enormously advanced the possibilities of thought and culture by way of printed books, including editions and translations of the Greek and Latin classics from which Italian Humanism drew the inspiration for its stimulating dialogue with antiquity.[1]

Some of the greatest painters of the Renaissance were at the peak of their production: Leonardo da Vinci, Botticelli, Giorgione, Correggio, and Michelangelo, whose "David" was completed at Florence in 1502, his paintings of the Sistine Chapel in 1508–1512, and the Medici Chapel, 1526–1534. Vernacular literature had made important strides with the work of Poliziano, recently dead (1494) only two years after the death of his great friend and patron, Lorenzo de' Medici, at Florence. Machiavelli, Ariosto, and Castiglione wrote their famous books in the early sixteenth century, while Aldus Manutius printed Greek texts from 1494 onward. The Renaissance popes, Alexander VI, Julius II, Leo X, and Clement VII (1492–1534) were active as patrons of art and learning. Savonarola's reforms in Florentine religious life took place between 1494 and 1498; in northern Europe the growth of Christian Humanism was contemporaneous with the advance of Biblical scholarship and the Lutheran revolt. The French invaded Italy under Charles VIII in 1494; Maximilian I, Holy Roman Emperor, campaigned against France; and Hapsburg expansion under Charles V led to the sack of Rome in 1527, a prime cause of the Counter-Reformation. Francis I of France had extended secular power over the Church by means of the Concordat of Bologna, 1516, and Henry VIII of England, who was to break with the Church in 1534, in the Treaty of Amiens (1527) which he made with Francis I against Charles V, helped for a time to revive the shattered Papacy. Vasco da Gama reached India in 1498, Columbus undertook his voyages between 1492 and 1504, and the papal line of demarcation between the Spanish and Portuguese spheres of conquest in the New World was made in 1493.

Spanish expansion in the New World had created competition in empire-building. One of the more sinister results was the spread of syphilis in Europe, which in turn raised one of the most controversial of contemporary medical questions. The growth of industrial capitalism and a steadily rising inflation of prices during the sixteenth century were promoted by wealth brought back from the New World, and this prosperity encouraged various expeditions to the Americas, resulting in the earliest printed maps of that region (1506–1508) and the beginnings of scientific interest in flora and fauna. Increasing wealth and power brought with them a realism in diplomacy that became frank cynicism on a grand scale in European affairs. It is interesting to note that it was Giulio, a member of the Medici family of bankers and patrons of art, who when he became Pope Clement VII in 1523 commissioned Machiavelli to write his history of Florence, a book which marked an epoch in the development of modern historiography. As Symonds has pointed out, Machiavelli was the first "to contemplate the life of a nation in its continuity, to trace the operation of political forces through successive generations, to contrast the action of individuals with the evolution of causes over which they had but little control. . . ."[2] The philosophy of history was enriched and deepened by this book, the predecessor of Guicciardini's greater work on the same subject. But in the light of later European history its consistent emphasis upon the masses of men as political instruments, not as thinking and feeling individuals, was to establish and encourage a way of thinking which would lead to Fascism.

The pre-Vesalian period was one in which the tradition founded by the Medicis became a constant practice among the rulers of men: the encouragement and subsidization of art, literature, philosophy, and science. It was in an atmosphere of sympathetic interest and active support unknown in Europe since the time of Frederick II of Sicily that these expressions of the human spirit began to flower into the greatest era of the Renaissance.

It is true that science had already found a haven and a base of operations in the universities while philosophy, apart from the court of Lorenzo de' Medici and the Platonic Academy at Florence (founded in 1459), had actually nowhere else to go. Hence we find Pomponazzi, the liberal Aristotelian, challenging in the *Studium* of Padua an anatomist and Averroist

* The introduction was delivered as the Clendening Lecture No. 13 on the History and Philosophy of Medicine in the Clendening Medical Library of the University of Kansas School of Medicine on January 16, 1966.

[1] Henry E. Sigerist, *The Conflict Between the 16th Century Physicians and Antiquity*; International Congress, London, 1922; published at Antwerp, 1923. Gernot Rath, "Pre-Vesalian Anatomy in the Light of Modern Research," *Bull. Hist. of Medicine* **35** (1961): pp. 142–148, is the sole account in English of the subject but is only a brief glimpse of a large field.

[2] John Addington Symonds, *The Renaissance in Italy* (Modern Library Edition, 1935) **1**: p. 167.

philosopher, Alessandro Achillini, in one of the most important encounters in the history of Paduan Aristotelianism during the years 1506–1508. Gabriele Zerbi, also an Aristotelian philosopher as well as an anatomist, found a post likewise at Padua, where Alessandro Benedetti, when he was not traveling either to Greece or the wars, was Zerbi's colleague. Berengario da Carpi became a long-time teacher of surgery at the University of Bologna (1502–1527) while maintaining a lucrative medical practice and traveling about Italy on important missions of healing for rich patrons.

Science was accompanied in that age by the pseudo-sciences, astrology, physiognomy, and chiromancy. Manfredi, Benedetti, and Achillini were interested in one or more of these.[3] Popes and princes, scholars, priests and laymen, burghers, farmers and artisans, all lived under the influence of specialized superstitions. Giovanni Pico della Mirandola's attack upon them in his *Disputationes adversus* (or *in*) *astrologos* (Bologna, 1495) or Symphorien Champier's several books in criticism of astrology and other occult arts and sciences during the pre-Vesalian period could not stem the tide of ignorance. The study of astrology was officially sanctioned at the University of Bologna; Bartolomeo della Rocca's (Cocles) treatise on palmistry, *Chyromantiae ac physionomiae anastasis,* one of the most influential books of its time, was dedicated to Alessandro Bentivoglio and published at Bologna in 1504, with the approval of Alessandro Achillini. The university lecturer on astrology was charged with the task of drawing up an almanac, or *taccuino,* in which were foretold on the basis of astrological calculations the harvests, weather, epidemics, and other important events of the coming year.

The mention above of Alessandro Bentivoglio brings up the subject of the patrons of learning and culture in the several cities where pre-Vesalian anatomy was practiced. The Bentivoglio were the most famous and the longest-lived of the rulers of Bologna and hence are closely linked to the fortunes of the oldest university in the world and to the sciences which it sheltered. The first Bentivoglio became the city's leader in 1401, and Giovanni II was the last ruler of the name; he left Bologna in 1506 before the advent of Pope Julius II, his great enemy. His sons attempted to rule for a few years afterward but the collapse of French power at the battle of Ravenna in 1512 at the hands of the Holy League brought about the final downfall of the Bentivoglio. The father, grandfather, and great-grandfather of Alessandro, the last of the line, had, however, established themselves as popular rulers and to the last the Bentivoglio had fostered the best interests of the city, including its ancient university with its great medical school.

Padua, the great rival of Bologna, was the next most significant center for anatomical studies and as the university center of the Venetian Republic provided an appropriate environment for Benedetti, Massa, and Zerbi, who were professors there. Achillini, the Bolognese, also taught there from 1506 to 1508. The *Studium* at Padua had, in fact, been established in 1222 by a migration of students from Bologna, which thus became its parent. In 1399 the *universitas artistarum* assumed its independence from the *universitas juristarum,* the law school with which the *studium* at Padua had begun. The territory of Padua was incorporated with that of Venice in 1406 and thereafter the republic of St. Mark became the arbiter of the destinies of learning at Padua. The Venetian Senate had made provision for regular teaching of the humanities by 1443.[4] For centuries Venice exercised a benign control over the fortunes of the university at Padua, keeping out rival schools such as those of the Jesuits, improving conditions for students both Catholic and Protestant, and hiring the best professors available. In 1594 a permanent anatomical theater was constructed at Padua, preceding by one year the one at Bologna, decided upon during the priorate of Ulisse Aldrovandi.

---

[3] Manfredi, who died in 1493 at Bologna where he had been a professor of medicine and philosophy from 1462 to 1463, drew up a horoscope or prognostication for the year of his death: see University of Bologna MS B. 2517, "Pronostico di Girolamo Manfredi per l'anno 1493," dated Feb. 8. He was in great demand at Bologna for such horoscopes although Pico della Mirandola made fun of him for his inability to predict his own death in the very year in which it occurred, as Charles Singer points out in "The Anothomia of Hieronymo Manfredi, etc.," *Studies in the History and Method of Science* (Oxford University Press, 1917), p. 98. His *Liber de homine,* etc. (Bologna, 1474) deals in Book II with physiognomy. Singer lists two other prognostications by Manfredi for 1478 and 1481 but overlooks the one for 1493. Other pseudo-scientific works by Manfredi mentioned by Gurlt, Hirsch, and Haller cannot be identified as separate books and are probably to be rejected.

Benedetti was asked to find an astrologically suitable day for completing negotiations between Count Niccolò Orsino Petiliano and the Venetian Senate: see his *De Bello Carolino* (ed. Giustiniani), p. 95, and my comments in the Introduction on the life and works of Benedetti. Achillini's discussion of physiognomy and chiromancy is contained in Escorial MS f. III, 2, of the sixteenth century, fols. 1 and 9, immediately followed by Pomponazzi's work *On Incantations*. None of these three scientists, however, brings any pseudo-science into their works on anatomy. De Laguna and Dryander also wrote on astrology.

[4] See, for useful accounts of education at Padua, Cesare Foligno, *The Story of Padua* (London, 1910); Francesco Maria Colle, *Storia scientifico-letteraria dello studio di Padova* (4 v., Padua, 1824–1825); for biography, G. Vedova, *Biografia degli scrittori padovani* (2 v., Padua, 1832); for Venice and its culture, Bruno Nardi, *La civiltà veneziana del quattrocento* (Florence, 1957), and for Bologna one or two items from many good available books: Francesco Cavazza, *Le scuole dell' antico studio bolognese* (Milan, 1896); Michele Medici, *Compendio storico della Scuola anatomica di Bologna* (Bologna, 1857). The *Atti del Sacro Collegio medico di Padova* are assembled in 76 volumes covering the years from 1367 to 1804 and its ancient archives are described by Giuseppe Giomo, *L'Archivio antico della università di Padua* (Venice, Fratelli Visentini, 1893). The Bolognese archives are listed as to their chief groups of documents in Giorgio Cencetti, *Gli Archivi dello studio bolognese* (Bologna, Pubblicazioni del R. Archivio di Stato in Bologna **3**, 1938).

II. THE CULTURAL BACKGROUND OF PRE-VESALIAN ANATOMY

The cultural background out of which pre-Vesalian anatomy grew was twofold: medieval scholasticism and Renaissance humanism. These merged their influence in the university training provided at the two great centers of medieval and Renaissance Italian university life, Bologna and Padua, where gradually more facilities for scientific investigation could be found, where both papal and Venetian liberalism had long ago removed restrictions upon dissection, and where the great medical minds of the time both in Italy and beyond the Alps gathered to establish a tradition of free inquiry and research.

Medieval scholasticism, the first and most persistent of the two sources of method in pre-Vesalian anatomy, was applied as early as the time of Pietro d'Abano, of Padua (1250–1316 A.D.) in the use of questions, solutions, doubts, and proofs. Dying ten years before Mundinus published his immensely influential anatomy, Pietro wrote another influential book in his *Conciliator Controversiarum quae inter philosophos et medicos versantur, Petro Abano, Patavino, philosopho ac medico clarissimo auctore* (first edition, Mantova, 1472; Venice, 1476). Thorndike calls this "the leading Latin medical text of the early fourteenth century,"[5] and Nardi says: "Pietro of Abano is perhaps the most characteristic scientist of that period in which Thomas Aquinas was the greatest theologian and Dante Alighieri the supreme poet."[6]

In the *Conciliator* Pietro d' Abano attempted by the process of question and answer accompanied by the syllogism to solve a large number of scientific and pseudo-scientific problems which are often reminiscent of the "Problem" literature that begins as early as the pseudo-Aristotelian *Problemata*; an example appears in my annotated edition of the Bologna Library manuscript, *Problemata Varia Anatomica* (1968). Nardi defends him against the charge of heresy arising from his Averroistic learnings. Pietro's chief purpose was to reconcile the Aristotelian and the Arabic sources of science and thus to harmonize Aristotle and Avicenna.

This purpose involved commentary on the *Canon* of Avicenna and not practical dissection. Other medical writers, who did not follow exactly the form of d'Abano's inquiries, were also commentators on Avicenna: Jacopo da Forlì, *Expositio in primum Avicenne Canonem* (Venice, 1508) and Ugo Benzi (Hugo of Siena) (1376–1439), *Commentaria super Quarta Primi Avicennae cum annotationibus quibusdam Jacobi de Partibus,* etc. (Venice, 1517).[7] Still others of the medieval scholastics commented on Galen (chiefly his *Microtegni,* also known as *Microtechne, Liber tegni, Ars medicinalis* and *Ars medica,* Galen's most popular work in the Middle Ages) or on Mesue the Younger: Pietro Torrigiano de' Torrigiani (Turisani Monaci Cartusiensis), *Plusquam Commentum in librum Galieni qui microtechni intitulatur. Cum questione de Ypostasi* (Venice, 1512), and Francis of Piedmont, *Complementum* to Mesue's *Grabadin* (Venice, 1471). Giovanni d'Arcolani of Verona, (1390–3–1458), wrote an *Expositio in nonum librum Almansoris* (Basel, 1540), a commentary on Rhazes which contains a rather exact description of the symptoms of delirium tremens. To this list of prominent medieval writers may be added Jacobus de Partibus (Jacques Despars) and Johannes Matthaeus Ferrarius de Gradibus, frequently quoted by Berengario da Carpi in his commentary on Mundinus (1521); the former was evidently so influential that like Zerbi he had his group of followers whom Berengario singles out as "Partists" along with the "Zerbists" (fol. 258 v).

These and other medical writers provided a method of commentary which was followed among the pre-Vesalian anatomists particularly by Gabriele Zerbi and Berengario da Carpi. The method aims at completeness of quotation from the Arabic scientists and includes a good deal of exclusively philosophic discussion as well as anatomical facts, especially in regard to the human soul, as, for example, in Pietro d'Abano's *Conciliator*. Zerbi's *Anatomiae Corporis Humani et Singulorum Illius Membrorum Liber* (Venice, 1502) includes references to many authorities among the Arabs and the Greeks; but Berengario's commentary on Mundinus is practically exhaustive in its abundance of quotation. A second motive in the medieval method of medical criticism is the harmonization of authorities, a tedious and confusing business as we see it in its most extensive form in Zerbi and Berengario.

It is, in fact, with a sense of relief that we see in Benedetti's *Anatomice* (1502) the first anatomy which abandons the medieval writers and the Arabs and thus returns to the Greek anatomical writers in the spirit of the humanists. This is not, of course, a wholly unmixed blessing when we realize that Benedetti has narrowed his sources of reference almost exclusively to Aristotle, Plato's *Timaeus,* and Galen for anatomical doctrine and to Pollux, Rufus, Aretaeus, and Paul of Aegina for terminology. His use of these authorities, moreover, is not critical;

---

[5] Lynn Thorndike, *A History of Magic and Experimental Science* (New York, Columbia University Press, 1941) **5**: p. 119.

[6] Bruno Nardi, *Saggi sull' Aristotelismo padovano dal secolo XIV al XVI* (Florence, 1958), p. 74. See also his second chapter, "Intorno alle dottrine filosofiche di Pietro d'Abano," pp. 19–74 (especially 27–38), where Nardi defends Pietro against the charges of error brought against him by Symphorien Champier, *Annotamenta, errata et castigationes in Petri Aponensis opera,* appended to the Giuntine edition of the *Conciliator* (Venice, 1520) and many later editions.

[7] Dean Putnam Lockwood, *Ugo Benzi—Medieval Philosopher and Physician 1376–1439* (University of Chicago Press, 1951), pp. 40–43 and appendix VIII has a good discussion of the medieval method of medical disputation.

he differs sharply with none of them and thus probably serves as a mentor in this respect for de Laguna, who actually boasts of his entire submission to his authorities. It seems that Benedetti, temperamentally oriented toward the Greeks by his long acquaintance with them in the Aegean region, may have sacrificed a wider and more firm basis of reference by thus arbitrarily ruling out the Arabs and the Italian scholastic anatomists. At most he was returning in the spirit of Italian humanism in general to the pure sources of medical knowledge which lay among the Greeks, beginning with Hippocrates. That he does not quote the latter more often (I find only five references to Hippocrates) is due to the paucity of strictly anatomical information given by the Father of Medicine.

The return to the Greeks, the abandonment of detailed reference to the Arabs and of the medieval scholastics as authorities, and a decisive reduction of the earlier attempts to harmonize conflicting doctrines marked the new anatomy which was to result in Vesalius's *Fabrica*. Achillini, Zerbi, and Berengario (who rarely quote Mundinus) are the chief pre-Vesalians to follow the method of multiple citation of both Greeks and Arabs as well as other medieval writers. Manfredi quotes few authorities, Massa depends chiefly upon Galen, as do de Laguna, Guinterius, Canano, and Estienne. With Galen anatomy began; and pre-Vesalian anatomy comes back to him full-circle as almost the one and only authority. The pre-Vesalian professors had come to rest at last, after centuries of fruitless wrestling with the conflicting testimony of Greek and Arabic texts, Latin translations, and interminable disputes over interpretation, in the bosom of another professor, Galen, the lecturer on anatomy at Rome in the time of Marcus Aurelius.

III. THE PRE-VESALIAN ANATOMISTS AND THEIR WORKS

The first important figures in medicine at Bologna, doubtless of Salernitan provenience and training, were those of the surgeons Ruggero da Palermo (or Parma) and Rolando da Parma, his student; some of their writings are now printed as well as preserved in manuscript. William of Saliceto, born around 1210, was a more eminent scholar; his work on surgery (1275) is still regarded by some as of an importance equal to that in anatomy by Mondino dei Luzzi (Mundinus) and certainly the best before him. He taught the suture of nerves, distinguished venous from arterial blood, and restored to use the knife which had been abandoned by the Arabs for the cautery. With William and his disciple, Lanfranco, begins the divorce of surgery from medicine. Other men, of whom little is known, appear in the *Studium* during the thirteenth century: Ugo da Lucca, the first to earn a salary as a medical teacher from the Comune of Bologna, and the first to practice anesthesia with belladonna, aconite, and henbane; Teodorico, his son, who died at ninety-three after producing his *Chirurgia Magna*; and Bruno di Longoburgo, who also taught at Padua and Vicenza. Lanfranco carried the teachings of Bologna to Paris, where he taught for one year, influencing French surgery in the person of Guy de Chauliac as well as Flemish surgery through Jan Yperman in the fourteenth century.

Throughout the Middle Ages the authority of Galen in anatomy had stood unquestioned. His works and those of the Arab physicians (both of them in Latin translations) were commented on and edited by many medical men, with slight additions or original contributions. The first European to dissect the human body publicly was Mondino dei Luzzi, who received his degree at Bologna in 1290 and died in 1326. His *Anatomy* (1316) became the standard textbook for the next two hundred years. Editions of it appeared as late as 1580, long after Berengario da Carpi and Andreas Vesalius had laid the basis for modern anatomy in their great works and after the errors of Galen had been disposed of by a series of excellent Italian anatomists. In fact, the history of anatomy, like that of certain other studies, can be called the history of its liberation from the domination of a book: Mondino's.

This is to praise Mondino, not to belittle him, for between the age of Galen (the second century A.D.) and that of the immediate pre-Vesalians (1490–1543) Mondino's *Anothomia* (as he preferred to spell the name)—the *editio princeps* is dated 1478 and its first association with the *Fasciculus Medicinae* in 1495—was the most complete description of the human body in the compact form of a manual and guide to dissection. He was the first medieval anatomist to give an adequate account of the preparation for and the process of human anatomical dissection as well as to provide a rudimentary knowledge of physiology and some information on surgical anatomy. More than thirty printed editions of his book appeared until well into the sixteenth century. Achillini and the other pre-Vesalians used Mondino carefully and with respect. Both Dryander and Berengario da Carpi edited the *Anothomia,* the latter with great success; he was the first to use tolerable illustrations, reproduced in the present book.

It is instructive to observe Mondino through the eyes of his best commentator, Berengario da Carpi, who in his *Commentaria* (1521) gives us the most thorough discussion and criticism of Mondino up to his own time. This work together with Charles Singer's translation of the *Anothomia* (Florence, R. Lier and Co., 1925) provides an almost exhaustive analysis of the text and thereby an impressive view of the importance of Mondino to the pre-Vesalian anatomists. Berengario does not attempt to equate Mondino with either Aristotle, Galen, or Avicenna, pointing out that Mondino did not have access to many works of Galen. He praises his predecessor chiefly for his concise and yet copious organization of material. He acknowledges both the faults and the good points of the *Anothomia* and his own debt to it. Although Mondino's work was dominated by authorities, his personal observations were considerable. It deserved its great popularity.

Contemporary with Mondino were Taddeo degli Alderotti, whose life was written by the chronicler Filippo Villani, and Bartolomeo da Varignana. Taddeo became famous enough to be mentioned by Dante (*Paradiso* 12. 82–85), if only as an example of a physician who loved money, while da Varignana gained high standing in legal medicine. He also rented rooms to students, taking them with him on his sick-bed rounds. A closer relationship between professor and student could scarcely be imagined. Other Bolognese anatomists worthy of mention in the fourteenth century are Bertuccio or Alberto Zancari, Mondino's pupil and continuator, Dino del Garbo, his son Tommaso, Gentile da Foligno, and Jacopo da Forlì. Tommaso Garbo was a friend of Petrarch and is recorded by both Villani and Sacchetti; his writings deal with fever and plague.

I pass over the first half of the fifteenth century entirely in this brief sketch of the rise of anatomy. The solid volumes of Sorbelli and Simeoni, of Malagola, Sarti Mauri, Fattorini Mauri, Medici, Cavazza, Martinotti, and others on the history of the University of Bologna do full justice to this less fruitful period. But like many a fallow upland in the varied landscape of history it was destined to give in its last years the nurture necessary for the most important age in the history of Bolognese anatomy and medicine—the sixteenth century.

The men who made this age what it was can be divided into two groups: the precursors and their successors, who built more extensively upon the foundation their predecessors made for them. The first group includes five men, not all Bolognese, who were born in the fifteenth century and thus passed the major part of their lives in the atmosphere of medieval anatomy, of Galen and Mondino, but give some indication of their impatience with authority. They are Hieronymo (or Girolamo) Manfredi (1430–1493), Alessandro Achillini (1463–1512), Alessandro Benedetti (1460–1525), Gabriele Zerbi (1468–1505), and Berengario da Carpi (1460–1530). The last named became more famous in later ages because his contributions to anatomy were both more considerable and point more directly toward Vesalius and an inevitable advance beyond Galenic anatomy. "In many ways the Danish astronomer's [Tycho Brahe's] role in the early history of modern astronomy is analogous to that of Vesalius in anatomy," says A. R. Hall in his book, *The Scientific Revolution*. In the light of this analogy Berengario da Carpi might then be called the Copernicus of anatomy; for he was the first truly original mind in anatomical research since Mondino, as Copernicus was in astronomy since Ptolemy.

I hope I have brought the life and works of Berengario da Carpi into a proper focus in the introduction to my translation of his "Short Introduction to Anatomy" (*Isagogae Breves*). As both surgeon and anatomist he is the most outstanding worker before Vesalius. His contributions to learning are considerable in the study of the brain, the sphenoid sinuses, the vermiform appendix, the tympanum, the thymus gland, the arytenoid cartilages, the action of the cardiac valves, the papillae of the kidneys, the chyliferous ducts, and the representation of the abdominal muscles. Not less significant is his first use of adequate anatomical illustrations, not mere schemata but drawn from the cadaver by a good artist.

Yet Berengario, for all his eminence, and his greater successor, Vesalius, have too long overshadowed their contemporaries at Bologna and elsewhere in Italy. Lynn Thorndike was the first to warn against such a distorted view of sixteenth-century anatomy:

> Therefore the publication in 1543 of the *De humani corporis fabrica* of Vesalius should not be regarded as a bolt from the blue. There was further marked activity and progress in anatomy during the sixteenth century, especially in Italy, where in addition to Achillini we have the names of Berengario da Carpi, Falloppia, Colombo, Eustachio and Aquapendente to set beside that of Vesalius. Between the works of Carpi and Vesalius, in the years from 1535 to 1541, appeared treatises by Andrés de Laguna, Nicolaus Massa, and John Dryander. These intermediary works help us to form some conception of the changing opinion and emphasis in anatomical studies and writings between 1521 and 1543.[8]

A second group of pre-Vesalian anatomists whose works were in most instances published a few years later than those of their precursors and contemporaries includes Antonio Benivieni (1450–1502), Andrés de Laguna (1499–1560), Niccolò Massa (1499–1569), Johannes Guenther of Andernach (1505–1574), Johannes Dryander (1500–1560), Giovanni Battista

[8] *Op. cit.* **5** (1941): p. 498. *cf.* Thorndike's comment quoted from p. 503 in my remarks at p. 10 and A. R. Hall, *The Scientific Revolution, 1500–1800* (Boston, Beacon Press, 1956), p. 49: "If Vesalius was pre-eminent among early anatomists he must not be made so at the expense of his contemporaries, who were also men of ability and precision, nor by neglecting his own faults and mistakes." W. P. D. Wightman, *Science and the Renaissance: an Introduction to the Study of the Emergence of the Sciences in the Sixteenth Century* (New York, Hafner Publ. Co., 1962) **1**: p. 230: "This brief survey of the works published or prepared before 1543 does nothing to impugn the priority of the *Fabrica* in respect both of its scope and quality; but it does reveal that far from pioneering an entirely new approach to the study of human anatomy it had only made a great leap forward on a road already trodden by several slightly earlier workers."

Canano (1515–1579), and Charles Estienne (1504–1564).

Of the pre-Vesalian anatomists Achillini, Berengario da Carpi, and Manfredi were Bolognese. Benedetti and Zerbi were Veronese but came within the orbit of Padua together with Massa, the Venetian. Benivieni was a Florentine, Canano a Ferrarese, Dryander and Guenther (or Guinter) Germans, Estienne a Frenchman, and de Laguna a Spaniard. Bologna could not therefore claim complete primacy in the field of gross anatomy even if we add the names of lesser Bolognese such as Lodovico Boccadiferro (1482–1545), Angelo Bolognini (1508–1536), Francesco de' Genoli (1498–1525), and Bartolommeo Maggi (1477–1516, or shortly thereafter), all of whom contributed little to the science of anatomy. But Bologna did have the first important men and one of the few leaders in the real sense of the word in Berengario da Carpi.

During the latter half of the sixteenth century other anatomists appeared at Bologna to carry on the tradition established by their predecessors: Andrea Albi (born *ca.* 1539–1540), Giulio Cesare Aranzi (1530–1589), Virgilio de' Bianchi (1522–1599?), Volcher Coiter (1534–1590 or 1600), Giovanni Battista Cortesi (1554–1634), Domenico Leoni (*ca.* 1559–1591), Bartolommeo Maggi (1477 or 1516–1552), Gerolamo Mercuriale (1530–1606), Francesco Muratori (1569–1630), Gian Francesco Rota (1520–1558), Carlo Ruini, the first great veterinary anatomist (*ca.* 1598), Gaspare Tagliacozzi, the pioneer in plastic surgery (1545–1599), and Costanzo Varolio (1543–1575). With the exception of Aranzio, Coiter (a Hollander who took his degree at Bologna), Mercuriale, Ruini, and Tagliacozzi, most of these men were minor figures.

I have provided in the following pages a brief sketch in chronological order of each of the important general anatomies produced by the pre-Vesalians, including those I have translated and in most instances provided with notes. Benivieni's book of case histories, forming the first text on morbid pathology, is included chiefly because it is based upon anatomical dissection and not because it represents any particular advance in the science of anatomy. There are other works of interest which might be added but since they tend to fall before the strictly pre-Vesalian period I have chosen to examine (1490–1543) I have paid little attention to them. Such works include, e.g., Giammatteo Ferrari da Grado (Matthaeus de Gradibus), *Practica*; Milan, 1471; *Expositiones super vigesimam secundam Fen tertii Canonis Avicennae*; Milan, 1494; Jacopo da Forlì, *Expositio Jacobi Forliuiensis in primum Auicenne Canonem,* etc.; Venice, 1508, the commentary of Johannes Arculanus (d. 1484) on the ninth book of Rhazes (publ. 1540), and a variety of other fifteenth-century commentaries on the works of Galen or the Arabs which do not imply actual practice of dissection. Many of these are quoted or referred to in Berengario da Carpi's commentary on Mundinus (1521), which is a veritable anthology of the anatomical literature prior to its publication but cites Zerbi most frequently of contemporary authors, if only to refute him.

In my survey of pre-Vesalian anatomy I have not included Jacobus Sylvius, *In Hippocratis et Galeni Physiologiae Partem Anatomicam Isagoge,* which, although it is said to have been written in 1542 (C. D. O'Malley, *Andreas Vesalius of Brussels 1514–1564*; University of California Press, 1964, p. 51), was published in 1555 and thus lies outside the chronological limits of my investigations. It probably should also be studied in the light of pre-Vesalian knowledge of anatomy and its use of Massa's *Liber Introductorius,* whom Sylvius quotes. I have also not included Johann Peyligk's *Compendiosa Capitis Physici Declaratio,* also known as *Anatomia Totius Corporis Humani,* first printed in 1499, with editions in 1503, 1509, 1510, 1513, 1515, 1516, 1518, which represents the very backward state of anatomical learning in Germany during this period. The situation was little better in England as can be seen in David Edwardes, *Introduction to Anatomy,* published in 1532: see the facsimile reproduction with English translation and an introductory essay on anatomical studies in Tudor England by C. D. O'Malley and K. F. Russell (Stanford University Press, 1961). The pioneering work of Leonardo da Vinci in anatomical drawing and description has been dealt with thoroughly elsewhere, and thus I do not include it in my account.

The Precursors

1. Hieronymo (Girolamo) Manfredi, *Anothomia* (1490). Manfredi, after a fervent apostrophe to Divine Majesty and a salute to Giovanni Bentivoglio, ruler of Bologna, writes to him:

> Hoc enim opusculum quantum melius potui ex variis antiquorum uoluminibus exserpsi ac id abreuiaui nec eumdem forte tenui ordinem et illi et ipsum materno composui sermone ut opus hoc delectabilius tuae sit magnificentiae. (I have excerpted this little work as best I could from various volumes of the ancients, abbreviated it without keeping the original arrangement of material, and written it in our mother tongue in order that it might be more enjoyable to your Magnificence.)

The work is thus described as a compilation from the ancients and written in Italian, presumably because Giovanni Bentivoglio felt more at ease in that language than in the standard language of scholarship, Latin. The *Anothomia* follows Mundinus closely, as may be seen in the three large excerpts presented by A. Mildred Westland in English translation: the head, the eye, and the heart. In the treatise as a whole there are not many borrowings from other authors apart from Mundinus, although Aristotle contributes the bone in the heart, which Mundinus did not take over in his description, and a few other items. Another term, grandid (from *grandineus,* like hail) is added for the crystalline humor of the eye, a term which scarcely occurs elsewhere. The description of the brain includes the ancient Augustinian division going back to the Greek philosopher Posidonius and followed by Mundinus, of the respective functions of the three ventricles: sensation and imagination, estimation and cognition, and memory, which, as Aristotle points out, man shares with the brutes. Achillini does not include this elaboration of the brain's function in his description nor does he follow the ten-layer account of the head as Mundinus and Manfredi do. At p. 140 (Singer's text of Manfredi) there appears what may be the source of Achillini's famous statement concerning Wharton's ducts of the submandibular gland, since it does not appear in Mundinus. The description is, in fact, more complete than Achillini's:

Circa [folio 15 *verso*] la radice de la lingua da ciascuno lato sono carne glandose facte acio che generasseno la humidita saliuale che hauesse a humetare la lengua acio che non se siccase per tanti mouimenti che ha in se: et in queste carne glandose sono dui buchi che poria intrare uno stile e per quilli buchi se distilla la humidita saliuale. Sotto la lingua sono doe uene grande uiride da le quale poi procedeno piu altre uene. (Around the root of the tongue on each side there are fleshy glands created in order that they may generate the salivary humidity needed to moisten the tongue so that it will not dry out due to its frequent movement, and in these fleshy glands there are two holes into which a stylus may enter and through these holes the salivary humidity is distilled. Under the tongue are two large green veins from which other veins proceed.)

The ultimate source is Avicenna, who adds the name "frog veins" but does not describe Wharton's ducts, which we may as well call Achillini's ducts in order to set the matter straight even at this late date.

Although he rarely mentions his sources, Manfredi, like all the early anatomists, borrows from both Greeks and Arabs what he does not draw from Mundinus. In this respect Achillini differs from him in usually naming his sources (as shown by my notes to the *Annotationes Anatomicae*) and in exhibiting much more scientific curiosity about his subject as well as more personal observation.

2. Alessandro Achillini, *Annotationes Anatomicae* (Bologna, 1520), a series of lecture notes probably not intended for publication and not published during Achillini's lifetime. They record anatomical observations made in 1502, 1503, and 1506, and follow the outline of Mundinus' text on which they are in fact a brief commentary. He quotes numerous authors and presents a broad view of the subject, avoiding certain dogmatic features of contemporary anatomy such as the doctrine of the humors or pseudo-scientific theories of astrology or physiognomy. He is credited with certain small discoveries or rediscoveries, as those of the fornix and infundibulum: his description of the brain is very good for that time. His work serves as an admirable short introduction to pre-Vesalian anatomy (translated herewith).

3. Alessandro Benedetti, *Anatomice, siue Historia Corporis Humani*; *eiusdem Collectiones Medicinales seu Aforismi* (Venice, 1502; Paris, 1514; Strassburg, 1528) (translated herewith). Benedetti is decidedly oriented toward the Greeks and away from the Arabs in contrast to Zerbi, who makes use of both sets of authorities and of the Latins as well. The *Anatomice* is not an original work but is well organized and clearly written except for a chapter on the nerves where the ancient confusion with sinews and tendons creates further confusion. It is, moreover, the work of a physician who had traveled widely and had learned his Greek in Crete, where he lived for some sixteen years toward the end of the fifteenth century; the book includes many first-hand accounts of medical practice and of anatomical anomalies.

4. Gabriele Zerbi, *Anatomiae Corporis Humani et Singulorum Illius Membrorum Liber* (Venice, 1502; 1533). This book, together with Berengario's *Commentary on Mundinus* (1521), is the most considerable pre-Vesalian anatomy but has been consistently the most ignored and maligned because it has not been read with care by medical historians. There are good reasons for this neglect and opprobrium. First, the wretched small Gothic type in which the book is printed together with many annoying abbreviations (only slightly improved in the second edition of 1533) makes reading it an infinitely tedious task. Second, the style with its endless *Additiones* and constantly recurring lists of features—*substantia, complexio, quantitas,* etc.—for each part of the body accumulates an overwhelming mountain of detail which at length wearies the most stalwart reader. I venture to conclude that few since the time of Berengario da Carpi have read the book through: but he read it with great care and attention, as his frequent references to it indicate.

As I point out in note 6 to the translation of Achillini's *Anatomical Notes,* Zerbi, unlike any other pre-Vesalian anatomist, divides his book on the basis of Aristotle, *De Incessu Animalium* 704b–705a, b and *De Caelo,* II, 284b, into three sections: I. *Anterior Parts,* fol. 5; II. *Posterior Parts,* fol. 136; III. *Lateral Parts,* vol. 152; see fols. 3v and 5v for his explanation of this unusual plan. Zerbi, like Achillini, quotes (at fol. 3r) Johannes Alexandrinus's definition of anatomy (see note 1 to the *Annotationes*). Zerbi's somewhat tedious procedure is to write a paragraph or two of *Textus* and then to add comments and quotations from the authors in corroboration of what he has written in the form of an *Additio.* He then discusses the nine features of each organ or part of the body including *substantia, numerus, complexio, figura, quantitas, situs, colligantia, iuvamentum,* and *passiones*; sometimes *color* and *operationes* are added. This process is repeated for each item described and seems to be drawn from the commentary of Johannes Alexandrinus on Galen's *De Sectis* (fol. 5r), upon which Achillini also drew for his basic principles of procedure without, however, subjecting every part of the body to the rigidly consistent treatment Zerbi employed.

Among the authors Zerbi quotes is Lactantius, usually by one of his other names, Firmianus (Lucius Caelius Firmianus Lactantius, who lived at the beginning of the fourth century A.D. and wrote between 305 and 311 a work entitled *De Opificio Dei*; Migne, *Patrologia Latina,* VII). Berengario among pre-Vesalian anatomists is the only other writer to quote Lactantius, surely a significant fact in view of his special relation to Zerbi and the inveterate habit of these sixteenth-century scientists of imitating each other. Zerbi, indeed, quotes more authors and adds more information not specifically anatomical than any other author since Mundinus except Berengario in his *Commentaria* on Mundinus. We may reasonably expect, therefore, that others beside Berengario would draw upon Zerbi for guidance, as Achillini seems to do.

Not only does he quote the standard Arabs—Avicenna, Averroes, Rhazes, and Haly Abbas—but the Greeks from Aristotle to Galen, including Rufus and Julius Pollux on terminology. The classical writers Lucretius and Pliny are quoted and the medieval authors Pietro d'Abano, Albertus Magnus, and Moses of Cordova. But in spite of extensive quotation from many authors, Zerbi's book is not a mere compilation nor another in a long series of commentaries on Mundinus, whom, by the way, he rarely quotes. His analyses and descriptions form the first systematic and sufficiently detailed examination of the human body since Mundinus and far outstrip the latter in scientific accuracy.

An indication of Zerbi's importance is Berengario's use of his work. In the 528 folios of the *Commentaria . . . super Anatomia Mundini* (1521) Berengario quotes and corrects or agrees with Zerbi (for he is not always in opposition to his predecessor) on a total of 123 folios, often several times on the same folio which actually increases the number of references. This is far more than Berengario quotes any contemporary anatomist; for example, he refers to Achillini only three times (fols. 348v, 496v, 517v). Many of his citations of Zerbi are used to corroborate his own views, and, although Berengario is chiefly hostile to Zerbi, he once calls him "noster" (fol. 391v), a great concession, and at fol. 301r speaks of the Zerbists (*Zerbistae*), evidently the followers of Zerbi. This implies students and other teachers and is a term of respect.

5. Berengario da Carpi, *Commentaria . . . super Anatomia Mundini . . .* (Bologna, 1521), is a significant work all too little known because of its unwieldy size. Thorndike said of it: "Carpi as the last commentator on Mundinus cited so extensively and exhaustively on all mooted points from the previous medieval medical literature that he left little to add in this respect and so may be said to have cleared the way for Vesalius and subsequent sixteenth century anatomists" (Thorndike, 1923–1941: **5**: p. 503). This is an eminently accurate statement and should long ago have led medical historians toward a closer examination of the *Commentaria* and its significance in the history of anatomy. It has, in fact, been suggested that the *Commentaria* was rarely consulted in the sixteenth century, in contrast to Berengario's shorter work, *Isagogae Breves* (1522), and thus had less influence on the advance of medicine than the contents deserved. If this statement is true we have an analogy here with many examples of the ancient handbook tradition: the larger work neglected for the simpler condensation or abridgment. This is not quite exact, of course; the *Isagogae Breves* is a combination dissecting manual and descriptive anatomy and has no such relation with the *Commentaria* as Vesalius's *Epitome* has to his *Fabrica*.

What is immediately apparent in the *Commentaria* is the rigorous criticism to which each statement quoted from the numerous authorities is subjected, in the light of actual observation. Berengario insists upon the precedence of sense perception over authority in the following passages:

> et cum isto textu [of Galen] concordat sensus: sic ergo sint cauti componentes libros de anatomia et non credant auctoritatibus sed sensui sicut nos facimus et faciemus (fol. 153v). (And let sense perception agree with this text and thus let those who write books on anatomy also not trust in the authorities but in their sense perception as I do and shall do.) G. cum suis sequacibus cuius opinionem semper tenemus nisi ubi discordat ab ipso sensus (fol. 412v). (Galen with his followers whose opinion I always maintain except when sense-perception disagrees with him.) Multi tamen aliter sentiunt, sed sensus in hoc est iudex (fol. 443r). (Many, however, believe otherwise, but sense perception is the judge in this matter.)

Berengario also frequently records his personal observations in anatomy both at Carpi and elsewhere in Italy in the attempt to arrive at the truth.

I have written at some length on the nature of the *Isagogae Breves* in my translation of it (University of Chicago Press, 1959; re-issued by Kraus Reprint Co., 1969); see the most thorough review of it by the late C. D. O'Malley in *Isis* **60** (1960): pp. 600–602. As to the illustrations of the various editions, I hope some day to publish an iconography of Berengario da Carpi, assembling the entire series of illustrations from all his works. When I stated on page 15 of my translation that Berengario quotes none of his contemporaries I did not imply, as O'Malley inferred, that the *Commentaria* was included in my remark; what I said was restricted to the *Isagogae Breves*. The latter continues the emphasis on observation and first-hand experience discovered in the *Commentaria* and is the best *brief* account of the human body between Mundinus (1316) and Vesalius (1543).

The Later Pre-Vesalians

1. Antonio Benivieni, *De Abditis Nonnullis ac Mirandis Morborum et Sanationum Causis* (Florence, 1507), which Streeter called "the earliest book on pathology", was written before the author's death in 1502 but published posthumously. Detailed comment on it is omitted here for the reasons given earlier. It is a series of autopsies which show considerable skill in dissection.

2. Andrés de Laguna, *Anatomica Methodus, seu De Sectione Humani Corporis Contemplatio* (Paris, 1535), is an example of the alternating reaction and progress which characterizes pre-Vesalian anatomy from 1490 onward. It is a brief handbook written in a fervent style full of conviction. Laguna deplores vehemently the state of medical knowledge in his day when it has fallen into "the hands of cobblers, weavers, and at last public wine sellers . . . turned over to the tender mercies of the lowest of muleteers." The *Anatomica Methodus,* written within three months, is somewhat ostentatiously described by its author as containing "nothing . . . which I could not also prove by the authority of Hippocrates, Galen, Cornelius [Celsus], Plato, Aristotle, Pliny, or, finally, Aphrodisiensis [Alexander of Aphrodisia]." His method is that of nature; he describes the human body beginning with the mouth, where food is taken in, and ending with "the brain, from which the spirits themselves receive their ultimate and most complete transmutation." He lashes out in his violent manner at Alessandro Benedetti for an error concerning the gall bladder (fol. 12v) and for his lack of method (fol. 27v) but does not criticize older authorities. His text is discursive and full of interesting although not specifically necessary information drawn chiefly from the Greek learning, of which he appears to be quite vain, carefully quoting Greek texts and then translating them into Latin. On the whole, however, his authorities are few and his contribution to anatomical progress negligible. Laguna's book is chiefly interesting for its reaction in favor of authority and its strong personal tone. Although a humanist he is not interested in terminology, even Greek terminology; he produces nothing new or noteworthy in either description or method and is perhaps to be classified with Benedetti, the sole contemporary anatomist to whom he refers, as derivative and conservative. He presents few personal observations, autopsies, or anomalous cases and seems to choose anatomy largely as a means of displaying his peculiarly Spanish-Latin rhetoric. As a stylist he is clearly the most interesting of the pre-Vesalian anatomists; but scientifically his book represents a backwash in the stream of tradition which was moving toward Vesalius. It contains one poor borrowed illustration of the muscles which might as well have been omitted (translated herewith).

3. Niccolò Massa, *Liber Introductorius Anatomiae siue Dissectionis Corporis Humani,* etc. (Venice, 1536). My translation of this book contains an introduction on Massa's life and works which includes the transcript of his last will and testament. Although the title-page bears the date 1559 both preface and colophon are dated 1536; hence it is clear that the date 1559 indicates a mere re-issue of the original work long after it was printed in 1536 by another bookseller. Massa is an indefatigably prolix writer whose garrulity is matched only by his considerable egotism. Like Benedetti, Massa addresses a number of his friends, whom he names, and seems most comfortable before an admiring audience. His impatience and contempt for amateur anatomists is almost as violent as de Laguna's. While Massa pretends too much about his "discoveries," such as the fleshy panniculus, which he admits was first described by Avicenna, or the two muscles in the anterior part of the throat which "not everyone has recognized," there is no need to take seriously any such claims. None of the pre-Vesalians except Achillini made any new discoveries in their general anatomical dissection. The book itself in spite of its verbosity is a remarkably clear account of the human body by a skilled dissector who was proud of his ability. The emphasis on personal observation and freedom from authority is stronger here than in other pre-Vesalian anatomies except those of Berengario. Massa follows Mundinus's plan, which in fact was always an excellent one that makes de Laguna's rejection of it seem ridiculous. Massa is careful to avoid any preoccupation with terminology and finds his precedent for this in Plato. In spite of his pomposity Massa displays common sense and a vigorous mind that thrusts aside all sham and ignorance: his attitude toward slavish adherence to authority is expressed in the following sentences:

> Let it not be ascribed to my arrogance if I point out parts of the body not seen by Aristotle and Galen and many others, since it is given to each person to seek better things to the extent of his ability. For questions and difficulties not solved by Averroes as well as other wise men have made me a sedulous investigator of the parts of the human body and of other things.

Massa is closer to Berengario in his point of view than any other pre-Vesalian anatomist and doubtless echoes the latter's forthright convictions to some extent, although he is careful never to quote or refer to any contemporary anatomist of any standing, anticipating Vesalius in this respect. Like Berengario and Vesalius, he has abandoned the old narrow practice of assembling authorities and choosing alternatives among conflicting opinions. His is one of the more modern sixteenth-century anatomies (translated herewith).

4. Johannes Guinterius, Andernacus (1487–1574 A.D.); *Anatomicarum Institutionum, ex Galeni sententia, Libri IIII per Ioannem Guinterium Andernacum Medicum. His accesserunt Theophili Protospatarii, de Corporis humani fabrica, Libri V. Item Hippocratis Coi de Medicamentis purgatoriis, libellus nunquam ante nostra tempora in lucem editus. Iunio Paulo Crasso Patauino interprete* (Apud Seb. Gryphium Lugduni, 1541). The first edition of this book was published at Paris and Basel in 1536; the second edition, revised by Andreas Vesalius, at Venice in 1538, and the third edition at Basel in 1539; the 1541 edition is a reprint of the 1539 form of the book. In Guinter's dedicatory letter to Jacob Ebulinus, physician to the bishop of Cologne, he outlines the contents and purposes of his volume. In the usual manner of the anatomists of his time Guinter was urged to publish his commentary on the body by his friends and students when he taught at Paris,

quod scilicet nullus recentior Anatomiae scriptor extaret, quem commode et tuto possis imitari. (because of course there is no recent writer on anatomy whom you may comfortably and safely imitate.)

This statement is, of course, sweepingly inaccurate in view of the existence at the time he wrote of several usable and useful introductions to anatomy by Italian writers; but Guinter may not have found copies of them available for his students at Paris and thus heavily underlines the lack of any good anatomical manual written by north Europeans. Guinter more vehemently than any other anatomist except de Laguna deplores the low state of anatomical knowledge in his time, when false "Rabini medicorum," who pretended to imitate Galen, Hippocrates, and Avicenna, set themselves up as physicians but were nothing but quacks and impostors, "rather delighting the palates of men [with their syrups and worthless potions] than healing their ills." Private citizens (*idiotae*) and even women practiced medicine in France and Germany, because of the lack of regulation and supervision by the proper authorities.

Guinter repeated, and added to, the familiar allusions to the great leaders of the past who had interested themselves in dissection and medicine: Alexander the Great, Mark Antony, the Egyptian kings, the Roman consuls Boethus and Paulus Sergius, who had attended the dissections and lectures of Galen at Rome. Galen had, however, written too diffusely in his works *On Anatomical Procedures* and *On the Use of the Parts*. Hence, Guinter adjusted his own book, a condensation in large measure of these two books of Galen, "ad minorum ingenia cum doctrinae compendio," "to the talents and abilities of lesser and younger students by means of a compendium or abridgment" of Galen's teachings.

In a straightforward and simple Latin style which is among the best of the pre-Vesalian anatomists Guinter then proceeded to discuss on the basis of Galen—"almost word for word" ("pene ad verbum descripta," p. 7)—and in four books the location, number, property of substance, size, shape, and connection of the parts of the body, their use and procedures (*administrationes*), and his method of dissection (Book I). In Book II he described the thorax and its inner and outer divisions; in Book III, the nature and processes or functions of the brain; and in Book IV the joints, bones, arteries, veins, and extremities. He mentions two assistants, Michael Villanova, who helped him in his dissections, and, far better known, the great Andreas Vesalius himself, who helped edit the Venetian edition of 1538: "primum, Andream Vesalium iuuenem mehercules in Anatomis diligentissimum." Vesalius is referred to once more in the book at p. 36 (the 1541 edition has 124 pages and no illustrations):

Hae venae seminariae deorsum membranulis dorso annexae feruntur, quibus iam in ilium regionibus arterias coniunctas reperies, longe aliter orientes, quam uenae a cava: quod antea a nullis Anatomicis scriptum reperi, nec animadversum opinor. Nuper autem opera Andreae VVesalii Imperatoris myropolae filii, mehercules iuuenis magnae expectationis, ac praeter singularem medicinae cognitionem, in utraque lingua etiam eruditi, in corporibusque dissecandis dexterrimi, post longam partium disquisitionem inuenimus. (These seminal veins are carried downward by little membranes attached to the back to which you will find arteries joined in the region of the ilium, arising in a far different manner from the veins that issue from the vena cava, a fact which I find has not been written down by any anatomist nor, I think, has it been noticed. But recently, through the efforts of Andreas Vesalius, son of the emperor's apothecary, a young man, by Hercules, whom I consider of great promise, gifted not only with an outstanding knowledge of medicine but erudite in both languages [Greek and Latin] and very skillful in dissection, we made this discovery after a long search.)

Guinter, with Vesalius's assistance, thus found the seminal ducts connected with small membranes to the back and joined with arteries in the iliac region which arose far otherwise than the veins arising from the vena cava. These ducts were hard to find since they were bloodless and white like the surrounding parts.

Guinter uses marginal rubrics which sometimes include a few Arabic and Greek terms. His book, which he frequently declares to be a summary of Galen (*e.g.*, p. 102, at the opening of Book IV: "Si Galeni, cuius doctrinam in hisce institutionibus *ex professo imitor* . . ." [italics mine]), represents no advance in pre-Vesalian anatomy and is typical of the state of anatomical learning at Paris and hence generally in northern Europe, not to mention England, where the feeble booklet of David Edwardes (1532) reveals how little progress had been made in anatomical studies in that country.

5. Johannes Dryander, *Anatomia Capitis Humani, in Marpurgensi Academia superiori anno, publice exhibita per Iohannem Dryandrum medicum.* Omnia

recens nata, Marpurgi ex officina Eucharum Cernicornum Agrippinatis anno 1536. Dryander (1500–1560 A.D.) also published an *Anatomiae, hoc est corporis humani dissectionis pars prior,* etc. (Marburg, 1537) which Singer in his *The Evolution of Anatomy* (re-issued as *A Short History of Anatomy from the Greeks to Harvey*; Dover Publications, Inc., New York, 1957) does not mention, although he reprints without comment Dryander's illustration of the heart (p. 100). Dryander's third and last work is an edition of Mundinus (Marburg, 1541). The first of these writings, his anatomy of the human head, is perhaps the most useful of the three. Its details are clear and, with Canano's book on the muscles, it forms one of the very first specialized anatomies devoted to single parts of the body to appear in the pre-Vesalian period. Had more anatomists followed this lead, that is, specialization in the study of restricted areas of the body, it is possible that more progress could have been made in anatomy before Vesalius; but almost all these scientists preferred to write general works which could not, in the nature of the case, become anything more than a base upon which Vesalius was to create the largest and most detailed general anatomy of the sixteenth century. At the end of his description of the human head Dryander refers to his somewhat larger general description of the body which was to appear in the following year, 1537. The illustrations of the head are fairly rudimentary. In one passage (letterpress to figure 6) Dryander quotes a description of the buttocks (*anchae*) of the brain from Berengario da Carpi, *Isagogae Breves* (see my translation, p. 143). Dryander's small work on the anatomy of the embryo continued a specialty begun by Gabriele Zerbi as early as 1502, in a brief work attached to his large work on anatomy (translated herewith).

6. Giovanni Battista Canano (1515–1579): *Musculorum Humani Corporis Picturata Dissectio per Ioannem Baptistam Cananum Ferrariensem Medicum, in Bartholomei Nigrisoli Ferrariensis Patritii Gratiam, Nunc Primum in Lucem Edita.* No date or publisher's name appears in this little book, but it was probably published at Ferrara in 1541 or thereabouts. It is dedicated to Bartholomew Nigrisoli, a friend. The twenty-seven copper-plate illustrations by Girolamo of Carpi showing the muscles of the body are printed on paper which has allowed both printing and figures to show through to the other side, thus making it quite difficult to obtain clear reproductions or a good facsimile; the annotated one by Harvey Cushing and E. C. Streeter (*Monumenta Medica* **4**, ed. by H. E. Sigerist; Florence, R. Lier and Co., 1925) is not satisfactory; the facsimile edition (with a faulty translation) produced by Giulio Muratori (Florence, Sansoni, 1962) is much clearer. Aldo Mieli[9] discusses the possibility of finding the entire manuscript one day of Canano's work on the muscles (the published volume I contains only those of the hands, arms, and shoulders) on the basis of information received from G. Agnelli, librarian of the Biblioteca Comunale of Ferrara, who reports that G. Faustini in his *Biblioteca de' scrittori ferraresi* had said that the entire manuscript had been deposited with a physician named Ignazio Vari, who had given it to Morgagni for correction. The manuscript then passed from Morgagni to a certain Barotti and finally to a Doctor Nicola Zaffarini, where the trail ends. That Canano was discouraged from further publication by the appearance in 1543 of Vesalius's *De Humani Corporis Fabrica* is a mere fancy which has been put forward almost as a fact; but it is not founded on any real evidence.

The book opens with a dedicatory letter whose recipient is named in the book-title, Bartolomeo Nigrisoli, and continues with the standard reasons for publishing pre-Vesalian anatomies: the urgent requests of friends to publish and the dearth of sound knowledge of anatomy, without reference to any of the contemporary works on the subject. A relative, Antonio Mario Canano, also a physician of Ferrara, assisted Canano and demonstrated the muscles as Canano dissected them. He writes to Nigrisoli that he is presenting the muscles of the arms and legs; but only those of the arms are included in this first volume. Probably those of the legs followed in the second volume, the first part of the "remaining books, already under the engraver's press" which were soon to be published, as Canano wrote (*mox edituri*). Canano's work is all the more valuable because of the fact that the muscles were most difficult to distinguish and dissect, especially in a dead body where they assumed a form different from those in a living body. For this reason he discusses vivisection and its impossibility in his own time and the information which can be derived from the dissection of animals. It is encouraging to note that Canano takes issue several times with Galen and uses those of his books which Guinter had summarized: *On Anatomical Procedures* I and *On the Use of the Parts* II. Canano also points out Galen's omissions (translated herewith).

7. Charles Estienne: *De dissectione partium corporis humani libri tres, a Carolo Stephano, doctore Medico, editi. Vna cum figuris, et incisionum declarationibus, a Stephano Riuerio Chirurgo compositis*; Parisiis. Apud Simonem Colinaeum. 1545. This is regarded as the most fully illustrated of the pre-Vesalian anatomies, with detailed systems of nerves, veins, arteries, and a good description of the joints. The canal in the spinal cord is observed by Estienne, and the glands and blood vessels are well described. His female figures are better than the male. Some of the plates are dated 1530, which places this book, although published in 1545, actually in the pre-Vesalian period. There is no doubt that Estienne worked independently of Vesalius and produced a highly artistic work which would clearly have been

[9] *Gli Scienziati Italiani* (Rome, 1921) **1**: p. 291.

an inspiration to the latter if it had been printed earlier than 1543. In the preface addressed to his students Estienne states that his woodcuts were almost entirely completed by 1539 and printed as far as the middle of Book III, when he was accused of plagiarizing Vesalius and sued. He in turn accused the Basle printers of plagiarizing his own work. His anatomy is the first to contain a thorough alphabetical index of the parts of the body. He uses the term *fabrica* (p. 3) which was to form part of Vesalius's famous title and reveals a gift for logical arrangement and a clear style. His preliminary remarks contain many of the familiar ideas which appear in the prefaces of other pre-Vesalian anatomies, but there is an engaging simplicity about his re-statement of them; it is well to remember that Charles Estienne was a Frenchman. His quotations of Galen and Hippocrates tend to be unobtrusive but there is no doubt that his work is thoroughly Galenic in conception and execution. He is well acquainted with a large number of Greek terms and uses them without the self-consciousness of earlier anatomists such as Benedetti or Zerbi. He begins with the bones in Book I and proceeds to cartilages, ligaments, nerves (a very detailed account), membranes, muscles, tendons, flesh, glands, "adenes," veins, arteries, fat and adipose matter, the skin, nails, and hair. Book II begins with a list of the external parts of the body and proceeds to a dissection of the lower venter with all of its parts, followed by that of the upper venter in which the term "valuulae cordis" appears (p. 219). The pictures which show the brain are most vividly posed, the cadaver lying upon tables or chairs in most realistic attitudes. Each book sums up briefly at the beginning what has been described in the previous book or books and no superfluous material or digressions are used. Estienne has abandoned all of the tedious paraphernalia of previous anatomists such as the number, shape, size, "colligantia," or remarks upon illnesses, categories, spirits, humors, and temperaments. Book III opens with the means of extracting a foetus from a dead mother, an early description of Caesarean section (p. 261) (see J. V. Ricci, *The Genealogy of Gynaecology* [Philadelphia, Pa; 1950], pp. 252–253). He goes on to deal with the problem of two such foetuses, one dead, the other alive within the mother's womb. Book III then proceeds to the description of the uterus, of the position of the foetus in it, of the seven-cell theory which Estienne accepts on the word of some French midwives, even going on to say that more than seven are sometimes seen although he himself has seen no more than three (p. 277), "hoc anno in nostra Lutetia" ("this year in our Paris"). He also retails the ancient idea that males are generated on the right, females on the left side of the uterus. The male organs of generation are treated much less thoroughly by Estienne, with half a page devoted to the testes in the general treatment of glands and three pages to seminal ducts, testicles, and penis in Book II (pp. 192–195). In Book III he continues with the eyes (including the first pictures of their muscles apart from Vesalius's work). Then he describes the muscles in general, with a great deal of detail. Next comes the spine, with pictures of its nerves. This description is followed by some observations on the process of dissection, the instruments used, and the theater in which dissection is to be carried on, with some remarks on the bad environment created by spectators. Characteristically, he mentions no contemporary dissectors or books on anatomy, contenting himself with modest references to Galen. He then devotes a few paragraphs in his closing pages to each of the major parts of the body and how they should be dissected, in a sort of extended summary. He ends with the extremities, as does Vesalius in the *Epitome* (1543), but details them more briefly. All told, Estienne's anatomy is the most modern before Vesalius and only the delay which caused it to be published two years after Vesalius's great work can account for its virtual eclipse. The history of culture can provide other examples of such chronological misfortunes.

IV. PRE-VESALIAN ANATOMY IN THE LIGHT OF VESALIUS

"Perhaps no other activity is as likely as anatomy to display so clearly and diagrammatically the rise of science as a self-conscious process." This judgment of the late Charles Singer[10] is perhaps more applicable to the period of pre-Vesalian anatomy (1490–1543) which I have chosen to study than it is to earlier periods in the history of this particular form of science. There is a rivalry and a sense of eager achievement at this time which cannot be matched in the comparatively dull and, from the point of view of the modern scholar, much less attractive years of the fourteenth and fifteenth centuries when anatomy seemed to slumber in the arms of scholasticism.

It is around the year 1500 when anatomy appears to spring once more to life as a form of research and publication, accompanied by a steadily increasing number of dissections, both public and private, until these attain their greatest brilliance under the dramatic guidance of that consummate showman, Vesalius. In 1502 Zerbi published his *Anatomy of the Human Body* and Benedetti his *Anatomice*; Achillini's *Anatomical Notes* must have been written shortly after this year, although it was posthumously published as late as 1520. In another spurt of activity Berengario da Carpi issued his *Commentary on Mundinus* and his *Short Introduction to Anatomy* in 1521 and 1522 respectively. A third phase of publication in pre-Vesalian anatomy begins with de Laguna's *Anatomical Method* in 1535,

[10] Charles Singer, "Brain Dissection Before Vesalius," *Jour. Hist. of Medicine* **11** (1956): p. 261.

Massa's *Introductory Book of Anatomy* (1536), Guinter's *Anatomical Institutions According to Galen* (1536), and Dryander's work on the human head (1537). These books, together with Benivieni's (written before he died in 1502, the banner year of pre-Vesalian anatomy), Canano's study of the muscles (1541), and lastly Charles Estienne's very respectable *Dissection of the Parts of the Human Body* (1545), represent a considerable body of published works and a devotion to science which is truly remarkable.

Why then did these writers and their books make so small an impact, relatively speaking, on the history of anatomy in comparison with the astonishing success of Vesalius and why has their prestige waned in comparison with his continuing and steadily increasing prestige from 1543 to our own day? These are undoubtedly among the most important questions we can ask concerning the subject, and I ask them here for the first time, so far as I know, with the realization that true scholarship is largely a matter of providing the correct answers to the correct questions.

The fundamental problem is, of course, the problem of genius in any age, of its inception, evolution, and achievement. One may say that the pre-Vesalian anatomists simply did not have the genius of Vesalius. Talented men they were: and it is in their talents that we can begin to see, paradoxically, the causes of their failure. The times after 1500 were ripe for the massive break-through toward the *Fabrica* (1543); interest in anatomy was at fever-pitch, but none of the pre-Vesalians, however much he may have furnished some impetus toward that break-through, could effect it. It is idle to speculate as to the possibilities latent in such men as Zerbi and Berengario da Carpi, to my mind the only two among the pre-Vesalians who could even begin to conceive of that encyclopedic approach to anatomy, free from a confrontation of authorities, which is Vesalius's special contribution to the subject.

It is to the practical world of fact and not to speculation that we must look for our answers and especially to the realm of biography. The subtitle of these studies on pre-Vesalian anatomy emphasizes one of the most important sources of our knowledge, in my estimation, and one which as I hope I have shown has not been hitherto sufficiently exploited. It is in the lives as well as works of the pre-Vesalian anatomists that we must first seek for guidelines in our study; and it is there that every fresh detail, each document and letter, each significant biographical fact will serve our purpose.

The pre-Vesalians were busy men, with many interests and duties, as much occupied in their age as the top physicians and medical research workers are today. I have reason to believe that Achillini, for example, one of the first pre-Vesalians and still one of the most interesting, literally worked himself to death with his arduous teaching, writing, and candidate-examining duties at Bologna, to say nothing of his strenuous two-year stay at Padua. Most of these men were professors of medicine or practicing physicians, or both, a combination of activities well known among modern physicians also. Benedetti was a military surgeon, actually a surgeon-general, as well as a greater traveler than all of them and a dweller in foreign parts. Massa was a chief public health inspector at Venice and risked his life in plagues to write in much detail upon their causes and cures. Berengario da Carpi served as a physician to a number of prominent people, journeying frequently to their bedsides in various parts of Italy, as did Zerbi, who was to lose his life in, admittedly, a not unselfish but obviously very dangerous trip to a foreign land in an attempt to save the life of a Turkish pasha. De Laguna was a court physician, like Vesalius; Dryander, Guinter, and Estienne were university teachers. No modern university professor worth his salt, be it noted, attains success without considerable labor in his profession and considerable publication; he will fully understand the busy lives of the pre-Vesalian anatomists.

Part of this absorbing activity arose from interests quite apart from medical science. These men were also philosophers. One of the most significant facts concerning Renaissance medical education is that students took their degrees in both medicine and philosophy, a standard and unchanged practice from the Middle Ages onward. The healing of the human body was from the times of Hippocrates and Galen inseparably associated with the healing of the human mind through philosophy; indeed, the function of astrology also was discussed in its relation to medicine.[11] Several of the pre-Vesalians, in fact, wrote on that subject: Achillini, de Laguna, and Dryander, although astronomy might be a more proper term for the latter's efforts in the field.

Not only were they philosophers: they were sometimes passionate contributors to philosophical literature and disputants in the lists of public argument. Achillini's production as a vigorous Averroist is, in fact, much more important to the modern scholar than his contributions to anatomy. His collected publications in philosophy fill a large and imposing volume, now read, it is true, only by experts in Renaissance thought. His *Anatomical Notes* are a slim and sketchy essay which was never included in his collected works and rarely mentioned in his time.

The Aristotelianism of Zerbi is one of the most remarkable features of his work. I have pointed out for the first time so far as I know the fact that his *Anatomy* (1502) is built entirely in its organization of

[11] Averroes had mentioned the relationship as debatable: "Adhuc inquiritur, si Astronomia [astrology] est necessaria scientiae medicinarum, an non." (*Colliget*, I, 2, Venice, Guinta, 1547, p. 2): "The question is still raised whether astrology is necessary to the science of medicine or not."

material upon Aristotle's division of the human body into anterior, posterior, and lateral parts. This is an organization which is not employed by any other pre-Vesalian anatomist and not consistently even by Aristotle himself, who merely states it in two separate works, *De Incessu Animalium* and *De Caelo,* but does not utilize it in his more important writings on animals and the human body. While Zerbi is the most attached to and the most consistent user of the scholastic methods among the pre-Vesalians, representing in this respect indeed a sort of throw-back to the fifteenth century, he is the first of them to perceive that what anatomy required at least by 1502 was greater, much greater detail; and in his treatise on the human embryo he exhibited a much-needed emphasis on specialized studies of separate features, a direction taken only in part and by few men, chiefly Zerbi, Benivieni, Canano, and Dryander. That their efforts were slight and abortive as far as changing the course of anatomical study is concerned is one of those accidents of history which later scholars are forever discovering. Zerbi's Aristotelianism took shape in his first published book, *Quaestiones Metaphysicae* (1482), which I have dealt with in more detail elsewhere. It is an original work and in no sense a mere commentary upon Aristotle. Zerbi examines the entire subject and quotes many other scholars, Arab, Greek, Latin, and Renaissance. In fact, next to Achillini he showed himself to be the most competent philosopher of the pre-Vesalian anatomists.

Benedetti probably cannot be rightly called a philosopher. Yet his primarily philological orientation is, like de Laguna's, toward Plato. In this sense Benedetti too is philosophically inclined; it is his philosophic pre-occupation which prevents his *Anatomice* from being anything more than a milestone (but important for this reason) in the development of anatomical terminology.[12]

Massa, who sometimes gives the impression that he is trying to ride off in all directions at once, promised to write special treatises on the muscles, the veins, at least a chapter on the ills of the body, wounds in the head (Berengario actually wrote *De Fractura Cranei*), and on anatomical nomenclature. These unrealized projects show at the least a great multiplicity of interests in a man who, true to the customary allegiance to philosophy on the part of Renaissance physicians, also wrote in Italian a treatise on logic which set out to teach in seven books "with admirable brevity and skill every sort of argument both probable and demonstrative. A book truly useful to all students not only of philosophy, medicine, and rhetoric but also of grammar, history and of every kind of learning."

Only Berengario among the more important pre-Vesalians betrayed no interest in philosophy, and in the light of an argument I shall shortly begin, he was wise to abjure it. He alone reveals a single-minded purpose which limits all of his published work to anatomy. Even there his specialized book on skull fractures was written primarily to show up the ignorance of his fellow physicians who were called in before him to attend Lorenzo de'Medici, nephew of Pope Leo X, wounded in the head on March 28, 1517, and hospitalized at Ancona.[13]

The besetting sins of the pre-Vesalians, as the reader may already have perceived, were philosophy and other non-anatomical interests. These can, of course, be considered sins only from the point of view, admittedly highly conjectural, that they might have written more and better anatomy books if they had been content, like the majority of modern scholars, to stick to a specialty. The chief lesson of modern scholarship—that one can become famous only as an expert and very seldom as a wide-ranging catholic mind—began to be taught in the first years of the sixteenth century, but the pre-Vesalians did not learn it. We may add that if they had learned it we would not have Zerbi's excellent book on medical ethics (*De cautela medicorum*) or the first work on gerontology or geriatrics, also his. We would not have Benedetti's book on the war with Charles VIII nor Massa's booklets on the plague and on venesection.

Another serious drawback apparently unnoticed by the pre-Vesalians was their lack of any real interest in illustrations in their books. Achillini, Zerbi, and Massa have none, nor seem to have felt any need for them. Berengario, it is true, presents the first usable illustrations of the sixteenth century, but they are crude and few in comparison with the wealth of pictures in the *Fabrica*. Almost a touchstone of their respective merits as scientists are the comparatively excellent illustrations in Canano's treatise on the muscles and de Laguna's single picture, lifted from the *Conciliator* (1496) of Peter of Abano. It has often been observed how little the world of art, so rich in these years especially in Italy, impinged upon the sphere of anatomy; the situation argues not only a lack of genius but of its prime element, imagination. While the pre-Vesalians had many friends among physicians, philosophers, and politicians they seem to have been

[12] See J. Steudel, "Vesals Reform der anatomischen Nomenklatur," *Zeitschrift der Anatomische Entwicklungsgeschichte* **112** (1943): pp. 675–681. In his simultaneous publication, "Der vorvesalische Beitrag zur anatomischen Nomenklatur," *Sudhoffs Archiv für Geschichte der Medizin und der Naturwissenschaften* **36** (1943): pp. 1–42, Steudel mentions only Berengario as a contributor to the immediately pre-Vesalian anatomical vocabulary and seems to know nothing of Benedetti's deep preoccupation with Greek terminology and Pollux. His students, however, are continuing Steudel's work on anatomical terminology: see their theses listed in Hermann Triepel, *Die anatomischen Namen . . . 27 Aufl., neu bearb. u. ergänzt* . . . von Robert Herrlinger (Munich, Bergmann, 1965).

[13] Berengario da Carpi, *A Short Introduction to Anatomy,* translated by L. R. Lind (University of Chicago Press, 1959), p. 7.

personally acquainted with few if any artists. Berengario and Canano are exceptions, but even they did not fully exploit the aid they could have received from the graphic arts. Nor were they better served by their printers, in view of the wretched typography of most of their books, beyond a doubt among the most poorly produced and printed since the invention of the printing press.

It was otherwise with Vesalius. Guided by the divine prescience and immortal common sense of genius, he did not diffuse his energies or squander his talents. Almost prophetically, his first publication, the *Tabulae Sex*, is as much a work of art as it is of science; at any rate, it pointed the way anatomy had to take—toward accurate detailed pictorial representation of the parts of the body—if it was to become anything more than description. None of his minor works was exhausting or even taxed his powers; they seem almost exercises in text-edition or the specialized essay beside the great monument of the *Fabrica*. Nor did Vesalius exhaust his strength in lecturing or in attendance upon the patient's bedside, although he did both brilliantly when he chose. While the pre-Vesalians busily lectured, traveled, and visited the sick, Vesalius, both before and after the wonderful years at Padua, husbanded his resources only for one supreme effort. With miraculous foresight, also, he contrived to produce his masterpiece while he was still in his twenties as though he knew that he would not live long afterwards. Yet, having produced it, he produced little more. We do not have the slightest new demonstration of his genius in print after 1543 beyond the second edition of the *Fabrica* (1555), as if he, like us, knew that year was an *annus mirabilis,* to occur but once in any life time. Few great men, whether deliberately or unconsciously, have had as much good sense as he and have rested permanently upon their laurels.

While the pre-Vesalians wrote general anatomy on a modest scale together with general medicine and assorted books on philosophy, history, logic, medical ethics, old age, plague, venesection, and one or two other subjects, Vesalius bent to a gigantic book which took five years to write and earned him immortality. While they prescribed and taught he carried on for the first time in anatomical history what may be called true research based upon intensive dissection. It was probably not the mere number of dissections in which he engaged that procured him his mastery but the infinite patience and detail with which he explored the human body and the marvellous powers of observation and memory which enabled him to record his explorations. No book as large and as complicated as the *Fabrica* could have been written without the preparation of huge masses of preliminary notes; but it is a mystery characteristic of Vesalius that not one page of these notes has come down to us to illustrate, however briefly, the manner in which he worked. The notes and drawings of a greater genius, Leonardo da Vinci, have been preserved and published; Vesalius, with a typical sense of drama, apparently destroyed all of his working papers together with proofs of both printed text and illustrations.

One of Vesalius's principal contributions to the further development of sound anatomical knowledge was his fairly consistent and thorough criticism of Galen. The incomplete index of the *Fabrica* lists at least 265 references to Galen, most of them involving corrections. Whether or not Vesalius should be charged with anti-Galenism is not really the question here, although his own age charged him and Sylvius, his teacher, attacked him for it.[14] Vesalius lived long enough to see himself vindicated in Leonhart Fuchs's *Epitome* of the books of Galen and Vesalius (1551), but he was forced to suffer at the hands of the conservative anatomists. In his criticism of Galen, however, he was not always without precedent in addition to his own firm dependence upon personal observation. For instance, O'Malley speaks of Vesalius's "denial of the belief in the liver's multiple, usually five, lobes."[15] There was, in fact, no real agreement on the matter among his immediate predecessors; Achillini insisted there were four lobes while Massa stated that there were not always five lobes since he had often seen only one division and sometimes two divisions in this organ. A faint air of skepticism was already blowing through the field of anatomy before 1543. Again, when Vesalius stated certain general principles, such as that on hardness and sensation in the bones or the earthiness of cartilage, he was not only following Galen but Galen's predecessors, Aristotle or Aristotle's imitator, Pliny. An important aspect of the problem of Vesalius's relation to Galen is the large number of instances in which he accepts Galenic (or Aristotelian or Arabic) doctrine without question.

In another area Vesalius also outstripped his contemporaries and predecessors. This was terminology.[16] He returned to the Greek vocabulary most frequently for his terms, as Benedetti had done before him in the *Anatomice*, which had so delighted the grandfather of Charles V (*Fabrica,* dedicatory letter, page 5, but unnumbered). Maximilian I had been Benedetti's patron as Charles V was Vesalius's dedicatee. In his dedicatory letter to the *Fabrica* Vesalius had complained of the ignorance of language which characterized the dissectors of his time. When he came to Italy, the birthplace of Renaissance science, he probably read the works of his Italian contemporaries

[14] See the interesting study of Vesalius's criticism of Galen in C. D. O'Malley, *Andreas Vesalius of Brussels 1514–1564* (University of California Press, 1964), chap. VIII, pp. 139–186, for more specific instances.

[15] *Ibid.*, p. 172.

[16] See note 12 and O'Malley, *op. cit.*, pp. 118–119, too brief for such an important topic. At p. 146 he identifies the reference to Benedetti's *Anatomice*.

(although he never mentions any anatomist among them by name) and absorbed some of their preoccupation with the lexicographical contributions of Julius Pollux, *Onomasticon,* Book II, in the field of medicine; however, his only references to that Greek author (*Fabrica* 65,66) are corrections of the latter's anatomical descriptions. The index of the *Fabrica* is dotted with 356 Greek terms for parts of the body, accompanied by their Latin equivalents. The Arabic terms are likewise faithfully cataloged and all three—Greek, Latin, and Arabic—are used by Vesalius, who seems to have striven for completeness in terminology.

A final, and most important, quality set off Vesalius from his immediate predecessors. This was courage. Although he never launched a full-scale campaign as such against the authority of the ancients, Hippocrates and Galen, he undermined their prestige sufficiently with his corrections and personal observations to begin the process of toppling them from their privileged position as the arbiters of anatomical accuracy. The attacks of Sylvius against Vesalius[17] were, on the other hand, only part of a consistent defense of Galen by Sylvius's contemporaries. De Laguna in *Anatomica Methodus* (Paris, 1535), the most colorful and picturesque of the pre-Vesalian anatomies I have translated, lashes out with typically Latin vehemence against some unnamed colleague (fol. 25 ff.; I have tentatively identified him with Benedetti) who dared to accuse Galen (without naming him) of denying that the bile had any nutritive effect among the humors other than the blood. The point at issue was a small one and hardly possible of proof either way, given the contemporary state of knowledge about nutrition; but it was enough to send the militant Galenist de Laguna into a rage. Indeed, a book could be written on the defense of Galen by sixteenth-century medical writers alone to illustrate the conservatism which dominated their thinking in terms of the past.

Vesalius is, of course, the most glaring light in which the efforts of the pre-Vesalians can be judged; the verdict is fairly given by this time, and it is unfortunate that they should be forced at this date to stand so damaging a comparison with genius. Vesalius and his fellow scientists in France, Italy, and Germany were not, however, the sole practitioners of the art of anatomy. It is perhaps idle to speculate at any length as to what students, who sat at the feet of these anatomists, might have produced in the field of dissection, for there is little evidence and few names we can cite. Dorothy Schullian[18] has published the eyewitness account of a dissection written by a student named Ippolito of Montereale in 1519, then living in Perugia with the anatomist who performed it, Giovanni Lorenzo of Sassoferrato, of whom little else is known. This brief but enthusiastic account contains nothing that can be called new; Ippolito mentions only the usual features of the cadaver dissected, including the *rete mirabile*. It is, however, a welcome piece of evidence of the eager interest in medicine and anatomy which some teachers could awaken in their pupils.

The most considerable such document is, of course, Ruben Eriksson's masterly edition and translation of Baldasar Heseler's eyewitness report of Andreas Vesalius's first public anatomy at Bologna, 1540.[19] This falls well into the pre-Vesalian period, before Vesalius had published the *Fabrica* and while he was engaged in writing it. Its value is enhanced by the fact that, as Eriksson writes, "although thousands of medical students must have attended Vesalius's lectures and demonstrations in anatomy during his career, no accounts written by actual members of his audiences have hitherto been known." This student account of a memorable experience in the dissecting room of Vesalius is remarkably detailed and useful because of its clear description as well as citation of authorities, for here as elsewhere, especially in the *Fabrica,* Vesalius referred to authorities in the traditional manner observed by all anatomists of his time. His references to Hippocrates, Galen, and Aristotle are normal enough, but it is surprising how often he refers to (if only to correct) Mundinus, who should have been obsolete as an authority at least to Vesalius and whom he cites only once in the *Fabrica,* published three years later than the time of his Bolognese lectures. These lectures constitute, in fact, an admirable introduction to the *Fabrica,* surpassing the *Epitome* itself for this purpose, containing also information on what we may now call physiology as well as anatomy and of a kind not to be found in the *Epitome*. It is strange that Vesalius himself did not realize that such a set of lecture notes as Heseler collected in Bologna would in itself have been a valuable publication, pending the appearance of the *Fabrica*.

In the last analysis, the charge that the pre-Vesalian anatomists failed to anticipate Vesalius and

---

[17] O'Malley, *op. cit.,* pp. 214–219, quoting in translation Vesalius's *Letter On the China Root*; see p. 219, where Vesalius writes of "the Italians who oppose Galen." There is no doubt that he profited from reading the works of Massa, Achillini, and Berengario da Carpi at least (*ibid.,* pp. 121–122), although he does not mention them by name in any of his works. Sylvius, who refers often to Massa's *Liber Introductorius,* pretended to Vesalius that the Italians (whom Vesalius respected: *ibid.,* p. 220) disliked Galen, a charge that Vesalius was quick to refute. Vesalius seems, in spite of his politeness toward his Italian contemporaries, to have regarded them as having small skill or experience in dissection and as usually in opposition to him, an impression not borne out in any of the works of the pre-Vesalians; the sole mention of him (not by name) by Massa conveys respect for the *Fabrica,* although like Sylvius (*ibid.,* p. 219) he had not read it.

[18] "An Anatomical Demonstration by Giovanni Lorenzo of Sassoferrato, 19 November 1519," in *Miscellanea di Scritti di Bibliografia ed Erudizione in Memoria di Luigi Ferrari* (Florence, L. S. Olschki, 1952), pp. 487–494.

[19] *Andreas Vesalius' First Public Anatomy at Bologna 1540. An eye-witness report by Baldasar Heseler, Medicinae Scolaris, together with his notes on Matthaeus Curtius' lectures on Anatomia Mundini,* edited, with an introduction, translation into English and notes by Ruben Eriksson (Uppsala and Stockholm, Almquist and Wiksell, 1959).

his work is to beg the question and to ignore the fact that Vesalius appears all the more brilliant because his older contemporaries were less so. They prepared the way for him as precursors have always done for the great genius who arrives to surpass them. An exceedingly close comparison of the *Fabrica* with the chief works of the pre-Vesalians (including both those translated in the present studies and the untranslated works of Zerbi and Berengario) might produce textual evidence that he had studied, and profited from their writings, since at the moment it remains only a plausible supposition that he did so. I am not certain that such a study would be worth the effort, for the course of Vesalius's investigations took him away from the pre-Vesalians and along other paths. Like other great innovators, he returned to the basic works available to all and extracted new knowledge from them by seeing them in a new light—his own. A sure instinct led him back to Galen, on whom his contemporaries had not improved in any appreciable degree. Starting from the works of the true father of Western medicine—for Hippocrates, or the writers of his corpus, did not produce anything like the bulk or the painstaking detail of Galen's contributions to medicine—Vesalius analyzed the Greco-Roman texts again with minute attention, and where his elder fellow anatomists had found only occasional cause for dissent he discovered that the foundations of anatomy, once believed to be firmly set forever by Galen, were all to be built anew—and by himself.

THE CHIEF PRE-VESALIAN ANATOMICAL WORKS IN CHRONOLOGICAL ORDER

1. Manfredi, Hieronymo, *Anothomia* (1490); Bodleian Library MS Canon. Ital. 237, Western 20287, written in Italian; printed with partial translation by Charles Singer. *Studies in the History and Method of Science* (Oxford, 1917), pp. 79–164.
2. Zerbi, Gabriele, *Anatomiae Corporis Humani et Singulorum Illius Membrorum Liber* (Venice, 1502; sec. ed. 1533).
3. Benedetti, Alessandro, *Anatomice, siue Historia Corporis Humani; eiusdem Collectiones Medicinales seu Aforismi* (Venice, 1502; Paris, 1514; Strassburg, 1528). Translated by L. R. Lind.
4. Benivieni, Antonio, *De Abditis Nonnullis ac Mirandis Morborum et Sanationum Causis* (Florence, 1507), translated by Charles Singer, with a reprint of the Latin text, C. C. Thomas (Springfield, Ill., 1954).
5. Achillini, Alessandro, *Annotationes Anatomicae* (Bologna, 1520). Translated by L. R. Lind.
6. Carpi, Berengario da, *Commentaria . . . super Anatomia Mundini . . .*; (Bologna, 1521).
7. *Idem, Isagogae Breves,* etc. (Bologna, 1522), translated by L. R. Lind, with anatomical notes by Paul G. Roofe (University of Chicago Press, 1959; reissued by Kraus Reprint Co., 1969).
8. Laguna, Andrés de, *Anatomica Methodus, seu De Sectione Humani Corporis Contemplatio* (Paris, 1535). Translated by L. R. Lind.
9. Massa, Nicolò, *Liber Introductorius Anatomiae siue Dissectionis Corporis Humani,* etc. (Venice, 1536 [colophon]; 1559 [title page]). Translated by L. R. Lind.
10. Guinterius, Johannes, *Anatomicarum Institutionum secundum Galeni . . . libros Quattuor* (Paris and Basel, 1536; Venice, 1538).
11. Dryander, Johannes, *Anatomiae . . . pars prior . . . quae ad caput spectant . . .* etc., (Marburg, 1537). Translated by L. R. Lind.
12. Canano, Giovanni Battista, *Musculorum Humani Corporis Picturata Dissectio* (Ferrara, 1541?). Translated by L. R. Lind. Facsimile edition by Harvey Cushing and Edward C. Streeter, *Monumenta Medica* **4**; (Florence, 1925).
13. Estienne, Charles, *De Dissectione Partium Corporis Humani Libri Tres,* etc. (Paris, 1545, but written fifteen years earlier in the pre-Vesalian period).

ANATOMISTS, CHIEFLY BOLOGNESE, OF THE FIFTEENTH AND SIXTEENTH CENTURIES

Achillini, Alessandro (1463–1512)
Albi, Giovanni Andrea (1525–1566)
Aranzio, Giulio Cesare (1530–1589)
Bartoletti, Fabrizio (1576–1630)
Benedetti, Alessandro (1450–1512)
Benivieni, Antonio (1450–1502)
Bianchi, Virgilio de' (1522–1599?)
Boccadiferro, Lodovico (1482–1545)
Bolognini, Angelo (1508–1536)
Canano, Giovanni Battista (1515–1579)
Carpi, Berengario da (1460–1530)
Coiter, Volcher (1534–*ca.* 1576)
Cortesi, Giovanni Battista (1554–1634)
Dryander, Johannes (1500–1560)
Estienne, Charles (1504–1564)
Genoli (Genuli), Chiaro Francesco de' (1498–1525)
Guenther, Johannes, of Andernach (1505–1574)
Laguna, Andrés de (1499–1560)
Leoni, Domenico (*ca.* 1537–1592)
Maggi, Bartolommeo (1477 or 1516–1552)
Manfredi, Hieronymo (1430–1493)
Massa, Niccolò (1499–1569)
Mercuriale, Gerolamo (1530–1606)
Muratori, Francesco (1569–1630)
Rota, Gian Francesco (*ca.* 1524–1558)
Ruini, Carlo (*ca.* 1598)
Sylvius, Jacob (1478–1555)
Tagliacozzi, Gaspare (1545–1599)
Varolio (Varoli), Costanzo (1543–1575)
Vittori da Faenza, Leonello dei (*ca.* 1450–1520)
Zerbi, Gabriele (1445–1505)

Alessandro Achillini

(1463–1512)

LIFE AND WORKS

THE most recent study of Alessandro Achillini, the Italian anatomist and philosopher, treats him as a philosopher and only incidentally as an anatomist.[1] This is not surprising since even in his own day he was best known as a philosopher; from the time of Berengario da Carpi to that of Osler and Singer his standing as an anatomist has not remained unchallenged.[2] Achillini's twin studies and contributions to knowledge have, it is true, produced a certain atmosphere of ambiguity in modern discussions of him; but it is quite possible to write about him without confusing the two subjects which, for several years (1500–1504), he taught conjointly at the University of Bologna with apparent success.

Since no record of his birth or baptism has been turned up at Bologna, where I have consulted the *Registri di battesimi* in the *Archivio del Battistero di San Pietro*, we must fall back upon two secondary sources of information: (1) the horoscope of Achillini, published among other horoscopes of sixteenth-century celebrities, by Luca Gaurico together with a statement by Francesco Giuntini;[3] and (2) the

[1] Bruno Nardi, *Saggi sull' Aristotelismo padovano dal secolo XIV al XVI* (Florence, G. C. Sansoni editore, 1958), VIII. I *Quolibeta de intelligentiis* di Alessandro Achillini, pp. 179–223; IX. Appunti sull' Averroista bolognese Alessandro Achillini, pp. 225–279. Both chapters have been reprinted and appeared first in 1945 and 1954 respectively. They constitute an important contribution both to the biography of Achillini and to the analysis of his thought. Nardi's recent article on Achillini in *Dizionario biografico degli italiani* (Istituto della Enciclopedia Italiana fondata da Giovanni Treccani, Rome, 1960) **1**, pp. 144–145, is also useful, although written largely from the point of view of Achillini's work as a strict Averroistic philosopher. Lynn Thorndike's excellent chapter III: "Achillini: Aristotelian and Anatomist," *A History of Magic and Experimental Science* (Columbia University Press, 1941) **5**: pp. 37–49, is a valuable pioneering work to which I am much indebted.

[2] Berengario da Carpi, *Commentaria . . . super anatomia Mundini*, etc. (Bologna, Hieronymus de Benedictis, 1521), fol. CCCCXCVI *verso*, speaks of Achillini *in sua confusa anatomia*. Berengario patronizes Achillini, calling him *noster* as he points out Achillini's error concerning the ostiola of the heart (fol. CCCXLVIII *verso*). They seem to have dissected together. Osler's judgment is painfully harsh: "To appreciate what he [Vesalius] did, compare the beggarly treatises of his predecessors in the sixteenth century, Achillinus, Berengar, Guinterius, Massa and the others, with the royal volume of the 'Fabrica' " (*Bibliotheca Osleriana*, etc. (Oxford University Press, 1929), p. 567. He is followed by Charles Singer, who is in addition sweepingly inaccurate: "Achillini's work [*Annotationes anatomiae*; the work was never "included in the 1502 edition of De Zerbis' *Liber Anatomiae*," as Singer insists it was] is but a slight advance on that of Gerbi. It is really little else than a notebook for students, and gives the baldest directions for dissection, accompanied by a few comments from Avicenna. [Achillini actually refers to thirteen authors in addition to several works of Galen and of Aristotle.] Achillini occasionally ventures to criticize Mondino, and his work has at least the advantage of brevity . . . Achillini, like Gerbi, was a windy and very 'scholastic' disputator. He was best known as a supporter of the philosophy of Averroes." ("A Study in Early Renaissance Anatomy with a new text: The Anothomia of Hieronymo Manfredi," 1490) in *Studies in the History and Method of Science* (Oxford University Press, 1917), p. 95; the paragraph on Achillini by Singer in his translation of Mundinus, *The Fasciculo di Medicina,* Venice, 1493 etc. (R. Lier, Florence, 1925), p. 40, is transferred *verbatim* from Singer's work on Manfredi above, with the deletion of "like Gerbi." In his treatment of Achillini, Singer also fails to realize that the "one Johannes Philotheus" who wrote the preface to Achillini's book is Achillini's own brother. In his last discussion of Achillini, in an essay on "Beginnings of Academic Practical Anatomy" contained in a reprint of Ludwig Choulant's *History and Bibliography of Anatomic Illustration,* etc., (New York, Hafner Publ. Co., 1962), pp. 21–22, Singer repeats both the error as to the printing of Achillini's *Annotationes anatomiae* in Zerbi's work of 1502 and the long-since exposed misconception of two (or more) anatomical writings by Achillini. He goes on to say with both exaggeration and inaccuracy: "Allessandro [*sic*] Achillini (1463–1512), like Gerbi, taught at both Padua and Bologna. By modern standards his writings are masses of frothy disputation. To glance at them is to raise uneasy reflections that cultured men once treated such windy nonsense with high respect. He left two [!] anatomical works as gaseously unreadable as his philosophical . . . The medical historian will always wish that Achillini had devoted himself exclusively to philosophy, for there are few writers whose works he would less readily peruse." The last sentence is a particularly gratuitous insult which comes at the end of what is essentially a third repetition of what Singer had already written in his study of Manfredi; but by this time he had at least learned who Philotheus was. Nardi's careful essays show Achillini as a very considerable philosopher; and no reputable medical historian except Singer has ever used such contemptuous language in writing of Achillini. Singer's scorn for Achillini's scholastic arguments is, of course, only proof of Singer's lack of acquaintance with medieval philosophy; but it has also obscured his estimation of Achillini's anatomical writings which, brief as they are, represent a respectable understanding of human anatomy as it was known *circa* 1502–1503 (see the three references to autopsies in these years, *Annotationes Anatomiae*, fols. Vv, XIIIIr). Achillini was an associate and sometimes an opponent in philosophical disputation of Renaissance thinkers like Pico della Mirandola, Pomponazzi, and Nifo; he was invited to occupy the chair of natural philosophy at Padua for two years (1506–1507) and was referred to by his colleagues as *citra contentionem philosophorum summus* or *philosophorum iubar*. Nothing is more jealously reserved among academic colleagues than such complete and unreserved admiration as Achillini enjoyed among his intellectual peers, without a single dissenting note. Only unknown inferiors seem to have expressed jealousy of him; see Giovanni Garzoni's letter to Achillini which mentions them only to dismiss them.

[3] *Lucae Gaurici Geophonensis Episcopi Civitatensis Tractatus Astrologicus* . . . Venetiis apud Curtium Troianum Nauò MDLII, fol. 58 v: Alexander Achillinus Bononiensis philosophus. The horoscope is drawn and in the center of it Gaurico adds: Anno 1463 Octobris D. H. M. Then below he gives a short biography of Achillini: "Erat venustus, altae staturae, sed bene proportionatus, letus, iucundus, ridens, affabilis, studuit Parisiis tres annos in philosophiae, evasitque dialeticus, et philosophus eminentissimus, disputator subtilissimus, et quasi semper erithimematice disputabat, edidit multos libros et potissimum de orbibus tractatum egregium. In legendo erat obscurus et perplexus, obiit Bononie, solo natali, ex febre acuta anno humanati verbi 1512 Augusti die 2 hora 19 circiter, suae vero aetatis 49 climacterico." (He was handsome, tall but well proportioned, cheerful, joyous, smiling, affable, he studied at Paris for three years, became distinguished in dialectics and philosophy, most subtle at discussion and almost always argued syllogistically, he published many

genealogy of the Achillini family included in the voluminous genealogical manuscripts of Count Baldassare Antonio Maria de' Carrati in the Archiginnasio Library at Bologna.[4] It does not seem possible at the present time to resolve the contradiction between Giuntini's "October 29" and Carrati's "October 20," and I follow Giuntini as the earlier and probably more correct source.

The first attempt to write a life of Achillini was made by my late friend, Dr. Ladislao Münster, professor of the history of medicine at the University of Ferrara and a practicing physician in Bologna. His article[5] has been of great value to me although he did not pretend to have given a complete picture of Achillini's life and works.[6] Dr. Münster's kindness to me while I worked in the libraries of Bologna for five months in 1962, his gift of reprints of his articles, and his generosity in supplying useful information have placed me under a considerable debt of gratitude. I have used his pioneering article as a basis for my own study and make full acknowledgments to him.

Early biographical accounts of Achillini include, beside Gaurico's brief description of the anatomist, the account by Paolo Giovio in his *Elogia Doctorum Virorum*[7] (my translation):

Alexander Achillinus, of Bologna, an accurate interpreter of Averroes, won renown for his firm and steadfast doctrine when he taught philosophy at Padua, to the point that Pomponazzi, his zealous rival, began by artful intrigue to rob him of his students. For Achillinus because of the complete simplicity of his nature was utterly unskilled in the arts of adulation and double-dealing to such a degree that the witty and impudent students, although they honored him for his learning, regarded him as an

---

books and especially an excellent one *On the Spheres*. In his lecturing he was obscure and difficult to follow; he died at Bologna, his native city, in his forty-ninth climacteric year from an acute fever on August 2, 1512, at 2:19 P.M.) It is, however, from Francesco Giuntini's *Commentaria in Quadripartiti Ptolemaei* (in his *Speculum Astrologiae*, etc., Lyon, 1583), Liber III, cap. xiv, page 558, that the date October 29 for his birth is derived: "Nascitur 1463, die 29 octobris, liv. 17, m. 20, post meridiem." Gauricus taught astronomy at the University of Bologna in 1506–1507 (U. Dallari, *I rotuli dei lettori, legisti e artisti dello Studio bolognese dal 1384 al 1799*, Bologna, Merlani, Vol. I, 1888, p. 195) and there is a treatise on astrology by him in a Berlin MS. 283 r (Lucas Gauricus, *Isagogicus tractatus in totam astrologiam* (exc.); see Valentin Rose's catalog of the Berlin Library, II, 3, pp. 1427–1428: *Verzeichnis der lateinischen Handschriften der kgl. Bibliothek zu Berlin* (Berlin, 1905)).

[4] *Genealogie di famiglie nobili bolognesi*; Bologna, Biblioteca del Archiginnasio, MS. B. 699, tavola 2: Alessandro Achillini, born October 20, 1463, was the son of Claudio, from Barberino in Val d' Elsa; he received the doctorate at Bologna University in 1484, was an Anziano in 1491, and died in 1512. His brother, Giovanni Filoteo, was also a celebrated professor of humane letters and an Anziano in 1537. The Achillini family was descended from Giovanni Acchillini, as the Carrati manuscript spells the name, whose son Claudio was the father of Alessandro, Giovanni Filoteo, and a sister Giovanna. Alessandro seems to have had no issue, if indeed he was ever married.

[5] Ladislao Münster, "Alessandro Achillini, anatomico e filosofo, professore dello studio di Bologna 1463–1512," *Rivista di storia delle scienze mediche e naturali* **24** (1933): pp. 7–22; 54–77; there is a bibliography at pp. 76–77.

[6] *Op. cit.*, p. 74. Herbert S. Matsen, Alessandro Achillini and His Doctrine of "Universals" and "Transcendentals"; Ph.D. diss., Columbia University, 1969, pp. 1–18 has a concise treatment of Achillini's life. He points out that there is no contemporary evidence for the date on which Achillini was awarded his degree in philosophy and medicine—September 7, 1484; the statement appears only in secondary sources such as Alidosi, Mazzetti, and J. B. Cavatius, *Catalogus omnium doctorum Collegiatorum* . . . (Bologna, 1664), p. 22. The first occurrence of Achillini's name in official university documents is in the *Minute dei Rotuli* 1384–1518 for September 30, 1484 (U. Dallari, *I Rotuli*, etc., I, 1888, pp. 124–125). His name also appears in the *Libri Partitorum* which authorized professors' salaries on April 7, 1486 (Vol. X: 1480–1489), carta 170 v and in the *Quartironi degli Stipendi*, or list of taxes to be paid by the professors, on March 29, 1485 (*Riformatori dello Studio*, Quartironi for 1485: *cf.* G. Cencetti, *Gli Archivi dello Studio Bolognese*, Bologna, 1938, pp. 83–84). A further appearance of his name is in the *Puntazioni dei Lettori* on October 15, 1485, in a document which recorded professors' teaching assignments and the number of students who listened to them; points (*puntationes*) were deducted from the professors' account for lack of scholars. Matsen's thesis is devoted to an analysis of Achillini's philosophical works primarily and does not deal except in passing with his work as an anatomist.

[7] *Opera*, Vol. II (Basle, P. Perna, 1578), 72. Nicolaus Papadopoulus, *Historia Gymnasii Patavini* . . . (Padua, 1776, I, 298), paraphrases Giovio's account and refers to the war of the League of Cambrai with Venice as the disturbance which drove Achillini back to Bologna. Papadopoulus's date for Achillini's death (1525) is, of course, an error connected with the death of Pietro Pomponazzi (1462–1525) in that year. There is an excellent brief account of Pomponazzi's position as an opponent of the dominant Averroism of his time by John Herman Randall, Jr., in *The Renaissance Philosophy of Man*, ed. by Ernst Cassirer, Paul Oskar Kristeller, and John Herman Randall, Jr. (University of Chicago Press, 1948), pp. 257–279, preceding a translation of Pomponazzi's famous essay "On the Immortality of the Soul." Randall summarizes Achillini's position as follows (pp. 264–265): "Alexander Achillini, who taught a vigorous and independent Averroism first at Bologna, from 1506 to 1508 at Padua, and from 1508 to his death in 1512 at Bologna, tried to find a place for human individuality within this framework. Man is a true substance, made single and individual by his cogitative soul in conjunction with the possible Intellect. Differences in intellectual power come from the body rather than from the intellect, from the senses and the bodily spirits, from the greater efforts of different cogitative souls. 'And thinking is in our own power, not only because the Intellect is our form, but also because the operation of the senses which thinking follows is in our power' [*De intelligentiis, Quod.* III, *Dubium* ii]. In the general introduction by Kristeller and Randall, Averroism is described as "This version of Aristotle without benefit of clergy is hence known as Latin Averroism. It had accompanied the introduction of Aristotle at Paris in the thirteenth century; when its spokesman, Siger de Brabant, was condemned in 1270 and 1277 and refuted by the more accommodating modernism of Thomas, it took refuge in the Italian medical schools." On the relations of science and philosophy at Padua in the sixteenth century see the stimulating analysis by Randall in *The School of Padua and the Emergence of Modern Science*; Editrice Antenore, Padua, 1961, which includes his earlier essay, "The Development of Scientific Method in the School of Padua," *Jour. History of Ideas* **1** (1940).

object of ridicule, especially when he walked about with an ungainly shamble, in an old-fashioned robe of scarlet, tight-sleeved and fringed with otter skin, without any folds billowing out behind him. His cheerful face, his slow, uncertain speech, betrayed the limitations of an unsophisticated and reflective spirit; but when it came to disputation among a group of professionals he used to gain the upper hand both by frequent sallies of wit and by his own unrivaled powers of mind. He had, indeed, some time before he came to Padua, published various books of Aristotelian doctrine, "On the Elements, On Intelligences, and On the Celestial Spheres," following the views of Averroes, in which the splendor of his entire talents shown forth. When the University of Padua was closed amid the alarms of a war that had broken out he returned to Bologna and departed this life, before he had quite completed his fiftieth year. His tomb can be seen in St. Martin's church with this poem by Janus Vitalis upon it:

Stranger, who seek Achillinus in this tomb,
You will not find him; joined to his Aristotle,
He dwells in Elysium. The natural causes which
He scarcely could find on earth he looks upon
Now with wide eyes. As in the blessed fields
That noble shade goes wandering, bid him there
A long and a perpetual farewell.

According to Gaurico's description (see Latin text in footnote 3) Achillini was

handsome, tall but well proportioned, cheerful, happy, smiling, affable, he studied at Paris[8] three years in philosophy, became a dialectician and a very eminent philosopher, a most subtle disputant, almost always arguing by syllogisms [reading *enthimematice* for *erithimematice* of the text], published many books and especially an excellent treatise on the celestial spheres. In his lecturing he was unclear and involved, he died in his native Bologna of an acute fever in the year of the incarnate word 1512 on the second of August around 2:19 in the climacteric year of his age forty-nine.

Nothing is known of Achillini's life at Bologna until 1484, when he received the doctorate in medicine and philosophy,[9] except a statement by Michele Medici repeated by Ladislao Münster that Floriano Cerioli, canon of San Pietro in Bologna, was his teacher,[10] the source of which I have not been able to trace. Achillini immediately began to teach at the University of Bologna,[11] continuing except for the years 1506–1508 when he was at Padua, although even then his name was carried on the books of the University in his absence. From 1484 to 1487 he taught logic in the morning, the first subject taught by academic fledglings; he was twenty-one when he began. From 1487 to 1490 he taught the extraordinary course (or held second place) in philosophy in the afternoon; from 1490 to 1494 the ordinary course (or first place) in philosophy in the afternoon. In 1495 he taught the ordinary morning course in medicine for the first time, continuing it until 1497 when he returned to the ordinary course in philosophy in the afternoon until 1500. In that year and 1501 he began the more arduous program of teaching both ordinary medicine in the morning and ordinary philosophy in the afternoon, continuing thus for the rest of his life down to 1512, with the single exception of 1507–1508 when he taught only ordinary medicine in the morning. No one in the records of the teachers at Bologna in medicine or philosophy seems to have taught as long as Achillini, a total of twenty-eight years including the two he taught at Padua, not even veterans such as Berengario da Carpi (1502–1527 in surgery) or Chiaro Francesco de' Genoli (1502–1525 in philosophy and medicine).[12]

[8] Ernest Wickersheimer in his brief but useful article on Achillini in *Dictionnaire biographique des médecins en france au moyen âge* **1** (Paris, Droz, 1936): p. 18, has not found his name listed in the faculty of medicine at Paris.

[9] "1484 * D. Alexr. Achillini Bonon. in U. C. 7 Septemb." This entry means that on September 7, 1484, Alexander Achillini of Bologna received the doctorate in both colleges (*in Utraque Censura*) of medicine and philosophy. It appears in a manuscript marked *Busta* (or envelope) 235, Archivio di Stato, Bologna (F. 3. 50), containing the *Nomina Doctorum Omnium, incipiendo ab anno 1480 usque ad 1800 laureatorum in medicina et philosophia*. This manuscript has been edited together with information from the *Libri segreti del collegio delle arti e di medicina* 217 (8), from A.D. 1481 to 1500, and 217 (10), A.D. 1504–1575, and the *Acta* (*Libri actorum utriusque collegii*, 1481–1604, 4 v., MSS 191–194, Archivio di Stato, Bologna) by Prof. Giovanni Bronzino, under the title *Notitia Doctorum-Universitatis Bononiensis Monumenta* (Soc. Tip. "Multa Paucis," Via G. Gozzi 29, Varese, Italy, 1964); I read part of the page proofs at Bologna and express my thanks to Professor Bronzino for the privilege. An earlier partial edition of the material Professor Bronzino has published is the *Catalogus Omnium Doctorum Collegiatorum in artibus liberalibus et in facultate medica incipiendo ab Anno Domini 1156*, etc.; Dicatus . . . Co. Federico Calderino . . .; Bononiae, Typis Iacobi Montii, 1664, bound with *Illustriss. Gabellae Syndicorum Nomenclatura*, etc.; Bologna, Typis Nicolae Tebaldini, 1641 (*Sindici Gabellae Grossae pro parte Collegii DD.Medicorum ab anno 1508 usque ad Annum 1641* inclusive) in the Biblioteca Gozzadini in the Archiginnasio Library at Bologna. The *Catalogus Omnium Doctorum*, etc., p. 22, records the fact that Achillini was twice an Anziano: "Fuit de numero DD. Ant. de anno 1491 Septembri, et Octobri, et de anno 1504 Novembri, et Decembri." See entry 48 of the list of references to Achillini in the *Libri Segreti* for the term *anziano*.

[10] Michele Medici, *Compendio storico della scuola anatomica di Bologna*, etc. (Bologna, Volpe e Sassi, 1857), p. 46; Münster, *op. cit.*, p. 10.

[11] U. Dallari, *I rotuli,* etc., for the years 1484 to 1512.

[12] An absence of Achillini at Padua in 1495 supposed by G. Facciolati, *Fasti Gymnasii Patavini* **2** (Padua, 1757): pp. 108, 112, is refuted by Münster, *op. cit.*, pp. 12–13 by reference to two important letters praising Achillini and requesting an increase in salary for him. Both letters are dated in 1495. The first, in Latin, is from the Council of Sixteen, which constituted in effect the Senate of Bologna, to Cardinal Ascanio Sforza, legate of Bologna (July 13) and the second, in Italian, to the Archbishop of Pesaro, also from the Council of Sixteen, mentioning the fact that Achillini had added the teaching of medicine in 1495 to that of philosophy "to satisfy the university and the students" and stating that he had taught more than ten years at a salary of 100 lire annually (Archivio di Stato, Bologna, *Litterarum* **5** [1491–1499]: carta 224–225); these manuscripts are listed in a separate hand-written "elenco" to be secured from the attendant and do not appear in G. Cencetti,

Achillini's labors included frequent attendance as a member of doctoral committees for the examination and approval of candidates. His name appears ninety times in the records of these proceedings, the famous *Libri Segreti delle Arti e di Medicina*.[13] These are the minutes of the doctoral examinations, or other business of the Faculty of Medicine such as the election of new members to the Company of Collegiate Doctors (*aggregatio*); the process is described by Gnudi and Webster in a book which contains much information about Bologna and its medical school, although it deals with a period long after Achillini's time.[14] Since the *Libri Segreti* have never been ransacked for biographical information on Achillini or other pre-Vesalian anatomists at Bologna, I have gathered for the first time a complete list of all references to him, with dates and other pertinent information, such as the names of colleagues with whom his name is joined.

The process of approving candidates for the doctorate was the following: John Doe was first *aprobatus*; his sponsors were then named. They were called *promotores*, usually three in number, members of the larger doctoral committee but probably supervisors of the candidate's studies and thus in a position to judge him ready for examination. Finally, after the committee vote by white and black beans, one of the *promotores*, possibly the teacher who was best acquainted with the candidate and his favorite professor, *dedit insignia*, or bestowed upon the candidate the tokens and habiliments of the degree granted. These were a cap or crown, an open and a closed book, and a gold ring (see Gnudi-Webster, *op. cit.*, p. 58 for their symbolism). The ninety references to Achillini follow in order: fol. 1, MS 218 (8) contains data from November 20, 1481 to March 21, 1484; fols. 2–18 are blank and must have contained the record of Achillini's own doctoral examination, known to us therefore only from the *Nomina Doctorum Omnium*; see footnote 9.

1. Fol. 19 r, April 3, 1487: a meeting of the college of medicine for an *aprobatio*.

2. Fol. 21 r, August 20, 1487: joined with Hieronymo Manfredi in an *aprobatio*.

3. Fol. 24 v, October 25, 1487: joined again with Manfredi, Egidio Antaldi, Giovanni Garzoni, Nestore Morandi, Lodovico Leoni, and Lorenzo Gozzadini in a meeting of the College of Medicine under the priorate of Andrea Crescimbene for an aggregation ceremony.

4. Fols. 26 v-35 r are blank until December, 1493, a gap of six years; the gap is partly filled with events recorded on folios 91 r-100 v, at the end of the manuscript. These are all in the same handwriting and are apparently a replacement for folios 26 v-35 r; the years included are 1492–1494 but not in sequence, shifting thus: 1492, 1494, 1495, 1488, 1490, 1491. In order not to disturb the sequence of folios I have not made any change in my recording of the references to Achillini; the dates of the 39th to the 46th reference range thus out of chronological order from 1500 to the disordered sequence of years given above: Fol. 35 v, Feb. 1, 1494: Achillini gave the insignia to Mario da Cesena.

5. Fol. 37 r, January 5, 1495: Achillini was prior of the college of medicine for the first time, with two *consiliarii*, one of them the previous prior, Andrea Crescimbene. Achillini was to hold the priorate at least six times during his lifetime. Sometimes a man was prior of both colleges of medicine and philosophy, as Achillini was in 1509; usually he was prior of only one college. The first priorate of Achillini seems to fall before January 22, 1490, since his name appears in a list of priors of the college of medicine in MS *Libri Actorum utriusque collegii*, 191–194, fol. 4, on that date; see fol. 7 also.

6. Fol. 39 r, April 1, 1495: *Alexander de Achillinis in prioratu* is crossed out because the scribe had forgotten that Lodovico Leoni had become prior for April, May, and June.

7. *Ibid.*, June 11: an *aprobatio* of Angelo de Allia and Dominico de Maecha.

8. Fol. 41 r, February 4, 1496: "aprobatus fuit in medicina (*in marg.*) magister Tiberius de Bazalerius nemine discrepante. Promotores eius fuerunt M. Johannes de Garzonibus M. Nicolaus de Saniis et M. Alexander de Achillinis et dictus magister Alexander dedit insignia." On the fortunes and misfortunes of this Tiberio Bacilieri, a student of Achillini's, see Nardi, *Saggi* etc., pp. 226–227, and *passim*. His *approbatio* in *artibus* is recorded at fol. 91 r, July 3, 1492. He was later (fol. 54 r, December 9, 1499) admitted to the college of arts and medicine as a supernumerary, with the right to vote but not discuss in the election of doctors. Because of misdeeds and insolent speech toward his colleagues he was suspended, with Achillini voting against him (fol. 57 r, July 9, 1500) for five years and was

Gli Archivi dello Studio Bolognese; Nicola Zanichelli Editore, Bologna, 1938, the printed guide to these archives. See also Bruno Nardi, *Saggi* etc., p. 226, who denies the Paduan absence without reference to these letters or to Münster's publication of them.

[13] Archivio di Stato, Bologna MS 217 (8), A.D. 1481–1500; 217 (10), A.D. 1504–1575; I have used photostats of these manuscripts down to Achillini's death in 1512; they are parchment and occasionally palimpsests; an earlier copy, a paper manuscript, is very difficult to read.

[14] Martha Teach Gnudi and Jerome Pierce Webster, *The Life and Times of Gaspare Tagliacozzi, Surgeon of Bologna, 1545–1599* (Milan, Hoepli; New York, H. Reichner, 1950); see especially chapters I, III, V: the *aggregatio* is discussed at pp. 83–84. The awarding of doctoral degrees in medicine in the Middle Ages is well described by Ruben Eriksson, *Andreas Vesalius' First Public Anatomy at Bologna*, etc. (Uppsala and Stockholm, 1959), p. 21.

reinstated November 8, 1505 (*Libri segreti* 217 (10), fol. 4 r) *sine partito sed viva voce tantum*, that is, without a vote but by acclamation only. Bacilieri meanwhile taught Averroist doctrine at Padua and Pavia.

9. Fol. 43 v, Sept. 22, 1497: the *aggregatio* of Paolo Antonio da Faenza *in artibus* with Achillini as a *promotor*.

10. *Ibid.*, October 7, 1497: Achillini served as a *consiliarius* with Nestore Morandi to Andrea Crescimbene, prior.

11. On a folio in a different hand intercalated between folios 43 v and 44 r (with a penciled note in Italian: *allegato a c. 43*), dated February 7, 1498, Achillini served as a *promotor* together with Leonello da Faenza. References 11 to 26 fall entirely during 1498, Achillini's most active year as a doctoral examiner and as a member of the college of medicine.

12. Fol. 44 r, January 8, 1498: Achillini was a *consiliarius*.

13. *Ibid.*, February 7, 1498: Achillini a *promotor* with Leonello da Faenza.

14. *Ibid.*, February 8, 1498: Achillini *dedit insignia* to Julius da Calzina.

15. Fol. 44 v, February 9, 1498: Achillini attended an *aggregatio* meeting for the admission of Hieronymo de Bombaxia and others.

16. Fol. 46 v, April 26, 1498: Achillini a *promotor* of Franciscus de Brissia.

17. *Ibid.*, May 4, 1498: the *aggregatio* of Julius da Calzina.

18. Fol. 48 v, May 14, 1498: Achillini *promotor* for Gregorio de Allia.

19. *Ibid.*, June 28, 1498: Achillini gave the insignia to Chiaro Francesco de Fontana (Genoli), who was to become a prominent teacher of medicine at Bologna.

20. *Ibid.*, July 1, 1498: Achillini prior for the second time.

21. Fol. 49 r, July 13, 1498: Achillini attended the *aggregatio* of Chiaro Francesco de Fontana (Genoli) with Leonello da Faenza.

22. Fol. 50 v, August 18, 1498: the *aggregatio in artibus* of Hieronymo da Imola.

23. *Ibid.*, October 2, 1498: a meeting of the college of medicine called by the prior, Achillini.

24. *Ibid.*, October 14, 1498: Alfonsus Hispanus failed to pay his fees for his bachelor's degree; Achillini makes note of the fact.

25. Fol. 51 r, November 3, 1498: the *aggregatio* of Bonifacius de Pixis *in medicina*; Achillini was in attendance at the ceremony.

26. *Ibid.*, December 19, 1498: Achillini a *consiliarius*, with Nestore Morandi as prior.

27. Fol. 52 v, June 10, 1499: *aprobatio* of Johannes Hispanus *in artibus*; Achillini *dedit insignia*.

28. Fol. 53 r, September 2, 1499: *aprobatio* of Johannes de Allia *in medicina*.

29. Fol. 54 v, February 9, 1500: Achillini prior for the third time.

30. *Ibid.*, February 11, 1500: Achillini gave the insignia to Christophorus Hispanus.

31. Fol. 56 r, May 15, 1500: Achillini served on two doctoral committees on the same day.

32. Fol. 57 r, July 9, 1500: Tiberio Bacilieri, or de Bazaleriis, was suspended from the college of medicine; *cf.* folio 59 v: "propter nonnulla demerita et facinora commissa suspensus fuit."

33. Fol. 58 v, same date as above: the proceedings for the suspension of Bacilieri, in which Achillini voted with the majority.

34. Fol. 59 r, July 14, 1500: the suspension proceedings continued.

35. Fol. 63 r, August 3, 1500: a meeting following the *aprobatio* of Jacobus de Sanisii in the sacristy of St. Peter's Cathedral.

36. Fol. 65 r, August 6, 1500: *aprobatio* of Maria de Boemia *in medicina*.

37. *Ibid.*, August 8, 1500: *aprobatio* of Andrea de Allia *in medicina*.

38. Fol. 65 v, September 1, 1500: a meeting of both colleges.

39. Folios 67 r-90 r are blank; 91 r begins in 1492 with a list of payments, June 26, 1492. At the bottom of the page is a reference to Tiberio Bacilieri, approved in medicine (with *artibus* written above *medicina* crossed out) under date of July 3. In the middle of the page appears the name of Achillini with the date June 26 and another *aprobatio*; he gave the insignia as a *promotor* with Lodovico de Leoni.

40. Fol. 91 v, September 26, 1492: an *aprobatio*, Achillini a *promotor* with Hieronymo Manfredi again.

41. Fol. 93 v, January 30, 1494: Achillini gave the insignia to Christophorus Hispanus.

42. Fol. 94 v, September 22, 1494: another *aprobatio* of a certain Fabiano de Cocagno.

43. Fol. 96 r, June 11, 1495: *aprobatio* of Angelo de Allia.

44. Fol. 97 v, May 17, 1488: *aprobatio* of Leonidas de Mirandola.

45. Fol. 98 v, January 9, 1490: three men approved in medicine, two of them from Poland; Achillini with Manfredi as a *promotor*.

46. Fol. 99 v, May 28, 1490: *aprobatio* of Antonio de Macenovo; Giovanni II Bentivoglio is mentioned at the bottom of the page.

47. Fol. 100 v, September 5, 1491: *aprobatio* of Hieronymo de Aprilis.

48. MS 217 (10), A.D. 1504–1575; fol. 1 r, some time before October, 1504: In the palace of the Anziani (honorary officers who had held power during the era of the Commune of Bologna, having been created in 1376 to govern the city) in a room called La Scaletta, where Achillini, then prior of the college of arts and an Anziano as he had been in 1491 (*Catalogus omnium doctorum*

etc., [Bologna, 1664], p. 22) was residing, Achillini presided over a meeting in which it was decided to set up two books of minutes, one for each college, of arts and medicine, in which to record their activities. These books were to be handed down from prior to prior. The presiding group of eleven men is listed; of them several are noted in the margin as *mortuus* by a hand that must be as late as 1527, when Lodovico Leoni and Galeotto Beccadelli died. Achillini (1512), Giovanni Garzoni, the Bolognese humanist and medical man (1505), Leonello dei Vittori da Faenza (1520), and Chiaro Francesco de Genoli (1524) are also listed as dead, as is Antongaleazzo Bentivoglio, apostolic protonotary (1472–1525), who attended this meeting in his capacity as chancellor of the University of Bologna. In 1504 Pietro Bernardino de Pizzani (died in 1505) was prior of the college of medicine; he uses the words *librum secretum* in recording, doubtless with his own hand, *omnia et quecunque gesta tempore prioratus mei*, as was agreed in the minutes of the meeting.

49. Fol. 2 r, November 6, 1504: Achillini a *compromotor* for Francisco Torres, a Spaniard; i.e., he is presented to the doctoral committee.

50. *Ibid.*, November 8, 1504: successful examination of Torres, attended by Achillini.

51. Fol. 2 r-v, November 12, 1504: Achillini met with the prior Pizzani and a doctor of laws, Jacopo Bonino, concerning the preparation of the secret books of the colleges.

52. Fol. 3 r, December 8, 1504: Achillini a member of the examining committee for Valerio de Monte Lupone. Chiaro Francesco de Genoli, prior for the first trimester of 1505, notes with his own hand the death of Giovanni Garzoni on January 28.

53. Fol. 3 v, June 14, 1505: Lodovico de Vitali *doctoratus fuit . . . insignitur ab excellentissimo patrono et magistro domino nostro Alexandro Achillino promotore ellecto*. At fol. 4 r the *liber secretus* is described as *librum secretum antiquum et corrosum in papiro scriptum* and its new parchment binding is noted. This is the first paper manuscript of the volume containing the proceedings from 1481 to 1500, preserved with the parchment copy in the Archivio di Stato, Bologna. On November 8, 1505, Tiberio Bacilieri was reinstated unanimously in the college of medicine.

54. Fol. 4 v, January 12, 1506: Alfonsus Hispanus was presented to the committee by Achillini, among others.

55. *Ibid.*, January 18, 1506: Lucas Poloniae (of Poland) was presented by Achillini.

56. Fol. 5 r, March 24, 1506: Andrea Mauro Turiensi was presented to the committee by Achillini.

57. *Ibid.*, March 28, 1506: the *puncta* (points for discussion in the examination) were given to the candidate above and, after he had done well in the questioning, Achillini gave him the insignia.

58. *Ibid.*, April 16, 1506: Agostino Palotio was presented to the prior for the examination in arts and examined by Achillini.

59. Fol. 5 v, April 17, 1506: Achillini presented Bartolomeo de Pontemilo for the doctorate in arts.

60. *Ibid.*, April 18, 1506: Achillini presented the *puncta* to Agostino Palotio.

61. *Ibid.*, April 18, 1506: Palotio was examined by Achillini.

62. Fol. 6 r, May 19, 1506: Achillini was absent in Rome to attend a disputation in the chapter of the Brothers Minor (Nardi, *Saggi*, pp 252–253; Münster, *op. cit.*, p. 75). In the words of Girolamo Bombaci, prior of the college of medicine for the second trimester (April, May, June):

Die decima octaua maii apresentatus mihi fuit Magister Gulielmus Spinola de Mutina qui per biennium fuerat rector studii doctorandus in utraque censura pro die sequenti, et quia optime se habuerat in suo officio ideo libenti animo admissi. Eadem vero die apresentatus mihi fuit Magister Guido Pisaurensis doctorandus in artibus qui anno MDV presentatus fuerat per magistrum rectorem praedictum ut apparet supra per notam Magistri Ludovici de Leonibus, et quia hic vir adeo erat doctus quod promeruerat lectura extraordinaria physice, ideo ilari uultu admisi supra. Die decima nona maii assignata fuerunt puncta praefatis viris qui in sua recitatione adeo docte et eleganter se habuerunt quod meruerunt approbari fuerantque unanimi cum sensu approbati et magnificus vicarius eos doctorauit, videlicet magistrum rectorem in utraque censura et magistrum Guidonem in artibus. Quibus doctoratis surexit magnificus rector et . . . [then follows the request for the doctoral insignia and their bestowal by Leonello dei Vittori, in his name and that of Alessandro Achillini who] tunc temporis Romam iuerat ut interesset disputationibus fiendis in capitulo generali fratrum minorum tam obseruantinorum quam conuentualium, gratia sui onoris studiique nostri ac almae ciuitatis bononiae nec non magistri Ludovici de Leonibus, qui etiam ob mortem excellentissimi utriusque iuris doctoris domini Floriani de Dulfis non interfuit ad laudem. Et eadem die habuimus opulentam colationem a doctoratis. (On May 18 Master William Spinola of Modena, who had been for two years Rector of the university, was presented to me for the doctoral examination in both subjects [philosophy and medicine] on the following day, and since he had conducted himself well in his office I accepted him readily. On the same day Master Guido of Pesaro was presented to me for the doctorate in arts; in 1505 he had been presented by the Master Rector above mentioned as appears earlier in the note written by Master Lodovico de Leoni, and because this man had shown himself learned in his extraordinary lectures on the *Physics* [of Aristotle] I admitted him with pleasure. On May 19 the aforesaid students sustained their examination on the points of discussion assigned to them and conducted themselves so learnedly and in such a gentlemanly fashion that they merited approbation and were passed unanimously. The Vicar Magnificent gave them the symbols of the doctorate, that is, to the Rector in both philosophy and medicine and to Master Guido in arts. When they had received the doctorate the Rector Magnificus and . . . who at that time had gone to Rome to attend philosophical discussions held in the Chapter General of the Brothers Minor both of the Observants and the Conventuals, for the sake of his honor and that of our University and the bountiful city of Bologna as well as that of Master Lodovico de Leoni, who also because of the death of the very excellent doctor of both laws [civil and canon] Florian de Dulfis was not present for the praising of the doctoral candidates. And on the same day we received a fine banquet at the expense of the newly created doctors.)

Achillini was prevented thus from attending the doctoral examination for Guglielmo Spinola da Modena, who had already been rector of the Studium for two years, and for Guido da Pesaro by his absence in Rome, where on June 6 he had engaged in a defense of his Averroistic doctrine in the house of Cardinal Domenico Grimani. Twelve years before (June 1, 1494) he had upheld the theses of his *Quolibeta de intelligentiis* in the capitulum generale of the Franciscans at Bologna, under the auspices of none other than Giovanni II Bentivoglio,[15] the ruler of the city.

63. *Ibid.*, June 8, 1506: magister Jacobus Nicolus de Baeza, who paid his fee on the sixth and was presented on the seventh, received the *puncta* and distinguished himself in the examination for the doctorate *in artibus et in sacratissima medicina*, giving an *orationem ornatissimam*.

64. Fol. 6 v, July 23, 1506: Vincenzo of Reggio *doctoratus fuit*; Achillini as a *promotor* "crowned" him with the doctor's cap.

65. Fol. 7 r, August 30, 1506; Alfonsus Hispanus *doctoratus fuit*, Achillini's last service as an examiner before leaving for Padua. Münster, *op. cit.*, note 6, p. 16, is in error.

66. *Ibid.*, November 7, 1506: Chiaro Francesco de Genoli succeeded Achillini (prior for the fifth time), who had gone to Padua to teach:

Ego clarus franciscus de genulis septima nouembris successi *priori medicorum sapientissimo uiro allexandro achillino citra contentionem philosophorum summo* [italics mine] qui ad studium patauinum ut ibi ere publico mereret concesserat. Erat enim tunc temporis uniuersa urbs in sagis ob terrorem summi pontificis qui magnis et gallorum et italorum copiis ad eam approperabat ut urbem suam liberam in liberiorem redigeret, quod sibi successit fuga optimatum bentiuolorum, qui tunc ei preerant, suscepta, in qua quattuor mensibus cum sacratissimorum heroum collegio commoratus est, quo tempore exiguum satis ac exile studium habuimus et scolares namque suspicione belli futuri et caritate annone cuiusuis non frumentarie alio secesserant. (I, Chiaro Francesco de Genoli, on Nov. 7 succeeded to the priorate in the place of the most wise physician, Alessandro Achillini, beyond doubt the chief of philosophers, who had gone to the University of Padua to teach for public hire. The entire city was at that time under arms in fear of the Pope who had approached with many troops both French and Italian in order to render his free city even freer, an opportunity which came to him because of the flight of the noble Bentivoglio family which was then in power. The Pope remained in Bologna four months, staying in the College of the most sacred heroes [of medicine]; at this time we had a quite small attendance at the University because through fear of the war to come and consequent scarcity of provisions the students went elsewhere.)

From this account we learn of the descent of Pope Julius II upon Bologna and the flight of its ruling family, the Bentivoglio. Achillini also fled, to Padua, where by November 10 he was hired to lecture on philosophy as ordinary professor of the second rank,[16] at a salary of 250 florins annually in the place of Antonio Fracanziano, with a subvention of 100 florins. A final entry in the record for September 3, 1508, shows a full salary to be paid him: "quod habeat integrum stipendium pro eius conducta." At Padua, Achillini found his rival, Pietro Pomponazzi, in the first chair of philosophy;[17] the clash of these two minds, one a strict follower of Averroes' interpretation of Aristotle, the other a critic who did not believe that Aristotle was a god or that he knew everything, must have been a stimulating experience for both.

67. Fol. 7 v, the first trimester of 1507: Lodovico Leoni was prior of the college of arts. ". . . absente praefato Alexandro achillino qui patauii ere publico ad philosophiam ordinariam legendum conductus est." During Leoni's regime and that of Jacopo Benacio as prior of the college of medicine the

[15] Nardi, *Saggi*, etc., p. 180, p. 252.

[16] This information appears from the entry in *Archivio antico dell' Università di Padova*, MS 649, Professori, Artisti e Legisti fino al 1509, under "profes. Philosophiae," fol. 272 r: "1506 vii. id. nov.

Alexr. Achillinus ad philos. ord. loci secundi flor. 250 loco Antonii Fracantiani cum subventione flor. 100.

1508 3 septembris quod habeat integrum stipendium pro eius conducta."

Nardi, *Saggi*, p. 259, refers to a letter of the Forty Reformers of the Bolognese Studium (September 11, 1507) published by B. Podestà, "Di alcuni documenti inediti riguardanti Pietro Pomponazzi," *Atti e Memorie della R. Deput. di Storia Patria per le provincie di Romagna* **6** (Bologna, 1868): p. 142, demanding that Achillini return from his unauthorized leave of absence under penalty of 500 gold ducats fine and other very grave penalties. It was only a year later, however, that Achillini returned: Archivio di Stato, Bologna, MS *Lib. Partitorum* 13, fol. 136 v (September 14, 1508), which records his re-engagement as professor of theory of medicine and ordinary in philosophy in the afternoon at a salary of 900 Bolognese lire. The vote of the Forty was a tie, 19 to 19; evidently the presiding officer decided the outcome. *Lib. Partitorum* 14, 47 records full payment of salary assigned to Achillini and 15, 27 the payment of remaining salary owed to his heirs under dates December 24, 1509, and December 11, 1514.

[17] Nardi, *Saggi*, p. 254, who cites Cod. Vat. Regin. Lat. 1279, fol. 3 r, concerning Peretto Mantovano's lectures in 1506–1507 at Padua, and the mention of Achillini's presence "ad concurentiam." See *Saggi*, p. 256, for the reference to Pomponazzi. Julius II entered Bologna November 11, 1506 "tanquam Dominus benemeritissimus"; the colleges of arts and medicine walked out in the Strada Maggiore to meet him, dressed in their collegiate garb, and accompanied "beatitudinem suam" as far as St. Peter's cathedral: *Libro segreto 217* (10), fol. 7 r, November 11, as described by the prior, Hieronymo de Bombaxia.

statutes of the college were revised and reformed; Achillini had been also charged with this task but did not carry it out with Leoni since he had fled to Padua. At fol. 9 v, as Chiaro Francesco de Genoli began his priorate for the trimester July, August, September, 1507, he noted that nothing worth remembering took place in the college because of the plague then raging. The Studium was closed for four months and the students went elsewhere.

68. Fol. 12 r, October 26, 1508: Achillini was once more a *promotor* at Bologna for Pietro Giorgio Bonvignati, having returned from Padua on September 14. Then in rapid succession he served at the following functions:

69. *Ibid.*, November 8: *aprobatio* of Giustiniano Fantini.

70. *Ibid.*, November 10: the doctoral examination of Giustiniano Fantini.

71. *Ibid.*, November 16: the *aggregatio* of Fantini.

72. Fol. 13 r, February 20, 1509: *promouetur in artibus et medicina magister* Alexander de Ripa Florentinus; Leonello dei Vittori was prior in medicine, Thomas Campeggio in arts, and Achillini was a *promotor*.

73. Fol. 13 v, February 20, 1509: Alexander de Ripa was "rigorously examined" and awarded the doctorate. Achillini ("philosophorum iubar") gave the insignia ("doctoralibus infulis insigniuit.")

74. At the bottom of fol. 13 v Achillini assumed the priorate of both colleges for the sixth time:

> Tempore prioratus Alexandri Achillini utriusque collegii prioris aprilis maii iunii" (In the time of the priorate of Alessandro Achillini, prior of both colleges, during April, May, June.) (1509).

Since the custom of writing the minutes had been established as the task of the prior in the meeting at the palace of the Anziani some time before October, 1504 (MS 217 (10), fol. 1 r), the minutes set down here (fol. 13 v) are undoubtedly the autograph of Achillini. Previous priors since 1504 had written the minutes in their own hand ("propria manu"): e. g., Leonello dei Vittori, Lodovico Vitale, Lodovico Rengherio, Jacopo Benaci, Chiaro Francesco de Genoli, Galeotto Becchadelli, Hieronymo de Bombaxia, Federico Gambalunga, Lodovico Leoni, and Benedetto di Campeggio. Achillini's handwriting here is practically identical with that of one of five short manuscripts attributed to him in the University of Bologna Library.[18] Achillini's minutes are only nine lines in length and record nothing unusual in the affairs of the colleges: payments of sums of money from Giovanni Battista Gualtini and Giovanni Andrea Garisendi (whose family name one of the two great towers at the end of the Via Rizzoli in Bologna bears to this day), arrangements for lectures by Antonio Menaliotto de Burgovalle and Giovanni Marcio in Monti Bianchi, between the dates April 19 and July 23. The priorate then passed to Leonello dei Vittori for the last trimester (October, November, December) of the year 1509.

75. *Ibid.*, May 22, 1509: Antonio Menaliotto (see above).

76. *Ibid.*, July 23, 1509: Giovanni Marcio (see above).

77. Fol. 14 r, December 17, 1509: Pietro Piemonte graduated in arts and medicine; Achillini "eum insigniuit."

78. *Ibid.*, February 21, 1510: Achillini "crowned" Bernardino Licinio of Cremona in medicine and arts, in the priorate of Benedetto di Campeggio. Achillini's name also appears as *promotor*.

79. Fol. 14 v, March 19: "Die 19. marcii panfilus de monte Bononiensis mihi presentatus fuit per magistrum Alexandrum de Achelinis et ita ego magister Benedictus de Campegio, prior colegii medicine, feci conuocari colegium ad dandum puncta in mane in die sequenti." This Pamphilo Monti was later the editor of Achillini's collected works (1545).

80. *Ibid.*, March 20, 1510: *aggregatio* of Pamphilo Monti on the same day he was awarded the doctorate.

81. Fol. 15 r, May 29, 1510: Achillini presented a certain Paulus for the doctorate and on the next day served as his *promotor*.

82. Fol. 16 r, July 16, 1510: a "magister-idaspus hispanus" was promoted *in artibus* by Achillini. On September 22 Pope Julius II entered Bologna, as he had done years before, with twenty-seven cardinals and an army to drive out Giovanni Bentivoglio II and his sons, according to the account by Chiaro Francesco de Genoli (fol. 7 r). This time he came to lay Duke Alfonso of Ferrara (who had been aided in his rebellion by King Louis of France) under an interdict and to drive him from his seat of power. The collegiate doctors again met Pope Julius at the city gates, dressed in their doctoral robes and accompanied by the doctors of civil and canon law.

83. Fol. 16 v, November 12, 1510: Achillini was *promotor* for George Schibtel of Hombach, Germany.

84. Fol. 17 r, January 16, 1511: Matteo dei Monti received the doctorate in arts and medicine with Achillini his *promotor*. The Pope with his army was ravaging almost all the territory of Ferrara, including the town of Mirandola.

85. *Ibid.*, April 12, 1511: Pamphilo Monti was elected to the college of medicine; Achillini gave him the insignia: "Et infulis doctoris a philosophorum monarcha d. m. Alexandro Achillinis insignitus."

86. Fol. 17 v, April 26, 1511: two Spaniards, Gundisalinus and Garsias, were unanimously awarded doctorates in arts and sacred medicine: "et a

[18] Expositio Alexandri Achilini super prima 4 i Avicene pulchra (September 7, 1509), listed in Lodovico Frati, "Indice dei codici latini conservati nella R. Biblioteca Universitaria di Bologna," in *Studi italiani di filologia clasica* 16–17 (1909), 110: p. 14 (10). The date 1509 for this manuscript coincides, moreover, with the year in which Achillini was prior.

philosophorum antiste distributis prius pecuniis utriusque collegii domino magistro Alexandro achilino laureatis."

87. Fol. 18 r, April 30, 1511:

Posito hoc die partito in utroque collegio cum fabis albis ac nigris electi fuere excellentissimi artium et medicinae doctores domini magistri Alexander achilinus ac hieronymus de bombice ad videndum et cognoscendum rationes et computus gabelle pro collegiis nostris una cum eximio doctore domino magistro Ludovico leone, ut ex nota nostra notus declaratus. (A vote having been taken this day in both colleges with white and black beans there were elected the most excellent doctors of arts and medicine master Alessandro Achillini and Hieronymo de Bombaxia to oversee and render the financial accounts and levy of tax for our colleges together with the outstanding doctor master Lodovico Leoni, as is indicated by my notes.)

Achillini, according to this report by the prior Giustiniano Fantini, was elected a syndic of the Gabella Grossa with the task of auditing its accounts and rendering a statement thereof.[19]

88. *Ibid.*, May 14, 1511: "Johannes Clementius Galus Reverendissimi Cardinalis de alebrhet medicus . . . in medicina . . . ab excellentissimo art. et med. doctore d. m. Alexandro achilino insignitus fuit, promotores domini magistri Beneventus Campegius et Justinianus Fantinus."

89. *Ibid.*, May 27, 1511: "Johannes kunza polonus . . . in sacra medicina . . . promotores d. m. Alexander achilinus . . . ab eodemque Alexandro achi. insignitus est." This is the last recorded service by Achillini on a doctoral committee.

90. Fol. 19 r, August 2, 1511 (1512). The death of Achillini occurred on August 2, 1512. His eulogy was written by Federico Gambalunga, prior of the college of medicine, and occupies the major part of fol. 19 r. Gambalunga failed to change the date 1511 at the top of the folio, where events of that year had been recorded; but the proper date 1512 is clear from fol. 18 v, where the heading lists Gambalunga as prior for the second trimester of 1512. This eulogy has been printed by both Münster and Nardi[20] but is presented here also in its correctly transcribed entirety as a testimony of the esteem in which Achillini was held by his colleagues and in fact the city of Bologna:

Egregius artium et medicinae doctor, D. M. r Alexander Achillinus obiit quarto nonas augusti: heu dies dies inquam [Achillinus moritur *in margine*] infausta nimium, non modo Bononie verum etiam uniuerse genti. ea enim talem urbs haec litterarijque omnes, quorum ingenia quasi phoebe ab illius uberrimi fontis splendore lumen recipiebant iacturam passi sunt qualem uix Roma Cesaris uel Ciceronis obitu uel maxime tunc quum apud Cannas infaeliciter pugnatum est, uix Graecia ex phylosophorum principis: uix Chous ex medicine parentis leto pertulit. Talis enim tantusque erat: ut cunctis uiuendi esset exemplar: unde non iniuria omnium oculi in ipsum converti erant. Omnes ipsum intuebantur. Omnes admirabantur, omnes denique tamquam deum colebant uenerabanturque. Deus deus inquam nobis. Nam ut de moribus taceam: quibus Laelium superabat, erat enim beneficus, comis, iucundus, hylaris amicis, inimicis autem non asper, nemini nocebat, tribuebat autem quod suum est unicuique, quod proprium est iustitie, utebatur pietate in parentes deosque mirifica, fauebat bonis, malos autem modeste incusabat. Quid de litteratura, qui quasi Crisippus in dyalectis, in physicis Aristoteles, in Metaphysicis Plato, quasi Galenus in medicina. Ita denique in omni scientiarum genere ualuit ut esset ex consensu eruditorum omnium extra omnem ingenii aleam positus. Fuit enim in ipso mira uigilantia, mira solertia, mirum exercitium. Nam die noctuque studebat, legebat, disputabat docebatque et equidem omnia docte, copiose subtiliterque quasi natura loqueretur. Ereptus est autem agens annum XXXXXI. Heu fortuna inuida, atrox, inconstans, indignorum fautrix. Heu mors crudelis, non potuisti huic uni indulgere, qui si uisisset [uixisset *Münster et Nardi*] archana dei mortalibus declarasset, nouissetque homines in uitam reuocare. Celebratum est autem eius funus tertio nonas augusti maximo doctorum scolarium ceterorumque comitatu. Astabant circa cadauer mulieres meste. Astabant affines egregiorumque doctorum nostra collegia. Discipuli ceterique, qui tamquam splendidissimum probitatis speculum litterarumque parentem flebant pullati omnes. Denique tota huius ciuitatis facies tristem sese praeferebat, quae semper gaudebit talem uirum habuisse dolebitque sibi fuisse abreptum donec funditus delebitur. Constitutum sibi fuit sepulcrum in aede diui Martini bononiensis ubi decantata fuere diuina offitia pro eo: cuius anima requiescat in pace Amen. (The excellent doctor of arts and medicine, Master Alessandro Achillini, died on the fourth day before the nones of August, alas a day most inauspicious not only for Bologna but all mankind. This city and all literate men whose talents as though from the sun had received light from the splendor of his most abundant source have suffered such a great loss as scarcely Rome suffered in the deaths of Caesar or Cicero, or especially when the Romans suffered defeat at Cannae, as scarcely Greece suffered from the loss of its chief philosopher, as scarcely Cos suffered with the death of the Father of Medicine. Such a one he was and so great that he was an example of life for all; thus the

---

[19] "The Gabella Grossa was an inclusive term applied to the body administering the various municipal customs-duties (the *octroi*) which were collected on fish, salt and other merchandise as well as on sales of property and dowries. After 1509 the members of this commission were professors of the Studium, and the funds were employed chiefly in paying the salaries of the professors" (Gnudi-Webster, *op. cit.*, pp. 35, 39, citing Guidicini and Mazzetti). The names of the syndics are listed in *Illustriss. Gabellae Syndicorum Nomenclatura pro parte Celeberrimi Collegii Medicae Facultatis ab anno Domini 1508 usque ad annum 1641*; by Barberi Jacobo and Ovidio Montalbani (Bologna, Niccolò Tebaldini, 1641), fol. A 2. Julius II had decreed in 1508 that six collegiate doctors should be elected to this commission by their colleagues.

[20] Münster, *op. cit.*, pp. 56–57: Nardi, *Saggi*, etc., pp. 264–265. The latter makes at least seven errors in his printed transcription of the eulogy. Gambalunga continued as prior in place of Leonello dei Vittori for the third trimester and thus to him fell the duty of recording the death of Achillini in a manner more handsome and extensive than had been employed for any other deceased member of the two colleges.

eyes of all were turned upon him in kindness. All gazed at him, all admired him, all in fact fostered and venerated him as a god. A god, a god, I say, he was for us. To say nothing of his character, in which he surpassed Laelius, he was kind, polite, jovial, pleasant to his friends, not harsh to his enemies, he injured no one, gave everyone his due, which is a trait of justice, he was most respectful toward relatives and gods, he favored the good, upbraided the wicked but in moderation. What shall I say of his accomplishments in literature, who was a very Chrysippus in dialectics, an Aristotle in physics, a Plato in metaphysics, a very Galen in medicine! He was so skilled in all sciences that he was by the unanimous opinion of all the learned placed beyond all hazard of genius. There was in him a singular vigilance, energy, activity. For he studied day and night, read, disputed, and taught, speaking learnedly, fully, and subtlely as though by nature itself. He was snatched away in his fifty-first year. Alas, envious fortune, fierce, inconstant, who favors the unworthy! Alas, cruel death, you could not indulge this man, who if he had seen them would have declared the secrets of god to mortals, he would have known how to recall men to life. His funeral was held on the third day before the nones of August [Aug. 3] by a great multitude of doctors and other scholars. Sad women stood about the corpse. Relatives stood about and our colleges of excellent doctors. His students and others wept, dressed in mourning, for such a shining mirror of probity and the parent of letters. Finally, the entire city presented itself as the picture of sadness, this city which will always rejoice to have had such a man and will grieve until it is completely destroyed because he was snatched away from it. His tomb has been erected in the church of St. Martin in Bologna, where divine offices were sung for him; may his soul rest in peace, Amen.)

From this eulogy it appears that Achillini's reputation among his colleagues was excellent, his qualities of mind and character highly respected. In its chief traits the eulogy is similar to the descriptions of Achillini by Luca Gauricus and Paulus Giovius. Ghiselli[21] is the authority for the statement that Achillini's skill in philosophical discussion gave rise to a proverb: "aut diabolus aut Achillinus." The phrase "quasi Galenus in medicina" is a tribute to his knowledge of medicine and to his teaching of that science.

Achillini could not, however mild and pleasant he was, have lacked those who envied his brilliance and success. This is clear from a letter by the Bolognese humanist, Giovanni Garzoni, whose works I am at present editing, who was a colleague of Achillini and died on January 28, 1505.[22] I present Garzoni's letter here with the variant readings included by Fassini in his copy in the selection of Garzoni's letters which he made in 1763 set off by parentheses:

IO. GAR. ALEXANDRO ACHILINO BONO. S. P. D.

(fol. 274 r) Dignitatem tuam, mi Alexander, diu statui mihi tecum habere communem; nulla enim pars corporis tui vacat officio. Addo, quam calles (calleas) philosophiam, que quantum tibi laudis afferat, res ipsa docet, ut ex omnibus philosophis quos mea videt aetas tibi principatus deferri possit. Sed cum (quum) invidia, teste M. Tullio [*Rhet. ad Herenn.* 4. 26. 36] (teste Marco Tullio) virtutis sit comes, nulla me admiratio tenuit (admirari tenet) si a quibusdam minus liberalis de te prolatus est sermo: satius illis fuisset si cogitationes suas in melius rettulissent nam quod nonnullos qui tibi virtute (virtutem) invident ingenio tuo et industria fregeris equo animo ferre non possunt. Flores tu eo tempore in quo videndo (videro) hominem philosophum habendum est loco miraculi. Non fero at (et) graviter exardeo cum (quum) quosdam animadvertam qui scientia, virtute, et (*om.* Fassini) tecum nolunt (volunt) in contentione (contentionem) venire honoris. Pridie apud legatum et elegans et gravis abs to habita est oratio. Admirabatur asellus quod inciperet [inciperes *scripsi*] cum timore; quotienscumque M. Tullius dicebat, toties ei videbatur venire in iudicium non ingenii solum sed virtutis et officii. [*Pro A. Cluentio* 51]. Eo fiebat ut magna sollicitudine animi et magno timore [*bis* MS 1896] diceret; nulli mirum videri debet si tibi, qui cum Cicerone minime conferendus es, dicendi eadem acciderunt. Sed relinquamus eos cum ignorantia sua. (fol. 274 v) Quantum in me fuerit non te deferam praesertim (*om.* Fassini) cum me optimum tibi esse amicum profitear. Non te fugiat, mi Alexander, me existimationis tuae procuratorem (procurationem) suscepisse. Vale. (I have long ago decided to hold your prestige, my Alessandro, in common with my own, for no part of your body is lacking in its duty. I add that the very fact that all philosophers yield the first place to you shows how well you understand philosophy, which brings you such high praise. But since envy, to quote Cicero, is the companion of virtue, I am not surprised that illiberal remarks have been made about you by some people; it would have been better for them if they had chastened their thoughts, for those who envy you will be crushed by your talent and industry: they cannot tolerate this. You flourish at a time in which the sight of a philosopher is to be regarded as a miracle. I cannot bear it, I am deeply angered when I notice some who wish to challenge you in knowledge and in virtue for the sake of honor. Yesterday you gave an elegant and serious oration in the presence of the papal legate. A little donkey present was amazed

---

[21] Ghiselli, Antonio Francesco, *Memorie antiche manuscritte di Bologna raccolte et accresciute sino a' tempi presenti* (1729) in University of Bologna Library Cod. 770, Vol. XII, 204.

[22] *Libri segreti,* MS 217 (10), fol. 3 r, as recorded by the prior Chiaro Francesco de Genoli, who wrote a brief eulogy for Garzoni. The letter to Achillini is contained in University of Bologna Library MS 1896, fol. 274 r, v. (IV, 67). There is no way to date it accurately, although the second letter in the collection after the one to Achillini is dated on the "vii kalendas novembres 1504." When Achillini gave the oration Garzoni mentions before the legate must also remain a matter for conjecture; but it was certainly before 1505. A second version of Garzoni's letter to Achillini appears in MS. Vat. Lat. 10686 as no. XXXIX, fol. 106, in a selection of Garzoni's letters made by Vincenzo Domenico Fassini, writing under the pseudonym of Dionisio Sandelli: *Joannis Garzonis Bononiensis Selectae Epistolae nunc primum e MS. Codice Bibliothecae S. Dominici erutae* [*editae*] *et auctoris Vita illustratae A F. Vincentio Domenico Fassini O. P. anno 1761* [1763]. I am editing the letters of Garzoni as contained in both collections, complete and selected, as well as his historical and philosophical essays.

that you began to speak with fear. Cicero said that as often as he spoke in public he also did so with dread; each time it seemed to him he was being judged not only as to his talent in speaking but as to his integrity and sense of duty. Hence he used to speak with great soul-searching and fear; no one ought to be surprised if you followed the same manner of speaking although you are by no means to be compared with Cicero. But let us leave them to their ignorance. As much as in me lies I shall not fail to second you especially since I profess that I am your best friend. Let it not escape your attention, my Alessandro, that I am the overseer of your esteem. Farewell.)

Achillini was buried on August 3, 1512, in St. Martin's Church. As Nardi gives the details on the basis of maps and documents,[23] the funeral cortège set out from Achillini's house at the corner of the Via degli Usberti and the Via de' Parisi or S. Columbano (probably the modern number 16). His tomb was doubtless situated in the Achillini family chapel, still in existence a short distance behind the main altar. Achillini's grand-nephew Claudio (1574–1640), one of whose sonnets is mentioned in Alessandro Manzoni's "I promessi sposi," was also buried here. The chapel was in later times sold and passed into other hands; today no trace of the tomb or its inscriptions remains although its location may be found in the *Monumenta* of St. Martin's Church by Pellegrino Antonio Orlandi (1723). I have translated the epigram on Achillini by Janus Vitalis[24] earlier in this introduction; it was never inscribed on his tomb, as some writers have mistakenly supposed. Cavalier Casio[25] wrote some time before 1525 another brief poem in Achillini's honor:

Del giovene Alessandro Achillino,
altro Aristotel, l'ossa son coperte,
ma l'opre stan tra Philosophi aperte.
Felsineo fu, anci pur fu divino.

A third epigram on Achillini is by Johannes Latomus Berganus (Iohan Steenhawer of Berg-op-Zoom), canon of Sant' Agostino. It is printed in Paulus Giovius' *Opera Omnia* (Basle, 1577), to accompany a medallion of Achillini:

Quisquis Averroem censes habere cerebrum,
falleris: haud est me passus habere meum.
Regula non fallit: propter quod quodlibet unum,
taliter affectum est, sed magis illud erat.

Other tributes to Achillini are printed by Alidosi.[26] Perhaps the most impressive salute to him was published by his younger brother, Giovanni Filoteo, in his terza rima poem *Viridario*, completed in 1504 but printed a year after Alessandro's death in 1513:[27]

Ne gli altri studii lo Achillino veggio,
che Theologia sparge in ogni zona.
lalta philosophia laudar non deggio,
che fama, e de laltre arti, il Mondo introna.
Me glorio, godo, e laudo il Creatore
che a questo unico son fratel minore.
Chi legge e intende lopre sue superne,
dove e insudato in la sua gioventute,
gli dara laudi gloriose e eterne.
Hor pensi, pervenendo a senettude,
le lucubration, calami e lucerne
seranno al letto et al lettor salute.
Di un lustro a punto il mezzo camin varca,
sel debito fara lhorrenda Parca.

This passage emphasizes Achillini's prominence in both theology and philosophy and in the next to last line indicates that he had passed the midpoint of life by a lustre: that is, thirty-five years (Dante's "nel mezzo del cammin di nostra vita") plus five, a total of forty, in 1504.

Certain portraits of Achillini exist. The first of these and perhaps the most authentic is a sketch by the Bolognese artist Francesco Francia done in 1486 when Achillini was twenty-three years old.[28] A second portrait appears in Paulus Giovius, *Elogia virorum literis illustrium* (Basle, 1577), 112, no. 363, several times reproduced, as by Münster, *op. cit.*, 15, fig. 3 and *Manoscritti e stampe venete dell' Aristotelismo e Averroismo* (secoli X-XVI)—Catalogo di mostra presso la Biblioteca Nazionale Marciana in occasione del XII Congresso Internazionale di Filosofia (Settembre 1958); Biblioteca Nazionale Marciana, 1958, fig. 18. A copy of this

[23] *Saggi*, etc., pp. 265–266. Ghiselli, *op. cit.* (August 2, 1512) has a long account of Achillini's death and burial.

[24] Printed in Ranuccio Ghero (Gherus), *Delitiae CC. Italorum Poetarum, pars altera* (1608), p. 1439.

[25] In *Libro intitulato Cronica, ove si tratta di Epitaphii, d' Amore e di Virtute, composto per il Magnifico* Hieronimo Casio de Medici, Cavaliero Laureato et del Felsineo Studio Reformatore. MDXXV, fol. 34 v.

[26] Giovanni Nicolò Pasquali Alidosi, *I dottori bolognesi di teologia, filosofia, medicina, e d' arti liberali dall' anno 1000 per tutto marzo del 1623* (Bologna, N. Tebaldini, 1623), p. 8. Münster, *op. cit.*, p. 54, lists several other writers of the sixteenth century who praise Achillini.

[27] (Hieronymo di Plato, Bologna, 1513), fol. 184 v-185 r. On Giovanni Filoteo (1466–1538) see the biographical article by Teresa Basini in *Dizionario biografico degli italiani* **1** (Rome, 1960): pp. 148–149, with bibliography, including T. Basini, "Spigolature e dipanature intorno alle opere di Giovanni Filoteo Achillini," *Paideia* **11** (1956): pp. 254–262.

[28] Nardi, *Saggi*, etc., p. 226; see also p. 258, note 88. Edward C. Streeter, "Francia and Achillini," *The Medical Pickwick: a monthly literary magazine of wit and wisdom* **1** (Saranac Lake, N. Y., 1915), p. 60, cites L. Olschki, *I Disegni della R. Galleria degli Uffizi, Serie Secondo, Fascicolo terzo, No. 2* (1913) for a reproduction of Francia's drawing: "a highly finished preliminary sketch for an oil portrait. Such a portrait was surely executed, according to G. C. Williamson, Francia's biographer, and proof of this is to be found in the Bolognese papers. Still better proof is this drawing, of which Williamson, writing in 1901, knew nothing." Streeter points out that Williamson believed that the figure in the Nativity scene (Misericordia altar-piece, Bologna Gallery No. 81 [Pinacoteca Nazionale di Bologna]) "which has masqueraded under the names of St. Francis or of Francia in *propria persona* for a century or more is none other than Achillini." This altar-piece was finished by Francia eight years after the sketch of Achillini.

portrait or possibly the original can be found in the gallery of the sala Fagnani in the Ambrosian Library at Milano. A third picture is the woodcut which forms the frontispiece of the "Annotationes Anatomicae" (1520), evidently a posthumous work. A fourth and final portrait hangs in a room in the University of Bologna Library, where it was pointed out to me in 1962 by the assistant librarian, Signor Luigi D'Aurizio, to whom I am grateful also for much useful and painstaking assistance in preparing the bibliography. This portrait forms the frontispiece to this book.

II

The first comprehensive list of Achillini's writings was made in the eighteenth century by Giammaria Mazzuchelli, who also wrote the first serviceable biographical article on him.[29] It remained for Claudius F. Mayer to draw up the most complete bibliography yet made, a remarkable undertaking that dates from 1941; historians of sixteenth-century medicine owe him a debt of gratitude that is not diminished by the fact that he, like Mazzuchelli, could not complete his work.[30] I have used his check-list but arranged my own in chronological order and in briefer style, adding comments here and there from other sources. The latest bibliography is Alberto Serra-Zanetti, *L'arte della stampa in Bologna nel primo ventennio del Cinquecento* (Biblioteca de "L'archiginnasio," Nuova serie, n. 1, Bologna, 1959), pp. 165–172. See also Index Aureliensis catalogus librorum sedecimo saeculo impressorum. Prima pars. Tomus A-, Volumen 1. Aureliae Aquensis, (1962): pp. 42–43.

A CHRONOLOGICAL LIST OF ACHILLINI'S WORKS

1. June 1, 1494: *De intelligentiis quodlibeta*, etc.; issued with no. 2 at Bologna, Benedicti Hectoris Faelli; other editions: *ibid.* 1506; collected editions: Venezia, Bonetus Locatellus for Octavianus Scotus, July, 1508; Venezia, Hieronymus Scotus, 1545, edited by Pamfilo Monti, professor of medicine at Padua; Venezia, Hieronymus Scotus, 1551; *ibid.*, 1568. See Nardi, *Saggi*, chap. VIII, pp. 179–223.

2. June 1, 1494: *De proportione motuum*; Bologna, Benedicti Hectoris Faelli; collected editions 1508, 1545, 1551, 1568 [see Mayer, *op. cit.*, p. 9]. There is a serious bibliographical problem in connection with this item. Hain, no. 71, who uses the title "De distributionibus ac proportione motuum," lists this work as published at Bologna "per Benedictum Hectoris" in 1494. The *Gesamtkatalog* I, 79, however, declares that its existence is "zweifelhaft." Nardi, *Saggi*, p. 263, insists, furthermore, that the publication has been "inventata," first, because within it the *De orbibus* of 1498 and the *De elementis* of 1505 are mentioned, and, second, because Giovanni Filoteo Achillini, who published the book posthumously (Alex. Achillini Bonon., *De proportionibus motuum*; Bologna, Girol. de' Benedetti, 1515, August), indicates its posthumous nature in his dedication to Pope Leo X: "Itaque Alexandri ipsius auctoris nomine (quando ipse funere praeventus acerbo non potuit) ea sanctitati tuae nuncupatim dico" (fol. 2 r). Mayer lists copies of the presumed 1494 edition at Krakow University and the National Library at Naples, printed together with the *Quodlibeta de intelligentiis* (Mayer, *op. cit.*, p. 13), and a second edition, Venezia, B. Locatellus, 1508, in the first collected edition of Achillini's works. T. M. Guarnaschelli and E. Valenziani, *Indice generale degli incunaboli delle Biblioteche d'Italia*, etc., I, A-B (Rome, 1943) list no such edition of 1494, although they record 7 copies of the *Quodlibeta* of 1494 and 11 copies of the *De Orbibus* of 1498. Mayer does not list the 1515 edition above to which Nardi refers in his footnote 99; I have examined a copy of it in the University Library at Pavia. Dott. Guerriera Guerrieri, directress of the Biblioteca Nazionale at Naples, informs me that the *De proportione motuum* does not appear in their copy of the *Quodlibeta de intelligentiis*, contrary to Mayer.

3. August 7, 1498: *De orbibus libri* 4; Bologna, Benedicti Hectoris Faelli, and collected editions. See Nardi, *Saggi*, pp. 228–238.

4. October 26, 1501: *De universalibus*, issued as part of *Septisegmentatum opus*, also known as Aristoteles, *Secreta Secretorum*, or Pseudo-Aristotle, *De secretis secretorum*, containing the following items: *De regum regimine, De sanitatis Conservatione, De physionomia, De signis tempestatum, ventorum et aquarum; De mineralibus;* Alexander of Aphrodisia, *De intellectu*, translated by Gerard of Cremona; Averroes, *De animae beatitudine*; Alessandro Achillini, *De universalibus*; Alexander the Great, *De mirabilibus Indiae ad Aristotelem* (spurious); Bologna, Bernard. de Vitalibus for Benedictus Hector, January, 1516; Paris, Du Pré, 1520; Lyon, A. Blanchard, March 23, 1528. Included in the *Septisegmentatum opus* is "the alchemical addendum to the Meteorology [of Aristotle] known as *De congelatione*" (Lynn Thorndike, *A History of Magic and Experimental Science* [Columbia University Press, 1941] **5**: chap.

[29] *Gli scrittori d' Italia cioè Notizie storiche e critiche intorno alle vite e agli scritti dei letterati italiani* **1**, parte 1 (Brescia, Giambattista Bossini, 1753), pp. 101–104. The facts contained here are correct and Mazzuchelli did not fall into the ridiculous error of assuming the existence of three different works on anatomy by Achillini as several later scholars did. It is unfortunate that Mazzuchelli never carried his work beyond the letters A and B; the plan was generous and masterful.

[30] Claudius F. Mayer, *Bio-bibliography of XVI. Century Medical Authors*; U. S. Army—Library of the Surgeon General, Index-Catalogue, Fourth Series, Third Supplement, **6**, G (U. S. Printing Office, Washington, D. C., 1941), pp. 7–10, *s. v.* Alessandro Achillini; p. 10, Giovanni Filoteo Achillini (1466–1538). Mayer's work goes down only through Alberti, Salomon.

III, "Achillini: Aristotelian and Anatomist," p. 48). There is a discussion of this work by E. J. Holmyard and D. C. Mandeville, *Avicennae De congelatione et conglutinatione lapidum*, [Paris, 1927], p. 14. It was an error in Panzer's bibliography (**10**: p. 62) which doubtless led Sarton to write: "In 1501 Alessandro Achillini of Bologna published a new Latin text which has been followed in all subsequent editions" (*Introduction to the History of Science* **4** [1927]: p. 557). See G. W. Panzer, *Annales Typographici,* etc. (Nuremberg, 1802). Nardi, *Saggi,* pp. 238–239, holds the *De Universalibus* was composed between 1484 and 1487.

5. August 23, 1502: *Examinatio figure quadrate et additio oblunge*; Bologna, Benedictus Hector. Printed with *De primo et ultimo instanti Gualterii Burlei*. Achillini's contribution to this 24-page, small double-columned booklet is a single page. The copy I have examined is in the Biblioteca Estense at Modena and bears the following information: "Magistri Alberti de Saxonia proportionum libellus finit foeliciter qui Bononie summa cum diligentia fuit impressus per Benedictum Hectoris. Die xxiii Aug. M.CCCCCII." Mayer, *op. cit.*, p. 9, lists one copy in Paris only. Nardi, *Saggi*, p. 239.

6. May 31, 1503: Augustinus Triumphus, de Ancona, *Opusculum perutile de cognitione animae et eius potentiis Augustini de Anchona cum quadam questione Prosperi de Regio*; also contains *Destructio arboris Porphyrii* by Triumphus (1243–1328), printed on July 10, 1503, both items "accuratissime revissum per A. Achilinum"; Bologna, Giovanni Antonio de' Benedetti (J. Jac. de Benedictis for A. de Placentia and J. de Ripis). Achillini added to this book in his capacity as editor a *Quaestio de sensibilibus interioribus* by Prospero da Reggio. Nardi, *Saggi*, p. 239.

7. 1503: *De Chyromantiae principiis et physionomiae*, preface to Bartholomaeus Cocles, *Chyromantie ac Physionomie Anastasis cum approbatione magistri Alexandri de Achillinis*; Bologna, Joannes Antonius de Benedictis, 1504. Achillini's work is also known as *Quaestio de subiecto physionomiae et chyromantiae*. Other editions: Venice, Bonetus Locatellus for Octavianus Scotus, 1508 (*Opera Omnia*); Pavia, Bernardinus de Garaldis, December 5, 1514; Bologna, Hier. de Benedictis, November, 1517; *ibid.*, 1518, 1523; collected editions 1545, 1551, 1568. The *De Chyromantiae* of Achillini begins "Quaeritur utrum"; the approbation begins "Venerandam sophiae." Cocles' name was really Bartolomeo della Rocca. Nardi, *Saggi*, p. 239; Thorndike, *op. cit.*, chap. IV, "Cocles and Chiromancy," pp. 50–68.

8. 1504: *De prima potestate syllogismi*; Bologna, Joh. Ant. de Benedictis; also in the collected editions.

9. 1504: *De subiecto medicinae*; Bologna, Joh. Ant. de Benedictis (issued with no. 8); also in the collected editions. This brief treatise is dedicated to Achillini's student, Virgilio Porto of Modena. Nardi, *Saggi*, p. 239, says he received his doctorate in 1505, although I can find no record of the event in the *Libro segreto* for that year. He taught theoretical medicine at Bologna from 1506 to 1525 and died August 6, 1527. Achillini addresses him affectionately in the following words as the collector of items 8 and 9:

Alexander Achillinus Virgilio Porto Mutinensi,
discipulo haud penitendo, foelicitatem.

Nostra quaedam fragmenta (ut moris eorum est), Virgilii mi amantissime, diligentem eorum collectorem adeunt. Tu enim urbanitate et virtutibus et doctrina is es, quem inter caeteros nobis dilectos elegi, apud quem aptissime reponantur: te enim semper cognovi nostri nominis studiosum. Logicalia quidem alios docebis; medicinalia vero exacte (ut assoles) contemplaberis: ex quibus non minus gloriae, Alexandro tuo aurigante, te iam comparaturum existimo, quam hactenus ex poeticis muneris (*read* numeris) adeptus sis. Haec igitur nostris aliis, quae apud te sunt, adiungas. Vale, et libenter res nostras perlege." (Alessandro Achillini to Virgilio Porto, of Modena, a pupil to be proud of, greetings: Certain fragments of my writings (as is their custom), my dearest Virgilio, are on their way to their diligent collector, for you are in your courtliness, virtues, and learning that person whom among other favorites of mine I have chosen in whose care these fragments of mine should most fittingly rest. I have always found you to be most zealous in my interest. You will indeed teach the works on logic to others; you will carefully contemplate (as you are accustomed to do) the writings on medicine, and from them all, with Alessandro as your charioteer, you will gain no less glory than you have gained thus far from your own poetry. Therefore add these fragments to the others in your possession. Farewell, and freely peruse my writings.)

10. September 11, 1505: *De elementis libri* 3; Bologna, J. A. de Benedictus; also in the collected editions. It contains an epigram by Virgilio Porto and a preface by Achillini addressed to Giovanni II Bentivoglio, important in establishing their firm relationship. Nardi, *Saggi*, p. 240, regards the *De elementis* as the third part of a more comprehensive work which embraces, together with the *De intelligentiis* (item 1) and *De orbibus* (item 3), the Aristotelic-Averroistic philosophy of nature. Porto's epigram hails Achillini as a new Aristotle:

Cum modo legisset titulum natura libelli
huius, Achillaeo est obvia facta seni,
atque ait: O nimium foelix hoc pignore, Claudi,
quam melius dici Nicomachus poteras.

Lodovico Boccadiferro (1482–1545), also a student of Achillini's and a professor of philosophy at Bologna from 1516 to 1531, having received his doctorate in 1516, also included an epigram in this publication to honor his teacher. He too taught the Averroistic interpretation of Aristotle: Nardi, *Saggi*, pp. 240–248.

11. 1506: *Quolibeta de intelligentiis* (sec. ed.); Bologna, Bened. Hectoris. This edition contains eighteen "doubts" raised by the Count Annibale

Rangoni, to whom it is dedicated. The work is also contained in the collected editions.

12. July, 1508: first edition of *Opera omnia philosophica* (none of which contains the *Annotationes anatomicae* or *anatomiae*); Venice, Bonetus Locatellus for Octavianus Scotus.

13. 1508: *Expositio primi physicorum*; Venice, Bonetus Locatellus, 1508 (contained in item 12); Bologna, Hier. de Benedictis, 1512; Bologna? publisher? 1518? with annotations by Francesco Mariano of Cremona (died 1528); and in Collected editions 1545, 1551, and 1568.

14. 1508: *De distinctionibus*; Venice, B. Locatellus for O. Scotus. (first *Opera Omnia*); Bologna, J. A. de Benedictis, 1510; Bologna, publisher? 1518, with annotations by Francesco Mariano; and in Collected editions 1545, 1551, 1568.

15. 1515: Aristoteles, *Rhetorica*; Venice, O. Scotus and G. Arrivabeni. Mayer, *op. cit.*, p. 8, observes: "Alessandro Achillini is mentioned as editor by the catalog of the Bibliothèque Nationale, Paris. How is this possible?" One can only suggest that Achillini edited this work (to whatever degree the word "edit" may be applied in this connection) before he died in 1512 and the book was delayed until 1515 in reaching print. Achillini is also listed as the editor of this work in *Index Aureliensis*, no. 107. 828.

16. I have left until the last in my chronological list of Achillini's works his poem, published by Giovanni Filoteo Achillini in "Collettanee Grece, Latine e Vulgari per diversi Auctori Moderni nella Morte de l'ardente Seraphino Aquilano; Bologna, C. Bazaleri, 1504." Serafino Ciminelli da L'Aquila died on August 10, 1500, and this collection of poems in 106 folios, octavo, was dedicated to Elisabetta, duchess of Urbino. Serra-Zanetti, no. 16, lists three copies, one each in the University Library, Bologna; in the Biblioteca Comunale dell'Archiginnasio, Bologna; and in the Biblioteca Colombina at Seville.

It would seem natural that so prolific a writer as Achillini should leave at his death a considerable amount of unpublished manuscript material. This is true; but only a small part of it can be identified as in his own hand. This material falls into two groups: (*a*) in the University of Bologna Library MS 14 (10), briefly listed in L. Frati, "Indice dei codici latini conservati nella R. Bibl. Universitaria di Bologna" (*Studi italiani di filologia classica* **16–17**, Florence, 1909), p. 110, of which I have photostats carefully prepared for me by Signor Luigi D'Aurizio; (*b*) MS Ambrosiana, Milan, A. 236, of which there is a microfilm in the National Library of Medicine at Bethesda, Md.

Group *a* consists chiefly of notes; perhaps only one item of the five contained in it can be called a consistently planned piece of writing. There are five paper manuscripts, each in a separate folder, with numbering from 1 to 5 and a title on the outside of the folder for each in the same hand but different from the hands in which the contents of the manuscripts themselves are written.

1. This item is entitled "Expositio Alexandri Achilini super prima quarti Avicene pulchra" and bears the date "septima septembris 1509." It contains on four and one-half folios, four of them written on both sides, a series of topics which probably served as lecture notes upon the first fen of Book 4 of Avicenna's *canon* the year after Achillini's return from Padua. They may well be an autograph since the handwriting is the same as that in *Libro segreto*, MS 217 (10), fol. 13 v, in Achillini's sixth priorate.

2. *Tabula Alexandri Achilini in medicina*. A series of questions or rubrics from various medical works: (*a*) *In afforismorum prima particula* (40 topics) from Hippocrates; (*b*) *in secunda particula* (43 topics); (*c*) *in tertia particula* (15 topics); (*d*) *in quarta particula* (12 topics); (*e*) *in quinta particula* (20 topics); (*f*) *in sexta particula* (one topic); (*g*) first fen first book of Avicenna's *Canon* (50 topics); (*h*) the first part of Galen's *Techne* or *Tegni* or *Ars Medica*, one of his most popular works (19 topics); (*i*) the second part of Galen's *Techne* (60 topics); (*j*) the third part (12 topics); (*k*) first fen fourth book of Avicenna's *Canon* (63 topics); (*l*) fourth fen first book of the *Canon* (63 topics). There are two and one-half folios in this item, again in a hand resembling Achillini's, if not identical with it.

3. *Quoddam consilium in medicina Alexandri Achilini pulchrum*, 5 folios. The first folio has the title: *Quoddam consilium de regimine itinerancium*. The handwriting is a diplomatic type of the fourteenth century and clearly not an autograph by Achillini. It is possible that the work bears some relationship to another one on a similar subject popular during the Middle Ages: the *Viaticum peregrinantis* (Traveler's Provision), "the most generally known of the works of Ibn al-Jazzar (d. *ca.* 1009). It was translated into Latin by Constantinus Africanus (d. 1087), and again by Stephen of Saragossa (*ca.* 1233). A Hebrew translation was made by Moses ibn Tibbon in 1259. See Sarton, *Introduction to the History of Science* I, p. 682; II, p. 65, p. 849, etc."[31]

4. *Autoritates Galeni colecte per alexandrum achilinum*, 10 folios written on both sides, the most extensive of the five manuscripts in group *a*. The first folio bears the same title as that on the outside of the folder and in addition the words at the beginning of the left-hand column "Liber de questionibus antecedit hunc librum." The manuscript, "quasi certamente autografe," (Nardi, *Saggi*, p. 269)

[31] Pearl Kibre, *The Library of Pico della Mirandola* (Columbia University Press, 1936), p. 47. Pico owned a copy of Achillini's *Quolibet de intelligentiis*; *op. cit.*, p. 253.

contains a summary of the first five books of Galen's *De simplicium medicamentorum temperamentis*, in eleven books.

5. *Multa ex Entisbari Sophista*, 4 folios, the last two only partially written upon. The title indicates a series of excerpts from the works of William of Heytesbury (*fl.* 1340–1372, fellow of Merton College, Oxford, and chancellor of the University), who wrote *Sophismata* (Pavia, 1481) (whence the epithet in the title of Achillini's manuscript), *De sensu composito et diviso* (Venice, 1494), and *Consequentiae subtiles* in R. Strode, *Consequentiae* (Venice, 1517).[32] Actually, item 5 has nothing to do with William of Heytesbury but consists of notes upon topics drawn from the *Expositio* of Paolo Veneto (P. Nicoletti da Udine, d. 1429) on Aristotle's *Posterior Analytics*. Paul of Venice, whose doctrines harmonize with those of Cajetan of Thiene (d. 1465) and Nicolettus Vernia (d. 1499), taught the Averroistic doctrines at Padua during the fifteenth century, bringing to bear what he had learned at Oxford and Paris.[33] As Nardi points out in his comments on Achillini's posthumously printed *De proportione motuum* (*Saggi*, 262), Achillini shows familiarity with the ideas of the mathematical calculators, Thomas Bradwardine, Nicola d'Oresme, R. Swineshead, William of Heytesbury, Albert of Saxony, Giovanni Marliani, and Paul of Venice, much discussed at Padua in their attempts to solve the problems raised by Aristotle in *Physics*, Book VII, 5.249b27-250b8, on the relations of force, resistance, and velocity. See also Thorndike, *op. cit.*, p. 41. Appropriately enough, the last folio of item 5 contains a number of arithmetical calculations.

Group *b* is listed by Lynn Thorndike,[34] who does not mention the University of Bologna manuscripts in group *a*. He calls it a late fifteenth-century manuscript. It was presented to the Ambrosian Library at Milan in 1673 by Giovanni Battista Capponi, doctor of medicine of Bologna, and contains the following title and table of contents: "Alexandri Achillini opuscula in Aristotelem numquam typis edita que dono dedit Ambrosiane Bibliothece vir doctissimus Jo. Baptista Capponius Bononiensis in Patrio Gymnasio Artis Medice doctor nonis Septembr. Anno 1673 Petro Paulo Bosea Biblioth. Prefecto." The table of contents follows:

"Quaedam Alexandri Achillini opuscula in libros Aristotelis
Praedicabilia Porfiri
Praedicamentorum Aristotelis
Peri Ermenias
Tractatus de Syllogismo tum probabili tum sophistico
Trac. de demonstratione
De consequentiis
fol. 155 r-178—Agregator plurium in logica Alexandri Achilini (in As. hand).
fol. 179 r—Nonnula circa Prohemium Aristotelis et Aver. in libris Phisicorum
Questio de subiecto philosophiae
188 r—(Albertus Magnus, cap. 3, I Physicorum)
193 r—Nonnule q. in philosophia naturali
fol. 195 r bears the date "1495, Julii 30."
200 r—Utrum ossa nutriantur medulla
203 r—Utrum hyle sit generabile aut corruptibile
207 r—Utrum proiectum moveatur a prohiciente post separationem ab eo.
208 r—Utrum elementa sint materia prima
214 r—Expositio dictorum Averrois in libris Phisicorum

The text ends at fol. 236 v. "The writing of the closing treatise is very abbreviated, and the other writing is very poor in places" (Thorndike, *op. cit.*, p. 49).

Thorndike's report on this manuscript is accurate, although the folio numbers he gives do not coincide with the results of my inspection of the writings. De Consequentiis Alexandri Achilini, according to my listing, begins at fol. 103 r; Agregator, etc., at fol. 107 r; fol. 132 v has the words "Alexander magnus" written large at the bottom; fol. 138 r is a page of Italian, evidently a letter without salutation or close; Albertus Magnus, cap. 3, I Physicorum begins at fol. 136 r; Nonnulla q. in philosophia naturali at fol. 139 r; fol. 141 r has the date "1495, Julii 30"; 146 r, Utrum ossa nutriantur medulla; fol. 154 r, Utrum hyle etc.; fol. 156 r, Utrum elementa etc.; fol. 162 r, Expositio dictorum etc. At fol. 169 r Achillini's name appears at the top of the page and "de anima" at the end of the line. Fols. 171 r-172 v are blank. After fol. 184 r, v there is a blank page and then a brief passage on an unnumbered folio with the last line thus: "Finis de intentione et remissione Alexandri Achilini Bononiensis."

[32] *The Cambridge Bibliography of English Literature,* edited by F. W. Bateson, **1**: p. 312, with references to F. W. Powicke, *Medieval Books of Merton College* (Oxford, 1913), pp. 25 f.; P. Duhem, "La dialectique d' Oxford et la scholastique italienne," *Bulletin italien* **12** (1912): pp. 8 ff. See *A Biographical Register of the University of Oxford* **2** (Oxford, 1958): pp. 927–928 for a biography.

[33] See Kristeller and Randall, *The Renaissance Philosophy of Man* (University of Chicago Press, 1948), p. 258, with bibliography cited.

[34] *Op. cit.*, **5**: pp. 48–49. *ibid.*, p. 49: "According to Alidosi Achillini left in manuscript a commentary on Averroes, *De substantia orbis*, a treatise *De mixtis* based on the twelfth book of the *Metaphysics*, a correction of the text of Aristotle's *Rhetoric*, and a *De anima* of 114 pages in his own handwriting." Nothing is known of these writings, although the "correction" of Aristotle's *Rhetoric* is probably a reference to item 15 in the chronological list of Achillini's works above. Thorndike, *op. cit.*, p. 43, finally, states that a copy of Achillini's *Questio* on physiognomy and chiromancy is also found in MS Escorial f. III. 2, at Madrid, fols. 1 and 9.

III

Achillini's *Annotationes Anatomicae* was first published posthumously by his brother, Giovanni Filoteo, in 1520. Various biographers and historians including Mazzuchelli, who was dubious about the matter, Fantuzzi, Hirsch, De Renzi, Münster, Capparoni, Pazzini, Giuseppe Favaro in the *Enciclopedia italiana s. v.* Achillini, Eycleshymer, and Singer have fallen into the error of taking the various titles under which the *Annotationes Anatomicae* were issued as referring to two or even three separate works on anatomy.

This misconception had been disposed of as long ago as 1844 by Francesco Mondini.[35] The facts are as follows. The first edition of this sole work on anatomy by Achillini (entitled "Annotationes Anatomiae," fol. II, but "Anotomicae annotationes," in the colophon, fol. XVIII) was published on September 24, 1520, at Bologna by Geronimo de' Benedetti (*per Hieronymum de Benedictis*) in a small format of eighteen folios, with a pair of poems of six and two lines each (elegiac distichs) by Annibale Camillo of Correggio to the reader beneath a woodcut portrait of Achillini (10 by 10 cm.) headed "Magnus Alexander Achillinus" and followed on the next page by a dedicatory letter dated September 13, 1520, from the editor, Giovanni Filoteo Achillini, to Panfilo Monti, doctor of arts and medicine and professor in ordinary in theory of medicine at Bologna, former student of Alessandro Achillini and later editor of the latter's second edition of collected works (1545). Giovanni emphasizes the compact nature of his brother's book as well as its great usefulness to students. He dedicates these "anatomica fragmenta" to Monti as the most redoubtable defender of Achillini's doctrines "contra rabidos oppugnantium ictus." The latter phrase recalls the main purpose of Giovanni Garzoni's letter to Achillini published earlier in this introduction: the defense of Achillini against his enemies. Fol. XVIII of this first edition contains twenty elegiac verses addressed to the Countess Veronica of Correggio by "Nicolai Camilli filius," evidently the same Annibale Camillo who wrote the verses on the title-page.

The second printing of the *Anatomicae Annotationes* appeared at Venice in January, 1521, at the press of J. Antonius et fratres de Sabio under the title *De humani corporis anatomia*, with the same dedication by Giovanni Filoteo to Panfilo Monti. I have not seen this book and know it only by its description in Nardi, *Saggi*, p. 272; Mayer, *op. cit.*, p. 8, lists copies only in London and Paris, but there is a copy in the New York Academy of Medicine.

The third printing is found as Tractate X in Johannes de Ketham, *Fasciculus medicie* [*sic*]. *Praxis tam chirurgis quam etiam physicis maxime necessaria*, etc.; Venice, Cesare Arrivabene, March 31, 1522, No. XIII in Charles Singer's list of editions of Ketham in his translation of Mundinus, p. 40. It bears the words "Incipit repertorium ex annotationibus anathomie Magni Alexandri Achillini bononiensis noviter excerptum" at fol. 47 r, the dedicatory letter from Giovanni Filoteo to Panfilo Monti, and at fol. 48 r the title "Annotationes anathomie magni Alexandri achilini bononiensis." The explicit appears at fol. 55 r. This book is somewhat better printed and punctuated than the first edition, although I have used both printings in making my translation.

Of other printings or editions earlier or later than the ones listed here no accurate account can be given since they are non-existent, in spite of references to them by certain authors; copies of such books have not been found anywhere.

The *Anatomical Notes* are clearly what Achillini meant by the title. More extensive and better organized than the excerpts and topics of the unpublished manuscripts, they are still lecture notes assembled in close adherence to the text of Mundinus.[36] Like the unpublished manuscripts, these notes were most probably regarded by their author as suitable at most for use in his lectures and not for publication; hence they were not included in the first (1508) collected edition of his writings printed during Achillini's lifetime nor were they published separately during his lifetime. The state of knowledge they represent recalls the years 1502, 1503, and 1506 from which he records anatomical autopsies (fols. V, *verso*, one in 1502, two in 1503; fol. XII, *verso*, one in 1506, when he failed to find the olfactory nerves in his cadavers; fol. XIII, *recto*, one in 1503; fol. XV, one in 1503; fol. XVI, one in 1502 and one in 1503). These eight instances form a rough method for dating the *Anatomical Notes*; they could not have been completed before 1506, the latest such autopsy recorded, and thus they must date at the latest from 1506 to 1512, during the last six years of his life. Their style in details is often difficult owing to their laconic and sometimes cryptic nature, but taken in the large they preserve a strict outline (Mundinus's text) and are remarkably consistent, as though they had been written in a short period and not over a long

[35] "De quodam codice Anatomiae Mundini, anno 300 exarato, Commentarium, quibusdam instructum Considerationibus ad historiam spectantibus Anatomiae: Memoria Posthuma (sermo habitus in conventu Academiae Scientiarum Instituti Bononiensis, die 7. Maii 1844)"; *Novi Commentarii Academiae Scientiarum Instituti Bononiensis Tomus Octavus* (1846), pp. 481–517 (especially pp. 488–489). Michele Medici, *Compendio storico della scuola anatomica di Bologna* (Bologna, 1857), p. 50, had called attention to Mondini's study but his efforts seem to have been in vain. Medici's review of the entire subject of pre-Vesalian anatomy is, by the way, still the best brief treatment available.

[36] See my remarks on the organization of Achillini's material in note 2 following the translation.

stretch of time which would encourage false starts, abrupt transitions, and unevenness of expression. They were at best the raw material for a book, lacking any formal dedication.

As to their scientific qualities, they represent a respectable and even a broad understanding of the state of knowledge of anatomy as it could be discovered in the writings of other anatomists within, let us say, the first decade of the sixteenth century. Far from being parochial and narrow like Benedetti, who quotes the Greeks almost entirely in his *Anatomice* (published in 1502, but from internal evidence clearly written in part at least as early as 1483; there is some evidence for an incunabulum of 1498), Achillini quotes a large number of authors and presents a wide perspective of views, some of which he disagrees with, introducing his own interpretation when his authorities fail to agree. I should also like to emphasize the fact that Achillini shows a remarkable progress in scientific thinking by avoiding entirely in this book any of the theories on humors, temperaments, complexions, influences, or virtues which vitiated the medical thinking of his time; nor does he use the physiognomonical or astrological theories evident in much of the work of his contemporaries and in certain other writings of his own. Finally, he betrays no narrow Aristotelianism in regard to categories for bodily members. In fact, we must conclude that he was seeking the truth within his lights without undue dependence upon authority and without fear of contradicting even so powerful an authority as Galen.

It is interesting to observe that much early anatomical dissection, far down even into the sixteenth century, had as its immediate object to prove that what Galen and Avicenna or Mundinus had said about a given part of the body was correct. That is, men did not dissect in order to discover new knowledge but simply to verify the old and to demonstrate its accuracy to their observers, students or laymen. If the anatomists disputed with each other it was a dispute over their colleague's interpretation of the authorities more than over the *original* observations of those colleagues. This state of mind was, of course, a hindrance upon scientific advances and an understandable cause for the slow progress of experimental work in medicine. Certainly Achillini cannot be singled out for blame on any of these counts since he shared the attitudes and preconceptions of his contemporaries.

Most of the *Anatomical Notes* deal with the standard descriptions of the parts of the body and the comparison of authorities for these descriptions. While Achillini follows Mundinus's order of presentation very closely his purpose is not primarily to comment on Mundinus's text, for he seldom mentions him. He is interested in commenting on conflicts among authorities and singling out the bare bones, as it were, the basic essentials, of an anatomical survey of the human body. Furthermore, the space he assigns to certain topics in proportion to the total length of his notes varies from the practice of Mundinus. I have introduced Roman numbers to indicate the three major divisions according to the bodily cavities involved since they are ignored in the printed texts.

Occasionally Achillini ventures what seems to be an original conclusion or at least what appears to him a more logical explanation of an anatomical problem. Thus at p. 49 he explains that the sperm for the generation of the female passes through the left hand emulgent vein full of watery blood before it is cleansed in the kidney, implying that it is the unclean nature of the sperm which results in a female embryo as contrasted with the clean blood from the right hand emulgent cleansed by the kidney which results in a male birth. At p. 51 he proves to his satisfaction that the sperm after exit from the preparatory vessels and before its entrance into the delatory vessels must lie in the substance of the testicle. At p. 54 he regards Aristotle's induction as to the three ventricles of the heart as more logical than that of Galen, Rhazes, or Averroes, who claimed there were only two. Berengario da Carpi's trivial criticism (*Commentaria*, etc., fol. 348 r) that Achillini was in error when he called ostiola certain bits of flesh contained in the heart since these are really pannicular in nature and shaped like the letter C, hardly shakes the essential truth of Achillini's observation. At p. 54 Achillini answers his own question about the artery above the vertebrae in the region of the heart by saying that it appears to have no connection with the heart because it is branched under the fork and hence concealed. At p. 56 he questions whether epileptics actually have green veins under their tongues called "frog veins" by Avicenna and whether they should be blood-let or not. It is at this point that later authors attribute to Achillini the discovery of the orifices of Wharton's ducts of the submandibular gland. What he says is as follows: "Two salival fonts into which a stylus can enter are visible; they open near the tongue, and pieces of glandular flesh exist in that place."

The attribution of certain discoveries to the early anatomists is a notoriously uncertain business. It is possible to say, however, that Achillini's description of the brain is excellent and his account of its dissection detailed enough to be helpful. He is said by various writers to have observed the course of the *ventriculi cerebri* into the inferior cornua, to have rediscovered the fornix and infundibulum, and to have shown that the tarsus consisted of seven bones. He described the midtrochlear or patheticus nerve, included by Vesalius, *Epitome*, with the second pair, and went on to mention an eighth pair of cranial nerves to add toward our modern total of twelve beyond the Galenic total of seven, although he adds: "Concerning this pair nothing has been discovered by learned men. I think that it gives motion to the eyebrows. It is immediately visible when the brain is lifted up." He was familiar with the ileocaecal valve and wrote a more satisfactory description of the duodenum, ileum, and colon than had been available.[37]

A small controversy was aroused in the late sixteenth century over another, and final, supposed discovery by Achillini: the malleus and incus bones of the ear. This was attributed to him by Alidosi,[38] who cited Eustachius and Casserius as his sources. Both of these men wrote after the first person who tried to date the discovery of these bones, Niccolò Massa, in whose *Epistolae Medicinales* (1550) the assertion first appears.[39] It should be noted that Massa does not attribute the discovery of the hammer and anvil bones to Achillini but to anatomists of Achillini's time. Falloppia in 1561 attributed their discovery to Berengario da Carpi, although Medici[40] insisted they had merely been more fully described but not discovered by him. Giulio Casseri (Julius Casserius) settled the matter by compromise, attributing the discovery to both Achillini and Berengario.[41] Lauth,[42] with Gallic simplicity, declared that these bones might have been discovered as they fell out of some skull after it had been boiled and cleaned; but he avoids naming the discoverer. What is chiefly clear from this survey of the problem is that Massa, the earliest known anatomist to refer to the problem of discovery, attributed it to Achillini's period and, if chronological priority can be pressed as an argument, Achillini was first before Berengario.

Of the Italian anatomists who cluster around the turn of the fifteenth to the sixteenth century and who publish their most significant (usually single) works on the subject of anatomy in that brief space Achillini is, of course, not the first to publish. Hieronymo Manfredi published his *Anothomia* in 1490, Gabriele Zerbi his *Anatomiae corporis humani et singulorum illius membrorum liber* in 1502, and Alessandro Benedetti his *Anatomice, siue Historia corporis humani* in the same year, if we ignore certain siren calls that speak of incunabulum ghost editions of the book from the 1490's. Achillini's *Annotationes*, although published in 1520, reflect the state of his knowledge of anatomy in the years 1502–1506 and this by an amazing coincidence causes it to fall, as far as much of its contents is concerned, within the same year when both Zerbi and Benedetti published their books, 1502, the banner year of pre-Vesalian anatomy.

Zerbi's book is too long, too scholastically oriented, of the three, to serve as a characteristic guide to anatomical knowledge except of a certain kind which still clung to the concepts and methods of medieval anatomy. Benedetti's book likewise is too exclusively oriented toward the Greeks and to Greek nomenclature to give us a complete view of the situation in 1502. Achillini's notes, while more brief than either of the other two books, does not cling too exclusively either to Arabs or Greeks and is the more balanced representative of the three for use as an introduction to the immediately pre-Vesalian phase of the history of anatomy—the somewhat

---

[37] On attributions to Achillini see Lynn Thorndike, *op. cit.*, **5**: pp. 43–47, who is wisely skeptical in his treatment; V. Portal, *Histoire de l'Anatomie et de la Chirurgie* (Paris, 1770–1773) **1**: p. 269; G. A. Brambilla, *Storia delle scoperte fisico-medico-anatomico-chirurgiche fatte dagli uomini illustri italiani* (Milan, 1780) **1**: pp. 165–167; Michele Medici, *Compendio storico della scuola anatomica di Bologna* (Bologna, 1857), pp. 46–47; Francesco Mondini, *De quodam codice Anatomiae Mundini*, etc. (Bologna, 1846), pp. 488–489 (see note 35); Carlo Calcaterra, *Alma Mater Studiorum: l'università di Bologna nella storia della cultura e della civiltà* (Bologna, 1948), p. 162; G. Zaccagnini, *Storia dello Studio di Bologna durante il Rinascimento* (Geneva, 1930), p. 262; Th. Puschmann, M. Neuburger, J. Pagel. *Handbuch der Geschichte der Medizin,* **2** (1903): p. 199. Thomas Lauth, *Histoire de l'Anatomie* **1** (Strasbourg, Levrault, 1815): p. 332, says: "Achillini decrit et prétend avoir découvert le nerf pathétique." But Achillini claims to have discovered no such thing, here or elsewhere in the *Annotationes*.

[38] Giovanni Nicolò Pasquale Alidosi, *I dottori bolognesi di teologia, filosofia, medicina e d'arti liberali dall' anno 1000 per tutto marzo del 1623* (Bologna, 1623), p. 8.

[39] Printed at Venice by F. Bindoni and M. Pasini; Epistle V to Antonio Fracanzano, on errors made by anatomists, fol. 55 v: "Haec vero ossicula anatomici, tempore Alexandri Achillini veri in omni genere scientiarum eminentissimi (ut ex eius scriptis clarissime uidere est) inuenerunt. Quare non ab istis sunt primo inuenta, nec ostensa, cum etiam Iacobus Carpensis loca istorum ossiculorum inuenire doceat." Just before this passage Massa had demonstrated the true function of these ear bones, "ut ego in meis anatomicis ostendi."

[40] Medici, *op. cit.*, p. 51.

[41] *Iulii Casseri Placentini philosophi, medici, et anatomici De Auris Auditus Organi, Historia Anatomica, Tractatus Secundus, Libri III* (Ferrara, 1601): *De auris auditus organi structura, Liber* I, De tribus ossiculis [he includes the stapes], cap. XII, p. 67: [incus] "Huius et mallei primi inventores Alexander Achillinus et Iacobus Carpensis." Casserius goes on to say: "Vesalio non simul cum malleo et incude stapes obvius fuit quia in cruda calvaria in qua humoribus quasi immersa ac abdita omnia latent, auditus organum . . . sicuti ipse in esam. obseru. Fallop. de se testatur. Ob eandem forte causam Alexandro Achillino Bononiensi, philosopho insigni, et Iacobo Carpensi, chirurgo et anatomico haud contemnendo (qui, ut de utroque Eustachius, de Jacobo vero Carpensi etiam Columbus et Fallopius attestantur, primi omnium in malleum et incudem inciderunt (p. 68), ipsaque promulgarunt) stapes sese non ostendit." The stapes bone was discovered by Ingrassia. Of Casserius, F. J. Cole, *A History of Comparative Anatomy* (London, 1944), p. 120, says: "We owe to Casserius the first detailed and comparative account of the auditory ossicles." See now for a thorough discussion of the entire problem C. D. O'Malley and Edwin Clarke, "The Discovery of the Auditory Ossicles," *Bull. Hist. of Medicine* **35** (1961): pp. 419–441, including the position of Achillini, Berengario da Carpi, Massa and other Italian anatomists both pre- and post-Vesalian in relation to its history. At p. 423 these writers say: ". . . it seems possible that they had been discovered by Alessandro Achillini of Bologna."

[42] *Op. cit.*, p. 320.

more than a generation, 1502 to 1543 (or 1490–1543, if we begin with Manfredi), which preceded the publication of Vesalius's *De Humani Corporis Fabrica*. The books published by Berengario da Carpi in 1521 and 1522 and by Massa in 1536 were the first to reveal the new emphasis upon personal observation and upon dissection conceived of as more than a means of confirming authority, an attitude which typifies the new anatomical science Vesalius was to bring to maturity.

A final word about my translation of Achillini is necessary here. Part directions for dissecting, part notes from the authorities, part commentary on Mundinus's text, and part the result of personal observation, the Latin style is difficult, at times elliptical and confusing, lacking transitions, connectives, even verbs. By means of bracketted words I have expanded and completed Achillini's thought as I saw it to be necessary. It was not possible even so to make a completely smooth translation but one can understand it better with the aid of my additions. Much labor has been expended on the notes and certain small discoveries should be emphasized here: the influence of Johannes Alexandrinus's commentary on Galen's *De Sectis*, the reference to Francis of Piedmont's presumable work *On Symptom*, and the six positions of anatomical procedure drawn from Aristotle but, as with the first of these items, without any acknowledgment. The notes should thus serve as a confirmation and corroboration of the text of the translation although I am not certain I have in every instance completely caught the meaning of Achillini's thoughts, and words, many of them, I conjecture, addressed to himself alone. Lastly, I have chosen to begin this series of studies with Achillini because of his compact text, his typical dependence upon authorities, much less evident in Massa, for example, and because he represents a midpoint between the more scholastic Zerbi and the already Vesalius-oriented Berengario da Carpi.

In the notes to the translation I have often quoted the Latin translation that accompanies the Greek text of Galen in Kühn's edition simply as a convenient guide for those who know Latin, although it is not always a completely accurate version; those who wish to check the Greek may do so in Kühn's text. To have quoted the Greek texts of Galen *in extenso* would have greatly increased the costs of publication. It should also be recalled that I have not tried to correct what the trained anatomist may regard as Achillini's errors but to translate his Latin text as faithfully as possible despite its obvious difficulties.

Finally, I have employed the terms of the *Nomina Anatomica* (1955) only where they seemed appropriate and where they could be safely and accurately used in the light of modern anatomical knowledge. There seemed to be no point in turning many commonly understood words such as "kidney" or "liver" back into their modern Latin anatomical equivalents or in turning Renaissance Latin into modern anatomical nomenclature with pedantic consistency. In fact, I have made my translations not for anatomists or for Latin scholars alone but for general students of the history of medicine and of science who may not be either Latinists or anatomists but are interested in both Latin and anatomy as important elements in the history of civilization.

ANATOMICAL NOTES BY THE GREAT ALEXANDER ACHILLINUS OF BOLOGNA

[I]

Anatomy is the skillful dissection of the members and is a clarification of those members hidden beneath the surface of the body.[1]

There is a threefold difference between human and animal anatomy: in the form or position of the members; in their nature or manner of use; and in certain parts.

By form or shape an erect stature is meant.

By nature is meant one that is gentle, simple, that is, naturally without art or skill, and civil.

Part refers to the fact that human beings are without horns, hard skin, long claws, beak, bristles, scales, and an external tail.[2]

Anatomical dissection is performed in reference to six features: number; the substance of the members; the location of the members, that is, internal or external, above, below, and so on; posture (*scema* [Greek σχῆμα]), that is, continuation or connection with each other; their use or operation in preserving the individual or the species; their particular illness.[3]

Actually there are seven features. Francis of Piedmont,[4] on the authority of Alexandrinus's com-

---

[1] The first sentence is borrowed with little change and without acknowledgment from the commentary on Galen's *De Sectis Ad Eos Qui Introducuntur* (ed. C. G. Kühn, I, [1821]: pp. 64–105) by Johannes Alexandrinus, to whom the introduction to a Latin translation of this work of Galen is attributed in the edition by Rusticus Placentinus (*Opera Omnia*, J. de Burgofranco, Pavia, 1515–1516), ff. 6r–12v, but whose date and general personality are thus far unknown (Owsei Temkin, "Studies on Late Alexandrian Medicine. I. Alexandrian Commentaries on Galen's *De Sectis ad Introducendos*," *Bulle. Inst. Hist. of Medicine* **3** (1935): pp. 405–430, especially pp. 407 and 421, with bibliography; see also, *idem*, "Geschichte des Hippokratismus im ausgehenden Altertum," *Kyklos: Jahrbuch für Geschichte und Philosopie der Medizin* **4** (1932): pp. 68, 71; the various and tentative identifications of Johannes Alexandrinus are discussed in detail at pp. 53–71). Since the Pavia edition of Galen is not at hand I have used the National Library of Medicine copy of the *Opera Galeni* (ed. by Diomedes Bonardus, 1490) which belonged to "Nicolaus Pol doctor" in 1494. The quotation from Johannes Alexandrinus follows (p. 4): Comentum: Anothomia est artificiosa incisio et clarificatio eorum quae in occulto ascondita sunt in corpore. Artificiosa incisio dicta est eo quod medici docte et articulate incidant quae foris sunt artarias [*sic*]. Clarificatio quae in occulto sunt bene dictum est quia in hoc non possumus scire nisi per anothomiam. Berengario da Carpi, *Commentaria . . . super anatomia Mundini . . .* (1521), VI, quotes Alexandrinus as follows: Quid rei sit anatomia. Noto primo quod anatomia capitur dupliciter: uno modo prout est artificiosa actualis incisio et clarificatio eorum quae in occulto sunt corpore abscondita. Istam diffinitionem dat Alexandrinus in libro de sectis et isto modo accepta anatomia non est scientia nec theorica nec pratica sed operatio quaedam etiam quod in vivo facto sit talis operatio manualis pro fine cuius intenditur forte sanitas, quia medicina nec quo ad practicam nec quo ad theoricam est operari sed scire operari, et isto modo etiam Medicina est scientia. Gabriele Zerbi, *Liber anathomie corporis humani et singulorum membrorum illius* (Venice, 1502), fol. 3 r, ends his quotation of Alexandrinus thus: Clarificatio que in occulto sunt bene dictum est quia hoc scire non possumus nisi per anathomiam *siue dissectionem et dilaniationem et diuisionem corporis* (italics mine). Mundinus does not quote the definition but merely lists the six features in essentially the Commentator's terms. For the history of ancient definitions of medicine to which this highly influential definition of anatomy leads us see L. Englert, *Untersuchungen zu Galens Schrift Thrasybulos* (Leipzig, 1929), pp. 4–31, cited by Temkin, *Studies*, etc., p. 417.

[2] In the preceding sentences Achillini follows the tenor of Mundinus's thought and expression closely, although in greatly reduced detail. His order of presentation also duplicates almost exactly that of Mundinus, with the exception of the eighteen sutures of the skull at the end of Achillini's work, drawn from Albertus Magnus. It is perhaps this close adherence to the order of detail in Mundinus that has given rise to the impression that Achillini wrote a commentary on Mundinus; it is rather a set of notes in which he follows Mundinus in outline but actually records his own departures and emendations together with abundant quotations and alternative views from many other authors. Mundinus, in fact, is rarely quoted by Achillini; I find only three instances, in two of which he disagrees with his predecessor. Man as a creature civil by nature is drawn from Aristotle, *Politics*, I, 2, perhaps influenced by Zerbi's reference to the passage in his *Liber Anathomie* (1502), fol. 4 v. *Ibid.*, fol. 5 r emphasizes man's lack of art by nature.

[3] The six features of anatomy grouped here also follow the list given by Johannes Alexandrinus, above, p. 4 (additions in square brackets are from the Pavia edition quoted by Temkin, *Studies*, etc., p. 411, note 30): In mortuis etiam fit anothomia propter vi occasiones: [scilicet] numerum [et] substantiam membrorum et positiones locorum et magnitudinem et scema et alterutrum continuatis [*sic*: the Pavia edition has *communicationes*]. In the 1490 *Opera Galeni* Galen says (f. 1 v): Medicus namque scire debet et positiones locorum et naturas. He goes on to explain them (f. 4 v): Propter positionem quia in dextra parte positum est epar et in sinistra splen. Propter magnitudinem quia maius est epar quam cor et renes. Propter scema quia epatis scema interius est concauo simile ut scemati lune. Propter alterutrum comunitatis quia vesica fellis communicat epati eo quod in scemate epatis posita sit et trahit ad se colericum humorem ut purget epar et sanguinem. He recurs to the six features at f. 5 r: Quibus respondemus quia nos non expectamus donec confusio fiat membrorum sed mox ut moriuntur incidimus ut inueniamus sex illa quae in superioribus diximus theuma numerum scema substantiam magnitudinem positionem et alterutrum comunitatis. *Theuma* may be an error for *scema*, which follows in the list, a Greek word used by Hippocrates, *On the Duties of Physicians* 11. The corresponding passage in Mundinus runs thus (italics mine: I use the text edited by Berengario da Carpi, chap. xxxxiii): De anathomia ventris inferioris. De membris autem officialibus sciendum est quod in quampluribus ipsorum quantum ad anathomiam factam in mortuis *ad minus* sex sunt videnda ut dicit commentator alexandrinus in commento libri sectarum scilicet quae ipsorum sit positio, quae sit eorum substantia et per consequens complexio et quae sit eorum quantitas, numerus, figura et continuitas eorum. Mundinus uses *figura* for *scema*, shape or posture, which is what the word seems to mean in the Commentator's description above (f. 4 v): he omits (as does the Commentator) Achillini's *use* and *illness*, which are added to the group also by other sixteenth-century anatomists. *Comunitas* or *continuitas* becomes the favorite *colligantia* in their works, the connection of organs with other organs or parts.

[4] Francis of Piedmont (Franciscus de Pedemontium or Pedemontanus, from Alife in the Terra di Lavoro near Capua, Italy) is for us a still obscure figure of the thirteenth century whose only printed work of consequence is his *Complementum*, a lengthy augmentation of the *Grabadin* or Handbook of Medicine by

mentary on [Galen's book] *On Sects*, enumerates six at the beginning of his commentary *On Symptom*: first, size, whether the member is small or large; second, number, whether one or two and so forth; third, the place where the member is located and its connection or relations (*colligantia*) with the other members; fourth, form or shape, whether concave, as the stomach; fifth, substance, whether thin or thick, as the slender and gross intestines; and sixth, complexion.

I reply that the subordination of connection (*colligantia*) among these features or properties reduces them to six; when it is listed separately then there are seven features. As to the second feature, under the term substance is understood thin or slender and so forth but also fleshy, pannicular or membranous, bony, and so on. Under complexion is understood that which is natural or unnatural [*praeter naturam*], but illness extends beyond complexion to embrace both composition and unity; thus the terms complexion and illness differ a little.[5]

There are six positions or locations: above, whence comes nourishment; the right, whence comes local motion; in front, whence comes sensation; below, to the left, and behind, which are opposed to the three former positions.[6]

The body cavities are three in number: of the animal members—the head; of the spiritual members—the chest; of the natural members—the lower cavity.[7]

The cavity of the natural members has two parts: those in front and those which are lateral. Those in front are the pomegranate or processus xyphoideus [shield cartilage]; the stomach; umbilicus; sumen [hypogastrium]; comb [regio pubica]; the fat under the sumen which is called the omentum in the *Anatomy of the Small Parts* [Pseudo-Galen, *De Anatomia Parva*].

The lateral parts are two: the hypochondrium; the ilium.

According to Avicenna, however, the mirach [abdomen, epigastrium] is made up of two parts: skin and the siphac [peritoneum]. Here Hugo[8] makes an

---

Mesue, called the younger (924 or 925–1015 A.D.), printed at Venice, 1471, by Giunta; 1549; 1561 (edited by Andr. Marinus and called "one of the best" editions by Th. Puschmann-M. Neuburger-J. Pagel *Handbuch der Geschichte der Medizin* **1** (Jena, 1902): p. 612. Francis lived at the court of Robert of Anjou as his physician and wrote what K. Sprengel, *Versuch einer pragmatischen Geschichte der Arzneykunde* (1823); **2**: p. 630, calls "the most complete practical compendium of this century." See also Heinrich Haeser, *Lehrbuch der Geschichte der Medicin und der Epidemischen Krankheiten* (Jena, third ed., 1875) **1**: pp. 706–707. The most extensive, although far from comprehensive, account of Francis is by Salvatore de Renzi, *Storia documentata della scuola medica di Salerno* (sec. ed., Naples, 1875): pp. 546–555. Giovanni Giacomo Bonino, *Biografia medica Piemontese* (Turin, 1824) **1**: pp. 24–25, lists as selections from the *Complementum* some *Excerpta de Balneis* printed in the large collection, *De Balneis Omnia etc.* (Venice, Giunta, 1553). George Sarton, *Introduction to the History of Science* (1947) **3**, Part 1: pp. 835–836, gives a brief life with bibliography. Francesco di Piedimonte (to give him his Italian name) died in 1319 A.D. I have used the *Complementum* in the 1497 edition of Mesue but, although he treats symptoms occasionally, I have found no separate work entitled *De Sinthomate*; it is possible from the wording of Achillini's text that a commentary by Francis on Galen's *De Differentiis et Causis Morborum Symptomatumque* is meant by Achillini's reference; but no work by Francis of Piedmont with such a title as *De Sinthomate* is known. It may be one of his lost works; in that case, Achillini is the sole author to mention it. I should point out that the Commentator on Galen's *De Sectis* does not include *complexio* in his six features; this is added by Mundinus, and according to Achillini involves illness also.

[5] "The *complexion* is the particular manner in which the four elements or four humours are mingled together and is to be distinguished from *composition*, which refers to the coarser structural nature": Charles Singer, *The Fasciculo di Medicina etc., Translation of the "Anathomia" of Mondino da Luzzi, etc.*, (Florence, R. Lier, 1925) in *Monumenta Medica* **2**: pp. 101–102. See Avicenna, *Canon* (transl. by Gerard of Cremona, 1595) fol. 11 for a detailed definition.

[6] The six *positions* (to be distinguished carefully from the six *features* of anatomy discussed above and of which they form only one) are drawn from Aristotle, *De Incessu Animalium* (translated by A. S. L. Farquharson, in *The Works of Aristotle Translated into English* under the editorship of J. A. Smith and W. D. Ross, Vol. 5): "Next we must take for granted the different species of dimensions which inhere in various things; of these there are three pairs of two each, superior and inferior, before and behind, to the right and to the left. (704 b) . . . Again, the boundaries by which living beings are naturally determined are six in number, superior and inferior, before and behind, right and left . . . The superior is that from which flows in each kind the distribution of nutriment and the process of growth; . . . The front is the part in which sense is innate, and whence each thing gets its sensations, . . . The right is that from which change of position naturally begins, the opposite which naturally depends upon this is the left . . . That the beginning of movement is on the right is indicated by the fact that all men carry burdens on the left shoulder; in this way they set free the side which initiates movement and enable the side which bears the weight to be moved" (705 a, b). See also *De Caelo* II, 284 b (translated by J. L. Stocks, Vol. 2): "The distinctions are three, namely, above and below, front and its opposite, right and left—all these three oppositions we expect to find in the perfect body—and each may be called a principle. Above is the principle of length, right of breadth, front of depth. Or again we may connect them with the various movements, taking principle to mean that part, in a thing capable of movement, from which movement first begins. Growth starts from above, locomotion from the right, sense-movement from in front (for front is simply the part to which the senses are directed)." Galen, in following Aristotle, gives a specifically anatomical meaning to the term position: *Hippocratis Aphorismi et Galeni in Eos Commentarii, Particula* V (Vol. XVII Kühn, 878): Posteriores namque partes nervosae sunt, anteriores autem venosae et arteriosae. Gabriele Zerbi, in fact, goes to the remarkable length of arranging his entire *Liber Anathomie Corporis Humani* etc. (Venice, 1502) along these Aristotelian lines; the three books into which his work is divided run thus: I. Anterior Parts, fol. 6r; II. Posterior Parts, fol. 135v; III. Lateral Parts, fol. 152v; see fol. 5 v: Summe itaque partes in quas primas totum diuiditur corpus secundum designationem Aristotelis sunt quas assignauimus de numero membrorum organicorum pauloante differentes, et he quidem communes sunt maris et femine, quarum situs quantum ad *superius et inferius, prius et posterius, dextrum et sinistrum* habeant pro corporis summa visu explorant (italics mine).

[7] These are the traditional divisions of the bodily cavities according to Galen's physiology; for a brief account see A. C. Crombie, *Medieval and Early Modern Science* (sec. ed., Anchor Books, 1959) **1**: pp. 162–172, with plate XIX. Pseudo-Galen, *De Anatomia Parva* is also known as *Anatomia Cophonis* or *Porci*.

[8] Hugo of Siena or Ugo Benzi: on him see E. Gurlt, *Geschichte der Chirurgie und ihrer Ausübung* (Berlin, 1898) **1**: pp. 866–868; D.

error [in their order of sequence] when he totals the five parts of the mirach. Avicenna, by the way, calls the siphac also beritheron.

The mirach is made up of five parts: skin; fat; a thin fleshy sheet; the muscles and their tendons; the siphac.

Hugo's argument runs counter to this listing: first, skin; second, fat; third, siphac; fourth, muscles; fifth, the thin fleshy sheet called beritheron by Avicenna, *Canon*, III, Fen 22, chap.1.[9]

Note the error of Jacopo of Forlì,[10] in the first chapter of his first book, where he says that the tendons of the muscles are called siphac by some and bertheron by others.

The order followed by Avicenna in the passage referred to above is as follows: the skin; under that the first panniculus and the combination of parts called the mirach; then the muscles; next, the beritheron; then the zirbus or omentum; then the intestines.

A discord exists here. Galen, in his book *On the Anatomy of the Uterus*,[11] says the umbilicus is nothing other than four vessels, that is, two arteries and two veins, which have the urachus in their midst. But according to Galen, *On the Use of the Parts*, XV, chap. 4,[12] two veins behind the umbilicus pass or blend into one, and thus we can say there are five vessels in the umbilicus and [by a different enumeration] four. The urachus is a duct which leads off the urine from the bladder of the foetus. The vein described above is implanted in the hollow of the liver. The artery or rather arteries of the aorta are implanted between the loins in the sides of the bladder.

On the contrary, the mirach [epigastrium] is made up of only two parts, that is, skin and siphac [peritoneum]; if it were otherwise, in the suture of a wound the muscles would be ligatured, which is inconvenient since the ligature impedes their operation.

There are three types of muscles, two longitudinal, two transverse, four oblique.

Galen in *On the Inner Parts*, V, Chap. 7[13] says eight muscles are placed beneath the skin, that is, four pairs, the first pair longitudinal and of the nature of flesh which descends from the bones of the ribs to the bones of the pubis. An abscess in it is a long one; in fact it occupies the lunar region of the liver.[14] Three other fleshy pairs stretch up to the longitudinal muscles but under them end in pellicular tendons. The second oblique pair descends from the chest, but not in a straight line, to the interior region in front. Another oblique pair begins from the lower parts, ascends also not in a straight line under the descending pair. The pair transverse to the oblique muscles touches the little skins of the cavity, that is, the siphac, or peritoneum.

The oblique muscles cross each other on the right and the left and thus make two crosses. The crossing appears in the substance of the muscles but not in that of the tendons, because the upper muscle has a tendon while the tendon of the lower muscles does not appear, with the exception of that of the transverse muscle.

Or the tendons of the upper muscles may make a third cross in the middle near the pubis.

Separate the tendon of the transverse muscles from the siphac [peritoneum] and demonstrate each of them.

Cut the siphac [peritoneum] or rather beritheron and demonstrate the natural location of the intestines.

The mirach [epigastrium] swells when the siphac [peritoneum] is burst.

Make a lateral incision in the peritoneum: as an example, on the left below, immediately above the bone of the ilium.

The protruding omentum [girbus] must be tied, or a

P. Lockwood, *Igo Benzi: Medieval Philosopher and Physician, 1376–1439* (University of Chicago Press, 1951). I cannot find the passage to which Achillini refers in Hugo Senensis, *Consilia Saluberrima ad Omnes Aegritudines* (Bologna, 1482; Venice, 1518).

[9] Avicenna, *Canon* III, Fen 22, Tractatus I (ed. Petrus Antonius Rusticus Placentinus, Venice, 1522) 299: De forma zirbi et duorum siphac. Cap. I: Oportet ut scias quod super ventrem post cutem sunt duo panniculi quorum unus nominatur siphac et comprehendit intestina et calefacit ea sua spissitudine et unctuositate et comprehendit lacertos et secundus est interior et nominatur beritero et nominatur rotundus quoniam quando singularis fit ab eo quod tegit ipsum est sicut spera super quam sit sumen et additiones molles et foramina et continuatur desuper velamen et processio eius est ex supremo et est subtilis sub cute ventris et panniculo eius et adherent ei duo ex lacertis ventris. . . . A much simpler version is that by Gerard of Cremona (Venice, Giunta, 1595) I, 961–962: Primum ergo, quod occurrit de uentre, est cutis, deinde sub ipsa est panniculus primus, et nominatur aggregatio eorum mirach: deinde sunt lacerti, postea beriterum, deinde zirbum, deinde intestina. . . . Oportet, ut scias, quod super ventrem post cutem sunt duo panniculi: quorum unus nominatur siphac. . . .

[10] *Expositio Jacobi Forliuiensis in primum Auicenne Canonem*, etc. (Venice, Jacopo Pentio de Leuco, 1508) fol. 44 v: Sciendum quarto ut patet in anothomia. Panniculus ventris est duplex. Quidam annexus musculis ventris. Alter autem immediate tegens intestina. Et secundum quosdam utrumque nominatur siphac. Secundum quosdam vero non. Sed primus secundum quosdam proprio nomine vocatur beliteron. Secundus siphac. Secundum vero quosdam econtra etc. . . . Sed qualiter hec vera sint videbitur in dubiis litteralibus. Jacopo of Forlì is speaking of panniculi, not of muscle tendons as Achillini does.

[11] Galen, *De Uteri Dissectione Liber* (Vol. II K., 907): Nihil enim aliud est umbilicus, nisi quatuor haec vasa in medio urachum habentia. For an English translation of this treatise see Charles M. Goss, "On the Anatomy of the Uterus," *Anatomical Record* **144** (1962): pp. 77–83.

[12] Galen, *De Usu Partium*, XV, 4 (Vol. IV K., 227): Si quidem venas videre est statim, ut umbilicum superarint, inter sese mutuo coire unamque effici . . .

[13] Galen, *De Interioribus*, V, 7 (*De Locis Affectis*) (Vol. VIII K., 346): . . . musculorum sub cute locatorum, qui octo numero sunt, quatuor contenti conjugationibus. . . . See also Mundinus, Singer's translation, p. 62.

[14] This sentence is clearly misplaced out of context and bears more relation to the passage referred to by note 47 than it does to the passage here.

ligature made in it; later it is cut off because it putrefies immediately upon contact with the air.[15]

The intestine is sewed up and placed back inside.

The mirach [epigastrium] is sewed up and so on. You do not sew up the muscles, for they do not form part of the mirach nor must they be impeded in their operation. Then the peritoneum is the last to be sewed up because it is above the muscles according to Avicenna[16] and not under them; but according to the first book of Almansor[17] the peritoneum is under the muscles. And thus a discord concerning terms exists between Avicenna and Almansor as to what you call the peritoneum.

The omentum [girbus] covers the stomach and all the intestines. It is of triple substance, with two thin pannicles containing flesh. The flesh is thick or fat with (*seposa*)[18] arteries and veins. However, since an artery differs from a vein the omentum is of fourfold substance and not only threefold.

Avicenna said in the *Canon,* Fen 22, Third Book, chapter 1,[19] that the arteries are very numerous and the veins are fewer.

The connection of the omentum is with the stomach, spleen, intestines, and especially the colon.

It arises from the fleshy pannicle [mesentery of small intestines] out of the back within the diaphragm, since that is the place where the extremities of the pannicles which clothe the peritoneum terminate. From the large vein and artery [therein] arise [the small veins and arteries] which cover [the stomach] and the mesentery, thick with arteries and veins (*seposum*).

The intestines are six in number: three slender ones above, three thick ones below.

According to Almansor[20] and Averroes' [*Colliget*],[21] the first book, the intestines are composed of two tunics not because of their contents but because of the danger of their excoriation. Inside them they have a clinging viscosity like a stagnation [mucous secretion]; there are fibers [villi] inside and outside of them.

When the straight intestine has been cleansed of the feces it should be tied up in two places and then cut between the ligatures.

Galen in his book *On the Anatomy of the Uterus*[22] says that the anus is said to comprise the entire rectum intestine, but the dilated lower part of the rectum at its end is called the sphincter. Note the great number of meseraic blood vessels which terminate at the intestines.

Galen in his book *On the Anatomy of the Living*[23] says the rectum has four muscles at its extremity by which it is squeezed. In order that the feces may be retained the rectum has a strong power of attraction. It does not have transverse fibers because its expulsion is from the inside toward the outside.

---

[15] Berengario da Carpi, *Commentaria* etc. (1521), CV (against Hippocrates and Galen) and CVI (against Niccolò Leoniceno), opposes the view that an extruded omentum putrefied instantly on contact with the air: Hippocrates, *Aphorismi* VI, 58: si omentum eruperit necessario putrescit; Galen, *Comment. Aphor. Hippocr.* VI, 58 (Vol. XVIII K., A. 96): omentum putrescit citius quam alia membra etc., where Galen contradicts Hippocrates; and see Aristotle, *Problemata*, XIII, 4, who also raises the question. I discuss the controversy between Berengario and Leoniceno in Jacopo Berengario da Carpi, *A Short Introduction to Anatomy* (*Isagogae Breves*), translated with an introduction and historical notes by L. R. Lind, with anatomical notes by Paul G. Roofe (University of Chicago Press, 1959), p. 23.

[16] See note 9 and the two versions of Avicenna's description of the mirach given there.

[17] Muhammad Ibn Zakarīyā, Abu Bakr, al-Rāzī, *Liber Rasis ad Almansorem*, etc. (Venice, Jacopo Pentio de Leuco, 1508), Tractatus Primus, fol. 6 v, cap. xxiii: Sub cute qua venter exterius cooperitus interius octo inueniuntur musculi quos nominauimus. Post quos interioribus est panniculus quidam rugosus qui vocatur siphac, post quem est girbus, post girbum vero sunt intestina. Ruptura vero que in mirach accidit est quando rumpitur hoc siphac.

[18] Charles Singer, translation of Mundinus, p. 102, note 24, is uncertain as to the meaning of the term *sepose* but on the basis of a study of Galen, *De Usu Partium*, IV, 2, V, 2, thinks that Mundinus refers to the pancreas by the phrase "sepose mesentery." I think the Latin word "saepe," "frequent," is the source of the adjective and that it means "thick, full of arteries or veins" in this passage and in the following (*seposum*).

[19] Avicenna, *Canon*, translated by Gerard of Cremona, with Johannes Costaeus's index; (Venice, Giunta, 1595), I: p. 961: . . . et zirbus, quidem est compositus ex duobus panniculis, quorum unus cooperit alium, inter quos sunt arteriae plurimae, et uenae pauciores eis. . . .

---

[20] Muhammad Ibn, etc., *op. cit.* fol. 5 v, cap. xvi: Intestina ex duabus tunicis componi reperiuntur. Super interiorem vero tunica viscositas quaedam adherens reperitur sicut stagnatio. . . .

[21] Averrois Cordubensis, *Colliget*, Libri VII (Venice, Giunta, 1574), I, chap. 26, fol. 11 v: Intestina composita sunt ex duabus tunicis.

[22] Galen, *De Uteri Dissectione Liber* (Vol. II K., 888): Sedem quidem totum rectum intestinum voco; laxam vero sedis partem inferiorem recti intestini partem, ubi maxime laxatur. Ad eius finem quoque est sphincter vocatus. . . . See the translation by C. M. Goss in *The Anatomical Record* **144** (1962): p. 78.

[23] The pseudo-Galenic *De Anatomia vivorum* is extant, according to C. G. Kühn (his edition of Galen, I, 1821, p. CLX), only in Chartier's edition (IV, 194) and in the Giuntine Galen (VII, 1576, ff. 43 v–56 v); but the excellent and very helpful list of editions by Richard J. Durling, "A Chronological Census of Renaissance Editions and Translations of Galen," *Jour. Warburg and Courtauld Institutes* **24**, 3–4 (1961): pp. 230–305, shows it also in the Pavia edition *Galeni Opera* (1515) and in the *Opera Omnia*, A 6–18 (*ibid.*, pp. 280, 283). This attributed work was published by Robert Ritter von Töply as the *Anatomia Richardi Anglici* (Vienna, 1902); see his *Studien zur Geschichte der Anatomia im Mittelalter* (Leipzig and Vienna, 1898), pp. 14–15, for a discussion of the work. George W. Corner, *Anatomical Texts of the Earlier Middle Ages* (Carnegie Institution of Washington, 1927) has an analysis (pp. 35–47) and an abbreviated translation with the Latin text (pp. 87–110). I quote the work in *Galeni Operum non Extantium Fragmenta quorum maior pars numquam prius edita*—prima editio (Venice, Giunta, 1576), fol. 50 v: Et in sua extremitate quatuor habet lacertos, quibus mediantibus anus clauditur ut teneatur faex et expellatur. Istud autem intestinum habet villos longitudinales, quibus mediantibus fortem habet attractionem. Sed quaeritur, quare non habeat villos latitudinales propter compressionem, quae ei necessaria esset. Ad hoc dicendum, quod ab extrinseco habet iuuamentum fortis expulsionis, quia in mirach pellicula inuoluente intestina sunt octo lacerti, quibus mediantibus fit expulsio fortis ab isto intestino.

The colon, according to Almansor,[24] begins near the ilia on the right. It is much involved or wound around the left kidney. It ascends, covers the spleen, rides above the stomach, and hence the material contained in it causes syncope. The omentum is closely bound to the colon. On the right it is covered over by the envelope [lobe] of the liver where the bile cyst is attached and thus colic pain grows when food is taken, but not kidney pain.

The monoculus [cecum or sack intestine], according to Mundinus,[25] has two orifices, but Almansor in the first book;[26] Avicenna, Fen 16, Third Book;[27] Averroes;[28] Albertus on the fourth book of Aristotle's *Meteorologia*;[29] and the Commentator on the *Book of Sects*[30] say it has only one orifice, serving for attraction and expulsion at two different times. It is located on the right near the hipbone, below the right kidney; however, it contains nothing in itself from the latter which is useful for nourishment.

The ileum intestine is thin; it lies near the ilia and is slender in substance. It slants down to the vertebrae on the left. It ends in many meseraic vessels; therefore it is rendered empty. The meseraics are branches of the portal vein, which is strongly embedded in the flesh of the mesentery.

Almansor[31] in the first book calls the ileum involute because it is involved in many turnings, but the jejunum is so called because it is empty, since the yellow bile falls into it from the end of the duodenum where it issues from the cyst, stimulating expulsion.

In the book *On the Anatomy of the Living*[32] the ileum is so called from [the Greek word] hyle because it is confused and twisted.

According to Almansor,[33] the duodenum and rectum extend lengthwise in the body so that matter falling into them may be expelled.

There are three causes for emptiness in the jejunum: the multitude of sucking meseraics; the freshness and large amount of yellow bile which falls into it and stimulates expulsion; its erect position.

Because the duodenum is twelve fingers [long] the Commentator on the *Book of Sects*[34] says every man has one as long as the measure of his fingers.

In the *Anatomy of the Living*[35] a nerve is said to dominate the duodenum throughout its length by whose action the intestine expels its contents; although it has a weak expulsive force its power of attraction is strong.

Do not break the canal of the black bile at the extreme end of the duodenum.

The intestines are attached to the back by means of the mesentery. It is composed of veins, tendons, panniculi, ligaments, a fat substance thick with veins and arteries (*seposa*), a glandular flesh which is thick in the upper part; within this [mesentery] is gathered the melancholia mirachia [black bile].

Inflate the stomach. Its place is the middle one of the six positions.[36] The heart is above it with the diaphragm between; below are the mesentery and the intestines, on the right the liver, on the left the spleen, in front the omentum; behind are the muscles of the back, the great vein and the artery, which descends to the left above the stomach; lower down it descends to the right.

The stomach is attached to the spleen by the vein which conveys the black bile to the mouth of the stomach and by the vein which carries nutriment to the left side of the stomach. It is attached to the liver by two meseraics which come to the bottom of the stomach. The latter is attached to the brain by a nerve which spreads through the upper part of the stomach near its mouth. It is attached to the heart by the great artery which is under the stomach.

The form of the stomach is circular, round, and ample like a twisted gourd. Those who have the rotund part integral with the oblong part, that is, a rotund stomach, breathe with difficulty after eating.

There are two tunics for the stomach. The exterior fleshy one is for digestion; it has transverse fibers. The inner membrane with sensation has longitudinal

---

[24] Muhammad Ibn, etc., *op. cit.*, fol. 5 v, cap. xvi: Post istud [i.e., monoculum] autem continuo sequitur intestinum quod colon vocatur, cuiusquidem principium in dextro existit latere, sed ipsum per latitudinem ventris usque ad latus sinistri porrigitur.

[25] Mundinus, Singer's translation, p. 66.

[26] Muhammad Ibn, etc., *op. cit.*, fol. 5 v: Post hoc [i.e., inuolutum] vero aliud sequitur intestinum quod monaculum [*sic*] vocatur. Hoc autem est intestinum amplum unum tantum habens foramen ac si esset saccus vel marsuppium. Hoc quoque os unum per quod illud quod hora una ingreditur alia egreditur hora quod est in dextro locatus latere.

[27] Avicenna, *op. cit.*, p. 808, col. 2: . . . et huic quidem intestino, sufficit orificum unum.

[28] Averroes, *op. cit.*, fol. 11 v: Et post ista est intestinum nominatum orbum et istud intestinum est latum non habens transitum nec iter sed simile est fisculo vel marsupio et quia non habet nisi unum os.

[29] Albertus Magnus, *Opera Omnia* . . . cura et labore Augusti Borgnet, IV (Paris, 1890): p. 744–745: intestinum quod dicitur monoculum, quia non habet nisi unum oculum. . . .

[30] Commentator on the *Book of Sects*; see note 1. *Op. cit.*, fol. 7 r: Et post hoc cecus intestinum et dicitur cecus quia unum habet porum unde ingreditur et egreditur et est positum in dextra parte.

[31] Muhammad Ibn, etc., *op. cit.* fol. 5 v: Precedentem vero intestinum sequitur quoddam aliud quod inuolutum vocatur quod multis inuolui inuolutionibus inuenitur.

[32] Pseudo-Galen, *De Anatomia Vivorum,* fol. 50 r: Hoc intestinum sequitur aliud, quod ileon appellatur, ab ile, quoniam confusum siue inuolutum appellatur. . . .

[33] Muhammad Ibn, etc., *op. cit.* fol. 5 v: Post hoc autem rectum sequitur intestinum in quo cum amplam habeat concauitatem colligitur stercus quemadmodum in vesica adunatur urina.

[34] Commentator on the *Book of Sects*, *op. cit.* 7 r: . . . post ieiunum intestinum est dodocadactilis, qui ideo sic dicitur quia omnis homo ad mensuram suorum digitorum eum longum habeat. Herophilus was the first to name the duodenum intestine: Galen, *De Anatomicis Administrationibus*, VI, 9 (Vol. II K., 572).

[35] Pseudo-Galen, *De Anatomia Vivorum*, fol. 50 r: Dominatur autem in hoc intestino neruus secundum latitudinem, quo mediante expellit contenta a se.

[36] See note 6.

and oblique fibers. The longitudinal fibers are inside, the oblique outside the first tunic and under the second.

Galen, however, *On the Use of the Parts*, IV,[37] says there are three tunics for the stomach. Others reply that Galen numbers the universal panniculus with the other members. On the contrary, Almansor in his first book[38] and [Averroes][39] in the first book, twenty-third chapter of the *Colliget,* say the stomach is composed of three tunics. One has longitudinal fibers, another has transverse fibers, and the third has oblique fibers; but if the tunic were common it would not have fibers. Nevertheless, Galen, *On the Use of the Parts,* XIV,[40] last chapter, [says] the stomach and the intestines have two tunics.

The mouth of the stomach is double, above and below. It is first attached to the thirteenth vertebra, which is the first of the alchatim, or the vertebrae of the region of the kidneys; these are twenty in number. [The number xx is omitted in the 1522 ed.] The diaphragm, however, is attached to the bone of the twelfth vertebra following that of the kidneys, that is, the universal vertebra which is the twenty-fourth [from the neck downward] but the eighteenth vertebra of the back.

The spleen appears when some ribs of the left side have been broken away; it clings along its concave side to the left wall of the stomach, along its hump [or convex side] to the back and to the peritoneum and the very thin pannicles arising from the latter. It is not situated so high up as the kidney; its shape is quadrangular and its substance thin and spongy.

The connections of the spleen are with the kidney, the mesentery, the omentum, and the stomach, since there issues a large vein from the gate [of the liver, i.e., the portal vein] whose first branch brings the more watery and humid blood to nourish the mesentery. The second branch passes near the spleen to nourish the left lower part of the stomach. Later this vein enters the spleen in whose hollow part it becomes bifurcated. The upper branch is split into other branches. One of these nourishes the upper left-hand side of the stomach. Another branch carries the black bile to the mouth of the stomach. This excites the appetite. The lower branch nourishes the omentum on the left. The spleen is attached to the heart by arteries.

According to Averroes in the first book of his *Colliget*, chapter six,[41] the portal vein is divided in the hollow of the liver into five branches which spread out to the capillaries and to the humped [convex] part of the liver where they meet the capillaries of the chilis vein [inferior vena cava] providing the material of the blood for them: Averroes in *Colliget*, Book I, chapter twenty-seven.[42]

Beyond the liver the portal vein is divided into eight parts. The first of the small veins [thus created] goes to the duodenum; the second, which is bifurcated, to the portanarius [pylorus] of the inner stomach; the third goes to the lower right hand part of the stomach to nourish it; the fourth, split into three, goes to the spleen. One of these three branches goes above to the left-hand part of the stomach; the second goes to the right-hand upper part, and the third to the stomach's humped or convex part. This third branch is also divided into two parts: one goes to the lower left-hand part of the stomach, the other to the mouth of the stomach for the black bile. The fifth of the small veins goes to the rectum intestine, the sixth to the omentum of the stomach, the seventh to the colon, and the eighth to the jejunum and the blind gut (*orbum*) [cecum].

It seems that the branch to the upper part of the stomach is divided into right and left parts, and the branch which descends to the lower region is also divided into right and left parts. Almansor in his first book[43] says that the portal vein does not carry blood; so does Averroes, *Colliget* I, chapter 27,[44] but in chapter 7[45] he had said a branch of the portal vein comes to nourish the upper part of the stomach in the right-hand region of the back. Therefore, either this vein carries blood or the stomach is nourished from outside but not with blood. Then Galen insisted that the spleen in some manner makes blood; therefore, the vein will carry impure blood from the spleen.[46]

---

[37] Galen, *De Usu Partium*, IV, 8 (Vol. III K., 282): Galen says there are only two tunics, however, in this passage: externa vero carnosior transversos [villos habet], oblique, not transverse fibers. Interna namque tunica in ventriculo simul ac stomacho membranosior existens villos rectos superne deorsum tendentes habet, a description which does not agree with Achillini's *longitudinal* and *oblique* fibers.

[38] Muhammad Ibn, etc., *op. cit.*, chap. xv: Stomachi quoque corpus ex tribus componitur tunicis, quarum una villos suos in longitudinem habet transeuntes.

[39] Averroes, *op. cit.*, I, 25 (not 23), fol. 11 r, v: Et corpus stomachi compositum est ex tribus tunicis, textura unius tendit in longitudinem et alia vadit per diametrum, et illa est intrinsecus, et est illa tunica, id est, neruosa et exterior est carnosa et textura ipsius tendit in latitudinem.

[40] Galen, *De Usu Partium*, XIV, 14 (Vol. IV K., 205): Cur autem intestina omnia et ventriculus ex duabus tunicis constiterint, matricibus vero una satis fuerit, quemadmodum et utrique vesicae. . . .

[41] Averroes, *op. cit.*, I, 6, fol. 6 r: Sed porta diuiditur in corpore hepatis in quinque partes. Et ramificantur hae quinque partes in ramos paruos, donec iungantur gibbositati hepatis.

[42] *Ibid.*, I, 27, fol. 11 v: Et istud canale, quod est in porta hepatis, diuiditur etiam in partibus interioribus ipsius hepatis ad partes minutas, scilicet ad venas capillares similes capillis.

[43] Muhammad Ibn, etc., *op. cit.*, chap. xvi: Ex huius [portae venae] autem concauitate canalis quaedam nascitur quae porta vocatur epatis, cuius figura vene tenet similitudinem, sanguinem vero non retinet (*cf.* also fol. 6 r, col. 1, middle).

[44] Averroes, *op. cit.*, chap. 27: forma ipsius est forma venae, nisi quia est non habens sanguinem.

[45] *Ibid.*, chap. 7 (actually 6): Sed una ex sex [partibus] tendit ad superficiem dorsi partis dextrae stomachi ad nutriendum ipsum.

[46] I can find no passage in Galen which says exactly this; but see *De Usu Partium*, V, 6 (Vol. III K., 372) where the spleen is said to

Others reply that the portal vein carries nutriment, not blood, through the meseraics from the stomach and intestines; but the portal vein sends the blood through veins transmitted to the members of the body. Yet what about the veins to the intestines?

Lift up the body which is being dissected. The liver on the right embraces the stomach with five lobes; sometimes it has four lobes. Almansor[47] in his first book says it has a crescent shape because it has both a hollow and a convex side; therefore, its abscess is crescent-shaped. On the inner side facing the stomach it is hollow. Its veins form a net; the empty spaces or interstices of the net are filled with flesh. The net is filled to its smallest division with food so that blood may be better made.

The liver is connected with the heart through the chilis vein [inferior vena cava] and the artery of the aorta, with the diaphragm by being suspended from it, and with the vertebrae; with the latter it is connected by a membrane. This membrane is double and surrounds the substance of the liver; by means of this membrane an extensive pain due to flatulence is felt. This membrane suspends and binds the liver to the diaphragm above, whence is felt a heavy pain from the weight of its humor; and thus it communicates with the brain.

Two other parts originate [here]: the veins; the gall bladder; two veins: chilis and portal. Aristotle[48] says the vena cava [chilis] originates from the heart; Galen[49] says it originates from the liver; and Averroes[50] agrees with Galen.

The substance of the liver is coagulated blood; it is completely like blood.

Its illnesses are dropsy, fleshy hyposarca, watery ascites [dropsy of the peritoneum], and windy tympanites [or emphysema, swelling of air in the abdomen].

The bile cyst is in the hollow of the liver in its middle lobe. It cleanses the yellow bile through a quite small branch so that the thinness of the humor may be dried out; this small branch enters the large neck of the cyst which extends to the duodenum. Thus on its way the thin branch is bifurcated near the bottom of the stomach, carrying yellow bile to stir up digestion but not to expel the digested contents. Because sometimes this branch is thick some people who have such a branch naturally throw up the yellow bile, and they are by nature unfortunate. The large branch extends to the end of the duodenum toward the jejunum in order to stimulate expulsion.

The substance of the bladder is pellicular or membranous. In the book *On the Anatomy of the Living*[51] it is said to be composed of longitudinal, transverse, and oblique fibers; to it come arteries, veins, and nerves.

Lift up the spleen and the liver but not entirely on account of the place of origin of the chilis vein [inferior vena cava] and the aorta artery, which is branched in the direction of the kidneys into two emulgent veins that draw the watery blood from the liver and enter the kidneys; the arteries enter with them. The orifice of the right-hand emulgent vein is commonly higher than that of the left-hand emulgent. Cut the kidney lengthwise and there will appear the colatorium [sieve] which separates the blood from the moisture of the urine. The colatorium is a thin membrane made of the emulgent vein when it becomes thinned out; from this sieve the urine is distilled as it passes to the orifice of the kidney where the urethral duct [ureter] carries the urine to the bladder. The illnesses of the kidney are stone, hair, etc.

In 1503 I saw a freak who had two urethral ducts [ureters] on the left and only one on the right.[52]

The kidneys have connections with the brain by nerves which come from the spinal cord to the membrane which covers the circumference of the kidney and through which the kidney has sensation; for the binding vestment is transmitted from the vertebrae. The connection of the kidney with the heart is by an artery that comes from the heart through the middle of the liver to the kidneys. With the liver, as in the *Anatomy of the Living*,[53] [connection is made] by

---

draw blood to itself (*cf.* also p. 320), and George W. Corner, *op. cit.*, p. 102 on the functions of the spleen in *De Anatomia Vivorum*. The idea probably originates with Aristotle, *De Partibus Animalium*, II, 7, 670b 5: "For the spleen attracts the residual humors from the stomach, and owing to its bloodlike character is enabled to assist in their concoction" (translated by D'Arcy Wentworth Thompson in the Oxford Aristotle, Vol. IV). Compare Hippocrates, *De Morbis*, IV, 9, and *De Morbis Mulierum*, I, 15, for the notion that the spleen attracts superfluous humors, and *Dictionnaire Encyclopédique des Sciences Médicales* **11** (1883): p. 529, *s.v.* splénotomie: La rate, en effet, est, de par la physiologie, un des principaux centres d' élaboration des globules blancs du sang. . . .

[47] Muhammad Ibn, etc., *op. cit.*, chap. xvii: Cuius quidem figura est lunaris quod interius a parte stomachi concauum cernitur. . . . A parte autem diafragmatis inuenitur eius gibbositas. . . . *Cf.* Singer's Mundinus, p. 71.

[48] Aristotle, *De Historia Animalium*, III, 513 a (Oxford Aristotle, **4** translated by D'A. W. Thompson): "These blood vessels have their origin in the heart, for they traverse the other viscera, in whatever direction they happen to run, without in any way losing their distinctive characters as blood-vessels, whereas the heart is as it were a part of them (and that too more in respect to the frontward and larger one of the two) [i. e., the *vena cava*] owing to the fact that these two veins are above and below, with the heart lying midway." Note 3 lists the following parallel references: *De Partibus Animalium*, III, 4. 665 b 16, 666 b 25, V, 667 b 15; *De Respiratione*, 14 (8). 474 b 7.

[49] Galen, *De Semine*, I, 8 (Vol. IV K., 541): spinalem namque medullam cerebrum velut truncum quendam producti, cor vero maximam arteriam, quam aortam Aristoteles appellat, hepar vero venam cavam. See also Galen, *De Placitis Hippocratis et Platonis*, VIII, 1 (Vol. V K., 657–659).

[50] Averroes, *op. cit.*, I, 6: Alia est vel exit a gibbositate hepatis . . . et nominatur concaua.

[51] Pseudo-Galen, *De Anatomia Vivorum*, fol. 51 r: contexta est autem exterior tunica ex filis longitudinalibus, et viget in ea magis attractiua. Interior autem tunica contexitur ex filis latitudinalibus et hoc propter comprehensionem, et ex transuersis, propter retentionem.

[52] Compare the somewhat similar anomaly reported by Berengario da Carpi, *A Short Introduction to Anatomy*, translated by L. R. Lind (Chicago, Ill., 1959): pp. 66–67.

[53] Pseudo-Galen, *De Anatomia Vivorum*, fol. 51 r: Item quaedam venae diriguntur ab hepate ad ipsos renes exterius, deferentes sanguinis nutrimentum ad exteriores partes renum. Interius enim non habent renes uenas sed nutriuntur a sanguine quam attrahunt per canales cum aquosa superfluitate.

veins which bring blood to nourish the exterior of the kidneys, for the interior is nourished with blood collected from the wateriness.

The substance of the kidneys is fleshy, solid, bloody, covered with a membrane: they have a colatorium, hollowness, oblong shape.

The seminal vessels are two, on right and left. The left-hand vessel arises from the left-hand emulgent [renal] vein in the middle between the chilis vein [inferior vena cava] and the kidney. The right-hand vessel arises within the right-hand emulgent vein above the lower part of the kidney. An artery originates from an artery, and thus the kidney is a bed for the vessels.

Avicenna,[54] however, said in his anatomy that two veins come from the left emulgent vein, and twice I saw them thus in 1502 and 1503. Galen[55] is doubtful about the matter in his *On the Use of the Parts*, XIV, 7; his solution is that the right-hand vessel is warmer and hence carries warmer material: chapter 8 for [the generation of the] male.

Why does the sperm for the generation of the female come from the left-hand seminal vessel? I reply that the left-hand vessel issues from the emulgent that is full of watery blood, that is, before the blood is cleansed through the kidney. The sperm for the generation of the male issues from the right-hand vessel because that vessel arises full of clean blood from the vena cava, that is, after the kidney sucks out the watery part, also because the branch of the artery is mingled with the right-hand vessel and thus carries more copious spirit. Hence it is apparent that the elevation of the left testicle after coitus is a sign that a female will be born because more of the matter at the left has issued forth, just as the elevation of the right testicle is a sign that a male child will be born because more matter passing through that vessel draws more of the matter of generation to itself.[56]

In the *Anatomy of the Living*[57] the preceding vessels are some veins which pass from the brain to the testicles and carry to them a whitened blood which has been used for the nourishment of the brain.

In the woman the seminal vessels descend to the uterus and outside it are woven together; near her testicles they are entwined in a texture or network filled with glandular flesh. In these pieces of glandular flesh there is generated the salival humidity which delights a woman in coitus. The vessels penetrate to the hollow of the uterus and are called cotyledons. By means of these the foetus is attached to the uterus, and the menstrual blood is carried by them. Some of them reach the mouth of the uterus to carry the humidity called salival. From these, two veins branch off on right and left, penetrating to the mirach [the abdomen or epigastrium]. They ascend to the skin and to the breasts, to which they come forth from the depths of the chest in the direction of the pomegranate [processus xiphoideus].

The cotyledons are two veins which ramify into many branches, some of them turning back to one trunk, again ramifying into many branches and again returning to a trunk. The same thing is true of two arteries.

The matrix or uterus is located in the concave region called the alchatim within the peritoneum and thus lies upon the rectum intestine and in front of it up to the femur. Galen in his book on the *Anatomy of the Uterus*[58] says the neck of the matrix [uterus] rests under the neck of the bladder, and in his book *On the Anatomy of the Living*[59] he says the neck of the matrix is a penis turned inside out.

The matrix [or uterus] is connected with the heart by arteries, with the liver by veins, with the brain by many nerves; by these nerves it is connected with the diaphragm, kidneys, abdomen, and especially the breasts, bladder, colon, and the slender intestines. It is attached to the hipbones and joints of the scia [ischium] by thick, strong ligaments near the womb but by thinner ligaments projecting near the hipbones like horns from the head of an animal. Therefore they are called the external horns, for the uterus has internal horns also to which the foetus is attached. They are like the head or nipples of the breasts. In the *Anatomy of the Living*[60] ligaments descend [from the uterus] to the knees and feet; on its upper part it is attached to the diaphragm. Ligaments ascend from it up to the breasts and eyes.

The form of the uterus is quadrangular but somewhat rounded; it has a long neck below. Mundinus said its size is moderate like that of the bladder.[61] Galen said in his book *On the Anatomy of the Uterus*[62] that it is commonly eleven fingers [7.7 inches in size]. Its upper part is near the umbilicus, the lower toward the vulva. Its substance is double.

[54] Avicenna, *op. cit.,* I, 69, col. 2: A sinistra autem venarum emulgentium duae procedunt venae.

[55] Galen, *De Usu Partium,* XIV, 7 (Vol. IV K., 175): Potest tamen accidere, ut interdum a caloris, qui semini inest, vi subacta masculum pro foemina foetum fieri permittat. This chapter and the next are devoted to the problem of generation by both male and female and the determination of sex by means of the right or left testicle. See p. 170 on the seminal vessels.

[56] Galen, *De Usu Partium,* XIV, 7 (Vol. IV K., 173–174).

[57] Pseudo-Galen, *De Anatomia Vivorum,* fol. 50 r: Vasa autem antecedentia sunt quaedam venae quae prodeunt a cerebro usque ad testiculos et deferunt sanguinem dealbatum, qui erat ad nutrimentum cerebri, ad ipsos testiculos, qui mediante sua virtute attractiua talem sanguinem attrahunt ad nutrimentum; residuum autem illius nutrimenti a testiculis transmittitur per seminaria vasa consequentia in ipsa virga virili et per illam in matricem ad generationem foetus.

[58] Galen, *De Uteri Dissectione,* I (Vol. II K., 887–890), gives a general description of the location of the uterus. C. M. Goss, *op. cit.,* p. 78 translates: "At the part toward the pudendum, the bladder is above the neck of the uterus."

[59] Pseudo-Galen, *De Anatomia Vivorum,* fol. 51 r: Instrumentum femine habet figuram inuersam et protensam, id est, fundatam intus; instrumentum hominis habet inuersam et protensam extra.

[60] Pseudo-Galen, *De Anatomia Vivorum,* fol. 51 v: Similiter eius ligamenta descendunt ad genua, a genibus usque ad pedes. Superius autem alligatur diaphragmati, cuius ligamenta ascendunt usque ad mammillas et oculos.

[61] Mundinus, Charles Singer's translation, p. 76.

[62] Galen, *De Uteri Dissectione* I (Vol. II K. 889): " . . . the interval is usually proportionate to 11 fingers." Translation by C. M. Goss, *op. cit.,* p. 78.

Galen,[63] however, in *On the Use of Parts*, XIV, last chapter, said that the stomach and intestines have a double tunic but that the uterus and the bladder have one tunic. But, given that there are two tunics, there is a disagreement here, for the tendinous tunic is external but the one with more veins in it is single and within the external tunic. The inner tunic is double, but in *On the Anatomy of the Living*[64] the external fleshy tunic is thicker, the inner sinewy one composed of longitudinal, transverse, and oblique fibers.

The exterior parts of the uterus are the sides to which the testicles are attached. The seminal vessels are entwined around the two horns on right and left. The extremity of the neck is the vulva. The neck is a palm in size. The side can be dilated; it is made of wrinkled skin. The extremity of the vulva is closed by two little skins; they correspond to the little skin which covers the prepuce [in the male]. It is written in *On the Anatomy of the Living*[65] that the neck of the uterus can be constricted because it alone has the urinal meatus to itself; but the mouth of the uterus inside is so compressed that a needle cannot enter it. Yet at the time of childbirth it opens to the extent of its entire cavity.[66]

Cut the uterus down the middle. In a virgin its mouth is covered by a thin membrane, but in a violated woman this is broken. The mouth of the uterus is like that of an old tench.[67]

There are three cells in the right side of the uterus, three on the left side, and one in the middle. A cell is a cavity in the uterus in which the sperm can coagulate the menstrual blood and bind it to the orifices of the veins.[68]

However, in the book *On the Anatomy of the Living*[69] some people who are led astray say that there are five cells; some say seven, but it must be admitted that more than one foetus can adhere to the same horn of the uterus.

Likewise, Galen *On the Use of the Parts*, XIV, chap. 4,[70] says that just as the entire body is divided into right and left sides so the sinus or fold of the uterus is divided because a multitude of sinuses is created for these uteruses, although in other animals who must bear many offspring this very need has created many sinuses, just as it has created two uteruses terminated at one neck and so on.

The seminal vessels stretch to the testicles, which are outside the body, causing slowness in coition.[71] They ride above the pubic bone. The testicles are covered by the didymus [tunica vaginalis], which is a membrane arising from the peritoneum, drawn tightly together at its place of origin but dilated in proportion to its contents.

At the top of the testicles the seminal vessels make a revolution around part of the testicle and reascend through the pubic bone to the neck of the bladder and when at that point they enter the penis as far as the testicles they are called preparatory vessels. But from the testicles to the penis they are called delatory [ductus deferens] or vessels which carry away, being more tendinous or sinewy than before.

There is a disagreement between Galen and Aristotle[72] as to the place where the sperm is generated, whether in those circumvoluted vessels, as it seems to Aristotle, or in the testicle, as Galen[73] believes.

In order to harmonize this disagreement some people say that the vessels in the place referred to are parts of the testicle. Therefore it is the same thing for the sperm to be completed in the vessels and in the testicles, and they agree that the preparatory vessels and those which carry away the sperm [the delatory]

---

[63] Galen, *De Usu Partium,* XIV, 14 (Vol. IV K., 205); see note 37.

[64] Pseudo-Galen, *De Anatomia Vivorum,* fol. 51 v: Componitur autem matrix ex duabus tunicis, quarum exterior magis carnosa est, interior vero magis neruosa. . . . Exterior autem spissior est . . . interior autem ligamentalis est, id est, ex ligamentis, et componitur ex omni genere neruorum, scilicet longitudinalibus, latitudinalibus, et transuersalibus.

[65] *Ibid.:* ita strictus est meatus colli matricis quod solummodo viam habet urina per ipsum.

[66] See the eloquent description of this marvel of nature by Galen, *De Locis Affectis,* VI, 6 (Vol. VIII K., 446).

[67] See Mundinus, Singer's translation, p. 76; Berengario da Carpi, *A Short Introduction to Anatomy,* translated by L. R. Lind (Chicago, Ill., 1959); p. 78; Aristotle seems to be the first to make the comparison.

[68] Charles Singer traces the seven-cell uterus to Michael Scot, *De Secretis Naturae,* 7, from whom Mundinus seems to have taken it although the idea goes back to the Hippocratic collection: Singer, translation of Mundinus, p. 104, note 61. Fridolf Kudlien, "The Seven Cells of the Uterus: the Doctrine and its Roots," *Bull. Hist. of Medicine* **39** (1965): pp. 415–423, traces the concept from early times and quotes Zerbi's "doxography" in *Liber anathomiae,* 1502, ff. 43v–44r.

[69] Pseudo-Galen, *De Anatomia Vivorum,* fol. 51 v: Unde quidam persimile dicebant decepti, in muliere esse quinque cellulas, quidam septem, quia tot foetus possunt simul esse in matrice.

[70] Galen, *De Usu Partium,* XIV, 4 (Vol. IV K., 150–151): . . . matricis sinus, alter quidem in dextris, alter vero in sinistris est locatus.

[71] Aristotle, *De Generatione Animalium* 717 b 11 (translated by Arthur Platt in the Oxford Aristotle, Vol. V): "When the testes are internal the act of copulation is quicker than when they are external, for even in the latter case the semen is not emitted before the testes are drawn up." Alessandro Benedetti, *De Re Medica,* XXIIII, p. 440 (1549 ed,) writes: "Seminarii praeterea meatus illi reuolutionem anfractusque habent ne libido uehemens crebraque citetur" (De pene et testiculis, ex Aristotele).

[72] Aristotle, *De Historia Animalium* III, 510 a 25: ". . . in the ducts that bend back towards the tube of the penis, the liquid is white-colored." *Idem, De Generatione Animalium* 717 a 30: "It is for this that the testes are contrived; for they make the movement of the spermatic secretion steadier, preserving the folding back of the passages (i. e., they keep the duct down in a bent and twisted state, thus retarding the passage of semen.") *Cf.* Galen, Vol. IV K., 575. The ducts are the epididymis and vasa deferentia in the vivipara, as horses and the like, and in man. (For details see the *De Historia Animalium* III, 1, which was illustrated by Aristotle with a diagram.) For the testes are no part of the ducts but are only attached to them . . ." (translated by D'A. W. Thompson in the Oxford Aristotle, **4**.

[73] Galen, *De Semine,* I, 16 (Vol. IV K., 582–583, 592).

are continuous. But it is not true that when the revolution or turning about of the preparatory vessels is lightly flayed or skinned including the revolution of the delatory vessels the ends of the preparatory vessels are found to be a third part of the testicle distant from the beginning of the delatory vessel. Thus the sperm after its exit from the preparatory vessels and before its entrance into the delatory vessels must lie in the substance of the testicle.

Castration: Pull the testicle high up near the pubic bone. Mark the place and cut, laying the testicle aside. Then tear the testicle away from the didymus and cut it off. Cut the spermatic vessel, making a ligature afterwards for the didymus and for the spermatic vessel. Then consolidate the two.

Lift up the kidney and skin the urethral duct [ureter]. You will see it ends at the middle of the bladder and does not perforate the latter with one large foramen but with small slanting foramina that stretch between the tunic of the bladder and its covering until they reach the neck of the bladder, where they perforate another tunic and fall into the hollow [of the bladder]: Rhazes in the first book of Almansor.[74] Cut the bladder. It is sinewy and hollow, but its neck is fleshy and muscular, with a thin foramen continuous with the prepuce.

The question arises: does the bladder have only one tunic? People reply that there is one tunic and one covering. Therefore it can be said that there is one tunic and that there are two tunics. The exterior one has longitudinal fibers; it is therefore proper to the bladder. Likewise, in the book *On the Anatomy of the Living*[75] the urine flows between the two tunics. The external tunic is woven with longitudinal fibers; the inner tunic, however, has transverse and oblique fibers. The inner tunic is doubled or folded toward the outer tunic.

Galen, however, *On the Use of the Parts*, XIV,[76] last chapter, says the bladder has only one tunic because it collects the superfluity.

Extraction of stone: with the patient bound and seated upon a perforated bench, thrust a finger into his anus. With one hand placed on the femur the stone can be drawn down to the neck of the bladder and thence extracted. Or it can be extracted by cutting if it is large or by drawing it out of the penis if it is small.

The penis is attached to the neck of the bladder with ligaments which arise from the pubic bone, and in the *Anatomy of the Living*[77] some ligaments also come from two broad bones with nerves that arise from the spinal cord, with large veins and arteries arising from the infurcation of the chilis vein (*vena cava*) and of the aorta artery near the hipbones. The [penis] has larger [veins and arteries] than any other member of the same size except the tongue.[78] The penis is fleshy and can be inflated with air generated in the arteries.

In the *Anatomy of the Living*[79] the penis has two pairs of muscles, one of longitudinal, the other of oblique fibers. The muscles of the penis issue through two ample spherical foramina in the thigh bone. Therefore the penis is said to be implanted in the bone by Galen, *On the Use of the Parts*.[80] There are two passages of the sperm and urine at the root of the penis; at its end toward the exit there is one passage in male and female.

The thigh bone is divided into right and left, but in the female when a light is brought close to it there appears a ligament. In the male there is no ligament. Yet by boiling the bone until the ligament is consumed or until the bones separate this duality appears in the bone of the male.

Break the thigh bone and cut the penis lengthwise to the canal. It is a palm in length. In the *Anatomy of the Living*[81] it has a natural length of from five to eleven fingers [3.5 to 7.7 inches]. Its substance is sinewy except for the prepuce.

Look at the four muscles of the anus which open and constrict the five hemorrhoidal veins. The anus is the entire rectum intestine whose extremity is called the sphincter. It does not have inner stagnation [mucous secretion] and is opposite to the other intestines in its position.

[II]

Skin the middle cavity up to the beginning of the neck; there examine the two forks [sternum and cartilage], an upper and a lower but laterally a right and left.

---

[74] Muhammad Ibn, etc., *op. cit.*, chap. xxi: Ad vesicam vero a renibus per duo colla profluit urina que emunctoria vocari diximus, quae cum ad vesicam perueniunt unam perforant tunicarum et postea inter duas ipsius transeunt tunicas donec ad collum perueniant vesicae, ubi aliam perforantia tunicam ad vesice descendunt concauitatem.

[75] Pseudo-Galen, *De Anatomia Vivorum*, fol. 51 r: et fluit urina inter duas tunicas.

[76] Galen, *De Usu Partium*, XIV, 14 (Vol. IV K., 205): . . . matricibus vero una satis fuerit, quemadmodum et utrique vesicae. . . . Vesicarum quidem corpora natura dura et vix patibilia comparavit, ut quae tantum excrementa essent recaptura. . . .

[77] Pseudo-Galen, *De Anatomia Vivorum*, fol. 52 v: Veniunt autem nerui ad eius compositionem ab inferiori parte spinalis medullae; veniunt etiam quaedam ligamenta a pubis ossibus.

[78] See the same statement in Berengario da Carpi, *A Short Introduction to Anatomy*, translated by L. R. Lind (Chicago, Ill., 1959), p. 72.

[79] Pseudo-Galen, *De Anatomia Vivorum*, fol. 52 v: Virga habet duo paria musculorum. . . .

[80] Galen, *De Usu Partium*, XV, 1–2 (Vol. IV K., 218–219).

[81] Pseudo-Galen, *De Anatomia Vivorum*, fol. 52 v: Longitudinem naturalem habet inter sex et nouem digitos. This statement does not agree with Achillini's five to eleven fingers; neither does his measurement of the uterus given in note 62, also as from *De Anatomia Vivorum;* possibly he used a manuscript of this text which differed in these two measurements from the text here quoted; see note 23 for details.

There are five inner parts: skin, fat, muscles of the breast, bones, cartilages.

In a woman the breast is like a round squash with a pinnule whence the milk is sucked; the pinnule may be a concentration of the veins which are scattered through the breast.

The substance of the breasts is glandular flesh, their number two, their location in the right and left parts of the chest.

The connection of the breasts is with the heart and liver through the ascending vein from which near the fork branches descend above the back and penetrate within the ribs to the breasts while the veins which have been mentioned make connection with the uterus.

You cannot see all the muscles. Some of them dilate only, others dilate and contract. There may be eight muscles above, below, and on right and left.

The muscles which dilate are those of the diaphragm; in the inner parts of the chest these open the large lower cavity.

Two muscles of the neck which send out tendons to the bones of the thorax open the small upper cavity.

There are muscles arising in the back near the origin of the first rib. There are also others, such as four between rib and rib [musculi intercostales sinister, dexter, internalis, externalis] according to Avicenna and Galen,[82] that is, two transverse and two oblique. There are therefore sixty-eight muscles according to Avicenna and Albertus Magnus, *On Animals* I,[83] because between the twelve ribs there are eleven spaces and each space has four muscles, that is, two inside and two outside since from the top of the ribs the fibers are raised more closely while at the root of the ribs the fibers are raised up at a greater distance.

The bones are double, both of the ribs and of the thorax. There are twelve ribs, seven true ones and five false. The true ribs are continuous with the thorax. The fork [sternum] is composed of bones and cartilages; at its extremity is the scutal cartilage or pomegranate [processus xiphoideus]. At the extremity of the false ribs there are cartilages.

The ribs of the chest are to be cut on right and left where the ribs are soft.

In the chest there are panniculi, the heart, the lung, and the aorta. There are three panniculi. The mediastinum divides the chest from front to back and separates the lung through the middle to right and left. It is not sinewy nor continuous nor actually a unity. It suspends the lung to the chest. Avicenna[84] in his *Anatomy* calls the diaphragm the mediastinum.

The pleura is a large hard sinewy panniculus which covers all the ribs on the inside.

Pleurisy is both true and false by reason of the place, the humor, or of the wind (air).

The diaphragm is a panniculus. At the end of the chest and the ribs it is attached to the twelfth vertebra from behind and to the cartilages of the ribs in front just as it is attached to the middle part which is sinewy and membranous with the lung so that it may move the latter in breathing by means of nerves which comes to it from the brain and from the spinal cord; for the diaphragm is a muscle and thus its motion is voluntary. This is the opinion of Galen.[85] The *Plusquam Commentator*[86] thinks its motion is natural [involuntary] from the ebullience of the blood but that the will can assist its movement. Averroes[87] says the lung has natural motion, the chest has voluntary motion. Natural motion comes from the diaphragm, and thus the motion is part natural, part voluntary, but more natural. If the diaphragm does not move naturally but voluntarily then its motion would be more voluntary because its voluntary movement comes from two sources, although only one of these is natural. in the *Anatomy of the Living*[88]

[82] Avicenna, *op. cit.*, p. 53, col. 2: Inter omnes enim duas costas sunt ex eis quatuor pro certo musculi, licet quidam existiment quod sit unus musculus. At. p. 54, col. 1, Avicenna counts 78 muscles of the chest; Achillini has only 68. Galen, *De Musculorum Dissectione ad Tirones* (Vol. XVIII K., B. 988–989) lists 22 intercostal muscles; see also *De Causis Respirationis Liber* (Vol. IV K., 467). Charles M. Goss, "On the Anatomy of Muscles for Beginners by Galen of Pergamon," *The Anatomical Record* **145** (1963): p. 492. translates this passage.

[83] Albertus Magnus actually lists 88 muscles: Albertus Magnus, *De Animalibus Libri XXVI nach der Cölner Urschrift* (herausgeg. von Hermann Stadler, 2 v., Münster i. W., 1916) (Beiträge zur Geschichte der Philosophie des Mittelalters—Texte und Untersuchungen XV), I, paragraph 325, p. 114; Omnium autem musculorum pectoris numerus est octoginta octo; hos tamen duo alii musculi iuuant, qui a furcula ad spatulae caput venientes ad primum eius latus continuantur et ipsum sursum levant et adiunant in dilatando pectus.

[84] This reference is a strange one; there is no separate *Anatomy* by Avicenna and Costaeus's index to the *Canon* (Venice, Giunta, 1595) does not list the word *mediastinus;* the references gathered under the word *diaphragm* say nothing of the sort as reported by Achillini. See also note 97. See R. Dunglison, *A Dictionary of Medical Science* (Philadelphia, 1860), *s.v.* pleurisy.

[85] Galen, *De Usu Partium,* XIII, 5 (Vol. IV K., 102–103); *De Anatomicis Administrationibus,* V, 7 (Vol. II K., 521, 525). Note that Achillini wavers between panniculus and muscle for the diaphragm. Galen calls it a muscle: *De A. A.,* V, 5.

[86] *Turisani (Trusiani) Monaci Cartusiensis plusquam commentum in librum Galieni qui microtechni intitulatur. Cum questione eiusdem de Ypostasi* (Venice, a Philippo Pincio Mantuano, 1512), Liber Secundus, Commentum 35: Digressio de respiratione, fol. 49 v: His expeditis, dicemus quod quamuis calidum intumescens ipsum quae infrigidatum detumescens sit principalis causa anhelitus habet tamen uoluntas ingressum in hunc motum sic ut dicebatur, non tamen sicut principale efficiens sed sicut adiuuans tamen motum diaphragmatis et pectoris quibus pulmo colligatur et dilatatur et contrahitur ipsis dilatatis et constrictis, est ergo hic motus commixtus ex voluntaria et naturali et non simpliciter voluntarius. . . .

[87] Averroes, *op. cit.,* II, chap. 19, fol 29 r, v: Averroes here summarizes the views of Galen on the process of breathing. At fol. 29 v, second column, below: Ergo melius quod dicere possumus ex his est ut sint sicut duo primi mobiles, scilicet, quod musculus fit primus in motu voluntario, et cor et pulmo primi in motu naturali. See note 85 for Galen's description of the diaphragm as a muscle.

[88] Pseudo-Galen, *De Anatomia Vivorum,* fol. 53 v: Cum autem motus dilatationis in circulo, a medio puncto incipiat et terminetur in extremitatibus et motus constrictionis incipiat ab extremitatibus

it appears that the motion of the diaphragm is voluntary because the nerve descends directly and perpendicularly through the midpoint of this panniculus, furnishing it with sensation and motion. Later this nerve branches and spreads throughout the entire diaphragm. Two other nerves come from the fifth and fourth pairs and in time of need the muscles of the chest assist the diaphragm itself. It is called metaphrenon [the back near the kidneys], as if it were the check rein (*frenum*) of the mind, since when it is injured the operations of the mind are impaired. Thus the check rein is that by which the heart rules the mind.

The shape of the diaphragm is round and oblong, its substance muscular and sinewy. It differs thus from the other muscles because it has the head of the muscle in its midst [reading *medio* for *numero*] and tendons round about it.

When you lift the panniculi you can see the lung. In the middle of these [reading *medio*] panniculi is the heart, veiled by the lobes of the lung. The root of the heart is at the right, its cuspis or point [apex cordis] to the left. It is proportionally larger in man than in any other animal because it has proportionally more heat for its size.[89]

The shape of the heart is that of a pine cone because a pyramid signifies heat; it is also triangular. On the other hand, there is that [view] of Plato, rejected by Aristotle in *De Caelo* III,[90] that if experience shows fire tends naturally to take the form of a pyramid this is an imperfect mixture which according to the nature of its lightness can be thus shaped by its motion. That which is opposed [to the pyramid] is said to be of a perfect mixture especially animated. Its shape is according to what its form demands for its work, and so on. Its point (*cuspis*) does not rise upward because of the weight of the blood in the right ventricle, and further because it could not then easily receive the blood. Outside the heart is the capsule [pericardium], sinewy and not continuous with the heart, containing the heart's aquosity. When this water dries out marasmus or atrophy results; but when this aquosity is excessive violent beating, tremor, and heart disease occur.

If you count the mediastinum there are three tunics for the capsule.

Around the heart there is fat on account of the great agitation of the blood, just as when you shake milk vigorously the result of that shaking is butter. The heart has auricles to hold its steam or spirit, such as the left auricle, or to hold the excess blood, such as the right auricle. Within the heart there are three ventricles, right, left, and middle. In the *Anatomy of the Living*[91] the heart has three chambers (*folliculos*). In the top one there is air, in the lowest one there is blood, and in the middle one there is a mixture of air with blood for the generation of the spirits. Therefore it is said to be the chamber of the spirits.

Aristotle[92] in the *History of Animals* III says that in large animals there appear three ventricles. In middle-sized animals the heart has no left ventricle; in small animals there is only a right ventricle.

Galen[93] in *On the Assistance of the Breath* scolds Aristotle concerning the number of the ventricles because Galen insists there are only two. Rhazes and Averroes name two.

---

et terminetur in medio, necesse fuit ut principium motus diaphragmatis esset in medio illius. *Phrenes* is the Greek work for diaphragm. It also means midriff, heart, and mind, among other meanings. Plato, *Timaeus* 70 A says the *phrenes* divides the mortal soul into upper and lower parts, the irascible soul from the appetitive. Frenzy was said to arise from the rupture of the diaphragm: see Hipp. *Acut.* 66; Rufus, *Onom.* 90; Julius Pollux, *Onomasticon* 2. 136.

[89] Aristotle, *De Partibus Animalium,* III, 4. 667 a 10–30.

[90] Aristotle, *De Caelo,* III, 304 a, b: "Some of them give fire a particular shape, like those who make it a pyramid, etc. . . . For it is a matter of observation that a natural body possesses a principle of movement. . . . (304 b) (2) If, on the other hand, a primary body is divisible, then (a) those who give fire a special shape will have to say that a part of fire is not fire, because a pyramid is not composed of pyramids, and also that not every body is either an element or composed of elements, since a part of fire will be neither fire nor any other element. (306 a) Because fire is mobile and productive of heat and combustion, some made it a sphere, others a pyramid" (translation in the Oxford Aristotle by J. L. Stocks, Vol. II). The description of the effects of change in the water of the heart's capsule is drawn from Mundinus, Singer's translation, p. 82, and from Galen, *De Locis Affectis,* V, 2 (Vol. VIII K., 303).

[91] Pseudo-Galen, *De Anatomia Vivorum,* fol. 52 v: Cor autem habet tres distinctiones siue folliculos, ad modum quarundam bursarum, in quarum parte suprema recipitur aer, in infima recipitur sanguis, in media, quae fovea appellatur, fit aeris et sanguinis commistio et spirituum generatio. Unde ille locus in corde dicitur minera spirituum. Berengario and Massa also use the term *minera*. Achillini does not.

[92] Aristotle, *De Historia Animalium,* I, 16. 496 2: "The heart has three cavities . . ."; *ibid.,* III, 3. 513 a 30: "but in the largest animals all three chambers are distinctly seen." Compare *De Historia Animalium* I, 17, 496 a 4; *De Partibus Animalium,* III, 4. 666 b 21, and Galen, *De Usu Partium* (Vol. III K., 480). The translation above is by D'Arcy Wentworth Thompson; see his long note. Achillini's statement, from "In middle-sized animals" to "right ventricle" does not appear in Aristotle, and Achillini is wrong in attributing such a description to him by garbling and misunderstanding *De Historia Animalium,* III, 513 a 25–30. The auricles were later distinguished as appendages to the two cavities of the heart recognized by the Hippocratics: Galen, *De Anatomicis Administrationibus,* VII, 11 (Vol. II K., 624).

[93] Galen, *De Utilitate Respirationis* (Galeni Opera, Venice, Giunta, 1556): fol. 64 r: Dicit enim tres esse ventres cordi cum sit patens visui esse tamen duos. This spurious work is not included in Kühn's edition. See also Galen, *De Usu Partium,* VI, 9 (Vol. III K., e 442): Non igitur recte Aristoteles numerum ventriculorum definivit, ad magnitudinem corporis et parvitatem multitudinem ipsorum referens. Neque enim maximis quibusque animalibus sunt tres, neque minimis est unicus. . . . See note 92 and Galen, *De Anatomicis Administrationibus,* VII, 10 (Vol. II K., 621) where he repeats his criticism. Galen chides Aristotle again in the later books of this work, which have survived only in an Arabic translation: *Galen on Anatomical Procedures—The Later Books,* translated by W. L. H. Duckworth, M. C. Lyons, and B. Towers; Cambridge University Press (1962) XIII, 9, p. 173. Muhammad Ibn, etc., *op. cit.,* chap. xiiii, fol. 5 r: Duos praeterea magnos etiam habet ventriculos, quorum unus in dextro est latere, alter in sinistro. Averroes, *op. cit.,* I, 24, fol. 11 r: Et habet cor concauitates duas magnas. . . .

Aristotle's induction is convincing to me, both because the right auricle is part of the heart and below it there is a hollow as well as because there is a middle separation between the part whence the air enters (the left part) and the aorta artery. Thus the separation of the heart into three ventricles is simple.

Cut the heart on the right side beginning with its point in such a way that you do not touch the heart's wall. Here there are two orifices, one toward the liver. The larger orifice is that of the inferior vena cava by which the heart sucks blood from the liver. Mundinus[94] said all the blood is digested in the heart, a statement not generally conceded; this appears from the exit of blood through the gate of the liver.

Haly [Rodan][95] errs in the second book of his *Tegni* when he says the heart touches the chest in its dilation. It does not touch along the side but along the cuspis when the heart contracts. When the heart is dilated the arteries are also dilated, a statement commonly denied although it is conceded by Haly and in *On the Anatomy of the Living*.[96]

Just as the inferior vena cava when it enters the heart sends a branch of itself around the heart so the artery when it enters the heart spreads a branch around the heart. When the artery is open three ostiola [semilunar valves] are evident in the aorta artery, dividing the same circumference into three parts. Under one of these is a foramen of the artery diffused through the substance of the heart. These ostiola open from the inside toward the outside, yielding to the spirit which issues forth. The inferior vena cava is closed imperfectly by three ostiola which open from the outside toward the inside, yielding to the blood as it enters the heart. This vein has branches in the heart. One enters the heart's cavity, another goes about and carries to it the blood which nourishes the heart. Correlatively, the lung is nourished by that blood which overflows from the heart and is not destined for its nourishment but for the nourishment of the spirits since the spirituous blood is drawn to the lung from the ventricle of the heart and not from the vein which surrounds the heart. Three threads [or fibers] cling to the middle wall of the heart [tricuspid and chordae] so that the inferior vena cava may be opened by the heart's dilation, for those threads are fixed in the gateway of the inferior vena cava.

There is another orifice of the arterial vein toward the lung [pulmonary artery and valve]; this carries spirituous blood from the heart to nourish the lung. Three ostiola of this arterial vein which close perfectly from the inside toward the outside I have never found. This vein is covered with a double tunic.

Cut the left ventricle of the heart. It has thick walls and two orifices around its root; one orifice is that of the aorta artery which is perfectly closed by three large ostiola from the inside toward the outside. Hence issue the spirit and arterial blood. At the entire gateway of the artery nature has laid down a lock or hindrance (*apodiamentum*) so that the artery may be closed when the heart is dilated.

The other orifice of the venal artery [pulmonary vein and mitral valve?] which has only one tunic is closed imperfectly by two ostiola, but they shut the wall of the heart until the spirit issues forth, lest it escape from the heart. This artery carries warm air to the lung and arterial blood to nourish the lung. In this digestion nothing superfluous is cut off, according to Avicenna in his *Anatomy*.[97] The extremity of this vein is pervious to air and impervious to blood; therefore the blood goes forth by sweating: Galen, *On the Use of the Parts*, VI.[98]

Why does not the artery which exists above the vertebrae in the direction of the heart [azygous vein?] appear to have a connection with the heart? One must say that the aorta artery is branched under the fork within, that is, by a branch bent toward the arm, a branch ascending to the head, and a branch descending upon the vertebrae. If you divide this artery which exists above the vertebrae much arterial blood appears. There also appears the foramina of the arteries, one pair of foramina or of arteries corresponding to each one of the vertebrae.

The middle ventricle is not one cavity but many small broad ones, more of them in the right than in the left side. But is there below the right auricle of the heart a middle cavity (*venter*) entering the heart?

There are four sources or places of origin springing from the heart: the chilis vein [vena cava], the arterial

[94] Mundinus, Singer's translation, p. 83.

[95] Haly Rodan, *Commentum Galenum* (commentary on Galen's *Techni*, with the *Expositio* and *Quaestiones* of Jacopo da Forlì on the same; Pavia, Michele and Bernardino de Garaldi, 1501): fol. 22 r: . . . quoniam pectus est creatus ut sit tutamen cordis et defensio ergo secundum dispositionem cordis est dispositio pectoris. Opus ergo ut ab illo mutetur figura pectoris ut in parvitate et magnitudine et secundum mutationem quantitatis capitis et quantitatis cordis. fol. 23 r: Et quando dilatatur cor dilatantur omnes artarie corporis simul et quando constringitur constringuntur omnes ipsae simul. Galen, *De Usu Partium*, VI, 16 (Vol. III K., 488) expressly says there is a sufficient interval between the pericardium and the heart in order to contain the latter when it dilates.

[96] Pseudo-Galen, *De Anatomia Vivorum*, fol. 55 v: De arteria. Arteria est membrum ortum habens a corde, concauum existens, ut per eius concauitatem concurrat spiritus vitalis ad omnes partes corporis. This is the closest corresponding statement I can find in *De Anatomia Vivorum*. See also *Turisani (Trusiani) Monaci Cartusiensis plusquam commentum* etc., who has a long discussion of the relation between the dilation of the heart and the arteries under *Commenta* 35, fol. 48 v, Utrum dilatatio arterie fit per influxum spiritus repletiui a corde an per ebullitionem sanguinis and 40, fol. 52 v: Utrum dilatatio chorde [*sic*] dilatentur similiter omnes arterie.

[97] See note 84.

[98] Galen, *De Usu Partium*, VI (Vol. III K., 445–446, 465, 469, 477, 503–504). All these passages deal with the general description here but do not mention *resudatio* of the blood. Note that Achillini follows Aristotle on the heart as the origin of the veins and arteries.

vein, the aorta artery, and the venal artery; add a fifth in the nerve that goes out from the heart and is spread through the diaphragm [vagus nerve?].

Notice whether in the interior of the artery there is a longitudinal or transverse fiber or whether the first has its purpose to receive and contain from the entire artery, the other to expel.

The lung is a triple organ of vessels, soft flesh, and membranes. The vessels, arterial vein, venal artery, and trachea, are divided to right and left, each of them into two branches which go above and below to the small capillaries woven together like a net. The concavities of the lung are filled with soft flesh joined triangularly. These three open foramina are led down to the exterior panniculus, although the larger is the aperture of the trachea and the panniculus is thin; because of this the corrupt blood or serous fluid is sucked from the chest by the trachea and not by other means.

There are five lobes of the lung, two on the left side, three on the right. The third is a cushion for the inferior vena cava and the artery in the direction of the heart. The lung may be inflated to show its size, location and shape.

Skin the branches of the trachea so that you may see the round rings and those not round which the sufferer from peripleumonia or consumption spits up.

Skin the gullet and you will see longitudinal muscles under the two guidez veins [jugular] which are called apoplectic and the veins of sleep and the deep veins, on the right and left. There are two tonsils on right and left, glandular flesh in the shape of almonds.

Galen, in *On the Assistance of Breath*[99] and on the authority of Aristotle, says the stoppage of these apoplectic veins is the cause of the sleep of apoplexy, of epilepsy, against the moderns (*Modernos*), who postulate stoppage in the first ventricle of the brain as the sole cause of epilepsy, etc. Therefore, Galen conceded to Aristotle that the apoplectic veins are the veins of sleep when their stoppage is light, and when it is heavy apoplexy is caused.

Cut the fork and take the ascending trunk of the aorta artery and of the vein which before it goes forth splits the fork into two branches, right and left. From each there descends a vein through the vertebrae of the back to nourish eight ribs and the upper part of the chest. Then the vein enters the subaxillary or armpit and makes the basilica vein. This is curved from the arm for bloodletting in pleurisy with pain ascending to the fork. The vein branches to the outer region of the arm, makes the cephalic vein in its curvature, and, bifurcated, ascends to the head; thus much about the arteries. The jugular veins are superficial; the apoplectic veins are deep, above the vertebrae of the neck and under the jugular superficial veins.

Behind the apoplectic veins there are two large nerves from the sixth pair [vagus] from which the reversive [recurrent] nerves of the voice branch out, the residuum of the nerve being joined with the right-hand nerve for the major part. And to the mouth of the stomach it is branched and proceeds up to the fundus of the stomach although some part of it with veins and arteries is joined in the fork and descends to the capsule. But the left-hand nerve for the major part is intertwined with the spiritual members as much as on the other side it [is entwined] with the natural members.

Lift the trachea artery or esophagus with the lower mandibles separated from the upper; lift the lips, lower and upper, with a nerve from the third pair, and with their flesh, skin, and muscles inseparably woven together in a marvelous composition. They are covered with a panniculus arising from the meri membrane [gullet or iugulum].

The teeth are thirty-two; two duals, two incisives, two canines, four maxillars, six molars, the same number also above the concave palate divided by a commissure from right to left and from front to back; the comissures cross each other near the straight rings. Two bones behind the comissure are quadrangular.

The pendent uvula is thin, spongy and cauterized with the cane [or cannula]. The fauces are wide and glandular; in them there occurs an abscess called

[99] Galen, *De Utilitate Respirationis* (*Galeni Opera,* Venice, Giunta, 1556): II, fol. 63 v: Dico autem ex hac oppilatione venularum fieri somnum, quandoque epilepsiam, et quandoque etiam apoplexiam. Fit autem apoplexia ex humore plurimo et crasso. Epilepsia ex humore non tanto, neque adeo spisso et quandoque ex solo fumo crasso, ascendente a stomacho, aut pede, aut aliquo membro alio; in his quidem videtur aura subtilis et frigida ascendere ad caput. . . . unde Aristoteles dicit epilepsiam esse somnum quendam. Aristotle, *De Historia Animalium* III, 514 a: " . . . if these veins are pressed externally, men, though not actually choked, become insensible, shut their eyes, and fall flat on the ground." Aristotle says nothing of apoplexy or epilepsy, however. Owsei Temkin, *The Falling Sickness,* etc., (Johns Hopkins Press, Baltimore, 1945): p. 33, cites the passage from Aristotle, *On Sleep and Waking,* 457 a, on the similarity of sleep and epilepsy. Temkin, who does not mention the theory of stoppage of the apoplectic (jugular, guidez) veins, collects the relevant passages from Galen and Aristotle as well as later authors. Gilbertus Anglicus, *Compendium Medicinae,* etc. (1510) is one of the "Moderni" who, according to Achillini, postulate stoppage in the first or principal ventricle of the brain (pp. 126–127). Temkin, *op. cit.,* p. 61: "Galen is quite emphatic about his contention that all epileptic attacks are due to affections of the brain." He cites Galen, *De Locis Affectis,* II, 11 (Vol. VIII K., 193; see Arthur J. Brock, *Greek Medicine;* New York, 1929, for a translation). The idea of apoplexy as a result of stoppage of the jugular veins seems to go back as far as Hippocrates: *Hippocratis De Acutorum Morborum Victu Liber et Galeni Commentarius* I (Vol. XV K., 775): Venarum interceptiones appellat venarum oppletiones a plenitudine obortas. . . . Quandoquidem hinc epilepsiae et apoplexiae et syncopae cardiacae oriuntur. Actually, pressure on the neck causes black-out from the carotid sinus or canal, not from the veins.

gotum.[100] The covering of the epiglottis is cartilaginous. Averroes[101] said it is marvelously composed of three cartilages which prevent the entrance of anything into the lung.

Lift the meri [iugulum, gullet] with the trachea. The substance of the meri is pellicular or of skins, yielding a passage for the food, and soft. The substance of the trachea which opens a way for the air is pellicular and cartilaginous, with many cartilages woven together from pellicular rings incomplete toward the gullet.

The substance of the meri is made firm by a double tunic; the inner one has longitudinal fibers, the outer one transverse fibers. Joannes Herculanus[102] said there are no oblique fibers therein. Others say the oblique fibers are few. In the book *On the Anatomy of the Living*[103] the statement is made that vomiting is difficult on account of the oblique fibers.

In the book *On the Anatomy of the Living* the meri proceeds directly to the fourth vertebra of the chest, then declines quite gradually to the right because to the left the cane or windpipe is longer for it extends to the diaphragm, below which it is continuous with the mouth of the stomach although they are of different composition; for the mouth of the stomach is more sinewy, mediating between the cane [trachea] and the vertebrae of the neck and the back.

Lift the meri from below carefully in order not to touch the reversive [recurrent] nerve. It is easily separated from the trachea up to the epiglottis, but at that point it is separated with difficulty because the tunic of the meri is dispersed in the epiglottis.

When the meri has been lifted the gross and solid substance of the epiglottis is revealed, composed of four skins, muscles, nerves, cartilages. Its muscles are twenty in number, twelve within and eight outside; they have their origin from the lower members.

There are two ascending nerves near the trachea, right and left; they are the reversives [recurrent nerves] which arise near the heart. They are the nerves of the voice and are bent back on the right of the heart by a second bend. The first bend is in the direction of the fork. By the incision of these nerves breathing with stroke or noise is lost. Galen, *On the Inner Parts*, XIV, chapter 6,[104] says there are five operations for these nerves: expiration, breath, breath with noise, voice, speech in sequence in the order mentioned.

The anterior cartilage [thyroid] is a shield or target for the others. The one which describes a circle has no name [cricoid, *innominata*]. The third is the cymbalaris [arytenoid] cartilage in the middle of which is the fistula [ventricle].

The tongue is soft white flesh, according to Averroes.[105] Avicenna,[106] *Canon*, Third Book, Sixth Fen, says small red veins cover this whiteness around the cane of the lung and the epiglottis. The tongue is fixed in or made firm by the lambda [hyoid] bone to which many large arteries come. Two pairs of nerves of motion go deeply into the root of the tongue. Two pairs of nerves of sensation are spread out in the tongue's surface and in its panniculus to give taste and touch.

On the other hand, Galen in *On Simple Medicine* IV, Third Distinction, chap. 4,[107] says that six nerves are connected with the tongue alone.

In the book *On the Anatomy of the Living*[108] the tongue has nine muscles.

Two salival fonts into which a stylus can enter are visible; they open near the tongue, and pieces of glandular flesh exist in that place. Do epileptics have green veins under their tongue which Avicenna[109] calls frog veins (*raninae*), *Canon*, Third Book, Sixth Fen? He called them green, without making himself entirely clear. And are they to be bloodlet or not?

---

[100] "squinancy, quinsey of the second kind, was called gotuni at Bologna": Mundinus, Singer's translation, p. 87 and notes 100 and 101.

[101] Averroes, *op cit.*, fol. 10 r: Et positum est supra viam istam cannae pulmonis coopertorium unum ad cooperiendum ipsam in hora deglutiendi, ut aliquid non possit subintrare de eo quod deglutitur. fol. 10 v: et ipsum est compositum ex tribus chartilaginibus.

[102] *Io. Arculani Omnes qui proximis seculis scripserunt medicos longe excellentis opera . . . Commentarii in Razis Arabis Nonum Librum ad Regem Almansorum,* etc. (Basle, Henricus Petrus, 1540), p. 460, chap. lxxvii, De anatomia meri et stomachi: In meri autem non fuerunt uilli transuersales, quia non ordinatur in retentionem. Quamuis aliqui dixerunt in ipso esse aliquos paucos transuersales. . . . See on Iohannes Arculanus, T. Puschmann, M. Neuburger, J. Pagel, *Handbuch der Geschichte der Medizin* **1** (Jena, 1902): p. 677; E. Gurlt, *Geschichte der Chirurgie,* etc. **1** (Berlin, 1898): pp. 884–894; and the recent article on Arcolani in *Dizionario biografico italiano* **1** (Rome, 1961).

[103] Pseudo-Galen, *De Anatomia Vivorum,* fol. 49 v: Huius interior panniculus componitur ex villis longitudinalibus, ex quibus habet vehementem attractum; exterior vero panniculus componitur ex villis latitudinalibus, ex quibus facit compressionem eorum quae recipit et componitur ex aliis villis transversoriis ex quibus habet aliquantulum retinere: propter hoc difficilis est vomitus. Procedit autem gula directe usque ad quartum spondylum pectoris. Achillini's "corrupte forte" which I translate "quite gradually" is not in the *De Anatomia Vivorum* and is probably a doubtful reading. Its meaning is certainly not clear.

[104] Galen, *De Interioribus,* XIV, 6 (= *De Locis Affectis,* IV, 9; Vol. VIII K., 270): His actionibus quinque inter se consequentia quadam convenientibus, expiratione, efflatione citra strepitum, strepenti effatione, voce, loquela.

[105] Averroes, *op. cit.,* I, 20, fol. 10 r: Lingua est caro mollis et alba.

[106] Avicenna, *op. cit.,* p. 586, col. 1: Et substantia quidem linguae est caro mollis, alba, quam cooperuerunt venae parue intrantes, sanguineae, rubeae ex colore eius, quare alie sunt uene et aliae sunt arteriae.

[107] Galen, *De Simplicium Medicamentorum Temperamentis ac Facultatibus,* IV, 15 (Vol. XI K., 669): In hanc [i.e., linguam] sex inseruntur nerui, quum aliae omnes aut unius tantum aut duorum sint participes, nec ita magnorum tamen.

[108] Pseudo-Galen, *De Anatomia Vivorum,* fol. 49 r: Habet autem lingua ix musculos in sui compositione.

[109] Avicenna, *op. cit., ibid.:* Et sub lingua sunt duae venae magnae, virides, ex quibus procedunt venae plures, et nominantur duae raninae.

[III]

The head at the top is bony rather than muscular; its form is that of a sphere a little compressed on the sides and tuberose in the occiput and sinciput.

The head has hair, skin, and flesh. In the *Anatomy of the Small Parts* [by Pseudo-Galen] it is said that under the skin there are other little skins (pellicule) which are called muscles. There is a panniculus on the outside of the cranium and two panniculi inside it, the brain, the marvelous net, and the basilar bone. Under the brain are two panniculi which go around the brain above and below it. There are eleven places of origin or layers.[110]

The external panniculus is generated from nerves and ligaments which arise from the dura mater and penetrate the commissures of the cranium. It is called the pericranium and the almuchatim.

The cranium is a large bone covering the brain on the outside; it is woven together with the junctions called adoreae, that is, sawlike, through which vapor or smoke goes forth and the sinewy fibers from the dura mater. The veins and arteries enter through them to the brain. Five bones are woven together with three commissures, coronal, sagittal, and lambdal; these bones are of the front, two parietal, the lambdal and basilar.

On the contrary, the cranium is made up of eight bones without the basilar because he [Mundinus?] does not place that one in the number, that is, of the two tables, the front, two lambdal, the petrous, the two bones of the pair. Then also Hugo of Siena[111] reprehends Almansor, Averroes, and Aristotle, who assert that there are six bones of the cranium because he regards the basilar as the seventh bone.

I answer that Mundinus speaks well if by basilar he understands the two petrous (rocklike) bones and the ossa paris and thus there are eight bones.

The other two commissures are of the petrous bones and are the false commissures, contrary to Hugo, because, as Joannes Herculanus[112] said, these commissures do not penetrate directly within the cranium; but nevertheless, according to Avicenna,[113] First Fen of the *First Book*, because they are petrous bones they are truly discontinuous from the cranium, yet are apposed to the cranium and placed above it. But according to Albertus Magnus in *On Animals* I[114] they are called false because they do not appear like the other commissures.

The bregma is the softer and thinner part of the cranium where two junctures are united. On the contrary, by Galen[115] the two parietal bones between which the sagittal commissure passes are called vervalia and these two bones are properly the cranium.

The inner panniculi are the dura mater and the pia mater. The dura mater penetrates the brain on right and left up to the anterior ventricle and divides the brain into front and rear parts. Similarly the pia mater is also divided through the cavities of the brain to bring nourishment to it. As it penetrates round about it covers the first and middle ventricle. The posterior ventricle does not need a panniculus because the brain is quite dry.

The brain is large, divided into anterior and posterior parts. The front part is divided into right and left; the division is apparent in the substance of the brain and in the ventricles.

Avicenna[116] in the third Fen of the fourth book, First Tractate, chapter 20, says both anterior ventricles have meatuses or nostrils.

Cut the brain lightly through the middle until you come to the large anterior ventricle. Some say the ventricle found by cutting along the sagittal commis-

---

[110] Mundinus, Singer's translation, p. 90, has these ten parts of the head on the authority of Avicenna, *Canon*, III, Fen 1, chap. 1. Achillini counts the two membranes as two and hence arrives at his eleven *orta* (or layers): see Mundinus, p. 108, note 109 (Singer): "The ten parts or layers of the head are a commonplace of mediaeval anatomy taken from Avicenna, whom the ensuing elaborate account of Mundinus closely follows. . . ."

[111] On Hugo of Siena see note 8.

[112] Iohannes Arculanus, *op. cit.* (note 102), p. 3: Sunt etiam et aliae duae commissurae non uerae lateribus dicte non uere, quia non penetrant uidelicet os utrinque, sed solum primam tabulam.

[113] Avicenna, *op. cit.*, p. 37, col. 2: Alii vero duo parietes, qui a dextra sunt et a sinistra sunt ossa in quibus sunt aures et vocantur petrosa propter sui duritiem quorum unumquodque terminatur superius ad suturam coronalem et inferius ad commissuram quae provenit a summitate suture lambda et protenditur usque ad coronalem et ab anteriori pars coronalis et posteriori pars suture lambda.

[114] Albertus Magnus, *De Animalibus*, etc.: after listing eighteen *divisions*, including three true *sutures*, of the head in paragraphs 183–184 Albertus proceeds in I, cap. v, paragraph 185 as follows: Istarum tamen divisionum quaedam sunt manifestae, quaedam autem occultae, et ideo vocantur mendosae. Manifestae quidem sunt suturae tres, quae sunt fere in omni capite hominis. Alia enim capita animalium suturas non habent. Occultae autem sunt ad modum linearum in compositione diversorum ossium, quae tres habent diversitates: ex duritia quidem maiori et minori, sicut ossa laterum capitis dividuntur ab ossibus cranei, et ossa petrosa aurium dividuntur ab ossibus aliis: et spissitudine et tenuitate, sicut dividitur os quod est fundamentum capitis ab ossibus superaedificatis: et situ, sicut dividuntur ossa in angulo convenientia ab invicem, sicut dividitur os frontis ab osse superciliorum.

[115] The name *vervalia* is an error for *nerualia:* see J. Hyrtl, *Das arabische und hebräische in der Anatomie* (Vienna, Braumüller, 1879), p. 195. Galen discusses the bregma in *In libros VI Hippocratis de Morbis Uulgaribus Com.* III (Venice, Giunta, Ed. III, 1556), fol. 169 v. He speaks of six bones of the head in *De Ossibus ad Tirones* (Vol. II K., 745: illa quidem sincipitis rarissima et infirmissima sunt) with the exception of the sphenoid (or basilar, as it is known to moderns) and in *Galeno Ascripta Introductio seu Medicus,* 12 (Vol. XIV K., 720): Ossa calvariae septem numerantur, occipitis unum, verticis duo, temporum item duo, frontis unum, multiformis unum. At p. 721 he discusses the *ossa jugalia:* Post alia duo ossa super foramina, per quae facultas audiendi est, emergant, jugalia appellata. Sub his rursus alia duo petrosa. See Hippocrates, *On Head Wounds* (ed. E. Littré, III, 1841, p. 190).

[116] Avicenna, *op. cit.*, II, 120, col. 2: (De exituris calidis) . . . eruptio eius que in cerebro ad duos ventriculos anteriores laudabilior est, quoniam ambobus sunt meatus, sicut nasus et aures.

sure is the middle ventricle because it was laterally divided before. Others reply that the cavities scattered laterally are the passages to the nostrils which purge superfluities.

Before you go deeper to the lacuna note that the ventricle is divided into right and left parts, also that there are walls which descend to the base. Divide the brain on the left and immediately there appears the amplitude or breadth of the two anterior ventricles.

Midway between the first and middle ventricles are three anchae (buttocks) [caudate nuclei?] which are the base of the anterior ventricle on right and left. They are of the substance of the brain. There is a long nerve bound with ligaments of a bloody substance, said Joannes Herculanus.[117] It is reddish and has little nerves; one of these on each side closes and opens the buttocks [colliculi]. Galen *On the Assistance* [*of the Parts*] IX, [118]last chapter, says a worm [vermis] lies above the buttocks; this worm when elongated contracts and when shortened it opens.

Are there three ventricles of the brain? I reply that Almansor[119] in the first book, on the authority of Galen, says there are four ventricles, two in the anterior part of the brain, one in the middle, and one in the posterior part of the brain. Averroes says the same thing: *Colliget* I, chap. 16.[120]

Albertus Magnus[121] in his book *On Respiration* says the breath (*spiritus*) enters the anterior cells from arteries through a certain duct or pore closed by the caruncula [fleshy body] of the little worm.

I open the brain through the middle where the conjunctiva separates the right from the left parts, going deeper until the white arch occurs. This arch is the tortoise shell (*testudo, fornix*) of the ventricles, the first and middle ones. The anterior concavity of this arch is the first ventricle. Behind it occur the buttocks; upon these like a muscle lies the worm; and upon the latter there is a vein from which the torcular is composed. Open the buttocks; in the middle of them there is the middle ventricle. Uncover the cuspis or point toward the posterior part of the brain; this is the top of the third ventricle. To the extent that the base of the cuspis is dilated just so broad becomes the third ventricle at its bottom.

Descend lower down to where the lacuna occurs. Its concavity is round and oblong. In the middle of it there is a foramen extending downward diagonally to the palate. By means of this foramen there is a direct path descending from the middle of the ventricle to the colatorium. The lacuna has large round eminences as supports for the veins and arteries which ascend from the marvelous net to the aforesaid ventricles. Thus there appear at the base [of the lacuna] prominent glands near the marvelous net.

The superfluities of the anterior parts [of the brain] are purged through the colatoria (or sieves) of the nostrils. The superfluities of the two ventricles are cleansed through the lacuna.

When these items are disposed of you can see the middle ventricle. As you proceed along the path from the anterior to the posterior part [of the brain] you see the posterior ventricle in the posterior part of the brain. It is covered and separated from the first ventricle by the two aforesaid panniculi, forming a solid place of origin for the spinal cord. Here originate many nerves of motion.

The posterior brain is pyramidal in shape because the ventricle located within it is pyramidal. Its base, through which it receives [impressions], is broad; its tip, by which it contains [impressions], is sharp.

Lift the brain gently in order not to break any of the nerves. Beginning from the anterior portion you can see the carunculae; according to Galen, *On the Assistance of the Breath*,[122] they are full of spongy flesh which arises from the medulla of the brain and are covered by the pia mater. They exist in the cavity of the emunctorii [or colatorium nasi] of the nostrils. They receive vapor through the porosities of the bone [lamina cribrosa of the ethmoid bone].

There are seven pairs of nerves. Galen, however, *On the Assistance*,[123] etc. X, chap. 5, says that anatomists do not count the two branches which pass to the nostrils. These provide the sense of smell; they are softer than the others and differ little from the substance of the brain. I have often exhibited this

[117] Iohannes Arculanus, *op. cit.* (see note 102), 4: et in fine huius primi uentriculi sunt duae substantie, una quae est sicut ancha, altera sicut est uermis subterraneus longa [*sic*], subtilis, rubea, contrahibilis et elongabilis, et ideo uermis dicta est. Iohannes Arculanus deals with the brain only on pages 1–6 and this is the sole statement which even approaches the reference to his work by Achillini.

[118] Galen, *De Usu Partium,* VIII, 14 (Vol. III K., 677–683). Chap. 14 is indeed the last chapter of this book. Achillini cites book IX instead of VIII and an old title of the work, *De Iuuamento Membrorum.*

[119] Muhammad Ibn, etc., *op. cit.,* chap. vii, fol. 4 v: cerebrum vero est solidum sed quasdam habet concauitates quae ipsius dicuntur ventriculi et sunt secundum Galenum iiii inter quos et quidam reperiuntur transitus quibus ab uno in alium perueniri possit, quorum duo in anteriori cerebri parte statuuntur et unus in medio, alius autem in occipitio sistitur.

[120] Averroes, *op. cit.,* I, 16, fol. 9 v: et nominantur cellulae cerebri et sunt duae existentes in parte anteriori cerebri, et una in medio, et alia in posteriori parte.

[121] Albertus Magnus, *De Animalibus,* etc., I, 538: Inter utrumque etiam oculorum sunt tres viae a cerebro et ad cerebrum: sed duae illarum procedunt ad occiput et una quidem illarum est quae est via spirituum per cellulas: et secunda quae magna est, est via carnis vermicularis: hae enim ambae ad occiput vadunt. Albertus has no separate book *De Respiratione.*

[122] Galen, *De Utilitate Respirationis* (Venice, Giunta, 1556), fol. 63 r: Et pelliculae plenae spongiosa carne locatae sunt intra nares, in quibus fit olfactus. Et omnes hae pelliculae originem habent a pelliculis cerebri. . . .

[123] Galen, *De Iuuamentis Membrorum,* X, 5 (Venice, edited by Diomedes Bonardus, 1490), fol. 31 r: Modo autem volumus narrare dispositionem alterius rami neruorum orientium ex cerebro, et est par quatuor parium neruorum quoniam anothomici non computant duos ramos qui ueniunt ad nares ex numero partium neruorum, quoniam isti duo rami non sunt nerui sicut alii nerui. . . .

pair, but in the year 1506 I did not find it. This was not strange because the cadaver I dissected was that of a freak; in fact, it did not have the mammillary carunculae. The nerves ought to have terminated at the carunculae, which was not the case [with my cadaver]. These nerves penetrate to the nostrils and pass under the carunculae.

In the *Anatomy of the Living*[124] the nerves are light to receive the spirit and thin in order to offer swift and easy passage to the spirit and flexible to serve the members.

Proceed to lift the brain and you will see two nerves larger than all the others; these are the optic nerves, which originate from the substance of the brain. In the bottom of the anterior ventricle outside the pia mater before they pass out from the brain they cross or are joined together. They are perforated at their point of contact in such a way that a cavity is formed therein. In the form of a cross they pass out of the cranium and each of them comes to its own eye.

According to Haly, Albertus,[125] and the Commentator[126] on the *Book of Sects* they form a cross thus: X. On the basis of experiment with a soldier wounded in the right temple [I discovered that] he lost the left eye.

Galen, Rhazes, Avicenna, and Averroes[127] say these nerves are joined by contact; according to experiment the wound touched the left nerve and not the right one.

However, we know by experience that when a living person has his right optic nerve touched he feels pain in his left eye, and when the left nerve is touched he feels pain in his right eye.

The second pair of hard thin nerves [oculomotor] moves the eyes voluntarily through the foramina which exist in the eye sockets. This pair issues from the cranium and is spread out through the muscles of the eye; it is quite large. William[128] [of Saliceto] said this pair gives the sensation of touch. Look at the pia mater which is collected there in the manner of a nerve, but of a reddish color, and enters the eye with these nerves.

The third pair of nerves [n. trigeminus] give sensation and voluntary movement to the face in one of their parts; in another part they are mingled with the fourth pair [sympathetic?] where the first body cavity reaches the second cavity. Avicenna[129] said this pair goes out between the anterior and posterior regions of the brain. Haly Abbas[130] in the second book of his *Theorica*, chap. 10, says that part [of this pair] carries taste to the tongue; some part of this pair is mingled at the temples with the fifth pair and at the places of mastication, the end of the nose, and the lips; part goes to the gums and another part brings feeling to the teeth.

Galen in his book *On the Assistance of the Members*[131] says this pair is spread out in the root of the nose among the hollow spaces therein. And Avicenna[132] said it is a branch of the third branch.

The fourth pair gives sensation to the diaphragm, viscera, and stomach, and from it come the reversive [recurrent] nerves. This [latter] pair is behind or comes after the third pair and declines more to the

[124] Pseudo-Galen, *De Anatomia Vivorum,* fol. 53 v: Item leues sunt nerui, rari, et flexuosi: leues, ut suscipiant spiritus; rari, ut praestent ei facilem motum, et directum et velocem; flexuosi, ut deseruiant membris ad regendum se faciliter versus quamlibet partem.

[125] Albertus Magnus, *De Animalibus,* etc., I, paragraph 358, p. 127: Statim autem post ortum suum ille nervus qui oritur a dextra parte, reflectitur versus sinistram: et contra ille qui oritur a sinistra, reflectitur ad dextram, donec in crucis modum sibi in medio obviant et se ibi secant. . . . Et sunt isti nervi concavi propter hoc ut spiritus visivus et formae visivae decurrant in ipsis, et vocantur in Graeco nervi optici, quod Latine sonat visivi.

[126] Commentator on the *Book of Sects* (see note 1): fol. 8 r: Oculi ipsi continentur a duobus neruulis qui oriuntur per duo foramina a cerebro venientia per que tenuissime vene a parte dextra in sinistram et a sinistra in dexteram oculorum implent orificia ipsorum nervorum unde quidem quia a dextera emanat in sinistra dirigitur, alter vero a sinistra in dextram et veluti duos nodulos faciunt et implent quedam concauum locum.

[127] Galen, *De Usu Partium,* X, 12 (Vol. III K., 813): *De Nervorum Dissectione Liber* (Vol. II K., 833); Muhammad Ibn, etc., *op. cit.,* chap. iiii, fol. 3 v: Primum vero horum septem parium neruorum qui a cerebro prodeunt duos habet neruos qui in anteriori cerebri parte exoriuntur et oculis directi sensum visus eis tribuunt, quorum unusquisque concauus est, qui cum aliquantulum ab eo elongati fuerint coniunguntur . . .; Avicenna, *op. cit.,* I, 60, col. 2: Et ille, qui a sinistra oritur, dextratur et ille, qui a dextra oritur, sinistratur, deinde sibi super sectionem cruceam obuiant . . .; Averroes, *op. cit.,* fol. 7 v: Par primum sunt nerui duo . . . et ipsi coniunguntur in oculis. Et isti duo nerui sunt concaui et quando elongantur a cerebro, tunc coniunguntur. . . . Et diuiduntur, donec sunt intus in cerebro et demum exeunt et tendunt unusquisque illorum ad oculum existentem in latere sui; demum coadunantur in humiditate vitrea ut deferatur oculo sensum visus. See Charles Mayo Goss, "On Anatomy of Nerves by Galen of Pergamon," *Amer. Jour. Anatomy* **118** (1966): pp. 327–335 for a careful translation.

[128] *Chirurgia Guilielmi De Saliceto Placentini, Medici celeberrimi, nunc primum sue integritati restituta*: in Guy de Chauliac, *Ars Chirurgica,* etc. (Venice, Giunta, 1546), fol. 352 v: Praeter hoc etiam ad oculos veniunt alii nerui a secundo pari neruorum cerebri per foramen cranei oculi, qui nerui dant oculis motum et sensum, per quos sentiunt res nociuas et iuuatiuas.

[129] Avicenna, *op. cit.,* I, 60, col. 1: Tertius vero paris origo est a termino communi inter anticessionem cerebri et ipsius postremum, scilicet, ab infimo basis cerebri. . . . The actual ramifications of this pair of nerves is, of course, much more complicated than Achillini's statement implies.

[130] Haly Filius Abbas, *Liber Totius Medicine,* etc. (Lyons, 1523): Theorice, Liber Secundus, cap. x, fol. 24 v: De neruorum assignatione. Tertii paris pars ad linguam vadit et praesunt ei gustus sensus quaedam autem ad timpora et ea quae masticant loca et nasi extremum labiaque. Pars ad gingiuas alia ad dentes sensum conferens illis tactus.

[131] Galen, *De Iuuamentis Membrorum,* X, 5 (Venice, edited by Diomedes Bonardus, 1490), fol. 31 r: Et inuenimus istum neruum esse diuisum in telam que induitur a naso et extenditur quousque perueniat ad superius palatum, quoniam ista tela est communicans palato et ori et copulat inter eos per foramina pertranseuntia ex naribus ad os, scilicet, foramina per quae fit hanelitus [anhelitus].

[132] Avicenna, *op. cit.,* I, 60, fol. 1 (see note 129).

base of the brain. Haly Abbas[133] says that it carries the sense of taste to the palate. Be careful with them because in that place there are two pairs of ligaments which pass from bones into the brain; these ligaments closely resemble nerves.

The fifth pair [n. vestibulo-cochlearis and n. facialis] comes to the petrous bone to weave a panniculus with its fibers; by means of one part of this pair the sense of hearing is created. In the *Anatomy of the Small Parts* the nerve which goes to the ears is called the posticus. By means of another part of this pair of nerves the muscles of the face are moved: the first book of Almansor. Haly Abbas[134] says part of the fifth pair gives motion to the muscle of the chest, part of it gives motion to the ear.

Avicenna[135] says that although it is unequal there is a fifth pair and that these nerves are very close to each other; therefore one pair is regarded as one nerve. Likewise, he said the fifth pair like the sixth is bound with panniculi and ligaments as if it were one nerve. Thus the sixth pair is slender with a small foramen a little above the fifth pair. Do there appear to be eight pairs? Yes, by doubling the fifth and not numbering the one which rides above.

The sixth pair [n. glossopharyngeus] which is slender carries sensation to the palate. Haly Abbas[136] says that part of this pair carries sensation to the viscera but part of it [n. vagus] brings motion to the muscles of the throat. Part of the sixth pair goes to throat and tongue, part [n. accessorius] to the muscles of the shoulder blades. Another part descends through the neck by many branches to the muscles of the epiglottis. When these reach the chest they are divided; some parts of them are turned upward as far as the epiglottis. Galen,[137] *On the Use of the Parts*, XVI, chap. 4, said: "For from the sixth pair it sends forth reversive nerves [rr. recurrentes nervorum vagorum], etc." Other parts are scattered through the capsule of the heart, through the gullet, through the lung and adjacent places. A larger part penetrates the diaphragm; another larger part is extended to the mouth of the stomach. Another part goes to the panniculus of the liver and is extended to the spleen and viscera to which also parts of the third pair [n. sympathicus; see Goss, *op. cit.*, p. 329] are extended.

The seventh pair carries sensation and motion to the tongue; it comes from the posterior part of the brain whence arises the spinal cord. Galen,[138] however, *On the Assistance* [*of the Members*], IX, chap. 2, says that soft nerves give sensation to the tongue and come from the anterior part of the brain. Hard nerves give motion and come from the hard [part of] the brain and are joined before to the lung. Thus Mundinus[139] made an error if he understood that one pair only comes to give sensation and motion.

There is another slender pair of nerves that issues from the posterior part of the brain and passes to the anterior parts above the location of the ears. Concerning this pair [n. trochlearis?] nothing has been discovered by learned men. I think that it gives motion to the eyebrows. It is immediately visible when the brain is lifted up.

Jacopo of Forlí[140] said that all seven pairs arise from the anterior part of the brain. But Avicenna[141] said four pairs arise from the anterior part of the brain and three pairs from the posterior part.

When the panniculi are lifted from the basilar bone the marvelous net is in the middle of the basilar in the direction of the colatorium [cribriform plate of the ethmoid]; it is very strongly woven of very slender arteries doubled by the apoplectic [jugular] arteries. The stoppage of this net is the cause of sleep. Note this because Aristotle[142] said that sleep is a passion (suffering) of the common sense. Two fleshy glands sustain this net; they also sustain two veins and two arteries [internal carotids] which ascend to the ventricles of the brain. Averroes,[143] *Colliget* I, chap. 16, says the vertebrae of the nape of the neck cover the net just as the cranium covers the brain. Or when the bone is broken is the marvelous net in the direction of

[133] Haly Filius Abbas, *op. cit.*, fol 24 v: Quartum par in superiori distribuitur palati, gustus illi sensus deferens. Duo autem quarti paris nerui post ortum tertii oriuntur paris coniunctique tertio pari ab eo diuiduntur distribuunturque in tunicam et operimentum quod palatum vestit eique sensum tactus perfert.

[134] *Ibid.:* Quinti deinde paris pars ad aures porrigitur ad auditus sensum conferendum. Quedam ad lacertum pectoris latum virtutem ad illud perferens motus.

[135] Avicenna, *op. cit.*, I, 61, col. 2: Sed unumquodque paris quinti impar cum duobus nascitur mediis, etc. The description by Achillini does not exactly coincide with Avicenna's.

[136] Haly Filius Abbas, *op. cit.*, *ibid.:* At vero sexti pars paris ad viscera dirigitur quibus etiam dat sensum. Quedam ad gutturis lacertos datque illis motum.

[137] Galen, *De Usu Partium*, XVI, 4 (Vol IV K., 279–280).

[138] Galen, *De Iuuamentis Membrorum*, IX, 2, fol. 28 v: Et nerui molles qui ueniunt ad linguam oriuntur ex inferiori ad interius anterioris cerebri. . . . Nerui ergo molles ueniunt ex cerebro ad linguam directe.

[139] Mundinus, Singer's translation 94, note 117. Mundinus follows Galen, *De Usu Partium*.

[140] Jacopo da Forlì, *op. cit.*, fol. 35 v–36 r: Sexto nota quod ut colligitur capitulo eodem et in sequenti ad illud nerui orientes a parte anteriori cerebri sunt valde pauciores quam dicantur esse orientes a posteriori parte eiusdem, nam a parte anteriori solum septem neruorum paria oriuntur quae in sentiendo et mouendo solummodo iuuant membra capitis faciei et viscerum anteriorum; a parte autem posteriori dicuntur omnes alii nostri corporis nerui ortum habere, qui sunt xxx et unum paria. Ultimo nota quod ut ait Auicenna eodem primo capitulo plurimi eorum neruorum qui tribuunt sensum oriuntur ab antecessione cerebri, a parte anteriori cerebri, et plurimi eorum neruorum qui praebent motum oriuntur a postremo, id est, a posteriori cerebri quae est nuca, ut dictum est.

[141] Avicenna, *op. cit.*, I, 60, col. 2: A cerebro septem neruorum paria oriuntur. This is all that Avicenna says on the matter.

[142] Aristotle, *De Somno et Vigilia* 1, 454 b 10: "For sleep is an affection of the organ of sense-perception—a sort of tie or inhibition of function imposed on it, so that every creature that sleeps must needs have the organ of sense-perception" (translation by J. I. Beare, Oxford Aristotle, 4).

[143] Averroes, *op. cit.*, *ibid.:* Sed nuche spondylia ipsam [retinam] cooperiunt. Achillini's doubt concerning the net indicates that he did not actually see it, obviously because it exists only in ungulates such as sheep and oxen.

the colatorium? I saw it thus in 1503, and there is a chamber in which it lies closed up with bones; this is what I think Averroes meant.

Or is there a double net, that is, the marvelous one, worthy of admiration, composed of arteries, and a secundinal net composed of veins scattered through the brain which make up its texture? Or is there only one net which is composed of arteries and veins, etc.?

The basilar bone, harder than the rest, is composed of five bones, two petrous [parietal], one that is cavernous and porous for the nostrils, and two bones for the ears [ossa paris].[144]

But why is not the basilar bone [composed] of the lambda [hyoid] bone, since it is bent back in the posterior part under the brain, returning to the anterior part? I answer that it is laid down under the spinal medulla. On the other hand, the posterior brain is immediately beneath the bone, and the spinal medulla originates in the middle of the brain. But if the basilar bone is included among these bones the doubt arises as to why it should be numbered among them and, secondly, how [does it result] that there are twenty-three bones in the head?[145] Likewise the frontal bone is bent back from the eyebrows and penetrates thus into the anterior substance of the brain and is therefore itself part of the basilar bone.

Cut the bone of the eye (roof of the orbit, the eye socket). You will see the place. Its connection is with the optic nerve and with the nerve of motion. This place is not too far outside nor too far inside. It holds the eyebrows and eyelids. The upper part of this place has muscles, the lower part does not; there is also the tuberosity of the jaw and the nose [maxilla].

The eye has seven tunics in name, but, in fact, according to Galen[146] in his *Anatomy of the Eyes,* it has four tunics and six muscles. The eye has three humors from front to back. The first tunic is the cornea, which is dried out and flattened in those who are dying. It is similar to horn in substance, hard and transparent. Its posterior part, which comes from the dura mater, is the sclerotic tunic. The cornea is composed of four tunics, although this can be perceived only with difficulty. The fact is apparent when the eye is thoroughly boiled. I knew this through my experience; then I found out about it in the book *On the Anatomy of the Living.* [147]

The second tunic is the conjunctiva, which arises from the almuchatim, the panniculus that covers the cranium [dura mater]. According to Galen[148] in his book *On the Assistance of the Breath* it is insensitive; thus it is clear why the eyes do not grow stiff. On the other hand, the conjunctiva is the first tunic because it is outside the cornea, is visible when the eye is completely boiled, and that is what the author of the *Anatomy of the Living*[149] said. The reason for this statement is clear from the fact that the conjunctiva arises from the pericranium, which is outside the cranium, and the cornea arises from the dura mater, which is inside the cranium. Galen called this tunic the circumossalis; it comes to the front through the sockets of the eyes, according to Galen, *On the Anatomy of the Eyes.*[150]

The third tunic is the uvea [choroid and iris], whose foramen is the pupil. Therefore this tunic is [also] called the pupil. Behind it is the secundina, nor does there seem to be a suture of the uvea with the secundina; by many veins which exist in the secundina it is separated from the front part of the uvea. The uvea opens in moderate light and contracts with excessive light in its foramen in order to repel what is harmful because of its shining nature. The uvea arises from the pia mater.

The fourth tunic is the cobweb (arachnoid, *aranea*), the second part of which is called the retina. It arises from the optic nerve and is of the utmost clarity and polish.

Galen, *On the Use of the Parts,* X, chap. 2,[151] says the beginning of the retina is a thin meninge that covers the brain and carries an artery and a vein when the nerves pass forth from the brain. Galen, *On*

---

[144] Mundinus, Singer's translation, p. 94.

[145] Compare the calculations given at note 165 to reach the total of 23 bones.

[146] *On the Anatomy of the Eyes:* Galen has no such separate work; Achillini is probably using this title for Book X of *De Usu Partium* (Vol. III K., 759 ff.) which Galen devotes entirely to a discussion of the eyes. He also has a good treatment of the relationship of tunics and muscles in *De Anatomicis Administrationibus,* X, translated by W. L. H. Duckworth, M. C. Lyons, B. Towers (Cambridge University Press [1962]), pp. 27–43; the word tunic is applied (p. 42) "not to a substance but to a relation of its surroundings to that which is surrounded." See also his *De Musculorum Dissectione* (Vol. XVIII K., 932–933; translated by C. M. Goss in *The Anatomical Record* **145** (1963): pp. 477–501. A reference to pseudo-Galen, *De Oculis,* may be intended here: see Richard J. Durling, "A Chronological Census of Renaissance Editions and Translations of Galen," *Jour. Warburg . . . Institutes* **24** (1961): p. 288, no. 74. For a text see *Collectio Ophthalmologica Veterum Auctorum Fasc.* VII: Constantini Monachi Montiscassini Liber de Oculis et Galieni Littere ad Corisium De Morbis Oculorum, etc. (Paris, Baillière et fils, 1909–1933).

[147] Pseudo-Galen, *De Anatomia Vivorum,* fol. 48 v: Et haec tunica [cornea] in sui constitutione quatuor habet cortices. Achillini does not distinguish between *tunica* and *cortex* here.

[148] Galen, *De Iuvamento Anhelitus* (Venice, Giunta, 1490), fol. 32 v–33 r; *De Utilitate Respirationis* (Venice, Giunta, 1556) Vol. II, fol. 63: . . . fitque visus non in crystallino humore oculi, ut Aristoteli visum est, sed in pelliculis oculi. Quarum duae procedunt a duabus pelliculis cerebri et tertia a neruo qui opticus, id est, visibilis dicitur praeceditque ad ipsis pelliculis cerebri. Est autem praeter has pelliculas tres, quaedam alia quarta, quae est in oculo, quae alba est et continet omnes alias praeceditque a pellicula quae tegit cranium, et est insensibilis.

[149] See note 147.

[150] Galen, *De Usu Partium,* X (Vol. III K., 792) on the periostium. See note 146.

[151] Galen, *De Usu Partium,* X, 2 (Vol. III K., 763): Principium autem huic quoque tunicae meninx est tenuis, quae cerebrum ambit quam paulo ante dicebamus simul cum neruis omnibus enatam venam atque arteriam secum afferre. (*paulo ante:* 762): Nam cum neruis omnibus, qui a cerebro proficiscuntur, portio quaedam meningis chorioidis enascitur, quae arteriam secum affert ac venam. . . .

the Assistance of the Breath,[152] says vision is not caused in the crystalline of the eye, as Aristotle thought, but in the little skins [that is, the membranes and tunics] of the eye, and in *On the Assistance of the Members,* X,[153] he says there is no true vision except in these nerves, that is, the optic nerves. And in *On the Use of the Parts,* X, chap. 2,[154] he says the first and greatest use of the retina is to perceive the alterations of the crystalloid humor. However, the retina perceives not by means of the optic nerves because it (the retina) is made up of arteries and veins. Or does the albugineus [aqueous] humor mediate between the aranea tunic?

A cataract results from vapor which descends from the head or ascends from the stomach; it prevents totally or partially the transit of forms of sight (*specierum*) to the crystalline humor, laterally or otherwise. A cataract can be driven downward by piercing the cornea outside the pupil of the eye; when it has deposited its cloud it should be perforated near the external region of the eye.

There are three humors: albugineus, crystalline, and vitreous. The albugineus precedes the crystalline and the vitreous follows the crystalline. Does the vitreous humor nourish the crystalline by its own substance or with the blood that passes through itself? I answer that the second [alternative] is true. Otherwise everything would appear reddish, since the blood would retain something of its own color. The vitreous humor surrounds the crystalline humor up to and a little beyond the largest circle of the crystalline humor.

The ear is on the side of the head, round in shape, cartilaginous, and with an aperture on both sides at whose extremity there is a membrane which closes off within the petrous bone the coronal air [which comes from the coronal suture], the jaws, etc.

Divide the entire body lengthwise into two halves from the neck to the tail. There are thirty vertebrae. But the round bone [atlas] upon which the head rests makes thirty-one when it is included in the number of the vertebrae. There are seven vertebrae of the neck; they are slender but have a larger cavity or aperture, however, and are hard and firmly joined to each other. Or does the tenth vertebra have two pieces or processes? According to Galen,[155] *On the Use of the Parts,* XIII, chap. 2, paragraph 18, you must understand here the middle vertebra not in number but in position in the length of the back; Albertus, *On Animals* I.[156] Or do the buccellae [processes] ascend above and descend below the tenth vertebra? Or does the tenth vertebra have two cavities and thus the vertebrae above the tenth have buccellae below them and the vertebrae below the tenth have buccellae above them, according to Avicenna, [*Canon*], I, 1, and Albertus, *On Animals* I?

There are twelve vertebrae of the ribs, seven true and five false. The five vertebrae of the kidneys are thick and large. Rhazes in the first book of Almansor calls them chatim. The alchatim is made up of three vertebrae; these are called the alhovius vertebrae [sacrum]. There are three vertebrae of the tail, called alhosos [coccyx]. All the vertebrae contain medulla. It is false, [however, to say] all unless you mean the vertebrae above those of the kidneys because in the vertebrae of the kidneys [lumbar vertebrae] are collected like a mass of threads all the nerves which descend to the lower muscles. I say that the nucha [spinal medulla] is more viscous and solid than the substance of the brain. It contributes sensation and motion to the entire body from the head downward. At its origination there appears a thread which does not penetrate; it divides the right from the left along the division of the brain. The nucha is covered with two panniculi proportional to the dura mater and to the pia.

From the nucha through each vertebra extends one pair of nerves. Galen[157] said in *On the Use of the Parts,* XIII, chap. 8, that through the same foramen through which the nerve goes out from the nucha there enter a vein and an artery. Between the spaces of

---

[152] Galen, *De Iuvamento Anhelitus* (Venice, edited by Diomedes Bonardus, printed by Filippo Pintio de Caneto, 1490), fol. 32 v–33 r: Dico etiam quia ista instrumenta omnia sensuum continentur fiuntque sensus in ipsis pelliculis ut est videre in oculo cuius pellicule plene sunt humore aqueo et originem habent a cerebri pelliculis fitque visus ut in cristallino humore oculi ut aristoteli visum est sed in pelliculis oculi quarum due praecedunt a duabus pelliculis cerebri et tertia a neruo qui obticus dicitur, id est, visibilis. . . . This statement represents a contradiction with Galen's view at *De Usu Partium,* X, 1 (Vol. III K., 760), where he insists that the crystalline humor is indeed "primum videndi instrumentum." The passage in the 1490 printing of *De Iuuamento Anhelitus* is influenced by Aristotle, *De Generatione Animalium* V, 1, 780 a 1–5, who says that "sight is the movement of this part [liquidity]. . . ."

[153] Galen, *De Iuuamentis Membrorum = De Usu Partium,* X, 14 (Vol. III K., 833).

[154] Galen, *ibid.,* 762: Porro usus est ipsius primus quidem ac maximus, propter quem superne fuit demissa, ut, quum crystallinus alteratur, id sentiat, praeterea ut vitreo humori alimentum advehat atque afferat, siquidem arteriis ac venis multo crebrioribus ac, quam suae moli conveniat, majoribus referta conspicitur. The following page in Galen's text is also part of Achillini's context.

[155] Galen, *De Usu Partium,* XIII, 2, paragraph 18 (Vol. IV K., 77): At qui in dorso est decimus, solus omnium spondylorum (ut diximus [76]) utriusque apophyseos fines mediocriter devexos adeptus invehitur adjacentibus, in cavitates quasdam superciliosas desinentibus. The bucellae (additamenta or processes) are the capita alarum spondylium, here equated with apophysis (pl., apophyseis).

[156] Albertus Magnus, *De Animalibus,* I, paragraph 271, p. 96: Decimae autem spondilis additamentum est rectum, et habet utriusque additamenta quae coniunctioni sunt apta, et sunt cava recipientia alia in se sicut bucellas. Earum autem quae sub decima sunt bucellae receptae, sunt ad superiora porrectae, et concavitates suae alias recipientes diriguntur ad inferiora, et ipsarum additamenta ad superius incurvantur. See also Galen, *De Usu Partium,* Vol. IV K., 76–77.

[157] Galen, *De Usu Partium,* XIII, 8 9 (Vol. IV K., 115): Nam foramine uno ex iis, quae antea fuerunt in neruorum expositione dicta, ad trium instrumentorum transitum est usa, nervum quidem intus foras, foris autem intro arteriam ac venam deducens.

the nerves there mediates a very slender artery. Nevertheless, you should know that, although there is one foramen on one side in the vertebra, nonetheless through that one foramen there sometimes comes out one nerve and sometimes one pair of nerves, sometimes more according to the need of the members, as I showed in the region of the arms in 1503. You should know that although there are thirty vertebrae there are nevertheless thirty-one pairs of nerves plus one nerve because the first pair issues from the upper first vertebra and the second pair from the vertebra below that one, similarly from the last of the alhovius and the first of the alhosos in the intermediary space there.

From the medulla of the neck come eight pairs of nerves, from the medulla of the chest twelve pairs of nerves, from the medulla of the kidneys five, from the medulla of the alhovious three, and from the medulla of the alhosos three with one-half of one [pair].

However, within the vertebrae of the kidneys there is no medulla nor below in any of the vertebrae but only above, for the medulla terminates at the vertebrae of the kidneys and thus it is necessary that, dried out by the dryness of the vertebrae of the kidneys, the nerves should be powerful and strong to cause motion of the tibiae.

Skin the arm lightly so that you may see the veins which penetrate through the subaxillary, through the inner region, and, appearing in the lower part in the curvature of the arm which is called the basilic [cubital fossa], come above the fork which branches off to the head and to the external region of the arm which is called cephalic. The branch from one vein to another [between the cephalic and the basilic veins] is called common. The vein between the auricular [little or ear finger] and annular [fourth or ring finger] is called seyles [sceiles], corresponding to the basilic through a branch which comes from below under the arm. The vein between the thumb and index finger is the salvatella; it is phlebotomized in the place of the cephalic vein.

In the *Anatomy of the Living*[158] the vein between the middle finger and the ear finger is called salvatella.

Or is it a branch of the common vein, according to Avicenna,[159] I, 1, or of the basilic vein, according to Almansor, I?[160] I reply that each alternative is true, whichever from the gathering of all those veins according to their branches (which in the middle is called the rope of the arm, *funis brachii*) holds the place of the common vein.

Do not seek to find the muscles and tendons.

Lift up the flesh from the bones. The adiutorium bone [humerus] is above the back. The spatula [scapula] has a cavity in which the adiutorium revolves with a round head in a box. It slants toward the inner region. The end of the adiutorium has two eminences joined to two fociles [condyles of the forearm] named rostrata, that is, embracing with a beak, and pyxis. Thus there are two fociles in the arm, the lower and the longer, thicker at the ends than in the middle.

The raseta [carpus, wrist] of the hand is made up of eight bones in two rows, four in a row. The comb (metacarpus, *pecten*) is made up of four bones. The thumb is not in its row. There are five fingers, fleshy inside and laterally but not on top (A. says "*post*" as the cadaver lies supine). Avicenna and Galen say that there are seven bones in the wrist, but the eighth bone is a protection for the nerve of the palm. Galen, *On the Use of the Parts,* II, chap. 2,[161] said the ninth bone is not numbered by anatomists. It is strongly attached to the bone for moving the wrist and thumb.

In the large hand [i.e., both limb and hand] there are thirty bones, the adiutorium, two fociles, eight bones of the wrist in two rows, four bones of the pecten, fifteen for the fingers at three bones for each finger. There would be thirty-one if the ninth of Galen is included. [The latter is, however, a monkey bone.] You should know that the nails are not reckoned among the bones of the hand. Look for the sesamoid bones between the loose joints, which thus far I have not been able to find.

The tendons are more distant or far apart and deeper. They do not ride above but pass through the middle.

Take the skin off the tibia very carefully in order to see the two descending veins which branch from the chilis vein [inferior vena cava] in the separation of the right buttock from the left. Each of the two trunks is branched into two large branches through the inner region up to the toes and is called the saphena; it empties from the natural members, the genitals, uterus, kidneys, and testicles. It appears above the knee and above the clavicle and in the heel in the pecten of the foot. Another branch slants and enters near the juncture of the scia (hip). Therefore it is called the sciatic vein and when incised it empties from the joints. It is found bifurcated under the ham of the knee and below around the clavicle.

---

[158] Pseudo-Galen, *De Anatomia Vivorum,* fol. 54 v: Sceiles siue saluatella quidem, quae in dextra, doloribus hepatis confert. . . .

[159] Avicenna, *op. cit.:* Costaeus's index gives uena saluatella under l. 79. b 71, but I cannot find such a description in Avicenna's text.

[160] Muhammad Ibn, etc., *op. cit.,* fol. 4 r, chap. v: Assellate vero pars altera quae est inferioris loci per partem brachii quae est in profundo transit quae usque adeo protenditur quousque ad caput inferioris focilis perueniat et fit ex uno suorum ramorum vena quae est inter medium et auricularem, quae saluatela vocatur.

[161] Galen, *De Usu Partium,* II, 2 (Vol. III K., 131): Videbitur itaque sic quidem carpus septem omnino ossibus constare. *Ibid.,* 134: . . . posuit et hic velut quoddam vallum praelongum os, intro nutans rectum; a quo alia, quae ibi sunt, muniuntur, et maxime nervus, qui a spinali medulla prodiens ad interiora manus disseminatur. Hoc est os carpi octavum, de cuius notabili generatione in superioribus disserere distuleram. *Ibid.,* 137: Nonum aliquis poterit hoc carpi os numerare, sed non est numeratum ab anatomicis, sicut nec aliquod eorum, quae sesamoidea vocantur.

The femur bone has vertebrae built above it and has a hollow [fossa acetabuli] beneath it, where the round bone of the cane of the coxa [acetabulum] is located. It is attached by a ligament in the middle. The joint is called scia; its pain is also sciatica. The cane of the coxa [femur] is larger than other bones and has a larger hollow. It is not entirely straight, but its ends lie toward the inner region. The middle part is convex near the exterior. Two fociles of tibia are bent in the knee. To the front is the patella made of ligaments bound into a knot. The larger focile is in the inner region and is thicker. The thinner focile is in the exterior and is smaller. It does not reach to the joint. The two fociles are united through the joint where the bone of the clavicle is depressed at a thick quadrangular bone called cayb [cahab] or achib, according to Almansor, I, 2.[162] The navicular bone is made in the shape of a little boat, quadrangular and oblong.

The raseta [tarsus] of the foot is composed of three bones, the pecten of five bones. The bones of the toes are fourteen. The bones of the large foot [entire limb] are twenty-eight, but according to Almansor I there are twenty-nine because the os calcaneum (heel bone) makes a division from the cahab bone.

I saw in 1502 five bones of the raseta of the foot.
Of the pecten five.
Of the toes fourteen.
Of the navicula one.
Of the calcaneum two.
Of the skleus [crea, shin, tibia] two fociles.
Of the knee one.
Of the coxa one.

In 1503 I saw seven bones in the raseta of the foot, three in the first row but four in the second row upon which the pecten was raised. And thus one ascends through a series of uneven numbers from one to three to four to five toes.

The tendons which extend the toes arise from the silvestris (inner) muscles of the tibia, but those which contract the toes arise from the muscles of the sole of the foot.

The muscles all told are seventy-nine.

According to Almansor I,[163] the bones are as Galen numbered them, two hundred forty-eight, with the exception of the bone of the epiglottis, the lambda bone, and the bone of the heart, called a cartilage by some, and the little bones by which the hollows of the joints are filled, that is, the sesamoid bones.

Bones of the head twenty-two.
Coronal one.
Lambda one.
Vervalia [nerualia] two.
Petrous bones two.
[Ossa] paris two.
Of the upper jaw twelve.
Of the lower jaw two.
The bones of the nostrils are lacking.

I have not named the basilar bone because he [Galen] does not place it in the number with the others since from the anterior part the coronal bone of the front is bent back underneath as it makes an arch for the eyes. The lambda bone is also bent under from the rear. The petrous bones and the ossa paris descending laterally are laid under the brain.

According to Albertus, *On Animals* I,[164] the bones of the head are twenty-seven, except for the teeth. The frontal bone is bipartite through the root of the hairs. I have never been able to find this commissure for in the root of the nose the frontal bone is separated from the bone of the nostrils. But in the temple the os paris is separated from the triangular bone. The lambda bone is bipartite through the curvature of the occiput, and this is not a real section or division. He names the basilar and it seems that he makes it bipartite. They do not place this [in the number of the bones of the head], as I said before, but taking four from twenty-seven there remain twenty-three.[165] Therefore if to the previously mentioned bones you add the bone of the colatorium you will have Avicenna's number, that is, twenty-three. But in regard to its substance the bone of the colatorium may be separated from the other bones because it is porous and spongy. Nevertheless, I have not been able to see the commissure which separates this bone from the lambda bone bent under from behind, for in it is the foramen of the neck joint. But no discontinuation appears in that place nor is it likely that nature made one there. If you add the round bone upon which the head revolves, it is clear that it is not the bone of the head although it is the support of the head. But if you count the middle bone common to the nostrils then the bones of the nostrils must be reckoned or the number of the bones of the upper jaw is increased to thirteen by the addition of one.

But if you number the bones of the head they are twenty-two.
Teeth thirty-two.
Vertebrae thirty-one.
Of the fork two.
Of the shoulder blade two.

---

[162] Muhammad Ibn, etc., *op. cit.*, fol. 3 r, chap. ii: . . . Calcaneo vero anterioribus os quod vocatur nauicula coniungitur. Inferius autem cuidam coniungitur ossi quod achib vocatur. *Ibid., infra:* Omnia igitur pedis ossa xxix sunt.

[163] *Ibid.:* Et cum omnia corporis ossa numerata fuerint prout Galenus numerauit ducenta et xlviii excepto osse quod est in epigloti quod laude grece littere os simile nuncupatur.

[164] Albertus Magnus, *De Animalibus,* I, paragraph 196, p. 71: Sic ergo omnia ossa capitis simul computata sunt quinquaginta novem, si dentes sunt triginta duo: si autem sunt viginti octo, tunc sunt quinquaginta quinque: et componitur ex eis caput secundum modum quem praediximus. Fifty-nine minus thirty-two gives twenty-seven, Achillini's number.

[165] I cannot find such a numbering in Avicenna's discussion, *De anatomia cranii, ibid.*, I, 36–39, although he totals the bones of the body at 248 (I, 48).

Ribs twenty-four.
Of the thorax seven.
Of the hips two.
Of the large hands [entire limb] sixty-two.
Of the large feet [entire limb] sixty-two.
This number differs from the number of Galen.
The eye sockets above, right and left, two.
The eye sockets below, right, left.
The cheek bones, right, left, two.
Right and left parts of the quadrangular bone under the nose, two.
The exterior, interior, right, and left parts of the triangular bone of the nose, four.
The middle, separating part of the triangular bones, one.
The bones of the upper jaw.
There are three parts in one eye.
With the upper let down, three.
Three in the other eye, three.
Three in the right part of the nose, three.
Three in the left, three.
The middle of the nose, one.
Galen in his book *On the Parts* or of the *Medical Art*.[166]
The artery has two tunics, but has three tunics, and appears to have four.[167]

One tunic has oblique fibres, but the other has straight fibres.

Albertus Magnus, *On Animals* I,[168] puts the commissures [or sutures] of the head at eighteen, apparent or not.

The first is the coronal, from temple to temple.

The second is the sagittal, from the coronal to the lambdal.

The third is the lambdal, from the top of the posterior part of the head to the sides.

The fourth commissure is equidistant to the coronal on the right.

The fifth corresponds to the fourth on the left side, one third of the petrous bone.

The sixth divides the jaws in the middle of the chin.

The seventh goes from the temple to the left eyebrow.

The eighth corresponds to the seventh between the frontal bone and the os paris.

The ninth goes from the ear through the right cheek.

The tenth corresponds to the ninth between the os paris and the petrous bone.

The eleventh division of the frontal bone [goes] transversely through the sinews [zirbus] of the eyebrows.

The twelfth is under the lambda bone and the basilar bone.

The thirteenth is between the sinciput and separates the occiput; it appears in the extremity of the palate from right to left.

The fourteenth from the head of the nose to the dual bones separates lengthwise between the roots of the dual bones.

The fifteenth goes from the root of the nose to the right canine tooth.

The sixteenth corresponds to the left side of the fifteenth.

The seventeenth goes under the head of the nose from right to left.

The eighteenth goes through the length of the nose between the triangular bones of the nose.

The eleventh and twelfth I have never found although I have searched for them through many heads.

The end of the Annotations of Anatomy of the great Alexander Achillinus of Bologna.

---

[166] This list of the bones may be best approached by way of Galen's *De Ossibus ad Tirones* (Vol. II K., 732–778). Achillini's reference to *De Usu Partium* is less precise; that to the *Ars Medica* (a very popular introductory work) is also unsatisfactory since the bones are not specifically treated or listed therein. I conclude that in this instance Achillini was citing Galen carelessly, an unusual lapse from his normal practice. Galen himself lists briefly the contents of *De Usu Partium* at the end of Book I: I. General Principles and Procedures; II. Parts of the Hands and Arms (continued from I); III. The Legs; IV, V. The Organs of Nutrition; VI, VII. The Lungs; VIII, IX. The Head and Its Parts; X. The Eyes; XI. The Face; XII, XIII. The Spine; XIV, XV. The Organs of Reproduction; XVI. The Vascular System; XVII. The Arrangement of the Parts of the Body and Their Uses. There is no separate book on the bones alone.

[167] This statement makes no sense as it stands and is doubtless corrupt. See Galen, *De Temperamentis,* II (Vol. I K., 602) and *De Anatomicis Administrationibus,* VII, 5 (Vol. II K., 601–603) on the tunics of the blood vessels. The latter passage in Singer's translation (*op. cit.*, p. 178) may help to explain Achillini's odd assertion: "The arteries have two intrinsic coats, the outer [*tunica adventitia*] like that of the vein, the inner [*tunica media*] about five times as thick and harder. It consists of transverse fibres. The outer coat, like that of the veins, has longitudinal fibres, some slightly oblique, but none transverse. The inner, thick, hard tunic of the arteries has a woven sort of membrane on its inner surface, which can be seen in the large vessels. Some regard it as a third coat [*tunica intima*]. There is no fourth intrinsic coat but, like certain of the veins, some arteries have attached to and round them in places a delicate membrane which guards or fixes them firmly or binds them to the neighbouring parts." It is probably this passage that Achillini has in mind as he writes; but he has misunderstood it.

[168] Albertus Magnus, *De Animalibus,* I, paragraphs 183–185, pp. 65–66 (see note 114).

Alessandro Benedetti
(1450?–1512)

LIFE AND WORKS

THE documentary materials attesting the dates of his birth, his doctorate, and his death are lacking for a biography of Alessandro Benedetti. It is thus fortunate that there are many other biographical details which may be gathered from his letters and scientific writings. The older biographers agree that he was born around the middle of the fifteenth century at Legnago Fortezza near Verona.[1] His father was named Lorenzo[2] and he had an uncle, Giovanni Giacomo Angelo,[3] who was also a physician. Benedetti's wife, as we learn from his last will and testament,[4] was named Lucia and he had a daughter named in the will as Julia; he died without legitimate sons as heirs and bequeathed his estate to his female relatives. He received his doctor's degree in medicine at Padua in 1475 where, according to Papadopoulus,[5] he was also a professor. There is no evidence that he ever taught at Bologna, although Facciolati makes that claim.[6] Benedetti's death date is given as October 31, 1512, by the famous Venetian diarist, Marino Sanuto.[7]

One of the chief facts in the life of Benedetti was his long stay in Greece both in the Morea (southern Greece) and in the Venetian domain of Candia in Crete. For sixteen years he practiced medicine in these regions, especially at Cydonia on the northern coast of the island, modern Chania. He must have gone there around 1474 or 1475, perhaps just after receiving his doctorate in medicine, probably at the age of twenty-four if he was born around 1450, as seems reasonable. He was about sixty-two when he died on the basis of this tentative calculation.

His stay in Cydonia, at the house of a certain Cornelia, was filled with interesting experiences, some of which he describes in his large work on medicine, known by a short title as *De Re Medica*.[8] This fascinating volume deals in thirty books with the diseases and ills of the head, eyes, ears, nose, mouth, tongue, teeth, throat, trachea, lungs, sides, heart, stomach (internal and external), liver, the mind, gall bladder, spleen, pericardium, mesentery, omentum, peritoneum, intestines, bile, with worms and insects in the intestines, with the kidneys, urinal vessel, genitals of both sexes, strangled uterus, menstruation, prolapse, inflammation, abscesses,

---

[1] Giammaria Mazzuchelli, *Gli scrittori d'Italia cioè Notizie storiche e critiche intorno alle vite e agli scritti dei letterati italiani*, **2**, 1 (Brescia, Giambattista Bossini, 1753): pp. 811–812, is the most complete in his information. Nicolaus Papadapoulus draws upon Pietro Castellano and Andrea Chiocco in his few remarks about Benedetti: *Nicolai Comneni Papadopuli Historia Gymnasii Patavini* **1** (Venice, 1726, apud Sebastianum Coleti): p. 291. Roberto Massalongo, "Alessandro Benedetti e la medicina veneta del Quattrocento," *Atti del Reale Istituto Veneto di Scienze, Lettere ed Arti* **76**, parte II, fasc. 1 (1916–1917): pp. 197–259, writes a discursive and not particularly informative essay based upon Mazzuchelli's account and citing a long list of early authors without actually using them for the purpose in hand. Massalongo is more interested in Benedetti's contributions than in the facts of his biography. Like Castiglioni and Pazzini, he has been misled by Benedetti's description of a proposed anatomical theater (see note 8 to my translation of the *Anatomice*) into concluding that Benedetti had such a structure actually built at Verona. It is true, however, that this description is the first in anatomical history and precedes the building of the first permanent theater at Padua (1594) by ninety-seven years.

[2] *De Re Medica,* XXIX, prooemium, p. 514: "Ex patre Laurentio audimus. . ." His father lived to the age of eighty without using spectacles: *ibid*. II, prooemium, p. 43. His son Cornelius seems to have died early in life: *De Re Medica*, X, 9, p. 193. G. Favaro, *Leonardo da Vinci-I medici e la medicina* (Rome, 1923), p. 19, discusses the question whether Leonardo knew Benedetti. Dr. Dorothy M. Schullian in her excellent edition and translation of Benedetti's *Diaria de Bello Carolino* (New York, Renaissance Society of America, 1967), p. 34, lists the principal biographical studies of Benedetti to date. She provides a full bibliography.

[3] See the dedicatory letter to Marco Sanuto, of Venice, which precedes the *Collectiones Medicinales* that follow the *Anatomice* in the edition I have used (Paris, 1514): "avunculo meo Ioanne Iacobo Angelo physico."

[4] Archivio di Stato, Venice, September 18, 1512, "nel protocollo del notaio Girolamo Bossi, busta 50, n. 178." This also, of course, mentions his father Laurentius. This will, of two pages only, was preceded by another of one page, dated July 1, 1511, contained in Busta 1236, n. 183, "protocollo del notaio Antonio Savina." The first will was witnessed by Pietro Antonio da Legnago, Barbero, and Antonio Agnello di Agnelli da Lonigo, the second will by the latter alone.

[5] As recorded under the year 1500 in MS Archivio Antico dell' Università di Padova, Vol. 649, fol. 223 v: "Alexander Benedetti Medicine professore apud Papadopoli p. 291" (Professori, artisti e legisti fino al 1509). Mazzuchelli, *op. cit.*, p. 811, points out, however, that there is no evidence for his teaching at Padua at this time; and, of course, the manuscript in the Archivio Antico cites a printed source of 1726.

[6] Massalongo, *op. cit.*, p. 226, note 1. Both Medici and Martinotti, trustworthy scholars, deny any stay at Bologna.

[7] *I diarii*, **15** (Venice, G. Berchet editore, 1886): p. 283: "Morite questa notte (cioè dal 30 venendo al 31 ottobre 1512) domino maestro Alexandro Veronese medico, qual corexe Plinio et fece la Diaria *De Bello Carolino*, et altre opere de observatione in pestilentia et altro che fo impresse. Questo è stato do mexi amalato è *tandem* è morto." His first book (*De observatione in pestilentia*, 1493) and his last (Plinius Secundus, C., *Historiae naturalis libri XXXVII ab Alexandro Benedicto emendatiores redditi,* 1507) to be published during his lifetime are mentioned in this death notice, which records the fact that he was ill for two months. His will left a bequest to the church of San Pantaleone in Venice, and Massalongo thinks he may have been buried there (*op. cit.*, p. 227).

[8] Marcus Hopper calls the book by this brief title in his dedicatory letter to Vesalius, dated at Basle, July 1, 1549, on p. 1 of Benedetti's *Opera Omnia* (Basle, Henrichus Petrus, 1549). The full title is formidable in the 1533 edition: "Habes lector studiose hoc volumine Alexandri Benedicti veronensis physici praestantissimi, singulis corporum morbis a capite ad pedes, generatim membratimque remedia, causas eorumque signa. XXXI. libris complexa, praeterea (there follows a list of his other works with which the *De Re Medica* is posthumously printed) . . . Venetiis in officina Lucae Antonii Iuntae mense Augusto anno MDXXXIII." I quote the Basle edition of 1549, whose title page varies somewhat from that of the 1533.

and ulcers of the uterus, ills of the back, spine, loins, hands, feet, and with certain secret mixtures for separate members of the body as a final category. He discovered, for example, that Greek women who suffered from sunstroke relieved it with cold water poured from a cup upon their heads (I, 2, p. 3). He saw a girl at Cydonia, not yet eight years old, who during an acute attack of fever was seized with epilepsy and, lying near death, was relieved of a mass of forty-two intestinal worms by a purgative of honey and acorns which her mother gave the girl; "smoky vapors caused by these creatures in the lower body venter had seized upon all the senses of the brain" (I, 26, p. 19; *cf.* V, 5, p. 117). Benedetti, in an interesting mental case history, described a religious young man of choleric temperament who, in a fit of madness, pursued all his servants with a sword and was finally restrained, bound hand and foot, and thrust into a cell (I, 28, p. 23). In the same passage he describes a naked madwoman he knew in Italy who, after sexual relations with fifteen men in one night, menstruated abundantly, although she had ceased to do so for many years, and was thus restored to sanity. He mentions the fact that his father Lorenzo lived without using spectacles to the age of eighty (II, prooemium, p. 43).

While Benedetti was writing the *De Re Medica*, a certain Giacomo Justiniano, in Padua, having taken too much opium for a toothache which persisted even after a molar had been pulled, was finally seized by the final sleep ("supremo somno correptus fuit") (II, 6, p. 47). In Greece, Benedetti discovered a great lack of oculists and surgeons as well as physicians as a result of the barbarity of the Turks, who had destroyed Greek education (II, 9, p. 51). Young men in Crete, he also noted, suffered from swollen hands and feet (XXIX, p. 514).

Occasionally Benedetti mentions cases and medical events in Italy, such as that of Francesco Viario, a Venetian nobleman who always trembled as he spoke (V, 15, p. 118) and the mule which gave birth while he was writing the *De Re Medica* in the year Julius II was warring against Ferrara and King Louis of France (XXV, 9, p. 467). This incident can be dated with some accuracy since we know that Pope Julius entered Bologna on September 22, 1510, and campaigned that autumn against Ferrara. By a curious coincidence Berengario da Carpi records such an unusual birth in the time of Leo X (1518–1521) at Rome: *Commentaria*, etc., fol. 270 r. At any rate, Benedetti was writing the 25th book of the *De Re Medica* in 1510. At VII, p. 128, he discusses the incidence of goiter among the inhabitants of Bergamo and Como (see note 104 to my translation of the *Anatomice*). In the same book, while describing the ills of the throat, he tells how a mother at Brescia in the act of forcing her son to take a pill clutched his windpipe too firmly and killed him accidentally.

He learned many things in his experience as a physician both in Italy and Crete: for example, that the Greeks regarded the smell of oysters as among the sweetest of fresh odors (X, 10, p. 194) and that a certain Venetian had fasted for forty-six continuous days while a nun of Benedetti's acquaintance, Columba by name, had become something of a miracle at Rome by her months-long fasting (XI, 10, p. 204). Benedetti was much interested in fasting and wrote a book on prodigious fasts: "De Prodigiosis Inaediis," edited by F. Boerner at Bern in 1604, with the variant title "Exsempla Prodigiosae Inediae." Only Mazzucchelli mentions this work and I have not seen a copy of it. There is none in the British Museum or in the Bibliothèque Nationale. Massalongo is clearly in error when he says this book appears in "molte delle edizioni delle sue opere."[9] The actual book referred to by Friedrich Boerner, *Noctes Guelphicae* (Rostock, 1755, ch. 1), is by Paulus Lentulus, *Historia admiranda de prodigiosa Apolloniae Schreierae, virginis . . . inedia . . .* (Bern, Joannes le Preux, 1604), and is listed with full details by C. F. Steiger, *Jean le Preux, Der erste obrigkeitliche Buchdrucker der Stadt Bern 1600–1614* (Bern, Verlag des Schweizerichen Gutenberg-museums, 1953, p. 26, no. 17). It contains only two pages (102–103) of Benedetti material, two case histories from *De Re Medica* XI (not XII, as Boerner has it) (1549, p. 204).

Benedetti mentions many odd facts in his account: the method of reducing the size of their breasts employed by Venetian women after giving birth: an application of honey, saffron, and turpentine (X, 10, p. 198); the two women he saw who swallowed hair pins, bits of broken glass, and hairs: one of them died because of this strange diet, while the other vomited up three huge dried nodes of a dog's tail and presumably survived (XI, 7, p. 203). He refers to the use of poisons frequently and at least twice mentions a book he had written on poisons and envenomed creatures (X, 10, p. 195; XII, 8, p. 219). He found a Cretan wine called monovaticum (probably named from Monemvasia, on the Peloponnesian mainland) which surpassed all beers and was exported at thirty gold pieces a barrel (XI, 10, p. 212).

His little daughter Cornelia suffered from worms which troubled her sleep (XII, 11, p. 221). In a discussion of dangerous remedies Benedetti cites the example of a certain priest who was afflicted with gout. Hearing of pills which contained *euphorbia* (whether *Euphorbia lathyris* or some other form of the plant is difficult to say) as beneficial to his illness, the priest took some of them in concentrated

[9] *Op. cit.*, p. 257.

form made by a dealer in spices as imprudent as himself, was seized with dysentery on that very day and was buried on the next (XII, 17, p. 225). A post mortem he performed on a young man at Rome who had died of syphilis showed that almost the entire tunic surrounding the liver had been eroded, although the patient had never complained of any pain in that region while alive (XIII, 23, p. 243). A certain noblewoman named Paula Auogaria who was fond of drinking vinegar became fatter thereby (XIV, 22, p. 262). At Cydonia a foolish mother thrust her consumptive son into an oven to make him perspire, having first taken out the bread that was baking there and prepared a plank for him to lie upon; the poor boy died (XV, 25, p. 285). However, heavy sweats such as were achieved in a cave on the island of Melos were often salutary in the cure of dropsy (*ibid.*). Benedetti remembered that Cassandra Trona, whose father was a prince, suffered from a large tumor on the leg. When she refused to have it cauterized blisters appeared above her toes and as these discharged their fluid she was in a marvelous way restored to health (XV, 40, p. 295). A Venetian matron named Alba, of the Cauchi family, suffered from great constipation while pregnant. After she gave birth she emptied her bowels of thick excrements a hundred times in succession and was thus cured in wondrous fashion (XXI, 47, p. 379). In the use of drugs and herbs Benedetti records many useful items. The herb *picris*, called "panella" by the Veronese people, was an herb of great bitterness which grew on Mount Baldo; it was helpful in expelling worms (XXI, 3, p. 385). In curing kidney ulcers the leaves of the rhamnos tree were eaten; it was common in Greece and Dalmatia. Some Italians called it a bush (XXII, 31, p. 410). The tart wild cherry, called fig by the common people, grew among the hills of Verona and when sweetened with sugar was good for ulcers of the bladder (XXIII, 10, p. 427). He refers to his *Collectiones Medicinales* (or *Medicinae*, as the early editions give the title), a set of medical aphorisms first published in 1493, in which he had written that those who go about barefooted are not inflamed by sexual desire since their bodies dry out and grow chill (XXIV, 4, p. 445). Benedetti gave a seventy-year-old Venetian nobleman some medicine to revive his flagging virility. Lame and paralyzed as he was on one side of his body, the old gentleman went out merrily looking for pleasure and, having found it among the women of the town, was stricken with a sudden diarrhoea, *stolidus homo* that he was (XXIIII, 18, p. 448). Under the heading of "Those Things Which Inflame Love in Women and Men" Benedetti describes the corpses lying on the battlefield on the banks of the Taro River at Fornovo, southwest of Parma, in 1495. Here the Italian allies led by the Holy Roman Emperor Maximilian I, Ferdinand, king of Aragon, the Pope, the Venetians, and Lodovico Sforza met Charles VIII of France and his troops on July 6, as he was returning from his conquest of Naples. The Italians were defeated with a loss of three thousand men. The corpses were lying naked on the ground with their private parts swollen by the sun's heat and the rain which had fallen the day before (XXIIII, 19, p. 450). Under the heading of "Excrements Which Flow Forth under the Stress of Venery, and Cures for the Same" (XIIII, 27, p. 455) Benedetti mentions a certain Andrea, of Dalmatia, who was thus affected during vehement pleasure. Another quack, a certain Spaniard, who offered a cure for intestinal hernia by a single incision, is mentioned at XXIIII, 27, p. 459. Unusually long pregnancies, one of a woman who bore a second twin in her nineteenth month, on the island of Cyprus, and another who gave birth after carrying her child in the womb for fourteen months are listed (XXV, 23, p. 475). The nineteen-month child had an emaciated sinciput and crept back through his mother's umbilicus after he was born; it was ordered that he be buried

FIG. 2. Alessandro Benedetti. Palazzo Comunale, Legnago. Courtesy of Cav. Gino Girardi, Sindaco, Comune di Legnago.

alive as a monstrous animal. At XXV, 29, p. 479, Benedetti refers to an earthquake which occurred on March 24, 1511, and shook all of Italy for forty days. This date, which appears under the rubric *De causis abortiendi*, shows that he was close to the end of his *De Re Medica* (in thirty books) in 1511, the year before he died. In 1510 he had dissected a monstrous polyp to which a Venetian noblewoman of his acquaintance, Dionora Georgia, had given birth and had found in its midst a seven-month male embryo. Benedetti noted that in regions where the wines were stronger than in others pains in the joints were prevalent, as in Crete.

His stay in Crete was ended by 1490. In that year an invitation from the Venetian Senate sent him by its representatives, Marco Trevisani and Luca Pisani, brought him to the University of Padua, where he assumed his duties as ordinary professor of practical medicine and anatomy, succeeding Marco Dotto Padovano in the post.[10] We know little of these years; and in July, 1495, he was called away to become surgeon general of the Venetian army, which was in league with other Italian forces to fight at Fornovo in that year and to lay siege to Novara. In his *Diaria de Bello Carolino* (Venice, Aldus Manutius, 1496) Benedetti described this military campaign by Charles VIII against the kingdom of Naples and the Italian resistance to his invasion.[11] His style is compact and terse, packed with details and names dealing with topography and the commanders who took part in the battle and showing a keen awareness of the political intrigues involved in achieving unified action by the various Italian contingents.

The *Diaria*, only 28 pages long in Giustiniani's edition, is divided into two equal parts: the battle at the Taro River (*De Tarrensi Pugna*) and the siege of Novara (*De Obsidione Urbis Novariae*). The latter describes the siege of a city thirty-one miles west of Milan where the duke of Orléans, left in command by Charles VIII, was attacked by the Italian allies while Charles dallied at Asti and failed to relieve him. In the battle at Fornovo, Benedetti says about one thousand Frenchmen fell in the space of an hour and two thousand Italians. Commines put the French loss at only one hundred. Obviously there is no agreement on this point.

Beyond an occasional mention of fever and diarrhoea among the troops Benedetti has something more extensive to say about the medical problems presented by these two military events and mentions himself once, in connection with the employment by the Venetian Senate of Count Niccolò Orsino Petiliano as a general, when the physician was charged with the task of finding an astrologically propitious day for completing negotiations in the matter between the Senate and Petiliano: "Comes Petilianus . . . statuit autem prius a me Alexandro Benedicto Physico pro accipienda conditione propitii sideris aspectu querere ut prosperam diem denunciarem." This same Petiliano was later struck under the right kidney by a lead bullet which became impacted near the left shoulder blade. Benedetti examined him carefully and found by urine analysis that the Count's kidneys and bladder were unharmed. He then proceeded with his cure in the company of other physicians although the wound healed somewhat slowly since a bronze probe a palm and a half long had been thrust into it. A wandering quack dared to promise speedy recovery if a grain of millet were introduced into the wound; it would emerge on the following day from its mouth. A certain medicated water was also to be administered to the patient, who was persuaded by the duke Lodovico Sforza to undertake the treatment recommended by the quack (*non medicus, sed circulator quidam*), being assured that no poison was involved. Thus two days were wasted. Benedetti, however, knowing the venality of the quack, convinced him, after a payment of fifteen gold pieces, to postpone his treatment of Petiliano until the following week. Surgeons were meanwhile summoned from Milan who denounced the faker.

After these five years of military medical service Benedetti returned to his professor's duties at Padua and to his writing. Apart from these more tranquil labors we know little of his life until the year of his death, 1512. A number of letters, both by him and to him, throw a little light upon his later career. These may be listed as follows, most of them having been printed chiefly in his own books:

THE LETTERS OF ALESSANDRO BENEDETTI

1. To Giacomo Contareno: "Iacobo Contareno Patricio Veneto Philosopho Iuris consultissimo Senatoriique ordinis Viro Integerrimo. S. P. D.," in *De Obseruatione in Pestilentia* (1493).

2. To Marco Sanuto: "Alexander Benedictus Veronensis Physicus Marco Sanuto Veneto patricio

---

[10] P. Tosoni, *Dell' anatomia degli antichi e della scuola anatomica padovana e serie cronografica degli anatomici dello studio di Padova*; Memoria di Pietro Tosoni pubblicata in occasione della sua laurea in medicina (Padua, Tipografia del Seminario, 1844), p. 65.

[11] See note 2 for Dorothy M. Schullian's translation of the *Diaria de Bello Carolino*, to which I refer readers for its extensive evaluation of the book in its historical and intellectual context. Benedetti's relations with the Venetian aristocracy are carefully detailed in the Introduction, pp. 3–45. Historical accounts of this event in Italian history are to be found in Francesco Guicciardini, *Storia d'Italia* (Pisa, Niccolò Capurro, 1822–1824) **1**: pp. 297 ff.; Philippe de Comines, *Mémoires* (English version by Thomas Danett, London, Samuel Mearne, 1674), pp. 286–290; H. Fr. Delaborde, *L' Expédition de Charles VIII en Italie* (Paris, Firmin-Didot, 1888), pp. 611–620; John Addington Symonds, *The Renaissance in Italy*, I, chapter X, "Charles VIII" (this can be found in the Modern Library Giants); Colonel G. F. Young, *The Medici* (Modern Library, 1930), chap. XI, pp. 244–249.

senatori optimo. S. D." In *Collectiones Medicinae* (1493).

3. From Jacobus Antiquarius, known also as Jacopo Giacomo Antiquari: "Alexandro Benedicto Veronensi Medico clarissimo Iacobus Antiquarius S. P. D." Dated 1494 in the 1502 edition of the *Anatomice*. Missing in the 1514 edition.

4. From Giorgio Valla:

Alexandro Benedicto Veronensi Georgius Valla S. D. P./Alexander mi benedicte sis abs deo benedictus ac salve. Habeo gratiam magnam tibi quod patriam nostram et valere et salvere juberis tantique mea de causa habueris. Quod autem scribis commentarios tuos planeque de rege gallorum historiam et legam et si quid videatur corrigam. Ego eis qui litteras attulerunt tuas respondi confestim videri mihi rem absolvendam vel saltem proximam pugnam quam cerno cruentissimam fore maximique omnium momenti; immo equidem statuum praecipueque mediolanensis ponderis gravissimi esse expectandam. Itaque cum responderim illis videri mihi hoc negotium perficiendum meam ipsi probantes sententiam tuos secum tulerunt commentarios. Nunc quod reliquum est tamquam amicissumus frater commonebo ut existimes imitatione aliqua te ista melius scripturum et quemadmodum Salustius suum bellum scripsit Iugurthinum ita tu Carolinum mentemque tuam ad istum aut alium similem dirigas imittandum. Deinde existimes fortunam laudare solam superbum esse. Consilia atque judicia apud sapientes solere vel reprehendi vel laudari; proinde non tam bellum referendum esse quam quo consilio regio necnon veneto fuerit administratum ut si quae legationes contionesque inter se fuerint ne quidem supprimantur. Dein quod ad doctrinam pertinere videbitur ut acutum tuum postulat ingenium fertque judicium singulare, inseras historiae quam equidem talem fore arbitror ut sit sibi aeternitate vindicatura. Tum si quid nos opis ferre poterimus non detrectabimus. Vale. Data Veneti quarto Idus Augustas 1495. (Giorgio Valla to Alessandro Benedetti of Verona, greetings: My Alessandro Benedetti, may you be blessed by the Lord [punning on his name Benedetti and *benedictus*] and continue in good health. I am very grateful to you because you desire that our country should be both strong and safe and because you have been so active in my behalf. As to the fact that you are writing down your notes, which are in fact a history concerning the king of France, I shall both read it and correct it where it seems necessary. To those who brought me your letter I replied at once that it seemed to me a task that should be completed and that at any rate as I see it the recent battle was the most bloody and its outcome of the greatest importance to all states, especially to Milan. Thus when I told them that I believed your undertaking should be carried out they approved of my opinion and brought your notes with them. The next thing to do is to advise you like a most friendly brother that you should consider for the better composition of this work the imitation of Sallust and his style in the *War with Jugurtha* as you compose your *War With Charles VIII* and turn your mind to this approach or something like it. Then you should consider that to praise fortune alone is an act of pride. Plans and judgments are wont to be blamed or praised among wise men; hence, not only the war itself should be discussed in your book but its planning and execution by the king and the forces of Venice so that whatever embassies and conferences were conducted among them should not be omitted in your account. This procedure will appear to be based upon accurate knowledge as your keen talents require and your outstanding judgment bears out; if you incorporate these into your history I think it will be one which will bring you immortality. Furthermore, if I can help you in any way to do this I shall not fail to do so. Farewell. Venice, August 10, 1495.)

This letter, which appears among the Tioli manuscripts at the University of Bologna,[12] was written on August 10 by the physician and writer, Giorgio Valla, of Piacenza (who lived from about 1430 to 1500), little more than a month after the battle of the Taro River (July 6). From Valla's remarks it appears that Benedetti set to work immediately after the event and sent some of his notes or first drafts to Valla, who in the manner and custom of the day, promised to correct them, urging Benedetti to complete his account of the battle, whose importance he clearly recognized, as quickly as possible. He also urged the imitation of Sallust, the Roman historian, or some other such reputable writer, with proper attention to the diplomatic and political aspects of the battle. Valla left it up to the talents and judgment of Benedetti as to how he should interpret the event *sub specie aeternitatis*, casting upon Benedetti's shoulders the illustrious mantle of a profound observer.

5. To Matteo Rufo: "Alexander Benedictus physicus Veronensis Reuerendo patri Domino Matthaeo Rufo conciui suo. Sal. P. D." This letter is prefixed to the edition of Pliny's *Natural History* edited by Johannes Britannicus and published at Brescia in 1496, to which Rufo had written an introduction. Benedetti was himself to publish an edition of Pliny in 1507 and in the present letter comments briefly on certain botanical terms in Pliny. He begins with the statement that he has just returned to Verona after fourteen years absence and finds the city entirely changed in customs, manners, and in public and private buildings. After vigorously supporting Rufo's introduction against all criticism he ends in great haste with the cryptic words "Alia forte in Graecia excitabis," which may mean that he intended to write Rufo again from Greece. Evidently another visit to Greece was being planned by Benedetti or already decided upon.

6. The two following letters, to Agostino Barbarico

[12] Notizie della vita e delle miscellanee di Monsignor Pietro Antonio Tioli, nato in Crevalcuore a'. XIX maggio. MDCCXII, defunto in Roma a'. XX Nov. MDCCXCVI, cameriere segreto di S. S. e segretario della S. C. de' confini dello Stato ecclesiastico raccolte da Francesco Cancellieri con i catalogi delle materie contenute in ciascuno de' XXXVI volumi lasciati alla biblioteca del SS. Salvatore de' canonici Lateranensi di Bologna; Pesaro, coi tipi di Annesio nobili nell' anno MDCCCXXVI: Vol. XIX, p. 91, University of Bologna Library Cons. ℒ139 (MS 2948). It is also published by J. L. Heiberg, *Beiträge* etc. p. 75 from Cod. Vat. Lat. 3537, fol. 165 r–166 r.

(Barbarigo) and to Sebastiano Badoer or Baduario and Gerolamo Bernardo, are dated in March and August, 1496, and are printed at the beginning and at the end of the *Diaria De Bello Carolino*: (*a*) "Alexander Benedictus Veronensis Physicus, Augustino Barbarico, Illustrissimo Venetorum Principi, S. perpetuam dicit." In this dedicatory letter, after explaining his theme with admirable brevity, not omitting a modest comparison between his history and that of Livy (both designed primarily as annals of daily events), Benedetti refers to other writers who have touched rather lightly upon the events which he records day by day and have written little truth. Whether he means the contemporaneous accounts of Marino Sanuto (see footnote 157 to my translation of the *Anatomice*) and Giovanni Garzoni or others is hard to say; but it is clear that the invasion of Italy by Charles VIII of France stirred the commentators and historians of Italy into almost instant action. (*b*) "Alexander Benedictus Veronensis Physicus Sebastiano Baduario equiti et Hieronymo Bernhardo Consiliariis Veneti Senatus clarissimis. S. P. D."

7. To Emperor Maximilian I: "Alexander Benedictus Veronensis Physicus Maximiliano Caesari Aug. Romanorum Regi Invictiss. Sal. Perpetuam Dicit." This letter bears the date Aug. 1, 1497, and is prefixed to the 1502 edition of the *Anatomice* and retained in subsequent editions such as that of 1514. It will be noticed that dedicatory letters by Benedetti or letters to him by other writers published in his books do not always bear the same date as that of publication of the book itself and vary as much as five years or more.

8. To Aloisio Trevisano: "Alexander Benedictus veronensis physicus magnifico et clarissimo domino Aloisio Trivisano, quondam magnifici domini Silvestri domino meo observando. In quodam chyrografo, oratores magnificae comunitatis Traguriensis pro rebus comunibus and pro salute totius Dalmatiae tibi commendo; coram tibi narrabunt, vale." This letter, written in haste from Zara (Zadar) in Dalmatia on the last day of June, 1499, is published in Marino Sanuto, *I Diarii* (G. Berchet, Venice, (1879), **2**: pp. 891–892), and describes a devastating attack by the Turks in the region from which Benedetti writes. The local Greek soldiers were helpless to prevent the slaughter of from four to six thousand people, the theft of farm animals, and the burning of villages. Benedetti describes his cure of Count Nicolas Benchovich from pleuritic fever as well as the alert administration of that part of Dalmatia by the local duke who was doing his best under difficulties to stem the Turkish incursions.

9. To Jacobus Antiquarius[13]:

Alexander Benedictus Jacobo Antiquario S. P. D. Georgius Valla[14] in omni genere litterarum doctissimus nostrique amantissimus opuscula nostra quae nuper praegustationis gratia edidimus ad te familiarissime misit, ductus nimia nostri affectione unde fortasse laxato paulum diligenti examine bona esse existimavit quae tu quoque vir praestantissime cum priscis qui de medendi ratione scripserunt certare affirmas. Non is sum profecto ex medicorum vulgo vel postremo qui hic audeam, sed temporum nostrorum maxima indignatio facit ut qui nihil sciam audeam scribere ut caeteri expergefacti nostris conatibus perficere idem audeant. Non ero primus qui medicinae rationem male prosequar, undique enim occurrit barbaries; hactenus milia errorum apparent, nec modus est. Nos forte domesticam medicinam aliis diligentius excolendam dabimus. Damnabantur enim qui vineam suam habuissent derelictui. Litterae tuae ad editionem anatomices sive historiae membrorum humani corporis ad Maximilianum Imperatorem quae libris quinque continetur me plurimum exagitatis [exagitant], et nisi praesto sit Georgius Valla vir aetate nostra doctissimus extra omnium aleam hic positus vehementer aborticum partum timeant; utriusque tamen periculo fuit: octo interea de omnibus morbis qui membratim hominem invadunt sub lima sunt. Sex de venenis et venenatis animalibus perfecti patrocinium quod amici maxime probaverint praestolabuntur. Triginta sex in summa volumina in plebem spargenda sunt, quae si priusquam edita sunt taxatorem invenerint maturius sese ad tempestivum partum conservabunt. Nec ita temere aliorum exemplo in lucem exibunt. Vale. Veneti Idibus Ianuarii. (Alessandro Benedetti to Jacopo Antiquari, greetings: Giorgio Valla, most learned in every form of literature and very fond of me, has in very friendly fashion sent you those small writings of mine which I recently published as a foretaste of better things to come. He was induced to do this by his excessive affection for me and hence perhaps relaxed somewhat his careful faculty of criticism in calling my writings good; you too, most estimable of men, declare that they rival in quality the ancient works on the art of healing. I am certainly not one of that mob of modern physicians who might dare to make such a statement, but it was my very great irritation with my own times that causes me, who know nothing, to dare write in order that others, stirred up by my attempts, would carry out the task more thoroughly. I shall not be the first to pursue the art of medicine badly, for barbarism meets us everywhere in the profession; up to the present day thousands of errors in its practice are apparent, nor is there any end of them in medical writing. Perhaps I devote myself to general

[13] Jacobus Antiquarius (Jacopo Giacomo Antiquari), of Perugia (1444–1512), held public office at Bologna and then went to Milan *circa* 1460, where he became secretary to Galeazzo and Lodovico Sforza, known as Il Moro, the Dukes of Milan. He was a friend of numerous famous men of his time, including Hermolaus Barbarus, Poliziano, Giorgio Valla, Marsilio Ficino, and Pico della Mirandola. This letter is also published by Heiberg, p. 88, who dates it on January 13, 1494. See further D. M. Schullian, *op cit.*, pp. 15–16.

[14] Giorgio Valla praised Benedetti as "eloquio et doctrina medicus primas tenens," comparing him with other famous Veronese citizens such as Catullus, Vitruvius, Pliny, and Domitius Calderinus in his edition of Nicephorus Blemmydes' *Logica*, with other tracts: (Venice, Simon Papiensis dictus Bevilaqua, 1498) Aristotelis Magnae Ethicae. Apostolo Zeno, *Dissertazioni Vossiane* **2** (Venice, G. Albrizzi, 1753): p. 43 says the name Paeantius was given Benedetti by Quinzio Emiliano Cimbriaco in his dedicatory verses to the *Diaria* and to B.'s edition of Pliny.

practice more diligently than others, for those who neglect their vineyard are condemned. Your letter concerning the publication of my *Anatomice or the History of the Parts of the Human Body*, dedicated to the Emperor Maximilian and contained in five books, stirs me up greatly, and if Giorgio Valla were not at hand, a man beyond all hazard the most learned of our time, the book would greatly fear an abortive birth, for there was danger of it; meanwhile eight books on poisons and envenomed animals are now completed and shall await that approval which friends especially are wont to bestow. Thirty-six books all told are to be published; but if before they appear they shall find someone to appraise them they shall more ripely preserve themselves for a timely birth. For not so rashly shall they issue forth into the light of day following the bad example of others. Farewell. At Venice, January on the Ides.)

The mention of Jacobus Antiquarius's letter of 1494 in the 1502 edition of the *Anatomice* places the date of the letter transcribed above as between 1494 and 1502; Antiquarius had urged the publication of the *Anatomice*. Eight books of the *De Re Medica* (published posthumously) are described as in process of revision (*sub lima sunt*). The work on poisons and envenomed creatures in six books (see note 158 to my translation of the *Anatomice*) is here declared to be completed and soon to be forwarded to Jacobus. Thirty-six books of writing are said to be on the point of publication, although only nineteen are referred to specifically. The sentence about those "qui vineam suam habuissent derelictui" is an allusion to St. Matthew 21:33.

10. To Paolo Trevisano: "Alexander Benedictus Veronensis Physicus: Paulo Trivisano Equiti Clarissimo Salodii Praefecto. S. D. P." This letter appears in a work without date or colophon inside a title page which, according to an early and inconvenient custom, serves also as the table of contents. The book is "Annotationes Ioannis Antonii Panthei Veronensis ex trium dierum confabulationibus Ad Andream Bandam Iuris consultum, in quo quidem eruditus lector multa cognoscet quae hactenus a doctis viris desiderata sunt." At the bottom of the same page are the words: "Alexandri Benedicti Veronensis physici atque oratoris clarissimi epistola ad magnificum atque insignem equitem dominum Paulum Trivisanum Salodii ac totius Riperiae praefectum." There is no publisher or place of publication recorded in this odd book, and the date of Benedetti's letter "Venetiis. Idibus Maiis. M. D. D." probably means May 15, 1505. The letter itself tells us that Giovanni Antonio Panteo, of Verona, who had written the series of conversations on baths, springs, and rivers (especially the baths of Calderiana near Verona), had died before he had prepared his book for publication. Pietro Banda, his friend, in consultation with Andrea Banda, archpresbyter and canon of All Saints' Church, Verona, to whom Panteo had addressed his book, prepared and published it. Margarita Banda had married Paolo Trevisano, the Venetian official to whom Benedetti addresses his letter; she and her husband had lived on the island of Cyprus and hence in the area of the Aegean where Benedetti had spent at least fourteen years. Trevisano himself had travelled extensively in the East and Near East, and had finally settled for some years on Cyprus, having studied the flora and fauna of the Nile and the Near East with some care. The small treatise bound with Panteo's book, "De laudibus Veronae," is dated 1483 at the end of it.

11. To the citizens of Verona: This three-page letter appears in Benedetti's edition of Pliny, *Natural History*, and begins as follows: "Alexander Benedictus Veronensis physicus clarissimis ciuibus suis Sa. perpetuam D." The title of the edition is: C. Plinii Secundi Veronensis historiae naturalis libri XXXVII ab Alexandro Benedicto Ve. physico emendatiores redditi (Venezia, Giovanni Rubeo e Bernardini fratelli Vercellenses, 1507). Pliny was a native of Verona, and there is an appropriateness which Benedetti and his townspeople could relish in the fact that Benedetti was also a Veronese. The edition itself is the usual sort of thing which in Pliny's case was multiplied dozens of times. Benedetti's interest in pharmacology results in many comments on plants and herbs in Pliny.

12. From Jacobus Antiquarius: "Iacobus Antiquarius Alexandro Benedicto Philosopho Clarissimo. S. P. D." This very brief letter, dated Nov. 10, 1509, at Milan, appears in the 1533 collected works as well as in the 1549 edition of the same. Both are posthumous, of course, since Benedetti died in 1512. Jacobus approves of the *Anatomice*, which he had exhorted Benedetti to publish (see letter 3, dated 1494) and refers to Benedetti's projected edition of Paul of Aegina, then in progress, an author neglected in Jacobus' opinion and little known except for his frequent citation of Galen.

Mazzuchelli (*Gli scrittori d'Italia, etc.*, II, p. 812) mentions a letter from Benedetti to Symphorien Champier in the collection entitled "Complures Illustrium Virorum Epistolae ad Camperium per Jo. Piroben et Jo. Divineur Alemanos, sumptibus Jacobi Franc. de Jonta Florent. Bibliopol. Veneti 1519" which I have not seen, as well as some letters among those of Leonardo Bruni (Aretino) in an edition published at Brescia in 1495. See D. M. Schullian, edition of Benedetti's *Diaria de Bello Carolino*, pp. 17, 41–42, for this letter and for one by Pietro Delfino.

THE BOOKS OF ALESSANDRO BENEDETTI

I am not able in this simple list to deal with the difficult problem of the incunabula editions of Benedetti's books, which have been listed in the *Gesamtkatalog der Wiegendrucke* **1** (1925), nos. 862–864 and accompanying remarks on ghost editions, by Frederick R. Goff, *Incunabula in American*

Libraries. A Third Census (New York, 1964), nos. A-388–390; and in *Indice generale degli incunaboli delle biblioteche d'Italia* **1**, nos. 1459–1461. Early writers mention various incunabula, but no library or collector seems to possess them; as a matter of conservative reason it is perhaps wiser to ignore them as unascertainable. The chronological list that follows is based upon information drawn from the catalogs of the British Museum, Bibliothèque Nationale, Library of Congress, New York Academy of Medicine, National Library of Medicine, Wellcome Historical Medical Library, Index Medicus, and other sources such as Mazzuchelli's list in his biography of Benedetti or Klebs' list of incunabula:

1. *De observatione in pestilentia*. Venice, Johannes and Gregorius de Gregoriis, July 29, 1493. This also appears as *De pestilenti febre liber unus, in quo sane compendio omnia quae de eodem morbo dici possunt doctiss. sunt tractata* in Joachim Schillerus ab Herderen, *De peste Brittanica commentariolus*, etc., ff. 25–60; Basle, 1531, and in Petrus de Abano, *De venenis eorumque remediis*, ff. 139–193; Basle, 1531? 1561? The BM Catalogue also lists Muhammad Ibn Zakarīyā (Abu Bakr) Al-Rāzī, etc. *Rhazes . . . de ratione curandi pestilentiam*, etc., Paris, 1528, among the printings. Loris Premuda, ed., *De Plinii in Medicina Erroribus* (Edizione de "Il Giardino di Esculapio," 1958, with biographical introduction on Leonicenus [Niccolò Leoniceno], notes, Latin text, and Italian translation of Leoniceno's book published at Ferrara by Jacopo Maziocho, 1509), contains among other items Benedetti's *De observatione in pestilentia* preceded by the letter to Jacopo Contareno, of Venice.

2. *Collectiones medicinae*; Venice, Johannes and Gregorius de Gregoriis, *ca*. 1493. Also printed by Symphorien Champier at Lyon in 1506. According to the type used this book was printed approximately at the same time as No. 1, *De observatione in pestilentia*: see the *Brit. Mus. Cat. of Books Printed in the XVth Century . . .*, Part V (1924), p. 344. H. P. Kraus, Cat. 126 (1971), p. 114, lists a copy.

3. *Diaria de bello Carolino*; Venice, Aldus Manutius, 1496. Italian translations: (1) *Il fatto d'arme del Tarro fra i principi italiani, et Carlo Ottavo re di Francia, insieme con l'assedio di Novara, di M. Alessandro Benedetti, tradotto per Messer Lodouico Domenichi* . . . Vinegia, G. Giolito de Ferrari, 1549; (2) second edition by Antonio Crosa and Carlo Moscotti, Novara, Italy, 1863.

The Latin text was printed twice in later publications: (1) *Pietro Giustiniani, Rerum Venetarum ab urbe condita ad annum MDLXXV historia Petri Justiniani*, etc.; Strassburg, Lazarus Zetzner, 1611; (2) J. G. Eccardus (Eckhart), *Corpus Historicum Medii Aevi*, II (Leipzig, J. F. Gleditsch, 1723), cols. 1577–1628.

4. *Historia corporis humani; sive Anatomice*; Venice, Bernardino Guerraldo, 1502. The National Library of Medicine catalogue card contains the following information: "The dedication is dated 1497. The following passage from book 3, chap. 18 indicates the time of writing: ". . . & si aetate nostra Venetiis cum haec conderemus anno a sal. universali MccccLxxxiii Lucretiae perquam familiari nostrae simile contigerit." Perhaps this date is incorrect and instead of 1483 we should read 1493, at which time the author was himself in Venice. (R. Massalongo, *Alessandro Benedetti* 1915, p. 29, states that the author was in Greece and Crete for about 16 years prior to 1490.) The prefixed epistle from Jacopo Antiquario, urging publication of the work, is dated 1494.

First edition? Various authorities list editions of 1493, 1496, 1497, and 1498, but bibliographies of incunabula record no edition of this work. G. Cervetto, *Di alcuni illustri anatomici italiani*, 1842, p. 86, n. 1, gives the dates 1493, 1496, and 1497. J. Astruc, *De morbis venereis libri novem, editio altera*, t. 2, p. 565, gives the date 1497, but states in a footnote that it is from the dedication prefixed to the 1514 edition. Mazzuchelli gives the date 1498 and cites Maffei, Verona illust. P. II. pag. 250. . . .

Benedetti "scrisse dottamente d'anotomia, e l'opera uscì nel 1498 con titolo d'Historia corporis humani, distinta in cinque libri, e ristampata poi più volte." —Maffei, Scipione, *Verona illustrata*, pt. 2, 1825: p. 244. The existence of such an incunabulum edition seems confirmed by the following passage by Giorgio Valla (d. 1500) here transcribed from his *Commodis et incommodis*, Argentine, n. d., p. 20: "At nostra tempestate . . . Alexander Benedictus . . . Anatomicen corporis humani tradidit, ad quam lectorem mittendum existimamus." Copinger, no. 377, records a Paris, 1497, edition of the *Anatomice*, allegedly printed by Henri Estienne, but as GW notes (col. 428) Henri Estienne was not active at this date. See Ph. Renouard, *Imprimeurs Parisiens*; Paris, 1898, p. 121: Estienne was active between 1502 and 1520. Copinger also recorded a copy in the Royal Library at Dresden, but Helmert Deckert's catalogue of the incunabula holdings of that library rightly omits all mention of it as a ghost edition (1957). What Copinger obviously referred to was the 1514 edition printed by Henri Estienne. Valla's remark, quoted above, may therefore merely mean that the book circulated in manuscript for some years before publication.

Other editions: Paris, Stephanus, 1514 with *Collectiones medicinales*; Cologne, Eucharius Cervicornus, 1527, with *Georgii Vallae Placentini de humanae faciei, corporis partibus, opusculum sane elegans et perutile*, pp. 24, based on Julius Pollux and Celsus; Paris, Simon Sylvius, 1527, with *Collectiones medicinales seu Aforismi*; Strasburg, J.

Hervagius, 1528, with *Aphorismorum liber, Aphorismi Damascoeni*, and *Hippocratis jusjurandum*, edited by O. Bumfelsius.

5. *Plinius Secundus, C., Historiae naturalis libri XXXVII ab Alexandro Benedicto emendatiores redditi*; Venice, Giovanni Rubeo and Bernardino brothers of Vercelli, 1507; the BM catalog lists three more editions: 1510, 1513, 1516; the BN catalog has one in 1512.

6. *Medicinalium observationum exempla rara:* see R. Dodoens, . . . *Medicinalium observationum exempla rara*; Cologne, Maternus Cholinus, 1581; this book contains in addition to Antonio Benivieni's *De abditis nonnullis ac mirandis morborum* etc., with notes by Rembert Dodoens (Dodonaeus), 1517–1585, "Medicinalium observationum exempla rara Valesci Tarantani et Alexandri Benedicti," pp. 289–306. These are excerpts from three works of Benedetti: *De curandis morbis; De observatione in pestilentia;* and *Historia corporis humani; sive Anatomice*.

7. *Habes lector studiose hoc volumine . . . singulis corporum morbis a capite ad pedes, generatim membratimque remedia, causas, eorumque signa XXXI libris complexa, praeterea Historiae corporis humani libros quinque, De pestilentia librum unum, et Collectionum medicinalium libellum . . .*; Venice, Lucantonio Giunta, 1533. This posthumous work contains the first edition of the general work on medicine called *De Re Medica* by Marcus Hopper in his dedicatory letter to Vesalius in the 1549 edition. The letter by Jacopo Antiquari in the 1533 edition, written while the work was in progress, is dated 1509. Benedetti on p. 405 (p. 479 of the 1549 edition) mentions as a current event an earthquake in the year 1511. Other editions, which differ in their contents, are: Basle, Henricus Petrus, 1539; 1549.

The *Anatomice* is a descriptive anatomy and not a dissecting manual. Benedetti begins with a dedicatory letter to Maximilian I in which, according to a fashion which begins as early as Zerbi's *Liber Anathomie corporis humani*, etc. (1502) and is preserved by Berengario da Carpi, de Laguna, Massa, Guinterius, Estienne, and Vesalius in the dedicatory letter to Philip II that precedes the text of the *Epitome* (1543), he deplores the barbarity of the age in which only an enlightened patron can understand a book on anatomy, when ignorant people practice dissection, and when one's students and friends demand the publication of the book in hand. There is also in his dedication a note struck by later anatomists such as Massa: the boast that the author has discovered features of the human body overlooked by previous investigators.

Benedetti's remarks about the Arabs and those who prefer to read them are sharp and uncompromising; he warns them not to read his book at all if they turn up their noses at it. He then proceeds to his preliminary chapter with the conventional comment on the ancient kings who dissected live criminals and the cruelty of such a practice. He mentions post-mortem examination in cases of syphilis in the manner of Galen and his famous ape, whose pericardium he found to be dried out. The first Greek he quotes is Aristotle, who with Plato in the *Timaeus* forms the chief basis of reference. Aristotelian teleology governs his investigation of the human body. The terms he uses are exclusively Greek, drawn largely from Julius Pollux, *Onomasticon*.

After listing the external parts of the body he takes up the qualities of the humors and of the members, beginning with the blood. Here the theory of dry, wet, cold, and hot categories is likewise Aristotelian. The doctrine of similar elements or parts, however, with which Benedetti opens Book I, chapter 5, is Galenic (ed. C. G. Kühn, **4**: p. 741). In describing the bones he gives the case of a woman who had died of syphilis and whom he found upon post-mortem examination to have her bones swollen with abscesses and suppurating as far as the bone-marrow.

The organization of the *Anatomice* (in five books) is as follows: I, general remarks on anatomy, choice of a cadaver, and the construction of a temporary dissecting theater; the shape and dignity of the body from a teleological point of view; the external parts; the qualities of the humors and of the members; the function of the members; the bones, cartilage, nerves, arteries, veins, muscles, membranes, marrow, skin, and nails. II, the stomach, umbilicus, sumen or abdomen, muscles of the lower cavity, peritoneum and omentum, intestines, mesentery, stomach, adipose matter and the pancreas, spleen, liver and its veins, gall bladder, kidneys, seminal veins, semen, testicles, scrotum, penis, bladder of the urine, seminal vessels in women, their sexual parts, the menstrual flow, and the anus. III, the spiritual members above the diaphragm, the diaphragm, the breasts, chest muscles, pleuritic membrane, lungs, trachea artery, veins and arteries of the lung, the heart, large hearts and the right and left lobes of the heart, the ventricles, right, middle, and left, the gullet, epiglottis and artery, tonsils, uvula, palate, tongue, teeth, lips, and gums. IV, the head, the divine intellect of man, the brain and its contents, the hair, the skin of the head, external membrane of the skull, the skull, its sutures, the inner membranes, the cavities of the brain, the sinus of the cerebellum, the pathway between each sinus or ventricle, other parts of the brain, the seven pairs of nerves, the nerves of the eyes according to Aristotle, the second pair, the third, fourth, fifth, sixth, seventh pairs, another observation concerning the nerve pairs, the marvelous net, on the strong bone of the head, the eyes, causes of their color, sharp vision, defects of the eyes, their membranes, cornea, white of the eye, second and third tunics of the eyes, another enumeration of the membranes of the eyes, the humors between the tunics, the hyaloid and crystalloid, cure of cataracts, the ears, nose, and lower jaw. V. The last book deals with the veins, muscles, bones, and nerves which pass through all parts of the body, the blood and spirit, the major vein and aorta and their branches, the vein of the middle of the forehead and of the temples, the common vein of the arm, the veins connected with the kidneys, the other veins, the aorta, arteries of the legs, veins and arteries corresponding with one another, the members to which only the blood veins pass, to which only the aorta passes, or to which both pass, the method of the fibers, origin of the sinews, the nerves which arise from the spinal medulla, the nerves of the neck, the loins, the muscles (additional remarks to those in the first book) of the thorax, arms, back, legs, the bones and vertebrae of the neck, the spine, the injured medulla and dislocated spine, hidden shoulder blades and the shoulder, the arm, the hand and its dislocation, the lower bones, the hip and pubic bone, the femur and its dislocation, the shins and knee cap, ankle bone or talus and its dislocation, the heel and the toes, and a final chapter on the praise of dissection in which he urges that a public dissection be held at least once a year for both beginners and veterans in the art.

The exaggerated Hellenism of Benedetti is reflected throughout his book and nowhere more than in his rather self-consciously Greek title, *Anatomice*, unique among pre-Vesalian works on anatomy. We must remember, of course, that he spent at least fourteen years in a Greek environment and doubtless knew Greek well; he could probably have written his anatomy in Greek had not Latin been the language of science and scholarship. There is, however, a greater clarity in a style which is not impeded by numerous references to many authors, Arab, Greek, and Latin; and what the *Anatomice* loses in richness and breadth of reference it gains in simplicity and forcefulness.

In his adherence to the Greeks, however, Benedetti is not quite slavish. He realizes, for instance, that "Aristotle has had so much authority for so many centuries that even those things which (physicians) have not seen they will affirm to exist, even without experiment" (Book III, 13). He points out in regard to the ventricles of the heart that Celsus said there were only two, right and left, although Aristotle insisted there were three. Since he said so, physicians continued to acknowledge that there were three ventricles. At Book I, 8, he states that Aristotle believed the nerves arise first from the heart; but Benedetti points out the fact that most of them originated from the brain.

It is doubtful whether one can speak positively about original contributions by Benedetti. Gurlt, Hirsch, and others[15] believe his chief service to science was to recognize the need of clinical and pathologic-anatomical observation as the single rational basis for any advance in medical learning; this conviction spurred him to write his medical encyclopedia, *De Re Medica*. He was an excellent surgeon especially in cases of hernia, bladder stone, gunshot wounds, and rhinoplasty; he writes as one of the very first physicians to do so on syphilis, examining the ravages of the former disease as far as the bone marrow. In pathological medicine he was a faithful follower of his teacher, Benivieni. His precautions against the plague were sound: he urged the isolation of the patient and the disinfection of his clothing, especially heavy woolen garments, to prevent spreading contagion.

Brambilla[16] points to Baverio Maghinardo de' Bonetti (1405 or 1406–1480), a physician of Imola, as a predecessor of Benedetti in the accurate anatomical description of the affected part before the description of wounds. Benedetti describes the secretion of moisture in the vagina during coitus and was the first since Hippocrates to trace apoplexy to the compression of the jugular veins, as he saw it in people who had been hanged, and the resultant overabundance of blood in the vessels of the brain (see note 99 to my translation of Achillini's *Annotationes Anatomicae*). Haller[17] believed that Benedetti was the first to describe stone in the bile duct. Portal[18] refers to his statement that the yellow bile empties from its duct into the duodenal intestine; Benedetti also knew the property of mercury for inducing salivation. Pazzini[19] declared that Benedetti was the first physician to attempt anatomical

---

[15] *Biographisches Lexikon der hervorragenden Aertze* (second ed., 5 v. and supplement, Berlin and Vienna, 1929–1935) *s. v.* Benedetti. "His contribution to the body of anatomical knowledge was nil." This is the rather too harsh judgment of C. D. O'Malley, *Andreas Vesalius of Brussels, 1514–1564* (Berkeley, Calif., 1964), p. 18.

[16] G. A. Brambilla, *Storia delle scoperte fisico-medico-anatomico-chirurgiche*, etc. **1** (Milano, 1780): p. 177. See the excellent monograph on Baverio by Ladislao Münster, "Baverio Maghinardo de' Bonetti, medico Imolese del Quattrocento: La vita, i tempi, il pensiero scientifico," *Atti dell' Associazione per Imola Storico-artista* **7** (Imola, 1956), 119 pages, with many documents, including Baverio's testament, and bibliography. Baverio's method of diagnosis is described on the basis of his *Consilia medica* (Bologna, Plato de Benedictis, 1489), at p. 65. Baverio wrote a treatise on the plague, *Reggimento in tempo di peste* (Bologna, Johann Schriver, 1479) published shortly before Benedetti's work on the same subject. A second edition of Baverio's *Reggimento* appeared at Brescia in 1493, the exact year in which Benedetti's *De observatione in pestilentia* was published at Venice.

[17] Albrecht von Haller, *Biblioteca anatomica* **1** (Zurich, 1774–1776): p. 166.

[18] Antoine Portal, *Historie de l'anatomie et de la chirurgie* **1** (Paris, Didot, 1770–1773): p. 245.

[19] A. Pazzini, *Storia della Medicina* **2** (Milan, 1947): p. 179.

preparations of muscles, vessels, and nerves by drying them. Gurlt[20] insisted that Benedetti's *Anatomice* was the best work of the pre-Vesalian period; but he was probably too much impressed with Benedetti's innovations in terminology and their unified effect without sufficiently comparing his actual contributions and total effect upon anatomy with those of Zerbi, Berengario da Carpi, and Massa. In dealing with all the pre-Vesalian anatomists it is perhaps wise to avoid such overenthusiastic estimates unless they can be substantiated with actual quotation from the texts and a comparison with the writings of all of them, a task which the reader can now carry out for himself, at least to a certain extent, on the basis of the translations I have provided for him.

This account of Benedetti's medical profession and writings should not omit some discussion of his second book, the brief and popular *Collectiones Medicinae, circa* 1493. These are a series of 399 aphorisms divided into fifteen very short books. They are of a practical nature and apparently the fruit of long personal experience in clinical medicine, although it is possible to discern that many of them are based upon earlier medical literature, without, however, quoting any author by name.

They begin with a few cautious remarks which fall into the category of legal medicine or a code of medical manners;[21] this aspect of the *Collectiones* is emphasized by a Latin translation of the Hippocratic oath from the hand of Niccolò Perotti which is printed on their last page. Examples are: "A good physician follows this primary rule: do not take many patients." "A prudent physician does not attend those who are incurable and desperately ill since he might be considered responsible for the death of a person who was about to die anyway, owing to his condition." The great majority of the aphorisms, however, are devoted to various aspects of clinical medicine and are of a non-ethical intent, divided generally into the principles of diagnosis, the causes and symptoms of disease, diseases connected with the blood, bile, and other humors, the use of food, treatment of fever, epilepsy, and similar ills, afflictions of eyes, ears, and throat, of the heart, the chest, and stomach, the handling of abscesses, fluxes, and dropsy as well as dysentery, colic, hemorrhoids, ulcers, stone, the variations of sexual appetite and behavior, diseases of the genitals, gout, suppurations, and pains in the joints. On the whole, most of these aphorisms are sound advice and characteristic of the best medical knowledge of the time. They complement Benedetti's achievements in the *De Re Medica*, where many of them are exemplified in what may be called detailed case histories, as well as his other works on plague and anatomy. They are the ripe fruits of a long practical

[20] E. Gurlt, *Geschichte der Chirurgie und ihrer Ausübung* **1** (Berlin, 1898): p. 954: "Benedetti's noch fast ganz Galenische Anatomie ist das beste Werk der vor-Vesalischen Periode und dadurch uns um Vieles näher stehend, als die Werke seiner Vorgänger, dass seine Nomenclatur zum grossen Theil bereits der heutigen entspricht und namentlich die bei den Früheren vielfach gebrauchten arabischen Benennungen fast vollständig verschwunden sind." It is interesting to add here the single adverse criticism which I have discovered against Benedetti's work, recorded by Lynn Thorndike, *op. cit.* **5** (New York, 1941): p. 501: ". . . it is deserving of note that much of the criticism of Mundinus in the first half of the sixteenth century was not the outcome of anatomical knowledge and research, but dictated by humanistic prejudice and a desire to rank ancient Greek medicine above that of the Arabs and the Medieval Latins. Thus Leonhart Fuchs had the effrontery to accuse Mundinus and Alessandro Benedetti of anatomical errors simply because they disagreed with Galen, although both had dissected a great deal more than Fuchs, whose attack upon them was based merely upon his reading of ancient authorities and not upon practical knowledge." Fuchs's book bears the title *Errata recentiorum medicorum LX numero adiectis eorum confutationibus* (Haguenau, Joh. Secerius, 1530). He was criticized in turn by Sebastian Montuus, in *Annotatiunculae . . . in errata recentiorum medicorum per Leonardum Fuchsium Germanum collecta* (Lyons, 1533). The Renaissance Humanist prejudice against Mundinus is, of course, part of the general Humanist hostility against all the products of the Middle Ages. Yet in other areas of learning than science it can be shown without much difficulty that the Humanists were well versed in, for example, the works of the Church Fathers and hence do not completely bear out the general impression given by most students of the Renaissance that it is an age of reaction against the Middle Ages. It is certain that Mundinus maintained his prestige among the scientists down to 1550 at least, when Matthaeus Curtius produced his excellent edition at Pavia (see Charles Singer, *The Fasciculo di Medicina*, etc. (Florence, 1925), p. 56).

[21] Ladislao Münster has a long and informative article on the subject which reviews the earliest samples of a medical deontology from the Salernitan school to the time of Gabriele Zerbi, who wrote the chief Renaissance work in this area of medicine: "Il tema di deontologia medica: Il 'De cautelis medicorum' di Gabriele Zerbi," *Rivista di storia delle scienze mediche e naturali* **47** (1956): pp. 60–83. He points out the quasi-Christian nature of the Hippocratic oath and the transformation of the physician in the late Middle Ages into an important member of a guild organized to protect his group interests. From the thirteenth-century physicians had also been employed by communities and thus entered their administrative life. This advanced status required a code of ethics, usually called "De cautelis medicorum," observed by the Colleges of Medicine in the university cities. The Salernitan code was attributed to Archimatteo and goes back to the eleventh century; it was entitled "De adventu medici ad aegrotum." The first code to bear the title "De cautelis medicorum" is attributed to Arnold of Villanova, toward the beginning of the fourteenth century. Another code was drawn up by the Bolognese doctor, Alberto de' Zancari; it has only recently been published by Riccardo Simonini. The fifteenth century produced another work of this kind in the "Introductorium sive janua ad omne opus practicum" by Cristoforo Barzizza of Bergamo, professor at Padua between 1434 and 1440. Münster mentions Benedetti's "Collectiones medicinae," although they are in only small part an example of medical deontology; he overlooks the fact that they were published *circa* 1493 and thus may antedate Zerbi's "De Cautelis medicorum" by two years. He then proceeds to give an analysis of the latter work. See further on the subject Friedrich Wilhelm Oskar Bandtlow, *Die Schrift des Gabriel de Zerbis: de cautelis medicorum* (Zeulenroda i. Thür., 1925; Inaugural Dissertation, Leipzig).

experience which remained consistently in the realm of reality and never reaches the heights of a philosophy.

THE LAST WILL AND TESTAMENT OF ALESSANDRO BENEDETTI

Benedetti's will, of which an accurate text is presented in Appendix I, provides details that give a clearer understanding of his life. The will was drawn up by the notary Gerolamo de Bosis and dated upon the Rialto at Venice, September 18, 1512; within it is incorporated (from the words "In nomine Dei eterni amen. Anno Domini MDXI die primo julij" to "Clausumque dedi Lucie uxori mee"; then there follows a paragraph from the beginning of the second will to the words "dare debeo ex causa mutui") a first will dated July 1, 1511, which itself makes mention of an even earlier will made at Iadre (Zara (Zadar) in Illyria, now Jugoslavia), rendered null and void by the later documents.

Benedetti names himself the son of Lorenzo, from the district of Saint Pantaleone at Venice. He appoints both Antonio Moreto, his sister's husband, and Vincenzo Saraceno, the ducal secretary, as his executors (*commissarii*) and makes provision for the usual masses to be said in the sacristy of Saint Pantaleone. He then indicates his wife Lucia as a third executrix and leaves six hundred ducats, some garments, and three pieces of land to his daughter Julia; the money is to be supplied by the executors from his property "in villa Sancti Bonifacij" near Verona. In referring once more to possessions which Julia is to inherit he mentions them as belonging formerly to Dionira, her mother, evidently his first wife (Lucia being his second wife at the time he made his will); she is mentioned in *De Re Medica* as "Dionora nostra" (XI, 10, p. 205; XIX, 33, p. 342; Dionora Georgia, in 1510: XXV, 29, p. 479). Benedetti mentions property at three villages, evidently nearby, as well as furniture and clothing to the amount of 600 ducats of gold. A house bought in the name of Julia's mother but completely paid for by him with improvements was to be set aside as part of Julia's dowry from her mother, with Benedetti's expenditures duly noted. Then, observing that he was without legitimate sons as heirs (his son Cornelius, mentioned in *De Re Medica*, X, 9, p. 193, having died: "ut olim Cornelius filius meus"), he mentions a certain Marin whom he had legitimized with the intention that Marin might wish to become a priest and leaves him 100 ducats. He also left small sums to the schools of San Marco, Spirito Santo, and Santa Maria Formosa, 25 ducats for his burial expenses, and 300 ducats to his wife Lucia from his property in San Bonifacio. He left money also to Cornelia, who is not designated in any other way but who was probably the little daughter who suffered from worms, as we learn from *De Re Medica* XII, 11. If Lucia, Cornelia, and Julia should die Marin was to receive 200 ducats beyond the 100 mentioned above. Messer Benedetto Calbo, son of Pietro, was appointed as executor in this part of Benedetti's will. Faustina, who might have been Lucia's sister, is mentioned as having died "in Ongaria"; in case her will did not leave a share of property to Julia Benedetti enjoins his executors to add a sum equivalent to that which Faustina would normally have bequeathed her to Julia's dowry.

In the next paragraph of the will, which shifts from Italian to Latin once more, Benedetti gives his full name as "Johannes Alexander Benedicti"; thus we may add Giovanni to the Italian form of his name. The will ends with a long statement in the appropriate legal language of the time which gives full authority to his wife Lucia to carry out his wishes in all respects. The witness to the will was "Antonio Agnello spicier quondam Ser Agnello da Lonigo." The notary signed as "Hieronymus de Bossis quondam domini Bartolomei Venetiarum notarius."

ANATOMICE OR THE HISTORY OF THE HUMAN BODY

Alessandro Benedetti, physician of Verona, sends eternal greetings to Maximilian Caesar Augustus, most invincible king of the Romans.

I have decided to dedicate to you, Maximilian Caesar Augustus,[1] the history of the human body, which the Greeks call anatomice and we call the dissection of the members, a work established by philosophy for the purpose of healing. When we examine the marvelous and divine handiwork of God the creator, that is, the body which is (according to Plato)[2] as it were the temporary vehicle of the soul, we also perceive whence arise those happy contemplations (which they call theories), images or imaginations, thoughts, memory, and reason for the sake of which nature cleverly disclosed the eyes, ears, and the other senses in the body so that by means of the approaches they provide knowledge might enter the mind. The entire theory and practice of medicine and the discipline of surgery depend upon anatomy. When certain illnesses attack many members or single members, if abscesses begin to grow, or an ulcer lurks within, when the body is wounded, or a member is torn loose or dislocated, or something similar happens, none of these can be properly understood without the dissection of the human body. Heavenly emperor, who are distinguished in your worship, moral qualities, and learning before that God who is the creator of the world and the founder of all things and who does nothing useless, nothing in vain,[3] you will come forth from this intimate contemplation (as it were) of nature and from the workshop of this private undertaking and be venerated more worshipfully for lingering there. You will more easily bring your gaze to bear upon the nature of the universe whose likeness is borne by insignificant man. You will recognize the power, moreover, one by one of all the bodily members and run over their names, both Greek and Latin, omitting those in the vernacular tongues, which Pollux,[4] the most excellent of authors, set down cursorily for the benefit of Commodus Caesar. If you should wish to see these sights (leaving to surgeons and physicians for the distasteful duty of dissection materials worthy of a theatrical spectacle) and more closely to scrutinize the particular force and various effects of each, it will be possible sometime for your majesty to judge at greater length concerning the function of nature. It was not inappropriate for Alexander the Great, drawn by a desire to understand the nature and properties of animals, to gain knowledge, among so many and such great triumphs during his career, of the advantages of dissections. What is more, in Egypt the kings themselves, as people have written, were accustomed to dissect the bodies of the dead in order to discover the nature of illnesses. This branch of philosophy which the art of medicine has received to itself as a related discipline I have decided to dedicate to you, for there seemed to be nothing greater which I was able to offer as worthy of you, Prince, thinking to present this gift in sacrifice which I have long since consecrated to you in solemn worship at your altars. For the common people and the nobility do not pray to God in the same fashion although the prayers of worshipers are received with equal grace, nor is a carefully composed invocation preferred to rustic babble. I do not do this as a mortal in order to render you immortal since you were, as a prince, born an immortal and most like the gods. It is not my task to furnish you with greater glory, you who already possess the greatest triumph through your own humanity, loyalty, self-control, liberality, and admirable justice or by the conquest of yourself. For these reasons, Caesar Augustus, you have always and long since been chosen and elected by the agreement of all as the parent of Christendom. Therefore accept

---

[1] Emperor Maximilian I (1459–1519) was himself a writer of sorts. He signed the League of Venice in 1495, two years before the publication of Benedetti's *Anatomice*. The drawing of him done by Albrecht Dürer, now in Vienna, is reproduced in Max H. Fisch, *Nicolaus Pol Doctor* (New York, Herbert Reichner, 1947), p. 18. Chapter 1 of the Anatomice has been translated, with slight omissions, by William E. Heckscher, *Rembrandt's Anatomy of Dr. Nicolaas Tulp* (New York University Press, 1958), pp. 182–183.

[2] Plato, *Timaeus* 69 C: "And they, imitating Him, on receiving the immortal principle of soul, framed around it a mortal body, and gave it all the body to be its vehicle . . ." (translated by R. G. Bury, Loeb Classical Library, 7: p. 18). Practically all of Benedetti's references to Plato are to the *Timaeus*.

[3] Aristotle, *De Generatione Animalium,* II, 6, 744 a 35: "But since Nature makes nothing superfluous or in vain . . ." (translated by Arthur Platt in Vol. 5 of W. D. Ross's edition of Aristotle in translation, Oxford University Press). See note 79 to my translation of Massa's *Introductory Book of Anatomy* for a complete list of repetitions of the same thought in Aristotle.

[4] Julius Pollux (*ca.* 158 A.D.), author of the *Onomasticon,* wrote at Rome under the Emperor Commodus. Book II of his work is a valuable and highly concentrated source for medical nomenclature which remains to be studied systematically for its influence upon the Renaissance medical vocabulary. Such a study should also be extended to Rufus of Ephesus, *Peri Onomasias,* ed. C. Daremberg and C. E. Ruelle (Paris, 1879) and to Pseudo-Galen's work on medical definitions (*Definitiones Medicae* (Vol. XIX K.)). On this work see Jutta Kollesch, "Zur Geschichte des medizinischen Lehrbuchs in der Antike," *Aktuelle Probleme aus der Gesch. d. Med. . . . Verhandlungen des XIX. Internat. Kongresses für Gesch. d. Med., Basel, 7–11 September 1964* (Basel, 1966), pp. 203–208. An introduction to the subject which mentions, however, only Berengario da Carpi among the anatomists immediately prior to Vesalius and in other respects is merely a sketch is by J. Steudel, "Der vorvesalische Beitrag zur anatomischen Nomenklatur," *Sudhoffs Archiv für Geschichte der Medizin und der Naturwissenschaften* **36** (1943): pp. 1–42. Steudel's work on anatomical terminology is, however, being continued by his pupils: see the list of their theses in Hermann Triepel, *Die anatomischen Namen* . . . 29 Aufl., neu bearb. u. erganzt . . . von Robert Herrlinger (Munich, Bergmann, 1965).

among your crowns of laurel, most humane Prince, these books on the history of the members of man which I have written in Latin to the best of my ability. In them, not without some labor, I have pointed out many things overlooked by the moderns who have disgraced the Latin language with foreign, nay, barbarian terms and various errors and have made thousands of wounds upon the studies of medicine through their ignorance, in which they have been followed by the younger generation, enveloped by so great a mist of wretched stupidity and driven headlong by the impulse of barbarism. Hence the name of medicine has for some time been transferred from the Greeks to the Arabs with such success that Arabian audacity has increased to the point where they dare openly to contend with the Greeks[5] concerning the origin of medicine. But theft and depredation in this art are readily apprehended. I know there are many who are accustomed to the former blight who turn up their noses at what I offer, as being inane and unheard of by them. Let such people as these know this, that if my work displeases them, it has been offered not to them but to you, most humane prince, and in the Latin language. Therefore may you read my trifles when leisure permits. If they have been composed more hastily than I wished it was because they were impatient of delay and, wishing to greet you the sooner, were not fearful of the danger of abortion. Farewell, divine Caesar. At Venice. 1497. August 1.

THE HISTORY OF THE HUMAN BODY BY ALESSANDRO BENEDETTI,—OF VERONA, PHYSICIAN

BOOK I

ON THE USEFULNESS OF ANATOMY, ON THE SELECTION OF A CADAVER, AND ON THE CONSTRUCTION OF A TEMPORARY DISSECTING THEATER.

Chapter I

Tradition holds that kings themselves, taking counsel for the public safety, have accepted criminals from prison and dissected them alive in order that while breath remained they might search out the secrets of nature, how she acts within herself with great ingenuity, and should note carefully the position of the members, their color, shape, size, order, progression, and recession, many of which are changed in dead bodies. The kings did this carefully and more than piously in order that when wounds had been inflicted it should be understood what was intact and what was damaged. But our religion forbids this procedure, since it is most cruel or full of the horror inspired by an executioner, lest those who are about to die amidst such torture should in wretched despair lose the hope of a future life. Let barbarians of a foreign rite do such things which they have devised, fond of these sacrifices and prodigies. But we, who more mercifully spare the living, shall investigate the inner secrets of nature upon the cadavers of criminals. Early physicians observed that if anyone died of unknown diseases and they dissected cadavers they might discover the hidden origins of diseases with equal advantage to the living. Galen was not ashamed to do the same thing with his ape[6] when the cause of its death was unknown, just as we have done in the case of the Gallic disease (syphilis). The pontifical regulations have for a long time permitted this form of dissection;[7] otherwise, it would be regarded as most

[5] Benedetti is decidedly oriented toward the Greeks in his medical learning and terminology and prefers to use Greek terms wherever he can, playing the Greeks against the Arabs. The majority of his references is to Plato, Aristotle, Galen, and Pollux, with an occasional glance at Hippocrates, Alexander of Aphrodisias, and Paul of Aegina, whose work he was translating when he died. The passage on vivisection by kings is borrowed from Celsus, *De Medicina,* I, 1.

[6] Galen, *De Locis Affectis,* V, 2 (Vol. VIII K., 303); the anatomists often refer to this famous passage.

[7] Pontifical regulations on dissection for post mortem examinations empowered the college of surgeons (*Medici Chirurghi*) to perform public anatomies at Venice by an act of the *Consiglio Maggiore* for a period of twenty-two days continously, beginning on May 27, 1368; such anatomies were thenceforth carried out annually in the Hospital of Saints Peter and Paul at Castello (*Statuta Collegii Physicorum,* cap. 31, 1507 A.D.) and in other places. Benedetti records at least three such anatomies performed by him, on the heart, on the brain, and on the veins, muscles, bones, and nerves: *Anatomice,* III, preface, p. 36; IV, pp. 49–50; V, p. 68. For an account of the establishment and history of the College of Medicine and Surgery at Venice see Francesco Bernardi M. F., *Prospetto storico-critico dell' origine, facoltà, diversi stati, progressi, e vicende del collegio medicochirurgico, e dell' arte chirurgica in Venezia,* etc. (Venice, Domenico Costantini, 1797), pp. 52–57; see also L. Nardo, *Dell' anatomia in Venezia* (Ateneo Veneto, Venice, 1897); G. Panzani, *De Venetae Anatomes Historia et Claris Venetiarum Anatomicis Prolusio Habita in Veneto Anatomico Theatro* (Venice, D. Deregni, 1783). A popular and quite brief statement is found in A. M. Lassek, *Human Dissection: Its Drama and Struggle* (Springfield, Ill., Charles C. Thomas, 1958), p. 37: "At Venice the council passed a law permitting dissection of a human as early as 1368 . . . Dissection was included as a definite part of the medical curriculum in 1433. Statutes were made regulating anatomy in

execrable and abominable or irreligious. Furthermore, ritual purifications of the physicians' souls take place and we propitiate their offense with prayers. Those who live in prison have sometimes asked to be handed over rather to the colleges of physicians rather than to be killed by the hand of the public executioner. Cadavers of this kind cannot be obtained except by papal consent. Thus by law only unknown and ignoble bodies can be sought for dissection, from distant regions, without injury to neighbors and relatives. Those who have been hanged are selected, of middle age, not thin nor obese, of taller stature, so that there may be available for spectators a more abundant and hence more visible material for dissection. For this a quite cold winter is required to keep the cadavers from rotting immediately. A temporary dissecting theater[8] must be constructed in an ample, airy place with seats placed in a hollow semicircle such as can be seen at Rome and Verona, of such a size as to accommodate the spectators and prevent them from disturbing the masters of the wounds, who are the dissectors. These must be skillful and such as have dissected frequently. A seating order according to dignity must be given out. For this purpose there must be an overseer who takes care of all such matters. There must be guards to prevent the eager public from entering. Two treasurers are to be chosen who will buy what is necessary from monies collected. The following are needed: razors, knives, hooks, drills, trepanning instruments (which the Greeks call chenicia), sponges with which to remove blood in dissection, paring knives, and basins. Torches must be also provided for the night. The cadaver is to be placed on a rather high bench in the middle of the theater in a lighted spot handy for the dissectors. A time for attendance is to be established, after which the gathering is dismissed so that the dissection may be completely carried out before the material putrefies. First let me run over briefly the form, shape, and dignity of the human body.

1495 . . . N. Massa . . . is said to have performed nine demonstrations on human cadavers between 1524 and 1536. An anatomical theater was built in 1552."

[8] Ruben Eriksson, *Andreas Vesalius' First Public Anatomy at Bologna 1540, An Eyewitness Report by Baldasar Heseler, Medicinae Scolaris, Together With His Notes on Matthaeus Curtius' Lectures on Anatomia Mundini,* edited, with an introduction, translation into English and notes (Uppsala and Stockholm, Almquist and Wiksell, 1959), p. 306, quotes Vesalius, *Fabrica* (1543), p. 548, for Vesalius's description of the manner in which to arrange an anatomical theater and points out that Benedetti's instructions imply an ancient Roman amphitheater such as those which exist at Rome and Verona. Admission was to be charged in order to defray expenses for equipment. It is probably this advice which has given rise to the certainly erroneous assumption that Benedetti was instrumental in building a permanent anatomical theater at Padua. Such an assumption is made by A. Castiglioni, *A History of Medicine* (New York, 1946), p. 370, and by A. Pazzini, *Storia della Medicina* **1** (Milano, 1947): p. 671.

Chapter II

On the Shape and Dignity of the Human Body

The human body was created for the sake of the soul and stands erect among other animals, as established by divine nature and reason so that it might look upward more comfortably, whence the Greeks gave it a name.[9] Man alone has the power of thought; likewise he alone remembers. Memory and self-control, however, (as Aristotle says)[10] he possesses in common with many animals. The members exist for the sake of their duties and functions, to which single features called dissimilars are accommodated. For in the human semen there is a heat with such great power and action that it can be gradually accommodated to any part of the body whatsoever.[11] The heart was first created since it contains the principle of life and sense.[12] Next came the brain and liver. Then nature, performing like a painter, sketched out the other members with a life-giving fluid; they gradually receive their colors from the blood, which is very abundant in man and stirs up very much heat. Hence the human body is more erect than others which would not be easy if the great bulk of the body were to sink down. Therefore man's head has very little flesh on it and holds the higher place in order that it may reflect that universal state in which the divine mind is considered. But the rest of the animals being heavier have their heads bent down to

[9] The Greek name Benedetti has in mind here is *anthropos,* from a fanciful etymology involving *anti,* "opposite, facing," and *tropos,* "turn or direction." See Mundinus, *Anathomia,* translated by Charles Singer, p. 59 and note 6, p. 101, Isidore of Seville, *Etymologies* XI, 1, 5, and W. D. Sharpe, "Isidore of Seville. The Medical Writings," *Trans. Amer. Philos. Soc.* **54**, 2 (1964): p. 38. Benedetti omits *plantenus,* given by Mundinus as a variant term for man. The upright nature of man became an anatomical commonplace before Mundinus and the "Problem" literature and onward to the sixteenth century.

[10] Aristotle, *De Memoria et Reminiscentia,* 1, 450 a 15: "Hence not only human beings and the beings which possess opinion or intelligence, but also other animals possess memory" (translated by J. I. Beare, Oxford Translation of Aristotle, **3**).

[11] Aristotle, *De Generatione Animalium,* II, 3, 736 b 29; in 737 a 6 he speaks also of the "spiritus conveying the principle of soul" (translation by A. Platt, Oxford Translation of Aristotle, **5**). The word dissimilars refers to an old classification of the parts of the body into similars and dissimilars. See Vesalius, *Epitome,* translated by L. R. Lind (New York, Macmillan, 1949), p. 1 and note. The *Collectio Salernitana* (ed. S. de Renzi, Naples, 1859) **5**: p. 43, lists "partes similares" as the following:

Nervus et arteria, cutis, os, caro, glandula, vena,
Pinguedo, cartillago, et membrana, tenantos [tendones]:
Haec sunt consimiles in nostro corpore partes.

[12] On the heart as the first organ created in the body see my note 47 to the translation of Massa, *Introductory Book of Anatomy,* and the references given there: Aristotle, *De Partibus Animalium,* III, 4, 666 a 10; *De Generatione Animalium,* II, 6, 741 b 15–20; Pliny, *Natural History,* XI, 59. 181, p. 545, Loeb Classical Library. Note the strongly teleological flavor of the description of the parts of the body that follows: each has its purpose.

the earth and to them nature allotted front legs in place of arms and hands. Arms and hands were provided for man as the most foresighted of all animals so that he might use them suitably for many crafts. The eyes in the highest part of the head are the watchmen of the entire body; the ears are their counterparts in sound. The head is covered with hair as a protection against cold and heat. Eyebrows were added against the flow of sweat and eyelashes against harmful objects which fall upon them. The nose provides breathing, judges food before it is eaten, while the bones of the cheeks project to guard the eyes. The lips are necessary for voice and teeth. The neck holds up the head for turning it around and is fitted with many vertebrae. The sides together with the thorax form the ribcage with many bones and as it were a wall for the heart, making it easier to breathe than if that cage were made of one bone; we see this also in the arrangement of the spine. But the stomach is without bone and hence more adapted for food and for the female uterus. The pubis has a bone wherein we carry the intestines and women the child they conceive. The genitals are made of nerves and are not bony, as in some animals such as ferrets and weasels.[13] The prepuce covers the glans so that it may not be injured by rubbing against the thigh. The Jews cut it off for the sake of cleanliness. The male genitals are external, the female genitals or vagina are located within the body; if they fall foward the male groin appears. The buttocks provide a cushion since it is not easy for man to continue standing erect but he desires rest in sitting. The arms and legs are bent in wondrous fashion, the legs inward, the arms outward, the latter for handcrafts and the former for walking. Nature also gave the fingers for seizing and pressing most suitably. The thumb, which is thicker and shorter, presses upward at the side in the lower region of the hand; the other fingers from above press downward. The thumb itself has powers equal to many needs, whence its name. The middle or long finger is like the middle of a boat. All fingers except the thumb are equal in grasping. The legs are the largest members of a man in proportion to his size because they preserve the entire body erect. The toes in the feet are short because they were given for no other use except for walking. I have, however, seen a woman born without arms who was skilled in using her feet instead of hands for spinning and sewing. Nails have been added on account of the weakness of the toes. But let me go over the external parts and their names in order. My good readers should recall that if ever I lack Latin words I have preferred to use Greek terms, rejecting the barbarian names.

Chapter III

On the Parts of the Body Externally Visible and Their Names

The principal parts of the body are the head, chest, stomach. The secondary parts are the neck, arms legs, and genitals. The exterior parts of the head are called by these names. The top of the head is called vertex; in some people it is found double. These are called long-lived and bivertices.[14] The bregma or sinciput slants toward the forehead and palpitates in infants. Next comes the forehead, and the occipitium opposite, the parting of the hair dividing the bregma at the top of the vertex. There are two temples, ears, eyes, exterior angles of the paropia (lesser canthus) and inner angles called the fountains (tear ducts) whence flow the tears. The eon of the eyes is their great circle. Then there are the eyebrows, whose middle part is called mesophrion by Pollux.[15] Then come the eyelids, in which there are the lines of wrinkling (*lineae rhitides*) and the eyelashes, then the cheeks, the cheek bones, nose, the ethmus (or sieve), the septum of the nostrils, the globus (or tip) of the nose, the wings or phenae, the hollows of the cheek, the mustache, lips, mouth, chin, all of which compose a face of great variety. The neck is made up of the upper throat, the gullet, and the outer part of the back of the neck from the stropheus to the atlas vertebra (first cervical vertebra). From this first cervical vertebra and the gullet to the genitals and the os sacrum extends the thorax, of which these are the parts: the collarbone on right and left verging toward the shoulders, the chest, breasts, nipples, sternum (*malum punicum*), the (xiphoid) cartilage, the hypochondrium, the mouth of the stomach, the girdle, the abdomen, the navel in the middle line of the body and in the middle of this the acromphalus. Around this and upon it is the "wrinkled old woman" (*vetula*). The epigastrium is near the liver and spleen. Beneath it is the hypogastrium. On both sides are the ilia, (also known as) lagona and ceneona. Then there are the sumen and pecten (comb), which is also used concerning the hair, and the genital member in either sex. The parts of the back are the scapulae or

[13] Aristotle, *De Historia Animalium,* II, 1, 500 b: ". . . in other cases it [the male organ] is bony, as with the fox, the wolf, the marten, and the weasel; for this organ in the weasel has a bone" (translated by D'Arcy Wentworth Thompson, Oxford Translation of Aristotle, 4). Benedetti connects the stem of *pollex,* thumb, with *poly,* much, mingling both Greek and Latin in the process. See Isidore of Seville XI, 1, 70, translated by W. D. Sharpe: "The first finger is called the thumb, pollex, because it surpasses, *pollere,* the others in strength and power."

[14] The terms *macrobii* and *bivertices* are not found either in Joseph Hyrtl, *Das Arabische und Hebraische in der Anatomia* (Vienna, 1879) or Adolf Fonahn, *Arabic and Latin Anatomical Terminology, Chiefly from the Middle Ages* (Oslo, Norwegian Academy of Sciences, II. Hist. Philosoph. Class, No. 7, 1922), the standard reference works.

[15] Pollux, *Onomasticon,* 2. 49. I use the edition by Wilhelm Dindorf (Leipzig, 1824, 5 v., including index).

metaphrenum (shoulder blades), table, and spine, in which are the cynolopha (spinous processes of the vertebrae), the sides, kidneys, lumbar region, os sacrum, buttocks, anus or seat; from the girdle to the buttocks is an area called osphys. The genitals are the penis, prepuce, and scortum (scrotum). The arms consist of shoulder, wings [arm], gibber [forearm], and flexus [wrist], that is elbow, and olecranon or cubitus. The flexus of the hand is called procarpion in Greek, the back of the hand is called opisthenar. Then there is the palm or vola. Finally there is the ir, which the Greeks call thenar (palm). The part from the index finger to the little finger is called hypothenar. The thumb, as stated above, is the first of the fingers. Then comes the index finger; the middle or shameful finger;[16] the one next to the little finger which some call medical, and finally the little finger. In the fingers there are internodes and the prominences of the fingers called condyles. There are three internodes or scytalides to each finger, two only in the thumb; then come the nails and the roots of the nails. The legs begin from the vertebrum or coxa, then come the femora and the inner thighs, the knees, the hams of the knees; the legs include the tibiae and calves from the knees to the feet; under the calf is the sphyron (*malleolus*). There are no heels in man although the lowest processes of the tibiae which are called malleoli have received the name of heels, hence talaria; there is a similar bone in a dog which is called talus by Marcus Varro.[17] There follow the neck of the foot or the mons flexus, the perna, whence comes the name perniones, calx (*os calcaneum*), the fleshy vestige, the sole, the toes, whose names correspond for the greater part to the names of the fingers, that is, pollex (big toe), little, middle. The rest are said to be nearest the little toe and the big toe. Finally come the toe nails, which are the nerve ends and given for the sake of a covering to guard the ends of the fingers. But I have spoken enough about these visible externals. Next I must speak of the qualities of the humors and of the members.

[16] The finger is so named from its use in obscene gestures. Berengario da Carpi, *Commentaria . . . super Anatomia Mundini,* etc. (1521), fol. XXXIII, quotes Giorgio Valla on the meaning of the term *impudicus:* Georgius Vallensis dicit quod impudicus dicitur quia subductis aliis utrinque digitis solo illo porrecto membri pudendi formam referat. Aliqui dicunt quia dicitur infamis eo quia cum opus est tractare uel matricem ut faciunt aliquae impudicae mulieres libidinis causa uel obstetrices necessitatis causa uel circa rectum intestinum propter aliquam eiusdem intestini aegritudinem uel propter calculum in uesica eo solo utuntur operatores digito, ideo notatur infamis et impudicus digitus propter loci impudicitiam et immunditiam. Isidore of Seville, Etymologies, XI, 1, 71, says: Tertius inpudicus, quod plerumque per eum probri insectatio exprimitur. Pollux, *Onomasticon* 2. 184 says the people of Attica considered the middle finger obscene, apparently because of a certain gesture made with it. Valla may have based his discussion on Pollux.

[17] I can find no passage in which Marcus Varro uses the word *talus,* either in his *De Lingua Latina* or *De Re Rustica.*

Chapter IIII

On the Qualities of the Humors and of the Members

The blood is the first of all four humors in the human body and is warmer and more humid, resembling the nature of air and containing the other humors mingled within it. The bile is yellow, warm and dry, resembling the property of fire since it is especially kindled by external heat, for it disturbs all with its vapor, as we see in a pernicious bilious remittent fever (endemic fever of the Levant: Hippocrates).[18] The phlegm is cold and moist, possessing the force of water. The black bile is dry and cold, containing the nature of earth. Individual receptacles contain these humors. The veins contain blood, the stomach contains phlegm (*pituita*). In the liver the bladder contains (yellow) bile, but the spleen contains the black bile, although the blood may contain these same humors which through all the small passages of the veins more or less in accordance with the condition of the body are scattered to its individual parts, as we understand with right reason. They furnish nourishment to the members and are so necessary that if a man is deprived of any one of them either by force or of his own will he is certain to be brought to destruction, as when all the black bile is rendered up. Nourishment sweating back thus through the veins and small passages (as water passes through a rather crude clay vessel when put into it) is finally united and (to use Galen's words) converted into flesh or some other similar member consistently hot or cold. Hence it comes to pass that flesh is dissolved by fire. For heat does not make flesh out of just anything or in whatever manner or time but from material adapted to the purpose so that it can be fitly accomplished. The nature of the flesh between skin and bones is readily split into threads like a vein which they call villus; these pass into veins and fibers whenever the living creature grows thin and slender. The flesh is likewise changed into fat whenever there is a supply of food and nourishment. Those who have much flesh have more restricted veins and redder blood, small abdomen and viscera. Those with little flesh have ampler veins, black blood, and larger abdomen and viscera. The arteries are the receptacle of the spirit, arising from the heart; they contain the spiritual blood which they call vital, for it breathes a particular life into all the members and also penetrates to the bones. From the arteries arises that pulse, the most certain indication of life, which by certain rhythms and laws of measured harmony moves in the members, whence the directions of life are obtained. In those who die of

[18] Hippocrates, *Coan Prognosis* (on causes; Gk. καῦσος); see *The Medical Works of Hippocrates,* translated by John Chadwick and W. N. Mann (Springfield, Ill., 1950), p. 227. Albertus Magnus, *De Secretis Mulierum* (Lyon, 1571), p. 54 also equates the four humors with the four elements and the four quarters of the moon.

heart trouble the veins collapse and sometimes by a natural fault in certain healthy individuals; heedless physicians are often deceived by such veins, concerning which I have now said enough. I shall in the last book speak at greater length concerning the spirit and the blood and their veins. Returning to the parts of the body, some are warm, some dry, others cold and some moist. Of these the heart is warm and the blood also for I believe the spirit is tempered, according to Aristotle;[19] then there are the liver and the flesh with its special quality, the muscles, kidneys, and spleen. But the hairs are dry in nature, the bones, cartilage, tendons, joints, veins, arteries, and nerves called vocal by physicians[20] are dry also. These parts, almost the same, are likewise cold, the hairs, bones, cartilage, every membrane, such as the one called hymen by the Greeks, the nerves, spinal medulla or marrow, and the brain. Moist are the fat or adeps, brain, spinal medulla, glands in the breasts and in other parts of the body (*adenes*). These are constituted according to a tempering or what the Greeks call mixture (*crasis*). One part only in the human body is especially sensitive, that is, the skin, particularly in the palm or at the tip of the index finger, to which a temperate mixture is given by nature among the qualities which exist in living bodies: dry, humid, cold, and hot. God, the Creator of the universe, has in His excellent reason joined with admirable union things which war among themselves, for by a certain relationship and conciliation single items adhere to single items, due to their own qualities.

Chapter V

On the Functions of the Members

The parts of the body are of two kinds. First, there are those with a similarity of composition, or homoiomereiai, as the Greeks call them. We, with a more recent term, have begun to call these similar parts, into which they can be divided; they are composed from both primordial elements and humors, such as flesh and bone. From these the second kind of parts is composed, that is, from dissimilars which can be divided into parts of dissimilar nature, such as the muscle and the face. From the first kind of parts, that is, from similars, the sensory parts are composed, such as from flesh, skin, nerves, since these are capable of individual sensibility. From the dissimilars arise actions and movements and various functions, since they are constituted from a varied nature of similars, whence we call these parts organic, functional, and instrumental. Of the members there are certain ones which at the same time receive and transmit, such as the gullet and pylorus. Others receive and contain, such as the gall bladder. Some are necessary to the body, others are superfluous. An example of the former is the stomach, of the latter the bowels and the bladder. Of these one holds dry excrement, the other holds wet excrement. Some likewise are necessary for the individual himself, others for the function of continuing the species; included in the former are the liver and stomach, in the latter the penis, testes, and vulva.

Chapter VI

On the Bones and Their Nature

The bones among the similar parts are of the hardest material, they sustain the entire weight of the body, are the least animated (as Plato says),[21] and are covered with the most and thickest flesh. Hence the head and chest are least fleshy, except for the tongue, since it is especially adapted for sensation. For the sake of the flesh alone the bones, nerves, skin, veins, arteries, nails, and hair were given by nature. This alone of all the parts possesses the sense of touch. The bones are extremely necessary for motion; among them there is the spine, like the keel of a ship; from it the ribs appear curved back and rising again to the sides. Some bones either protect the body, such as the skull, ribs, radius, or ward off objects that strike, such as the spinous processes of the vertebrae (*cynolopha*), whence the name projecting canines, or fill out empty spaces, as the bone which they call the patella in the knee. Others are more solid and most strong, some are perforated such as the cranium, some are full of marrow for nourishment. The stronger bones are the femora and tibiae. All the bones are covered with a membrane or periosteum. When this is torn away they are straightway destroyed, nor do they have sensation without it, since they are cold (as has often been stated) and dry of nature for they are hardened by cold but are boiled to pieces under heat; although their humor dries they cannot be destroyed by fire. Except for the skull all the bones arise from the spine. At the first constitution of the parts the bones spring from the seminal excrement; when the human being grows they take their increase from external nourishment coming either from the female matrix or from outside. Thus the bones like the other members are marked off by a definite term of growth and as in the animals they have a limit to their size. The nerves are midway between flesh and bone; cartilage likewise is made of a substance between that of the nerves and that of the bones.[22] As one may easily see, the bones in man number two hundred forty-eight. It

[19] Aristotle, *De Partibus Animalium,* II, 3, 649 b-650 a; *De Generatione Animalium,* II, 6, 743 a; V, 4, 784 b.

[20] These are the recurrent laryngeal nerves; see Galen, *De Usu Partium,* VII, 14 (Vol. III K., 577–581).

[21] Plato, *Timaeus* 64 C: " . . . the bones and the hair and all our other parts that are mainly earthy" (translated by R. G. Bury, Loeb Classical Library 1961), p. 165.

[22] Galen, *De Usu Partium,* VII, 3 (Vol. III K., 519); Avicenna, *Canon,* I, First Fen, Doctrina 5, chap. 1 (translated by Gerard of Cremona, Venice, 1595), p. 29.

is not unusual for them to be broken, but for an abscess to result in them is strange, as I recently saw when I dissected a woman who had died of syphilis in order to discover the cause of the disease. Her bones under their intact periostea or membranes were swollen and were suppurating as far as the bone-marrow simply because bones are nourished and for this reason are affected by abscesses.

Chapter VII

On the Cartilage

The cartilage or chondrus is of the same nature as the nerves and bones, according to Pollux and Galen.[23] Almost all the bones terminate in cartilage to prevent injury from rubbing or striking. Broken cartilage almost never is consolidated again. It is inserted as a cushion at points where bending occurs.

Chapter VIII

On the Nerves

Aristotle believes that the nerves first arise from the heart,[24] in which they appear before the other members are formed, but almost all of them (as is more evidently established) are perceived to originate in large part from the brain or from its substitute, the spinal medulla or marrow, whence also the animal spirits and the spirits of sensation have their paths. The nerves are white, flexible, and diffuse the powers of sense and movement. When cut they are not consolidated again, when wounded they feel the greatest pain, but when cut they feel none. Those members called tendons are round and adapted to muscles in arms and legs. The inner nerves conduct the members, those on the surface recall them; this is understood from dissected bodies. Furthermore the nature of nerves, as well as of [reading *necnon*] the material of veins, proceeds continuously from the same beginning. Thus also the bonds annexed to bones and nodes of bodies which are called joints have the nature of nerves. Around the nerves is a certain sticky mucous substance which is given them by nature for nourishment. But I shall speak of the origin, nature, and number of nerves in their place.[25]

Chapter IX

On the Nature of Arteries and Veins

The arteries and veins, which are made of the material of nerves, arise from the heart, with a thicker stock or trunk near their place of origin, are cold and dry of nature, they are broken apart by heat and brought more compactly together by cold. The arteries, which are the receptacles of the spirit, are composed of two membranes; they continually dilate and contract by turns; whether they do this of their own accord or by the motion of the spirit will be discussed elsewhere. Hence arises the government of life. The veins are the receptacles of the blood, made of one membrane only, and motionless. All these members are irrigated with little streamlets for the most part so very slender that they can scarcely be discerned. What are called the fibers of the aorta are made entirely of nerve substance; those which are distinguished by no hollow space within stretch in the manner of nerves; where they end they are attached to the slippery joints of the bones, according to Aristotle.[26]

Chapter X

On the Muscles or Lacerti

The muscles are counted among the dissimilar parts which have the function of causing voluntary motion. They are composed of sinew, flesh, membrane and surrounding covering, and of the nerves [tendons] which adhere to the ends of the bones. They are thicker and more ample in the middle, fleshy within, of a sinewy material running here and there. On account of the fact that they end in a slenderness at both ends they are called muscles [little mice]. They exercise the function of stretching or reaching out. In those who are wounded muscle stretchings are made not without peril of life; the Greeks call these stretchings spasms, the Latins call them distentions. They feel weariness in labor. They vary in many ways in the body, both in shape and power, since some retain, others expel; they are stimulated by heat, rendered sluggish by drugs, their power for holding is injured by ointments, they are restored by astringents. Some are oblong, others compact, some rather broad, some so mixed that they can scarcely be torn apart. They are regarded and understood by their various motions and number, as is particularly comprehensible in the forehead and lips. Some muscles vary as the body's shape varies. More than five hundred have been gathered by very diligent authors. Their order and location will be discussed separately in the last book.

Chapter XI

On Membranes

Membranes or tunics follow. These are of sinewy nature, such as those which contain the brain and the

---

[23] Pollux, *Onomasticon,* 2. 32; Galen, *De Usu Partium,* VII, 3 (Vol. III K., 519).

[24] Aristotle, *De Partibus Animalium,* III, 4, 666 b 15; *De Historia Animalium,* III, 5, 515 a 30. Benedetti breaks here with the Aristotelian view of sense perception as a function of the heart in favor of the true view, the brain as the source of sensation.

[25] In Book V, chap. 14.

[26] Aristotle, *De Historia Animalium,* III, 5, 515 a 30. Note that Benedetti counts five hundred muscles while Achillini has only seventy-nine. In Book V, chap. 22, he gives the number five hundred twenty-six.

marrow. They connect members with nerves or cover them, as the liver and kidneys, and through the nerves they have sensation. In other respects they have no feeling.

Chapter XII

On the Adipose Matter and the Marrow

Adipose matter is found in man, for it grows around membranes. The fat is under the skin. Seuum (hard fat) is found around the kidneys in horned animals. Fat also warms the membrane, for in a temperate body the overflow of the blood is gathered from the members by saturation; otherwise the senses are oppressed by a great weight of fat and often the members die. Lard in a pig is regarded as free from sensation. The marrow in hollow bones has no sensation; it is covered, however, by a very thin surrounding membrane which is reddish in youth, whitish in old age and less around the heads of the bones, where it is seen to be bloody and less fat. Broken bones are not glued back together, due to the fact that the marrow flows away from the point of the breakage.

Chapter XIII

On the Skin

The skin is formed by the drying flesh as (according to Aristotle)[27] the crust is formed in polenta (pearl barley). For when the sticky matter cannot evaporate it settles down since it is fat and viscous. Hence the skin consists of the same sort of sticky material and is a similar part since it is created in the uterus. It is very thin in man, opposed to the air around us, lightly perforated (with pores) whence the vapors issue forth. This fact is quite evident from the cruelty of tyrants when the skin of a wretched criminal is torn off and dried out amid the greatest tortures. Then the foramina of the skin are seen to have the size of holes in a wheat sieve. The sweat exudes from the skin, all forms of touch are carried out by it. By means of it as a temperament we judge with the sense of touch; therefore, it is itself without sensation. Under it there is a mucous humor. Wherever the skin is without flesh and is injured it does not heal, as in the jaws, cheeks, prepuce and parts similar to these.[28]

Chapter XIIII

On the Nails

Nails arise from the earthy part of man and have little humor or heat. When they grow cold the humor or moisture and heat evaporate into the air and the nails take the hard form of earth, like hoof or horn. Thus they are softened by fire. They possess a nature between bone and nerve. I saw in Crete while I was writing this book a knee, wounded by an arrow, which grew out similar to the black horn on the head of a mountain goat, and the material which was to have been converted into the substance of bone easily changed into the nature of horn by the breath of the air and immediately grew hard like gum. This process is very similar in plants, whose parts when they are covered by the earth have the nature of a root, but when the air blows on them they bring forth leaves. Those who teach how to propagate by slips from the branches of trees have observed this. For they transfer slips from the tops of the trees and coax the roots from the branches passed through clay vessels and having thus obtained the roots with this coaxing they plant them on another tree high up from the earth between the fruit and the top of the tree. The nails in the hands grow more since (as Alexander of Aphrodisias says)[29] an increase upward is carried on more. Let these words be uttered while the body to be dissected is washed and the razors readied according to custom and necessary preparations for dissection made. Now let us pass to the inner parts.

---

[27] Aristotle, *De Generatione Animalium,* II, 6, 743 b 5: "The skin, again, is formed by the drying of the flesh, like the scum upon boiled substances; it is so formed not only because it is on the outside, but also because what is glutinous, being unable to evaporate, remains on the surface" (translated by Arthur Platt, Oxford Translation of Aristotle, 5). See Plato, *Timaeus* 76, A.D.

[28] Aristotle, *De Historia Animalium,* III, 11, 518 a. For the nails as earthy see Plato, *Timaeus* 64 C.

[29] *Problemata Alexandri Aphrodisei Georgio Valla interprete* (Venice, Albertinus Vercellensis, 1501), fol. 6 v: Cur qui in manibus sunt ungues citius crescunt quam qui in pedibus, aut quia sursum uersus actio magis scandit quo circa et pili in mento potius crescunt quam in pube, aut quod in calore quidem manus, pedes uero in frigore capiunt incrementum etc.

BOOK II

OF THE ANATOMICE OF ALESSANDRO BENEDETTI, PHYSICIAN, ON THE NATURAL AND GENITAL MEMBERS CONTAINED UNDER THE DIAPHRAGM

Preface to Friends

Christopher Sorovenstanus,[30] most skilled orator of our Caesar, Plato advises people not to appear ungrateful toward their friends and particularly not toward their fatherland.[31] I am doing one as well as the other when I call you away from affairs of court to this dissecting theater with Christopher Lamphranchus, Justus Lelianus, with my comrades Andreas Peregrinus and Dionysius Cepius, men of Verona, and lawyers, among whom also my Ruffus and Jacobus Mapheus I wish to be present, who shall not unfittingly be first to view these parts of the human body with the most serene king Maximilian Augustus, since I know that all of you are most temperate in both food and drink, often checking the importunity of the stomach and good and keen at warding off its intemperance. This intemperance often troubles that portion of us which is divine and unjustly pushes away our spirit, as it advises us, with its insatiable importunity. Our Justus therefore will be present so that this intemperance may leave us in tranquil ease. You should not be displeased if I wish you to observe human entrails since they say nothing exists in nature that is not admirable. For Heraclitus[32] also when he chanced once to be sitting in a bakery for the sake of the heat bade some people to approach with confidence when he saw them hanging back. "Don't you know the immortal gods are present here?" he said. Thus in all things we are certain to find the power of nature and all things are full of soul. Among the entrails Plato[33] especially commended the nature of the liver for divination and for likenesses (used in divination) because like a looking glass it was suited for reflecting images and hence the function of the genital members was likewise suited for perpetuating the human species through individuals in whom, as you know, God has placed the marvelous desire and incredible love of giving birth to their kind. These members are by no means to be despised although often in women they fatally hinder the conceived creature from being born. When we have examined these parts we shall condemn those who pursue the wantonness of the stomach and the violence of lust since (as Plato tells)[34] we see how they demonstrate that such people not unseasonably degenerate into the forms of various beasts.

Chapter II

On the Stomach and the Umbilicus

The human body to be dissected is divided into three parts, upper, middle, and lower. The first contains the head and is the part which creates sensation. The middle is the spiritual part. The lower, in which the stomach lies, is called the natural part. In the first part the nobility of the senses has its place. In the second the source of life and the reason have their place. The lower part is the workshop of nature; from it we must begin to dissect since the intestines rot more swiftly than other portions of the body. We must begin therefore with the worst of all bad parts, the stomach, for the sake of which the greater share of mortals serves forever nor do they fear its vileness because of the foul work of its excrement since many people undergo the almost infinite perils of sea and land on its account. I add the

[30] Christopher Sorovenstanus (or Scrovenstano; so Massalongo p. 248) was apparently a member of the court of Maximilian I as a spokesman of some sort, but I have been unable to find any information about him.

[31] In his salute to friends Benedetti is illustrating the words of Galen (who follows Plato) in *Methodus Medendi,* VII, 1 (Vol. X K., 456): Tu enim mihi conscius es neque hoc me opus neque aliud ullum popularis aurae studio fuisse aggressum, sed quo vel amicis gratificarer, vel me ipsum simul utilissima ratione ad rem propositam exercitarem, simul ad oblivionem senii, ut Plato inquit, commentarios mihi reponerem. Charles Singer translates the passage in his translation of Mundinus, p. 59: "a work in any science or art is published for three reasons: first, for the satisfying of friends, second, for the useful exercise of the faculties, and third, as a remedy for the forgetfulness which doth come with lapse of time."

Concerning the friends mentioned here I have been able to find a few fragments of information. Lamphranchus is probably Christophorus Lafranchinus (Cristoforo Lafranchini), who lived in the fifteenth century, was a friend of Marsilio Ficino, and addressed a poem to Marcus Aurelius. Dionisio Cepio (Cippico), also of the same century, was a friend of Muretus and wrote an epigram contained in Raphael Regius's edition of Terence. Ruffus is Matthaeus Ruffus (Matteo Ruffo), of Verona, a member of the Roman Academy of Pomponius; to Ruffus, Benedetti addressed a brief letter printed in the edition of Pliny, *Natural History,* edited by Johannes Britannicus (Brescia, 1496). Ruffus also wrote an *Epistola de C. Plinii Secundi patria* (Brescia, 1496) and epigrams in Greek and Latin.

[32] Aristotle, *De Partibus Animalium,* I, 5, 645 a 15–20 tells the story about Heraclitus. Its moral is eloquent: "as even in that kitchen divinities were present, so we should venture on the study of every kind of animal without distaste; for each and all will reveal to us something natural and something beautiful. Absence of haphazard and conduciveness of everything to an end are to be found in Nature's works in the highest degree, and the resultant end of her generations and combinations is a form of the beautiful" (translated by William Ogle in Vol. 5 of the Oxford Translation of Aristotle).

[33] Plato, *Timaeus* 71 B: ". . . the power of thoughts which proceed from the mind, moving in the liver as in a mirror which receives impressions and provides visible images, should frighten this part of the soul . . ." (72 B): "For these reasons, then, the nature of the liver is such as we have stated and situated in the region we have described, for the sake of divination" (translated by R. G. Bury, Loeb Classical Library, pp. 185, 189).

[34] Plato, *Timaeus* 73A.

fact that like a most insistent creditor it presses upon all people nor does it allow, the worst vessel of all the members, a day to pass without making its appeal. This it is which many have in common with the life of brutes and it is lowest and has a hidden place in the body because of its uncleanness, as though nature had spared the principal members and had relegated the stomach or bowels farther away from the site of reason and of the mind and fenced it off with the diaphragm in order not to disturb the rational part of the mind with its importunity. Here as in a cook shop is the location of the dregs and excrement just as in buildings the architects avert from the eyes and nostrils of their masters whatever is necessarily regarded as foul and unclean. These members serve the higher ones. Some of them concoct the food into juice, others digest it into various humors, some expel the superfluity. I must first describe the external lower body cavity. This is located below the lowest point of the sternum within the ilia and groin. Part of it is the sumen. The cavity lacks bone because it increases with food, as has been said, and decreases with hunger, which also happens to women who are pregnant. In the middle of the cavity is the umbilicus (navel), embracing two arteries and two veins, with a kind of shell surrounding it for each of the veins is surrounded by its small membranes. The four vessels from the cups of the uterus, that is, its cotyledons [fimbriae or fringed extremities of the Fallopian tubes] furnish food for the embryo through the secundae [secundinae] which correspond to the veins and arteries of the uterus, with the orifices of the veins clinging among each other, of which I shall shortly speak. The umbilicus reaches to the gates of the liver with two veins that come together in one [round ligament] and with twin arteries [umbilical] near the aorta where it is split from the region of the bladder and two arteries are made from one. At a more advanced age the umbilicus now unfit for service and hardened by many years appears bloodless and slender like a nerve to such a point that the veins and arteries cannot be discerned in any manner. Dissectors should take care not to cut the umbilicus carelessly, a precaution I advise in respect to other details also.

Chapter III

Of the Sumen or Abdomen

The part below the umbilicus is called the sumen; by the ancients it is called abdomen. It has very slender veins scattered under the skin by which the embryo in the uterus expels water, as we understand more clearly in an abortion. Below the sumen is the comb or pubis, occupying the lowest region facing toward the genitals. The lower cavity consists of these items (in order that we may now begin dissection): skin, fat, fleshy membrane, muscles, nerves, and the inner membrane joined to the omentum which the Greeks call the peritoneon. All these inner parts are covered from the ilia and pubis upward toward the diaphragm. With the skin and umbilicus thus intact cut the entire area along transverse lines from the top of the stomach to the pubis and from the left to right ilium (or flank) in the manner of a cross. Under the skin is fat, after this a fleshy membrane [fascia], then nerves and muscles, and the very strong covering just mentioned of the membrane [peritoneum] of the inner abdomen, wrapping all the intestines.

Chapter IIII

On the Muscles of the Lower Cavity

Eight muscles are found in the lower cavity, according to Galen.[35] Two of these from the lowest part of the chest are more fleshy and are joined to the pecten or pubic bone [rectus abdominis]. In these there are very long stamens or filaments which can be split; they can be seen proceeding up or down. Under these are two muscles which near the first two muscles on both sides originate from the more slender and softer ribs which the Greeks call rhoai and end in sinews and are joined transversely to the pubis [external oblique]. Two other muscles opposite from the pubis tend upward to the diaphragm in the form of the Greek letter chi (X) [internal oblique]. Finally, two other muscles under the previous ones pass at more of a right angle across to each side of the lumbar region (the loins), where they are attached [transverse abdominis]. However, some people think that the sinews of the muscles by which they are attached come together above the larger muscles. Others think they are joined under the larger muscles since they come together in one from different places. This has been accomplished by nature so that the intestines might be more strongly contained and the food more easily concocted by their reflected heat. For the muscles contract, dilate, call back, accept, retain, expel and relax when there is need, protected by the nature and location of the sinews with the aid of the transverse septum pressing above; the first muscles attract, the last compress, but the middle ones contain. When any of them are injured, however, the bowels are compressed with difficulty. The upper and more ample muscles contain the food and keep it from flowing backwards. The lowest muscles, however, contain the food before there is need to send it forth through the bowels and are so constituted that they can draw off any superfluous moisture. These muscles are not to be cast aside after being observed. Dropsy renders the lower cavity unsightly; in this form it is called ascites. The method of drawing off water in cases of dropsy will be discussed in other books of my history. We choose a spot toward the ilia or groin where there is almost

[35] Galen, *De Anatomicis Administrationibus,* V, 7 (Vol. II K., 511–520); *De Usu Partium,* V, 14 (Vol. III K., 393).

complete bareness of the peritoneum, for the thick part of the sumen is by all means to be avoided. The skin must here be moved aside with a knife that is introduced at this point as far as the omentum, which occurs as the first member below the peritoneum. This must be penetrated (it is called paracentesis by Paul of Aegina)[36] [tapping for dropsy] but not too boldly lest the intestines be injured by rash handling. Only so much liquid is to be collected as to prevent the strength from collapsing, for the liquid contains vital spirit. Finally, put back the skin and the wound heals by itself. Thus gradually day by day the water is drawn off. In this operation physicians keep up the patient's strength with suitable food.

Chapter V

On the Peritoneum and the Omentum

When these parts are moved aside except for the umbilicus, the inner membrane of the abdomen becomes visible. It is very strong in order to prevent rupture by a heavy load or by spirit retained therein (as happens quite often), for this rupture occurs above the groin where the membrane is thinner. If it is ruptured it takes the form of a tumor called bubonocoele [hernia of the groin]. Many surgeons have been deceived by this tumor and have inserted a knife in it, thinking it to have pus, and have injured the intestines, whence has immediately arisen a foul odor and in a few days the patient has died. This ill happens more frequently to women since it is necessary to strain this membrane for the expulsion of excrement and flatulence. This membrane is wonderfully extended in pregnant women. It is attached to the back. To it pass certain veins [inferior epigastric] which carry blood from the uterus to the breasts for lactation immediately upon parturition. We apply cupping-glasses to the breasts when we wish to check the flow of milk but under the breasts when we fear an abundance of the menstrual flow. When this membrane is moved aside the omentum lies beneath (for so it is more particularly called); it covers almost the entire intestines and the lower part of the stomach, for it arises from the middle of this body cavity and depends upon it whence like a cushion it assists the concoction of food. People have been found who lived without an omentum, but they were more inept at digestion, according to Galen.[37] The omentum is composed of very thin little membranes on either side with adipose tissue attached in the middle, with arteries and slender veins, smooth on the inner side, softer on the upper side, lacking sensation completely. It is joined to the fleshy membrane of the back along the diaphragm.

Chapter VI

On the Intestines and First On the Bowels

When you have laid aside the omentum the longest intestines, wound in coils, become visible. In some animals these proceed straight from the lower cavity and expel the excrement through the bowels by means of the straight or shortest intestine, as lynxes and certain waterfowl. Hence these are regarded as insatiable and very often pass excrement from their bowels. This would be a very great hindrance to human activity and unseasonable hunger would result. The intestines receive excrement, that is, superfluous material; they act at different times and places, as Aristotle believes.[38] Their material is sinewy for receiving, containing, and expelling and they use up completely whatever excess moisture the stomach has. They have scallops or notches by means of which they concoct and change the moisture with stamens and fibers of different kinds; they contract and relax when there is need. The intestines do not possess a manifest power of attraction since they do not have stamens drawn out lengthwise. But we must begin from the lower intestine, that is, the bowel. This part thus specially named is similar to pork meat; by means of it the excrement is expelled. By some it is called the longanon, by others the straight intestine (*rectum*). It is not twisted into coils. A part of the thicker intestine falls forward in straining at stool; hence the more reverent ancients called it prolapse of the seat. The furthest part of the bowel is called anus or seat. Galen calls it sphincter. The bowel extends upward to the ilia [descending colon] and is bent back to the left kidney, then to the stomach; hence it stretches to the liver [transverse colon]. It is less fleshy than the rest of the intestines. Certain veins come to this part from the mesentery which suck out the final portion of the moisture. By means of these veins all patients receive nourishment if any is administered to them through an enema; this is not a new discovery, as some people think it is.

Chapter VII

On the Colon and the Blind Intestine

Above the bowel is located the colon, a thicker intestine. It is pervious and spread out widely, with many little cells among its windings in which the excrement remains for a longer time and receives a certain shape. The Greeks call it catocoelia. I know it is also called bowel by some people. It is attached in a more involved form to the left kidney; hence kidney pains and colic pains have the same sensation for the

---

[36] Paulus Aegineta, ed. I. L. Heiberg (Leipzig, Teubner, 1929), III, 48, 5, p. 258.

[37] Galen, *De Usu Partium,* IV, 9 (Vol. III K., 286): . . . in quibus per vulnus excidit omentum, deinde lividum factum cogit medicos partem laesam adimere. Omnes enim hi frigidiorem sentiunt ventriculum, minusque concoquunt. . . .

[38] Aristotle, *De Partibus Animalium,* III, 14, 675 a, b; Pliny, *Natural History,* 8. 22. 34. 84 on the lynx (lupi or lynx cervarii); *ibid.,* 11. 37. 68. 80 on scallops (*crena,* a corrupt word).

patient. These pains can be distinguished from each other, among other known characteristics, by fasting. Food helps kidney pains but aggravates colic pains. Then this intestine reaches the spleen, climbs up toward the right side of the stomach, and on account of this the noxious humor of this intestine affects the mouth of the stomach and the diaphragm. Thus patients often fall into syncope (lipothymia, or fainting spells), purging the bowels on the stool. Not without reason is this intestine included among the lower organs although it seems to hold a higher position. Likewise it is rightly very close to the liver in order to receive through a duct from the latter the bile from the gallbladder. The biting effect of the bile causes the excrement to be expelled; hence that part to which the bile duct is attached is yellow and bitter to the taste. The blind intestine [cecum] is joined to the colon and located on the right side under the kidney near the coxa. It is called blind because it does not have two orifices at opposite ends like the other intestines, for it expels at the same orifice where it receives since it is considered to be of the same nature as the previous intestine [ascending colon] and is troubled by the same ills.

Chapter VIII

On the Three More Slender Intestines

Next, the more slender intestines are attached to the stomach. In man they are called lactes. The first of these intestines, called the more slender one, is much entwined in tortuous coils which are connected to that intestine which the Latins call middle and the Greeks call mesenterion. It turns more toward the right side and ends in the region of the right coxa but fills the upper region more. Because it is located near the ilia it has received the name ileon (ileum) intestine and its pain is iliac pain. A little above, toward the stomach on the more left hand side of the spine there is the intestine called jejunum, not entwined so much as the other intestines are. Its emptiness has given it its name for it never contains what it receives but immediately transmits it to the lower region and is always found to be empty. It is called hira and hilla by some people. The reason why food slips out of it so quickly is the fact that the bile has a biting effect upon its contents. To this intestine, finally, is attached that small juncture from almost the middle of the stomach, turned a little to the right side, which the Greeks call pyloros. Since in the manner of a gate [duodenum] it sends forth into the lower parts the matter which we are about to excrete, it is narrower than the other intestines. A duct [common bile] extends to this gate of the bile for the same reason. This gate is twelve fingers long. During dissection that duct must be handled carefully and must be shown intact.

Chapter IX

On the Mesentery and Its Veins

The mesentery (which the Latins call medium or middle and median) is sinewy, filled with many small membranes and glands, which Galen calls adenes. It is curled or wrinkled everywhere. The lactes adhere to it in tortuous entwined coils, as I have said, and the nearness of the member has confused some people who have often mistaken the lactes for the mesentery. In it the fibers of vein are constituted by nature in such a way that they stretch from the intestines to the larger vein and aorta. Without the mesentery the moisture would otherwise be uncomfortably attracted from the intestines (for the veins have something to which they may adhere); the latter end like roots in the material of the intestines and do not pierce them. The physicians call these mesenteriac veins, often the origin of obstruction. The mesentery is attached to the diaphragm and to the back or depends from the larger vein and the aorta, as Aristotle prefers,[39] whence it is certain the blood and spirit are carried down. The veins themselves are solid and hard; they take nourishment from the stomach and from the intestines as do the roots of trees which draw moisture from the earth and nourish the trunk, for the intestines and lower cavity of animals are similar in function to the earth. There are some people who think the orifices of these veins are among the broader intestines in the living person and lie hidden in the cadaver. Such people are in serious error for they think that bits of food can be drawn through these same very slender fibers. They also believe the following story. A young girl in our day at Venice while I was writing this book carelessly swallowed in her sleep a hairpin four fingers long which she held between her teeth when she was about to go to bed. Ten months later she passed the hairpin in her urine amidst great suffering since the viscous humors collected in her bladder had made a rounded stone around it the size of a hen's egg. These people insist the hairpin was first swallowed by the meseraic veins and that they know the passages which nature found for this purpose. These fibers in the intestines end in very slender sinews and become blotted out in the material of the intestines. Here they are more numerous and thicker than those which lead to the liver and thus carry more food; they more quickly assist in need. But if these veins, all or in large part, grow weak or are obstructed they deform the person with an induced flux to the point of wretched emaciation. When you have examined the mesentery blow up the stomach with a bellows inserted at the pylorus and thus you will more clearly reproduce the similar position of the living stomach.

---

[39] Aristotle, *De Partibus Animalium,* IV, 4, 678 a: *De Historia Animalium,* III, 4, 514 b 25.

Chapter X

On the Stomach

The passsage which stretches from the mouth of the jaws to the stomach is called the gullet [esophagus]. It is composed of sinew and flesh and through it food and drink are brought to the stomach. For it is called the gateway of the food. It is not attached to the middle of the stomach but at the side to keep the food from easily flowing back when the head is tilted. Thus the other entrance to the stomach which I called the pylorus a short time ago is not joined to the bottom of the stomach in its middle but a little higher to keep what is eaten undigested in the bowel from slipping out more swiftly than is necessary. For this reason the transverse septum or diaphragm is tightly squeezed in drunkenness to keep any of the food from being vomited up. These twin gateways differ in location and size, for the upper one is broader, the lower one is narrower. For this reason, because the stomach must transmit that which is more ground up by digestion the beginning of the intestines is like that of a dog; although it is not much wider than the intestine it seems similar to a wider intestine, according to Aristotle,[40] under the diaphragm near the region of the heart with its narrower part. For what is called the mouth of the stomach is found where the thirteenth spinal vertebra is located. It digests and changes food and is called the father of the family since it alone controls the entire living creature. If it becomes ill life hangs in the balance. Between the spleen and the lobes of the liver it stretches from side to side; above and below between the diaphragm and the intestines it holds the middle position. The omentum protects it in the anterior region; the large veins foster it from the rear. In its upper part it slants to the left, in the lower part to the right. Thus the reason for its orifices seems to be explained for it receives the black bile running down in its upper region through a certain duct from the spleen but in the lowest region takes the yellow bile carried through another duct from the bile sack. The stomach swells and hangs in the middle; it is connected by means of slender membranes to the diaphragm, spleen, liver, heart, and to the brain by veins, arteries and nerves; hence when the head is struck the stomach is often seriously disturbed. It has two tunics; the inner one is thick and sinewy, the outer one thin and fleshy. The inner tunic has two kinds of threadlike filaments, which the Greeks call ines. One of these stretches inward and lengthwise, attracting downward in times of appetite. The other, on its exterior surface, has a retentive force, called cathectica by the Greeks and a very necessary one, which operates through transverse (or oblique) filaments. The outer fleshy tunic of the stomach has a broad threadlike vein it extends forward by means of which the food is expelled as well as the superfluity of digestion; when this tunic contracts it is said to excrete. Apocritica is the name of this power of secretion or excretion. The inner tunic has the sense of touch, whence proceeds the appetite for food. The outer tunic modifies food and drink and mixes them; this power the Greeks call alloeotica, to which the other powers of the stomach submit.

The stomach on the inside is denticulated or corrugated [rugae] with thick skin in the manner of a blackberry for the purpose of breaking up food, with the scallops or wrinkles decreasing toward the orifices of the stomach; they have the fresh roughness of a carpenter's rasp. From these wrinkles the fibers arise; they are thus called from their similarity to the meseraics. They have a nature between that of a vein and a nerve. By them (as I said shortly before) the moisture is attracted and transmitted hence through larger veins to the liver. While the stomach is digesting it is conscious of itself since in hunger and thirst it continuously attracts moisture and is the first to be refreshed.

Chapter XI

On the Adipose Matter and the Pancreas

Before we pass to the spleen a few words must be said about the adipose matter. This lies under the thin membrane of the abdomen and is more ample in an obese body. It begins from the ilia and tends downward toward the kidneys. On the inner side it is held in by a membrane which is joined to the omentum. It is not concocted nor broken by any dryness like suet (*sevum*) since its nature is less earthy. It is without sensation for it lacks nerve, artery, and vein. Therefore, it has been said that when the adipose matter is removed from many people and the body relieved of its immovable burden they have lived. The pancreon [pancreas] also has a nature between that of flesh and of glands. It is attached to the spleen, stomach, and liver. It takes its name from flesh. To it extends a vein common with the omentum which arises from a larger vein.

Chapter XII

On the Spleen

The spleen is located a little lower in the left-hand region toward the liver; it is similar to the fat of pigs.

[40] Aristotle, *De Partibus Animalium,* III, 14, 675 b. The ines, or fibrous connective tissue mentioned below, is explained as a material between a sinew and a vein by Aristotle, *De Historia Animalium,* III, 6, 515 b 30.

It is the receptacle for the purging of the blood from that black humor which we call bile and the Greeks call melancholia. It is not joined, as the liver is, to the diaphragm but to the left side of the stomach and to the omentum. Its humped part is attached to the back with little membranes; in order to make it more easily visible the lower ribs which are more slender and under which the spleen is located must be opened. Its material is soft, thin, and fistular (or full of abnormal passages), of moderate length and thickness. In splenetic patients it swells so much due to the collection of moisture in it that it seems to girdle a man as with a belt. Since (as Plato says)[41] the liver abounds with impurities the thinness of the spleen purges them and absorbs them since it is concave and bloodless. Thus it grows with the increase of purged material, a particular impediment (as they say) to runners. Hence sometimes in those who are suffering the spleen is burned or cauterized. It is not reasonable that it can be removed from a man without peril of life. Intemperance is attributed to this member in proportion to the size of its location. No vein comes to the spleen nor to the liver from the aorta. I see the crowd of commentators aroused in many places [passages] against Aristotle and stirred to quarrels or disputes, as I shall shortly state concerning the vulva, since they judge that these members cannot possibly exist without spirituous blood, for they are nourished with blood by the smaller vein which is not without spirit, just as the vulva is without blood. The larger vein by a direct passage from the gate of the hollow of the liver is led to the spleen [splenic]. If the latter is carefully uncovered there appears a certain small branch (*rivulus*) from the middle course of the same vein which leads more diluted blood downward toward the mesentery to nourish it [inferior mesenteric]. Again the same rather large vein is divided into a fork near the spleen. One branch leads to the bottom of the stomach for nourishment [short gastric], the other proceeds to the hollow of the spleen, inside which it is spread out into two branches, passing up and down. The lower of these irrigates the omentum with blood; the upper one, however, from the hollow of the spleen is again divided into two branches, one of which tends toward the more left-hand and upper region of the stomach; the other tends upward to the orifice of the stomach and pours in the black bile by which the mouth of the stomach is contracted and a burning sensation arises. The remaining portion of the vein is absorbed in the spleen itself except for that share of it which the large vein transmits to similar members.

[41] Plato, *Timaeus* 72 C. The comparison of the stomach's bottom to a rasp is borrowed by Benedetti from Pliny, *Natural History,* 11. 67.180, p. 545 (Loeb Classical Library).

Chapter XIII

On the Liver

Plato thought the liver was the noblest member.[42] He placed in it the virtue of divination. When it is removed no clear sign of fortune-telling results. Later physicians, however, attribute to it the origin of the blood. If the liver is strong the members seem to grow with increased blood; when it is weak there is no doubt the members decrease. When the members have become diminished they are readily filled out as the liver convalesces. The liver is proportionately very large among the inner organs. It is suspended under the diaphragm at the right side, filled with blood and natural spirit, hollow inside and humped on the outside, projecting gently toward the stomach from the region of the spleen. It fosters the stomach and is more pendent in the living body. Its powers are multiple for it attracts, retains, concocts, digests, and expels. It digests all juice [chyle] into bloody mater, [yellow] bile, and black bile. These powers by means of the temperate heat of the heart serve the nutritive function which is born with the individual, for Averroes[43] has no doubt that the power which nourishes and that which causes growth are one and the same. The liver is divided into five lobes; it is without sensation inside, fistular, spongy, and perforated, with very small veins in knotty union scattered throughout the entire member. When the chyle is changed into blood or bloody matter by the second digestion it is then carried forth to various parts, for it is warm and humid. The liver is covered by a very thin membrane. It is attached to the diaphragm, to the spine, and to the stomach. The major vein [inferior vena cava] is more closely joined with the heart than with the other members. The aorta does not stretch hither since it would be more visible at first entrance to the liver and would have to be distributed to all the parts, as is apparent in the lungs, for from the heart a vital force is poured through the larger vein into the liver, as has been said; the same passage receives and administers the humor. The gall bladder hangs from the lower part of the hollow of the liver. In this hollow is an assemblage of veins from different places; it is called the gate of the liver [portal vein] and forms a sort of crossroads of ducts and passages; through these pass that chyle (*succus*) by which we are nourished. This vein is immediately spread out inside the liver into five branches and into very slender veins which are then dispersed throughout the entire substance of the

[42] Plato, *Timaeus* 71 E.

[43] Averroes, *Colliget,* II (Venice, Giunta, 1574), fol. 21 r, v, speaks of the nutritive function of the heart which gives all other virtues (or powers) to the members; among these is growth.

liver. These dwindle into a minute fineness and are at last obliterated in the liver itself. By means of these very subtle fibers the bloody matter of the chyle is concocted and turned into blood (as many insist) by the ministering heat of the liver; the blood is finally completed entirely in the heart.

Chapter XIIII

On the Veins of the Liver

At the gates of the liver eight veins are branched off from the portal vein. Two of these are small: one stretches to the pylorus [pyloric] and to the mesentery, the other is joined to the lower stomach near the pylorus [coronary]. Of the remaining six veins one turns to the right-hand part of the stomach [right gastro-epiploic]. A second larger one stretches to the spleen [splenic or lienal]. A third is joined to the rectum intestine [inferior mesenteric] for it runs down to the left side. A fourth goes to the right-hand side of the stomach, where it is divided into slender fibers [short gastric]. The fifth vein has two offshoots; one of these is extended to the omentum and is scattered out in the region of that branch which is seen to arise from the spleen [pancreatic]. The second offshoot is joined to the thicker intestine; by means of this offshoot the remains of the chyle are attracted. The last vein stretches to the jejunum intestine [superior mesenteric] and to that which I have called the blind intestine, to which many veins clearly pass for the purpose of drawing off the chyle.

Chapter XV

On the Sack of the Yellow Bile

All animals do not have a gall bladder. There is none in the horse, mule, or deer, and of the same kind are those animals which partly possess one and partly do not, such as mice, sheep, and men. Those who consider that the nature of the gall bladder (says Plato)[44] exists for the sake of some sense perception do not think rightly. Thus they insist the gall exists to excite with its biting quality that part of the soul located in the liver and by relaxation promotes rejoicing. The sack of the yellow bile hangs from the lower part of the hollow of the liver; it is called choledochus by the Greeks. In it there is a yellow humor, sometimes seen as pale by Galen or of a bluish color (green) as of the herb isatis[45] according to various dispositions of the body. In the cadaver it changes color entirely. It is yellow in the living body. It arises from the liver to purify the blood. There are two ducts [hepatic] to the gall bladder. One arises from the middle of the liver and by means of it bile is attracted or sometimes poured forth. The other is more visible and leads to the pylorus; sometimes stone is found in this duct. From the same place another third duct[46] is said to arise which passes to the lowest part of the stomach; other people deny this. But this is not constant in its appearance in the body. The largest duct [common bile] of all goes to the intestines which we call lactes. The health of those who have no gall bladder is considered more firm and their life longer.[47] When the passage of the bile is blocked the king's disease (which is called icteritia (jaundice) or by others, arquatus) results. The entire body is suffused with a yellow color and the natural color removed; we first perceive that what ought to be white in such patients has become yellow. Fevers also accompany this condition. The intestines have thus far in large part been described. Thus everything except the humped part of the liver because it is annexed to the large vein is to be cast aside, leaving the stomach.

Chapter XVI

On the Kidneys

These items examined, we see that two large veins pass down from the highest region along the spine. I shall discuss these in the last book. The large vein [vena cava, chilis] sends two branches toward the kidneys [renal] opposite in the form of the Greek letter lambda, one to the right kidney, the other to the left kidney, at equal lengths. The other large vein, called the aorta, does the same thing. It is annexed to the spine. Since the right kidney is less fat it is higher than the left kidney, which happened for a similar reason in the eyebrows. The branches of the major vein [vena cava or chilis] do not pass into the folds of the kidneys but rather adhere with their heads and do not enter the hollow part but disappear in the body of the kidneys, just as the aorta vein does. From these kidneys the genital[48] semen is generated in large part.

---

[44] Plato, *Timaeus* 83 C, 87 A; the entire passage on the action of the bile is diffuse and complicated, involving its action on both body and soul.

[45] On isatis see Pliny, *Natural History*, 20. 7. 25. 59, p. 37 (Loeb Classical Library): "Its leaves pounded up with pearl-barley are good for wounds" (translated by W. H. S. Jones).

[46] Compare my translation of Massa, *An Introductory Book of Anatomy*, p. 193; Berengario da Carpi, *A Short Introduction to Anatomy*, translated by L. R. Lind (University of Chicago Press, 1959), p. 62.

[47] Berengario da Carpi, *loc. cit.*, says, however: "Sometimes a man lacks a gall bladder; he is then of infirm health and shorter life." Benedetti follows Pliny, *Natural History*, 11. 74. 192, quoted also by de Laguna (note 28 to my translation of his *Anatomical Procedure*).

[48] The origin of the semen from the kidneys is a strange idea. Neither Aristotle (*De Generatione Animalium*, I, 17–18, 721 b–726 a 30) nor Galen (*De Semine*, Vol. IV K., 521–651) mention such a

They are oblong like the kidneys of oxen, as though they were composed of many kidneys, bent back in the interior, rather rounded on the outside toward the intestines. Their flesh is hard and firm, without sensation, and from each of them fat comes forth from the middle. The kidneys are to be cut lengthwise and, when a probe has been inserted, the method in which they are joined must be observed. In the middle there is a very slender little skin by means of which the liquid only without blood distils through ducts, or white veins, called ureters. These are hollow, sinewy, and near the spine by means of veins they stretch to the ilia; soon by a narrow passage they are placed in each coxa and disappear; then thrust forward once more they appear in the buttocks in such a way that they come to an end in the male near the bladder and genitalia, in the female near the vulva. These veins which come forth from the middle of the kidneys have nothing in common with the major vein [chilis or inferior vena cava]. The kidneys are covered with little membranes by means of which they have sensation. They are troubled with stones since many kidneys seem to be afflicted with them. They are double in order, according to Galen, that they might offer an equal balance of the body to living creatures. From superabundant heat, incontinence, and the dripping of urine, sand and stones are generated in their passageways.

Chapter XVII

On the Seminal Veins

Two seminal [spermatic] veins come from the kidneys to the heads of the testicles. They are called ducts, or pores, and are bloody. Two bloodless arteries [spermatic] stretch forth from the aorta vein, for in coitus spirit precedes the semen; by means of the spirit the semen is propelled in spurts.[49] From the head of the testicles a duct [vas deferens] receives it, thicker than the previous one above and more sinewy. Having progressed through the testis it bends and seeks the head of the testis again. Then the ducts come together again and in the prior region enter the genitalia, called parastatae, and are received into a common canal and are covered by a continuous membrane. Then they settle among the testicles and seem to be one duct unless you carefully draw away the membrane. The duct contains a bloody humor which settles there but less than in those [ducts mentioned above?]. Those which turn back and pass under the neck (*cervicem*) have a whitish genital urine; the duct is also extended from the bladder and reaches the neck, around which there goes a covering or shell with a channel in it which we call the penis.

Chapter XVIII

On the Semen

The semen is a superfluous nourishment of the body, a material pure and separate from the principal members necessary for generation. The greater quantity of the material of generation, it is believed on the authority of Galen,[50] is drawn from the brain. For this reason a certain very lecherous man whose head was dissected after his death was found to have a very little brain. The veins which pass behind the ears, if cauterized or cut, induce sterility [51] through a marvelous and secret limitation of nature, as I shall soon state, which I learned by chance. The same thing happens also to other members, for many people insist on the basis of various arguments that those who castrate roosters by burning their genitals or legs with a red-hot iron deprive the entire body of the power of procreation.[52] This view seems to be bolstered principally by the fact that in children are reborn a distinguishing mark, mole, or scar similar to that of a parent except that grandsons resemble in appearance and distinguishing marks grandparents from whom no semen has been withdrawn [to create the grandsons]. A woman of Elis (*Elide*) who had lain with an Ethiopian did not give birth to an Ethiopian daughter but the son of that daughter was an Ethiopian.[53] We have seen this also in plants for there

theory. Aristotle, however, in *De Historia Animalium,* III, 1, 510 a 15, speaking of the spermatic arteries and veins, says that two of them extending from the kidneys are supplied with blood. This statement may be the source of Benedetti's unexplained statement.

[49] That the semen is propelled by the spirit or air is a Hippocratic idea: Hippocrates, *Opera,* III, 748 (ed. C. G. Kuehn, *Medicorum Graecorum Opera Quae Exstant* **28** [1827]). The passage is cited by Arthur Platt in his translation of Aristotle, *De Generatione Animalium*, I, 20, 728 a 10. See *De Historia Animalium*, 7. 7. 586 a 15.

[50] Pseudo-Galen, *Definitiones Medicae* (Vol. XIX K., 449) quotes Plato and Diocles on the origin of the semen from the brain: CDXXXIX. Semen, ut Plato ac Diocles autumant, ex cerebro et dorsi medulla excernitur: ut autem Praxagoras atque Democritus praeterea et Hippocrates censent ex toto corpore. The latter view is carefully refuted by Aristotle, *De Generatione Animalium,* I, 17–18, 721 b–726 a 30.

[51] That sterility is induced by cutting the nerves behind the ears is an idea associated with Hippocrates' statement about the treatment of varicose veins among the Scythians in *On Airs, Waters, Places* 22: ". . . vessels behind the ears which if cut, cause impotence . . ." (translated by John Chadwick and W. N. Mann, *The Medical Works of Hippocrates* [Oxford, 1950], p. 108).

[52] The castration of roosters by cautery of their genitals is discussed in my translation of Ulisse Aldrovandi's famous chapter on chickens (*Ornithologia,* Vol. II, Book xiv, 1600) in *Aldrovandi on Chickens,* translated by L. R. Lind (Norman, Oklahoma, University of Oklahoma Press. 1963), pp. 408–411. Aristotle, *De Historia Animalium,* IX, 50, 631 b 25, mentions the practice.

[53] Aristotle, *De Historia Animalium,* VII, 6, 586 a; *De Generatione Animalium,* I, 18, 722 a 9–10; II, 6, 743 a 10, on nutriment oozing through blood vessels like water through unbaked pottery. *Meteorologica,* II, 3, 359 a: "Make a vessel of wax and put it in the sea, fastening its mouth in such a way as to prevent any water getting in. Then the water that percolates through the wax sides of the vessel is sweet, the earthy stuff, the admixture of which makes the water salt, being separated off as it were by a

are many quince trees which bear fruit similar to a priapus. Some seeds of the latter preserve the same form, others are mixed. Seeds sown from these in turn show the same deformed result born from the parent fruit. The semen of man in his youth is strong; in old age, ill-health, and lechery it becomes impure. It comes forth from a warm place within since by a mixture of spirit and water it is liquefied from its state of thickness and cold; because of this it is rendered white and spirituous, which qualities do not come together (or coalesce) since spirit is included. From the heart (the semen) passes through arteries, from the liver through veins, and from veins and arteries descending from the brain to the smallest veins and arteries in the loins. In these receptacles it is changed at the proper time into the substance of blood. Then it is carried to larger and ampler veins so that from both one vessel full of spirits might be constituted for each testicle. In this vessel it receives that power of generation which other people attribute to the testicles from which another creature similar to the one who generates is produced. Some have asserted that by the natural warmth of the generative semen matter is gradually collected like sweat among these ducts, as in the uterus, just as empty perforated earthenware vessels let down into the sea receive sweet water within themselves, or as cathartics applied to the stomach or intestines or as from the lung without a certain duct of the veins sweating back the matter within evoke or draw away evil humors by the marvelous work of nature.

Chapter XIX

On the Testicles

The testicles which are called didymi hang in a man in such a position as to sustain more strongly the impetus of sexual relations and to incline less forward and outward. There is a great deal of pleasure in them, as Alexander of Aphrodisias declares,[54] since they are much pained when struck a blow. When they are cut off the masculine form and behavior is almost completely changed and become feminine, for men lose their strength, boldness, habits, voices, and beard. The seminal vessels pass around these (testicles) and do not enter them, as some people falsely state, but (as Aristotle thinks)[55] they increase the firmness of movement of the genital excrement, since nature creates nothing without great reasons; thus they only preserve their winding back upon themselves by this argument. Once, as it is reported, a castrated bull coupled directly after castration with a cow when the ducts had not been retracted as yet and procreated offspring.

Chapter XX

On the Scrotum

The proper covering and protection of the didymi is the scortum (scrotum), which the Greeks call oscheum. It hangs from the loins, is wrinkled, and has little veins scattered through it. In its lowest part there is a suture, which is lacking, however, in some men. If the scrotum swells in an unsightly manner through injury they call it a hernia; I have discussed this in the books which treat of all the ills which afflict the members.[56] Inside there are tunics which cover the testicles hanging by single sinews; the Greeks call the latter cremasters. With each of them two veins and arteries descend; I indicated these a short time before. The tunics are thin, sinewy, white, and bloodless. Upon them is another stronger membrane which inwardly clings strongly to their lowest part. The Greeks call it edarion. Finally, many little membranes cover these veins, arteries, and sinews.

Chapter XXI

On the Penis

The stalk itself (*coles, penis*) hangs from the groin as if connected with skin. It has more nerves than the other members, with more arteries and veins. It is joined to the pubic bone with harder nerves and nervelike connections [suspensory ligament]. Other nerves descend to it from the marrow of the lowest part of the spine, but otherwise this member is less sensitive than others. I once watched a certain impious surgeon of the Jewish race extracting one of these nerves on account of its inflammation, for he pulled it back little by little day after day from the

filter." See also *De Historia Animalium,* VIII, 2, 590 a 24. The translator, E. W. Webster, points out in Vol. 3 of the Oxford Translation of Aristotle, that the facts do not support Aristotle's statement.

[54] Alexander Aphrodisiensis, *op. cit.,* fol. xx: Cur magis dolemus in testibus percussi ? quod colei plurimum sub sensum cadant praeterea natura quoque pudendis suam praebuit uoluptatem coitus causa neruosa enim pars est atqui plurima uoluptas a multo gignitur sensu testiculi igitur cognatione et affinitate uoluptaria affecti multi sensus ipsi quoque particeps sunt. Ubi sensus inquam multus ibi ingens oritur uoluptas; ubi porro ingens uoluptas, illic etiam ingens dolor maximo siquidem bono, maximum ut plurimum opponitur malum etc.

[55] Aristotle, *De Historia Animalium,* III, 1, 510 a, 510 b, describes the seminal ducts; see the diagram in the Oxford Translation of Aristotle, Vol. 4. On the bull which caused a cow to conceive after it was castrated see *De Generatione Animalium,* I, 717 b; *De Historia Animalium,* III, 510 b. See note 79 to the translation of Massa's Anatomy for the phrase "Nature creates nothing . . ."

[56] Alexander Benedictus, *De Re Medica,* or *De Omnium a Vertice ad Plantam Morborum Signis, Causis, Differentiis, etc. . . . libri* XXX (Basle, 1549). The later reference to the advent of syphilis in Italy points to the date 1493–1494.

root; finally he drew it out entirely, with fatal results. The extreme end of the member is the glans (acorn), covered with very thin skin which has sensation from the upper nerves of the spine. It is the seat of venereal pleasure, whence the seminal veins are shaken with a peculiar tickling. In the middle of the member is a certain spongy hollow. When the spirit of desire enters this it causes a stiffness; authors report that the member is also excited even in a dead body by a spider's sting. The other duct likewise is uncertain as to its purpose. In puberty, as some believe, a certain viscosity is transmitted from it as the rudiment or beginning of the semen for procreation. The glans surrounds the end of the penis. From its top to the bottom a ligament (*frenum*) comes down from the upper orifice or fissure which they call the urethra, where the urine and the semen come forth. The covering of the glans is called the prepuce. The Jewish race and almost all the rest of mankind except the Christians now for many ages circumcise the prepuce by law for the sake of cleanliness, as they say, a practice later referred to religious rites. The skin which covers the penis is not simple but sinewy in nature and possesses a great sense of touch. For this reason through venereal contact the new Gallic disease, or at least unknown to previous physicians, has crept upon us from the west under the aspect of a pestilential star while I was publishing this book, with such disfigurement and torture to the members especially at night that it even surpasses incurable leprosy or elephantiasis in horror, to say nothing of its danger to human life. This pestilence has already infested the rest of the provinces and even recurs again in those who have been once healed of it, to the great perplexity of medical men. Empirical physicians whose medicine is learned from experience only have come from the west to travel about to their great gain among our cities, professing to heal this disease alone.

Chapter XXII

On the Bladder and Its Ducts

White veins called urethrae pass (as I mentioned above) to the bladder in that part of it which consists of a double membrane. One of these veins is pervious, and between these tunics the urine flows past until it reaches the foramen of the inner membrane and falls into the bladder. These ducts are not set opposite each other since the urine would sometimes be forced to flow back to the kidneys when the bladder is full; but the more the latter becomes filled the more the membranes mentioned are compressed and thus they contain this liquid refuse. Inexperienced people think there is only one duct at the place where we separate them since they believe the bladder is simple, or of one piece, in all of its extent. The bladder is sinewy and located differently in men and women. In men it is near the rectum intestine inclined somewhat to the left. In women it lies above the genital mouth and when it has slipped from above it is sustained by the vulva itself. Two strong and permanent ducts (*meatus*) proceed from the aorta to the bladder, and no vessel comes to it from the major vein (*vena cava, chilis*). Those vessels which come from the hollow of the kidneys do not have a material continuous with the major vein, as I said above. The cervix of the bladder is fleshy. It is filled with muscles by means of which the liquid is contained and released when necessary. The neck alone coalesces if cut; the rest of the bladder cannot do so, which, as Aristotle[57] explains, is due to the difference between the two parts. In males the urinary passages [urethra] are longer and more compressed; they descend to the end of the penis. In the female, however, these passages are shorter and more ample. They are revealed in a common pathway to the neck of the vulva. Thus women are less afflicted by stones in the bladder; however, hairy threads, which cause great pain grow in the mucous membrane. Whatever the method for extracting stones the place must be opened with a knife and care taken lest the surgeon carelessly and fatally wound the patient by mistaking the proper area for incision. Certain other things grow in some peoples' bladders which resemble shellfish or oysters. But now let us pass to the female seminal vessels.

Chapter XXIII

On the Seminal Vessels of Women

This is the place to speak about the genital members and seminal veins of women. These veins [ovarian], which many physicians declare to be prolific in humor, have almost the same origin in the female as I described above in men and pass in winding coils to the testicles. Thence they pass to the mouth of the vulva, whence the semen bursts forth with such great pleasure, something which is vain and useless according to the statement of Aristotle,[58] chief of philosophers, with whom the Arab Averroes agrees. However, I acknowledge that from the kidneys, spine, coxae, and loins many veins and arteries which are scattered forth pass into very slender fibers through the vulva and from these the genital humor necessary for conception, thick, white and similar to egg white, is collected within (as bile is daily collected in the stomach), just as happens in eggs in which the white and the yolk produce

---

[57] Aristotle, *De Historia Animalium,* III, 15, 519 b 15: "The bladder, like ordinary membrane, if cut asunder will not grow together again, unless the section be just at the commencement of the urethra: except indeed in very rare cases, for instances of healing have been known to occur" (translated by D'Arcy Wentworth Thompson, Oxford Translation of Aristotle, **4**).

[58] Aristotle, *De Generatione Animalium,* II, 4, 739 a 20.

offspring from the female, with a constituent supplied in the seed from the male. From this material members which we call spermatic[59] are generated and from the menstrual flow they become red as flesh for nourishing. Aristotle[60] reports this concerning fish: when the egg is broken the embryo comes into being like a nucleus, at first accepting no food but growing by absorbing the liquid of the egg. Later the fish are fed in the fresh waters of rivers until they grow large enough. The same author does not say whether it is true or not that all their nourishment comes from outside[61] but from the first just as there is something inside called milky as in the seeds of plants, so in the material of animals there is an excrement of the constitution which is nourishment and the foetus grows through the umbilicus in the same way plants do by means of their roots and as happens also to animals themselves when they have been completely formed and released from that aliment in which they are contained. But that which bursts forth from women in the act of sex does not issue in spurts as from a man since any such thing is useless and vitiated, differing completely from that which overflows within the vulva without any lust. The members of the foetus are formed from both the male and the female and the spiritual life is produced from both, but the principal members are constituted from the male seed and the other more ignoble ones from the female seed, as if from purer matter, just as the spiritual vigor is created. For the male semen is collected in the midst of the genital female whiteness so that it may spread out its power of generation to all matter, as Galen believes;[62] hence the members are formed of menstrual blood and thus far are not in need of nourishment until, the larger amount of matter becoming exhausted, the body requires nourishment which is later sought through the umbilicus. From conception as far as the seventh day the flow of the prolific semen appears simple, at which time the umbilicus and the secundinae seem to come into being. On the tenth day the bubble of the heart swells with the boiling spirit within it. With the drying heat outside it the material of the secundinae is made firm since both plant and shell or covering are generated at the same time. After these days the other members are sketched out; the principal members are not generated at one time since they are different in nature. The heart is created first, then the rest of the members are delineated and finally completed as if in a painting. The male members are formed within thirty days, according to Hippocrates.[63] Others, however, among them Aristotle,[64] declare that the male form is completed in forty days, the female within three months or later, since the form in humidity does not remain after abortion but is immediately dissolved. If the masculine form is spread out in cold water it is compressed and becomes consistent in the membrane. When this is broken the foetus itself appears to be as large as a rather big ant. The members are already separate, both genitals and the others, and the eyes are quite large as in other animals. For this reason I once dissected the uterus of a deer in which there was the foetus of a fawn. It was as long as my finger and formed in all parts corresponding to those in a larger deer, except for the eyes which were rather large. The vulva was full of a white viscous liquid about the size of a goose egg surrounding the very slender members of the foetus. The latter hung high in the midst of this liquid. From its umbilicus a bloody thin threadlike connection seemed to reach the surrounding skin, the earliest source of nourishment being that whiteness which I would contend is the female semen not unnaturally contained in the uterus without a sign of menstrual blood. This does its duty first to make the foetus larger. The testicles in women are bent back on either side like those of a rabbit. From them (as others believe) other passages go out to the hollow of the vulva which are suited for generation. Dissection easily reveals the truth of this belief.

Chapter XXIIII

On the Sexual Parts of Women

We see that the entrance of the vulva in woman as in other animals is called the nature and the genital mouth. It appears in the middle with a little tunic on both sides called nympha and clitoris. This nature is guarded within like the glans which is guarded by the skin of the prepuce in males in order to keep out the air by this primary protection. The inner part is called the cervix. In almost its middle part there is a thin nervosity or sinewiness sprinkled with tiny veins which presents the proof of virginity. It is called eugion or hymen for it is first split in coitus. The natural parts are sometimes so closely compacted by nature that they require the services of a surgeon.

[59] Avicenna, in *Canon,* I, Fen 1, Doctrina 5, chap. 1 (translated by Gerard of Cremona, Venice, Giunta, 1595), 31, defines spermatic member thus: Membra autem, quae ex spermate sunt creata, quum solutionem continuitate patiuntur certa continuitate non restaurantur nisi pauca ex eis et in paucis habitudinibus et in etate pueritiae, sicut ossa et rami venarum parui. . . .

[60] Aristotle, *De Historia Animalium,* VI, 10, 564 b 25–30.

[61] Aristotle, *ibid.*

[62] Galen, *De Semine,* I, 7 (Vol. IV K., 535–536); II, 4 (*ibid.,* 622–625).

[63] Hippocrates, *De Natura Pueri* (Vol. I K., 392): Iam vero genitus est infans eoque pervenit ut foemina quidem primam concretionem duobus et quadraginta diebus, cum longissime, accipiat, mas vero triginta diebus, quod longissimum.

[64] Aristotle, *De Historia Animalium,* VII, 3, 583 b: "In the case of male children the first movement usually occurs on the right-hand side of the womb and about the fortieth day, but if the child be a female then on the left-hand side and about the ninetieth day" (translated by D'Arcy Wentworth Thompson, Oxford Translation of Aristotle, 4).

The inner part we call matrix [uterus] and loci (places). This often descends in desire of coitus; in it there are the acetabula or cotyledons. By means of these through the umbilicus the foetus is caused to grow. The vulva has no power for changing the semen but only the power to contain it. The inner entrance (*vagina*) is glandular, shaped to a sword point like the male glans with its spindle projecting forward into the cervix. In deer there is a triple entrance;[65] for this reason pregnant deer refuse coitus because it hurts them, while women like it and hence sometimes superfetate, or conceive again while still pregnant. Not all animals have complete cotyledons of the vulva. Horn-bearing animals have those with one side dentated while they bear their young. Rabbits, mice, and bats have cotyledons dentated on both sides. The pig has a thin vulva and the embryo does not adhere to its cotyledons; thus the semen is drawn away from the places of the uterus and thus many pigs are generated in proportion to the number of places and of cotyledons for often twins arise in the same spot in the uterus. The foetus clings more strongly from the cotyledons of the vulva in that region where there is a certain confused mass of fibers. The orifices of the umbilicus correspond to the orifices of the cotyledons by which, as I said, nourishment is obtained, for the roughness of the fibers and the orifices on both sides make for tenacity as with walls which workmen first scrape before they put cement on them in order to make it cling more strongly. The umbilicus with its involucrum, that is, the secundinae, or part of the placenta, when its function is accomplished is released from the cotyledons like the stem of an apple already ripe falling of its own accord from the branch. The menses flow from these cotyledons at parturition. There is a similar material in males which is turned into the genital semen by concoction. The menses, of which I shall soon speak, flow for the most part each month at the moon's phase; by means of it virgins change voice as males do at the first coitus; then the breasts swell. In man the umbilicus is larger than it is in other animals while the child is being born. At this time it does not furnish nourishment; hence the veins and arteries around which there is a tunic compress themselves and thus birth takes place at the proper time, for the umbilicus must be tied off since the offspring is born rather imperfect. But brutes are born more robust; they leap forth immediately. In conception the entrance of the uterus is opened and then closed. It does not open further unless the woman superfetates or is troubled with discharges or if the menses flow. This mouth is closed so tightly that the air cannot enter. The position of the uterus is between the lowest part of the spine which we call the os sacrum, above the bladder opposite the middle of the bowel, turned a little to the right coxa. Later proceeding to the ilia above the rectum intestine its sides connect with the thighs. The cervix is continuous toward the bladder; we call this the canal; it is placed upon the neck of the bladder and is fleshy and quite cartilaginous. The cervix is contracted in the pregnant woman. The vulva in a virgin is small. In a woman, unless she is pregnant, it is not much larger than that which can be grasped in the hand. It is joined to the heart by arteries, to the brain and spinal medulla by nerves and likewise to the diaphragm, breasts, kidneys, stomach, ilia, coxae, and pubis; all of these connections are perceptible in strangulations of the vulva or suffocations of the womb by which the breath is withdrawn, dizzy spells ensue, and women think the hair on the top of their heads is being pulled. They also feel pain in the loins, femora, groin, and stomach. The throat also swells greatly. These connections are more apparent in the living body, but collapse in the dead person. This condition others attribute to remaining vapors which the wounded vulva immediately betrays. In this state the eyes give great pain, women grow mute, some lose their minds. The uterus is certainly joined to the ilia with certain small nerves twisted back which are extended in pregnant women. No vein stretches directly to the vulva from the major [chilis] vein, but many veins come from the aorta. Its material is sinewy, wrinkled, pellicular, as thick as the thumb; at conception it can stretch to a great size. Those who investigate the cotyledons of the vulva at this time to the great desire of the crowd are deservedly frustrated, for very little of them can be seen, nor at the beginning of conception, since the involucra which they call secundinae cling more strongly to the vulva. At that time the membranes are joined more firmly through the roots of the umbilicus to the cotyledons, as the Greeks call them, such as we see in the arms of polyps which hold on to anything they grasp by suction. These membranes are conspicuous while the foetus is growing; after the birth they diminish and are finally abolished. They are joined to the convex part of the uterus and there, according to Aristotle,[66] are converted into the foetus. In this covering something fleshy from the secundinae, full of fibers, is affixed to the smallest cotyledons which suck out blood and spirit just as the fibers of roots do and carry it through the veins. The blood is brought to the umbilicus and the spirit through which the foetus draws nourishment. There are two sinuses of the matrix (or womb), gently divided from each side (in the right side they say males are conceived, females

[65] Pliny, *Natural History,* 7. 11. 9. 48: few animals except women ever have sexual intercourse when pregnant; 8. 55. 81. 219: the hare practices superfoetation; 10. 63. 83. 179: rabbits superfoetate. I have not been able to find the source of the idea of three entrances in deer; it does not appear in Aristotle or Pliny.

[66] Aristotle, *De Generatione Animalium,* II, 7, 745 b 20–746 a 10.

on the left side),[67] corresponding to the twin testicles in the male. Like the horns on a cow these are quite far apart.

Chapter XXV

On the Menstrual Flow

A few words remain to be said about the menstrual flow of women and its origin. Aristotle[68] calls it nothing but semen that has not been concocted. From the regions above and below the kidneys and from the loins the menses come forth in very small veins to the cotyledons of the vulva by which the foetus is nourished in the uterus (as has often been said) or at childbirth near the mouth of the vagina from two veins following the pubic bone near the place where the urine issues. From this part the non-prolific semen of women is poured forth in coitus, as I have said, with so great an impetus in many that it spurts out farther like the semen of males, of a different nature at any rate from that prolific semen which overflows in the hollow of the vulva, as I described at greater length above.

Chapter XXVI

On the Anus

The exit of the bowel is called the anus. There are muscles [sphincter] here which contact and relax when necessary. Within the anus are mouths of veins which rise as if from some small heads which often pour forth offensive blood similar to that of nose bleed. The Greeks call them hemorrhoidal veins. Their flow is used instead of purging nor do they become much weaker when so used. Certain patients who had no exit for the blood were seized with sudden and very severe illnesses when that poisonous matter reached their internal organs. In that region which is called the crown (*corona*) and seat knots or tumors and tubercles, alternately hard and soft, arise from an inflammation. If they become old and chronic they are sometimes burned out with caustic medication. Mariscae (hemorrhoids) and rhagades (fistulae) infest the anus in the skin which is cut around it. A lascivious lechery and a burning lust cause these ills, not without injury to nature and the divine majesty. Lust, the inventer of vices, neglecting nature,[69] treads upon both civil and canon law and seeks low retreats; this is the way to ruin the human race. Would that the parents of the human race had known only that love and not these painful results! The anus suffers prolapse as the vulva (uterus) does; we call this the falling forward or a turning around of the seat. We bring it back in place by the use of astringents. Ulcers and fungi and such things are created in the anus. Thus far I have described the internal organs under the diaphragm. Now let us pass to those members known as spiritual.

PREFACE
TO THE THIRD BOOK OF THE ANATOMICE BY ALESSANDRO BENEDETTI, PHYSICIAN

Chapter I

On the Spiritual Members Which Lie Above the Diaphragm

Nature has located the heart and lungs above the diaphragm, where (as Plato says)[70] boldness, dread, flight, and counsellors of madness at one time or another made their abode, with rage that is often implacable. Love, likewise, the fearless invader, has pitched his camp there, with gentle hope his companion. If these have been evil they have accustomed pleasure, the seed bed of evils, and sorrow to be their equals. But if fortitude, on the contrary, the daughter of a good mind and the despiser of all, has scorned the commands of desire and pleasure, when the mind has started back in pale terror at terrifying sights, the blood and spirit of the veins as with a closely packed legion come suddenly running as at a given order, then fortitude, driving back fear, makes use of the power of the mind. In a similar fashion if the force of anger suddenly bursting forth cleverly stirs up flames, reason straightway checks them by frequent cooling of the heart with the breath of spirit from the hollow fistulae of the lungs. Thus anger, otherwise implacable, is easily calmed. In those who are wise power is given to reason against the enemies which are pleasures and other perturbations of the mind. The wise alertly receive with fortitude and patience of body the

---

[67] Aristotle, *De Generatione Animalium,* 763 b 35; Galen, *Hippocr. Aphor.* V, 38 (*Galeni in Eos Commentarius*), Vol. XVII K., 829; *idem, De Semine,* II, 5, Vol. IV K., 633; Aristotle, *De Generatione Animalium,* IV, 1, 763 b 20–765 b, discusses this idea at length.

[68] Aristotle, *De Generatione Animalium,* I, 19, 726 b 30 and the following passages in which Aristotle proves that women do not emit semen as men do; 728 a 25: "For the catamenia are semen not in a pure state but in need of working up . . ." (translated by Arthur Platt, Oxford Translation of Aristotle, 5).

[69] These sentences are borrowed without acknowledgment to Benedetti by Berengario da Carpi, *A Short Introduction to Anatomy,* translated by L. R. Lind (University of Chicago Press, 1959), p. 76. They bear some resemblance to Plato, *Timaeus* 91 C on the effects of sexual desire; compare 86 C, 73 C on the marrow as "seed-stuff."

[70] Plato, *Timaeus* 69 D.

attack of war set raging with leveled spear; hand to hand they meet the heavy blows of wrath and hatred against the secret chambers of the heart and drive them off without harm. All philosophy dwells herein and the entire doctrine of the good life has its abode in this place. To it happiness owes its origin, by it the terrors of death are driven away, and the sublime intellect penetrates to the divine sanctuary. To the dissection of these spiritual parts tonight in this theater I invite my wise senators from the patrician class of Venice, Bernard Bembo, Antonio Boldù, Knights, likewise Antonio Calvo the triumvir, and Pietro Priuli,[71] senator, that they may observe with me the divine workshop of the heart and search out the secrets of nature, now that they have weighed and transacted the business of the republic, and may then go home tired but happy.

Chapter II

On the Spiritual Members

The spiritual members are situated between the head and stomach, for the heart (thanks to which the whole body exists) is placed in their midst since nature is accustomed to establish that which is noble in more noble regions, that is, rather higher than lower and in front rather than behind.[72] The principal member of the viscera is not unreasonably placed rather to the left than to the right, as is evident in all animals. For the heart is the laboratory of the heat, the origin of life and of the mind, the beginning of the blood and the spirit. It is the first of all the parts and appears as a spot of blood about the size of the chick embryo in eggs which have been incubated for three days.[73] When the body is formed the heart is felt to palpitate as if there were another animal within. The upper part of the thorax is called the chest. It is situated between the upper arms in man and forms a broad ribcage as a protection for the heart and the breasts. But I shall name the parts beginning with that member, the transverse septum (diaphragm), by which the less noble parts are separated from those which are more noble. Finally the members contained in the mouth itself (because they are accommodated to it and to the stomach) will be given in order.

[71] Bernardo Bembo (1433–1518 or 1519) was the father of Pietro Bembo, an ambassador, a friend of Ficino, Bracciolini, and others; he also paid for the restoration of Dante's tomb at Ravenna. Antonio Boldù was a president of the Council of Ten at Venice and shares the dedication of Benedetti's *Diaria de Bello Carolino* (1496) with the two other presidents. Antonio Calvo was also a senator of Venice, a friend of Hermolaus Barbarus, and a collector of manuscripts. Pietro Priuli, son of Marco, was a friend of the Naldus Naldius to whom George Merula dedicated his editio princeps of the *Scriptores Rei Rusticae* (Venice, 1472).

[72] This idea comes from both Plato and Aristotle and is sedulously repeated by the sixteenth-century anatomists.

[73] Aristotle, *De Partibus Animalium,* III, 4, 65 a 35; 666 a 10–20.

Chapter III

On the Diaphragm

The diaphragm is located under the lung. Others call it cinctus (girdle) and the ancients called it praecordia since it is stretched in front of the heart; still others call it the diaphragm. Plato calls it phrenes as if it were the participant of reason and joined to the mind;[74] hence some have though that the mind received its subtlety from the diaphragm. Thus there is no flesh in it but a strong sinewy thinness especially in the middle. It is attached to the lower ribs and the spine at the twelfth vertebra. It separates the seat of the heart from the stomach and the intestines so that the origin of life may be preserved without offense from the vicinity of the bowels and the excrement and so that the mind should not be weighed down with the vapor of food. For this reason the intervention of the diaphragm cuts off as if with a wall the nobler from the less noble part of the body. If it were more full of flesh it would hold and attract a greater amount of vapor. The diaphragm is the source of the sensation of tickling, which man alone feels among the animals; tickling takes place because of the thinness of skin in the diaphragm. Aristotle[75] bears witness to this fact: laughter is due to the midriff (diaphragm), which enters the armpits with some nerves. When the diaphragm is gently warmed it moves the person to involuntary laughter; for this reason gladiators have been seen to die in the arena with a laugh when they were transfixed through the diaphragm with a weapon.[76] The diaphragm is set apart from the liver with certain small membranes downward in the direction of the spleen, but upward toward the middle it also sets apart and sustains the lungs. It also assists in breathing and has the power of relieving the stomach of its contents; this power is called apocritice (excrementory). When the diaphragm is injured

[74] Plato, *Timaeus* 70 E.

[75] Aristotle, *De Partibus Animalium,* III, 10, 673 a 10: "It is said also that when men in battle are wounded anywhere near the midriff, they are seen to laugh, owing to the heat produced by the wound: this may possibly be the case . . . for no animal but man ever laughs" (translated by William Ogle in the Oxford Translation of Aristotle, **5**).

[76] The reference given by Ogle (see above) in his note to *Dictionnaire des Sciences Médicales,* ix. 214 is wrong; xxix. 66–122 deals with the diaphragm. See Pliny, *Natural History,* 2. 77. 198, p. 557 (Loeb Classical Library, translated by H. Rackham): "To this membrane unquestionably is due the subtilty of the intellect; it consequently has no flesh, but is of a spare sinewy substance. In it also is the chief seat of merriment, a fact that is gathered chiefly from tickling the arm-pits to which it rises, as nowhere else is the human skin thinner, and consequently the pleasure of scratching is closest there. On this account there have been cases in battle and in gladiatorial shows of death caused by piercing the diaphragm that has been accompanied by laughter." Pliny here adds his first-hand knowledge of gladiatorial combats at Rome to Aristotle's statement and Benedetti has combined information from both Aristotle and Pliny.

the bowel stands firm, but the mind is at once completely shaken by inflammation, as if the brain were affected. Hence arises the word phrenetic, which is applied to such patients.

Chapter IIII

On the Breasts

The middle of the breast must be opened, in order to continue the order of dissection, from the breast-bone to the collarbone. The skin must be cut with a knife downward to the sides because of the breasts, for these parts in a woman must be handled in their own place. The breasts project on either side of the chest. In their middle is the nipple adapted for feeding babies. Its texture is looser, containing glands and carunculae of a fungous nature and perforated; this is more clearly seen in a carefully preserved cadaver. In males the flesh is thicker; in them something similar to milk has been found while I was writing this book. M. Maripetrus, of the Sacred Order of Knights, relates that a certain Syrian, whose wife had died leaving an infant son, often presented his breasts to the child to calm him as he cried in hunger and after frequent sucking milk trickled out of the nipple by which the baby was fed, to the great wonderment of the entire city. Aristotle,[77] likewise, in discussing milk in males says there is a tradition to the effect that through much and frequent sucking milk has come forth from some men. I also saw a woman who, while a virgin and hence still menstruating, always had milk in her breasts. The same author (Aristotle) says that if hairs are found by chance on the breasts of women and one is placed in a cup and drunk this is traditionally called "depilating the illness," nor will the pain be soothed until the hair is pulled forth or is sucked out with the milk. Others deny this since hairs seem to grow at random on the body, which the insane credulity of the mob believes is caused by bad medicines; for, as Galen says,[78] when abscesses grow chronic on account of various qualities of the humors, it is proven that their material is converted into stone, sand, wood, coals, dregs (of oil), feces, pig bristles, and into other forms of this kind. Another ill peculiar to the breasts is called colostratio; aloe in rose oil frequently smeared on inhibits this ill for a month before parturition. Those veins which ascend from the uterus (as Galen writes) run throughout the uneven texture of the breasts so that by its slower movement the blood may be more suitably prepared; it grows white at last through the power of the glands.

Chapter V

On the Chest Muscles

There are many muscles around the breasts which cannot be seen. By some of them the chest is lifted on intake of breath; by some the chest is depressed and expanded again around the diaphragm. Muscles are more necessary in the upper part of the chest near the neck, for when they are injured the chest can scarcely be raised. Others deny this because hindrance does not prove power exists in a member; the muscles can hinder, but by themselves they cannot institute motion since that is the function either of the lung or of the diaphragm, according to Averroes.[79] Other muscles begin from the back, still other very small ones are scattered here and there, some between single ribs which now raise, now depress the chest. They contract the chest with stamens woven into their length and breadth. Around these parts arises the very acute disease called pleurisy, from the word for side in Greek (*pleura*). The front part of the chest, which is joined with much cartilage to the ribs, is located immediately under the throat. There are seven ribs joined to each other. They are rather long and bent in crescent shape on each side. The five other ribs are shorter; they are called false ribs and rhoai. Highest of all is the collarbone, hard, strong, and curved backward. With its curved heads it makes a gap above which is the actual collarbone; these heads terminate near the shoulders, are moved a bit with the movement of the arms, and are connected with the broad bone of the scapular (shoulder blade) below the skull with nerves and ligaments. The ribs under the armpits are twelve; they stretch to the lowest region of the diaphragm and ilia. With their first parts they are round and easily joined to the transverse processes of the vertebrae and little by little inserted there. They are visible on the exterior; gradually curved back, they disappear in cartilage. That part of the ribs toward the exterior is gently curved and in their lower part they are joined to the chest which is nothing but cartilage, at the termination of the breast bone. Thus the first parts of the thorax, that is, the chest, sides, and breast bone, contain the spiritual members. This part is not made of one bone so that man may bend to necessary work by means of its flexibility and breathe more easily. When this region has been examined let the dissectors cut the middle of the chest as far as the collarbone.

---

[77] Aristotle, *De Historia Animalium,* I, 12, 493 a 15; III, 20, 522 a 20.

[78] Galen, *De Tumoribus,* chap. 4 (Vol. VII K., 718): Etenim coeno, urinae, grumo, melleo mucosque succo, ossibus, lapidibus, callis obduratis, unguibus, et pilis similia corpora in abscessibus reperiuntur; quin et multoties inventa sunt animalcula, prope omnibus, quae ex putredine generationem habent, simillima. Another list of materials found in abscesses which is closer to that given by Benedetti is Galen, *Methodus Medendi,* II, 9 (Vol. XI K., 116): Etenim lapidibus, arenis, testis, lignis, carbonibus, luto, strigmento, amurcae, feci, multisque id genus aliis similia corpora saepe in abscessibus contenta deprehenduntur.

[79] Averroes, *Colliget,* II (Venice, Giunta, 1574), folios 26 v–30 r discusses voluntary motion in breathing.

When the ribcage is opened the lungs become visible. But first the pleuritic membrane, which receives its name from the sides rather than from the chest, must be discussed.

Chapter VI

On the Pleuritic Membrane

Under the ribs there is a stronger membrane similar to the peritoneum which is called pleuritic. The single ribs are covered with its small folds. To this another membrane [mediastinum] is connected, which divides the lungs through the middle and is joined to the chest and spine so that if one lung is injured the other may be preserved unharmed. The illness called vomica (encysted tumor or abscess), rarely cured by physicians, clearly fills the hollow of this membrane. But we must turn to the lungs, which form a cushion to the heart[80] and with their cool air temper the heat of the heart.

Chapter VII

On the Lungs

The lungs are divided into two lobes by an artery [pulmonary] that separates into two parts and by a medial membrane called the disseptum or praecordia [mediastinum]. The function of breathing is given to the lungs, continuously drawing in and expelling the breath. Otherwise man is immediately destroyed since breathing results from a voluntary and natural power through the function of the muscles of the heart and lung. The lung changes the breath, as the liver changes the chyle, into food for the vital spirit; the rest of the smoky or noxious air, especially the vapors of fevers, is expelled. The lung contains the heart within its sinus or fold and cherishes it with continual rekindling. It is spongy, thin, hollow with empty fistulae in addition to veins and arteries. It holds the spirit and contains much blood. Some people believe it is bloodless in the cadaver, deceived by the fact that the blood immediately flows out of the lung and it becomes greatly changed, bearing little resemblance to the inflated lung in a living person. It hangs from the neck with its dividing membrane through the middle. A very thin membrane covers the lung, very much perforated in cats and domestic animals. The lung begins at the last vertebrae of the neck. It has the shape of an oven or, as they say, a cow's hoof. It ends at the thirteenth spinal vertebra, reaching out further toward the diaphragm, in such a way that it fills the entire cavity of the chest.

Chapter VIII

On the Trachea Artery

Two little canals come together from the throat. One is shorter and exterior; it is called trachea, or rough, artery or pharynx. The Latins call it guttur or spiritual pipe. The second canal is longer and on the interior. The Greeks call it esophagus and stomach, the Latins gullet and cibal fistula (pertaining to food), of which I shall speak shortly. The first canal brings air only from the throat to the lungs; in addition, if anything else falls unexpectedly into it this is at once expelled from the canal. The second canal sends food and drink to the stomach, although Plato,[81] who is in other respects the first of philosophers, thinks that drink flows to the lung and then to the bladder, an opinion for which he is justly taken to task by the physician Erasistratus. I shall subjoin the words of Plato: "They have applied or extended the covering of the lungs to the heart, soft at first and bloodless, then distinguished by hollow fistulae within like those of a sponge so that with the breath and the drink which the lung takes up it may cool the heat of the heart by a respiration of this sort. The drink or liquid runs elsewhere, flowing from where the lungs are under the kidneys to the bladder." These are the words of Plato, who has undoubtedly fallen into error for lack of anatomical investigations. Through the other canal, that is, the gullet, both food and drink slip into the stomach which is called the catocoelia. Through the trachea artery only the breath goes back and forth from mouth and nostrils. This artery is hard and rough with denticulations, clinging to the gullet; with its other parts it stretches to the lung. It is made of flesh and cartilaginous semicircular rings like the letter *C* only in that part where it is joined to the gullet. In the remaining part of the artery the rings are complete. This has been purposefully contrived by nature to keep any foods which are thicker than they should be from slipping downward. This part is distinguished by slender little veins; hence there is little blood in it. Above, it is joined to the throat and supported by the foramina of the nostrils. They call it the bronchus; it stretches to the mouth. Therefore, when drink has gone astray into the pathway of the lungs it flows out of the nostrils. The artery is cartilaginous and the cause of the voice, as Aristotle[82] reports. For that which sounds ought to be light and solid. It is placed in front of the gullet since the heart is located in the front region.

[80] See Mundinus, *Anathomia,* Charles Singer's translation, p. 106, note 92, for a discussion of the cushion of the lungs. Singer thinks the thick posterior border of the lower lobe of the lung is meant. See the next note, where Plato speaks of the lung as a cushion of the heart.

[81] Plato, *Timaeus* 70 C, D. The last sentence of Benedetti's quotation is not in Plato. Despite his consistent adherence to him, Benedetti recognizes Plato's shortcomings as an anatomist.

[82] Aristotle, *De Generatione Animalium,* V, 7, 788 a 25.

Chapter IX

On the Veins and Arteries in the Lung

From the single artery many arteries are divided on both sides of the lung and soon becoming smaller and smaller within its paths these are scattered to the smallest lobes, filling the entire member. When these arteries are inflated the lungs are lifted up. The right side of the lung is divided into three smaller lobes; the left side is separted into two lobes only. On the right side, which is larger, the lung is attached to the major vein (*vena cava, chilis*). From this vein almost infinite branches are scattered to all parts of the member. The left side, which is the smaller, is joined to the aorta artery and its branches similarly scattered through the lung. Thus because the branches or rivulets of blood, spirit, and air are not at all continuous or run into each other while the final diminutions escape observation on account of their smallness, if any of these rivulets are crushed when the lung is too greatly distended, blood bursts forth from the mouth, or pus, foul blood, or gore, or anything else which flows into the chest penetrates the membrane by which, through the marvelous work of nature, the lung is covered. It enters the smallest fistulae and proceeds thence to the larger ones as far as the trachea from which purulent, bloody excretions are emitted; this happens in cases of pain in the sides. From the lung arises consumption and peripneumonia and asthma. But in this part of the body one may also contemplate the majesty of nature or of God the Creator, Who has daily shown us miracles in the human body. For in the illness called pleurisy a bloody pus sometimes is accustomed to be passed through the urinary duct, as happened to Helius Bassus, the son of my friend Cymbriacus,[83] in the greatest crisis of this disease. Sometimes the pus is sent through the bowel as I have seen. This is an incident even more marvelous: once the blade of a scalpel was broken while a careless bloodletter was cutting some one's vein. The blade, or point, went through various passages and stopped behind the ear, where it was observed by its almost constant pricking; this person often begged me to relieve him of this annoyance. Unbelievable also is something I saw while I was in Crete. A certain country man had been wounded during a riot by an arrow which was impacted in his back. When the arrow was extracted the iron arrowhead was left in the thorax. The wound was healed prematurely while the physician probed in vain for the arrowhead. After two years the latter came forth from the lower regions (bowels) and miraculously exhibited the shape of a bearded figure two fingers broad. The weapon had penetrated point forward along the diaphragm where it is more fleshy, then descended along the intestine and gradually near the bowel cut the anus and made a way for itself to the exit, half-corroded. There was a great discussion among the physicians about this unusual occurrence. But I think I should speak about the heart, for whose sake the lung was created.

[83] Helius Bassus was the son of Quinzio Emiliano Cimbriaco, who wrote a dedicatory poem to Benedetti's *Diaria de bello Carolino,* etc. (1496) and is prominent elsewhere in the dedicatory material of Benedetti's writings.

Chapter X

On the Pre-eminence of the Heart

The heart is the source and beginning of the blood, which, according to Averroes,[84] is of a hot and humid nature, or of a temperate nature, if we believe Aristotle.[85] In man it faces toward the front with its cone-shaped point below the left nipple; its highest part faces to the right. In the cadaver it is very often moved out of position. I have seen a certain person who was killed by a dagger thrust; when his chest was dissected and the dagger shoved into the wound the point neither of the heart nor of the lung corresponded or was lined up with the wound, so great is the difference in position between cadavers and living persons in regard to their organs. In the heart is the beginning of the vital heat and nourishment and the seat of anger. The heart is covered with a strong membrane (or capsule) like a sheath [pericardium]. This is fat bodied and sinewy, full of water so that the heart may be continually moist and so this covering may not be a hindrance to the heart's motion, allowing it to beat freely and more easily. The lung surrounds the heart as a cushion for it,[86] so great is the reverence shown to this principal organ of the viscera. The heart alone among all the members is not wasted away by ills nor does it sustain the more serious punishments of life.[87] As Aristotle[88] says, no bile can approach the heart since it suffers no violent

[84] Exactly where Averroes may say this is difficult to point out since there is no index to his *Colliget*. The statement is one which accords, however, with that given by Aristotle in the next note. Why Benedetti should attempt to make a distinction between the two is puzzling.

[85] Aristotle, *De Historia Animalium,* III, 19, 520 b: "Blood . . . is neither very thick nor very thin. In the living animal it is always liquid and warm. . . ." See the entire discussion of blood in chap. 19 and *De Generatione Animalium,* IV, i, 765 b: "It is true that blood is hot" (translation by D'Arcy Wentworth Thompson, Oxford Translation of Aristotle, **4**). Aristotle does not seem to bear out Benedetti's statement as to the temperate nature of blood.

[86] See note 80.

[87] This sentence is also used by Berengario da Carpi, *A Short Introduction to Anatomy,* translated by L. R. Lind (University of Chicago Press, 1959), p. 93. It is based upon Aristotle, *De Partibus Animalium,* III, 4, 667 a 30: "The heart again is the only one of the viscera, and indeed the only part of the body, that is unable to tolerate any serious affection" (translation by W. Ogle, Oxford Translation of Aristotle, **5**).

[88] Aristotle, *De Partibus Animalium,* III, 4, 667 b.

disposition for in its entrance no evil disposition has ever been seen. It dies at once when injured. When the other parts of the body are corrupted the vitality whose seat is in the heart still endures. When the water of its capsule is exhausted by some ill the heart beats more than usual. This is called heart disease (*cardiacus morbus*) and it brings a man to a decline, as Galen saw when he dissected his ape.[89] The heart itself is fleshy and muscular in nature, fat at the top, without bone, although that is found in some animals such as in the hearts of deer.[90] From the heart as if from a parent trunk all the major veins are seen to arise because in it there are no veins. The rest of the members have no blood except in veins like receptacles. The blood comes from the heart to the veins and thence to the members. In the lobes of the liver there appears a rather watery, bloody substance which does not have the nature of blood until it has received this in the heart. Arteries do not reach the liver since they are peculiar to the heart.

Chapter XI

In Whom the Hearts Are Large

The vessels of the heart, about which I shall soon speak, seem to have a certain articulation similar to the crevices and sutures of the head, although this is not true because the heart is made up of many parts but reveals this articulation only in some details. Some people think that those animals in which the heart is more articulated are more capable of sensation; those in which there is less articulation are less able to use their senses. The soft vessels are full of sensation, such as the flesh has; on the contrary, the hard and thick vessels imply less sensation. Larger vessels indicate timorous animals; smaller or middle-sized vessels indicate the opposite.[91] In the larger vessels there is a small portion of heat which disappears in the large space and makes the blood colder. The same thing happens in the ventricles of the heart and in the veins, for the same amount of fire does not warm equally in a small as well as a large dwelling place. There is less heat in a large space; hence in these vessels the heat is not equally distributed.

Chapter XII

Why the Right Hand is More Adaptable Than the Left

The fibers (or lobes) of the heart in its ampler part slant to the right because they pour a greater heat into it. Hence the great share of men use their right hands more capably. If the larger part of the heart slanted to the left, the opposite would be true about men's hands, and if the larger part were in the middle of the chest, equally balanced, this would make both hands equally adaptable for use. Aristotle[92] calls such people equal-handed, ambidextrous. The Latins call them scaevae, as Ulpian[93] indicates.

Chapter XIII

On the Ventricles of the Heart

According to Celsus,[94] the heart has a double sinus, both right and left; he is followed in this matter by many of our physicians. But Aristotle [who says there are three sinuses] has had so much authority for many centuries that even those things which the physicians have not seen they will affirm to exist, even without experiment.[95] And if they judge they have seen only two sinuses, nevertheless they acknowledge that there are three in the heart. This matter is of great concern in all of medicine since these physicians according to Aristotle's view are greatly in error as to the number of ventricles, of passages or ducts, and of locations. Among these men Galen,[96] in other respects the most learned of

---

[89] Galen, *De Locis Affectis,* V, 2 (Vol. VIII K., 303) for this much-quoted account of the postmortem examination by Galen of an ape which died from loss of water in the heart's capsule.

[90] Aristotle, *De Partibus Animalium,* III, 4, 666 b 20, where only the horse and a certain kind of ox are mentioned as having a bone in their hearts.

[91] This view as to the relation between the relative size of an animal's heart and its courage comes from Aristotle, *De Partibus Animalium,* III, 4, 667 a 20 and is repeated by anatomists after him, e. g. Berengario da Carpi, *op cit.,* p. 93. An opposing view, however, is taken by Pseudo-Galen, *De Anatomia Vivorum,* fol. 53 (see also fol. 46 v): Non enim magnitudo cordis causa est audaciae neque paruitas est causa timiditatis, quoniam lepus et ceruus magnum cor habent secundum mensuralem sibi proportionem cum timidi sunt; quaedam similiter animalia paruum habent cor, sed calidum, et propter hoc audacia sunt, sicut serpens, thirus, mustela etc.

[92] Aristotle, *De Historia Animalium,* II, 1, 497 b 33; *Ethica Nicomachea,* V, 10, 1134 b 34; *Magna Moralia,* I 34, 1194 b 30; *Politica,* II, 12, 1274 b 13.

[93] Ulpian, *Digesta* 21. 1. 12: Item sciendum est scaevam non esse morbosum vel vitiosum, praeterquam si inbecillitate dextrae validius sinistra utitur: sed hunc non scaevam, sed mancum esse (ed. Theodorus Mommsen [Berlin, Weidmann, 1905], p. 271).

[94] Celsus, *De Medicina,* IV, 1. 4: Huic [pulmoni cor] adnexum est, natura musculosum, in pectore sub sinistriore mamma situm; duosque quasi ventriculos habet.

[95] This salutary observation would carry more weight if Benedetti himself had followed its implications more consistently. There is a curious indecision and timidity in even the boldest pre-Vesalian anatomists when they are faced with the necessity for breaking with tradition and contradicting the medical authorities of the past. Aristotle, of course, held that the heart had three ventricles, a conviction which was regarded as the truth until the time of Massa (1534): see note 70 to my translation of Massa's *Introductory Book of Anatomy.*

[96] Galen, *De Anatomicis Administrationibus,* VII, 10 (Vol. II K., 621); the later books translated by W. L. H. Duckworth, *et alii* (Cambridge University Press, 1962), XIII, 9, p. 173; and Galen, *De Usu Partium,* VI, 9 (Vol. III K., 422). But in the first of these passages Galen had written: ". . . he [Aristotle] deserved to be forgiven," because he lacked anatomical experience.

men, attacks Aristotle with too little justice, a man most praised for so many centuries. For that author[97] [Aristotle] speaks thus concerning the cavities of the heart: "The heart has a triple sinus, quite large on the right side, smallest on the left side, in the middle moderate in size and location between the right and left sides. All (the sinuses) are pervious to the lung." In single hollows or cavities the heart contains blood without veins and these hollows are divided by septa (membranes). Of the hollows the right one contains most of the hot blood because this right part of the body is warmer. The left sinus has little blood but most of the cold spirit, for it is certain that air passes to the heart, attracted from the trachea artery. The middle sinus, however, middling as to both blood supply and heat but very thin and temperate (as he says), provides the spirit. There is no doubt that the aorta artery originates from this sinus. For it is proven by argument that this sinus is the beginning of the heart since that which is the beginning of anything must be the most quiet; such it will be if its blood is pure and it is mediocre both in suppply of blood and of heat. The position of this sinus is perceived to be in the middle in the living person. The left venter is more swollen, not flaccid of air as it is seen in the cadaver, for in the dead body everything loses its firmness and is displaced. The venters or sinuses are tighter, more compressed, in the smallest animals so that one which is larger can scarcely be seen and the rest (the smaller ones) escape notice entirely. The second sinus is conspicuous in middle-sized animals, but in the largest animals all the sinuses are visible. But the sinus containing the colder spirit which is connected with the lung, the receptacle of the air which Pollux[98] names the left ear (auricle), people do not enumerate among the cavities of the heart, as Celsus[99] and other later people, among them our Galen who rails most constantly against Aristotle and more severely than is right. Yet Aristotle, not ignorant of nature, saw only two sinuses in the point of the heart but contemplated the nature of the heart more loftily and wished to make no mention of the heart's auricles. Those who have placed auricles in the heart have made their conclusions only from the posture of the cadaver. But Aristotle, more cautious and clever, paid attention to the function of nature in living bodies, something which can easily be observed in living brutes and larger animals. The philosopher (Aristotle), however, stated that the sinuses are triple: the right one of the collection is the source of blood and the largest of all; the left one is the smallest and contains the air drawn from the lung (called the left auricle by the inexperienced, as I have said); the middle sinus is middling both in location, size, and temperament; it is the receptacle of the blood whence it is certain the aorta vein draws its origin. All the sinuses have passages leading to the lungs since the middle sinus has access to the left lung by means of the venal artery. These passages are conspicuous in the larger animals. But since two auricles are visible in dead bodies, why should we not believe that thus there are four auricles? One can readily meet this objection, for there cannot be a sinus beside the right sinus since it is part of the major (*vena cava, chilis*) vein which issues from the heart as from a trunk or stock and contains the same material of the major vein although not constantly since it is created only to contain a superabundant supply of blood. Otherwise it settles back in living bodies and has no septum which divides it from the larger cavity, where the material of the blood is proven to be different, all of which can be seen to exist in the left follicle (bag) or cavity. The right ear of the major vein is rather a wide space, created not without reason by nature. But those who use the term auricles for sinus are ignorant of their order, number, and purpose; hence Galen, in other respects the greatest of physicians, and Avicenna, his counterpart among the Arabs, have been forced into a shameless controversy. I subjoin the words of Aristotle;[100] he makes no mention of auricles: "The major vein arises," he says, "from the broadest and highest sinus; then, since this ventricle of the heart is a part of the vein in which the blood overflows, and from the same, the major vein drives its root into the heart when it thrusts itself forward." This is what can be said briefly as far as the condition of the site is concerned about the number, nature, and position of the ventricles. The veins and arteries are carried around the heart from the right and from the middle cavity; they stretch with their ducts to each ventricle. Hence come the spirit and the blood in their mutual turns; from them it is not possible for blood to be created without spirit nor spirit without blood.

Chapter XIIII

The Contents of the Right Ventricle and On Its Veins

On the right side of the heart which is the dwelling place of the blood there is a wide part similar to an ear which is located there for the purpose of receiving at times the superfluous blood as was mentioned above. When not in use it lies flaccid and pressed

[97] Aristotle, *De Historia Animalium*, III, 513 a 30.
[98] Pollux, *Onomasticon*, 2. 219.
[99] Celsus, *De Medicina*, IV, 4, p. 357, Loeb Classical Library: duosque quasi ventriculos habet. See note 94.

[100] Aristotle, *De Historia Animalium*, III, 3, 513 b: "The great blood-vessel, then, is attached to the biggest of the three chambers, the one that lies uppermost and on the right side; it then extends right through the chamber, coming out as blood-vessel again, just as though the cavity of the heart were a part of the vessel, in which the blood broadens its channel as a river that widens out in a lake" (translated by D'Arcy Wentworth Thompson, Oxford Translation of Aristotle, **4**).

down and hence is not without reason considered a neighbor which receives the over-abundant blood when there is need. In the right ventricle (for it must be dissected from the heart's point) two veins appear inside. One of these, the larger, stretches to the liver; this is called the big vein [inferior vena cava]. The heart receives the imperfected blood by means of this vein with its large duct; the word receive is believed to be a better term than contribute since in its orifice three valves [tricuspid] are purposefully placed by nature like movable gates which by turns when the heart is contracted in emitting blood do not completely shut off its passage, for these valves close inward. But when the heart swells (in diastole) the blood is attracted upward. This large sinus contains rather fleshy little nerves that run among each other so that the middle sinus may not have the firm structure of its sides torn away by too much motion. The same large vein is ampler near the heart as a tree is near its root and is mingled on the inside with the material of the heart. Near the gate (or entrance) a certain branch from this vein [coronary sinus] is carried around to the fibers (or lobes) of the heart. From this branch other more slender branches are scattered into the substance of the heart. The other orifice of the arterial vein sends a branch to the lung [pulmonary artery] so that the lungs may abound with a great deal of blood. This vein is composed of a double membrane (whence its name) lest it be easily split by the constant motion of the lung. If it is ever injured excretions of blood come out of it. Before the ostium or opening of this vein is a triple valve [semilunar] that closes easily; it opens outward and closes inward almost completely, by turns. It is created to pour the blood alternately to the lungs while the heart reciprocates with its open valves; when these are closed and the blood taken away the heart receives nothing from it.

Chapter XV

On the Middle Domicile or Dwelling Place of the Heart

In the middle dwelling place [left ventricle] of the heart they say the vital spirit has its abode; hence this, according to right reason, is more firm and strong, for, on the same account, all the arteries are double around the heart's fibers. From this cavity two mouths of the veins are visible. On of these, which is large [ascending aorta], is joined to the aorta; the other, more slender, is sent to the lungs through the left sinus. The larger one scatters the vital spirit to all the members when the heart is contracted, even to the finger tips. In the orifice of this vein there are three valves [semilunar] of the same construction and purpose; they are stronger, with small nerves within and membranes, which shut off almost everything so that the vein may be capable of returning spirit but not vice versa since the valves are closed on the outside. Near the root there are a small vein and an artery which carry blood and spirit to and from the right sinus to the middle sinus, as is apparent in dissections, according to what was said a short while ago.

Chapter XVI

On the Left Ventricle

The left ventricle, which the inexperienced call the left auricle [atrium] and the Greeks call ota, is wrinkled, cavernous, and has less fleshy material than the other sinuses. To it an artery [pulmonary veins] sends air. It is attached to the lung. From this ventricle the venal artery stretches to the lung since it has a thinner tunic than a vein. By this means the heart's heat is cooled while we breathe and the superfluous smoky air borne from the heart is transmitted from this region. Between the confines of this sinus and the middle one two large valves, the strongest of all of them, attached to the walls of the heart with numerous hard little sinews shut up the entire orifice, for it is moved inward. By means of this bivalve [mitral] the middle cavity is divided from the left hand upper one and sends two veins to the two lungs; these veins descend to the smallest lobes. In the veins the air after it is prepared is carried to the left cavity of the heart, for the air growing white within is perceived by its mixture of spirituous blood. For this reason the valves face inward so that they can better receive than send back, for in the body coolness itself is considered necessary for tempering its natural heat. When the lungs and heart are removed at the seventh vertebra there appears a caruncula similar to a gland; it is called the thymus and holds evil and superfluous humor.

Chapter XVII

On the Gullet

Now the gullet must be described. It holds a posterior position behind the trachea artery. It consists of sinew and flesh so that it can expand when food is carried through it and be flexible to avoid injury from the roughness of food as it descends. From the throat it hangs down with a continuous connection of both spine and artery and passing through the diaphragm comes to an end in the stomach; it has the appearance of flesh. It is named for its narrowness and length, very similar to a straight trumpet (*tuba*, tube) or as to width and length, as Pliny[101] reports, poured out like a bottle. In

[101] Pliny, *Natural History*, 11. 68. 179, p. 545, Loeb Classical Library: "This name denotes the cavity attached to the spine below the fleshy part of the windpipe, bulging out lengthwise and breadthwise like a flagon" (translated by H. Rackham).

the throat the gullet dilates to its greatest extent. The part of it which reaches the stomach is very sensitive in the nature of nerves, by which retchings are induced. That part which is attracted to the bronchus is more fleshy; some call it the minister of the food, others the dispenser. For this reason Philoxenus Eryxius[102] prayed to the gods that he might have the neck of a crane. There are also other Philoxeni who cherish the gullet beyond measure in their lives and attribute the greatest good to this part of their bodies, the followers of the prodigal Apicius[103] who flung away the most ample possessions and huge patrimonies into the abyss of their gluttony. Hence even now there have arisen rigid laws among the Venetians forbidding solemn banquets and the flesh of wild game except that of the flamingo. The gullet consists of a twofold membrane with fibers of various kinds. One membrane with stamens proceeding crosswise expels the food and yields without injury when the food is attracted. The other inner membrane draws the food downward with stamens running lengthwise; this power is called cathelctice and is peculiar to the throat, through whose narrow passage we stuff the thick, congested material when we eat. If this power fails with acute disease, death threatens the patient. It is sometimes so stricken with angina that it can swallow neither food nor drink or can send down only food or only drink, for in eating the throat draws that part of it called gurgulio (windpipe) up toward the gullet. In some places in Italy peoples' throats swell so horribly due to defect in the waters they drink that they are regarded as monstrosities;[104] but on the other hand newcomers to these places contract this disease in a short time. The Greeks call it bronchocoele [goiter, enlargement of the thyroid gland]. Dissectors tear away the skin from the neck so that the longer muscles may become visible. Rather large veins adhere to these muscles whose power must be considered; when the chin has been smashed the jaw bones must be dislocated so that what is contained in the mouth and throat may be seen.

Chapter XVIII

On the Epiglottis and Artery

The epiglottis is adjusted to the throat in its upper part as a sort of cover attached to the root of the tongue to keep any food or drink from descending

---

[102] Aristotle, *Problemata*, XXVIII, 7, 950 a: ". . . Philoxenus longed for the throat of a crane" (translated by E. S. Forster, Oxford Translation of Aristotle, 7). See *Ethica Eudemia*, 1231 a 17; *Ethica Nicomachea*, III, 1118 a 32; Aelian, *Varia Historia*, X, 9, for accounts of Philoxenus Eryxius.

[103] M. Gavius Apicius, who lived in the time of Tiberius, is called by Pliny, *Natural History*, 10. 68. 133: nepotum omnium altissimus gurges, which H. Rackham translates "the most gluttonous gorger of all spendthrifts." See Tacitus, *Annals*, IV. 1; Seneca, *Ad Helviam*, 10. 8, and the scholiast to Juvenal, *Satires*, IV. 23 for further references. To Apicius is attributed the most famous of ancient cookbooks, *De Re Coquinaria*, a compilation of the fourth century A.D.

[104] Pliny, *Natural History*, 11. 68, pp. 544–545, Loeb Classical Library, seems to be the first to connect goiter, or at least swollen throat, to some defect in drinking water: Guttur homini tantum et subus intumescit aquarum quae potantur plerumque vitio. H. Rackham translates: "Man and swine alone suffer from swollen throat, usually due to bad drinking water." Soranus, *Gynecology*, translated by Owsei Temkin (Baltimore, Johns Hopkins University Press, 1956), II, 6 [70 b], p. 74, uses the word bronchocele as tumor of the bronchus, as does Pseudo-Galen, *Definitiones* Medicae (Vol. XIX K., 443): Bronchocele tumor est gutturi adnascens differtque ab eo qui in scroto progignitur. Benedetti, *De Re Medica*, VII, p. 128 recurs to the subject of goiter in his discussion of throat diseases (de malis faucium): Guttura quoque si intumescunt, quod aquarum saepe uitio euenit, ἐντερόκηλαι a Graecis dicuntur, uocis ac sermonis usum, ac anhelitum impediunt, quae in Bergomensi ac Nouocomensi agro monstrifica sunt, quibus plerique strangulati sunt; quibusdam impune adeo dependent ut ab humeris gestentur modice infestant nostros castro Iulienses ad Natissam. According to this account the people of Bergamo and Como were especially afflicted with goiter even to the point of strangulation, while those who lived at Cividale on the Natiso River in the region of Friuli, north of Aquileia and below Mount Pedril, had such huge goiters that these were carried slung over their shoulders. My friend, Richard J. Durling, has called my attention, by way of the German translation by August Bürck (Leipzig, 1895), to Marco Polo's *Travels* (translated and edited by Sir Henry Yule, third edition, London, John Murray, 1903, Vol. I, chap. XXXV, p. 187) for Marco Polo's remarks about the people of Yarcan or Karkan: "The inhabitants are also great craftsmen, but a large proportion of them have swoln legs, and great crops at the throat, which arises from some quality in their drinking-water." Yule's note on this passage states that according to Dr. Sven Hedin three-fourths of the population of Yarkand suffered from goiter. Earlier travelers confirm this situation and attribute it to the bad quality of the city's drinking water. Dr. Isidor Greenwald, of the New York University College of Medicine, has very kindly answered my inquiry concerning the passage in Benedetti's *Anatomice* as follows: "I believe that Boussingault was the first to hold that gotrogenic waters lacked something (*Annales de chimie et de physique* (2) 48, 41–69, 1831; 54, 163–177, 1833). The lack was in dissolved air. This is overlooked or disregarded by those writers on goiter who mention, and emphasize, his finding of iodine in certain salts in Colombia. I believe that it was exceedingly improbable that anyone, even as late as 1500, believed that any water lacked something that protected against goiter." It was obviously impossible before the age of sound chemical analysis to determine the relationship between the incidence of goiter and the lack of iodine in drinking water but the fact remains that from Pliny onward scientists did attribute goiter to some defect in drinking water, to some fault or flaw which only modern science could isolate. I should add that the term goiter does not, according to a recent article on the subject (*McGraw-Hill Encyclopedia of Science and Technology* **6** [1960]: p. 229), necessarily mean thyroid disfunction; the ancient term *tumor* covered any swelling. Jacques de Vitry, who died in 1241 A.D., is the first medieval source on the history of goiter. In his *Historia Orientalis et Occidentalis*, written at Damietta, Egypt, in 1220 A.D., he describes goiter in people who lived on the border of Burgundy and Switzerland near the Alps: In quibusdam regionibus et maxime in extremis Burgundiae circa Alpes quaedam sunt mulieres guttur magnum usque ad ventrem protensum tanquam amphoram seu cucurbitam amplam habentes (ed. Jacques Bongars in *Gesta Dei per Francos*, I, 1047–1145 (Douai, Beller, 1597). The passage is discussed by Dr. F. Merke in his forthcoming history and iconography of endemic goiter.

along the wrong path since they would close the air passage, as happened to Anacreon the poet, who was killed by a raisin seed and to Fabius the praetor,[105] who was strangled by a hair swallowed in milk. Likewise, in our own time at Venice while I was writing this book in the year 1483 the same thing happened to Lucretia (Lucrezia), a very close acquaintance. While she was breaking a hazel nut with her teeth she burst into laughter. The nut was immediately swallowed into the air passage and her life was endangered. While in peril of strangulation when the nut became lodged near the inner parts of the lung she began to suffer less; however, her breathing was heavier and more labored since the passage was obstructed. But we began to have some hope for her life when some small veins were ruptured and an abscess resulted in her chest from which she was finally cured. The hoarse sound of the air going in and out of her blocked passages is even now perceptible. This covering (the epiglottis) is wonderfully constructed. Galen[106] calls it the lesser tongue because it is situated close to two pipes so as to open and close them by turns. It is cartilaginous and joined to the root of the tongue so that it cannot be seen. The bronchus [larynx] is situated in the highest part of the throat. It is a pellicular member, muscular, sinewy, and cartilaginous so that it may be more clear-sounding. By means of this member the breath, sound, and the voice itself when uttered can be more perfectly emitted.

Chapter XIX

On the Tonsils and Uvula

At the end of the palate and far within the mouth there are two pieces of flesh which our physicians call tonsils and mala, the Greeks antiades and paristhmia; we call them glands in a pig. They are of almost the same nature as the other glands. When they swell, their illness is called by the same name (tonsillitis). They have a constant and very strong force of saliva which helps to prepare the food and moisten the artery. They are attached to the throat with certain thin little membranes. Under them there is a bone called hyoid which looks like the Greek letter upsilon; to this bone the head of the bronchus is attached. Between the tonsils is the uvula at the end of the palate; it hangs there in man alone of the animals. It is called columella by some. If it is inflamed or swells it is called uva. It is of a fungous material, adapted to receive distillations from the head; it is necessary to the voice. If it is cut off the voice loses its brilliance. Because of this fact it is usually cauterized for it is healed with difficulty if cut off. From this condition arises a fetid smell in the mouth and finally consumption or wasting disease, as Alexander of Aphrodisias reports.[107] This part of the body first tempers the spirit to keep it from descending raw to the lungs.

Chapter XX

On the Palate

The palate is necessary both for the food and for the voice. It is called the sky (*caelum*). In it there exists the consciousness of taste, as also in the tongue. When the food is properly prepared by the teeth it is revolved under the hollow of the palate which has frequent folds or wrinkles which help prepare the food. In fish the palate is so fleshy that it is considered a tongue by many people. A nerve which descends from the brain furnishes the sense of taste.

Chapter XXI

On the Tongue

The tongue is the messenger of the mind and of the will. For this reason Plato[108] asserts that a vein (which we would more correctly call a nerve) passes from the tongue to the seat of the heart. This nerve is the moderator of the articulated voice, since the epiglottis is the instrument of simple sound. The tongue judges taste by means of nerves of sensation from the sixth pair (*syzygia*)[109] descending from the brain, for the power of movement proceeds from the posterior cerebellum. Taste is a sort of touch.[110] The tongue's fissure is not clearly distinguished but is recognized by any misfortune which befalls it because each part of the two divisions of the tongue is

---

[105] For the deaths of Anacreon and Fabius see Pliny, *Natural History*, 7. 7. 44, Loeb Classical Library, p. 535 and Valerius Maximus, *Facta et Dicta Memorabilia*, ed. Carolus Kempf [sec. ed., Leipzig, Teubner, 1888], p. 35). Berengario, *Introduction*, etc., p. 134, also cites the classical instances of Anacreon and Fabius.

[106] Galen, *De Usu Partium*, VIII, 16 (Vol. III K., 587 and *Galeno Ascripta Introductio seu Medicus* (Vol. XIV K., 716) are the two major passages of description concerned but in neither does Galen call the epiglottis a second tongue.

[107] Alexander of Aphrodisias, *op. cit.*, fol. viii v: Cur qui radicitus uuam subsecuerunt tabidi euadunt ? See Aristotle, *De Partibus Animalium*, II, 17, 660 b 35 on the fleshy roof of the mouth in fish.

[108] Plato, *Timaeus* 65 D. Benedetti once more corrects Plato; compare note 81.

[109] Berengario da Carpi, *A Short Introduction to Anatomy*, translated by L. R. Lind (University of Chicago Press, 1959), p. 121, says the nerves of taste descend from the third pair.

[110] Aristotle, *De Partibus Animalium*, II, 17, 660 a; *De Anima*, II, 421 a 15; 422 a 10; 492 b 30 on the size of the tongue (see Benedetti below): "The tongue is sometimes broad, sometimes narrow, and sometimes of medium width; the last kind is the best and the clearest in its discrimination of taste" (translated by D'Arcy Wentworth Thompson in the Oxford Translation of Aristotle, 4). Below, "when it [the tongue] is constricted"; see Aristotle, *De Historia Animalium*, I, 11, 492 b 30.

endangered by the disease of apoplexy and in the uvula grows black with ills; when this happened once I saw the other part of the tongue remain uninjured for days on end. When the tongue is dissected lengthwise many veins and arteries can be seen in it. It is moderate in size, not too broad nor too narrow, and considered at its best when it is most flattened out. When it is constricted it causes stuttering, lisping, and stumbling of speech (*antipos*). Its nature is thin, fungous, and loose. It clings to the bone we called hyoid by which it is sustained as by a prop, for its motion is usually carried out above an immobile object. Under it there is a frenum or a ligament to which silence and loquacity are attributed, whence comes the reproach uttered against those who are unbridled in speech.

Chapter XXII

On the Teeth

There are various teeth in the mouth, first, those in front which are broad and sharp to cut the food, then two in the back call maxillars which grind up the food. Those teeth which separate the food are called canines; they are partly sharp and partly broad or flat, for in the middle they partake of both characteristics. The latest teeth to grow in the mouth are called genuine (*genuini*, or wisdom teeth) which come in around the twentieth year. For many people they appear in the eightieth year, which has happened in my time; in some people they appear in the palate. The son of Sir Antonio Boldù chanced to lose his front teeth in childhood due to rheumatism of the jaw but nature restored them in a short time. Men have more teeth than women have.[111] Of these, the four front teeth are called quaterni (temnici by the Greeks) and gelasini (dimple teeth) [incisors]; these are surrounded on each side by the four canines. Beyond the latter on each side of the mouth both above and below four maxillars [premolar and molar] are placed. Beyond those in turn the last teeth are those which I said grow in late; because of this they are called sophronisteres (wisdom teeth) and cranteres. The front teeth as well as those called canines send down single roots; some maxillars are fixed in the jaw with two, three, or four roots: their roots are longer than the teeth. Straight teeth have straight roots; the root is bent for curved teeth. All the teeth imitate the nature of bone; they develop cavities, however, due to the action of sharp phlegm. They alone have sensation, for they suffer with the cold and pungency makes them numb. The front teeth are established for the guidance of the voice. When Alexander Merula,[112] my teacher, lost these teeth he was obliged to enunciate by means of a golden thread or filament.[113] Those who have fewer teeth and farther apart, according to Aristotle,[114] generally have a shorter life.

Chapter XXIII

On the Lips and Gums

The lips are the farthest forward part of the mouth and necessary for handling drink and food, for expressing words, and for protecting the teeth. They utter sounds, for they can be extended, contracted, and dilated by a varied texture of muscles. For the purposes of all motion they are enlarged on the outside with plainly visible skin and covered on the inside with a reddish cuticle which surrounds the inner mouth with its rosy color. This color vanishes in dying people (except those with liver trouble)[115] and changes quickly from purple to a pallor. The lips are connected with the throat by some nerves, of which I shall shortly speak. Hence the lips, especially

[111] Aristotle, *De Historia Animalium*, II, 3, 501 b 20: "Males have more teeth than females in the case of men, sheep, goats, and swine . . ." The Italian name of Boldus was Antonio Boldù; see note 71. On the wisdom teeth see also *De Historia Animalium*, II, 3, 501 b 25; Pollux, *Onomasticon*, 2. 93.

[112] Alexander Merula (Giorgio Alessandro Merula), from Alessandria, was born in 1424, 1430, or 1431 and died on March 9, 18, or 19, 1494, at Milan. He was a teacher at Venice and at Milan, whither he was called by Lodovico il Moro to write the history of the Visconti (*Antiquitatis Vicecomitum libri* X); in Milan he discovered a manuscript of Ausonius. He served as a corrector for Speyer's press at Venice, engaged in polemical writing, and edited classical authors, Pliny, Quintilian, and others, including the *editio princeps* of Plautus. His *Commentarii in Iuvenalem* appeared at Venice in 1474 and 1478. Among his students, in addition to Benedetti, was Johannes Britannicus, an editor of Pliny. See for further information Ferdinando Gabotto ed Angelo Badini Confalonieri, *Vita di Giorgio Merula* (Estratto dalla *Rivista di Storia, Arte, Archeologia della provincia di Alessandria*, Anno II, Fasc. III, Alessandria, Italia, 1893); Remigio Sabbadini, "Giorgio Merula," in *Giornale storico letterario italiano* **47** (1906): pp. 25–40; also Mario Cosenza, *op. cit. s. v.* Merula.

[113] This is an interesting example of early orthodontics which might well be included in a history of dentistry. Vincenzo Guerini, *A History of Dentistry* (Philadelphia, Lea and Febiger, 1909), pp. 157–159, devotes more than a page to Benedetti's contributions but does not mention this item. Guerini's references are exclusively to Benedetti, *De Re Medica*, VI. *De affectibus dentium*, pp. 119–127, and deal with diagnosis before extraction of teeth, the evil effects of mercury on the teeth, dental worms, the use of opium in dentistry, and the story of the slave Benedetti did not buy because the man's teeth were like those of a wild beast; this was considered a bad omen by Benedetti. Benedetti did not write a treatise in 1460, when he was barely ten years old: this error appears in J. -Léonard André-Bonnet, *Histoire générale de la chirurgie dentaire* (Lyon, 1955), p. 79. There is nothing useful on Benedetti, although his *De re medica* is mentioned, in Karl Sudhoff, *Geschichte der Zahnheilkunde* (Leipzig, sec. ed., 1926). Bernhard Wolf Weinberger, *An Introduction to the History of Dentistry* **1** (St. Louis, C. V. Mosby, 1948): p. 240, mentions the Ethiopian slave boy Benedetti saw and did not buy, "all of whose teeth were sawed off, just as if he had canine teeth."

[114] Aristotle, *De Historia Animalium*, II, 3, 501 b 20.

[115] This seems a personal observation.

the lower one, tremble during periods of vomiting. We yawn with the lips; if we do so excessively it becomes an illness called oscedo. The place where the teeth are fastened in is called the gum; it is reddish and has many clefts. I have spoken thus much about the parts of the mouth which are adapted to the spiritual and natural members. Let us pass now to the brain, the principal member of the body, which has been left until the last according to the procedure followed in dissection.

BOOK IV OF THE ANATOMICE OF ALESSANDRO BENEDETTI, PHYSICIAN, ON THE ANIMAL MEMBER CONTAINED IN THE HEAD

Chapter I

On the Head

The head which is composed everywhere of a little hard bone, flesh, and sinew (as Plato says)[116] seems rather feeble in man since in other brutes it is constructed so as to come more to a point with more foresight as to its use. At the extreme end of the head it joins nerves to the strongest part of the neck and sends nerves or tendons or what they call connections to all the members and remaining joints. Hence it was not to be left a thing of mere bone, on account of excess heat or cold, nor on the other hand of flesh alone lest because of the heavy weight it should be rendered dull of sensation, but dried skin around the bone made it firm and light.[117] In other respects those who have a head that is fat, sinewy, and fleshy are stronger and longer lived, as he says. Besides, when nature was making up her mind whether the human race should live longer and worse or shorter and better she decided to give it a short and better life, in overwhelming contrast to a long and worse one. Everywhere in the head the natural heat has pierced the skin and opened passages whence the viscose humor drawn outward and extended little by little in length and rendered equal or even by these punctures has spread it out into the slenderness of the hair and the humor turned back under the skin has put forth roots. For these reasons therefore the hairs grow in the skin; on account of coagulation due to cold they have become harder than the skin and denser in substance. This covering has been contrived as a protection to the brain; the skull itself like a bony sphere made on a lathe fences the brain around and covers its powers as the celestial sphere covers the very similar divine souls of the planets. Hence the smooth medulla which is a kind of proxy or substitute (of the brain) descends by means of the rotating hinges of the neck and vertebrae to the lowest part of the spine, the assistant of the human semen (as the same author [Plato] says), [118] and is so necessary that it exists with a wall of protection that goes round it with a stony and articular circuit suitable for motion and for bending. For the contemplation of this citadel of the mind our Hermolaus Barbarus,[119] who is considered first among the Latins, and Antonius Cornelius, first of philosophers among us, have approached, since in a very spirited fashion like the immortal sky-dwellers they always cultivate the mind itself and are most shining of spirit beyond the meanness of this mortal life.

Chapter II

On the Divine Intellect of Man

The neck sustains the head, in which is contained the brain, the most excellent of the members and the nearest to the sky. Hence the intellect quite clearly holds the highest position as does the human reason, which God the Author and Founder of the world made as close as possible to Himself through His Son. From the brain issue ideas and thoughts which are also accustomed to share the divine counsel which is granted (although rarely) to holy men and by which through a sure divination mortals sometimes have a presentiment of what will happen. Also in sleep visions appear most clearly as heralds of what is

---

[116] Plato, *Timaeus* 75 C; flesh and sinews are explicitly absent, however, in Plato's account. See also Aristotle, *De Historia Animalium*, III, 5, 515 b: ". . . in the head there is no sinew."

[117] Aristotle, *De Partibus Animalium*, II, 10, 656 a 15–656 b 15, discusses the relation of flesh to the head and to sensation. For him, of course, the heart was the center of sensation; the reasons why he refused to believe that the brain was the sensory center are given in note 3, *De Partibus Animalium*, II, 10, 656 a, by the translator of the Oxford Aristotle, Vol. 5, William Ogle.

[118] Plato, *Timaeus* 73 C; 86 C; 91 B.

[119] Hermolaus Barbarus, from Como, was born in Venice May 21, 1453 (or in 1454 or 1464) and died in Rome June 14, 1493, 1494, 1495 or 1498. He was actually the son of Zaccaria di Francesco but lived with Bishop Hermolaus Barbarus at Verona (known as Maior, died 1471), from whom he received his name, with the addition of the word Minor. He was poet laureate in 1468, a member of the *Maggior Consiglio* in 1471, taught at Padua (1474–1475), became a senator of Venice in 1483. He was created Patriarch of Aquileia in 1491; this made him at once an exile at Rome according to Venetian law. He wrote a *Compendium Galeni* and other scientific and literary works. See Ermolao Barbaro, *Epistolae, Orationes et Carmina*, ed. V. Branca (2 v., Florence, 1943); T. Stickney, *De Hermolai Barbari vita atque ingenio* (Paris, 1913); A. Ferriguto, *Almoro Barbaro* (Venice, 1922); George Sarton, *Appreciation of Ancient and Medieval Science During the Renaissance (1450–1600)* (New York, A. S. Barnes, 1961), pp. 81–82, for a brief account.

to come; owing to these I do not doubt that the soul is immortal since it is judged to have no share in bodily action.[120] These visions do not have the character of dreams which occur in the appearance of false images to those who are distended with wine and food. The brain, which is without sensation,[121] is nevertheless held to be the instrument of the power for causing sensation or feeling. In the brain sleep takes place; it bears the appearance of death when the limbs are stretched out without sensation. Those who are drowsy and who do not limit their nights to slumber have less abundance of the sweetness of life. The brain is the cause of sleep since it is cold and moist and the vapors which flow into it are at once congested and obstruct the passages of the powers of sense. The members of the body lie without feeling and hence draw strength if proper measure of sleep is preserved. The brain is the coldest[122] of all the members in proportion to the temperament of the heart. For nature is always accustomed to devise aid by means of the conjunction of opposites against any sort of exaggerated preeminence and checks one excess by another. For in the entire body cold was given by nature for tempering the natural heat, as is perceived by touch and understood by the reason. The brain has no blood but in order to bring a moderate heat to that part of the body veins pass to the membranes which envelope it from the major vein and from the aorta. And to prevent the heat from becoming troublesome not a few ample veins but many thin veins go about through the windings of the brain so that a blood not copiose and thick but thin and pure may flow to it. The brain consists of an earthy and watery substance because when it is cooked it grows hard and dry just as a liquid when boiled leaves behind an earthy portion. For this reason it is not marrow but a member in its own right[123] and of its own kind. If it grows moist or dry beyond measure the mind is affected or destroyed or disturbed. If it becomes rather cold owing to the north wind distillations flow from it after the south-wind blows on it. For a lofty vapor seeks the brain which, chilled because of the latter's frigidity, is changed into humor, as usually happens when the rain falls since vapors have collected in the upper air, if one may compare something great with something small. For when from earth or water the vapor resulting from heat is borne upward to the highest level of air which is cold the vapor stops and is converted once more into water on account of refrigeration and flows at last to earth.

Chapter III

That Which is Contained in the Brain

The nerves of sensation from the brain are regarded as the interpreters and messengers of the objects of perception; they are located in large part in the head as in a citadel. Thus I see that the head is called caput because sensation takes its beginning from the head (*initium capit*). It is the largest member in proportion to man's size among the animals.[124] Besides the covering of hair the head consists of skin, membrane, the skull, a double panniculus [meninges], the brain, the ventricles, and the marvelous net of mingled veins and arteries, and other windings of veins with nerves and other dissimilar parts such as the eyes, ears, and such items, as I shall relate in their proper place.

Chapter IIII

On the Hair

Man's hair is abundant, harder or softer according to difference of the bodily regions. It is a protection against the cold. The material of the hair is made from the excrementitious humor and smoky vapor carried from the entire body to the skin. In proportion to the quality of the skin the hair is now hard, now soft, and now white, now black according to the color of the skin, as in leprosy. The hair grows gray when covered rather than when allowed to blow in the breeze. The hair begins to grow gray under certain circumstances. The pubic hair grows white last of all. The hairs on the head, the eyebrows, and eyelashes are present at birth. After birth the pubic hair grows first, then the hair in the armpits, lastly the beard. The hair falls out but never in the occiput. The eyebrows also fall out from a disease for which no name has been discovered. Women are not troubled with baldness nor are boys and eunuchs. The hairs which grow after puberty, except for the pubis, only fall out when they are cut. There is a story told about a Greek peasant. He suddenly and with laughter asked a certain philosopher why the beard existed. The astonished philosopher made no reply. But the peasant said, with a jeer, "I who abide in the country

---

[120] Dreams as a proof of the immortality of the soul is an ancient classical idea which occurs from Homer onward in Greek and Roman thought and literature. Figures appearing in dreams were considered to inhabit Hades or Elysium; see Homer, *Odyssey*, Book 11; Vergil, *Aeneid*, Book 6, and elsewhere. See the article *After-Life* in the Oxford Classical Dictionary (Oxford, 1949).

[121] Aristotle, *De Partibus Animalium*, II, 7, 652 b 5. Benedetti's description of the brain is thoroughly Aristotelian. See also Pliny, *Natural History*, 11. 49. 135, p. 517, Loeb Classical Library, on the brain as the source of sleep.

[122] *Ibid.*, 652 a 35; the view that the brain is cold, bloodless, and fluid is partly Hippocratic in origin; see also Pliny, *Natural History*, 11. 49. 133, p. 515, Loeb Classical Library: "the brain is the largest in man in proportion to his size; it is also most moist and coldest."

[123] *Ibid.*, II, 7, 652 a 25, where Aristotle points out the error of confusing the brain with the spinal marrow (called nucha by the sixteenth-century anatomists.)

[124] *Ibid.*, II, 7, 653 a 25–30.

shall tell you readily and properly. Since it is contained in the testicles so the chin also is adorned with Greek majesty."[125] All the hairs except the eyelashes grow out and swiftly turn gray. In some people the beard is rather thin but the mustache area is never naked. Hairs grow in bodies that are emaciated but the hairs of the head fall out. They sometimes grow in dead bodies, as do the nails. The congenital hairs of those people who are worn out with sex drop out more early; the hairs which grow after birth are produced more swiftly. Those who have dilated veins are less bald. Those blind from birth do not grow bald since (as Alexander of Aphrodisias says)[126] their heat is transferred to the nearest place.

Chapter V

On the Skin of the Head

All the skin lacks sensation, especially that which covers the head since there it adheres to the bone without any intervening flesh. Likewise, it is thicker in nature and thus the hairs rest upon it with a larger root. Under the skin there is a certain tough substance more viscose than elsewhere. In some people this skin is movable from the forehead to the occiput and back again, with some nerves penetrating the top of the skull to the skin.

Chapter VI

On the Exterior Membrane of the Skull

Under the skin there is a membrane [periosteum] which covers the skull externally to which another membrane is connected from the inner region by means of the sutures of the head to keep the brain from pressure from the adherence of the skin. Thus nature has designed this first membrane as a covering so that the skull may be preserved uninjured by means of yet another membrane, for it bears a close relationship to the inner membrane.

Chapter VII

On the Skull

When the skin is torn off along transverse lines from the occiput through the vertex as far as the lowest part of the forehead the skull appears, called testa or pot by the ancient Latins. It is concave within and globular on the outside, and hence called spheroid by the Greeks. The hemispherical parts, smooth and flat on the temples and double on the forehead, are thus the strongest; their bones are hard in the exterior portions and soft on the inside. Veins pass through these parts which constitute a perforated area without marrow; it is believed that these bones are nourished by the veins. There is an uncertain number of sutures in these bones; not even their location is certain, according to Hippocrates.[127] At a dissection in Padua I saw a solid skull without sutures. Such heads are called canine from the similarity; they are considered to be very strong and very well protected from pain. Celsus[128] reports that fewer junctures in the head contribute to its strength. Fractured bones in the head are made whole by some and removed by others; in their place a hard body and a scar appear. In infants the sinciput grows hard after some time.

Chapter VIII

On the Sutures of the Skull

Three sutures ascend to the top of the skull and come together in the shape of a triangle. The head is formed from six bones, according to some of eight, two of which are the bones of the upper and lower jaw. The latter is movable. The upper jawbone divides the middle foramina of the eyes and the nose and sustains or supports the brain with its inner process by means of the upper row of teeth. The upper jaw is closed with one suture. The bones of the cheeks [zygomatic] which are covered by the upper part of the face are contained in the upper jaw on both sides and are bounded by sutures near the two very small bones of the ears. These are lightly divided and held by one suture. From the front of the forehead bone, however, the bones seem rather to rest lightly in place than to be joined by any structure or connection. Behind the ears the bones are very thick and sustain the heaviest blows. They send a process downward. At the temples the bones of the skull are very thin and therefore liable to injury. The frontal bone (forehead) ends above the ears and divides the eyes with a thin suture. This bone is conterminous with the upper jawbone. It passes from the ears upward toward the vertex with saw-toothed joinings [coronal suture] called stephanitides; its hardness is so remarkable that fire can be struck from it as from a

---

[125] I have not been able to trace this story to its origin.

[126] Alexander of Aphrodisias, *op. cit.*, fol. VI v: Cur qui nati sunt calui non euadunt ? Utrum quia caliditas mutilis oculis se in proximum transferat locum ? nunquid non frigefit locus, sed semper nutrimentum habet. Caluitium autem a frigiditate oritur aut quia cum locus caliditatem admittit ut nutrimentum absumat non fit in sincipite ulla superfluitas. Luminibus autem orbatus frigefactus est, proinde et ibi multa redundantia est atqui crines ab superfluitatibus exoriuntur. On the congenital hairs which drop out due to sexual excess see Pliny, *Natural History*, 11. 56. 154, p. 529, Loeb Classical Library.

[127] Hippocrates, *De Capitis Vulneribus* (Vol. III I., 346).

[128] Celsus, *De Medicina*, VIII, 1, p. 475, Loeb Classical Library: "It is rare for the skull to be solid without sutures; in hot countries, however, this is more easily found; and that kind of head is the firmest and safest from head-aches. As for the rest, the fewer the sutures, the better for the heads; and there is no certainty as to the number, or even as to the position of the sutures."

flint. In the eyebrows and above the nose where it projects a little it is double; in some people it is single. The occipital bone which is the sixth in my enumeration is closed by sutures in the shape of the Greek letter lambda; hence the suture is called lambdoid. It covers the posterior part of the cerebellum with a suture [sagittal] that is sent upward toward the vertex of the skull. This part is also considered weak. The remaining two bones which together compose the eighth in number are above the ears and are called the sides of the head, a little bent in from the forehead and occiput; they [parietal] are joined to the vertex as well as to the occiput and frontal bone. The skull itself consists of only four bones, the frontal and the occiput, which is called epicranis by Pollux,[129] and the two side bones above the ears. The sutures of these bones are quite visible; the other sutures are called deceptive (*fallaces*). I have shown that the base of the brain is part of those bones I have described. This is the location and the number of the sutures and bones. Thus from this description we have an indication as to the method physicians are accustomed to use either in cauterizing or scraping the sinciput when necessary. They move the lowest part of the hand at the top of the nose to the space between the eyebrows (*intercilium*) and extend it to the bregma. This is the sinciput. The middle fingers indicate the top of the head, with a thread divided off on each side. They observe what lies between, for each one makes his own measurement and by this method they think the crossroads of the three sutures can be found. Their conjecture is sometimes fallacious; hence Hippocrates[130] openly confessed that he had sometimes been deceived (in using this procedure for identifying the sutures of the skull.)

Chapter IX

On the Inner Membranes

When you have examined these items in part and recognized them in part from my description alone I must show you how to remove a piece of the skull when the head is injured by a blow or a fall. I shall do this in the presence of surgeons, who must be at hand in our dissecting theater so that the blood or pus, if any, may be drawn forth. Otherwise abscesses, fever, and distentions of the nerves, which are called spasms, and delirium finally result. At the cost of great labor they scrape or cut the bone with a circular saw. When these procedures have been demonstrated they cut the entire skull from the eyebrows to the ears in the occiput with a saw. When this part of the skull is removed the size of the brain which is incredible in proportion to the members, especially in men, must be examined. It is covered by two tunics; the rupture of either of them is fatal. The upper tunic is stronger and attached to the sutures of the skull; it is called the custodian of the brain [dura mater]. The other tunic [pia mater] envelopes the brain itself, very thin and less strong than the first one. Pollux[131] calls it the ilamide (eilamides). This membrane contains slender veins which come together in a sort of coiling; they are scattered everywhere above in the brain. They do not pass through but above the substance of the brain. For this reason the brain lacks blood for the coolness of the brain has the function of tempering the spirit which comes from the heart in order to establish its power of causing sensation. These membranes serve as covering; they lack flesh in order to keep from holding or attracting vapors. The thicker membrane is attached to the skull so that no weight may be brought to it. The membranes divide the anterior brain into two parts, right and left [falx cerebri]; hence they have the name heterocrania. They also divide the anterior brain from the posterior cerebellum, for thus it is called [falx cerebri]. It is smaller and harder. Between these parts of the brain a certain unskilled surgeon thrust a probe to the deepest region while he was caring for a patient who had fallen from a height. As ignorant people do who make diligent search for vain purposes he thought the brain was perforated. When his error was recognized by a physician he realized that he had placed his probe between the two membranes.

Chapter X

On the Cavities of the Brain

There are also three hollows or sinuses in the brain, two in the fore part of it with a very thin membrane surrounding them on the inside. The hollow of the rear part of the brain is without a membrane but is preserved by its hardness only. The first two hollows [lateral ventricles] are constructed like many other members of the body such as the kidneys, lungs, ears, eyes so that if one of the two is injured the other one can continue to function. For the vital spirit which is spread upward by the arteries from the heart is tempered in the brain and thus receives its power for creating sensation (which they call animal, or animate perception). The arteries administer the spirits or spiramenta from these hollows or caverns, distributing movement and various powers through the members by means of the senses, nerves, and muscles. If these caverns are disturbed by spontaneous dizziness or through the effort of turning the body around on horseback or on shipboard or by some illness as occurs with those who suffer from dizzy spells, they upset the humors and spirits at the same time. When the caverns are thus deprived of their functions, hearing, vision, and movement of the bodily members collapse in ruin with the entire body. The material of the brain must be cut gradually so

[129] Pollux, *Onomasticon*, 2. 45.

[130] Hippocrates, *De Capitis Vulneribus* (Vol. III K., 561).

[131] Pollux, *Onomasticon*, 2. 44.

that the inner caverns may appear; these are rather large in living bodies, that is, in both right and left sides, for each is of the same construction. Philosophers, however, differ as to whether each sinus of the brain presents twin images or forms (Cicero[132] calls them spectra) since they occupy different places and do not come together or focus in a third place, as happens with the eyes which, although two in number, nevertheless present one form or shape to the vision. For in these sinuses the powers of the most noble senses are contained; from them is acquired the power of thought, judgment, and intellect. In them is found that which is called common sense and the grasp of sensible objects as the form of things, various and different images are formed here, and from them proceed vision, hearing, smell and the other senses along the pathways of nerves or of membranes. Images themselves according to their nature occur by involuntary action nor are they always subject to human will. This power of imagination we call the fancy; certain later thinkers have falsely judged the fancy to have different powers. Nicolaus Leonicenus and Laurentius Laurentianus[133] do not refute this view, for the common sense brings judgment to bear from external objects. No one except more recent philosophers has dared to mention anything concerning the location of these powers since if parts of these sinuses of the brain are affected the reason and the mind itself seem to become confused for such patients see, hear, think, and imagine empty falsehoods and make wrong judgments. Thus they become giddy, mad, phrenetic, delirious, or stupefied; others fall in epileptic fits and some grow suddenly speechless.

Chapter XI

On the Sinus of the Cerebellum

In the posterior part of the brain, the most visible instrument of sense, there is a larger cavern in which the faculty of memory is located and maintains its vigor. Here others place the faculty of reminiscense, the continued conservation of images. This cavern is sometimes cut off or rendered useless. Images or phantoms are subservient to these two faculties. Others contribute that power of recall, as I term it, to a middle pathway. The sinus of the cerebellum [fourth ventricle] is broad at the top with its point verging toward the spine, in the shape of an inverted obelisk. When this part is injured lethargic illness ensues, during which some people forget both how to read and the names of their relatives. One man, owing to greenish bile and the evil vapors produced by it, forgot the names of the books which were at hand, under the influence of a sudden illness, which soon after were restored to him and his memory likewise. From this sinus arises the spinal medulla, the manifest origin of most of the nerves. It is of a different nature from that of the brain itself, for it is fat and made softer by cooking. This is called the base of the brain or paracephalis or epicranis by the Greeks. Its color is yellower than that of the brain.

Chapter XII

On the Middle Pathway Between Each Sinus

Between the anterior ventricles and the posterior one there is a path attached to each by which images, as they say, are brought from each of the ventricles and from both sides [third ventricle]. Here it is held that what we call decisions and judgments are made. Here acts of understanding are recognized, for along

---

[132] Cicero, *Ad Familiares*, 15. 16. 1; 15. 19. 1.

[133] Both Niccolò Leoniceno (1428–1524), the famous philologist and editor of Pliny, and Lorenzo Laurenziani (died 1502; see Leon Dorez, "Le portrait de Lorenzo Lorenzano . . . par Sandro Botticelli," *Bull. Soc. franc. Hist. Med.* **6** (1907): pp. 235–238) were associated in the publication of a book printed at Venice by Jacopo Pentio de Leuco in 1508; it is a translation by Leoniceno of Galen's *Ars Medicinalis* and *De Tribus Doctrinis* with another by Laurenziani of Galen's *De Differentiis Febrium*. The book is entitled *Nicolai Leoniceni in libros galeni e greca in latinam linguam a se translatos prefatio communis*, but it contains no discussion of the fancy (*fantasia*) or imagination. What is involved in Benedetti's statement here is the theory of the three parts of the brain: reason, memory, and fancy, discussed by the Commentator on Galen, *De Sectis* (edited by Diomedes Bonardus, Venice, 1490) and by later anatomists: fol. 1 v: Ergo quia in ipso videmus aliquas operationes fieri consideremus per quas virtutes fiunt; tres namque sunt virtutes in corpore nostro: animalis, vitalis et naturalis. Animalis est per quam videmus, audimus et ambulamus et operamur. Animalis sunt tres species: rationalis, sensibilis, et mobilis. Mobilis est que corpora nostra movet. Sensibilis quoque [quinque sunt *ex marg.*] ut sensus attribuit: visum, auditum, gustum, odoratum, et tactum. Rationales sunt tres: fantasia, legismos [logismos] animae, et memoria. Fantasia est in anteriori parte cerebri; legismos anime in medio cerebri, que discernit bonum a malo. In posteriori parte cerebri est memoria. Vitalis est virtus per quam viuimus et statum accipimus et reddimus et diversitates pulsuum et alia similia. Naturales sunt vii, tres secundum genus, et quatuor secundum speciem. Secundum genus sunt genetike, rethike, antypike. Genetica dicitur per quam nascitur homo, rethiche qua nutritur, antypiche qua crescit. Istis serviunt quatuor ille que secundum speciem sunt electica, crastica, alliotica et apropitica. Electica est que trahit, crastica que tenet, alliotica que resoluit, apropitica que expellit. The four qualities described in the preceding sentence are based upon the account of the four digestions by Macrobius, *Saturnalia* 7. 4. 9–19. This is perhaps the earliest and most concise statement of the entire subject of the parts of the brain and their functions according to the Renaissance anatomists; it is interesting to compare it with the modern description of the location of these functions also. Walter Sudhoff, "Die Lehre von den Hirnventrikeln in textlicher und graphischer Tradition des Altertums und Mittelalters," *Arch. Gesch. Med.* **7** (1914): pp. 149–205, shows that the trichotomy φανταστικόν, λογιστικόν, μνημονευτικόν goes back to Posidonius. See also D. H. M. Woollams, "The Historical Significance of the Cerebrospinal Fluid," *Med. Hist.* **1** (1957): pp. 91–114, based on Sudhoff's study for the earlier period.

this pathway come all images of memory and thence it brings back the forces of recollection. Its power is voluntary and accomplished by man's will. By means of this power we think, remember, and come to agreement. We perceive with the accompaniment of reason itself; by this power we compose and release the discourse of our minds. For this reason this path to the heart as if to the beginning of all things was established above. It is quite similar to an earthworm for it is reddish in color and contracts and stretches out in an arch like a worm when necessary. But which man, I ask, has explored this pathway, what skillful talent has pursued it thoroughly? Life will decide whether what I say about it is true or false.

Chapter XIII

On the Other Parts of the Brain

Before you cast them aside you should examine with no less care three items which are handled in their investigations by physicians. Between the anterior ventricles of the brain and the middle pathway there are certain parts called supports made of the same material as that of the brain [choroid plexus]. They lie between the ventricles as earthworms lie intertwined, joined on both sides by small nerves. When they are stretched out at length they close the passages from the first ventricles which go to the middle pathway. But if they are contracted it is believed that then these ducts are opened, and thus the spirit runs to and from where the images are carried. Under these supports (or worms) there appears a certain shell-like cavity in the middle of which there is an aperture or foramen which descends to the farthest excretory pathway of the palate whence the phlegm flows out from the middle brain and sneezes burst forth. This lacuna has round processes on each side by which the veins and arteries are supported that pass from the marvelous net of arteries to the aforesaid ventricles. Following this net there are certain prominent glands able to hold superfluous fluid.

Chapter XIIII

On the Seven Pairs of Nerves According to the Ancient Physicians

The brain must at last be completely removed together with its two membranes to keep the nerves which arise from it from being dislodged. Seven pairs of nerves arise from the brain. We call them a pair (syzygia) when two nerves emerge together. They transmit the sense-producing breath (*spiramentum*) to certain places of the members so that they may pour in sensation to neighboring as well as distantly placed members along the pathway of the spirit. The first pair [optic] goes to the eyes; some people call them veins. By the nerves of this pair we recognize colors and distinguish between light and shadows. The second pair spreads itself to the ears, [facial, auditory], contrary to the opinion of Aristotle.[134] Through this pair arises the awareness of sounds. The third pair [olfactory tract] is inserted into the nostrils and furnishes the sense of smell. The fourth pair holds the palate [maxillary], the fifth proceeds to the tongue and to the face [mandibular]. The sixth seeks the stomach for which a great deal of sensation is necessary in order that it may have appetite and reject what is superfluous [glossopharyngeal, vagus, accessory]. The seventh pair of nerves pours sensation to the spinal medulla [hypoglossal]; this is compact and descends to the lowest part of the spine. But I shall add some more truthful details about these nerve pairs. Earlier physicians have paid more attention to their place of origin rather than to the place where they end, but later physicians have considered the matter concisely or in greater detail. The first pair which goes to the eyes from the lowest part of the anterior (or first) ventricle of the brain they call optic nerves. These are first joined in a cross [optic chiasma] on account of the more slender membrane of the brain; thus when forms are received by both eyes one single image is presented. But when the eyes are distended (dilated) beyond their normal custom and the superior nerves are sent down to join transversely one object appears double. From these nerves there is another pair [oculomotor], slender and of harder substance; these bring the power of motion. But Aristotle assigns them another number; his opinion must now be quoted.

Chapter XV

On the Nerves of the Eyes According to Aristotle

This is what Aristotle says about the nerves of the eyes: "Nature has established three nerves stretching from the eyes to the brain; just as the largest and the middle-sized nerves tend toward the cerebellum so the smallest pass to the brain itself. These latter nerves arise nearer the nostrils. The largest nerves are equidistant from each other and do not come together; the middle-sized nerves do come together as

[134] Aristotle's theory of sense perception does not involve the nerves, which, furthermore, he confuses with sinews (*neura*) but is based upon sensoria dependent on movement or affection from without. Hence, for him "the organ of hearing is physically united with air, and because it is in air, the air inside is moved concurrently with the air outside" (*De Anima*, II, 8, 420 a). See also *De Historia Animalium*, I, 11, 492 a 15: "This receptacle [the ear] does not communicate by any passage with the brain. . . ." Thus nerves are not participants in the function of hearing. See *De Historia Animalium*, III, 5, 515 a-b for the sinews (nerves) which originate from the heart, the source of sense perception in Aristotle's theory.

is apparent especially in fish, and are closer to the brain than the largest nerves. The smallest nerves are distant from each other by a very wide interval and can therefore never come together."[135] This is what Aristotle says. The membranes of the eyes have a connection with those nerves which pass to the stomach, since no eye is plucked out without causing a rising of the stomach.

Chapter XVI

The Second Combination (Coniugatio) *of Nerves*

The second combination ends at the ears, which is something Aristotle[136] firmly denies. From this combination the sense of hearing originates; I shall speak of this in its place. If any connection appears between these it will be judged by the entire audience in the dissecting theater since such a connection never appears.

Chapter XVII

The Third Pair of Nerves

This pair extends to the mamillary carunculae (*tubercula papillaria*) [olfactory bulb] of the nostrils. It is soft and serves a sensory function, coming from the nearby material of the brain which is covered by a thin membrane. These nerves sense odors since the organ of sensation is temperate; sneezing is motivated by them. In some people this power of smell is taken away as it was in Faleria Contarena, a friend born of a noble family. Those whose nostrils stretch out forward have a strong sense of smell, just as those who shade their foreheads with the hand can see farther. Those who find heavy smells pleasing have this faculty owing to the bad temperament (*discrassia*-[*dyskrasia*]) of the tubercles (*mamillary carunculae*).

Chapter XVIII

The Fourth Pair

The fourth pair passes through the palate, disappears, and stops there. Others think that the nerves of the sixth combination which provide the sense of taste to the tongue also extend to the palate.

[135] Aristotle, *De Historia Animalium*, I, 16, 495 a 10: "From the eye there go three ducts to the brain: the largest and the medium-sized to the cerebellum, the least to the brain itself; and the least is the one situated nearest to the nostril. The two largest ones, then, run side by side and do not meet; the medium-sized ones meet—and this is particularly visible in fishes—for they lie nearer than the large ones to the brain; the smallest pair are the most widely separate from one another, and do not meet" (translated by D'Arcy Wentworth Thompson in the Oxford Aristotle, **4**).

[136] See Aristotle, *De Historia Animalium*, I, 11, 492 a 15 and note 134. The Greek term *dyskrasia* used below means hard to mix or temper.

Chapter XIX

The Fifth Pair

The fifth pair extends partly to the tongue and moves the cheeks, and is partly connected to the seventh combination of nerves; hence those who touch the tip of the tongue incite vomiting for from the same combination those nerves stretch to the stomach; these are also joined to the membranes of the eyes.

Chapter XX

The Sixth Pair of Nerves

The sixth pair (as I related) is simply the medulla descending from the posterior part of the brain; I shall speak of it in the following book (the fifth). From the medulla originate more than thirty combinations of nerves. I should prefer to call the last pair the seventh in order since it issues in the last place.

Chapter XXI

The Seventh Pair

The last pair or combination of nerves [glossopharyngeal, vagus, accessory], which others call the sixth, is spread out from the posterior part of the brain to the stomach, diaphragm, epiglottis, and to both sides of the throat. From this pair are constituted the nerves which are called conversive (reversive) and vocal, or of the voice. They pass upward again to the epiglottis. By some people they are called *toni,* necessary for the voice and heralds or messengers of the heart (that is, of the will). The rest of this combination of nerves is joined with the gullet for the greater part and is dispersed to the lowest part of the stomach and ends there. Another part of this combination of nerves is connected with the involucrum of the heart; the nerve which is located in the left-hand side is joined to almost all the members which are located above the diaphragm. From the same foramen of the skull (in order to make the matter more easy to recognize) three separate nerves run forth. The first of these [glossopharyngeal] passes to the muscles of the neck and to the farther end of the tongue. The second [accessory], which is larger, goes to the muscles of the shoulder blade and to the regions nearer to it and ends there. The last of the three nerves [vagus] extends to the spiritual viscera (as I have said) by a fixed pathway near the veins which we call sphagitides (jugulars) and there it is connected. From the region of the top of the throat the same branch sends out a shoot which contributes the power to move the throat and its parts. The remaining part of the same nerve is carried back to the diaphragm and thus we call them conversive nerves; when these are cut a man immediately becomes mute, and when they are inflamed the voice is affected. These nerves are not directly established by nature from the brain for the function of the voice because there would be no convenient pathway such as would be necessary to lead downward in the upper region. Thus it is better that these nerves proceed upward again before they are carried down to the diaphragm. For this reason they attain (as Galen bears witness)[137] a greater hardness and are made stronger the farther they are from their beginning in the brain; hence also the shorter they are the softer the nerves become. They are made harder by the heat of the heart and rendered more suitable for the voice; otherwise they would easily be oppressed by the speech which restrains them. Not without justice are they connected with the brain and with the heart, since from the former in which resides the power of sensation and of thought and from the latter in which the mind itself and the soul dwell, these functions may appear by conceptions of the will and mind without any mediation between the two members.

Chapter XXII

Another Observation Concerning the Nerve Pairs

Some people have observed the nerve pairs in another manner. To sum up their views briefly, two pairs of nerves are carried to the eyes, one to the mamillary carunculae of the nostrils, the other to the palate; this pair ends on each side near the teeth and is double in some animals. In people whose jaws are rather long another pair goes to the cheeks and lips and eyes; it comes out from between the eyes; one pair goes to the tongue. To the ears nothing comes from the substance of the brain but only from its membrane. Another pair flows between through the spine. Another pair furnishes movement to the lower jaw. Another pair near the origin of the cervix (its posterior part) passes to the strong part of the neck. In addition to these nerves some pass to the highest part of the skull; they are few and move the skin of the head. Some nerves go to the ears in order to move them and some to both the eyebrows. But these nerves are not consistent in their occurrence, for their combinations proceed in this manner as far as their origins are concerned although they are carried down to many members in their reversion, to the stomach, tongue, diaphragm, and epiglottis.

Chapter XXIII

On the Marvelous Net

Each membrane near the brain goes around it. Then the bone of the skull lies beneath; this is fungous in order to be lighter in weight. In the middle of it there is an excretory foramen. Here veins and arteries stretch to the brain; they are carried near the ears and spread out into very small branches and pass under the membrane which is the custodian of the brain, forming a trellis in the shape of a net woven together from very slender intertwined fibers. In this net (as I have said) the vital spirit is diffused; there is no doubt among our physicians that the vital spirit is changed by the power of the brain into spirit which causes sensation. Others think this change is effected in the hollows of the brain. From these veins and arteries proceed what we call sphagitides or jugulars and parotids and are scattered to the thinner membrane of the brain since the latter is entirely bloodless and no vein ends there.

Chapter XXIIII

On the Very Strong Bone of the Head

This bone is very strong because it is the support of the brain. It extends to the location of the sense of smell where the folds of the eyebrows are or the foramina which are made of double or twofold bone. These descend toward the nostrils and above them the frontal bone ends. It is supported by the bones of the upper jaw. Combinations of nerves penetrate the lowest part of this bone and it is hollowed out with a double cavity within the temples. There is a ridge [petrous] projecting in the middle which is a process of the occiput above the foramen of the spinal medulla. On each side there is a very hard stony bone called the yoke-bone because, according to the Greek term for it, zingoides, it is supported by the lower bones [*arcus zygomaticus,* which connects the cranial with the facial bones]. Likewise it is divided into both the bones of the nose and of the eyes where the eyebrows are. From the region of the eyebrows where the nose begins within the frontal bone a bony point [crista galli] sticks out which divides the mamillary carunculae and from the side of the point there lies adjacent a very thin perforated bone which sticks out through which the humor of phlegm distils to the nostrils (*lamina cribrosa*). At this place I shall not omit an admirable example of nature. A certain Greek named Suirus, very well known to me, was wounded in the temple at the siege of Chalcis and taken prisoner by the Turks.[138] Somehow his wound was healed and he

---

[137] Galen, *De Usu Partium*, VII, 14 (Vol. III K., 581): . . . quin potius contra in primis quidem productionibus nervi omnes sunt molles et cerebro ipsi assimiles, progredientes vero magis magisque se ipsis efficiuntur duriores. See also Charles Singer, translation of Mundinus, note 104, p. 108.

[138] The siege of Chalcis began in June, 1470, when the Turks struck both by land and sea at the stronghold of Venetian power on the Euripus in Greece; within four weeks the Turkish leader, Mohammad, took the city (July 12), executing the Italian survivors and enslaving the Greeks: *The Cambridge Modern History*, **1**, The Renaissance (Cambridge University Press, 1934), chap. 3, The Ottoman Conquest, by J. B. Bury, p. 79. Adding twenty years that Suirus spent as a slave to the five he spent at Cydon in Crete before his strange experience occurs brings the event to the year 1495, when Benedetti was already in Crete.

spent twenty years of his life as a slave. Finally freed, he came to Cydon [on the island of Crete]. Five years later, while he was washing his face with cold water in the summer time, he fell into a fit of strong sneezing, accompanied by a great itching, and after the most strenuous efforts a piece of the arrow which had wounded him at Chalcis leaped forth from his nostrils. It was as long as his middle finger, with a steel arrowhead, and it left no other sign of the wound. While I was relating this event among the marvels of nature to the physicians of Venice and while some expressed disbelief, others (as usually happens) began to discuss the case and to attempt an explanation, suddenly and unexpectedly my friend Suirus, the Greek, arrived in Venice. Summoned by the discussion, he showed the arrow to those who disbelieved the tale and presented a most truthful testimony to what he regarded as a miracle. Since he had not come to the city within ten years before and came only by chance during the discussion of the case and the truth of it, he convinced the skeptics that what he said was true.

Chapter XXV

On the Eyes

Cicero and Galen[139] considered the very great utility of the eyes as a gift of the gods to the human race and that the head was created for their sake. First, the eyelids cover them; those which grow on each cheek (the lower eyelids) are also called eyelids. These are very soft to keep any contact from injuring the vision; they are equipped with a row of hairs for protection against flying objects. Both upper and lower eyelids open and close with wonderful swiftness. In sleep or in drunkenness they wink and close involuntarily. When the eyes are rubbed sneezing is checked; if anything falls on the eye either cheek droops away. The lightly prominent cheeks are placed under the eyes; they are called hypopiae by Pollux. Among the Sarmatians the cheeks project immoderately as a deformity in conjunction with a very small nose; in other people the nose is like a wall lying between the cheeks. The eyes are sunken, not without a useful purpose for they are hedged about by projecting areas on all sides. The suture which runs from the temples must finally be opened; this is the largest foramen or huge sinus of the head, called conchos by the Greeks. There is no larger; next comes that of the nostrils, and the smallest of all is that of the ears.

[139] Cicero, *De Natura Deorum*, 2. 57. 142, contains a discussion of the eyes. See also Benedetti, *De Re Medica*, II, 1, Prooemium, p. 43: Oculorum sensum diuinum Cicero uocavit quoniam nec tempus nec terminum habeat. . . . Galen, *De Usu Partium*, X, 1 (Vol. III K., 759–760). hypopiae: Pollux, II, 52, 53; X, 108.

Chapter XXVI

On the Causes of Colors in the Eyes

In man the eyes have more or less humor. Those eyes which have much humor are black since a large supply of humor cannot be seen through satisfactorily. Colors vary in the eyes in proportion to the quantity of their humor, as can very readily be recognized in sea water, for when it is sufficiently transparent it appears bluish gray. When it is less transparent it appears blue, and when on account of a gulf its depth cannot be determined the water seems dark and black in that spot.[140] Those eyes which have colors between these (or shades of them) differ in the quantity of humor and for the same reason bluish gray eyes cannot sometimes see sharply even in the daytime; black eyes cannot see sharply at night.[141] The former eyes on account of their slight amount of humor are less moved by the light and by visible objects since they are so humid and translucent, although vision, which is dependent upon the motion of the humor of the eyes, is due to its lucidity, not its humidity. Black eyes are moved less by their supply of humor, for the light is faint at night, and at the same time any humor whatever is more difficult to move at night. It is suitable for the humor to be moved where there is very much light but not to be moved more than is proper, for a gentler motion is displaced by a more violent one. Thus those people who have turned their gaze away from a stronger color or from the sun to the darkness do not see. For a motion already strong within it prevents that motion which takes place externally. Vision, finally, whether strong or weak, cannot see shining things since its humor is affected and moved beyond measure. Illnesses give an indication as to the color of the eyes, for glaucoma occurs rather in black eyes and purblindness in bluish gray eyes. The former is due to abundance of humor, the latter to dryness. That vision, lastly, is the best which contains moderate humor.

Chapter XXVII

Those Eyes Which Have Sharp Vision

In the middle of the eyes is a pupil by means of which we see. Its narrowness does not allow that

[140] Anyone who has seen the beautiful variations of color in the Mediterranean waters off the shore of southern Sicily, for example, can well understand Benedetti's analogy from nature. It is drawn from Aristotle, *De Generatione Animalium*, V, 1, 779 b 25–35.

[141] Aristotle, *De Generatione Animalium*, I, 780 a 5: ". . . they see less well in the dusk, for the nocturnal light is weak . . ."; 780 a 15: "what is called 'nyctalopia' [attacks] the dark-eyed" (translated by Arthur Platt, Oxford Translation of Aristotle, 5); see in addition Benedetti, *De Re Medica* (Venice, Giunta, 1549), II, 31, p. 73.

uncertain gaze to wander but directs it as by a channel; for this reason things seen at an angle easily avoid one's sight. The narrower the pupil the more keenly it sees, just as those who gaze during the daytime at the stars from pits or wells since the motion of sight is not dissipated in the vastness of space but is shot forth in a straight path, like those eyes which we call deep set. But those eyes are more dim which project too much; they are called *emisitii* (prying).[142]

Chapter XXVIII

The Defects of Eyes

To turn to the defects of eyes, man of all the animals is the only one in whom eyes of different colors can be seen.[143] In other animals the eyes are the same, except in horses, according to their species. Man alone suffers distortion of the eyes, whence arise the terms cross-eyed and blinking with a cast. These flaws are often contracted through the carelessness of nurses. Medium-sized eyes are considered the best by all. The eye is composed of thick and thin membranes and of hard, firm material against cold and heat. There is a continual humid saliva in them since they must be slippery and movable on account of things that appear to them. A supply of tears constantly drips from them, whether by force or involuntary action or through cold or due to a blow. Around the pupil the blackness varies; hence this part is called the iris by some people. It is seen to be dark or black, gray, blue or blue gray; in some it is red, yellow, or goatlike, which indicates the best manners and is believed to give a clearer vision.[144] A moderate location [of the eyes] also is an indication of the best manners, as with those who wink moderately. But those eyes which are rigid signify impudence; those which wink too much indicate inconstancy. The eyes grow stiff with illnesses, as in those who are terrified. In the eyes the soul itself is believed to dwell; for this reason those who kiss them believe that they are reaching the soul itself.[145] Some eyes see only at a distance, others only when objects are moved close to them. Still others, on the other hand, who have dim vision see better than other people at night. These are called purblind or nyctalopes. My friend Hermolaus [Barbarus] suffered from this defect. When we had something to read during the daytime he handed over the part to be read by daylight to me and he listened as I read. When evening came, however, and I could scarcely see anything he took the book and read quite accurately; he used to say to me: "My Alexander, bad things sometimes give way to good." Some people when awakened at night gaze at something for a little while in the manner of a cat; but gradually as the shadows draw away they can discern nothing. In many people when suddenly awakened the eyes appear to sparkle.

Chapter XXIX

On the Membranes of the Eyes

Modern physicians believe the eyes are covered with seven membranes, although the ancient doctors taught that there are four or only three. The former are certainly in error when they conclude that the exterior and interior half-tunics or semicircles are of a different nature, for they cut the eye first into prior and posterior parts. In the former they unfold four membranes, in the latter three only which are connected with the first four.

Chapter XXX

On the Cornea of the Eye

The first membrane of the eye is called the cornea because it is like translucent horn. It is pervious, very bright, hard, sinewy, and thick and does not perform its function in the depths of the eye but is only attached to the other inner tunics. It is clear and almost colorless so as to receive the forms or images of visible objects.

Chapter XXXI

On the Sling or Hollow of the Eye (Funda)

After the cornea there follows a part [sclera] that is hard, external, and white, in the shape of a sling, whence comes its name. Sometimes it is suffused with bloody veins or with bile, the first indication of the king's disease [usually indicating scrofula; but the

[142] Pliny, *Natural History*, 11. 37. 53. 141; Plautus, *Aulularia*, 1. 1. 2. Benedetti says his own father lived to be eighty without using spectacles: *De Re Medica*, II, 1. pits or wells: Aristotle, *De Generatione Animalium*, V, 1, 780 b 20.

[143] Aristotle, *De Historia Animalium*, II, 492 a 5–15.

[144] Aristotle, *De Historia Animalium*, II, 492 a 11; Pliny, *Natural History*, 8. 50. 76. 203; 28. 11. 47. 170.

[145] Pliny, *Natural History*, 2. 54. 146, p. 523, Loeb Classical Library: hos cum exosculamur animum ipsum videmur attingere . . . animo autem videmus, animo cernimus. . . . This is a reference not mentioned by Sir Stephen Gaselee in an article in which he traces the entire idea briefly through Greek, Latin, and English literature: "The Soul in the Kiss," *The Criterion* **2** (1924): pp. 349–359; see also A. J. Festugière in *Revue des études grecques* **65** (1952): p. 260; A. D. Nock in *Gnomon* **29** (1957): p. 526; and Georg Luck, *The Latin Love Elegy* (New York, Barnes and Noble, 1960), p. 40. There is a verse translation of the anonymous poem on the soul in the kiss in Aulus Gellius printed in my anthology, *Latin Poetry in Verse Translation* (Boston, Houghton, Mifflin Co., 1957), p. 306. See now for a general survey of the subject Nicolas J. Perella, *The Kiss Sacred and Profane: an Interpretive History of Kiss Symbolism and Related Religio-Erotic Themes* (Berkeley and Los Angeles, University of California Press, 1969).

symptom here is that of jaundice]. It is joined to the fleshy corners of the eyes and is called albugo and the white of the eye. By others it is called rhagoides, rhogoides, haemathoides, charoides, for under it is a humor very similar to water. To this the first of the interior membranes is connected; it veils the eye in the middle in its inner part and is similar in nature to the previous membrane; some people call it the sclerotica. The third of the exterior membranes which lies beneath the cornea is called the phacoides or lenticular membrane [iris]. The pupil forms a window in the middle of this membrane to which images are presented. In this membrane a great diversity and difference of colors is recorded as with a varied disk seen through by means of the cornea.

Chapter XXXII

The Second and Third Tunics of the Eyes

To this tunic another is attached, opposite to it and from the interior. This is called the secunda [sclera]; by others it is called thogoides [rhogoides], the same as the previous tunic. There is another very thin membrane likewise which is number four among those thus far enumerated. This is called arachnoides or amphiblestroides [choroid]; it has the same nature as the tunic posterior to it, to which the arachnoides is attached. People think this latter membrane is called only the reticularis [*retina*].

Chapter XXXIII

Another Numbering of the Membranes of the Eye

Others, however, of the most learned authorities have dealt briefly with these tunics. They have called the outermost tunic ceratoides or cornea; this one completely veils the entire eye. Under this cornea in the front part of the eye where the pupil is located is that humor similar to water. To this tunic the middle one [iris] is joined. It has a small foramen like a hollow window, circular and thin, with various colors seen through the middle of it from the cornea. It is hidden by the more distant parts and is somewhat more full; certain people call it rhagoides; it encircles the entire eye on the inside. Under these tunics is the very slender one called arachnoid by Herophilus, similar to a spider web [choroid]. Some of the more recent physicians insist that the cornea is attached to the membrane opposite it which we call the sclerotic.

Chapter XXXIIII

On the Humors Between the Tunics

Between the tunics are the humors. Between the cornea and the second tunic is the white humor [aqueous] through which images pass to the crystalloid [lens]. By means of this humor the straining or stretched cornea is lifted up, and when this humor is destroyed as happens in dead bodies the cornea is compressed or contracted. The pupil is contracted when this humor is dried out just as it is dilated when the humor is at its fullest amount or excess. Hence arise cloudings, obscurations, suffusions, and glaucoma of the eye which attack the region of the pupil. The angles of the eyes are troubled by epinyctides, pterygia, epilope. What they call mydriasis[146] (dilatation of the pupil) attacks the vision of the eye. Epiphore or persistent flow of tears affects the entire eye.

Chapter XXXV

On the Hyaloid Humor

The very thin membrane of the eye called the arachnoides contains a humor which from its similarity to glass is called hyaloides (or glassy) by the Greeks [vitreous]. This is neither liquid nor dry but a sort of hardened humor no different from the white of an egg. From this humor comes the color of the eye, black, gray, reddish, or bluish gray, since the outermost tunic is white.

Chapter XXXVI

On the Crystalloid (Humor)

Under these is a hardened drop of humor [lens] like the white of an egg from which arises the faculty of sight; it is called the crystalloid. On its account Galen calls the eye divine. The drop itself slants more toward the front; it is round in the rear, smooth and flat in front. Its power of vision is first located in the brain or in the heart as a place of chief origin. Since many men sleep with their eyes open the Greeks call it corybantia,[147] just as those whose eyes are open in fits of epilepsy see nothing while their souls are clouded. Thus we discern with the soul and see with the soul, as Plato[148] says, speaking of vision: "But if anyone shall approach the discussion of colors with earnestness he will appear ignorant of the great distance between human and divine nature. As-

---

[146] Galen, *Methodus Medendi*, III, 2 (Vol. X K., 171); *Galeno Ascripta Introductio seu Medicus* (Vol. XIV K., 769); Celsus, *De Medicina*, VI, 6, 37; Caelius Aurelianus, *On Chronic Diseases*, translated by I. E. Drabkin, **2** (University of Chicago Press, 1950): p. 569, Alessandro Benedetti, *De Re Medica* (Venice, Giunta, 1549), II, 32, p. 73 all discuss mydriasis.

[147] Pliny, *Natural History*, 11. 54. 147, p. 525, Loeb Classical Library: "Moreover, hares sleep with the eyes wide open, and so do many human beings while in the condition which the Greeks term 'corybantic' " (translated by H. Rackham). See also Benedetti, *De Re Medica*, II, 29, p. 72.

[148] See note 145 and Plato, *Timaeus* 45 D; 68 D; Pliny, *Natural History*, 11. 54. 145–146, p. 523, Loeb Classical Library: in oculis animus habitat.

suredly God is adequate to collect many things into one and conversely to disperse one into many, for he both knows how and is able to do it." For at almost the same moment the power of vision appears to take place in the heart and the brain just as with the reason, mind, and intellect and the other powers of the soul since these seem to take place in the site of principal origin, the heart. In large part the indications of the soul appear in the eyes, in happiness, love, and hate, when the eyes are gentle or stern, just as they are fierce or ardent in a state of anger.

Chapter XXXVII

How Suffusions (Cataracts) Are Cured By Hand

Since I have mentioned suffusions, I will say that they are called hypochysis (cataract)[149] by the Greeks, and the oculists speak of them in the eyes alone. How suffusions must be cured can be demonstrated in the eye of a cadaver. There are various kinds of suffusions. If they are small and motionless, of the color of sea water or shining steel and leave some sense of vision at or from the side in the form of light that reaches the eye, there is hope for their cure. If they are large or if they are black and the part of the eye that is lost (or obscured) changes from its natural shape to another, if they are blue or like gold, if they slide and move about here and there, there is little hope for a cure. Nevertheless, well-nigh worse are the eye troubles which come from a graver ill and arise from greater pains of the head or from a rather strong blow; there is no complete cure if the patient is aged since he has dim sight even without any new defect. Youth is not a suitable age for the cure of cataract but middle age is; if the eye is not small or concave (hollow) it is quite suitable and if its suffusion or cataract has a certain ripeness. Then one must wait until it no longer flows with liquid but appears to have a certain hardness. The physicians of our time offer the patient food of bad quality in order to hasten the maturation of the cataract since the maturation of the ill is hastened by evil vapors from the food. For three days before the operation for cataract the patient must eat and drink moderately and on the day before abstain entirely from both food and drink. Next the patient must be seated in a lighted place with the illumination coming from a direction opposite so that the physician will sit across from the patient and a little higher. An attendant will sit behind the patient and hold the latter's head in a motionless position, for a light movement (during the operation) can deprive him of sight forever. In fact, to render immobile the eye that is to be cured a cloth of cotton or wool can be laid on or bound upon the other eye. The left eye is to be operated on with the right hand and the right eye with the left hand, and both eyes by a person who is ambidextrous. Then the needle must be quite strong and grasped by its thicker part to prevent it from slipping; it must next be lowered (upon the eye). It must be held straight when the outermost tunics from the middle between the black of the eye and the corner nearer to the temple are affected by cataract, so as not to injure the vein. One must not, however, lower the needle too gingerly into living bodies since it is received in an empty place (that is, without bone or flesh) where not even the moderately skilled operator can make a mistake for there is no resistance to the pressure of the needle. When contact is made with the cataract the needle must be slanted, gently turned, and slipped beneath the area of the pupil. The point of the needle can then be seen through the cornea as if through glass. When the needle moves the cataract more vigorously the operator must press downward until the cataract rests on the lower part of the eye. If it clings there the operation is completed. But let us pass to the remaining parts of the head.

[149] Celsus, *De Medicina*, VI, 6, 35; VII, 7, 13–15, on the operation for cataract.

Chapter XXXVIII

On the Ears

The ears are always open since the sense of hearing is also necessary for those who are sleeping, in order that we may awaken at a sound. Those parts which stick out widely are called cypseles[150] or hollows of the ear [pinna] by the Greeks. Our doctors speak of the many folds of the ears called cochleae (snails), which are designed to prevent sounds from slipping out of them. Therefore the ears have hard, flexible entrances [external auditory meatus], which are called scaphi (boats or skiffs). Sound is amplified by repercussion in these entrances. The highest part of the ears is the alula or little wing; the lowest part is the lobe. In man the ears are so situated that they receive the air in the upper wing. They are cartilaginous so as to project; otherwise they would be flaccid and hamper the intake of sound. In proportion to his size man of almost all animals can hear least from a distance, although he is best of all at distinguishing between sounds. The reason for this superiority is his purity of sense, which is least terrestrial and corpulent. Sound, as Plato[151] believes, penetrates through the ears, brain, and blood as far as the soul, into the seat of the liver. Aristotle,[152] however, says that from the ears passageways penetrate to part of the occiput

[150] Pollux, *Onomasticon*, 2. 85.

[151] Plato, *Timaeus* 67 B.

[152] Aristotle, *De Partibus Animalium*, II, 10, 656 b 15–20, on the channels (poroi), not nerves, with William Ogle's footnote 2 in his translation; *De Generatione Animalium*, II, 6, 743 b 36; 744 a 1; V, 1, 781 a 20.

just as the passages of the eyes stretch to the veins scattered around the brain. Elsewhere he says the passages of smell and hearing reach the external air, and are full of native spirit; originating from the heart they extend to and end at the veinlets joined to the brain. Again in another place he says the passages of all the organs of sense extend to the heart. The indication of the differences both of sound and of odor lies in the purity or wholeness of the outermost membrane of the organ of sense; in these membranes rather bitter impurities are concocted. Those who hasten to purge the bile by this part of the body are easily overcome because of the passages. Entrance for little beasties or insects who wish to break in is made more difficult by these membranes, and they are also restrained by the viscosity of the membranes. In the depth of the ear's convolutions there is a membrane called the eardrum (meninge). The impact of the air ends at this point and is then carried to the brain nearby since this membrane comes from the brain and by its connection is joined to the tongue. Thus words are learned which are heard before and then pronounced by the mouth; for whatever movement passes through this organ of sense, such a movement, as if copied from the same likeness, is acted upon by the voice, so that what you hear you may also say. Aristotle,[153] followed by Averroes, declared that these passages of the senses extend to the heart because those who yawn, or scratch their heads, or breathe with great labor, hear less well, which does not happen to the other senses since the extremity of the organ of hearing terminates at the spiritual part and is shaken and moved together with the spiritual instrument whenever it moves the spirit; and important proof of this is that anyone who coughs to relieve himself shakes his ears as sneezes shake the nostrils. If the membrane of hearing is affected by humor or spirit this brings deafness, or heaviness; murmurs, whistling, ringing, and other uncertain sounds cause deception upon it. Under the ears tumors grow and take root violently; they often break out again with an offshoot in spite of remedies and after the tumors the parotid glands swell up.

Chapter XXXIX

On the Nose

The more prominent part of the face is the nose. It is cartilaginous and begins from the bone between the eyebrows which we call ethmus. The nose ends in a globule and in man is moved by the motion of the upper lips. In dogs the lips are moved and contracted by their own pair of nerves. We draw in and expel air by means of the nose, and here sneezing is incited since the nose has a direct pathway to the brain. It is properly situated in the upper part of the body since every odor is carried to the upper regions and makes the first indication of food and drink. For this reason the nose is in the vicinity of the mouth. In man the sense of smell is weaker than in other animals and for this reason is extremely attractive, as when we smell rose water. In this organ of sense we draw a mucous humor which is not completely purged from the brain. The nose in both sexes has hairs which are useful for catching or repelling dust and other objects. That which divides the nostrils we call the interseptum. The internal foramina of the nostrils are called thalamae by some people; the phlegm distills from these. They are also called lacus or torcular. Thence to the end of the nose the foramina are simple; they are again divided into two passages pervious to the throat and make exchange of the spirits. I shall speak of this in its place. The highest narrownesses of the bones are called isthmoides. In them occurs ozena,[154] and ulcer of foul humor which distills into them (ozaina, a fetid polyp). Polyps and carcinomata infest the nostrils. If the vein of the nostrils is cut it pours out blood sometimes so unceasingly that there is no way of stopping it. Nowadays there are ingenious men who have shown how to correct a nasal deformity. Frequently this method is to mold a flap cut from the patient's arm to the shape of a nose and apply it to the stump. For they excise the topmost skin of the arm with a scalpel. Having made a raw surface if the nose has to be freshened, or if the nose was recently cut off, they bind the arm to the head so that wound adheres to wound. When the wounds have conglutinated, they take from the arm with a knife as much as is required for the restoration of a nose. For the connected blood vessels of the nose offer nourishment to the flap, and finally a covering of skin is produced, with hairs sometimes growing there after the nature of the arm. And thus in a most remarkable manner they form a new nose. They make nasal passages, with bold resourcefulness having nature do their will. This supplement scarcely endures the severity of winter and during the early part of the treatment I advise not taking hold of the nose for fear it may come off.[155]

---

[153] Aristotle, *De Generatione Animalium*, V, 1, 781 a 30: Aristotle does not mention head-scratching, however, as a deterrent to hearing. On yawning as a deterrent see *Problems*, XXXII, 13.

[154] Benedetti, *Opera Omnia* (Venice, Giunta, 1533), *De Re Medica*, 99, describes ozena as a purulent ulcer; see also Pollux, *Onomasticon*, 4. 204; Galen (Vol. XII K., 678).

[155] The passage beginning with "Nowadays there are ingenious men . . ." is quoted from the translation by Martha Teach Gnudi and Jerome Pierce Webster, *The Life and Times of Gaspare Tagliacozzi, Surgeon of Bologna 1545–1599*, etc. (Milan, Ulrico Hoepli; New York, Herbert Reichner, 1950), pp. 112–113; they write as follows: "The fourth early reference to the Brancas is that of *the first medical writer since ancient times to mention this type of surgery* [italics mine]. Alessandro Benedetti (1460–1525), a native of Legnago who taught surgery and anatomy

Chapter XL

On the Lower Jaw

The forward part of the lower jaw is called the chin or the posterior mandible. Ancient suppliants grasped it as they begged. The ends of the mandibles are as if twin-horned; one process [coronoid] is called proboles by the Greeks. It is wider within and becomes slender at the vertex. Proceeding farther it passes under the yokebone (hyoid) and above it is attached to the muscles of the temples. The other process [condyloid] is shorter and rounder. In the sinus which is near the foramina of the ear it clings in the manner of a hinge; there, inclining this way and that, it offers the faculty of motion to the jaw. The jaws are covered by the cheeks, which indicate hilarity and laughter since they are made quite mobile by their muscles. The art of dissection shows that teachers or demonstrators of it must be warned about wounds in this region, as when the lips, cheeks, forehead, and eyebrows are by some chance to be dissected. In the case of the first two, section must be made from the nose downward; in the case of the last two, section is made from the vertex to the eyes. The jaw is often propelled or pushed forward, now on one side, now on the other, now on both sides, so that the chin is inclined to the opposite side by dislocation and the teeth do not respond to their equals in both jaws but the canines come under the incisors. If the dislocation is on both sides the chin is moved forward and the lower teeth far exceed the upper teeth in being thrust to the front and the muscles appear tense above them. The method of forcing the jaw bone back into place must be sought from Celsus[156] and the later anatomists. Now we must pass to the veins and arteries.

BOOK FIVE OF THE ANATOMICE OF ALESSANDRO BENEDETTI, PHYSICIAN

Chapter I

On the Veins, Muscles, Bones, and Nerves Which Pass Through All the Parts of the Body

Since all the members of the body are in constant need of nourishment the Lord has established channels of spirit and blood for irrigating the members by passageways hidden under the skin throughout the entire body just as men make water ditches that lead off to irrigate their fields. By means of nerves likewise, although they do not proceed from one stock or source to all parts of the body, the pathways of the senses are distributed like the roots and fibers of a tree. The bones of the arms and legs and the ribs arise from the spine, hard, stable, heavy, as the firm foundation of the entire body. For this reason they are stiff and dry, without sensation, since otherwise they would be continually subject to pain. Their ends cohere to each other, one with a hollow end, another with a round end, or each one hollow as in the middle bone of the heel. All of the bones are attached with sinews just as a ship is constructed from many pieces of wood and nails fastened together. Therefore, Giacomo Cavallo, uncle Angelo, Antonio of Faenza, Gerolamo Cordis, Domizio Gaverde and Luca Donato, son of Andrea, and Marino Sanuti, son of Leonardo,[157] both of noble family, in every natural object there is something marvelous to see, and since the greater part of the weary crowd of onlookers who thronged our dissecting theather both day and night have left vacant its seats and steps we shall examine in a manner more leisurely the nature of the veins, arteries, muscles, nerves, and bones, unhampered as in previous public dissections by eager youth. Together with a few onlookers we shall consider more closely those things which others do not wish to gaze upon.

at Padua, described the principal points of the procedure in more detail than Fazio had done. He, too, specified the 'top skin' as the material of the flap. He did not name the surgeons who performed it nor the place where they worked, but the phrase *aetate nostra* (his book was published about 1497) would indicate that the Brancas were meant." Since we know that Benedetti was writing in 1483 this remarkable description of rhinoplasty can be referred to an even earlier date than Gnudi and Webster realized was possible. See Ladislao Münster, "Alcune considerazioni e precisazioni a proposito di un lavoro su Alessandro Benedetti, con riguardo per la rinoplastica," *Rivista di Storia delle Scienze Mediche e Naturali* **47** (1955): pp. 1–6, on Gaspare Speranza Manzoli of Bologna as a predecessor of Tagliacozzi and (p. 2) Benedetti as "precursore dell'anatomia patologica moderna," suggesting a detailed comparison of the *Anatomice* and the *De Re Medica* of Benedetti for a full understanding of his work in pathology.

[156] Celsus, *De Medicina*, VIII, 12, on the method of reducing a dislocated jaw.

[157] Marino Sanuto, il Giovane, of Venice, (June 22, 1466–1535), son of the patrician Leonardo and a cousin of Marco, was a senator of Venice, a *Savio agli Ordini* and quaestor of Verona, an early member of the Aldine Academy, collector of books, manuscripts, and inscriptions, and to whom Aldus Manutius dedicated editions of Catullus, Horace, Ovid, Propertius, Tibullus, and Politian, wrote an important set of *Diarii* in many volumes (according to different editions, 56, 58, or 59), covering the years 1496 to 1533. Among his other works is a *De Adventu Caroli Regis Francorum in Italiam, Adversus Regem Neapolitanum* A. D. MCCCCXCIV which may have stirred Benedetti to write his own *Diaria de Bello Carolino* (1496). The relation between the two books should be explored. Giovanni Garzoni (1419–1505) also wrote on the subject: Tioli MSS, Bologna, VI, 262.

Chapter II

On the Blood and Spirit

The blood of man among the animals is particularly abundant. Foolish antiquity believed the soul had its habitation in the blood and that when it was spilled forth a man had to die immediately. The blood is contained in the heart as in a spring and in veins like rivulets which arise from the heart. It is generated by concoction or preparation from bloody matter (*sanies*). The spirit likewise originates from the heart and the aorta artery is its receptacle. But it is proved that neither blood without spirit nor spirit without blood can be found. The heart palpitates with spirit except in case of defect or failure. And in man alone of all the animals the heart is stirred by hope and expectation of what is to come in the future. The spirit contains the purest blood which is received by the middle ventricle (*domicilium*) of the heart through the surrounding vein; however, it is digested before in the material or substance of the heart. The veins of blood never move but the aorta palpitates or pulsates in the same manner as the heart does. By means of this pulsation and bubbling the spirits hasten to the extremities of the body, just as the movements of the sea which are observed near the shores or just as a pot sends up bubbles while it is boiling. Others attribute these movements to the arteries, that is, the movements of contraction and dilation, calling the former systole and the latter diastole. A proof of this bubbling or boiling is the fact that when the aorta is injured or cut the spirit spurts forth with the impulse of dilation. In the same manner a man's semen spurts forth in coitus, a certain proof of the aforementioned bubbling. But later physicians think that the spirit is contracted and expanded and have added intervals between two contrary movements, which is necessarily understood only for the movements of the receptacles. Whole blood is red, sweet, thin, pure, if the body is in good health and middle age. When it is impure it is black or watery or too full of bile or too thick or of bad flavor. In very fat people there is less blood than in thin people. A thin blood is emitted in place of sweat when someone is bitten by the haemorrhois snake, as I wrote in my book on envenomed creatures.[158] And from Christ our God when He was exhausted with hunger, sorrow, suffering, and sweat, the blood flowed forth.[159] In those who are sleeping the blood deserts the extremities and seeks the inner regions; this also happens to those who are cold or seized with fear. It is useless to perform blood-letting upon them. In the meseraics bloody matter is prepared and draws its diluted color from the chyle (*succo*) of the stomach and intestines. If the blood putrefies it is changed into pus and bad humors and, growing warm, is transmuted into a porous substance (*poroides*). From good blood fat is created; from corrupt blood skin eruptions, the itch, leprosy, and warts are generated. In the liver and lungs it grows hard like a stone. In young people the blood is thin and pure, thicker and blacker in old age. It is viscose like glue and is neither thin nor thick in middle age.

Chapter III

On the Major Vein and the Aorta

When all the intestines and spiritual members are removed two large veins appear along the spine. One is larger. The other is smaller and posterior; it is called the aorta. Both clearly arise from the heart where the fifth spinal vertebra is situated, for they proceed from the heart as if from a root stock with their thicker stems. In the liver, however, and the lung the veins, as I have said, are more slender and pass through the material of each, for they are offshoots of the two veins mentioned above. The veins receive the blood and spirit from the heart. Their nature and location can scarcely be observed in living bodies nor their ultimate beginnings in cadavers. The aorta is so called because its sinewy part is more evident to view in dead bodies.[160] The veins are

[158] On the haemorrhois see Pliny, *Natural History*, 20. 20. 81. 210; 23. 1. 23. 43, and on poisons Paul of Aegina, V, 16. Benedetti, *Opera Omnia* (Venice, 1533), VIII, 26, p. 131 (there is an error in book numbers; actually it is Book VII), also mentions "in primo venenatorum libro." Another reference to this work in six books appears in the *Opera Omnia* (Basle, Petrus, 1549), VII, 26, in Benedetti's preface to Maximilian I: ob eas res, inuictissime Imperator, libros sex de uenenis ac uenenatis animalibus tibi dedicaturi sumus. . . . G. Mazzuchelli, *Gli Scrittori d'Italia*, etc. (1760), II, 2, 812, lists this work among the writings of Benedetti: "anche i sei libri *De Venenis et venenatis animantibus* che andava scrivendo, la quale però non è mai venuta alla luce (*cf.* Zeno, *Dissert. Voss* II, 45)." This book, as to whose state of completion at the time Benedetti wrote his dedication to Maximilian I we know nothing, was, like his edition of Paul of Aegina on which he was working when he died, never published. The book on poisons is mentioned as early as 1493 in the dedicatory letter to Benedetti's *Collectiones Medicinae*.

[159] Berengario da Carpi, *Commentaria*, etc. (1521), fol. CCCXXXVI, discusses the blood of Christ upon the cross in connection with the fluid in the pericardium and points out the view of "aliqui proterui catholicae fidei" who declared that if this blood is natural in man the water mingled with blood which issued from Christ's side at the thrust of the soldier's spear was also natural and not miraculous. Berengario denies this rationalizing explanation of the phenomenon by some Renan born four hundred years too soon, insisting on the miraculous nature of the effusion since ordinarily the water is indistinguishable from the blood in color or consistency as it issues from the pericardium: the water which flowed from Christ's body was *separate* from the blood, a veritable miracle. Berengario was right, of course, as he so often was in a matter of controversy. Note Vesalius's refusal to enter the argument in *Fabrica*, VI, p. 585.

[160] Aristotle, *De Partibus Animalium*, III, 5, 667 b discusses the vena cava and the aorta in their relation to the heart. The etymology of the word aorta is "lifted above" the vena cava, hence more visible in dead bodies. The heart is also abundantly supplied with sinews: *ibid.* III, 4, 666 b 15; see *ibid.* II, 651 a 4 on the fibers in the blood.

so arranged that one is above, the other below. The larger vein holds a position rather on the right, the smaller vein holds one on the left; it issues from the middle sinus of the heart with a more narrow pipe from the side of the latter. The former, larger vein arises from the largest and uppermost sinus and holds a position at the right. It receives the shape of a vein from the middle sinus; one might rather say this ventricle of the heart is a part of the major vein which with its root passes through the heart when it comes forward. This vein is made of membrane and skin. But the aorta appears to be particularly sinewy so that by its nature it distributes the streams of blood and spirit throughout the entire body. This is more clearly observed in thin and emaciated bodies, in which many veins, and these the larger ones, are visible, as in the drying leaves of trees in which only the veins are left. The cause of this phenomenon is the blood and because the vein has a comparative relation in size to the blood, since the vein is potentially body and flesh, or because the vein corresponds in size to body and flesh. Likewise, the ampler veins of the body endure or hold out but the very small ones are made into flesh by their action, although the power may be nonetheless in the vein itself since it is also flesh. Our physicians, however, assign other humors between the flesh and the blood; thus the blood is contained in every part of the flesh and flows forth when a wound is inflicted. Therefore, the flesh cannot exist without a vein but no vein is manifest. From these veinlets a sweat like dew is elicited or a yellow bile of a fetid odor or lentil-shaped spots are scattered in fevers or warts appear from the black bile. Furthermore, in the substance of the blood there is a certain kind of fibers, which I prefer to call stamen or filament, called ines by the Greeks.[161] This does not appear in all animals but in those only in which it coalesces with a firmer combination. For if this fiber is removed, as in stags and deer, then blood does not coalesce except in a fluid combination as milk grows thick without coagulation; but this combination is observed in the blood if it is collected into hot water from the vein.

Chapter IIII

On the Branches of the Major Vein and of the Aorta

The major vein as it ascends is split into two branches, to complete the entire account of the sequence of veins. One of these [pulmonary] passes to the lungs with a bipartite branch, since the lung is twofold, as I said when I described the lung in its proper place. The second of the two branches descends through the spine [inferior vena cava] and is stretched upward to the neck where it [superior vena cava] is extended with a rather larger double offshoot [brachiocephalic] to right and left. Both of these, before they enter the armpits, send out single veins down from each side of the spine [internal thoracic] to each of the ribs, which as I indicated are seven in number, and the same number to the vertebrae. These also nourish the chest. In smaller animals, as we noticed, there is only one vein which stretches along the spine from the right side downward and nevertheless nourishes the sequence of ribs on both sides. The major vein is split in two near the vertebra above the kidneys but again with the upper part which proceeds from the heart it is split into a quite numerous set of offshoots and reaches two places since it sends some offshoots to the sides and shoulders which then are carried through the armpits to the hands [subclavian]. From these same branches where they are split above the gullet the veins called jugulars or sphagitides are led upward on each side of the throat; when these are seized on the outside they cause a man to sink to the ground senseless. Thus extended and embracing the artery they are carried to the interval of the ears where the jaws come together with the upper part of the throat. Once more they are split into four veins, of which one passes through the neck and shoulder and descends and with the previous offshoot is joined to the curve of the arm. This vein is called the cephalic. The second vein ends at the palm and fingers, the third stretches to the brain from each region of the ears and is scattered into many very slender branches and passes to the membrane which is the custodian of the brain, which later physicians have called the marvelous net. The remaining branches go partly around the head and partly to the region of the senses, to the palate, tongue, and teeth and terminate in a slender series of veins. Part of the major vein is sent to the heart; this is carried higher, as I said, and passing through the diaphragm it is connected with the liver. Thence, from it as it passes through the liver two branches extend; one of these ends at the diaphragm and midriff, the other seeks the upper regions and led off through the armpit to the muscle of the right arm it reaches out and joins itself to the other veins near the inner bend of the arm. Hence when blood is let from this point almost all liver pains and afflictions, such as obstructions, abscesses, and the like, can be relieved. It is called the regal vein [basilica]. Another part of the same vein, short indeed but thick, seeks the spleen and disappears in it. From the same larger vein in similar fashion an offshoot is sent to the muscle of the left arm, but the offshoot which proceeds to the right arm also stretches to the liver. Two other offshoots are propagated from the major vein; one of these goes to the omentum, the other to the pancreas from which a great number of veins pass through the mesentery. All these end in one large vein stretched out through the entire intestinal region and abdomen

[161] The ines are described by Aristotle, *De Partibus Animalium*, II, 4, 651 a 3; *De Historia Animalium*, I, 17, 497 a 21; and Pollux, *Onomasticon*, 2. 39.

up as far as the gullet. This vein scatters out many branches from those regions through which it crawls. Both the major vein and the aorta stretch together and continuously to the kidneys, as I said. Here they adhere with their heads to the kidneys and are split in two like the Greek letter lambda turned upside down. The major vein slants more than the aorta into the posterior part of the body. The remaining offshoots have been for the most part mentioned in the discussion of the internal members. It might not be unsuitable to run over them in order again.

Chapter V

On the Vein of the Middle of the Forehead and of the Temples

The forehead vein, which is also called middle, is seen to be swollen in many people. In those who are insane or stupefied its illness is a chronic headache that seems to be located in the occiput, for this vein draws matter from a higher place. In apoplexy it is quite effective to let blood from the sphagitides (jugulars) which are sinuous in the temples; in illnesses of the eyes they are easily cut. Some people apply cauteries to the parotids when a cataract begins to form, just as they do to the temples. After this, dissectors tear the entire skin of the arm in order to display the rest of the veins.

Chapter VI

On the Common Vein of the Arm

The liver vein goes to the liver, the cephalic vein to the head, as I have said. To these is attached a vein called common. The first two veins go to the very end of the hand and the fingers. Under the liver vein there extends an artery which makes incision of the vein a danger; this artery climbs strongly along with the vein, as is observed by its pulsation. The cephalic vein ends between thumb and index fingers. Among the smallest fingers is the location of the splenary vein (as they wish to call it) and the substitute of the liver vein.

Chapter VII

On the Veins Which Are Connected With the Kidneys

The major vein is connected with the liver, spleen, and kidneys in which it is stabilized as though by anchors. Just as with the aorta in its members, so the substance for generation is prepared in the kidneys, whence blood flows in those who are about to give birth. I have already spoken about this.

Chapter VIII

On the Other Veins

In the nape of the neck there is a vein which is incised in those who suffer from headache. The veins of the neck and those behind the ears are also incised for bloodletting. In the nostrils, under the tongue, there are veins cut in cases of angina, and in the uvula, under the lip and iugulum, the right-hand side of the diaphragm and on the left also, there are veins which are bloodlet in illnesses of the liver and spleen. From the ham of the knee there are veins for bloodletting in sciatica, and from the ankle there is one which is bloodlet in illnesses of the vulva. Above the heel in the farthest part of the foot is the sciatic vein, or between the little toe and the one next to it. All these veins must be demonstrated by skilled dissectors after the skin has been removed. I should say, once for all at this point, that the location of the veins, just as the location of the suture of the bones when I spoke of them, is not always the same for many of them vary their position. But now I must run through the main features of the aorta.

Chapter IX

On the Aorta

The aorta vein [artery] is scattered in almost the same manner into branches which follow the rivulets of the major vein. The latter is almost always stretched out above the aorta and seems to be the guardian of the other vessel except in the upper arms and in the thighs where the artery and vein change places, the one on the right side climbing to the left and the other on the left climbing to the right, for nature always avoids perils and accidents and always presents the more ignoble part to dangers. The branches of the aorta are everywhere smaller but they have stronger tunics than those branches of the major vein. It is also connected with the spine by means of two slender veinlets for each interval between vertebrae and with almost the same number of veinlets for the vertebrae. It is likewise attached to the sinews of the heart from which a prominent vein is extended. It sends branches to the mesentery just as the major vein does, but they are much smaller and thinner, quite similar to fibers; I have pointed out that these branches, thinned out, end in hollow fibers. No aorta vein extends to the liver but only as far as the kidneys, just as the major vein is forked into the shape of the Greek letter lambda and is absorbed in the substance of the kidneys. Veins which pass out from the kidneys themselves and not from the major vein are carried into both coxae or hips. They become hidden there and then once more thrust forth they pass (as I have said) to the genital parts. Many well-filled veins pass from the aorta to the uterus as do also branches which reach to the rest of the intestines. The arteries ascending in the same way around the region of the throat are called parotids [carotids]; these proceed upwards and are extended beyond the ears. When these arteries are cut sterility results. They pass on to the brain and its protecting membrane, constituting that network [the marvelous net] with the accompanying branches of the major vein.

Chapter X

On the Arteries of the Legs

From the aorta near the groin two branches [femoral] descend to the legs, through the thighs and legs into the soles of the feet and the toes, where they end, in the same way as the arteries of the arms.

Chapter XI

On the Veins and Arteries Which Correspond to One Another

By the marvelous proof of experience it is agreed among authors that certain offshoots of the veins correspond among themselves. Galen[162] was admonished in his dreams that for a gravely diseased liver the vein which is located between the thumb and index finger should be incised for bloodletting. He obeyed the admonition of his vision and the patient was saved. This experience likewise is not to be scorned: Diogenes[163] the physician reported, and I confirmed myself in a case of very great pain in the stomach, that it is of very great assistance to let blood from the liver vein or splenetic vein. Aristotle[164] scolds the physician Polybius because the latter reported that a vein stretched from the forehead through the neck to the sides of the spine, then to the buttocks, thighs, and tibiae to the exterior malleoli or ankle bones and to the smallest toes. If this vein were cut the pains of headache were alleviated. Aristotle did not disprove this experience by Polybius but only the exact course of the vein which the latter had so rashly believed was the true one. And if all veins come together as one, for the same reason antiquity persuades us to rub the ham of the knee for relief of pain in the nape of the neck. In cases of nosebleed it is helpful to place cold compresses on the testes or to squeeze the extremities. A cupping glass placed under the breasts checks the flow of the menses. It is held that when the big toe is tied firmly to the next toe it will soothe a tumor of the groin. When the two middle fingers of the right hand are lightly bound together with a linen thread distillations and bleariness of the eyes are driven away. In cases of hemorrhoids the black bile is purged in a straight line or path from the spleen. Other similar remedies are associated with hidden offshoots of the veins, but this is sufficient concerning them. The same thing is to be observed in nerves since itching of the nostrils is an indication that the gullet is disturbed by animal spirits (?) from the stomach through the connection of the nerve pairs, as has been said. I have also noticed that coughing is provoked when the ears are cleaned on the inside.

Chapter XII

On the Members to Which Only the Blood Veins Pass, To Which only the Aorta Passes, Or to Which Both Pass

Both veins stretch to the brain as to the kidneys, stomach, omentum, mesentery, pancreas, tongue, eyes, lungs, and diaphragm. In the liver only veins of blood appear, none from the aorta since by the ministry of the liver the blood is prepared and transmitted to the sinus of the heart. The blood received by the heart from the liver does not lack vital spirit and has been prepared in the lobes of the liver, for to the heart the lobes of the humped part of the liver correspond in the great slenderness of its veins. This quality is called alloeotice.[165] The blood is prepared not in the fibers but in the substance of the heart for the generation of the spirit. But in the lung the fibers or lobes of blood and spirit by the marvelous work of nature do not correspond to the fistulous passages of air; this must be examined in the very great thinness of the fistulae. Nothing comes to the vulva from the vein, as has been said, since it requires little blood but from the aorta many veins assemble there.

Chapter XIII

On the Method of the Fibers

Since I have mentioned fibers I think something must be said about them here. The fibers have a nature between that of nerves and veins. Some of them contain a humor called sanies or bloody matter. They are not nerves since they attract the blood, nor are they veins since they are hard and not hollow and contain bloody matter as veins contain blood. The

[162] Galen, *Hippocratis De Humoribus Liber et Galeni in Eum Commentarii Tres* II (Vol. XVI K., 222).

[163] Aristotle, *De Historia Animalium*, III, 2, 512 a 30 on Diogenes.

[164] *Ibid.*, III, 3, 512 b-513 a 10.

[165] Aristotle, *De Sensu*, 4, 441 b 21; *Physica*, VIII, 5, 257 a 24; Galen, *De Usu Partium*, IV, 7 (Vol. III K., 275); *De Simplicium Medicamentorum Temperamentis ac Facultatibus*, I (Vol. XI K., 380), and the Commentator on Galen, *De Sectis*, quoted *in extenso* in note 133 on the term alloeotice.

fibers of the arteries do not contain bloody matter but thin spirit and they end in a noteworthy thinness in the substance of the member. They stretch from the nerves to the veins and thence again to the nerves. None of them are found in the kidneys, few in the heart, but very many in the lung and the liver. The fibers in animals are covered partly with skin, partly with the material of the members. They are evident in the roots of plants. The fibers of the umbilicus in the uterus are very conspicuous because with their force of attraction they pour the blood through the veins of the secundinae as nourishment for the foetus. The common defect of all the fibers is obstruction with a thick humor, which we call oppilation; due to this, the fibers grow feeble and the patient is quickly debilitated, for the members gradually dry up like the branches of trees when their roots have been torn away since the stomach and intestines are like the earth, and the food and drink furnish nourishment. Those who think the material of the food enters intact are convinced by the analogy of plants, since the hairy thinness of the fibers disproves their view and reveals their ignorance. A certain man who was considered among the great physicians declared that he had seen a piece of mushroom emitted through the bladder but the example of that needle I mentioned earlier which was passed through the urine excites many people ignorant of dissection who falsely suppose that it passed through the veins of the mesentery from the stomach to the liver and through the major vein to the narrow parts of the kidneys and thence penetrated to the bladder, believing that the mouths of the fibers in living people are more wide open to receive both the needle and the piece of mushroom. For the needle with its point little by little and over the space of much time penetrated the intestines, making a way for itself as its own nature propelled it, and was delivered to the fleshy part of the bladder and thence was passed through the urine, just as also an arrow which in two years pierced the bowel with its point, a fact proven by wonderful examples.

Chapter XIIII

On the Origin of the Sinews

The system of the sinews in dissections is less clearly evident. Those who wish to contemplate more closely their origin, offshoots, and interlacing must do so in intact bodies, tied with various ligatures, that have been placed in a stream for maceration when the entrails have been removed and the skin pulled off. For the flesh and everything else that is soft in the body quickly decays. The sinews, however, and their interlacings cling stubbornly to the bones and reveal their various contexts when they have been sufficiently dried out in the sunshine, but the hideous sight of their tightened contraction deters physicians from them. The sinews originate in the heart[166] (since it is the first of all the members to be generated) without the assistance of the brain; among these is the sinewy aorta. In other respects the nature of the sinews is more clearly perceived to take its origin for the most part from the brain and its substitute, the medulla; the farther they stretch from the brain and the medulla the harder they become, such as those nerves we call vocal. Some are accommodated to the muscles, some to the bends and folds of the members. Those which lack a name are called nexus by some people. The round sinews are called cords. The flat shoulder sinew called platys by the Greeks makes for strength. The neck and shoulder blades are particularly held together by those sinews called tentondes (tendons) for the head is sustained by straight and powerful sinews on both sides. During almost all the flexures of these sinews one is always taut in order not to allow the upper regions to fall forward. There is another broad sinew, called the nerve of the poples, or ham of the knee, which is joined to the heel and renders the member suitable for jumping and, passing under the calves, for walking; it is necessary for running. I have spoken about the sinews of the head. The feet and hands abound in a supply of sinews woven around the joints thereof; it is more convenient to specify these sinews by letters of the alphabet since dissectors must point them out. There remain the nerves which arise from the spinal medulla.

Chapter XV

On the Nerves Which Arise From the Spinal Medulla

The spinal medulla descends from the brain. It is covered by a very thin membrane. If it is cut death results; hence it is called the custodian of the brain. From the medulla arise many kinds of nerves, for on both sides through the foramina of the vertebrae from the neck to the os sacrum thirty-one pairs of nerves extend. Two nerves go forth from each vertebra. From them power is furnished to the hands, feet, and other parts of the body by which necessarily and fitly life is maintained since the nerves provide the ability to feel and to move. From the lowest part of the spine

[166] Benedetti is confusing (and is perhaps himself confused) because of the ancient confusion of nerves and sinews going back to Aristotle, *De Historia Animalium*, III, 6, 515 b 15–20. Galen also uses the same word *neuron* for both nerve and tendon: see Charles Singer, Galen, *On Anatomical Procedures*, translation, pp. xix and 243, note 57. Perhaps it would be more advisable to use the word neuron throughout this chapter and to allow the medically trained reader to decide for himself whether in each instance Benedetti is speaking of sinews, nerves, or tendons. I have chosen to employ the less precise term sinews, which is itself a synonym for tendon. Note that Benedetti first gives Aristotle's view of the place of origin for the nerves and then the true origin, the brain.

nerves arise on both sides; passing under the hips and coming together they send single nerves to the knees which are called teretes or smooth. For the succession also of the species of both sexes they proceed to the genital members to furnish them with power in order that these may fulfill their functions. Thus almost no part of the body is without exhalation from the brain or the benefit of the spinal medulla. Hence it deserves to be called the long brain by physicians, extending upward like a thread lengthwise, lightly indicating the divided substance of the medulla. When the membrane is removed the brain is seen to be of almost the same material as the medulla, as is observed by the evidence of paralysis. For when one side or the other of the body is affected by apoplexy, only that side is endangered, the other remains uninjured. This is called hemiplexia, when the nerves which are the messengers of movement and sensation are impaired. By the nerves the sense of touch necessary to animation is diffused to all the members. There is no certain continuation of the nerves as in veins and arteries. All the bones are covered by sinewy membranes of the muscle and within the muscles there also runs a sinewiness. The bones are also joined with many sinews drawn around the articulations and bound with interlacings passing in and out and around and through the structure. Here are smooth sinews, there flat ones, elsewhere thicker ones, together with sinews of a membranous nature, each type where it is needed for the individual purpose. Some sinews mingled with flesh are more adapted for feeling; others which are smooth and simple cover the joints. All of them drip with continual moisture so that the members may be properly lubricated for swift action.

Chapter XVI

On the Nerves of the Neck (Cervix)

I have indicated that the nerves from the spine pass through foramina. But at the neck they issue from the joining of the vertebrae in order to avoid the necessity for frequent apertures which would weaken the body at this point. The neck contains many nerves (sinews), of which the platys is the largest. The jugular veins cling there to two huge muscles. The neck is subject to grave and perilous ills and to some which are extremely acute. By the strength of the nerves (sinews) the head is joined to the shoulder blades and the chin to the chest while the muscles hold the neck upright and firm. The Greeks called the front muscle the opisthotonos, the rear muscle the emprosthotonos.[167] They call the last of the three tetanos. The illnesses of the neck originate with catarrhs and descend also to the spine, for it is bent backward like a Scythian bow with a very acute disease, which often kills within four days. Among examples of it, Sanctius Pollanus, of a Cydonian noble family in Canea, Crete, suffered from this ill for forty days. When the material which caused it was digested and passed through the urine in the form of a milky fluid he was relieved; the use of castoreum [a secretion from beavers] was especially effective in achieving this relief.

Chapter XVII

On the Loins

The parts of the back in the region of the stomach are called the loins. Aristotle[168] reports that they are free of flesh so that they can be bent, for the joints everywhere are fleshless. The loins grow weary with venery and are often tormented by pain. Those who are weakened in the loins by a blow or a fall are in great danger. From the loins proceed the buttocks which are given the body as a cushion on which to rest; therefore those who are emaciated by illness cannot sit upon them. Cupping glasses applied to the buttocks are a good substitute for phlebotomy of the basilic vein. Now let us pass to the muscles of the body, whose arrangement I have in large part indicated in reference to the quality of the members.

Chapter XVIII

On the Muscles

I have said a few words on the muscles in the first book. Antiquity passed over their enumeration as a fruitless labor but paid attention only to those which were prominent in the members and those which were lethal when they were struck blows, for these muscles are more easily healed when they are cut or torn; the other muscles are of less account and less subject to fatality. All the muscles are composed of flesh, nerve, membrane, and sinew. When the muscles contract they extend the members; when they relax they draw back the members. They are confused and more perceptible by their motion than by their dissection in the forehead, head, eyebrows, lips, and cheeks. There are two in the upper jaw, and the same number in the cheeks and nostrils. In the lower jaw there are twelve, about twenty-three around the neck, thirty-two in the epiglottis, throat, and gullet, thirteen from the neck to the shoulder blades, and nine in the tongue.

---

[167] Aretaeus, *On the Causes and Symptoms of Acute Diseases*, I, 6 (translated by Francis Adams, Sydenham Society, London, 1856), p. 246. See also Plato, *Timaeus* 84 E for the terms tetanus and opisthotonus. Note that Benedetti has erred in stating the positions of the opisthotonus and emprosthotonus muscles by exactly reversing them; I have preserved his error in my translation.

[168] Aristotle, *De Partibus Animalium*, II, 9, 654 b 10–30; III, 9, 672 a 20.

Chapter XIX

On the Muscles of the Thorax

In the chest there are eight muscles which expand it and seven which contract it. Between each rib there are four so that the filaments of the four muscles which entwine the member are seen to be mingled with flesh. Around each shoulder above twenty-eight muscles are enumerated; some of these are split into two heads although they begin from a single one.

Chapter XX

On the Muscles of the Arms

In the arms there are twenty-four larger muscles. In the upper arms and elbows, in the palm and the hand when turned away and the fingers there are believed to be twenty-eight muscles to produce movement; most of these are so confused that they cannot be separated in any manner.

Chapter XXI

On the Muscles of the Back

There are eighty-eight muscles which bend and straighten the back. There are eight muscles of the stomach, concerning which I spoke in the first book. There are four for the male testes and single ones for the female testes, one only for the neck of the bladder, four for the male member, and the same number in the anus. One of these contracts and contains by voluntary action; of the rest one is connected with the male member; when this is titillated pathics or sodomists by the unspeakable vice of their nature burst forth in lust. For the same reason when those muscles are worn out with venery they allow the excrement to flow forth involuntarily.

Chapter XXII

On the Muscles of the Legs

Twenty-six muscles, almost the largest of all, are found in the leg. Two of them are quite prominent; one contracts, the other stretches the leg. From a lumbar vertebra it is joined by a nerve to the big toe for walking. The muscles which extend the leg are considered more noble; they are especially fatigued by running, walking, or standing. There is one muscle around each hip which is recognized as triceps [three headed] at the uppermost part and terminated in a double head at the tail, where it is connected. Some muscles are connected to the loins, others to the ilia. In each femur there are forty. Twenty-eight are joined in the knee and around the tibia. Around the bending of the foot twenty-two muscles for each foot are located, of which one sends out two cords from itself. Around the toes there are two muscles for each which send out two cords from each. Under the sole of the foot there are five muscles, of which some are so mingled with one another that they cannot be separated; when one of them is injured the rest are affected. The total number of muscles (as I have said) is reckoned to be around five hundred twenty-six.

Chapter XXIII

On the Bones and First On the Vertebrae of the Neck

I have said something about the bones in general in the first book, and the bones of the head, chest, and sides have been described in more detail in their special place. But now the nature of the rest of the bones must be investigated not only on the basis of authorities but more carefully (as Galen reports)[169] from cadavers. Some dissectors remove the muscles and flesh from half a cadaver and they boil it in a huge caldron. The boiling is completed when the flesh falls off the bones easily. They stir what is left with a laurel branch, carefully remove it, and then pick the bones slowly. During these dissections a certain physician ate some cooked human flesh in my presence and then spat out that which remained, so that he might thus indicate its flavor to those who asked him what it was like; he said it tasted like the flesh of a gazelle. When the bones are picked clean they are easily distinguished. Some people examine cadavers in tombs not without nausea because they are filthy. Cadavers washed in a stream or dissolved in lime are preferred. Physicians who treat wounds should often inspect bones of this kind of cadaver; those also who attend patients at their bedside ought not to neglect them. It happened once when a dissection had been completed at the University of Padua that a certain student kept the bones for his own use. Halfway on his journey to Venice at night he left his boat to dine and went to an inn with his companions. Along came a most impudent squad of tax collectors looking for contraband merchandise and found the box of bones the student had left behind in the boat. When they asked to whom these belonged and received no reply, the tax collectors carried off the bones and opened the box next morning in the presence of their overseers. When they found the bones, picked clean of flesh and shining white, lying among odorous herbs, they began to worship them with bared heads as though they were the relics of some saints. Then they brought the box to the highest magistrate. In the crowd that assembled

[169] Galen, *De Anatomicis Administrationibus*, I, 2 (Vol. II K., 220–221). In the story below about the student's box of bones Francesco is the son of Marino Sanuto, the Venetian diarist; see note 157.

before him Francesco Sanuti, a man of high reputation and standing in the legal profession, revealed that the bones were the remains of an anatomical dissection and ordered them to be restored to the student of medicine, who had by this time lodged a complaint about this stolen box. Everyone laughed loudly to see the frustrated greed of the tax collectors. But to return to the vertebrae of the neck. These consist of seven orbiculated bones, or bones that have a round shape. They are perforated in the middle by the medulla which descends from the brain, where the medulla is more abundant than in the spine. These vertebrae sustain the head; the first is called stropheus and epistropheus. Hippocrates calls it the odonta. The second is the axis since it sustains its burden like a porter. But Pliny[170] changes the order and makes the first vertebra joined to the spine the atlas; the last is the epistropheus by which the head is bent to left and right; between this vertebra and the following axis we bend the head to the shoulders and the chin to the chest. If the head therefore which is inserted by means of two processes [condyles] into two sinuses or folds of the highest vertebra falls forward so that the sinew under the occiput is extended and the chin is fastened closely to the chest then the individual can neither drink nor talk and sometimes the genital semen is involuntary emitted. Such people die very quickly.

Chapter XXIIII

On the Spine

The spine, which is the origin of the bones, is connected to the neck, according to Aristotle.[171] This spine is a round structure with foramina in its middle through which the medulla descends. It is called holy reed by the Greeks and sacred fistula by us. It consists of nodes which are called spondyles or vertebrae corresponding to each of the ribs, that is, the true and the false. Some people attach to the last of the ribs, that is, to the ilia the termination of the thorax. The first of the twelve nodes or vertebrae is called asphalia, the next the maschalister, so named from the arm pits, the rest pleuritae, from the sides. The twelfth is called diazoster from the waist, from which, according to Pliny,[172] a kind of sacred fire which goes around a man's waist is named. After these there are five nodes opposite the kidneys stretching out to the hips, round, more firm, short on each side. They send out two processes which extend from the spine; these are called cynolopha from their resemblance to dogs.[173] Muscles fill the loins; they are called psoae or ulpes by Cleanthus.[174] All the vertebrae except the three uppermost have sinuses cut in a little from their higher part in their processes [articular]. From their lower part they send out other processes facing downward. The lower vertebrae are equipped with sinuses on both sides into which the upper vertebrae are received. They are held together by many sinews and much cartilage so that with a moderate bend a man may stoop to his necessary tasks. After these vertebrae the bones of the lowest part of the spine constitute the os sacrum or holy bone. Paul of Aegina [175] calls it oropygium in birds. The os sacrum is formed of six vertebrae, smaller than the rest and more compact. They also call it spondylion, where the upper vertebrae leave off; these bones of the os sacrum are moved apart in childbirth by the marvelous work of nature. Galen[176] established four parts of the spine: the cervical region, metaphrenum [thoracic] (the broad part of the back behind the midriff), the lumbar region, and that which they call the os sacrum or others call the side. The cervical region has seven vertebrae, the metaphrenum twelve, the lumbar region five, and the os sacrum is composed of the remaining vertebrae. All told there are thirty vertebrae; each of these has foramina on both sides from which issue the same number of pairs of nerves, plus one more nerve. These engirdle the body and furnish motion and sensation.

Chapter XXV

On the Injured Medulla and the Dislocated Spine

If the medulla is injured the nerves are destroyed or distended and sensation dies in some member or other. Either one leg is weakened by some injury or the semen or excrement flow forth involuntarily. The same thing happens to those who suffer a dislocated spine, which cannot take place unless the medulla is bruised or the nerves severed. If the vertebrae are dislocated above the diaphragm the hands are enervated (or paralyzed), vomiting follows, and breathing becomes difficult. But if this dislocation takes place

---

[170] Pliny, *Natural History*, 28. 8. 24. 99; he uses the term "nodum Atlantion."

[171] Aristotle, *De Historia Animalium*, III, 7, 516 a 10.

[172] Pliny, *Natural History*, 25. 74. 121. p. 357, Loeb Classical Library: erysipelas, of which herpes or shingles is a variety.

[173] Pollux, *Onomasticon*, 2. 180.

[174] Cleanthus may be an error for Clearchus, ed. T. Kock, *Comicorum Atticorum Fragmenta* II, 72, who does use the word; I cannot find any reference to Cleanthus. See Hippocrates, *Mochlicus seu Vectiarius* (Vol. III K., 272); and Jones-Withington's Loeb Classical Library edition, Vol. 3, pp. 398–449; Aretaeus, *On the Causes and Symptoms of Chronic Diseases*, II, 3 (translation by Francis Adams, Sydenham Society, London, 1856), p. 341.

[175] It is difficult to locate the reference in the Greek text of Paul of Aegina since no index to it exists and the text is quite extensive. See Aristotle, *De Historia Animalium*, 504 a 32; 525 a 12; 618 b 33; *De Partibus Animalium*, IV, 13, 697 b 11.

[176] Galen, *De Ossibus* 7 (Vol. II K., 755); *De Usu Partium*, XII, 12 (Vol. IV K., 50).

below the diaphragm the legs are enfeebled, the urine is suppressed, or voided involuntarily. Galen[177] mentions how he cured a patient whose fingers the other physicians had treated with medicaments when they were rendered useless by a blow on the spine. Galen placed the same medicament on the vertebra instead of on the fingers.

Chapter XXVI

On the Hidden Shoulder Blades and the Shoulder

Backward from the neck two broad bones stretch on both sides to the scapulae (shoulder blades), which we call the hidden scoptula and the Greeks name omoplatae. These bones are curved at their highest vertexes; from them triangular bones which gradually become hidden extend to the spine and the broader they are the thicker they become. They are quite cartilaginous in the posterior part where they seem to swim since they cling to no bone except at their very top. But at that point they are held firmly by strong muscles and sinews. From the highest rib and a little inward where its midpoint lies a bone grows out, quite thin in that place but becoming more substantial as it approaches the broad bone of the scapulae. It also becomes broader and is gradually curved outward. Near the other side of the vertex it swells moderately and supports the collarbone [clavicle], which I said is moon-shaped and curved backward; it is fixed with one head in that place I described above and the other in the small hollow of the pectoral bone structure. It is connected by its sinews and cartilage with the broad bone of the shoulder blades. Here the shoulder [humerus] begins, swollen with its extreme heads on both sides, one soft, without medulla, cartilaginous, round in the middle, and the other hard, marrowy, gently humped toward the front and exterior.[178] The front part extends from the chest, the posterior part from the shoulder blades, the inner part to the side and the exterior recedes from the side. It will be shown to extend to all the joints in its more remote parts. The upper head of the shoulder is more rounded than the other bones of which I have spoken thus far. It is inserted into the verticulated bones of the shoulder blades and for its larger part it is bound fast by the passage of nerves which we call conjunctions. But the lower head has two processes; between these the middle one is more curved in its extreme parts and provides a seat or resting place for the arm, for I am speaking of the bone. The shoulder slips out from its seat, that is, from its omocytele [glenoid fossa],[179] now into the front part, now into the armpit. If it has slipped out in the latter place, in addition to other characteristics the shoulder which is dislocated has a longer arm than the other shoulder. If it is dislocated in front the top of the arm is extended. The elbow is stretched out less naturally and with more difficulty in front than behind. The methods of caring for dislocated members of this sort do not pertain to the present subject.

Chapter XXVII

On the Arm

The arm fits into the shoulder. The place where it bends both ways is called ancon (the elbow). This consists of two bones. The radius, which the Greeks call cercis, is the superior and shorter one; thinner at first, with its round and slightly hollow head it receives the small tubercle of the shoulder and in that place it is held by sinews and cartilage. The cubitus [ulna] is lower and longer; at its beginning it is more full in its uppermost head with two processes like vertexes which exist in the sinus of the shoulder and is inserted there between the two processes. The bones first joined to the two radii of the arm are gradually torn away and later come together again to the hand, or near it, with an altered thickness; where the radius is more full the cubitus is quite thin. Then the radius rises into a cartilaginous head and is bent into its vertex. The cubitus is round in its extremity; with its other part it proceeds somewhat. Celsus,[180] whom I follow in my account of the bones, observes that many of them end in cartilage, as do the ribs of which I spoke; but no joint ends thus. If the elbow is dislocated the injury is more serious than the dislocation of the shoulder, for in the elbow the bones of the humerus, radius, and cubitus come together, but the cubitus, which is attached to the humerus, is dislocated from the latter. The radius, which is an adjunct, is sometimes drawn out of place, sometimes remains firm. The elbow can be dislocated in all four directions. If its prolapse is forward the arm is extended but not curved or bent; if it is dislocated toward the rear the arm is curved but not extended and is shorter than the other arm. Sometimes the bile causes fever and vomiting if the arm is stretched out in front or backwards, but it recedes a little curved back, in that direction from which it ought to bend away. Whatever dislocation it suffers there is one method for placing it back in position which is considered very difficult both by recent as well as ancient physicians.

---

[177] Galen, *De Locis Affectis*, I, 6 (Vol. VIII K., 56–57).
[178] Celsus, *De Medicina*, VIII. 1. 102.
[179] Pollux, *Onomasticon*, 2. 137.

[180] Celsus, *De Medicina*, VIII, discusses the bones with their fractures and dislocations; this observation, if not as explicit as Benedetti indicates, is inherent in Celsus's treatment, as at 8 D 2.

Chapter XXVIII

On the Hand and Its Dislocation

The palm of the hand consists of many small bones, whose number was not certain among the ancient physicians. More recent physicians, however, count twenty-six, twelve in the hand and fourteen in the finger joints. All of them are attached to each other by a certain triangulated structure since by turn the upper angle of one is the flat side or level surface of another. Hence they present the appearance of one concave bone a short distance toward the interior; in fact, however, two small processes are brought together in the sinus of the radius while five bones [metacarpal] extending to the fingers from the other straight side fill out the palm. From these bones the fingers themselves arise; they are composed of three bones each except for the thumb, all in the same manner. The inner bone is bent toward the vertex and receives the small tubercle of the exterior bone. Sinews hold the bones together. The nails that arise from these bones grow hard and thus cling with their roots not to the bone but rather to flesh. The upper parts of the hand are arranged as follows. It can be bent in all directions. It is worthy of note that the fingers cannot be stretched to the rear. If the hand is dislocated in the front part they are not bent but the hand can be bent to either side on the contrary, either to the thumb or to the little finger. In the palm also the bones are dislocated toward the front or toward the rear only but not in the other two directions. A common indication of this dislocation is a swelling in that direction into which the bone has been forced and a hollow in that direction from which it has receded. The same thing happens to the fingers and the same indications of dislocation are found in them as in the hands themselves.

Chapter XXIX

On the Lower Bones

The legs correspond to the arms, the omokotylae to the coxae, the femora to the shoulders, the elbow to the knee or olecranon, the shins or tibiae to the armbones, the radius to the radius, and the ham of the knee to the bends of the elbow. The malleolus and calx (or ankle and heel bones) are peculiar to the foot. The toes correspond to the fingers and the soles of the feet to the palms of the hands. The toenails are similar to the fingernails.

Chapter XXX

On the Hip and Pubic Bone

The lowest part of the spine settles down upon the hip bone which is transverse in its position and very strong, guarding the vulva, bladder, and straight intestine (rectum). On its exterior the hip bone is humped and bent backwards to the spine. At the sides, that is, in the hips themselves, the bone has round hollows from which arise the pubic bone. It is solid and forms a transverse cavity above the intestines and under the pubis, straighter in men but more curved back to the exterior in women in order not to prevent childbirth. This bone is falsely called the holybone (*os sacrum*). It divides itself by its own effort in the middle and I see that some people have believed that it separated thus rather by divine power; otherwise it cannot be broken even with heavy blows. But this natural force is understood as something other than divine, for at the time of childbirth the lowest part of the spine which we call the os sacrum and the comb (*pecten*) open of their own accord when the foetus is larger and consent to the parturition of the infant. Then after childbirth little by little these bones return to their natural position. When one or other of these bones refuses to open a difficult childbirth results although conditions in other respects may be normal. When a pubic bone is felt to be divided it does not remain firm in childbirth. Aristotle[181] says that women in childbirth feel pain in many parts of the body, numerous women in either femur, some in the loins and others in the lower body cavity.

Chapter XXXI

On the Femur and Its Dislocation

Among the largest bones the femur has rounder heads than the humerus (shoulder bone) but below it has two processes from the front and rear. In the middle it is hard and marrowy, humped in its external region, and lightly hollowed in its interior. It swells out once more in the lower region in the knee, The upper part of the femur is brought into the hollow of the hip bone [acetabulum] just as the humerus is brought into the bones of the scapulae; then it extends inside the hip bone in order to support the members above it more evenly. The hollows or sinuses of the hip bone are called cotylae by the Greeks and acetabula by our physicians. The parts which fill these sinuses and around which rotation occurs are called coxendices by us and ischiae by the Greeks; hence those who suffer from pains in the hip are called ischiadici. The sinew by which the coxendix is joined to the acetabulum is also called ischion. The similarity of the bones indicates that they are also subject to similar accidents; the humerus is similar to the femur, for example, in its capacity for dislocation in all four directions. Very often it is dislocated to the inner region and then one leg is

[181] Aristotle, *De Historia Animalium*, VII, 9, 586 b 25–30.

longer and larger than the other and the end of the foot faces outward to the side. If the dislocation is to the outer region the leg is shorter and bent inward; the heel does not touch the ground in walking but only the end of the sole does so. The leg carries the upper members better in this form of dislocation than in the previous case and less requires the use of a cane. But if in the previous case the leg is extended it cannot be united or joined; it is similar to the other leg as far as the heel is concerned but the sole of the foot can be bent less in the forward direction. There is great pain in this type of dislocation and the urine greatly suppressed. When the pain subsides with the inflammation patients can walk comfortably. But if the dislocation is toward the rear the leg cannot be extended and it is shorter. Where the leg stands the heel also does not touch the ground; there is great danger to the femur and its dislocation reduced either with difficulty, or, when reduced, the bone slips out and is dislocated once more. One can see daily the mistakes of those who have undertaken to reduce such dislocations, for this department of surgical medicine has been taken over from us and has passed into the hands of hired servants, workmen, and peasants. Hence they think that a dislocated femur is a permanent injury and that no remedy of it can be found, a great disgrace to our Hippocrates,[182] who tells us that dislocated femora can be completely restored. Furthermore, it is clear that various worthless implements for extending and reducing the dislocated femur have been invented by these journeymen hacks.

Chapter XXXII

On the Legs (Shins) and the Knee Cap

The lower heads of the femora are hollowed or bent in the middle so that they may be more easily received by the legs. A small, soft, cartilaginous, round bone called the patella covers the joining together of these heads. It swims above their combination or structure and clings to no bone but is bound together with sinews and flesh. Tending a little more toward the femur bone it protects that joint among all the flexures of the legs. The leg itself is composed of two bones similar to those in the arm in order to be lighter and stronger. Hence the beauty and appearance of one of these members in living persons can be recognized from the other for those who have small legs also have small hands. The other bone from the exterior is located above and is also called the radius [fibula] as its corresponding bone is called in the arm. This bone is shorter and thinner above and swells out toward the ankle bones; some people call it the sura (or calf). The other bone is located in the front part of the leg and is named the tibia. It is longer and fuller in the upper part and is joined to the femur only in the lower part of the latter bone, just as the cubitus is joined to the humerus. These bones are joined above and below but gape apart from each other in the middle as the bones of the arm do. In the knee there is a double joint in the front part on right and left where there is a hollowness[183] as of the cheeks; from it the spirit escapes when it is punctured as though its bonds had been loosed. The knee also can be dislocated in the exterior, anterior, and posterior regions; it is rarely or never dislocated in the prior (front) region because the patella opposite prevents dislocation.

Chapter XXXIII

On the Ankle Bone or Talus and Its Dislocation

The leg in the lower region where the ankle bones or tali are located is received by a transverse bone [talus] which is located above the heel bone and is bent in a certain direction and in another direction has deviations. It also receives projections from the malleolus (ankle bone) and is inserted into the hollow of the latter. It is hard, without marrow, and projects more into the posterior direction and there presents a smooth round appearance. The talus, which they call also the astragalus, does not exist in man (according to the opinion of Aristotle); it can be dislocated in all directions.[184] When it is burst forward into the inner region then the sole of the foot is turned to the exterior direction and when the talus is contrary to this position a contrary indication is revealed. If it is dislocated in front then the broad sinew from the rear becomes hard; if it is dislocated in the posterior region the heel is barely attached and the injury is greater. When the dislocated talus is reduced and the sinews have grown firm again there is little danger that it will become dislocated once more.

Chapter XXXIIII

On the Heel and the Toes

Under the talus is located the heel [calcaneus], thicker and quadrangular in shape. Its flat surface treads the earth. Its callous substance is not filled with nerves but is without sensation; in infants, however, it is painful since the callous has not

---

[182] Hippocrates, *Mochlicus seu Vectiarius* (Vol. III K., 288; *cf.* 258, 250, and *passim* in this book); *De Articulis* (Vol. III K., 143).

[183] Pliny, *Natural History*, 11. 45. 103. 250, p. 589 (Loeb Classical Library).

[184] Aristotle, *De Historia Animalium*, II, 1,499 b 20: "of the many-fingered or many-toed, no single one has been observed to have a huckle-bone, none of the others any more than man" (translated by D'Arcy Wentworth Thompson, Vol. 4 of the Oxford Translation of Aristotle; his footnote 7 is puzzling).

become hard. The hill of the foot (its upper convex part) consists of three bones only; it is also called the grandineus. Five bones fill out the sole of the foot [metatarsals]. The toes consist of fourteen little bones. All told, the foot contains twenty-three bones. The bones of the sole of the foot appear in the same manner as those in the hand; the toes also have the same arrangement. Hippocrates[185] reported that dread of death exists in the humerus and femur bones; in other words, their damage is very serious. Dangerous also is any injury to the legs and elbows. Nothing in the way of dislocations is more safely reduced than in the sole, the hands, and the fingers in which (as some people insist) there are certain small bones between the joints called sesamoid bones, overlooked by the ancients and which have likewise not been found by our modern anatomists.

Chapter XXXV

On the Praise of Dissection, the Last Chapter

But now I shall put an end to this dissection; at last the theater of spectators must be dismissed since all the parts of the human body have been usefully rather than wordily handled; it has not been unsuitable to run through their history in a sort of summary. I urge all physicians or surgeons, both beginner and veteran practitioners, to hold a dissection in a theater of this kind at least once each year since in it we see the truth and contemplate its revelations as the works of nature lie under our eyes as though they were alive. In other respects writing is quite similar to a picture which often jogs the memory and drives away the darkness from the mind. But those who trust only the monuments of literature (as Plato says)[186] without an observation of things themselves do not reflect upon the things which are clearly visible. They are often deceived and entrust opinion rather than truth to their minds. Thus it happens to those who newly begin[187] to make sea voyages on painted maps where islands, bays, inlets, and promontories do not fully correspond to the true geographical features which lie before their eyes. But pictures of this kind recall to mind things which are well known and, as it were, bring back the images of living people. For I deliver the living and breathing discourse of one who knows, and I shall call its written form, without any offense, a kind of image.

The End

[185] Hippocrates, *Mochlicus, etc*. (Vol. III K., 291); he does not speak of these dislocations as fatal, however. Vesalius demonstrated and described the sesamoid bones in the *Fabrica*.

[186] Plato, *Republic* 7, 533 E; *Sophist* 244; *Statesman* 261 E; see the quotations of this observation by Galen, *De Anatomicis Administrationibus*, VI, 13 (vol. II K., 581), *Methodus Medendi*, XI, 12 (Vol. X K., 772); *De Usu Partium*, IV, 13 (Vol. III K., 309), and note 12 to my translation of Massa's *Introductory Book of Anatomy*.

[187] In the phrase "sicut nouiter nauigare incipientibus in depictis chartis contingit" Benedetti seems to be speaking of the contemporary activity in geographical exploration and map-making.

Gabriele Zerbi

(1485–1569)

LIFE AND WORKS

I

THE life of Zerbi has been discussed in some detail by Ladislao Münster, who points out the difficulties and lacunae of information which it presents and attempts to rescue Zerbi from the calumnies of Berengario da Carpi.[1] The variants of Zerbi's name in the records and the literature vary from Zerbus, Zerbo, and Zerbis to Gerbo, Gerbi, Gerbus, although the Paduan documents consistently use Zerbi while the Bolognese *Rotuli* and *Libri Segreti* have the variants Gerbi, Gerbus, or De Gerbis; the latter variants undoubtedly result from the tendency in the Venetian dialect to corrupt *Z* into *G*.

He was born at Verona in 1445 of an old, rich, and aristocratic family with possessions at Pigozzo nearby. The Zerbis lived during the sixteenth century in the now vanished Via Binastrova, as may be seen from statements made in Antonio Torresani,/ *Elogiarum Historicarum Veronae Propagium*, MS N. 808, Biblioteca Civica, Verona, completed in 1656. Gabriele's father Francesco is listed in documents in the Archivio di Stato at Verona in various capacities of civic responsibility: *consiliarius*, 1446; *aedilis Domus Mercatorum*, 1467 and 1473; *consul Domus Mercatorum*, 1470; *ratiocinator publicus*, 1458, 1460; and *provisor lanificij*, 1449, 1452, 1460. He seems therefore to have been a person of importance in the financial affairs of the community, in the guild of the wool merchants especially. Other members of the Zerbi family—Paolo, Niccolò, and Antonio—also held prominent positions in the business world of Verona. In the section of the Veronese Archives entitled *Reparto delle Anagrafi (Ponte Pietra*, 1456) there appears the name of "Paganinus, master of the arts of medicine" and that of his brother Franciscus (Francesco), with the names of their sons; Gabriele is listed as eleven years old in 1456, although the record does not clearly distinguish to which brother all the children named belonged. Zerbi's will, however, which I publish in Appendix III for the first time, names Benedetto and Giovanni as his brothers; the Taddea named in the Veronese records may have been his twin sister, as Münster suggests, since he named his own daughter (evidently the only one) Taddea. The Angela of the records (misspelled Anela in Münster, *op. cit.*, p. 67) is also listed as his sister in his will (see *infra*). The entry, however, serves to distinguish clearly Zerbi's year of birth. Since he was born in 1445 and died in 1505 he lived sixty years. The names of female relatives of Francesco and Paganino are given as: Alanadora, their mother, aged seventy-eight; Valeria, wife of Paganino, aged twenty-three; and Paula, wife of Francesco, aged forty. The latter was, of course, Gabriele Zerbi's mother; her Italian name was Paola.

It is not known where Gabriele obtained his university degree. According to a passage in his *Quaestiones metaphysicae*, fol. 277 b, he spent some time during his youth at Venice. He probably studied in the University of Padua, where at the age of twenty-two in 1467 he held a teaching post, filling the place of Conte Facino for a four-year period in philosophy. He continued thereafter at Padua for three more years until 1475, when he was thirty years old. From 1475 to 1483 Zerbi taught medicine and logic at the University of Bologna,[2] that is, from his thirtieth to thirty-eighth year. He achieved this extraordinary honor without the possession of Bolognese citizenship only because the University Statutes had been revised in 1479 to allow the appointment at the Studium of Bologna of specially qualified non-Bolognese; evidently the rule against them had been relaxed even four years earlier. The fact that Zerbi had distinguished himself as a professor at Padua was of course much in his favor. The letter which somewhat peremptorily demanded Zerbi's presence at Bologna when classes had already begun and he was lingering at Padua was sent by the Bolognese Senate through the Sixteen Anziani, November 6, 1475 (Archivio di Stato, Litterarum). His name appears for the first time in the *Rotuli* on September 25 of the same year. Among his contemporaries at Bologna were Baverio Bonetti, physician to Nicolas V and Callistus III, Girolamo Manfredi, Alessandro Achillini, Giovanni Garzoni, Benedetto and Leonello Vittori da Faenza, Galeotto Marzio, and Aristotile Fioravanti, who as an architect restored the royal castle at Buda, built cathedrals

[1] "Studi e ricerche su Gabriele Zerbi: Nota I, Nuovi contributi biografici: la sua figura morale," *Rivista di storia delle scienze mediche e naturali* **41** (1950): pp. 64–83, based upon researches in the Archivio di Stato, Bologna, the Biblioteca Universitaria of Padova, and the Civica Biblioteca and Archivio di Stato of Verona. Of the critical discussions of Zerbi's works promised by Münster at the end of this article (p. 82) only the following have appeared: "Il primo trattato pratico compiuto sui problemi della vecchiaia: la *Gerentocomia* di Gabriele Zerbi," *Rivista di Gerontologia e Geriatria* **1** (1951): pp. 38–54, and "Il tema di deontologia medica: Il "De cautelis medicorum" di Gabriele Zerbi," *Rivista di storia delle scienze mediche e naturali* **47** (1956): pp. 60–83.

[2] Umberto Dallari, *I rotuli dei lettori, legisti e artisti dello studio bolognese dal 1384 al 1799* **1** (Bologna, Merlani, 1888): Artisti 1475–1476 Ad lecturam Medicine in nonis D. M. Gabriel de Gerbis veronensis; 1476–1477 Ad medicinam in nonis D. M. G. de G. veronensis; 1476–1477 Ad logicam de mane D. M. G. de G. veronensis; 1477–1478 Ad medicinam in nonis D. M. G. de G. veronensis; *idem*, 1478–1482; 1482–1483 Ad philosophiam ordinariam de sero (when he exchanged the teaching assignment with Leonello dei Vittori da Faenza and ended his teaching career at Bologna). Dallari, Vol. I, prefazione V–XVI, gives a good account of the *rotuli* or teaching records, the courses taught, when they began (or became *ordinariae*), and other facts.

in Russia, and built a wing of the Kremlin at Moscow. This was indeed a galaxy of brilliant scientists and humanists among whom Zerbi must have shone with distinction.

On May 13, 1478, a notice in the manuscripts of the Archivio di Stato at Bologna known as *Partitorum* (Vol. 8, fol. 139a) records the fact that Zerbi obtained a leave of absence from his teaching duties to visit his mother, who had fallen ill at Verona. By a unanimous vote of the authorities he was not to be docked in pay for his absence nor was he obliged to supply a substitute lecturer. His mother was sixty-two years old at this time and Zerbi thirty-three. On December 9, 1479, his salary was raised from 150 lire to 200 lire annually (*Partitorum*, Vol. 8, fol. 198 b):

Item cum Magister Gabriel Gerbus veronensis artium et medicinae doctor impresentiarum legens in Studio Bononiense sit vir non mediocris doctrinae et acuti ingenij ac ipso in Studio utilis, pro omnes fabas albas ei constituerunt pro suo salario libras ducentas bononienses singulo anno quas integras percepire debeat et si tamquam privilegiato integre persolventur cum affectu L. 200. (Furthermore, since Messer Gabriel Zerbi of Verona, doctor of arts and medicine, at present lecturing in the University of Bologna is a man of no small learning and sharp talent and useful to the said University, it has been decided by a unanimous vote that he should receive two hundred Bolognese pounds annually as a salary without deductions and under complete privilege the said salary shall be paid intact.)

Just one year later the Sixteen Anziani, who were the chief administrators of the University of Bologna, appointed Zerbi to the post of ordinary professor of philosophy "de sero" to replace Nestore Morandi, who had accepted an appointment at Pisa. Zerbi's salary was increased by 50 lire; he continued his teaching until 1483.[3] In 1482 he had published his first book, *Quaestiones metaphysicae*, a commentary on Aristotle's *Metaphysics*, stimulated no doubt by his recent concern with philosophy as a teacher of it.

His service at the University of Bologna now at an end, Zerbi next appears at Rome. Berengario da Carpi,[4] in a famous passage of his *Commentary* on Mundinus, makes a vicious attack upon the moral character not only of Zerbi himself but of his sons, presumably while they were at Rome. Beginning with Zerbi's use of Aristotle, *History of Animals*, I, 8, 491b, as authority for some general statements concerning physiognomy (a passage which helped form the foundation for the entire edifice of this pseudo-science), Carpi proceeds to insult Zerbi for further invoking the authority of Pietro d'Abano's *Conciliator*, who used the Bolognese as examples of those people whose foreheads were stretched and polished and thus were tricky and deceitful. Carpi disregards the fact that Zerbi, in the passage to which Carpi takes such violent exception (*Liber anathomie corporis humani et singulorum membrorum illius*, etc., 1502, fol. 99v), merely cites Aristotle (see above) and Pietro d'Abano (who makes the original comparison with the Bolognese as tricky and deceitful) without indicating in the slightest his approval of d'Abano's statement. At fol. XXXIIr Carpi continues his baseless charges by attacking Zerbi's etymology of *carpus* as from "carpendo, apprehendo" on the ground that no Greek etymology is derived from Latin; he takes the opportunity of sneering at Zerbi as accustomed to "carpere, rapere, siue furari," that is, to thieve, repeating his defense also of his countrymen, the Bolognese. All of Carpi's insults were without any basis of evidence, and Zerbi's reputation so far as is known was spotless. The etymological pun, moreover, fits Carpi better than Zerbi.

Zerbi's residence at Rome, apart from this outrageous fabrication by Carpi, is not well documented except for a letter to Zerbi from Pope Innocent VIII, dated March 11, 1490,[5] which contains these words: "Virtutum merita et doctrina, alia multiplicia dona, quibus personam tuam fidedignorum testimoniis ornari accepimus, merito nos inducunt ut te specialibus gratiis et favoribus prosequamus." The Pope praises Zerbi for his virtues and learning together with many other gifts, attested by people whose opinions of him were trustworthy, and proceeds to show him special grace and favor.

---

[3] *Partitorum*, Vol. 10, fol. 38 a (December 9, 1480); *ibid.*, fol. 86 a.

[4] *Commentaria cum amplissimis additionibus super anatomia Mundini una cum textu eiusdem in pristinum et verum nitorem redacto* (Bologna, Hieronymo de Benedetti, 1521), fol. 17v, with the heading "Contra Zerbum in defensionem Bononiensium." Münster, *op. cit.*, pp. 75–79, effectively disposes of Carpi's charges, citing the fact that Carpi himself engaged in criminal acts (see my translation of his *Short Introduction to Anatomy*, University of Chicago Press, 1959, introduction, 5, 7, 8), and that he speaks of Zerbi as at Rome, committing the theft of two silver cups from a bishop whose illness he treated and being forced to flee to escape crucifixion, while his two sons were hanged as thieves during the pontificate of Julius II, as Carpi claims he saw with his own eyes. But Julius did not become Pope until 1503, when Zerbi had long since left Rome. Nor did Zerbi stab himself to death after his sons were hanged; his death was quite otherwise, and even more tragic, as we shall see. The two silver vases mentioned by Carpi present a peculiar coincidence with those which Benvenuto Cellini (*Autobiography* [Everyman Edition, 1907], p. 291) says he made for Carpi at Rome, where the great anatomist stayed six months in 1526, and for which Carpi paid Cellini wretchedly.

[5] Münster, *op. cit.*, p. 77. Lynn Thorndike, *A History of Magic and Experimental Science* **5** (Columbia University Press, 1941): p. 505, cites P. Capparoni, "Les maîtres d'anatomie à l'Athenée romaine de la Sapienza pendant le XVIe siècle," *Ve Cong. Internat. d'Hist. de la Médicine*, 1926: p. 97, as saying "that a brief of Innocent VIII shows that Zerbi taught at Rome after 1480 until a larger salary attracted him to Padua. But perhaps the 1480 is a slip or misprint for a later date." Thorndike is justified by Münster in this conjecture. The letter from Pope Innocent is printed in Gaetano Marini, *Degli archiatri pontifici*, **2** (Roma, 1784): pp. 238–239, dated March 11, 1490.

His Roman sojourn lasted more than ten years, from the second half of 1483 to the beginning of 1494. Zerbi had given a parchment copy of his *Quaestiones metaphysicae* (a unique example) to Pope Sixtus IV, who died in 1484, two years after its publication. The copy is now in the Vatican Library and bears a beautifully illuminated first page which shows the pope seated on his throne, wearing his tiara and surrounded by cardinals. Two figures kneel before him; the one on the right, shown in profile offering a copy of the book to the pope, is Zerbi himself; the figure on the left may be the artist who made the picture.

When at the end of the year 1490 the two chairs of theoretical medicine at Padua fell vacant the Senate offered the post of lecturer "de sero" to Zerbi with a two-year contract at 400 florins a year. Girolamo Donato, ambassador of the Senate at Rome, was charged with persuading Zerbi to accept the offer.[6] The details of the transaction were recorded in a Senate decision preserved in the University archives at Padua.[7] Zerbi was to remain at Padua for a second period, this time of eleven years, from the date of the last entry of renewal of his appointment, May 6, 1494, until his death in 1505. Hence his life as an active teacher and scientist may be summed up as follows: I. First Period at Padua, 1467–1475; II. Stay at Bologna, 1475–1483; III. Stay at Rome, 1483–1494; IV. Second Period at Padua, 1494–1505.

From this last period of his life date five letters from Benedetto Rizzoni to Zerbi[8] which I publish here for the first time (Appendix II) as the only correspondence concerned with Zerbi thus far known to me,[9] although I have made efforts to discover others.

In these letters Rizzoni expresses his gratitude and friendship for Zerbi in the elaborate terms of typical Renaissance Humanist prose. His style, while prolix and mannered, does not, however, conceal his sincere affection for his correspondent. Letter 1 conveys his thanks to Zerbi for medical care rendered to Rizzoni while Zerbi was at Rome in a way which reveals that they had been friends for some time and that Zerbi had rescued him from a serious

[6] *Archivio Antico dell' Università di Padova,* Vol. 662, fol. 192 v: "1491, 2 gen. (January 2). Gabriel Zerbi Veronese residente a Roma con f. (florins) 400. 1493, 17 ottobre, proposto di nuovo se vuol venire a leggere la teoria di medicina in questo luogo con f. 600." Another entry for May 6, 1494, in the same record of professors of the theory of medicine at Padua (actually the Archbishop's records) reports that Zerbi was once more teaching there. The final entry for Zerbi is in 1498 and then at last the words "1505 mori" (Vol. 649, fol. 220 v has "obiit"). See Giuseppe Giomo, *L' Archivio antico della Università di Padua* (Venice, fratelli Visentini, 1893).

[7] *Archivio Antico dell' Università di Padova,* Vol. 648, fol. 334, "Decreti del Senato per la nomina dei Professori, Artisti, e Giuristi," contains the notarial minute: 1491. Die 3 Januarij. In Rogatis. Viget fama universalis et in Gymnasio nostro Patavino et ubique de Magistri Gabrielis Zerbi Veronensis, qui impraesentiarum reperitur in Roma; et quoniam eius persona non tantum futuro est utilis verum etiam necessaria predicto Gymnasio nostro Patavino, cum vacent nonnullae lectiones in medicina quae maximopere exigunt eius doctrinam et virtutem. Vadit pars quod scribatur et mandetur Nobili Viro Hieronijmo Donato Doctori Oratori nostro in Curia Romana ut practicare debeat cum praedicto Magistro Gabriele et concludere quod velit venire ad praedictum Gymnasium Paduanum ad legendum in Medicina illam lectionem quae sibi per dominium nostrum deputabitur, cum salario florenorum quadrigentorum in anno, pro annos duos de firmo, et de vespera, in libertate nostri Dominij. Et de successu nobis det de suis litteris notitiam. F. Giovanni degli Agostini, *Notizie Istorico-critiche intorno la vita, e le opere degli Scrittori Viniziani,* etc. (Venice, Simone Occhi, 1754) **2**: p. 206, speaking of the consecration of S. Maria de' Servi at Venice, reports that the Prior, Girolamo de' Franceschi, was certainly absent from the ceremony because he was in Rome on January 3, 1492, trying to persuade Zerbi to accept the chair of theory of medicine at Padua: A codesta sacra funsione Girolamo [de' Franceschi, priore] certamente non intervenne, poiche a' iii di Gennajo dell' anno 1492 appresso fu in Roma a lui scritto, perche insinuasse a Gabbriello Zerbo medico Veronese di salire la cattedra teorica in Padova, con l'onorario di cccc fiorini. [Ex Regesto apud Excell. Reformatores Studii Patavini, Vol. II, pag. 75]. Evidently the efforts of Girolamo Donato, ambassador of the Senate of Padua to the Vatican, were reinforced and renewed in the following year by those of the Prior Girolamo de' Franceschi, since Zerbi did not accede to the request made in 1491, as appears from other documents in the Antico Archivio dell' Università di Padua. It was not, in fact, until October 17, 1493, that he seems to have agreed to come back to Padua, and on May 6, 1494, he was once more at his post in that city three years after the first attempt had been made to draw him away from Rome. Facciolati is explicit on this point: *Fasti Gymnasii Patavini Jacobi Facciolati studio atque opera collecti Patavii, Typis Seminarii MDCCLVII apud Joannem Manfré,* I, 134: *Gabriel Zerbus* Veronensis, qui decem ante annis Philosophiam in Gymnasio tradiderat, hoc anno Romae cum esset iii. non. jan. invitatus est ad Medicinam profitendam, oblatis argenteis quadringenis: sed conditionem non accepit. *Triennio post sexcentenis invitatus accessit* [italics mine] ad regendam scholam primam Theoricae Ordinariae. Letho indignissimo extinctus est anno MDV. De illo eiusque fato Jovius in Elogio M. Antoni Turriani, et Maphejus in Ver. illustr. 1. 3. p. 128.

[8] Benedetto Rizzoni is identifiable in none of the sources I have checked. Mario E. Cosenza, *Biographical and Bibliographical Dictionary of the Italian Humanists etc., 1300–1800* (Boston, G. K. Hall, 1962) **5**, Synopsis and Bibliography, lists two other Rizzonis, Martin and Jacob, but gives almost no information on them either. Thus far at least he appears in history only from these letters, although it is of course possible that research in the archives at Verona or Rome might turn up more information on him.

[9] These letters are listed by Giuseppe Biadego, *Catalogo descrittivo dei manoscritti della biblioteca comunale di Verona* (Verona, G. Civelli, 1892), p. 150: "MS 233 [1467] Rizzoni Benedetto.—[Epistolae]. Cart. degli anni 1480–1501, di 233 carte, m. 0.29 × 0.21, in carattere corsivo. Ha parecchie macchie d' umidità; qualche pagina fu rapezzata. Legatura moderna in mezza pergamena. Fu venduto a q. B. il 28 aprile 1873 da certo Luigi Rossini." The catalog references are p. 155, c. 134 r (1); p. 156, c. 141 r (2); c. 141 v, 142 r (3); p. 158, c. 174 v, 175 r (4); p. 159, c. 200 r (5). I have corrected Biadego's folio numbering for letter 3 according to the microfilm reproduction of the manuscript from which I have transcribed the letters.

Other letters, in Italian, are mentioned by Michele Medici, *Compendio storico della Scuola anatomica di Bologna* (Bologna, 1857), p. 42, but de Renzi thought they were by another Zerbi, son or nephew of our Gabriele. One should add, of course, Zerbi's Latin dedicatory letter to Marino Brocardo and Brocardo's reply in the front matter of Zerbi's *Liber Anathomie* to complete the list.

illness the nature of which is not, however, disclosed; Zerbi's fame as a physician at the Papal court during the reign of Innocent VIII (1484–1492) is also heralded. Rizzoni closes with the reflection that he has had it in mind to leave Rome and come to Padua both to avoid the dangers of life in Rome at that particular time (perhaps disease was increasing there temporarily) and to take a vacation where he might visit Zerbi. The letter is dated at Rome on July 3, 1495, shortly after Zerbi had returned to Padua.

Letter 2 describes Rizzoni's anticipation of Zerbi's return to Rome which Zerbi had mentioned in a letter to Rizzoni, enclosing a regimen of health for Rizzoni's benefit. Vincenzo of Cremona, a mutual friend, had been for some days the guest of Rizzoni at Rome; he evidently carried Rizzoni's letter to Zerbi, who is invited to ask Vincenzo for further details about Rizzoni and affairs at Rome. Rizzoni again expresses his appreciation for Zerbi's good medical care of him although he mentions the fact that he has had a slight fever that summer. He ends with a cryptic statement which is difficult to determine without further information of the facts surrounding Zerbi's departure from Rome; some element in the situation which Rizzoni regarded as a cause for Zerbi's decision to leave is still in force and a detriment to the practice of medicine at the Papal Court as well as an expense to the courtiers of the pope and all others, from which Rizzoni wishes they were relieved. This letter is dated at Rome, October 18, 1495, with the final news that the pope has gone for rest and recreation to his retreat at Rignano, eighteen miles from Florence.

In letter 3 Rizzoni refers again to Zerbi's return to the University of Padua and regrets that he had not visited his medical friend more often while the latter was still in Rome. He reiterates his sincerity and faithfulness of friendship, free from adulation and vanity, recalling the precepts of his father by which Rizzoni has tried to mold his life. He has sought as friends only those who were worthy of friendship, according to Cicero's saying (in *Laelius De Amicitia* 21. 79): "those are worthy of friendship in whom there is a reason why they should be loved." Since Zerbi and Rizzoni have proved their friendship frequently the anatomist may depend upon faithful performance on Rizzoni's part of any favor, duty, or task which may be required of the latter, to the best of his ability. There is no date but the letter presumably comes from Rome not long after the previous one.

Letter 4, a brief one dated from Rome on March 25, 1498, repeats the familiar theme of preoccupation with business, apology for delay in writing, and the usual protestations of friendship; he has not forgotten Zerbi, he has been too much taken up with affairs. He then mentions a third party to whom Zerbi had written to reprove him for an ill-considered transaction (*sui muneris uenditionem*) but who has now striven to mend his ways and to compose his spirit for a better life. Rizzoni too had suffered from the consequences of this rash act, although he refuses to give Zerbi any details. He presumes that some of the facts of the affair have already reached Zerbi's ears, such as Rizzoni's attempts to console the afflicted wife of this third, unnamed party.

Letter 5, undated, begins once more with the trite apologies for not writing sooner. Rizzoni refers once more to the mutual friend involved in letter 4 and his ill-advised sale (probably of some property of Zerbi's at Rome, which Zerbi had perhaps commissioned him to sell). The friend's desire to reform his methods of behavior are referred to in the same sentence with Rizzoni's assurance to Zerbi's question that he is well and happy and in no need at the moment of Zerbi's ministrations nor those of another physician, returning his best wishes for his friend's health at Padua. Rizzoni then asks Zerbi whether, without discommoding himself, he might wish to sell his property (*aedes . . . tuas*) in Rome at a fair price; Rizzoni assures him that he could make a satisfactory profit and urges him to accede to the request of the third party involved, who evidently wishes to buy Zerbi's possessions; thus Rizzoni may not appear to have written his letter in vain.

A number of details on Zerbi's later life may be added at this point. In an essay on the Paduan philosopher and physician Nicoletto Vernia, Bruno Nardi[10] points out that according to a record in the Venetian archives Gabriele Zerbi served as a "compromotore" on the doctoral committee for Girolamo Avanzo on July 29, 1494 (Archivio della Curia Vescovile, Vol. 44, f. 247 r). The Venetian archives of state (Senato terra, Registro 12, f. 109 v, dated September 17, 1495) also record the fact that his colleague in theory of medicine at that time was Pietro Trapolin:

> Dominus autem petrus trapolinus qui legit extraordinariam Theoricae medicinae ad secundum locum in concurrentiam D. Gabrielis Çerbo, cum summa satisfactione totius studij patavini. . . . (Messer Pietro Trapolin, who is lecturer in extraordinary of the theory of medicine in the second place concurrent with Messer Gabriele Zerbi, to the complete satisfaction of the entire University of Padua. . . .)

There is a further reference to Zerbi, although not by name, in a minute which describes the deliberations of the Venetian Senate (Senato terra, Reg. 13, f. 97 r) on October 31, 1499, mentioning the fact that Trapolin, then old and with many children, was teaching alongside a colleague who was paid 600 florins a year while Trapolin received only 250.

---

[10] *Saggi sull' Aristotelismo padovano dal secolo XIV al XVI* (Florence, G. C. Sansoni, 1958), pp. 157, 159, 166.

The famous diaries of Marino Sanuto,[11] a prime source on Venetian history during the fifteenth and sixteenth centuries, have a number of passages which concern Zerbi in the last six years of his life. On January 2, 1499 (II, 314) the first entry concerning Zerbi states that he, with other Paduan teachers of medicine, John of Aquila, Nicoletto Vernia, and Hieronimo da Verona, came to Venice to practice medicine during the vacation, where they did not pay the taxes which the Venetian physicians paid.[12] In June of the same year (p. 802) Zerbi accompanied the rector of the College of Arts at Padua to Venice to obtain the services of a Paduan doctor of arts then teaching at Ferrara to lecture on philosophy at Padua for 120 florins a year. Zerbi is here called Marco Gabriel, a name he bears nowhere else. Further details of this mission (p. 845) reveal that the teacher at Ferrara was Antonio Fracanzano, of Vicenza, who wished to have no professor in competition with him (Lat. *concurrentia*, Ital. *concorrente*, *concorrenza*). It was the place of this very Fracanzano which Alessandro Achillini had taken in November, 1506, when he came from Bologna to teach at Padua. The name of Fracanzano appears in the records of the teachers at Padua as late as 1538 and thereafter from 1545 to October, 1564 (*Archivio antico,* Vol. 663, fol. 2 r); minutes in regard to his salary increases in 1566 and 1567, after thirty-two years of service and with a family of fourteen children, are contained in the same volume, ff. 32 r, 33 r, v. In earlier years he had taught also at Bologna; since we hear of him first in 1502 at Padua this remarkable man must have been active as a teacher for more than sixty years.[13]

The entry for August, 1500 (Vol. III, 654) gives the teaching record of Zerbi at the head of a list as follows: "Dominus magister Gabriel Zerbus, ad ordinariam theoricae medicinae." When in May, 1503, Lorenzino de' Medici lay ill and by the mediation of Marco, a Florentine citizen, called upon the medical faculty at Padua to send a physician to Florence to heal him, the choice lay between Zerbi and Hieronimo da Verona, an indication of the high repute in which Zerbi was held (V, 30). In July, 1503, letters arrived from Corfu, where the captain-general was suffering from a flux; his nephew, Piero da Pesaro, accompanied by Piero Zustignan and Gabriele Zerbi, went to attend him, Zerbi having been given a leave of absence from his teaching post by the Signoria and a payment of 130 ducats a month. Unfortunately, the captain-general died of a fistula in the spine after Zerbi and the others had reached Corfu (V, 67). On January 25, 1504, the son of Gabriele Zerbi, the rector of the college of artists at Padua, arrived in Venice to deliver a lecture. We find this Paolo Zerbi mentioned also as rector in *Archivio di Curia Vescovile,* Padova, Vol. 47, f. 169 r, and present at the doctoral examination of Andreas Mocenigo on August 12, 1503: Nardi, *op. cit.,* pp. 167–168 (V, 759, 766) (See footnote 10).

On October 15, 1504, there arrived "con jubillo et campano" a galley bearing a message from Skander Pasha to the Signoria of Venice for a physician to attend him in his illness. Zerbi was allowed to go for two months, again suspending his teaching duties. He set out for Bosnia at the request of Andrea Gritti, then Venetian Consul at Constantinople and later Doge of Venice, with one of his sons, clearly not Paolo, or Hieronymo but Marco or Giovanni Alvisio, all of them named in his will, the first two as adult (see *infra*). On November 5 it was learned at Venice and in the college at Padua by private letter from Giacomo di Giuliano da Ragusi to Andrea Gritti that Skander Pasha had died; Niccolò Aurelio, secretary, halted his proposed visit to the Turkish court in Bosnia (Bossina), whither Zerbi had gone for the high pay of 300 ducats a month. The news was confirmed in January, 1505, with the additional tragic word that Zerbi had been sawn in half between two planks and his son likewise. In the same month Sier Hieronimo Contarini, in charge of provisions for the Venetian fleet, reported that Skander Pasha had died on November 26 (VI, 77, 101, 120, 122, with a "Soneto per la morte di domino Cabriel Zerbo per se," i.e., Hieronimo Capello, an administrator of the Venetian arsenal, which contains the lines:

> questo huomo, excelente
> l'era in effecto senza alcun errore,
> splendor e gloria dil secul presente . . .

The final details of the murder of both Zerbi and his son are contained in Sanuto's *Diary,* VI, 124, 277; the last entry (January 1506, recording an event of October 28, 1505) notes the engagement of Antonio da Faenza as replacement for Zerbi at Padua for 400 florins a year; this excellent doctor, however, became insane and later regained his reason.

The ghastly story is told again by Giovanni Piero Valeriano in his famous book on the unhappy careers of certain literary men.[14] I translate it here as the best and most complete single version:

---

[11] *I Diarii di Marino Sanuto* (Venice, a spese degli editori, pubblicato per cura di G. Berchet, 1879).

[12] See also Nardi's remarks on the reasons why Vernia joined his colleagues at Venice, *op. cit.*, p. 125.

[13] See Fr. Franceschetti, *La famiglia dei conti Fracanzani di Verona, Vicenza ed Este con notizie dei loro antenati,* etc. (Bari, La direzione del Giorn. Araldico, 1896), pp. 30–31.

[14] Joannes Pierius Valerianus, *De litteratorum infelicitate libri duo* (Venice, Jacob Sarzina, 1620; a new edition appeared at Geneva in 1821: *editio nova curante Domino Egerton Brydges. Genevae, typis Gul. Fick*), pp. 38–40. Another account of Zerbi's death is found in Neander, *Syntagma de Medicinae Originibus,* according to Thomas Bartholin, *On Medical Travel,* translated by Charles D. O'Malley (Lawrence, Kansas, University of Kansas Library Series No. 9, 1961), p. 52. See *Antiquissimae et Nobilissimae Medicinae Natalitia . . . auctore Johanne Neandro Bremano* (Bremen, Johannes Wesel, 1963), p. 151.

But these men may perhaps appear to have suffered a fate common to many and probably not to be numbered among the wretched compared with the most calamitous death of Gabriele Zerbi, a physician of Verona, than which no death is more cruel among the deaths of long ago, nor more inhumane. He was a quite famous professor in the University of Padua; his ever-ready talent and fearless confidence based on training were even then well known to you Romans since in the most crowded gathering of philosophers and theologians, where matters of extremely serious importance were being discussed, he once dared to expose the ignorance of Pope Sixtus in the course of the disputation. The Pope was much offended at the man's lack of discipline in thus holding his papal majesty in contempt. Learning of the Pope's plan to punish him, Zerbi fled at once and in secret from the city and thereafter taught for many years at Padua. It happened, however, that Skander, chief minister of the king of the Turks, lay ill of a most severe dysentery and asked Andrea Gritti, a man very well known to him and now the Doge of Venice, to send a physician from Italy as quickly as possible, who should receive a huge reward for his journey, his labors, and his cure. Zerbi undertook the mission and, eagerly pursuing the hope of a huge weight of gold, set out for Skander, bringing with him his son, a mere boy. He cured the prince carefully and restored him to health. He was thereupon loaded with much gold, raiment, gems, and a mass of silverware and other precious things; if he had been permitted to carry it all home he could by means of this wealth have rivaled any king in Europe. The entire affair had gone well, he had been assured by the Turk that safety and security were his due to the successful cure, and after giving his patient directions and a regimen for maintaining his health, Zerbi had received his passport to return to Italy. With his loaded mules he had reached an outlying Turkish fortress, whence the road led into Dalmatia; here he stayed a few days until a convoy of Christians could be assembled and he could travel under the safe-conduct the Turk had given. It came to pass, however, that Skander, a lustful man by nature, forgot the warnings Zerbi had uttered and in pursuit of his desires took a turn for the worse; in a short time he was dead. His sons, in order to get their hands on the precious gifts their father had given to Zerbi, accused him falsely of having given him a slow poison. They sent speedy runners in pursuit to bring back the fugitive, as they called him. They caught Zerbi and his little son and, having charged him with murder, they punished him as follows. First, they took his innocent boy and, before his father's very eyes, placed him between planks of wood and sawed him in two. Then they put Zerbi to death by the same most horrible torture.

I come now to the last item of biographical importance: Zerbi's last will and testament. It is dated on October 13, 1504, in the very month when Zerbi, according to the entry in Sanuto's *Diary,* left for Skander Pasha's court (VI, 77). The galley which brought the request for a physician from Skander to Venice arrived there on the fifteenth of the month. Perhaps it was with some presentiment of events to come that Zerbi had made his will two days before. The transcription printed in Appendix III is published for the first time.

Zerbi's will gives us a number of biographical facts. It was made at Padua, where he wished his body to be buried in the church of St. Francis if he died in that city and a suitably elevated monument erected in his memory for which a sum of sixty ducats was provided. If, however, he died elsewhere (here a note of pathetic prophecy enters the matter-of-fact notarial language) he invoked God's mercy upon his soul and body. He left twelve gold ducats to Battista Barzizzi, thirty-eight to fifty ducats to his nieces, the daughters of his sister Catherine, wife of Giovanni Sorii, when they came to marry; provision at the discretion of his executors was also to be made for some one of his more distant relatives in the same situation. To his wife Helena de' Metaselimi, of Bologna, he left the usufruct of all his property with all the rights of principal executrix. His property after her death was to go to his sons Paolo, Hieronymo or Girolamo, Marco, and Giovanni Aloisio on the condition that if Paolo and Girolamo, his adult sons, dissipated these goods and failed to live properly with their equal portions they should receive only a part of the income. The same condition applied to Marco and Giovanni Aloisio, one of whom, by the way, accompanied his father on his final tragic journey into Bosnia: these were evidently both under age. Zerbi wished his property to be held by his male descendants in perpetuity and kept in the family without limits and without the possibility of pledging it or disposing of it in any way, with reference to the Roman law of inheritance according to the injunctions of Trebellianus and Falcidius, the Roman lawyers.

Zerbi chose Pietro of Mantua, whom he calls the godfather of his children, and his wife Helena as executors of his estate. As long as his mother lived (and at the time of his will she must have been at least eighty-eight years old) she was to receive a part of the income from his property. To his daughter Taddea he bequeathed an income to be determined by the executors. To his daughter Hermodoria, a nun in the convent of San Lodovico at Bologna, he left twelve Bolognese pounds annually as long as she lived, a sum which her mother was to supplement annually as far as she could. To his sister Clara, a nun in the convent of Santa Clara at Verona, he left similar alms, and finally to his sister Angela he left three gold ducats in testimony of his love as well as four ducats to his brother Giovanni, canon of Belluno, and the same to his brother Benedetto. To Francesco, son of Benedetto, Zerbi left three ducats.

II

In addition to his contributions as physician, anatomist, professor of philosophy and medicine, and writer of medical works, Gabriele Zerbi was a collector of manuscripts and is credited with the discovery and preservation of the following scientific works of the Middle Ages:[15]

Pseudo-Apuleius, *De herbarum medicaminibus liber,* published in 1788 by J. C. G. Ackermann, *Parabilium Medicamentorum Scriptores* (Nürnberg and Altdorf), 127–380.

Pseudo-Antonius Musa, *De herba uettonica* (on the uses of betony in medicine).

Sextus Placitus Papyriensis, *De medicamentis ex animalibus pecoribus et bestiis uel auibus.*

Pseudo-Dioscorides, *De herbis.*

These works date from between the fourth and sixth centuries A.D.

Zerbi's own books are the following, three of them incunabula:

1. *Quaestionum Metaphysicarum libri XII ad Sixtum IV*; Bologna, Iohannes de Noerdlingen and Henricus de Harlem Socii, Dec. 1, 1482. Goff No. Z–27.

2. *Ad Innocentium VIII. Pon. Max. Gerentocomia feliciter incipit*; Rome, Eucharius Silber alias Franck, Nov. 27, 1489. Goff No. Z–26.

3. *Opus perutile de cautelis medicorum editum a clarissimo philosopho ac medico Magistro Gabriele Zerbò veronense theorice medicine ordinariam studii patauini publice legente,* etc.; Padua, Dominicus Berthonus [an error for Bertochus], 1495, 1503; Pavia, 1508, 1517; Lyon, 1525, 1582; Pavia, 1528, 1598. The incunabulum in the Yale University Library was printed by Christophorus de Pensis but not before 1495. The book was reprinted in a collection by Pantaleone da Confienza, *Pillularium omnibus medicis* . . . Pavia, Jacob de Burgofranco, 1508, and Lugduni, Per Antonium Blanchard, 1525, fols. 28 v–38 r. Goff No. Z–25.

4. *Liber anathomie corporis humani et singulorum membrorum illius, editus per excellentissimum philosophum ac medicum D. Gabrielem de Zerbis Veronensem, cum gratia*; Venice, printed by Boneto Locatello of Bergamo and published by the heirs of Ottaviano Scoto of Monza, Dec. 22, 1502; fols. 1, 2–134, 121–184, 1–20, 1–16 (Books I, II, III); Venice, O. Scoto, 1533.

5. *Anathomia matricis pregnantis et est sermo de anathomia et generatione embrionis.* fols. 1–14 v; printed at the end of no. 4, 1502 edition; Marburg, 1537, with Dryander's *Anatomy,* and at Venice, 1592.

6. *Libellus de Preseruatione Corporum a Passione Calculosa*; MS Biblioteca Comunale di Verona, No. 775, Cl. Medic., 91, 1, Busta A. I am in process of editing this work which has not hitherto been known and which now must be added to the list of Zerbi's works. He does not mention it in the appropriate part of *Liber anathomie,* on the illnesses of the kidneys, ff. 36 r, v, although he quotes the same authors as in the *Libellus.* He makes use of it, however, in the dietary portion of the *Gerontocomia*; see below. The catalog reference for this manuscript is as follows: Giuseppe Biadego, *Catalogo descrittivo dei manoscritti della biblioteca comunale di Verona* (Verona, G. Civelli, 1892), p. 314: "MS 618 775 Zerbi Gabriele.—Gabrielis Herbi veron. libellus de preservatione corporum a passione calculosa. Ad Reverendissimum in Christo patrem et dominum dominum Gabrielem Cardinalem Aggriens. Cart. del sec. XVI, di 69 carte, C. 20 14; legatura originale in cuoio. Fu acquistato dal libraio Robolotti." The entire work consists of 130 folios.

Cardinal Gabriele Rangone, of Verona, to whom Zerbi dedicated this work, belonged to the order of Minor Friars and became bishop of the diocese of Erlau, or Eger (Agrien) in Hungary, suffragan of Strigonien (Esztergom), Transylvania, in 1475. Under Pope Sixtus IV on December 10, 1477, he was promoted to the rank of cardinal, retaining his bishopric of Agrien and receiving the titulus of saints Sergius and Bachus. He held the degree of master of sacred theology and died on September 27, 1486.[16] Since Zerbi speaks to the cardinal as still alive at the time of his dedication the *Libellus* probably precedes the *Gerontocomia* in order of composition and thus may be his second book after the *Quaestiones metaphysicae.* Whether or not it was intended for publication and why it was not printed during Zerbi's lifetime are questions difficult to answer. It is reasonable to suppose that Zerbi met the cardinal during his stay at Rome in 1483–1494 but before 1486. The *Libellus* may then well have been written between 1483 and 1486, just after the publication of the *Quaestiones.*

1. The *Quaestiones Metaphysicae* is a large work in twelve books and 512 folios, in an ugly, small Gothic print with many abbreviations as in the *Liber anathomie*; Zerbi was surely unfortunate in his choice of printers. The book is not a mere commentary on any single work but an original and

[15] See David Diringer, *The Illuminated Book, Its History and Production* (London, Faber and Faber, 1958), pp. 47–48. The miscellaneous manuscript which contains these medical writings is Laurentianus Pluteus 73. 41, with figures 1–19c. It was written in the early ninth century. There are many drawings of human beings, animals, birds, and sixteen scenes showing surgical operations (ff. 122 r–129 v). The manuscript belonged to Zerbi in 1474 and is edited in *Corpus Medicorum Latinorum* IV (Leipzig, B. G. Teubner, 1927). On pseudo-Apuleius see M. Schanz, C. Hosius, G. Krüger, *Geschichte der römischen Literatur* **8**, 3 (Munich, C. H. Beck, 1922): pp. 130–131; H. J. Rose, *A Handbook of Latin Literature* (New York, E. P. Dutton paperback D-67, 1960), pp. 428–429, 523. The first mention of Zerbi's books in the Laurentian Library appears in Gaetano Marini, *Degli archiatri pontifici* **1** (Rome, 1784): p. 310, who also spoke of another copy of the *Gerontocomia* known to him in the monastery of San Martino in Palermo, Sicily.

[16] Conrad Eubel, *Hierarchia Catholica Medii Aevi,* etc., **2** (Münster, 1901): pp. 19, 82, 93, 279.

independent production based upon many authorities ancient, medieval, and Renaissance. Zerbi's painstaking thoroughness is shown in the long list of authorities he quotes; they include Pythagoras, Hippocrates, Plato, Aristotle (and commentators on him such as Themistius and Porphyrius), Euclid, Galen, Hermes Trismegistus, Boethius, St. Augustine, St. Gregory, Lactantius, Anselm, Algazel, Scotus, Avicenna, Averroes, Serapion, Damascenus, Haly Rodoan, Pietro d'Abano, Trusianus, Albertus Magnus, St. Thomas Aquinas, a certain "Alexander peripateticus," "Gualterius anglicus in ultima parte tractatus sui te [de] intensione et remissione," "David in libro de causis," "Henricus de gandavo," and even Horace's lost philosopher, Archytas of Tarentum (*Odes* I. 28), as well as the philosophers John of Jandun, Paolo Veneto, Niccolò Boneto, and John of Buridan. Zerbi knows the works of these men intimately and quotes them with sophistication and intelligence.

His book is a study of theology and philosophy with the emphasis on both characteristic of the Middle Ages. He uses the scholastic method of argument in a masterly fashion and is completely familiar, for example, with the work of St. Thomas Aquinas. This aspect of the *Quaestiones* is referred to in a laudatory epigram printed in the book by Petrus Almadianus Viterbiensis:

> Si physico forsan queris transcendere sensus
> Hoc eme Zerbeum sedule lector opus.
> Que Scotus Thomas ve sacer sensere volumen
> Explicat et partes conciliantis agit.

Zerbi is here described as harmonizing the ideas and arguments of the scholastics, especially Scotus and St. Thomas. There is much of value in the book, which has been completely ignored by the chief modern students of medieval and Renaissance philosophy; they do not even mention him.

2. The *Gerontocomia* (the title is misspelled in the book as *Gerentocomia*) is the first printed treatise on geriatrics and thus of particular interest to our times as it was to the Renaissance, which occupied itself in both art and medicine with the aesthetic and physical qualities of the human body.[17] The quarto volume consists of 135 unnumbered folios in double-column Gothic print with many abbreviations. The book begins with a prologue to Pope Innocent VIII, followed by a table of contents (3 folios) and 57 chapters. The first 17 chapters are devoted to general topics such as the nature, causes, and signs of old age; its uncertain terminus; its proper care and the functions of those in charge of the aged; the climate most suitable for old age; the garments, beds, exercise, and baths which old people should have. Chapters 18–38 deal with the diet of the aged, with great attention to specific foods. Chapters 39–57 discuss the physical condition of the aged and their medical care; the last six chapters describe certain medicines appropriate to aged patients. I find verbal evidence for my belief that in the chapters on diet Zerbi made use of his unpublished treatise on stone in the kidneys and bladder (see no. 6 in the list of his works), which deals largely with diet.

After observing that however miserable men still desire to remain alive and that however wretched old age may be the aged stand nevertheless in dread of death Zerbi proceeds to list for his patron the Pope some of the three hundred ills which are said to attack the elderly. Nonetheless, there are compensations in a long life well spent and one should so live that he may consider he was not born in vain, since nature, as Aristotle says, does nothing in vain. Wishing a long life to the Pope Zerbi then turns to his task.

Galen had defined old age as, first, the pathway to death, and, second, an animal disposition joined with a complexion caused by the innate warm action in a humid being during the period of decline of life. The first stage in old age began from the thirtieth to the thirty-fifth year or even the fortieth and extended to the fiftieth or sixtieth year. Its beginning was imperceptible and during it no clear defect in quality or quantity of substance or members could be perceived. Hence [Isaac] Judeus called this the "unrecognized" old age and Ovid [in the *Fasti* 6. 771], although Zerbi did not name the poet he was quoting somewhat inaccurately, said:

> Labimur occulte tacitisque senescimus annis.

In fact, Zerbi's frequent quotation of the Roman poets (whom he often fails to identify) is one of the many delights of his treatise on old age, together with his faithful citation of the astrologers.

The causes of old age are both extrinsic and intrinsic. The astrologers attribute the former to the impression of the planets upon the human body and the influence of Jupiter in "prima senectute" and of Saturn in "postrema senectute." Crates of Pergamon called those who live longer than one hundred years by the name gymnestes. One of the intrinsic causes of old age is the diminution of

[17] Ladislao Münster has published a brief useful summary of this book in his essay "Il primo trattato pratico compiuto sui problemi della vecchiaia: la "Gerontocomia" di Gabriele Zerbi;" *Rivista di Gerontologia e Geriatria* **1** (Roma, 1951): pp. 38–54. He includes a facsimile of the first page of the prologue and a brief bibliography of writings on old age from Cicero to Roger and Francis Bacon. Frederic D. Zeman, "The Gerontocomia of Gabriele Zerbi—A Fifteenth Century Manual of Hygiene for the Aged," *Journal of the Mt. Sinai Hospital* **10** (1943): pp. 710–716, gives a short discussion, pointing out that Zerbi's book was published before Marsilio Ficino's *De triplici vita* (n.p.n.d.) and that among his sources are Avicenna, *Canon* I, and Roger Bacon's *Care of Old Age*, etc. (translated in 1683). He also cites L. Crummer's translation of Zerbi's *Anatomia Infantis* in *American Journal of Obstetrics and Gynecology* **13** (1927).

youthful heat by the moisture in the body and the transformation of its heat into humidity and coldness, which is death, caused by consumption of the body's substance and warmth.

Zerbi returns to the effects of the signs of the zodiac upon human life in chapter 3. These are definitely accepted by him, contrary to Münster's view that he did not give much importance to them. Longevity is aided by the proper use of the six non-naturals (*cf.*, for example, Massa, *Epistolae Medicinales* II, 13, fol. 109v), air (or respiration), sleep, food (and drink), motion, bowel movements, and the affections of the mind; their decline is typical of old age. The discussion of physiological and pathological degeneration emphasizes the fact that the former has been studied and the latter neglected by those concerned with the subject. Diet is, for example, in part responsible for the weakness of old age since a change in diet is necessary as one grows old. Deficiencies in digestion, changes in the humors and temperature of the body, loss of mobility, gray hair, wrinkles, and baldness are among the inevitable signs of old age and partly its causes. The gradual loss of body heat is especially important. Those who are longest-lived, in fact, have the most abundant heat and humidity and are neither too fat nor too thin, according to Galen. Averroes, however, thought that muscular people with sufficient fat were long-lived also. Other psychosomatic characteristics conductive to longevity are a large chest, a full set of teeth, long life-lines in the palm, and the length of the fingers: the last two items arise from the lore of chiromancy.

Chapter 6 is entirely occupied with quotations from the sixth-century medieval poet Maximianus, who wrote a long elegiac poem on old age.[18] Zerbi does not name him, but he was a popular medieval school author whose description is extremely vivid and detailed, presenting almost all the chief physical and mental characteristics of old age. Zerbi's quotation of Maximian is the most extensive of any medieval Latin poet by the pre-Vesalian anatomists and is one more indication of his wide reading and thorough scholarship.

After quoting Ptolemy to the effect that the lot of human birth depends upon the planets and their signs as iron clings to a lode stone Zerbi continues his discussion of old age in terms of complexions, humors, and the other Medieval elements of physiology. His immediate purpose is to present the evidence for both brevity and longevity in terms of life expectancy. Aulus Gellius and Xenophon are among the authors he quotes; it should be said that the *Gerontocomia* is not packed with references to earlier writers as is the *Liber anathomie*, although Maximianus is once more quoted (lines 171–174, 221–222). Part of Zerbi's theme is the inevitability of old age. Longevity depends upon the complexion of both the species and the individual; in the latter it can be modified within certain limits and is thus a specific determinant for his length of life. After an analysis of the uncertain limits of life the author considers old age as a certain natural disposition beyond nature and comes at length to the function of those who have the aged in their care. It should be observed that the care of the aged involves not the cure of its numerous ills but the conservation of the health of those who are growing old and the prevention of as many of their ills as possible. There are three methods to be pursued: 1. conservation of those forces still active in the aged; 2. the resumption or renewal of these forces and 3. the reduction of inequalities of certain humors by restoring their equilibrium.

The gerontocomos or caretaker of the aged must have the qualities of a good physician or be one himself. He must oversee the environment of his charges, providing good food, quiet, a comfortable bed, and keeping watch over their physiological needs and variations from normal conditions. The climate and surroundings are important, such as hills, streams, or seashore, together with gardens and flowers, wide space and good air. In such an ideal setting the morose and difficult dispositions of the aged can be rendered more tranquil and a revival of pleasure in life among them promoted. Particularly important is the proper dress of old people for each season and occasion to avoid either excessive sweating or dryness of the body. The caretaker must be kind and patient, understanding and merciful, cheerful, moderate, clean in dress and free from offensive odors. The assistants of the gerontocomos must also be chosen with care since they should preserve a harmonious life for the aged; they ought to be chaste, obedient, and hard-working, the first to rise to their duties and the last to go to bed. They should be sober and silent, without any vices. Not every race produces good servants of the aged. The English are too proud, the Swiss suspicious, the Illyrians foul-mouthed, the Hungarians hostile to the Italians. The best servants are Bretons, Germans, some French, those Spaniards who are most similar to the Italians, and best of all the Italians are the Lombards. These assistants of the gerontocomos perform better under his watchful eye.

[18] See Aemilius Baehrens, *Poetae Latini Minores* **5** (Leipzig, Teubner, 1883): pp. 316–329, a total of 292 lines. Zerbi quotes the following passages in this order: lines 139–142, 163, 154–156, 165–166, 167–168, 131–144, 249–255, 157–162, 175, 119–121, 145–150, 137–138, 123–126, 105, 245–246, 195–208, 241–244, 209–225, interlarded with his brief comments. Among his other favorite unnamed poets are Vergil and Juvenal. See E. R. Curtius, *European Literature and the Latin Middle Ages* (London, Routledge and Kegan Paul, 1953), p. 50, on Maximianus as a popular school author and expert on the description of old age, as well as Max Neuburger, "The Latin Poet Maximianus on the Miseries of Old Age," *Bull. Hist. Med.* **21** (1947): pp. 113–119.

The purest air is ideal for the aged, far from sewers or stagnant waters. The proper location for the rest home is not necessarily that which faces south; even a northern exposure is advantageous for some people. An eastern exposure is generally better than any other because of its more salubrious air and more temperate climate; such an exposure facing somewhat toward north and south is even more suitable. A western exposure is the worst of all because of its more impure air, thicker and more moist, which dulls the nature and clouds over one's judgment, corrupting thoughts and warping the affections. At all events, the proper setting must take into consideration the effects of heat and cold at all seasons. This portion of Zerbi's discussion gives him the opportunity to describe (from his reading) southern climes such as Arabia and exotic regions where crops grow twice both winter and summer and where a certain people of India lives two hundred years with white hair in their youth that turns black in old age. There is little evidence here of any information brought back to Europe by recent explorers of the New World or of Africa, however; Zerbi's sources are Pliny, St. Jerome, Isigonus (probably Hesychius), Pietro d'Abano, the Old Testament, and Lactantius, among others. He offers a number of examples of longevity and slow aging as at Parma, Piacenza, Faenza, Bologna and their hill regions, as well as the Lago di Garda near Verona, and Bergamo. At Tivoli also men have lived one hundred and fifty years.

He continues with remarks on proper bedrooms, beds, and gardens nearby with flowers to purify the air and to stimulate the senses. The rooms should be heated when cold weather comes and should be kept clean particularly of pigeon dung, should these birds dwell near. The changes of weather and seasons are to be carefully observed and guarded against to prevent sudden shocks from temperature changes. The winds are also of special importance in their effects upon the health of the aged and should be guarded against. Zerbi contrasts the different influences of the south wind at Rome, where it brings rainy weather and illness, and among the Scots, Dutch, and Norwegians, to whom it brings salubrious bright weather. His etymology of *auster,* south wind, connects it, as Isidore of Seville does, with *haurio,* draw, that is, drawing winds, clouds, and water; but that of *ventus,* from *venio,* come, is not Isidore's: "dictus autem ventus quod sit vehemens et violentus" (*Etymologies* 13. 11. 1).

Proper clothing must be provided for the aged; that is discussed in detail in terms of materials, including wool, silk, linen, and skins or furs. Each has its property for preserving the wearer from diseases that afflict various organs of the body. The beds used by old people are also described in reference to the material of coverings, their weight, kind, and quality: silk sheets are mentioned among other items. Zerbi quotes Querolus (ed. R. Peiper, 1875), a fourth-century author who wrote a reworking of Plautus's *Aulularia*: "Mollia fulcra thori duris sunt cotibus equa," indicating that even soft bed clothing may be hard to some testy old people.

The value of exercise, moderate as to extent and time; the use of baths and their therapeutic effects (authors such as Albucasis and Theodorus Priscianus being quoted here); and the care of nails and beard are then discussed, leading at last to the major part of the *Gerontocomia,* the discussion of diet for the aged; this covers folios 49 to 134, much more than the third which Münster calculates, and is discussed in too great detail to be taken up in a short summary. Diet must be adapted to individual needs, be easy to digest, have sufficient viscosity and fat content so that the chyle may be properly carried through the capillaries after the food has been prepared for the blood. Ease of mastication must also be observed. Some old men clasp boys to their stomachs to promote warmth in digestion; Zerbi approves of this practice for improving natural heat. On the use of wine he quotes a certain Tilephus Grammaticus, *Methodus Sanitatis.* He also quotes the *Libellus de Morte et Pomo* sometimes printed in texts of the pseudo-Aristotelian *Problems* and Johannes Alexandrinus, the commentator on Galen *De Sectis* (see note 1 to my translation of Achillini), in reference to the good odor of food which makes it more palatable. Aristotle is supposed to have sat sniffing an apple before he died, discoursing on philosophy the while. Democritus the philosopher was requested not to die by the inhabitants of the town where he lived until they could complete a three- or four-day festival; he therefore sat for four days sniffing a jar of Attic honey until he died.

Soft foods except for sweets are preferable for the aged; soups and semiliquids are advisable. Eggs, butter, cream, milk, fowl, wine in moderation, fish, and game are among other foods recommended. Naps before dinner but not at other times are helpful; bloodletting should be rarely performed upon the aged. The astringent wood of the date palm or cypress is to be used as a toothpick for the care of teeth and gums. Coitus in advanced age is damaging to the body, especially dangerous to sanguine temperaments and to be avoided in the summer. Those who are accustomed to it, however, should not interrupt the practice. Aphrodisiacs are to be avoided since the gradual extinction of sexual potency is in the order of nature; music and agreeable conversation should form a substitute for diminished sexual activity.

The psychic health of the aged is of particular importance. Depression and irascibility become practically habitual states of mind; sadness, fear, envy and other forms of intemperate behavior increase with age. Worry becomes chronic and must be combated with pleasant surroundings, music, conversation, and other distractions. The final portion of the *Geronto-*

comia deals with specific medicines for the aged. Some of their ingredients are sublimated human blood, powdered precious stones, and potable gold, all of which are believed to be very helpful.

It can be seen from this necessarily brief summary that Zerbi's purpose is not philosophical but practical in this first general treatise on geriatrics; it presents a direct contrast in this respect to its predecessor, the *Quaestiones Metaphysicae*. The book is a guide to the proper hygiene, physical and mental, as well as particularly to the diet of the aged. Its physiological principles are, of course, those which spring from the current doctrines of the humors, temperaments, and qualities. The book represents a long step forward in the enlightened treatment of the handicapped, and its essential humanity and good sense are remarkable for an age in which cruelty and barbarism were still characteristic of the care of orphans and other unfortunates. The authors cited, chiefly classical and medieval Latin, with a few Greeks and Arabs, include all those who had contributed in any way to the treatment of the aged. There is nothing polemical or doctrinaire about Zerbi's discussion; even his references to astrology are in keeping with the practice of his times. He mentioned no contemporary author, a trait familiar among the pre-Vesalian scientists. A curious feature of the *Gerontocomia* is its complete neglect of women among the aged, although what he writes about old men may in large part be accepted as applying to women also. The entire subject of the menopause, however, and its disturbing effects upon the aging female is thus ignored. The fact that women generally lived longer than men was well known, moreover, from antiquity. Zerbi also neglects any discussion of the medico-legal aspects of old age, as Münster points out. Nonetheless, with all its defects the *Gerontocomia* is the most complete and satisfactory treatise on geriatrics to reach us from either the Middle Ages or the Renaissance and as such remains the classic Latin work on the subject, still unfortunately untranslated into any modern tongue. Zerbi remains the founder of the science of geriatrics.

3. *De Cautelis Medicorum,* a much shorter work, is in its turn the first practical treatise on medical ethics. I have pointed out in note 21 to my biographical essay on Benedetti in a summary of Ladislao Münster's article on the subject[19] the relation of the medical guild to a code of ethics and the earlier attempts at providing one by the physicians of Salerno, by Arnold of Villanova, Alberto de' Zancari, Cristoforo Barzizza, and Alessandro Benedetti. Zerbi is the next author on the subject.

The book, in 16 folios and 6 chapters, begins with a prologue in which Zerbi quotes Aristotle, *Metaphysics,* Haly Abbas, Mesue, Damascenus "in afforismis," Avicenna's *Canon,* the New Testament, Galen, and Haly Rodoan, on the relationship of the physician both to the Lord and to his human patients, emphasizing the loftiness of his calling and its great moral responsibility. Medicine is a career fraught with perils and temptation, fraud and error, which the true physician must avoid. *Cautela* means both caution and precaution and, in legal language, security; Zerbi defines it as "the avoidance with diligent attention of deception, fraud, and delusion, infamy, ignominy, and shame which [can] occur to the physician in his operative activity upon the human body while his practical reason directs him to preserve the honor and usefulness of his profession," thus expressing all material, formal, efficient, and final causes of *cautela.* The ideal conduct of the physician, his duties toward his patients, their relatives, and his own colleagues, while evading the perverse intentions of society in general is the subject of Zerbi's opening remarks. The following discourse is divided into six topics according to the means or attitudes by which the physician must seek to preserve himself from danger.

The first of these is derived from the nature or character of the physician and his physical appearance as well as his training. The second is his attitude toward God. The third is his attitude toward himself and the fourth his behavior toward his patient. The fifth is his relation with those present in the sickroom, the women, his disciples, ordinary folk, and the druggists. The sixth deals with the image he presents to the world at large outside the sickroom. It will be seen at once that Zerbi's preoccupation is with a series of problems in public relations well known to all modern physicians and, more recently, to the American Medical Association. Zerbi, in fact, presents what he calls a compendium on the subject well worth reading today.

Natural talent is a great asset to a physician; it should be combined with patient learning and reading as well as attendance at celebrated universities where excellent professors reside. Constant exercise of his

[19] "Il tema di deontologia medica: Il "De cautelis medicorum" di Gabriele Zerbi," *Rivista di storia delle scienze mediche e naturali* **47** (1956): pp. 60–83. Münster is in error in asserting that Zerbi's book was published ten years before Benedetti's *Collectiones Medicinae,* which appeared *circa* 1493, only two years *before* Zerbi's *De cautelis medicorum.* Benedetti must thus be regarded as Zerbi's most immediate predecessor; but the medico-ethical content of the *Collectiones Medicinae* is so slight as to be negligible in comparison with Zerbi's much more detailed work. It may be added that the "De Cautelis Medicorum" exists in an anonymous and undated Italian translation recently published by Clodomiro Mancini and probably done by him (see p. 11), "Un codice deontologico del secolo XV (Il "De cautelis medicorum" di Gabriele de Zerbi): *Scientia Veterum—Collana di Studi di Storia della Medicina diretta e curata da G. Del Guerra,* No. 44 (Pisa, Casa Editrice Giardini, 1963), pp. 5–74. It was discovered by the editor in a miscellany in the Biblioteca Canevari in Genoa and has been thus far unknown to scholars. This publication is the sole translation into any modern language and was called to my attention by Mr. Richard J. Durling, of the National Library of Medicine.

knowledge at the bedside is another prime requisite for the practitioner: as Galen said in the eighth book of his *Therapeutice* (or *Methodus Medendi*): "Medicus non potest esse perfectus nisi exercitatus." In physical appearance the physician should be neither gigantic nor a dwarf, as the *Conciliator* (Pietro d'Abano) says. He should not be either too handsome lest he incite his women patients to become amorous nor so ugly that all will avoid him. Nor should he have a chill breath or cold sweat, neither be goatish and oily in bodily odor nor have polyps in his nose.

The physician must be a god-fearing man since from the Lord comes all healing, for He has created medicine and holy things which are not revealed to the wicked. The physician should speak to his patient in the name of the Lord and regard his cure as attributable to divine authority and not to his own learning or skill. The physician ought to pray for heavenly guidance in his work and when administering antidotes he should say: "If God is willing." In Zerbi's largely borrowed doctrine on the physician's attitude toward God there is, of course, an obvious parallel with Hippocrates' teaching that the healer's soul must be pure, for without purity of soul there is no health in the body.

The fame of the physician creates confidence in the patient and increases his psychological impetus to become well. There is a certain difficulty in the astrological fact that while good doctors and special curative powers are found under the domination of Scorpio and Mars these signs also portend bad moral habits. Although medical science does not have as its province acquired virtues and habits but only natural ones (as appears from the second book of Galen's *Tegni*) nevertheless by his own efforts the doctor can overcome these disadvantages of his nativity and attain laudable habits. Zerbi then proceeds to list a good physician's qualities in the negative. Let him not be ambitious, or a lover of honors, neither blasphemous nor double or triple-tongued, a deceiver, avaricious, thieving, drunken, gluttonous, proud, nor talkative, litigious, nor prone to argument. Let him not be a liar, envious, bearing grudges, or vengeful but astute, sagacious, provident, serious, humble, honest, tireless, gentle, chaste, watchful, abstemious, slow to speak. However, when he does speak he should be affable and use proverbs and parables which present images and similitudes since much medical knowledge can be imparted by no other means. He should be also discreet and not speak to others about his patients' ills nor reveal secrets, as Hippocrates insists in the *Oath of the Physician*. Haly Abbas, Aristotle, and Galen in certain of their works are drawn upon for further ethical guidance in the patient-physician relationship.

Since reading and meditation are the keys to wisdom the physician should be a diligent student of the best authors. If he finds something false in the books he reads he should attribute this not to the respected author but to some other person, such as the translator. With Galen in his *De Ingenio Sanitatis* he may say to himself: "Experiment which I have joined to reason will comfort me and give me great boldness." The physician must, finally, take care of his own health lest he fall ill and suffer the taunt: "Physician, heal thyself."

In his behavior toward his patients the physician should avoid all falsity or love of adulation, setting an example of modesty. He should be a physician in fact and not name only, bringing pleasure to his patients and thus increasing their desire to grow well. On the way to visit him the doctor can inquire from his guide the condition, symptoms, appetite, bowel movements, and other facts concerning his patient in order to facilitate his eventual diagnosis. He will thus make an excellent impression when he arrives. He should then inquire whether the patient has made his confession and exact a promise to do so if the latter has not. The physician needs to rest after climbing stairs, for instance, and to allow the patient's pulse to grow normal after his initial reaction of joy or shame or worry for the fee he must pay has passed. Zerbi describes the method of taking a patient's pulse beat, of examining tongue and teeth, smelling the patient's breath, and analyzing urine. The physician must be cautious about expressing his opinions aloud concerning pulse and urine since the patients are all ears to hear him. This paucity of expressed medical information is still characteristic of the medical craft, as every patient knows, but its origin is obvious.

Returning to his patient, however, after making his urinoscopy the physician should promise him good health and should not fear even to lie, for his "white" lies will be courteous and helpful. The physician's interrogation should be thorough and in detail, concerning past and present phases of the illness involved. Zerbi's sources on these matters are Galen's *De Ratione Curandi ad Glauconem, De Interioribus,* and *On Hippocrates' Prognostica*; Damascenus on the *Aphorisms* of Hippocrates; Mesue; and a certain Nicolus, whose work is not named.

As Galen suggests, the physician should not walk about in public voicing his diagnoses or his cures lest his colleagues hate him and call him a diviner and mage. Diagnosis and prognosis are not to be stated in absolute terms since there are many factors which may cause change in the course of the disease. A positive response to many questions concerning a given illness is not therefore possible and silence is often the best policy; it is safer to depart without committing one's self. The physician's chief care will be for the patient's diet rather than for medicines or surgery. If medicines are needed he should use milder and more familiar ones in order to soothe the body. The physician should not return to a patient he has found on the brink of death but send a servant or a messenger to inquire as to his condition. He should avoid those who are about to die lest the death be imputed to his negligence; he should not be present at the death of a patient. Clearly Zerbi has in mind here the danger of what is now called a suit for malpractice and in his own case was actual murder visited upon himself, although he could not have foreseen it. The kinds of cases from which *infamia* or disrepute could most readily arise were infant diseases, pregnancies, and eye troubles. The physician would be especially wary of giving any assistance toward abortion nor present any drugs or potions to a pregnant woman except those familiar to those about her.

Although the physician ought not to be intent upon gain, the laborer is worthy of his hire. The poor are not to be neglected by the physician; kindness toward them will have its hundredfold reward in the love of the Lord. For other patients the fee should be within reason and they should not be given cause to see in the physician's visits anything but natural kindness and helpfulness. A cure should not be prolonged for the sake of larger fees. Zerbi is well aware that some members of his profession demand more than is just:

> plerumque enim medici ita modum excedunt in petendo ut ars nostra vilescat. (For most physicians are so excessive in their fees that our profession is becoming a bad one.)

The result was often litigation and unpleasant public relations. Discussion of fees with slow-paying clients should be carried on through a third party, a mutual friend or a pharmacist.

The fifth chapter deals with relations between the physician and the family and friends of the patient. His care should be entrusted to dependable people with exact instructions and without unnecessary discussion of the case. The physician may sometimes have to play upon the good instincts of a lazy nurse in order to obtain more diligent service for the patient. It is better for nurse and physician to remain friendly and cooperative with each other. It is equally important for the physician to maintain good relations with the patient's family, praising their solicitude and care and meeting all hostility with conciliatory words, modifying his prescriptions within reason and thus making any critical relative a kind of partner in the cure.

Of great interest are Zerbi's remarks on the physician's behavior toward the women folk of his patients. He should never in word, deed, or glance create any situation of an amorous nature and avoid all appearance of questionable relations with these women. He should have no secrets with them except those which may be safely discussed with an aged female relative. Zerbi refers to the Hippocratic oath in the matter of illicit sexual relations.

Relations with consulting colleagues are also highly important and to be conducted with great care, tact, and modesty. The physician in charge should select colleagues who are not jealous, malicious, or vainglorious, but older than himself and known to the patient. No dissension or unprofessional behavior should be allowed to arise among the consultants, no invidious comparisons made, and full cooperation should be required. Avicenna's *Canon* is here used as a guide. No one brought in as a consultant should express disagreement with the chief physician's handling of the case. He should, on the contrary, freely share his skill and knowledge as a matter of professional pride as well as ethics. All discussion of the case should be conducted out of the patient's earshot. Each colleague must realize that he has his companion's reputation in his care and can by a mere jealous word undermine it: each has a sword at his comrade's throat. Modesty is best in such a situation; the physician should recall Horace's phrase: "Est modus in rebus, sunt certi denique fines/quos ultra citraque nequit consistere rectum" (*Satires* 1. 1. 106–107).

The teaching physician bears a special relationship toward his students. It is his duty to impart his knowledge both practical and theoretical without thought of gain and with a generous spirit, cautious of errors and as clearly as possible. "Unhappy is he who knows much and scorns to teach it," says Zerbi. Effective teaching attracts students to the profession of medicine, but they should not waste their time wholly with theory nor grow old upon the Siren cliffs of meandering speculations by modern disputants. Yet theory must accompany practice lest the latter be lost in error.

The physician should choose a competent druggist to supply him with medicines at a reasonable price. These must meet the proper specifications in every respect in order to avoid any mistake in their use, particularly those which are opiates. The prescriptions must therefore be quite explicit and transmitted secretly to the druggist.

The common people are ignorant of medicine and it is impossible to please them all. Thus the physician needs to be diplomatic with them and to avoid familiarity or quarrels, presenting himself and his remedies affably to all yet with dignity. Let his speech be in parables like those of Jesus to the folk and let him seek no popular praise, for the "vulgus ignobile" cannot discern the skilled from the unskilled. Let him not reject honors or reverence shown him, however, whether from a knight in armor or a man on foot; these he should return politely. Medicines or prescriptions are not to be given on the spur of the moment but on the morrow after a specimen of urine has been delivered and examined.

The physician's home should not be poor but one known to all for easy access. His conduct must be circumspect, avoiding festivities, singing, and dancing; he should not visit houses of prostitution or mingle with sinners, criminals, or murderers. He should avoid public affairs and business, buy no household goods or groceries personally, go hunting or engage in alchemical experiments, in which one grew wealthy only in hope, soot, and foolishness. He should not engage in farming but devote all his energies to medicine. Thus he will be loved and honored, of good repute and favored by the Lord. If he should be invited to a banquet he should not occupy the best seat. He should praise all the food but be moderate in eating and drinking. In the home of the patient he should observe the same abstemiousness. Thus he will be fortunate and happy, a prince among men. One may well ask how many modern physicians observe all these rules and standards of conduct on the very highest level of morality.

Zerbi's medical ethics embrace, as Münster points out, three levels: (1) that of the ancient Hippocratic Oath; (2) that of precepts which are included neither in the Oath nor in other books of the Hippocratic corpus; (3) practical hints and suggestions which arise from Zerbi's personal experience and represent a wealth of practical wisdom, understanding of human psychology and behavior, and the basic elements of public relations. *Cautela* in this latter sense is a much more realistic concept than it appeared in Zerbi's initial definition, embracing the physician's moral and physical qualities, his professional duties and responsibilities, and his daily conduct. Eight editions of the "De Cautelis Medicorum" attest its popularity and prestige in an age when medical standards were never at the high level generally of Zerbi's injunctions; they represented, however, an indispensable ideal all the more attractive for their forceful, simple style and for the genuine moral earnestness they reveal.

4. I have elsewhere commented briefly on the organization and method of Zerbi's *Liber Anathomie Corporis Humani* on Aristotelian lines (note 6 to my translation of Achillini's *Annotationes Anatomicae*) into I. Anterior Parts; II. Posterior Parts; and III. Lateral Parts (*cf.* Aristotle, *De Incessu Animalium* 704b–705b; *De Caelo* II. 284b). These involve the six Aristotelian positions also: above, below, front, behind, right, left. The scholastic method of the book includes the interminable repetition of recurring features: substance, complexion, quantity, form, number, location, and connection (*colligantia*), in which Zerbi was to be followed by his most important successor before Vesalius—Berengario da Carpi. This fact does not predispose the reasonable student of pre-Vesalian anatomy against Zerbi's approach to his subject; these features provided and still provide, apart from purely physiological considerations, a tolerably accurate description of the parts of the body. When Zerbi adds, as he occasionally does after presenting a *Textus* with *Additiones,* the further features *iuvamentum* (assistance or use), *passiones* (illnesses), color, and operations, his description becomes quite complete. Although cumbersome, the scholastic method was thorough; this trait apparently accounts for its maintenance down to as late as 1502 by Zerbi and in modified form to 1521 by Berengario da Carpi. Only Vesalius was to abandon it entirely.

One of the most remarkable aspects of the *Liber Anathomia* is the fact that Berengario da Carpi quotes Zerbi (and no other contemporary) so often in his *Commentary on Mundinus.* This can mean only one thing: for Berengario the book was the sole really comprehensive anatomy available to him. This is all the more remarkable in view of Berengario's open hostility to Zerbi on grounds that must have been personal more than professional, since he is sometimes willing to praise him for his accuracy.

The book is both formidable and forbidding. Its wretched small Gothic type is a continual torture to the eyes in the 1502 edition; the type of the 1533 edition is larger and more comfortable but both editions are very rare and one is more likely to find the first edition. Zerbi opens the work with brief letters to Cardinal Domenico Grimano and to the physician Marino Brocardo; the latter's more lengthy reply is also printed. It contains among other laudatory remarks the statement that there was nothing more accurate *at the time* than Zerbi's knowledge of anatomy: "qua nihil est hodie accuratius in terris." In 1502 this was in fact the truth and no mere complimentary hyperbole; there was no other rival except Benedetti for such praise and his *Anatomice* is much less detailed. The *Liber Anathomie* is the most complete and scholarly work on the subject up to that year.

It is impossible to do more in a brief summary than to give a general idea of its contents. The book extends from folio 1 to 184; then, at the beginning of Book III, the numbering starts afresh and extends from 1 to 20v. This book is followed in turn by a new numbering for the brief essay *De generatione embrionis,* which proceeds to fol. 14v and gives way to an index of one and one-half folios. This rather confusing arrangement is explained at fol. 3v. There is, first, a prologue with general remarks on anatomy and its uses which begins at fol. 2r; Book I, the members in general, begins at fol. 3v. Book II begins at fol. 5v and ends at 184v. It is divided into three tractates dealing with the anterior parts, beginning at fol. 6r; the posterior parts, beginning at fol. 135v; and the lateral parts, beginning at fol. 152v. I am indebted to Signor Luigi D'Aurizio, of the Biblioteca Universitaria of Bologna, for a careful handwritten statement of the erratic signatures and foliation of the 1533 edition, which contains a total of 233 folios as against the 219½ of the 1502 edition: the former is more readable since it is in larger print.

It is perhaps most practical to present excerpts in translation as illustrations of Zerbi's general procedure. After defining and defending anatomy, listing briefly the Alexandrian anatomists and the Roman Celsus, and discussing the remarks of Galen, Moses of Cordoba, the pseudo-Galenic *De Anatomia Vivorum,* Pietro d'Abano, Johannes Alexandrinus's commentary on Galen's *De Sectis,* Gentilis, Aristotle,

Firmianus Lactantius (a favorite with both Zerbi and Berengario da Carpi but with no other pre-Vesalian), Isaac Israeli, John of Damascus, and Varro within the space of three folios, Zerbi takes up the proportions of the human body (fol. 5r and 87r: compare Erwin Panofsky's brilliant essay "The History of the Theory of Human Proportions as a Reflection of the History of Styles" for a brief survey of the subject)[20] and then settles down to his major task.

I shall begin with the heart at fol. 62. Zerbi first describes the adjacent parts: the diaphragm and the pericardium, the heart's fat and the auricles, each of them in detail. Then comes the following "Additio":

The anatomy of the heart which the Greeks call cardia. The heart is among all the viscera and members of the body the first root of the innate and vital heat of life, by means of which as if from a mine or a fountain the living creature dispenses that heat, according to Galen, *De Usu Partium* VI. The peripatetics have given it even greater dignity by teaching that the heart holds in the human body a position analogous to that of the first intelligence in the world. No blooded animal lacks a heart, that is, an animal which has blood by its nature, according to Aristotle, *De Partibus* [*Animalium,* III. 4. 665b5]. The viscera appear more bloodlike and larger in the newly born and not without reason since in their first formation there appears a supply of blood as material. Bloodless animals have none of it and no heart or viscera while the blooded have all the viscera. Hence Democritus [*ibid.* 665a30] does not seem to have discussed this matter accurately when he thought the viscera were not visible in bloodless animals because the latter were so small, for as soon as blooded animals are formed their heart and liver can be discerned unmistakably although they are quite small. But I must omit discussion of the heart in other animals and return to the human heart alone. Its articulation, fashioning, and organization when carefully examined exceed those of the hearts of any other animals, for the hearts of animals of keen sense are more articulated than those of duller sensation, such as pigs, according to Aristotle [*ibid.* III. 4. 667a5], and all the differences or distinctions of the heart as to size, smallness, hardness, softness pertain to some extent to its temperament, for those whose sense is dull have a hard, dense heart, but those with keen sense have softer hearts. Animals whose heart is larger are timorous, and those with a small or smaller heart are more bold and confident. For the affection produced by motion is uppermost in these. Their heart is of a thin substance and texture, indicative of the more subtle operations of the soul, but a heart of dense substance and texture is dull in its operations. They say that some men have a hairy and bristly heart and hence are very powerful and active, as Aristomenus the Messenian; but this view must be regarded as completely fanciful.

There follows a "Textus":

The heart is a fleshy member produced from arterial blood as the liver is from venal blood, for the uterus at the beginning of conception attracts at the same time the thinner and warmer blood with spirit through the arteries and venal blood through the veins. From these two kinds of blood the uterus creates the warmest of the viscera. From the thick blood it makes the substance of the liver, said Galen, *De Semine* I. 7, confirmed by Haly Abbas, the more dependable commentator of Galen. Hence the liver is the heaviest of all the body's fleshy members, as is also demonstrated by the senses, said the author of the *Anathomia.*

The last clause is probably a reference to Mundinus, who speaks of the very great size of the liver and its weight on account of its density.

Most of the above translation is drawn from Aristotle, and there is no denying the derivative nature of much of Zerbi's anatomy. But as one proceeds through such a characteristic passage, which details further aspects of the heart in the form of an "Additio" Zerbi's own observations based on dissection come to the fore; although quotation from authorities is not entirely abandoned it becomes much reduced.

The meaning of Zerbi's terms—anterior, posterior, lateral parts—should be clarified. The anterior parts include those of the lower body cavity, with the mirach (hypogastrion or abdomen), the siphac (peritoneum), and the zirbus (omentum); Zerbi uses the Arabic terms, with their Latin equivalents only for the zirbus. Berengario da Carpi also used Arabic terms. The muscles and vascular system for these parts are discussed, all in great detail. The intestines, stomach, spleen, kidneys, sexual organs, bladder, heart, lungs, throat, mouth, tongue, lips, larynx, the head, the brain, the torcular of Herophilus, the rete mirabile, the nerves, eyes, ears, and nose constitute the anterior parts. The posterior parts include the neck, the muscles of the head, the spine and its vertebrae, and the nerves of the spine. The lateral parts are the shoulder blades, the veins of the arms and hands with their bones and muscles, the coxae, legs, and feet with their bones and muscles. Zerbi then devotes 19 chapters to separate discussion of the skin, fat, flesh, muscles, fibers, tendons, nerves, ligaments, veins, arteries, cartilage, bones, joints, and medulla.

The impression Zerbi's *Anathomia* presents is one of great learning but perhaps too little evidence of original

---

[20] In his *Meaning in the Visual Arts* (New York, Doubleday, 1955), pp. 55–107. Among the anatomists Panofsky quotes only Galen, *De Placitis Hippocratis et Platonis,* V. 3. Zerbi is the sole pre-Vesalian I know who mentions the subject, which goes back to Polyclitus, "the father, or at least the formulator, of classical Greek anthropometry" (Panofsky, *op. cit.,* p. 64). Evidently Zerbi draws his view of comparative measurements from Vitruvius, who specified the navel as the center of the circle into which the body fits when spread-eagled, a theory of cosmological origin underlined by Zerbi's remark: "et propter hoc physici hominem minorem mundum iudicauerunt." It may be noted that Vitruvius, *De Architectura,* I, 2, 3, while listing the forms of knowledge which contribute to shape the ideal architect, says: "medicinae non sit ignarus." He is speaking, however, of that branch of medicine which Hippocrates dealt with in his *On Airs, Waters, and Places,* on endemic diseases and the choice of healthy or unhealthy localities.

work although occasionally he mentions a dissection, such as that of a pig which he conducted at Rome (fol. 7v). He is critical of Mundinus but not of other previous writers. He quotes contemporaries or refers to their efforts as, for example, Guy de Chauliac's measurement of the stomach (fol 20v), which Zerbi read in French:

Quantitate stomaci hominis conatus metiri cauliacus sermone barbaro dixit ipsum continere duos vel tres pancherios vini; nos tamen putamus ipsum esse insignis continentie. (Having tried to measure the size of the human stomach Chauliac wrote in French that it contained two or three measures [*pancherios*] of wine; I however think it is of considerable contents.)

The abundant detail, the numerous authorities quoted, and the attempt to cover the subject of anatomy with encyclopedic thoroughness but with chief reliance upon Galen make the *Anathomia* the superior work it is for its time. It is the last medieval scholastic anatomy and the most complete of its kind, as Vesalius's *Fabrica*, published forty-one years later, was to be the most complete of its new and Renaissance kind.

5. *De Generatione Embrionis* is Zerbi's last published essay, attached at the end of the *Liber Anathomie* and printed by Dryander in 1537 as part of his *Anatomiae, hoc est, Corporis Humani Dissectionis Pars Prior*. A variant title is *Anathomia matricis praegnantis*. It is one of the few specialized anatomical treatises produced by the pre-Vesalians and thus of more than intrinsic interest. Zerbi begins by stating that his investigations of the pregnant female have been carried out by dissection of monkeys, pigs, and other female animals but not of humans. Hippocrates, Galen, and occasionally Haly Abbas, Pietro d'Abano, Avicenna, and Aristotle are his chief authorities. The various stages of development of the embryo are first sketched and then the theories of the semen and its function in generation are outlined on the basis of earlier writers, chiefly Aristotle, Galen, and Avicenna. The formation of the secundina is the main feature of this *sermo* or discussion, with much use of analogy from the baking of bread in an oven. The nine characteristics of the bodily organs employed in the *Liber Anathomie* are also used in the description of the embryo.

The next portion of the work deals with the formation of the three principal members—the heart, liver, brain—and the umbilicus, together with its veins and those of the secundina. The connections or *colligantia* are next discussed. Two other panniculi or coverings of the fetus are then described, with their various names, although it is impossible to learn anything from the authorities about their relative origin. The formation of the members of the fetus itself can be determined from abortions; it is clear that they are created at different periods of development. The heart is the first created; a long discussion amplifies Aristotle's well-known reasons for asserting its primacy of origin. Then the liver and the brain come into being.

The progress of the first twenty-seven days of the existence of the fetus is thus recorded; the developments within the period from the twenty-seventh to the thirty-sixth or fortieth day are then taken up. Within this time-scale sex is determined in the fetus; a problem arises as to the point when the soul enters the living embryo together with reason. Various theories on this matter are presented from Pietro d'Abano, Moses of Cordoba (a favorite authority with Zerbi), Averroes, and Haly Abbas in addition to Aristotle. The time at which the embryo first begins to move in the uterus is the next item for discussion as well as when it is born; according to Avicenna, movement begins at the end of three and one-half months after conception, but this statement is compared with other views on the matter.

Next to be handled is the anatomy of the complete and animate infant in the womb; eight of the total nine features usually employed are brought into this description. At this stage the fetus has been heard to cry in the uterus, a sign of misfortune to come, despite the denial of Albertus Magnus, *De Animalibus* IX that children do not cry before they are born. The fetus according to Aristotle, *De Generatione Animalium,* who is Zerbi's basic but by no means sole source, maintains a state midway between sleep and waking during this period of lethargy.

Natural birth and its manner are next on the list of events in the life of the fetus. Pliny reports that at Saguntum in the year of its destruction by Hannibal an infant which was being born crept back into the uterus, a manifest sign of disaster. Various conditions govern birth in addition to the normal one of the completed cycle of time: nutrition, the phases of the moon, injury to the mother, etc. Contraction and expansion of the uterus indicate the approach of childbirth: that is, labor pains. Zerbi ends his essay with a list of the signs of conception and an analysis of the indications of sex in the fetus: when milk exudes from the right nipple, for example, the child is a boy.[21]

---

[21] As a final note on Zerbi's scientific contributions I may add that William S. Heckscher, *Rembrandt's Anatomy of Dr. Nicolaas Tulp* (New York University Press, 1958), pp. 173–174, says that supposedly the first scientific observations on the property of ethereal oils and their use in the preservation of corpses is found in print in Zerbi's *Anatomy* of 1502. He gives no exact citation but may be referring to Book III, fol. 3 v, where the preservation of body fat with camphorated oil is mentioned.

Berengario da Carpi
(1470–1530)

COMMENTARY ON MUNDINUS (1521)

THIS book, like Zerbi's *Liber Anathomie,* is an unexplored wilderness that requires a more complete map than I can supply in these pages: in fact, Berengario's commentary needs a commentary of its own. The last of the variorum anatomies, it assembles as great an array of authorities as Zerbi does and is in this respect dependent upon and inspired by Zerbi. It is also the last attempt of the pre-Vesalians to allay and reconcile the disagreements of earlier anatomists, with Mundinus as a pretext. This is clearly the intent of Berengario's opening words.

The praise he gives Mundinus is for his brevity, a virtue neither Zerbi nor Berengario has imitated. That brevity, in fact, gives Berengario an opportunity to "add something worthy of note not without its usefulness for younger men." He complains that special works by Galen available to his own time are only two: on the anatomy of the eye and of the uterus, and that Mundinus had not seen any of Galen's forty other treatises before he wrote his own small book. It is not Berengario's purpose, however, to exalt Mundinus above Aristotle, Galen, or Avicenna for any talent except that of conciseness. Those who criticize Mundinus forget the paucity of anatomical books in his day; no wonder he is occasionally found wanting in some respects, as in accepting the erroneous view of the seven-celled uterus. Here Berengario employs the famous image of the epigoni who are like children seated upon the shoulders of giants and thus able to see farther than they.[1] In further defense of Mundinus's shortcomings Berengario then quotes both Vergil, *Eclogues* 8. 63: "non omnia possumus omnes" and Horace, *Epistles* 2. 3. 359: "quandoque bonus dormitat Homerus," without naming either author or identifying either passage. In his quotation of the Latin poets Berengario is again at one with Zerbi: the two have certain habits they share. Lack of books accounts for Mundinus's failings, for which Gentilis, Nicolus, Arcolanus, Ugo, Jacopo da Forlì, and others chide him. Berengario insists that Mundinus is his own first guide and deserves homage as first among the Latin anatomists.

In discussing the advantages of vivisection over dissection of the cadaver Berengario points out that he had frequently discovered skull injuries apparent only from the livid color of the exterior, as he had described them in his *De Fractura Cranei.* Yet, dissection being the chief aid in discovering inner ills the physician who did not know anatomy was a blind leader of the blind. In defining anatomy he employs, as Zerbi and Achillini had, the words of Johannes Alexandrinus's commentary on Galen, *De Sectis.* The utility of anatomy was clear to both physician and philosopher (the well-known combination) in their investigation of nature. Whether, finally, anatomy is an art or a science is a question Berengario left to the sophists, whose books were full of its discussion.

The trained anatomist must read Aristotle, Galen, Avicenna, and others on animals both live and dead; Berengario is the first and only pre-Vesalian to insist that he must also read the *iuniores,* the later writers. Dissections must be performed on various species, both sexes, different ages, pregnant and non-pregnant, and both live and dead creatures; the fetus is also to be dissected. He had himself learned much from dissections of the latter. Sight and feeling or touch are also necessary to accompany wide reading; nor is one dissection sufficient. "One must not believe other

FIG. 3. Jacopo Berengario da Carpi. Museo Municipale, Carpi. Regarded by Putti as a purely imaginary representation. Courtesy of National Library of Medicine.

[1] Compare John of Salisbury, *Metalogicus* III, 4, ed. Clement J. Webb (Oxford, 1929), p. 136: "Dicebat Bernardus Carnotensis nos esse quasi nanos gigantium humeris insidentes ut possimus plura eis et remotiora videre, non utique proprii visus acumine aut eminentia corporis, sed quia in altum subvehimur et extollimur magnitudine gigantea." Lane Cooper, in *Evolution and Repentance* (Cornell University Press, 1935), p. 182, writes: "Newton said that he stood upon the shoulders of giants." If so, he was echoing Bernard of Chartres—and Berengario da Carpi.

authorities when experience and sense-perception run counter to them'' is Berengario's motto (fol. VII).

Various divisions of the cadaver for purposes of dissection are listed and that of Mundinus into the natural, spiritual or vital, animal parts, and extremities is praised for its simplicity. Whether a body which has died through hanging, decapitation, or some other form of death should be chosen is a question settled in favor of strangulation, although some say that strangled corpses displace the eyes. Diseased members should be chosen for the knowledge of that disease which dissection will produce: i.e., eyes with cataracts will teach one about this infirmity, or a pleura afflicted with pleurisy or a bladder with stone. The upright nature of man—a commonplace—is adorned by reference to Plato, who said that man must look at the heavens with inner eyes, the intellect. Only man is able to sit down properly; as Galen, *De Usu Partium* IV. 19, said, a monkey does so only to mock men. The eyes are situated close to the brain so that the latter may transmit the visual spirits to the eyes in greater quantity and more swiftly. Examples of keen sight are adduced from the eagle and the lynx as well as from authors who describe carvings so delicate that they can scarcely be perceived.

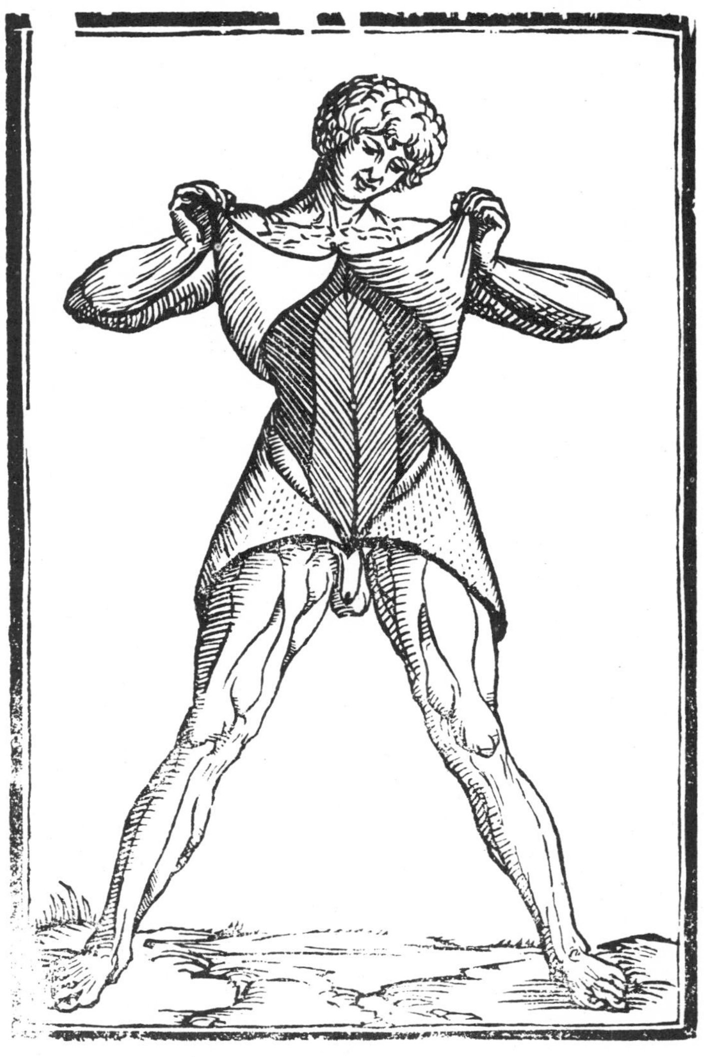

Figs. 4–11. Figures from Berengario da Carpi, *Commentary on Mundinus* (1521).

Man differs in arts, habits, and instinct from the other animals although some say he would lack free will if he had natural instinct. Berengario refutes this statement by pointing out the way a child at the breast sucks milk as soon as the nipple is presented to him as if he had learned the process long ago. Since man has such natural arts and inclinations parents should encourage those for which their children show aptitude. As Galen said (*De Usu Partium* I. 3), man does not possess arts because he has hands nor is he wise on account of them: because of his wisdom he has hands, for reason teaches the arts and hands are the organs of the arts. Nature, however, rather than reason dominates among brute animals, according to Hippocrates. Some people say centaurs have hands; but Berengario calls this view fanciful (*poeticum*), because it is impossible for the sperm of man and horse to mingle and form a centaur (Galen, *ibid.* IV. 9). He adds many differences unmentioned by Mundinus between man and the other animals, some of them concerned with habits of coitus but chiefly in anatomical features. He cites Boccaccio, *De Genealogia Deorum* IV. 68 on the remains of a giant found in Sicily who was more than two hundred cubits long, although this is very difficult to believe, as he adds.

After giving a catalog of the names of the external parts Berengario proceeds with the commentary on Mundinus in a series of forty sections (fol. 44V). It is impossible to give in a brief space an adequate view of this huge work in 528 folios, although these are small in size. It may be well first to list his sources and in this way to show how much of his work is borrowed from others. He quotes all the usual authorities, Greek, Latin, Arabic, and in addition a number who are not drawn upon by any other pre-Vesalian anatomists; among his contemporaries he refers to or quotes Zerbi at least 123 times and Achillini 3 times, a testimonial to his judgment of their relative importance. Lactantius, *De opificio Dei* (Migne, *P. L.* VII), Nonius Marcellus, Varro, Ambrosio Calepino's dictionary, Venerable Bede's commentary on the Gospel of Mark, Albertus Magnus, Plautus, Terence, Martial, Apuleius, Ennius, the Laws of the Twelve Tables, Persius, Homer, Juvenal, Vergil, Giorgio Valla, Suetonius, Alpharabius, various experts on chiromancy, Aulus Gellius, Macrobius's *Saturnalia,* Isidore of Seville, Horus, Pausanias, Rufus, Niccolò Leoniceno, Guy de Chauliac, Johannes Matthaeus Ferrarius de Gradibus, Nicolus Aggregator, Oribasius, Jacques Despars (Jacobus de

Partibus), Cato, *De Re Rustica,* Gentilis, Peter of Spain, Giovanni Baldù of Florence on the third book of Avicenna's *Canon,* Rabbi Moyses, Ugo Benzi, Columella, Gilbert the Englishman, Trusianus, Thomas del Garbo as well as Dino del Garbo, represent a partial list of the various sources used by Berengario. He is also the sole pre-Vesalian to quote foreign contemporaries such as Jacques Despars and Guy de Chauliac.

Berengario's frequent references to Zerbi prove, on close examination, to be more frequently than not simple corroborations of Berengario's own views or those of authorities of whom he approves. They form an imposing testimonial to the respect in which he held his colleague. Occasionally he emphasizes Zerbi's correct views by such a word as "recte." Whatever the source of his personal animosity toward Zerbi, as illustrated by two insulting passages (fols. 17r, 31v), Berengario's estimation of Zerbi's professional standing is impartial and determined only by scientific standards.

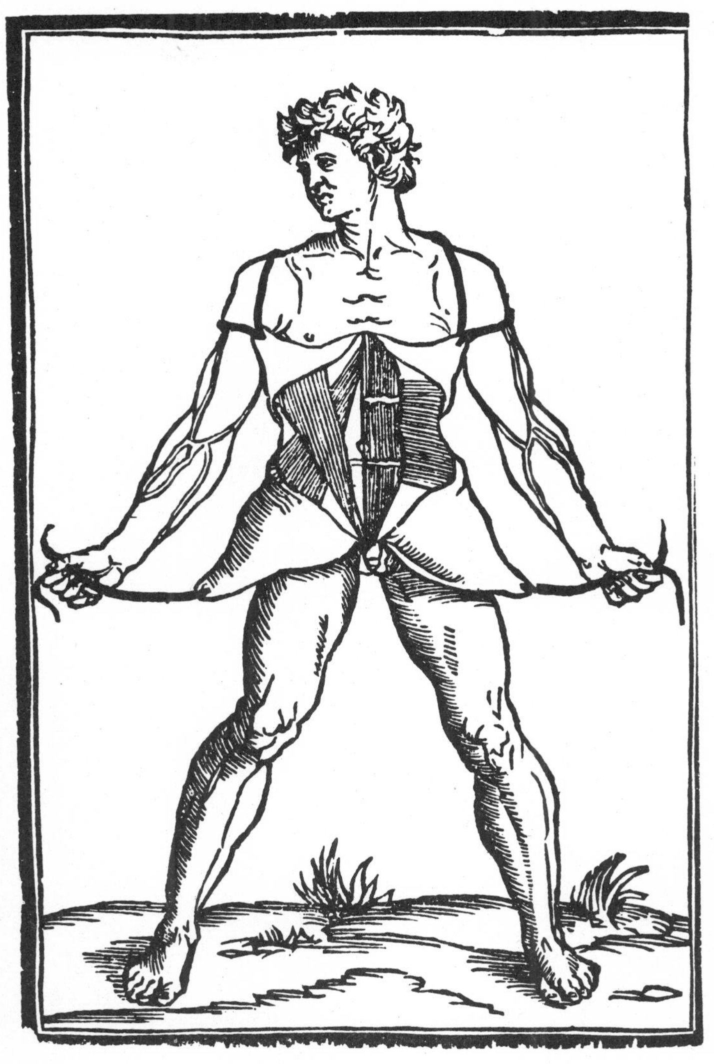

Fig. 6.

Fig. 5.

Berengario, in addition to an exhaustive quotation of authorities, is comprehensive in assessing the various terminology of anatomy, giving the Greek, Arabic, and Latin as well as occasionally the current vernacular names in Italian for the parts of the body. Pollux, Rufus, and other philological medical writers are cited by him. He also at times engages in textual criticism or discussion of works attributable, for instance, to Galen, as at fol. 89v where he points out that the *De iuvamentis membrorum* is not a different text from the *De Usu* (or *utilitate,* as he puts it) *Partium* (or *Particularum*) but is one and the same book. Translations and their merits are likewise discussed by Berengario, who thus proves himself the best philological anatomist of the pre-Vesalians, far superior, for example, to Benedetti, who greatly prided himself on his care with nomenclature, especially Greek.

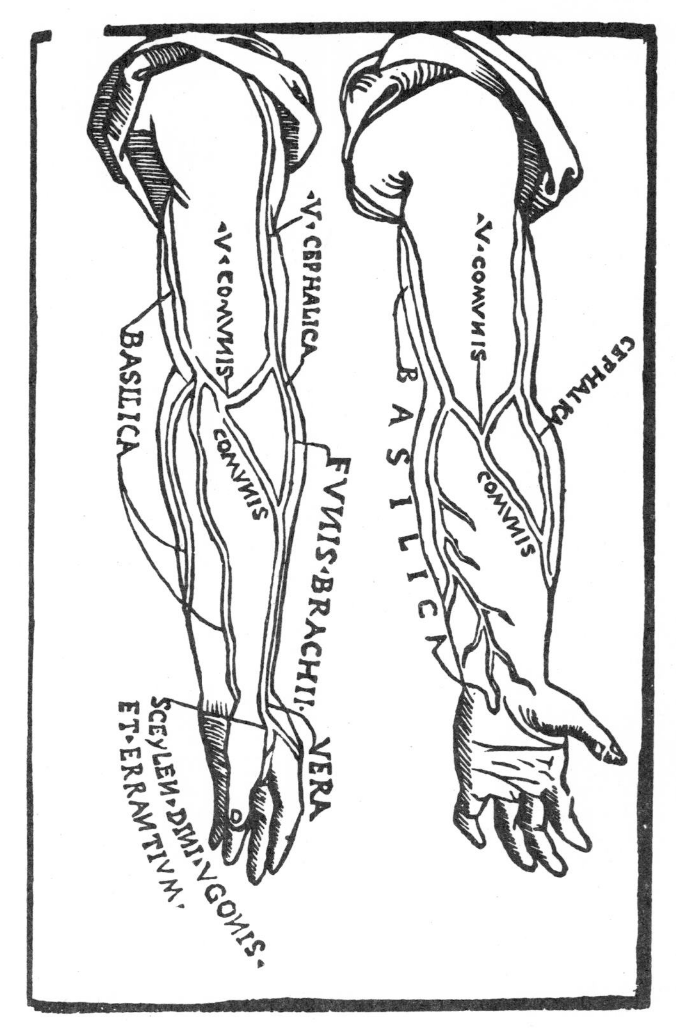

FIG. 7.

He supplies us with a number of personal observations from his medical practice and dissections. He experimented with the emulgent vein of the kidney, forcing warm water from a syringe through the vein to discover whether the water emerged to the urinary duct. He found that it did not do so; the kidney merely filled with water and when its surface was ruptured the water poured out. Berengario drew certain conclusions about kidney stone from this experiment.

On other occasions he gives case histories of a limited sort concerning his patients, such as the Marchese Galeazzo Pallavicini, whom he attended for kidney disease together with the physicians Francesco da Bobio, Homobono da Cremona, Hieronymo Caranzono, and Giorgio da Cremona. He mentions an expensive mule of his which grew very ardent when he sensed the presence of a mare or a donkey and refers to a female mule born at Rome in the time of Leo X. He publicly dissected a young and beautiful pregnant woman who had been hanged at Bologna; for four days before her execution she had had intercourse and he found intact sperm in her matrix, proving that it opened in coition. He states that the sex of a newly elected pope is verified by the college of cardinals as a matter of historical tradition and cites Cardinal Zabarella on conception without a ruptured hymen, adding a case of it vouched for in France by Matthew of Gradi. He objects to modern anatomists who follow mere authority and are like cattle who follow each other. In one instance he attended a woman with a large tumor in her uterus as a colleague of the prominent physician, Leonello dei Vittori da Faenza, and mentions the fact that he (Berengario) dissected many non-pregnant female cadavers. He details frequent operations which he performed upon noble Bolognese girls and upon two- or three-day old infants

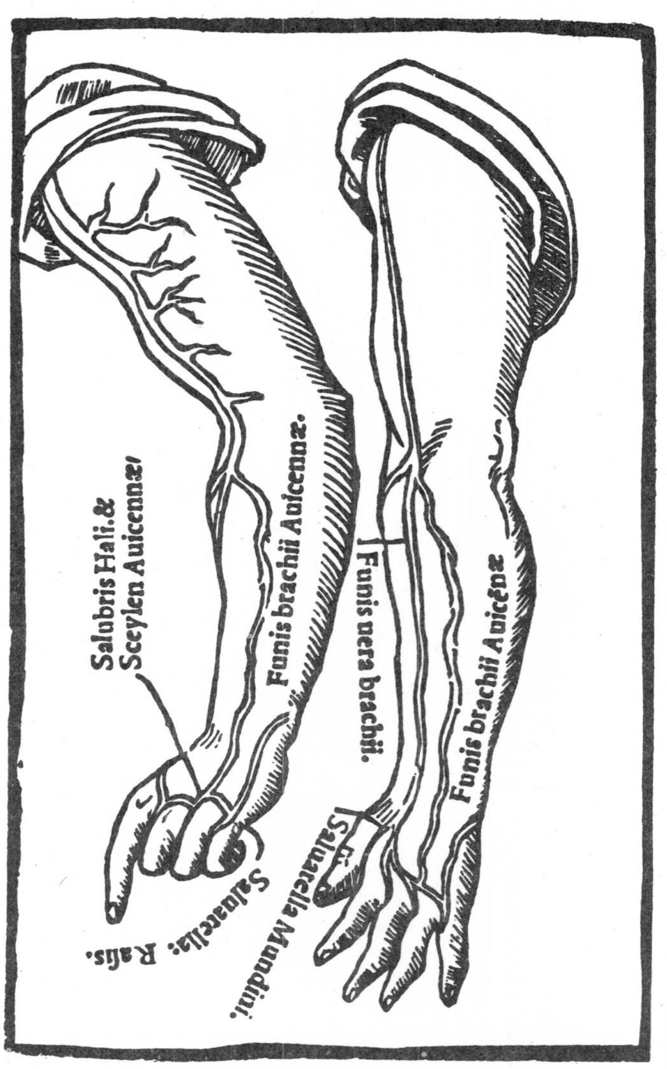

FIG. 8.

because they had membranes covering the urinary and vaginal orifices, as well as the rectum, nostrils, and ears. On the theory of the seven-celled uterus he refers to the *Enneas Muliebris*, a book written by Lodovico Bonaciolo and dedicated to Lucrezia Borgia, duchess of Ferrara, which contained a long discussion of the subject. Some said that the seven cells had ten folds or wrinkles each; thus the woman could bear seventy children. Apparently this theory had never been demonstrated in practice. Berengario regards the entire matter as pure mendacity. On the topic of multiple births, however, he declared that he knew at Bologna the sole survivor of seven children born at the same time; his mother was the sister of Floriano de Dulphi, a relative of Berengario. He had also seen quintuplets born at Carpi. An event attested by Alessandro de Campo Fregoso, bishop of Ventimiglia, who swore his episcopal oath upon it, was the birth of sixteen human foetuses by Giovanna de Egregia, of the Boccanegra family; they were born alive. These were marvels that had to be believed because they were related by historians —"quia historiographi multa dicunt miranda de monstris"—a statement somewhat damaging to Berengario's generally respectable scientific reputation.

Two cases of prolapse of the uterus were treated by him at Carpi, both of them successfully cured:

(fol. 225r) However, I saw two remarkable cases of this sort which were healed. One was a woman I saw at Carpi. She was called Euphemia and was thirty years of age. My father attended her and I accompanied him at the bedside. She was and is the wife of Alessandro Michele, a goldsmith of Carpi. Her uterus was completely extruded beyond the vulva and putrescent. My father cut if off with a razor and healed it. Later she was in very good health and carried on her household duties. Her husband still has intercourse with her because the neck of the uterus remained in its position and was not cut off with the remainder of that organ. She, however, has no pleasure in the act. Also in May, 1507, I was called by a certain tanner, who lived at Bologna in the Saragoza district to see his wife who lay very ill with fever and jaundice. She had the body of her uterus extruded beyond the vulva like a large purse turned inside out; it was black and fetid with gangrene. The uterus had fallen forward during a difficult delivery and the midwives could not by any method return it to its proper place; hence it had become gangrenous. I tied it off near the first orifice of the neck with a thread quite twisted and thick and suddenly cut it off with a razor and it fell off. I ordered the spot to be bathed in wine in which honey and aloes had been boiled and when two days had passed the fever left her together with certain pains of back and head that had accompanied the gangrenous state of the uterus. Later I cured the woman's jaundice with the proper remedies and restored her to her former health. She lived safe and sound for a long time thereafter.

The pioneering investigation of Berengario's *Commentary* was made by Lynn Thorndike in his invaluable work, *A History of Magic and Experimental Science*.[2] As I point out in my translation of Berengario's *Isagogae Breves*[3] Thorndike defended Berengario as a commentator on Mundinus and as a follower of authority against Moritz Roth's unwarranted attacks in his life of Vesalius (1892), using the same source, the *Commentary*. Berengario followed Galen except where his own observation led him to differ from the latter, pointed out errors in Mundinus and other anatomists, and attempted to substitute first-hand experience for authority where he could. As Thorndike says, he relied "first and foremost upon his own observations and experience." He took no credit for any discoveries and is innocent of the charge of vanity brought against him by Roth, who wished to exalt Vesalius by deprecating the accomplishments of the pre-Vesalians. Berengario allowed himself only one statement which reflects pride in his achievement; its truth cannot be denied in view of Roth's open hostility: "On account of this book [the *Commentary*] envy will love me after death" (*Isagogae Breves*, fol. 8v). Thorndike also emphasizes Berengario's interest in textual criticism of his sources and his wide knowledge of the literature. In fact, one may say that Berengario as the last pre-Vesalian of real eminence combines reliance upon personal observation rather than authority—a position toward which Massa had also moved—with a philological criticism of texts to a degree reached by none of his predecessors, pointing out the double path which his successors were to follow.

Berengario comments sensibly on the treatment of wounds, the fat of human hearts, operation for bladder stone, unskillful tampering with saliva outlets, half-dead foetuses, monstrous births, medical versus astrological explanations of certain phenomena, the supposed indication—from nodes on the umbilicus of the first-born—of how many children the mother would have, the greater thickness of skull in those who went about hatless, dolichocephalic and brachiocephalic people, etc. He was highly skeptical of such views as the one held by Zerbi that there was a sieve in the kidneys, of those who believed the *os sacrum* was in the breast, and that the uterus was divided into many cells.

His criticism of authorities led to his controversy with Niccolò Leoniceno (1428–1524), the teacher of Paracelsus and according to Castiglioni, "the best critic of his period." Leoniceno's Latin translation of Galen's comments on Hippocrates' *Aphorisms* had been adversely critized by both a former student

[2] 5 (New York, Columbia University Press, 1941), chap. 23, "Anatomy from Carpi to Vesalius," pp. 498–531.

[3] Jacopo Berengario da Carpi, *A Short Introduction to Anatomy (Isagogae Breves)*, translated with an Introduction and Historical Notes by L. R. Lind, with Anatomical Notes by Paul G. Roofe (University of Chicago Press, 1959), pp. 21–23.

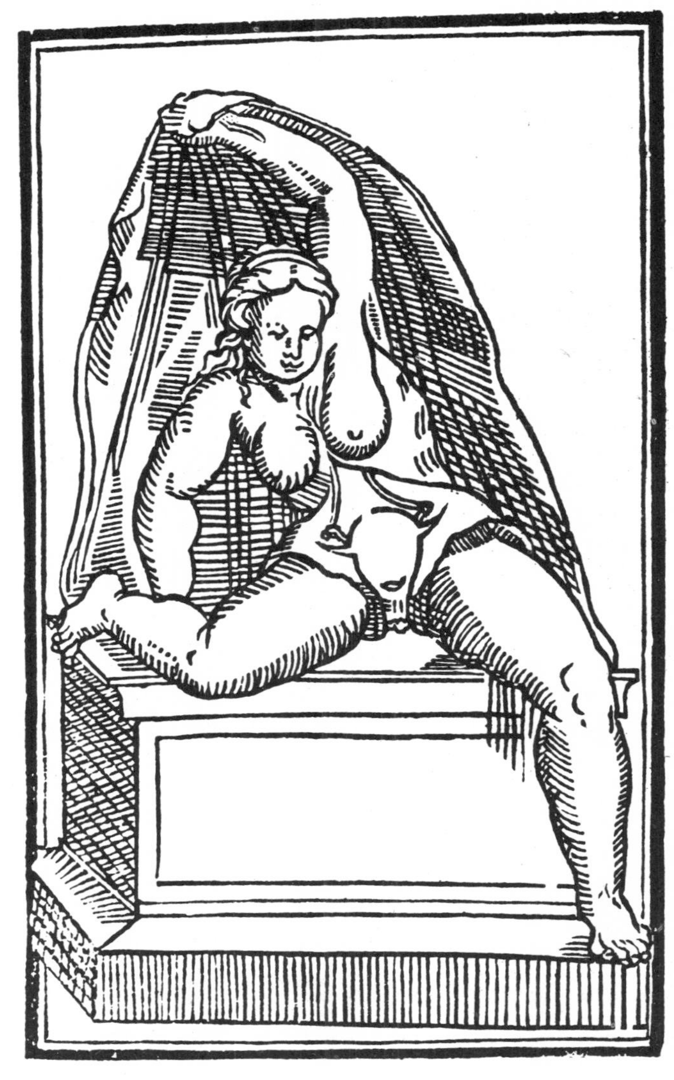

FIG. 9.

of Leoniceno and by Berengario (*Commentary,* fols. 89v, 105v, 116v, 361v–377v, 388v) on the subjects of attribution of certain texts to Galen, the condition of the omentum in a state of hernia, and Leoniceno's discussion of error in Pliny. The affair resulted in the publication of a small work now quite rare entitled *Nicolai Leoniceni medici clarissimi contra obtrectatores apologia* (Venice, Jacopo Pentio de Leuco, 1522). This book is an indication of the vigorous and often acrimonious differences of opinion that made the study of anatomy so exciting in the Renaissance, in this field as in all others the age of intellectual controversy. In this aspect, too, of the culture of his time Berengario played a distinguished role.[4]

He was bold enough to contradict even St. Thomas Aquinas and to hold with the "moderni" on the subject of red sperm in the female, seeking with them to harmonize the discord between Galen and Aristotle as to its exact nature, whether semen or menstrual blood. In a long discussion of the topic when the soul enters the human body he passed in review the opinions of others and stated his objections to their conclusions. The soul enters the male more quickly than into the female, according to Berengario. The entire topic is concerned with the development of the embryo; even the Salernitan medical verses, outdated as they were, are deemed worthy of quotation in this regard.

At fol. 208v one of his references to Zerbi runs as follows: "and Zerbi among others says many things about this subject; he is in truth a very great collector of anatomical authorities and worthy of praise although he frequently departs from accuracy." The word used for collector is "aggregator," a name not

FIG. 10.

[4] The preceding two paragraphs are quoted with some changes from my translation referred to in note 3 above.

unknown elsewhere among the medieval anatomists, and indicates a position of prominence among physicians.

Berengario dissected many foetuses of dogs (fol. 260r) in which he sought the passage by which the urine passed from the bladder to the umbilicus and thence to the elancoidea, a membrane immediately within the secundina; he had also demonstrated this membrane in the dissection of a woman dead in the ninth month of pregnancy, performed in his own home on May 17, 1520. He did not find this passage, proving that the foetus urinated through the usual organ, not through the umbilicus, as was generally believed.

He once saw a donkey giving birth on the road; when its navel cord was cut it bled to death. The implications of this incident for human childbirth were carefully set down. The various positions of the foetus at birth were of particular interest to Berengario, and in connection with them he discussed the theories as to the positions of both men and women who had drowned, the man face down, the woman face up. He adds that he had not verified this statement from personal experience. In the long account of the development of the embryo he quotes, among others, a poem by Lanfranc the surgeon on the sequence of time involved (fol. 267r). He notes an hereditary birthmark in Count Ercole Bentivoglio of Bologna, a rough blemish under the right eye which became red and painful when rainy weather was at hand. In connection with the resemblance of children to their fathers he comments on the frequency of adultery at Florence and Bologna and in the next breath admits that "in my judgment all Italy is Bologna and Florence and Venus and Mars are everywhere" (fol. 277v).

Berengario designed his commentary for students and for those inexperienced in anatomy: "hoc nostrum opus non scribitur nisi scolaribus et aliis in anatomia inexpertis" (fol. 287v), thus excusing his pedagogical prolixity. He warned the envious to keep silence since his book, which brought the truth of personal observation to bear upon anatomy, would, in the place of the magical alexipharmic called bezoar, drive away the poisonous barking of his enemies.

At one point he said he would write further on hernia in his "chirurgia," a work on surgery which he did not complete, if it was ever begun (fol. 294v). Even his copious *Commentary*, it seems, could not contain all he wished to write. He mentions the recent advent of syphilis, calling it the French disease and promising to discuss it further in the promised work on surgery (fol. 308v). He was the first to show the attending physicians at Florence, where a noblewoman was suffering from a fistula in the mediastinum that the serous membrane was separate from other membranes and did not touch the lungs (fol. 326v). The

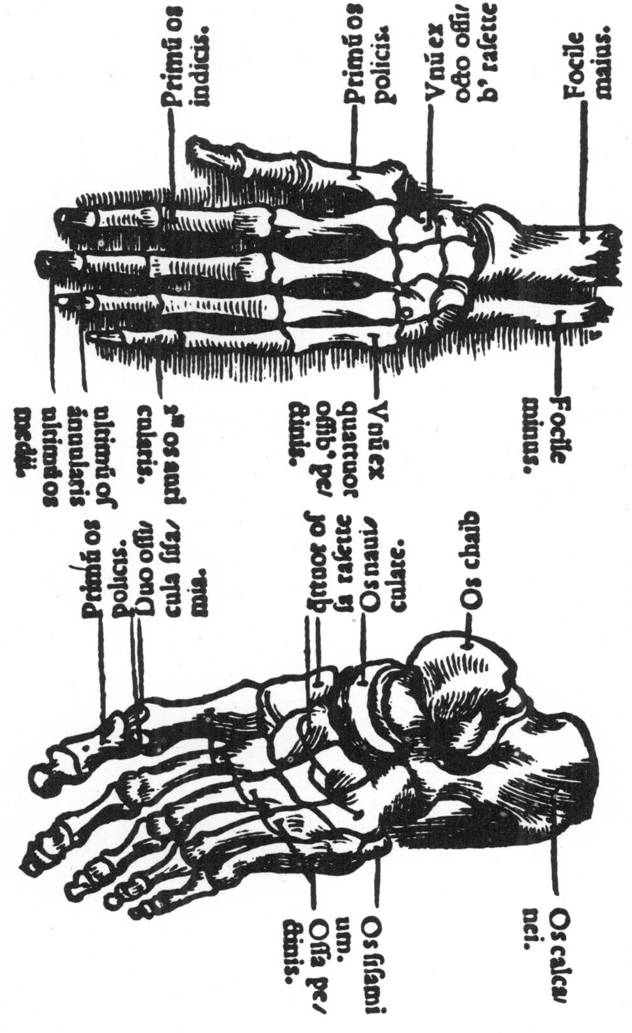

Fig. 11.

attachments of the various membranes of the chest cavity, including the pericardium, puzzled the anatomists of his time.

On the whole, Berengario wrote more completely on the organs of the body than did his predecessors, treating them with more detail than anyone except Zerbi, his redoubtable and respected contemporary. His acute observations resulted in many departures from authority and in many differences of opinion, although even he could not break away from antiquity in every respect. His attitude toward Galen, however, is refreshingly independent when we compare it with that of de Laguna, who wrote on anatomy fourteen years later than Berengario and represents sheer reaction. The student of pre-Vesalian anatomy who wishes to discover the broadest view of the subject will find it in Berengario's *Commentary*.

Niccolò Massa

(1485–1569)

LIFE AND WORKS

I

FROM his last will and testament,[1] a transcription of which I am publishing here for the first time (Appendix IV), we learn that Nicolo (as he spells the name) Massa was the son of Appollonio and the grandson of Thomaso and that he lived at the time he drew up his will in the district of San Giovanni Nuovo ("san Zaninuovo") at Venice. As his heiress he named his daughter Maria, whom he legitimized in her marriage contract with the elder son Francesco of Giovanni Grifalconi. His nephew Appollonio and another nephew, Lorenzo Caresini, son of his sister Paula, together with the physician Giovanni Gratasuol, were appointed his executors; he writes that he loved the latter like his son. Both Appollonio and Lorenzo were also listed as heirs, although he specifically lists most of his possessions as the legacy of his daughter. Niccolò Massa had two sisters, Paula and Lucrezia, no sons, and only one daughter. He had possessions at Mestre, not far from Venice, at Maran, San Piero di Castello near the monastery of San Domenico, and at Pesegia, as well as bank credits and revenues. To his three heirs and executors he left his books, both in Latin and in the vulgar tongue, although Maria was to have first choice of them; his medical books, however, Massa left to his friend Dr. Gratasuol. The books left to the former three were first to be examined for any that might be prohibited "by the sacred orders of the council" and these were to be burned. His daughter Maria was enjoined to allow three poor widows (who were not to have sons of twelve years of age or more) to live from time to time in his two houses in the district of San Piero di Castello; she was also to provide 25 ducats for a silk dress to be given to the daughter of his nephew Appollonio when she married, "although I know she has no need of it, but as a sign of my love and relationship." To his nephew Lorenzo he left his marten-lined cloak. Massa further ordered that his burial should take place in the monastery of San Domenico da Castello. Recalling that he suffered from dizzy spells, he took the precaution of insisting that his body, when presumed dead, should be left for two days upon the earth before burial to avoid the possibility of being buried alive. Masses of the Madonna were to be said for him by the father Agostino de San Zanipolo, his confessor. Appropriate alms were left to the sacristy of both San Zaninuovo (San Giovanni Nuovo) and San Domenico (Domenego) and to the nuns thereof. To his school of San Marco he left the interest of his possessions when there remained no more heirs descendant from his three beneficiaries and executors. Witnesses to his will were Giovanni Battista Peranda and Angelo Benedetti.

This summary of Massa's will gives us a few biographical details which cannot be gathered from any other source and indicates the importance of this kind of document in connection with the pre-Vesalian anatomists. Other facts are, unfortunately, rare, and documents lacking. E. A. Cigogna[2] in his catalog and description of Venetian inscriptions presents both Massa's funerary inscriptions and some facts about his life and works.

Niccolò Massa was born at Venice, although the year of his birth is unknown except by reckoning backwards from his necrology in the parish records of San Giovanni in Oleo, preserved in San Zaccaria:

---

[1] Archivio di Stato, Venezia, Busta 196, no. 870, protocollo del notaio Marcantonio Cavagnis: July 28, 1569.

[2] Emmanuele Antonio Cigogna, *Delle Inscrizione Veneziane raccolte ed illustrate da E.A.C., cittadino Veneto* (Venice, Giuseppe Orlandelli Editore, 1824) **1**: pp. 113–114; 131, nos. 41 and 42. The inscriptions in the church of San Domenico di Castello include one by Niccolò Massa to his father, brothers, and sisters, dated 1520, and another to his father Apollonio and his mother Francesca as well as to his brothers, Tommasso, Domenico, and Lodovico, dated 1530; Antonio, a fourth brother named in this inscription, was treasurer of the Republic of Venice. Massa's own epitaph runs as follows: Nicolai Massae Magni Phil. Ac Medici Ossa. Maria F. P. MDLXIX. It was set up by his daughter Maria and records simply the fact that her father was a philosopher and a physician. The epitaph of Massa's favorite nephew and heir, Appollonio, son of his sister Lucrezia, is at p. 115, no. 3. He was also the object of attention by Elizabeth, princess of Saxony: see Niccolò's dedicatory letter of June 28, 1540, in his *De Febre Pestilentiali.* The bust of Niccolò Massa done by Alessandro Vittoria was set up above his tomb in San Domenico di Castello.

The slight information on Massa presented by Luigi Nardo in his article "Dell' anatomia in Venezia," *L'Ateneo Veneto-Rivista bimestrale di scienze lettere ed arti,* Anno XX, Vol. I, Fascicolo 2 (Venice, 1897): pp. 152–157, is based almost entirely on Cigogna (*op. cit.*) and represents no advance in knowledge, although it is utilized by C. D. O'Malley, *Andreas Vesalius of Brussels 1514–1564* (University of California Press, 1964), pp. 438–439, who gives a good summary of the special items of anatomy which Massa seems to have been the first to emphasize properly: the "panniculus carnosus"; the intestinal canal; the variability in size of the spleen; the division of the portal vein; the study of the liver by the technique of maceration; the fact that the cavity of the renal veins is not continuous with that of the sinus of the kidney; the first adequate description of the prostate gland, although Berengario da Carpi had preceded Massa in an attempt to describe it; the denial of the existence of a third ventricle in the heart and of pores in the cardiac septum; skepticism as to the existence of the rete mirabile; the description of the malleus and incus, after Berengario da Carpi. One should also note pages 122–123 with their emphasis on Massa's influence upon Jacobus Sylvius, the Parisian anatomist, who refers frequently to Massa's *Liber Introductorius* in his own *Isagogae* (1555). I do not find any admission of indebtedness to Zerbi and Berengario in Massa's *Liber Introductorius,* as O'Malley seems to discover (p. 122); Massa merely mentions these anatomists to point out that they had not recognized the fleshy panniculus except in name. It is a pity that O'Malley did not indicate exactly which remark in the *Fabrica* he considered an oblique reference to Massa by Vesalius (p. 123).

"adi 27 agosto 1569 lo eccelente ms. Nicolo Masa medico de ani 84 in circha e sta amallato mesi 4 da frieve [*sic*]." I conclude therefore that he was born in 1485. He became blind in his eightieth year as appears from his words in a dialogue on blindness published by Luigi Luisini, of Udine: *Le cecità dell' ecc. medico m. Luigi Luisini da Udine* (Venice, per Giorgio de' Cavalli, 1569), where Massa is an interlocutor. The same Luisini dedicated to Massa his work *De Morbo Gallico* (Venice, 1566, 1567) and praised him for his skill in fighting this "pertinacem efferamque aegritudinem."

He took his degree at Padua, where he learned philosophy from Sebastiano Foscarini (*Epistolae Medicinales,* I, Venice, 1550, no. 3) and Greek from Giovanni Bernardo Feliciano, as Massa himself reports in Epistle 29 (*Epistolae Medicinales,* II). His work on logic in seven books (Venice, 1559) probably arose out of his philosophic studies at Padua. Massa was enrolled in the college of medicine at Venice in 1521, according to Cigogna, and practiced medicine, surgery, and anatomy there, sewing up intestines in a new manner, trying to cure syphilis, and writing books.

He was praised highly by contemporaries such as Luisini and Marco Guazzo in his *Cronaca* (Venice, 1553, fol. 429 *verso*) who says of him: "da Pontefici e da Re questo Huomo degno fu dimandato, et egli come buon filosofo contendandosi cio negando serve nella sua patria et molto onoratamente vi vive." (This worthy man was in demand by popes and kings but contenting himself like a good philosopher and refusing their requests he serves his native city and lives there with great honor.) Although at the age of eighty he lost his eyesight he continued to practice medicine with the help of his son-in-law, Francesco Grifalconi, who read to him and took dictation, and others. In his last years he was writing a treatise *De Partu Hominis* (On Childbirth) which he left in manuscript, but which is not extant. Luisini wrote his dialogue on blindness to comfort Massa in his infirmity.

FIG. 12. Niccolò Massa. Courtesy of National Library of Medicine.

II

The writings of Massa include a variety of subject-matter more extensive than that of any of his predecessors except Benedetti. It is characteristic, in fact, of the Renaissance scientists that their range of scholarship widens increasingly until by the end of the sixteenth century it becomes actually encyclopedic in the work of Ulisse Aldrovandi and Conrad Gesner. It is also true, of course, that perception and penetration become thinner in the case of those men whose output is large.

I have made a simple check-list in chronological order here of Massa's books and shall comment upon them separately in more detail.

1. De morbo gallico (with the occasional variant title "De morbo Neapolitano, etc."); with A. Bolognini, De cura ulcerum, etc.; Venice, Francesco Bindoni and Maffeo Pasini, 1532; the Wellcome Historical Medical Library has a copy published by Bindoni and Pasini, Venice, in 1507, with a note in its catalog: "F. Bindoni was not printing in 1507 and the first edition of this work is usually dated 1532." H. P. Kraus, the New York book dealer, has recently acquired a copy printed at Parma by Francesco Ugoleto and Antonio Vioto whose colophon gives the date 1507. Its type is larger than that of the Venice, 1532, edition, hitherto regarded as the first, and it contains 44 leaves instead of 40, with some variations in spelling. However, Lucas Panaetius's edition of Galen's *Therapeutica* (Venice, 1527) contains in its dedication to Massa a reference to the latter's *De morbo gallico* as being under the press in that year. Richard J. Durling, after comparing the National Library of Medicine copy of the Venice, 1532, edition, points out that it contains a mention of a dissection in 1524 on fol. A 7r; the same statement appears in the so-called Parma edition of 1507 on fol. A 4v; hence the date 1507 is either a misprint for 1527 or a deliberate falsification, most probably the latter. The newly discovered copy then appeared perhaps after 1524 and before January 15, 1527, the date of Panaetius's edition of Galen. It is evidently an unrecorded first edition. Liber de morbo neapolitano, etc., Lyon, 1534; with an epistle to Thomas Cademustus, Belgian from Liège, physician of Pope Paul III: see Gaetano Marini, *Degli Archiatri Pontificii* **2** (Rome, 1784): p. 149; Venice, Bindoni and Pasini, 1536, 1559; another edition at Basle, 1536 (*Morbi Gallici,* etc., pp. 103–217; the first such collection

appeared without place of publication in 1532; see also the printing by Scipio de Gabiano et fratres, Lyon, 1536); Venice, Ziletti, 1563; J. G. Schenck, *Biblia Iatrica,* etc. (Frankfurt, 1609), lists a Basle edition by Bebel, also 1563; Il libro del mal francese (Italian translation); Venice, Ziletti, 1566.

2. Epistola XXVI (dated 1524), De balneis Calderianis ad doctissimum Dominum Vincentium Riccium Illustrissimi Domini Veneti a secretis: in Aleardus de Pindemontibus (Minardo), ff. 299 b-302 b; Venice, Giunta, 1553; also 1689; extracts in Ventura Minardo's collection, Venice, 1571.

3. Liber introductorius anatomiae (dated 1559 on title page but 1536 in the colophon); Venice, Bindoni and Pasini. This is a remaindered edition reissued by Giordano Zilleti with merely the first four leaves re-set. BN has 1559 as a second entry but it is the same book of 1536.

4. Liber de febre pestilentiali, etc.; Venice, Bindoni and Pasini, 1540; Venice, A. Arrivabeni, 1555, 1556; with notes by Diomede Amico and Doctor Mead; London, G. Mears, 1721.

5. La loica, divisa in sette libri: ne 'quali con mirabile brevita e agevolezza s' insegna ogni sorte de argomentare, cosi probabile come demostrativo. Libro veramente utile a' tutti li studiosi, non solo di filosofia, medicina, e retorica ma anchora di grammatica, historia, e di qualunque altra sorte di scienza; Venice, Ziletti, 1559 (but the copy in the Biblioteca Malatestiana at Cesena has a colophon which reads: Stampata in Vinegia per Francesco Bindoni e Mapheo Pasini compagni nello Anno del Signore MDXXXXX, 167 folios). This is another remaindered edition reissued by Zilleti.

6. Epistolae medicinales, et philosophicae, etc., Vol. I, containing 34 letters chiefly written in the 1540's; Venice, Bindoni and Pasini, 1550; *idem,* Vols. I and II; Venice, Ziletti, 1558; with this edition is also printed an "Opusculum de pestilentia urbem venetam vexante anno 1556," in the copy I have examined in the Biblioteca Estense at Modena. See also P. Tingus, *Epistolae medicinales,* etc. 1556; 1557.

7. Avicenna, Liber Canonis; contains Massa's Life of Avicenna; Venice, Giunta, 1544; other editions: 1555, 1562, 1582, 1595, 1608.

8. Ragionamento . . . sopra le infermità, che vengono dall' aere pestilentiale del presente anno MDLV; Venice, Giovan. Griffio, 1556.

9. Examen de venae sectione et sanguinis missione in febribus ex humorum putredine ortis, ac in aliis praeter naturam affectibus; Venice, Ziletti, 1568.

One of the most serious drawbacks to the acquisition of an understanding of pre-Vesalian anatomy beyond the scanty and often ill-advised statements that now circulate on the subject is the scarcity and unavailability of the Latin texts themselves, the sole source of accurate knowledge of what these anatomists actually said. These texts are often exceedingly rare and copies can be found in only a few American libraries. None of them has been reprinted in modern times or edited with the notes they require, nor are they ever likely to be reprinted or edited. There are two ways in which to overcome these obstacles: one is to provide the modern reader with full-length translations, plus notes and introductions, as I have tried to do in a limited number of instances, although we are probably always to be denied translations of the longer works, such as Zerbi's *Anatomy,* Berengario da Carpi's *Commentary on Mundinus,* and, above all, Vesalius's *Fabrica.*

The second method is to furnish the reader with adequate summaries of the Latin works, and this I have followed in connection with the other works of Massa exclusive of his *Liber Introductorius.*

De morbo gallico (or *neapolitano*) is an essay in 6 books and 115 pages in the edition of Basle, 1536. It begins with an expression of pity toward those who are ill which echoes Galen, *De Morbis Curandis* (also known as *De ingenio sanitatis* and later as *Methodus Medendi*). Massa speaks of the Neapolitan disease (syphilis) as "nova" and describes its symptoms in detail, including pustules and ulcers; its history begins, according to him, in 1494 when the king of France occupied Naples with his troops (see my note 56 to the translation of Benedetti's *Anatomice*). Syphilis is contracted not only by sexual intercourse but by ordinary contact with others who are infected. It is interesting to note how the ancients and medievals, Galen, Aristotle, Avicenna, and Peter of Abano are brought into the discussion although the disease Massa writes of was unknown to them, another indication of the reluctance to abandon authority which dogged medieval and Renaissance medicine. Even children, who could not have contracted the disease in the usual manner, were infected. Massa cites three, giving their age and sex, whom he had cured (or at least treated). He then proceeds to discuss the relation of the liver and the humors to syphilis, together with its causes—including the conjunction of the zodiacal signs Saturn, Mars, and Venus in Scorpio in the year of its inception at Naples. Those who are cured, he notes, often become obese or hoarse or stiff in some member.

The treatment of syphilis involves the environmental conditions of air, moisture, temperature, and weather; the marshy conditions of Venice, Ferrara and other locations were of adverse influence upon the disease. The regimen of food and drink, the efficacy of sleep and exercise, and the regulation of sexual and other passions are discussed together with the use of drugs, bloodletting, leeches, "Indian wood" (described carefully at p. 148, this is called guaiac by the Spanish, as he notes, or "holy wood" by others: Indian wood, however, is its most frequent appellation). Quicksilver is also described as a medicament and other ointments are discussed; their usefulness is firmly vouched for by Massa and his anointed patients favorably compared with the unanointed patients of other less resourceful physicians. Ulcers of the mouth and gums are particularly characteristic of syphilis; Massa gives many recipes for their treatment. Bathing, both natural and artificial, is also a mode of cure; sulphur baths consistently employed were most efficacious for one of Massa's friends, who also used sulphur internally. The treatment of carbuncles by oil of vitriol and other remedies is discussed, with many detailed recipes.

The shorter essay on pestilential fever (*Liber de Febre Pestilentiali ac de Pestichiis, Morbillis, Variolis, et Apostematibus Pestilentialibus*) is dedicated to Elizabeth, duchess of Hesse and princess of Saxony, largely because of her kindness toward Massa's favorite nephew, Appollonio. This work discusses the nature and treatment of plague, fever, and smallpox as well as their after-effects in the form of pock-marks and other disfigurements. Case histories are presented, such as those of Niccolò Bianchi, a seventy-year-old sailor, or Maestro Francesco Marino, a cleric of St. Anthony of Padua. Massa's own father Apollonio, "eques aequoris," died of fever (fol. 12v) when Massa was fourteen years old (i.e. in 1499); his case history is given in detail. The tragic deaths of the years 1527 and 1528 by floods and subsequent plague in Italy are referred to from an eyewitness point of view. Massa's brother Thomas also died of fever, "quo tempore nondum eram medicus," "at a time when I was not yet a physician" (fol. 13r), i.e., before 1521, when Massa was studying medicine at Venice. In 1530 a heat wave drove farmers from the fields, killing many. The inhalation of noxious vapors as a cause of disease was illustrated by the fate of those who descended to clean an old well at Venice, witnessed by Massa. Public health and the means of preserving it seem to have concerned Massa deeply, to judge from his remarks on street and canal sanitation at Venice (fol. 24v): "Civitas ab omni sorditie et immundicie purgetur." Not only the city but its immediate environs should be cleaned. Hospitals also were to be built in healthful locations.

His friends, Girolamo Marcello and Vincenzo Ricci, addressed in his *Introduction to Anatomy,* are mentioned again at the opening of the second tractate. Here he deals in greater detail with the means of preserving an entire community from pestilence, as, e.g., preventing those infected from moving to or from the city and by strict isolation. Many foods and drugs are mentioned in his recipes, including the juice of lemons and oranges. It is in fact remarkable how much space Massa employs in the discussion of proper foods. The proper use of sexual relations, sleep, and exercise are also carefully discussed, together with the proper control of one's feelings and passions. The case history of Gaspare Butafuoco, of the district of St. Peter, who recovered from an attack of fever in the unhappy year 1527, is related; that year saw all Italy afflicted by plague. The even more detailed history of a twenty-five-year-old virgin who fell ill at the end of August, 1527, shows Massa's ingenuity in improvising treatment. Finally, his sister Vincenza and his nephew Appollonio were stricken in 1527 with pestilential fever, but recovered to live into their old age, saved by Massa's skill.

The *Epistolae Medicinales,* of which the first letter in volume one is addressed to his favorite nephew and heir, Appollonio, and the first letter of volume two to his nephew Lorenzo, have already been discussed to some extent elsewhere (notes to translation of Massa's *Introductory Book of Anatomy*). I shall deal with their information, chiefly biographical, more exhaustively here. These letters occupy to some degree the same position in the corpus of Massa's writing as the *De Re Medica* does in that of Benedetti; that is, they form a discursive work on medicine distinct from anatomy, although they are far less extensive and systematic than Benedetti's treatise. General principles are Massa's chief preoccupation in the *Epistolae*, with the usual heavy foundation of quotation from authorities. Many of these letters are actually *consilia,* or written consultations addressed to the patients themselves. The title page mentions both theory and practice of medicine; the strictly medical discussion is augmented by letters 33 and 34 in Vol. I, on the creation of the world and the immortality of the soul, a reminder to us that the practice of medicine was as a matter of tradition accompanied by the study of philosophy among the Renaissance physicians. Letter 35, Vol. I, is a return to Massa's serious interest in public health: "Concerning the Means of Freeing a City from Pestilence." Letter 29, the last of Vol. II, is a discussion of the generation of man and of his nobility, a general theme favored by the Renaissance thinkers.

Massa's nephew Appollonio was a student of Heinrich Auerbach at Leipzig in 1541 as well as of Philip Melancthon, to both of whom Massa sent his respects in Vol. I, Letter 1, addressed to his nephew. To Sebastiano Foscarini, senator of Venice, is addressed letter 3, full of information on food, especially Italian fish and where they may be caught. Letter 5 deals with anatomy and was sent to Antonio Fracanzano, professor of medicine at Padua (see my translation of Achillini, p. 29); it refers to the discovery of the malleus and incus in the ear in the time of Alessandro Achillini: "Haec vero ossicula anatomici, tempore Alexandri Achillini viri in omni genere scientiarum eminentissimi (ut ex eius scriptis clarissime videre est) invenerunt. Quare non ab istis sunt primo inventa, nec ostensa: cum etiam Iacobus Carpensis loca istorum ossiculorum invenire doceat." Massa divides the honor of their discovery with Berengario da Carpi but does not specifically attribute Achillini with any part in the event (fol. 55v).

In letter 6, also to Fracanzano, Massa mentions his *Introduction to Anatomy* as not a complete treatise but the opening of a path for modern readers to the subject. He also complains that other students, having learned from him, have become so conceited as to think themselves more accomplished than the ancients: "ut vidistis in eo volumine quod de fabrica corporis humani inscribitur" (fol. 58r). This is clearly a reference to Vesalius's *Fabrica,* which appeared in the same year as this letter (dated January 23, 1543), in a criticism of the latter's description of the nerves as differing from Galen's. Since the *Fabrica* did not appear until June, 1543, it would be puzzling that Massa should have been already aware of part at least of its contents if Roth, *Andreas Vesalius Bruxellensis* (Berlin, 1892) p. 183, had not pointed out that by Venetian reckoning the year of Massa's letter is 1544. It is, further, an historical irony that Massa should have been the person to whom the dedication of the Luisinus edition of Vesalius's *China-Root Epistle* (Venice, 1566; Cushing, *Bio-bibliography of Andreas Vesalius,* p. 166) was made. This is the only passage in the works of the pre-Vesalian anatomists in which mention is made of the great man who was to eclipse their work; and it is not beyond conjecture that Vesalius had indeed profited from Massa's teaching and writing, without which, as the latter declares, these imitators "nihil potuissent scribere, aut spectantibus ostendere, praeter illa, quae vulgaria in omnibus gimnasiis erant ex codicibus quibusdam accepta, et ex sensatis male demostrata (*sic*) et declarata ut vidistis in eo volumine quod de fabrica corporis humani inscribitur." As Massa explains further on (fol. 59r) the ungrateful moderns against whom he inveighs (including Vesalius) "are very frequently accustomed [to reprehend their elders and betters] in order thereby to acquire more authority for themselves."

Letter 5 has likewise a rather disingenuous disclaimer by Massa in regard to Vesalius, who is not named. In reply to the questions of his contemporaries as to his opinion of the *Fabrica,* Massa says of that "huge volume" (*de ingenti volumine*): "I could not by any means read big books, let alone even browse through them, busy as I am with many duties toward my patients. I have heard many, however, who praise the work highly, as I must also, considering the very great task of composing it, which could not have been accomplished except by wakeful study and toil." Massa goes on shrewdly to question the degree of originality in the books of his contemporaries, in spite of their distaste for Galen and Aristotle. In his defense of these ancients Massa insists that both of them dissected human cadavers, quoting Galen, *On Anatomical Procedures,* I, 2. 223 (Charles Singer's translation, Oxford Univ. Press, 1956).

Letter 7 is addressed to his friend, Hieronimo (or Girolamo) Marcello, familiar also from the *Liber Introductorius Anatomiae.* It is one of a series in which the treatment of fever is discussed in great detail; this was one of Massa's overwhelming preoccupations. Paolo Magnolo of Brescia is the recipient of several letters; he seems to have been a special friend. Among other addressees are Ulrich Steydle (or Steidle), Giovanni Francesco Marino, of the minor order of St. Francis at Padua, Giovanni Maria Plebano of Arquà, Giovanni de Ulmo, and others, whose symptoms he discusses and for whom he prescribes remedies, by mail, so to speak. It is hard to distinguish some letters from what may be called actual *consilia,* or physician's consultations. In letter I, 23, to his nephew Appollonio on dislocation of the intestines toward the pubis and tumor in the lower stomach he refers to a discussion of hernia of the scrotum by Alessandro Benedetti, his famous contemporary, whom some scholar had criticized in this regard:

> Quaerit deinde vir doctissimus numquid in rupturis in quibus scrotum durius tractu temporis factum fuerit, Medicus studens sanare hunc affectum per incisionem, una cum ipsa indurata cute, testiculum etiam amovere debeat, necne, et carpit curam Alexandri Benedicti. Dico, (quicquid dixerit Alexander vir alioqui doctissimus: non est autem meum taxare mortuos) quod, ut superius dixi, non debet Medicus in similibus agere, nisi quantum ad amotionem rei praeter naturam, sine spe, ut rectificari possit, facit. (The very learned man then inquires whether in hernial ruptures in which the scrotum has become somewhat hardened due to the passage of time the physician, desiring to heal this condition by surgery, ought also to remove the testicle itself together with its hardened skin or not and criticizes the method of Alessandro Benedetti. I say (whatever Alessandro, a very learned man in other respects, said, since it is not for me to reproach the dead) that, as I said above, the physician in similar cases ought not to operate unless he does as much as he can to heal the organ, except when there is no hope of remedy without its removal.)

Massa shows here an exemplary respect toward a dead colleague (fol. 122r). In this letter he also refers to Johannes Dryander, who had sent these questions on matters of medicine through Appollonio to his uncle. Dryander later relayed a question on superfoetation to Massa by the same means (letter 25); Massa refers to his work "in libello de sectione corporis humani" and the quotation from Galen therein on the closing of the uterus upon conception so tightly that not even a needle might enter. The work is, of course, his *Liber Introductorius Anatomiae.* Letter 26, to Vincenzo Ricci, at Venice, is the first edition of "De balneis calderianis," on the baths "in vico Gauderii in agro Veronensi." In letters 27 and 28 Massa returns to another favorite subject: the Gallic disease (syphilis). The long letter 32, which might rather be called an essay or lecture, is devoted to the topic of rhetoric and the uses of language and grammar. Its treatment of logic is still medieval, with the antique terms for the syllogism. Letters 33 and 34 of Volume I, upon philosophic subjects, are followed by 35, on the nature, causes and cure of the pestilence at Venice in 1556, addressed to Lorenzo Prioli, lord of Venice; they contain references to previous such outbreaks in Italy as described by Boccaccio and Marsilio Ficino.

Vol. II is also dedicated to Prioli, although letter 1 is addressed to lovers of the fine arts and of truth and wisdom in general. It affords Massa an opportunity to pass in review the names of many philosophers and medical writers from previous ages and to comment upon their virtues. Letter 2 is a general discussion of the use of medications for the benefit of his nephew Lorenzo. In it Massa demonstrates his quite considerable knowledge of simple medicine and in letters 4 to 7 continues his assiduous instruction of his nephew in a discussion of fever. Letter 10 is written at the request of some of Massa's students and deals with the signs of fever. Letter 11, on wounds of the head, chest, and lower body, and of the nerves was written to a certain Pasinus of Belgrade. Letter 12 is designated as a *consilium* or consultation on headache to the presbyter Vincenzo Mauro, and letter 13 to Cristoforo Valabi of Brescia discusses the latter's ulcerated right jaw, examined by Massa at Venice. Letter 14 is a similar discussion directed to Giovanni de Ulmo, also of Brescia. Epistle 16 is addressed to no less than Albertus Corvinus, duke of Poland, on his stomach disorder and difficulty in passing urine. Several letters on syphilis are sent to reverend theologians and churchmen.

Letter 22 concerns the syphilis from which the dear wife of Giovanni Romano, doctor of medicine, was suffering. Diagnosed by Massa in consultation with doctors Donato de Mutis and Giovanni Battista of Reggio, there was no mistaking her malady: she had contracted it, of course, from her husband, with all its characteristic pustules, ulcers, and tumors. In addition to the usual regimen of drugs, baths, and diet, Massa treated both husband and wife, linked in their misery, with guaiac ("Indian wood"), recommending that the doctor read Massa's treatise on syphilis, the first edition of which had appeared in the same year as this letter (1532).

Epistle 23 was sent to Prince Ladislaw Orza, in far-off Poland, also dealing with syphilis and with ulcers and varicose veins in the legs. It is interesting to note that throughout these prescriptions on syphilis sexual relations are strictly forbidden to the patient. Epistle 24 prescribes for a syphilitic who had recovered for some years and then had noted the appearance of hard swellings in the groin. Another Pole, the physician Valentine Podlodosky, is addressed in letter 25, also on his syphilitic condition. In fact, Massa seems to have been particularly sought out in correspondence by Polish medical men for consultation; his fame in their country must have been widespread indeed.

Letter 27 prescribes for a woman named Elena Centani, who had slipped and fallen down an ice-covered stairway in February (no year is given) and in her obesity contracted a tumor in the knee and consequent lack of mobility. The patella had been seriously affected. Massa's analysis of her case is ingenious and involves anatomical knowledge, a rare occurrence in these letters. The twenty-ninth and last letter discusses the generation of man and his nobility, a typical Renaissance theme, with his good friend Bernardino Feliciano, whose paternal uncle, Giovanni Bernardo, had taught Massa Greek, presumably at Padua, "tot annos antea," "so many years before" (the letter is dated January, 1556). One of the most interesting passages in this letter is an account of a phenomenal patient Massa attended at Venice while he was a young man. She was around sixty, her husband about seventy, and she had conceived a child, although Massa, unaware of this remarkable fact, had treated her for what he thought was dropsy. In the fifteenth month of pregnancy this woman bore a girl child, without eyes or hands, who lived five months. Massa attributed her prolonged pregnancy to the particularly frigid uterus of an elderly person which required an unusually long time to create the material for the members of the foetus and even so produced a monstrosity.

The *Ragionamento dello eccellentiss. M. Nicolo Massa sopra le infermità, che vengono dall' aere pestilentiale del presente Anno MDLV* (Venice, 1556) is dedicated to S. Francesco Veniera, prince of Venice. It describes in brief chapters called "parts" the pest or giandussa, as it was known to the people of Venice, which afflicted the city in 1555, dwelling upon the illnesses it caused, their source and signs, and the exact manner in which the infected atmosphere created a dangerous problem of public health. Massa also discusses the fact that the pest did not always attack those who nursed the stricken. He compared the present plague with the very deadly one of 1527 and 1528; that of 1555 was less severe. He insisted upon cleanliness in the care of those infected and serious attachment to duty among the nurses, who should be older people. He noted that the rich, sustained by better food, tended to survive while the poor died. He also discussed the probable duration of the epidemic, depending upon the change of weather when the warmer spring days should arrive. Precautions against infection were also set down, although Venice with its watery environs was a difficult place in which to escape the dread disease. Constant measures were to be taken to keep the streets and canals clean, especially in the quarters where dwelt the poor. Diet, exercise, and other regimens were to be observed carefully, as well as proper ventilation and dress.

Certain remedies were prescribed, and people were advised to read Massa's book on pestilential fever (1540) for further details.

The *Examen de Venae Sectione,* written in Massa's eightieth year, is a typical example of this kind of specialized medical writing; Vesalius and others were to follow Massa's lead in elaborating on the subject. An index at the front of the book facilitates reference to the appropriate disease for which blood was to be let. Massa dedicated the book to his nephew Niccolò Crasso, a jurist. He counsels experience, not dependence upon authority alone, as a guide in blood-letting. Galen and Avicenna are quoted, but sparingly, Galen in *Methodus Medendi,* Books XI, XII, XIII chiefly. Vincenzo Trincavella, who died in 1563 (see Bernardi, *Prospecto* etc., p. 38), is mentioned in fond memory (p. 29). A useful set of Canons to be observed completes the little book (pp. 33–44) with a defense of specific Galenic references (pp. 45–51).

Massa's *Life of Avicenna* presented a puzzle to George Sarton. He wrote in a fascinating and most useful book (*The Appreciation of Ancient and Medieval Science During the Renaissance 1450–1600*, 1955, p. 43) as follows:

> The Juntine edition of 1544 [of Avicenna's *Canon*] contained the Latin translation of an Arabic biography of Avicenna by Niccolò Massa. This is puzzling: Massa was an outstanding anatomist, professor in Padova, discoverer of the prostate (d. 1569). Did he know Arabic? Did he know it well enough to translate an Arabic text from MS? Of course, the translation might have been made for him or he might have been helped by an Arabic-speaking man of whom there may have been quite a few in the Venetian district. Massa's translation was reprinted in many other editions (1556, 1582, 1595, 1608).

Had Sarton turned to page 24 of the Juntine edition he mentions he would have discovered that the life of Avicenna written in Arabic by Sorsanus, one of the great man's pupils, had been turned into Italian at Massa's request by Marco Fadella Damasceno, an interpreter for the Venetian merchants at Damascus, and that it was this Italian version which Massa in his turn translated into Latin: "cum invenissem Marcum Fadellam Damascenum, Venetorum mercatorum interpretem, atque operam dedissem ut is verba Sorsani vulgari sermone exponeret, latinis ea mandare literis decreui. . . ." Massa dedicated his translation to Thomas Cademustus, physician to Pope Paul III; in the Juntine edition of 1595 it appears with a dedication to Cardinal Carlo Borromeo, dated by Massa in 1562.

The life is brief and laudatory, written in simple, factual Latin. The corruption of Avicenna's name from Ibn Sina is pointed out, his parents and descent carefully set forth. Avicenna's education and passionate dedication to learning are emphasized, even to his habit of drinking a measure of wine in order to revive his flagging energies for study, as a modern student might take drugs to keep awake. The books which influenced him, the friends who helped him, the travels he undertook are described, together with his first essays in writing scientific books; his works, including the famous *Canon,* are then listed. His attempts to ward off sexual lust by avoiding all laxatives brought him into a state of colic from which he rescued himself by enemas which caused an ulcerated colon. This distressing condition was increased by his servants who had placed not two but five drams of parsley seed, intended to dispel the pain, in the enema water. Further carelessness on their part caused Avicenna to scold them severely, whereupon they were angered against him and plotted his death. He finally died of his ills at fifty-eight and was buried in Chemedan, where his monument stands, in the year of the advent of Mohammed, 428. Massa emphasizes his own faithful adherence to Sorsanus's text, once removed, of course, by Fadella's Italian working version.

INTRODUCTORY BOOK OF ANATOMY

*in which as many parts, functions, and uses as possible of the human body which had been thus far overlooked by others both ancient and more recent are now for the first time made manifest; a work quite useful indeed to all students of medicine and philosophy; Venice, from the Printing Shop of Stella Giordano Zilletti, 1559**

To Lord Pope Paul the Third, Most Holy and Great, Nicolaus Massa wishes the best of health and a lasting happiness

When many good husbandmen with a deep and careful interest in agriculture bring their master some of their finest selected products they are accustomed to prove to him by the beauty and delightful quality of these fruits how much effort and toil they have expended in the cultivation and planting of the field. Pleased with their diligence and sense of duty, the master usually gives them more freely and liberally the wherewithal for tilling the soil so that his farmers may more conveniently sow the fields and grow better and more abundant crops for themselves and theirs. I thought I should do the same in my literary cultivation since I have for a long time devoted my efforts to medicine and tilled literary fields, sowing in them sweet, tasty, and odorous fruits which are, unless I am mistaken, far better than those which are gathered[1] in other literary fields not only of the moderns but of the ancients. For while in earlier years I dwelt in Venice, although it was a very busy city, nevertheless I took note of anatomical dissections and by inspection of the parts of the human body not only among as many physicians as possible but among devotees of other fine arts. Among these I had always as my faithful Achates that fine man and most learned philosopher, Gerolamo Marcello,[2] to whom I promised I would write a small introduction to anatomical dissections, since I have often revealed many members of the body unknown to the ancients, and not only members but many operations of the members overlooked and unknown to them. I decided I should direct these revelations, destined to be no less pleasant than necessary for the human race, to you, most Holy Father, so that as you excel and are beneficial to the entire human race by your own virtues and wisdom, if I have benefited the world in any way through my labors, as I believe I have, that fact would become known to mortals through your name and protection. Accept therefore, most Blessed Father, with cheerful countenance these my labors so that by your nod I may be strengthened for literary culture, and that you may see me hereafter offering you better and more abundant fruits. A long farewell.

At Venice, 1536, on the Ides of November

AN INTRODUCTORY BOOK OF ANATOMY

or

Of the dissection of the human body

by

Nicolaus Massa of Venice,

doctor of arts and medicine, now first edited by the author himself; in it are revealed very many members overlooked by both ancients and moderns, as will appear to studious readers.

Chapter I

Introduction to the Entire Work

I must offer deep gratitude to God, the Best and Greatest, to Him Who aids and protects me, and so should philosophers and especially those who profess medicine, if they recognize all the particles of the human body, in no wise possible (as they say) for me to tell with tongue or to write with reed. You should not wonder, great and most learned Gerolamo, if I begin my discourse to you with such words as these, when there are very few at this time who are students

* Title page: The first edition of the *Liber Introductorius* was published by Francesco Bindoni and Maffeo Pasini at Venice in 1536, with this date on the title page and the names of the printers in the colophon. Zilletti later obtained some unsold copies and brought them out in 1559 with a new title page but a colophon which retains the original date of publication, 1536. The Latin text is the same in both issues of the book.

[1] "gathered" involves a pun on the double meaning of the Latin word *legere*, "to read" and "to gather or collect."

[2] On the basis of the reference to Gerolamo Ferro at p. 181 (apparently a relative of Massa) I conclude that Gerolamo Marcello is the same man and that Massa is here addressing him by his first and second names. His dates are 1503–1546 and his correspondence with Benedetto Ramberti is mentioned by F. Giovanni degli Agostini, *Notizie istorico-critiche intorno la vita e le opere degli scrittori viniziani*, etc. (Venice, S. Occhi, 1752) **2**: p. 571. To him Marcus Antonius Muretus dedicated his *Orationes Tres De Studiis Literarum*; see Mario Emilio Cosenza, *Biographical and Bibliographical Dictionary of the Italian Humanists and of the World of Classical Scholarship in Italy*, 1300–1800, expecially Vol. V, *Synopsis and Bibliography* (Boston, G. K. Hall, 1962), from whose great book I have drawn some details concerning other friends mentioned by Massa. Cosenza, however, writes that Ferro died of the pest in 1561 at Constantinople while Agostini (above) gives 1546 as his death date. It is notoriously difficult to date accurately many of the Italian Humanists or other figures of the Renaissance. Achates: the faithful companion of Aeneas in Vergil's *Aeneid*.

of anatomy, however useful and necessary it is to philosophers and likewise to physicians. Why, it would be a most handsome attainment even for private individuals to know themselves [i.e. their own bodies] since man by his nature shows that Nature's ultimate intention is his own ultimate perfection. Therefore that wise Greek spoke shrewdly when he said, "Know Thyself." Do not cease therefore to give thanks with me to Blessed God, by whose mercy you have made yourself so learned a philosopher under my leadership concerning the parts of the human body, to such an extent that you need envy no one of our age; rather you have become the keenest censor of the worthlessness of many recent writers on anatomy read here and there in schools and universities. They are read and regarded as oracles. You know, however, that they present either a curtailed knowledge of the subject or know nothing at all, having transcribed the writings of others imperfectly understood by themselves and decked out with certain sophistical arguments, solving questions between Aristotle and Galen by glosses and distinctions, and have thus clouded the minds of young men, so that you see there is no one at all who perfectly understands anatomy learned through the senses. I am silent about those you know who are each day most highly praised by the common folk, when they know nothing; possessing only the formulas of philosophy and medicine, they are considered by men to be most wise. I protest to God I am ashamed to say how great is the ignorance of certain moderns not only in the discussion but in the actual sense-perceptions of anatomy, always swearing as they do by Aristotle, Hippocrates, Galen, and Avicenna, but knowing nothing for certain in fact itself; they are like little birds who utter in song the things they hear and by their chirping delight the ears of certain people. Listen to what often happens to me. A few months ago I attended a meeting of many physicians, for at Venice are to be found men who like yourself are more learned and skilled than those in the entire world. Over the cadaver of a certain quite troublesome and dangerous man [evidently a criminal] I was explaining to them the function of the muscles of the lower body cavity and their anatomy. An elderly person who is regarded among the first people of this noble city said that the muscles of that region do not have broad tendons which go around the longitudinal muscles and also a large part of the cavity, for he said all that region was fleshy. But the others around me, who had seen with me the sensible truth of anatomy, laughed and said nothing to him. I also recall an occasion when I was summoned to a certain man who had been wounded deeply near the false ribs and when there arose serious complications many physicians were summoned. One of them, considered by all the others both quite learned and skilled, was not ashamed to say that in the lower cavity there are no muscles. I shall not tell you how many other foolish remarks I hear daily from such men lest I delay you more than I should with such things. It is sufficient, most learned Marcello, to see how great is the worthlessness of certain men in a matter so clearly evident who (as you have often said to me in surprise) are believed according to their own estimation to be most expert and skillful investigators of nature and of men. I do not wish you to believe that I am attacking the ancients, by whom I was trained, with calumny; nay, rather, I always praise them since even those are worthy of praise who handed down to us imperfect principles or parts of doctrine since they gave us something of a handle for adding [knowledge] afterwards. I cannot refrain, however, from lamenting with Galen in the second book of his *Anatomical Procedures*,[3] since now the science of anatomy after Galen has been handed over to oblivion nor are there men who care about making actual incision itself. Yet nonetheless they have tried to write about those things which they have neither seen with their own eyes nor touched with their hands and have clouded the brightest light of the sun with manuscripts accepted from others, corrupted by age and by the carelessness of printed books. I am sorry that I have promised to write for you in these times when the very wise Arabs and many others have been pursued with gratuitous insults, for I do not wish to appear guilty of the same offense. I have written this book not in order to make accusations but rather to excite the spirits of the noble. It is a great evil to attack the highly learned from whom come all good writings even if they too have made some statements which are either obscure or not approved by all. Indeed, I praise Plato, Aristotle, Hippocrates, Galen, Avicenna, Averroes and other very learned men who have been most helpful to posterity with their labors. But I am not the kind of man who would say that nature created men without errors,[4] for since men

[3] Galen, *De Anatomicis Administrationibus*, II, 280–292, translated by Charles Singer (Oxford University Press, 1956), pp. 30–36. Galen here explains that handbooks of anatomy were not written in earlier days because the practice of dissection was handed down in the families of the professional physicians. But when instruction in the subject was given to those outside these families the practice deteriorated and a demand arose for manuals, which were severely criticized by Galen for their careless inaccuracy. Such handbooks were needed, however, when the oral tradition died out. See especially the various purposes of anatomical study outlined at pages 33–34 of Singer's text.

[4] W. P. D. Wightman, *Science and the Renaissance: an Introduction to the study* [*sic*] *of the Emergence of the Sciences in the Sixteenth Century* (Edinburgh and London, Oliver and Boyd; New York, Hafner Publishing Co., 1962) **1**: pp. 231–232, makes the following absurd translation of this passage because he mistakes the subject of the infinitive (it is true that Massa's Latin is not always entirely clear): "I am not one to claim that men have reproduced nature without error." Wightman's remarks on Massa's position in the history of anatomy are, however, quite sound: "What I am claiming for this book [Massa's *Liber*

have existed all of them have been capable of mistakes. They have also erred in certain matters of fact, for all of us cannot do all things.[5] Therefore let those remain silent who attempt open warfare with the dead and have thus impudently inveighed against them, seeking to adorn their own silly prattle by injuring such great men and appearing to exclaim that no one except those whom they have praised should be read. But since I now appear to be exceeding the bounds of an introduction, let my readers be indulgent and lay the blame upon the condition of the times. I decided to write this introduction to anatomy for no other reason than charity toward the ignorant and compassion toward those who are ill and often perish through the lack of physicians who know anatomy which they have learned through their senses. In order that men might not lose a form of knowledge so useful and necessary which can be learned by daily reading of the ancients and by laborious exercise in the dissection of bodies I decided to show others after me the method of investigating these matters. If I do not thoroughly accomplish this task in every respect those writers who follow me will have at hand the material for adding to and perfecting what I have written; I earnestly request them to do so. Therefore, concluding these prefatory remarks, I shall turn completely to the task itself, praying that the omnipotence of God may honor my undertaking with His aid and protection. Let me say then (according to the authority of Hippocrates in his book *On the Nature of Women*)[6] that there exists in man a certain divine power. Thus he who intends to write about the parts of man must take his beginning first from God Himself, from Whom come all things and without Whom nothing exists. And let this be your preparation before you approach the dissection of the human body itself for the reason that God gave that body a noble substantial form, capable as it is of a beatitude both natural and supernatural; this is recorded in Holy Scripture: "Thou hast made him but little lower than the angels" (Psalms 8:5, which in the Revised Version reads "God" for "angels"). From this substantial form necessarily proceed the other features in which man surpasses other living creasures, according to Averroes' *Commentary* VIII, 31, on the *Physics* [of Aristotle] where he says: "From substantial form necessarily follow the accidents," among which is man's upright gait. Thus that [poet, Ovid] sang in the *Metamorphoses* I, 85–86: "He gave a sublime face to man and ordered him to look at the sky and to lift an upright countenance to the stars."[7] You therefore see that man has an erect stature. I do not think the definition of the word anatomy is necessary since dissection itself seems to be the most important objective, that is, the knowledge which is obtained by cutting. Therefore let me say this much by way of definition: anatomy is the knowledge of the members gained through skillful dissection. And lest I appear to abandon the method of dissection, although (as Galen taught in his *Anatomical Procedures*) there are many methods,[8] eager for the truth and not for the statements of others, I follow the method handed down by Mundinus,[9] a man most famous for dissection. Nor will you see me defending the declarations of the most weighty philosophers since, as I said, they sometimes made mistakes and were unable to know everything. Thus when I do not agree with the ancients it is because the fact of truth differs from their statement about it. And although I love the truth more than the authorities of men the statements of the wise must be glossed or explained. Thus you will nonetheless frequently find me differing from them in matters of sense which do not coincide with reason; but in other respects you must understand that I always praise these men highly, since I was taught by them. Since all the members in one body cannot be observed in dissection because occasionally some members must be cut off and thrown away so that we

Introductorius] is not that it 'anticipated' the *Fabrica*, but that it is an important document in the still incomplete investigation of the degree to which Vesalius was not so much an innovator as the Man born in Due Season, sensitive to every indication of change, and able by his enthusiasm and urbanity to unite these diverse forces—critical, scholarly, artistic and instrumental—into a movement which ultimately swept everything before it."

[5] Massa is quoting Vergil, *Eclogues* 8. 63: non omnia possumus omnes.

[6] Hippocrates, *On the Nature of Women*, ed. E. Littré (Paris, 1851), 7: pp. 313: *μάλιστα μὲν τὸ θεῖον ἐν τοῖσιν ἀνθρώποισιν αἴτιον εἶναι*: le divin est chez les humains la principale cause.

[7] Berengario da Carpi, *Commentaria . . . super anatomia Mundini*, etc., XXIIII, also quotes the same passage from Ovid. That man by nature turns his face upward is likewise Aristotle's view: *De Respiratione*, 13, 477 a 21; *De Partibus Animalium*, 2. 11. 656 a 15; 3. 6. 669 b 5; *De Iuuentute et Senectute*, 1. 467 b. Compare Psalms 8: 5. The question why man does so is discussed at length as the first *problema* in *Problemata Varia Anatomica* (anon.), University of Bologna MS 1165 (2327), edited by L. R. Lind (Lawrence, Kansas, 1968). The author begins with Galen's remarks on the subject and proceeds to produce ten reasons drawn from Scripture, Aristotle, and Boethius. The Renaissance Humanists, especially Pico della Mirandola in his oration on the dignity of man, also make much of this erect attitude of mankind, in contrast, of course, to the creeping beasts.

[8] Galen, *De Anatomicis Administrationibus, op. cit.*, p. 35. Massa is probably referring to the schools of medicine discussed briefly by Singer in his introduction (xv–xvii) or at least to those who were content to learn anatomy by chance, i.e., by the examination of wounds as opposed to those who insisted upon systematic dissection, whether of animals or humans. It is useful to point out that Singer (xxii–xxiii) was convinced that Galen was acquainted with human dissection at least by way of post-mortem examination; the most telling evidence for this view is *De Usu Partium*, VI, 4 (Vol. III K., 423): "If death come not to me too soon, I shall some day explain construction in animals too, dissecting them in detail, just as I have done for man."

[9] That is, in the order in which he takes up the bodily cavities for explanation and discussion. Massa mentions Mundinus rarely, however, preferring to strike out on his own in matters in detail while preserving the general outline followed by his famous predecessor.

may perfectly recognize the other members it is thus also convenient to know that I have not intended to write about everything which can be observed in the human body in this book but only about those features which can be comfortably examined in one body. Furthermore, I have decided to reveal by dissection the facts necessary for those who heal the body and as an introduction to the remaining parts which must be examined. If God aids me with His mercy (as He has done) I shall write about the remaining parts of the body in another large volume.[10] You should not expect from me a kind of bench of preparation or a seat where the crowd of spectators may look upon you while you dissect or demonstrate. Do not expect any such ridiculous thing since I do not wish to display myself in my writings as an ignorant man and not a philosopher. If fortune is fair—and it is reasonable to suppose that it may be—who could have any doubts about what kind of body ought to be used for dissection? I say that whatever the bodies may be they differ either in sex, age, complexion, or size. The physician must inspect all bodies carefully without distinction so that he may understand their manner of creation in regard to their differences of nature in order that he may know the variation of the bodily members in reference to sex, age, and other characteristics. Although many have written statements which are at variance with my own on this point I do not know how to dispute with them since according to my purpose it is the spectators who will gain the greatest and most numerous advantages from my method of procedure. Therefore, gather around the human body, of whatever age or sex, that body from which the soul has departed. You will, however, find a more perfect form or shape and size of the members of the entire body in a young person or in one with a temperate and equable complexion or of a similar temperament. You should also know that for the recognition of muscles, veins, and nerves the better bodies are those without fat, that is to say, thin bodies, which are clean of dirt and hairs. If you remove the hairs as when you excoriate the skin of the pannicle which covers the entire body you will find during that inspection and testing of the skin by seizing and pulling it how the hairs are attached to the skin and that the skin is double, that is, both inner [superficial fascia] and outer. In the latter layer the hairs are fixed in the exterior, as you may see in animals. The outer skin is also double, as you may discover by a closer inspection: that is, there is a thin superficial skin [epidermis], which everyone can see in bladders and in parts of the body which have reddish eruptions upon them. The other [inner] part of the skin [dermis] is thicker than the outer skin; you can inspect it by separating the entire skin [by layers] with a razor. Although it is difficult to do this nevertheless the skin can be separated. It is thick and whiter on the inside; the layers are not attached to each other, as some people think they are, but each layer is continuous in its extent over the area of the body. However, the layers of skin differ in substance, color, and consistency from each other. There is no point in denying this threefold series of differentiae. I certainly should not wish anyone to conclude that I am describing the human body in a curtailed or abbreviated form just because I am trying to be brief, but I shall assume that an inspection of the body's exterior is unnecessary here since it is well known to everyone. Nevertheless, I think there are bits of useful information to be learned even from that exterior. Note, therefore, the parts of the human body as other anatomists have examined them. Look first at the head with its parts in which the animal faculties appear; they call this the upper cavity. The head has many members as you will see when you dissect it. After you have examined the head look at the middle cavity from which arises the first source of life.[11] This part of the body is called the chest. It terminates at the false ribs and is composed of many members which also protect it. Its terminus is the diaphragm or transverse septum. The region from the diaphragm to the lowest extremity of the loins is called the lower cavity. From this cavity proceed the natural faculties. It is also made up, as you will see, of many members both internal and external. There remain the other parts of the body, such as the neck, arms, legs, and so forth. These include the testicles, the penis or genitals, which observers should examine immediately since they are available at first glance, for I do not wish to appear excessively scrupulous in describing such well-known features. I do not intend, before I come to the actual task of dissection, to

[10] This large volume was never completed, if, indeed, it was ever undertaken. In at least seven passages (pp. 184, 209, 221, 222, 225, 242, and 247 of this publication) Massa also mentions a book on the muscles which he intended to write and once refers to a similar book on the veins; neither was ever written. Giovanni Battista Canano's little book, *Musculorum humani corporis picturata dissectio*, which I have translated herewith, appeared at Ferrara around 1541; a new facsimile edition with an Italian translation has recently been published by Giulio Muratori (Florence, Sansoni, 1962); the introduction adds little or nothing to our knowledge about Canano or the circumstances of the publication of his book. Massa may well have been vying with Canano, as his frequent references to this projected book imply; it is entirely possible that he was aware of Canano's interest in the specialized subject of the muscles and dropped his plans for proceeding with his own book when he learned that Canano was preparing one, as similarly it is held that Canano ceased to write further on anatomy when Vesalius' *Fabrica* appeared. Thomas Lauth, *Histoire de l'Anatomie* (Strasbourg, Levrault, 1815), pp. 328, 336, emphasizes Massa's bent toward myology.

[11] Massa does not name the spiritual virtues residing in the middle cavity, but these are implied, according to the standard description of the three virtues of spirits: animal, vital, natural. See Lynn Thorndike, *Science and Thought in the Fifteenth Century* (Columbia University Press, 1929), p. 75, note 30, on Franz Joseph Gall's influence in his *Anatomie et Physiologie du Système Nerveux* (1810–1820) in abolishing the speculations about the vital spirits in the ventricles of the brain.

dismiss those parts I think worthy of notice which have also been described by others and which exist in the lower cavity. From this cavity dissection must be begun, and all moderns take it up first in cutting lest the entire cadaver grow rotten under the effect of the putrid feces in the intestines and thus prevent dissection of the remaining parts. Therefore these external parts of the lower cavity must be noted. First, in the middle under the ribs lies the scutal cartilage [xiphoid process] called the pomegranate. This part they call stomachal [epigastric] since it lies above the ventriculus or stomach, as they commonly say. After this part inspect also the rest of the parts, which are diverse both in name and location. Under the ribs at right and left lie those places which are seen first. They call them hypochondria; under them and inside are the liver at the right, the milt or spleen at the left. They call them hypochondria from a Greek compound, that is, hypo, which means under, and kondrion [chondrion], which is a piece of something, as if you said under the ribs which are not complete but simply pieces in comparison with the complete ribs. Then under the hypochondria on one side and the other are those places called ilia because the bones of the ilium lie under those parts. In the middle of the cavity under the stomachal part is the umbilicus, so called either because it is in the middle of the cavity or because it contains in itself all that is collected to it. But since I must treat the facts themselves and not names lest I appear to be rather a grammarian than a physician, I shall dismiss the etymology and derivation of names as unnecessary, as Galen often did, especially in the sixth book of his *Anatomical Procedures* and in his little book *On Affected Places*. He thought it was more satisfactory to dispute matters of utility and to seek the truth and usefulness of facts rather than to lose time in quarreling over names, as he said in his little book *On the Differences of Symptoms:*[12] "If names are not available we must make up ordinary, suitable, and native words nor (as some do) seek them from abroad." But I return to the point whence I departed. Below the umbilicus to the pubis extends the part which is called sumen. Below the sumen is the last part of the cavity, which is called the pubis; but the groin seems to be part of the hips. These are the external parts of the lower cavity. I shall name the remaining external parts of the body according to the dissection of the members adjacent to them. If you find any members left undescribed for any reason, whether due to its freakishness or my ignorance or the dicta or others, do not call me a liar. I have proposed to write here the truth of sense-perception for those who wish to become skilled in dissections through continuous efforts and by no means for those who are unwilling to work frequently at dissections, since these do not comprehend the parts before their eyes with only a limited practice in cutting. Therefore in order to understand all things you must carry out dissections very frequently in the manner to be described below, or in another manner if you find a better one.

Chapter II

Concerning the Method of Dissecting the Lower Cavity

Begin therefore from the scutal cartilage [processus xiphoideus] called the pomegranate, above the ventriculus, which is also called stomach. Cut the skin lightly with your knife or razor along a straight line up to the pubis, but do not touch the umbilicus. Cut circularly around that part. The Latins call it vetula [old woman], the Greeks graia because it is wrinkled. Cut it out entirely and lay it aside. Then near the umbilicus cut the skin again diametrically from the right hypochondrion or ileum to the left, proceeding gently in the form of a cross. So that you may comfortably see the parts which constitute the umbilicus in their place, tie up the umbilicus itself with a thread where you cut it in circular manner, then proceed to remove the skin gently from every part until it is entirely removed from the cavity itself. This skin, as I said, is double; the inner one is whiter and softer. The outer skin is harder and also double, as it

[12] Galen, *De Anatomicis Administrationibus*, VI, 13 (Vol. II K., 581): Tu autem, si mihi et Platoni auscultas, nominibus semper neglectis, ad rerum scientiam primum et praecipue incumbes. . . . Singer (*op. cit.*, p. 168) translated: "But follow Plato and me in taking little account of names and seeking first and foremost facts and next clarity in exposition." He adds: "The allusion to Plato is either to the *Republic*, 7. 533 E, *Sophist* 244, or *Statesman* 261 E." Galen, *Methodus Medendi*, XI, 12 (Vol. X K., 772) refers again to the same statement by Plato; see also *De Usu Partium*, IV, 13 (Vol. III K., 309): Sed hoc quidem in omni re servare oportet, Platonici praecepti semper memores, *nos ditiores sapientia ad senectutem perventuros, si nomina neglexerimus*. Another reference in *De Anatomicis Administrationibus*, XII, 2 (translated by W. L. H. Duckworth, *et alii* [Cambridge University Press, 1962], p. 115). The "little book" *On Affected Places* (*De Locis Affectis*) runs to 452 pages in C. G. Kühn's edition. It contains a careful discussion of the terminology used by Archigenes (Vol. VIII K., 97–120) which tends to offset his undeniable impatience with terminological niceties. Furthermore, the pseudo-Galenic *Medical Definitions* (Vol. XIX K., 346–462) are carefully composed with due observance of philological propriety. Galen is by no means scornful of exact signification; see, for example, *De Locis Affectis*, V (Vol. VIII K., 328) on terms for the diaphragm. There seems to be no passage in the latter work to justify Massa's statement. In fact, Galen goes so far as to say in his book *On the Differences of Symptoms*, I, 2 (Vol. VII K., 10; cf. 65, 66): Non enim absurdum eos, qui dilucidiores doctrinae gratia haec vocabula distinxerunt, imitari. He is everywhere most careful to discuss his terms while noting that some learned men have so wasted their lives at such discussions that they never arrived at the end for which they strove in their art. The sentence Massa quotes from the latter work resembles this one: deinde nomina iis imponere potissimumque si possit Graecis usitata; sed si haec ignoraverit, propria fingere. (Vol. VII K., 45).

appears in bladders and other parts which have the upper surface skinned away. This part lacks sensation. The other part can be separated with a razor from that soft inner part, not without labor since it is continuous but not attached [to the other layers]. There are some who say in their delusion that if there were many parts in the skin they would be separate. But because the two parts of the skin are united therefore they form one skin only. If those people had looked at Aristotle, *Physics,* VI,[13] they would have seen him say that that which has its ends joined is continuous. They would have known that one part can be united with another and can thus create one thing in continuation, though the parts are different in substance and location. I always separate the skin with a razor and show it to those who are standing by, for not always can parts which are different in substance be separated in bodies with one method or instrument, as may be seen in the composition of the lips, whose muscles and skin cannot be skinned off from them, but are well separated only by boiling in some cooking-pot. But here it is sufficient to recognize the parts of the skin by their diversity of substance, color, and hardness. In the skin there are small veins [which function] through slender capillaries for its nourishment, for under this aggregated skin you will see immediately after it is flayed a notable amount of fat; but there is none or little in those newly born or in some cadavers emaciated with hunger or some other cause (as could be seen in that most unhappy year of hard times, 1527) or bodies consumed by illness or age. You must remove this fat with the utmost diligence if you wish to find the fleshy panniculus of Avicenna which surrounds the entire body. Galen called it the membranous muscle in *On Anatomical Procedures,* Books I and III, although in the third book he calls it a membrane.[14] In the removal of this fat, however, you will preserve two veins [superficial epigastric] within it which ascend to the breasts particularly in women so that you can observe their course from the womb to the breasts. You will see their progress in the dissection of the womb and breasts; therefore they should be kept safely until you dissect the womb or uterus.

Chapter III

On the Description of the Fleshy Pannicle [or Fascia or Sheath] or of the Membranous Muscle or the Covering (Involucrum) of the Entire Body

Having raised the fat, if there is any, since little appears in thin bodies and none at all in the newly born, you will see the panniculus [fascia] which covers the mirach or abdomen as you call it. This panniculus [fascia] is not white like the siphac or peritoneum, but tends toward redness, nor is it so thin as the peritoneum. It is thicker with its own notable thickness; its substance seems to be sinewy or similar to many broad membranous tendons of some muscles in our body. Here and there it has very thin fleshy parts scattered through it; because of this composition the very learned Galen calls it a membranous muscle in the first book of his *Anatomical Procedures*. This panniculus [fascia] has its origin or place of greater adherence in the back from the spine; in this place you will see many nerves coming to it from the spine; these are slender. Here the panniculus [fascia] proceeding toward the armpit seems to have a more manifest fleshy part than in other regions, but not so manifest and broad as in animals. Here the panniculus [fascia] in its progress sends out some sinewy parts toward the arms which pass under the armpits to the bones of the humerus and to the muscles both of the arm and to those which cover the ribs. These tendons also pass between the muscles and seem to have separate terminations; but in the substance of their muscle or panniculus [fascia] they are not separated. Note, however, that although the fleshy parts around the ribs under the armpits in this panniculus or muscle are larger in size, this part is not distinct from the panniculus as a whole as a separate muscle is. That is because in this region the fleshy parts are thicker than they are in other regions and more apparent. Nor does this fleshy part proceed very clearly up to the back but grows less and near the back is almost membranous, that is to say, in adults, since in the newly born the entire substance of it is fleshy also where it proceeds toward the back. Therefore those who say that in kids there is found a membranous muscle beyond the one mentioned which moved the shoulder blade (spatula) are deceived by that more fleshy part of the aforesaid muscle; this part is not so apparent in an old animal. Therefore, according to Galen, who says all of this membranous muscle arises from the back,[15] one ought to say that all this is a membranous muscle and that it

---

[13] Aristotle, *Physics*, VI, 231 a: "Now if the terms 'continuous,' 'in contact,' and 'in succession' are understood as defined above—things being 'continuous' if their extremities are one . . . " (translated by R. P. Hardie and R. K. Gaye in the Oxford Translation of Aristotle edited by W. D. Ross, Vol. 2).

[14] The fleshy panniculus of Avicenna is described in *Canon*, III, Fen 13, Tractate 1, Chapter 1, p. 691, in the translation by Gerard of Cremona, published at Venice by Giunta, 1595. Galen, *De Anatomicis Administrationibus*, I, 233, Singer, *op. cit.*, p. 8: "a thin membranous muscle [*panniculus carnosus*]. This was continuous with the covering of the spinal muscles at the loins as a fascia (syndesmon) from the bone of the spine." *Ibid.*, III, 348, p. 63: "take away the membrane [fascia] beneath. . . . " See Berengario da Carpi, *Commentaria*, etc. (1521), fol. LXIIII for a discussion in some detail of the fleshy panniculus.

[15] See note 14, first passage from Galen, *De Anatomicis Administrationibus*. Massa carries out his promise to "dispute this matter in another place" (see p. 180) and emphasizes the dissimilarity of human and animal anatomy by another reference to Galen at note 88.

is identical with the fleshy panniculus and that this muscle or panniculus is one member and not many members. Galen has spoken of it in many passages, as is apparent from the first and third books of his treatise on *Anatomical Procedures*. In these passages you will clearly perceive that he is speaking of this [muscle] and not of any other. But the text is sometimes confused since Galen calls this part a muscle and sometimes a membrane; hence pay attention because you will see him speaking of one member (by the testimony of sense) for if the others are membranous muscles he has not excepted this one although he has spoken of the others, as you clearly see. I shall dispute this matter in another place, and we shall also see that in no wise are the members of an ape completely similar to the members of man, as some have believed. Nor could the fashioning of the members of brutes ever deceive me so that I should desire to see a completely similar fashioning in men, although there are many brute animals who are quite similar in some of their members to men. But when you carefully consider them they are not completely alike; for the members of a stag differ from the members of a lion since the spirit or soul of a stag differs from the spirit of a lion, according to Aristotle, *On the Parts of Animals* I.[16] Averroes recorded this opinion in his commentary on Aristotle's *De Anima I*, chap. 53,[17] and emphasized the difference by which the members vary both in operations and movements. You will find this panniculus or muscle also in brutes but not completely similar to the panniculus or muscle in man. This panniculus or membranous muscle proceeds throughout the entire lower cavity and the middle cavity called the chest, surrounding all the muscles of both cavities and the remaining parts of the body such as the head, arms, and legs. And in those newly born this muscle or panniculus is most manifestly apparent and distinct. In children it has the substance and color of flesh and has an especial thickness. However, as it proceeds toward the armpits it grows thinner and similarly in the other extremities such as the legs and arms. But when it is dried out in adults it is not of such thickness in all parts of the body as it is in children. Galen in *On Anatomical Procedures* I[18] promised to write about the services of this membranous muscle or fleshy panniculus, but I do not remember that I have read about them in his volumes, which I have in my possession, since its service is not only to support the veins, as some say on the basis of the third book of Galen's *Anatomical Procedures*.[19] Therefore I shall try my best to describe its services. The first of these is to protect the muscles of the entire body from external injuries such as of heat, cold, and the rest. Since the skin is vulnerable to external impressions, when it is injured the muscles which are the soul's instruments in voluntary motion would also be injured unless this panniculus served as an intermediary. By reason of their composition these muscles are adapted for inducing pain when their quality is changed; then there follows lesion of voluntary motion, or its absence, by necessity. Secondly, when the skin is broken it can be healed more easily by the assistance of this muscle or panniculus because it is fleshy. For the skin is a spermatic member[20] and therefore not perfectly and easily healed except by something else which has some fleshiness, such as this panniculus which has a substance midway between that of a muscle and of a membrane although in adults it is more similar to a membrane. I know that that which grows can also be healed either by means of itself or

---

[16] Aristotle, *De Partibus Animalium*, I, has no such statement. Massa has transferred it from Averroes' comment on *De Anima*, 407 b 20, 24, given in the following note.

[17] *Averrois Cordvbensis Commentarium Magnum in Aristotelis De Anima Libros*: (Corpus Commentariorum Averrois in Aristotelem; Mediaeval Academy of America, Cambridge, Mass., 1953) **6**, 1: p. 75: Membra enim leonis non differunt a membris cervi nisi propter diversitatem anime cervi ab anima leonis.

[18] Galen, *De Anatomicis Administrationibus*, I, 3 (Vol. II K., 233), Singer, *op. cit.*, p. 8, found the membranous muscle in his ape and wrote: "the nature of which will be fully and duly explained," a promise Massa points out which was never fulfilled.

[19] *Ibid.*, III, 1, Singer, *op. cit.*, p. 60; 2, Singer, pp. 63–64 (Vol. II K., 340–353), probably contains the passage to which Massa refers, especially "the membrane [fascia] beneath through which the nutrient veins reach it [the skin]" (Singer, *op. cit.*, p. 63).

[20] See Mundinus, *Anatomia*, ed. Berengario da Carpi (Bolgna, 1521), fol. XXXXVII: alia est cutis quae dicitur uera cutis, quae est alba immediate sequens corium, et est spermatica, ideo si depredatur, non potest restaurari. This passage is missing from Singer's translation of Mundinus: *The "Fasciculo di Medicina," Venice, 1493: A Translation of the "Anathomia" of Mondino da Luzzi* (Monumenta Medica, Vol. II; Milan, R. Lier, 1924). He used the 1495 Latin Ketham edited by Pietro Andrea Morsiano of Imola as the basis for his work, a curious fact in view of his high praise for both Berengario's edition and that of Matthaeus Curtius (Pavia, 1550), which he calls the best editions "from the point of view of textual study" (p. 56). Berengario in his *Commentaria*, fol. LII, discusses the "true " skin (the lower layer, not the epidermis, also called a panniculus) on the basis of Galen's references and Aristotle, *Historia Animalium*, 3. 11, 518 a: "And wherever the skin is quite by itself, if it be cut asunder, it does not grow together again, as is seen in the thin part of the jaw, in the prepuce, and eyelid. In all animals the skin is one of the parts that extends continuous and unbroken, and it comes to a stop only where the natural ducts pour out their contents, and at the mouth and nails" (translated by D'Arcy Wentworth Thompson, Vol. 4, Oxford Aristotle, 1910, reprinted 1949, 1956). As to Aristotle's use of the term spermatic see *De Generatione Animalium*, 4. 5, 774 a 20 (Vol. 5, Oxford Aristotle, translated by Arthur Platt, note 1): "By 'spermatic' Aristotle means that both sexes produce large quantities of residual matter which is worked up into semen and remains in a sanguineous condition in the female." Furthermore, "spermatic" animals are able to superfoetate. Avicenna's definition of a spermatic member appears at *Canon*, I, Fen1, Doctrina 5, chap. 1 (translated by Gerard of Cremona, Venice, Giunta, 1595), p. 31: Membra autem, quae ex spermate sunt creata, quum solutionem continuitatis patiuntur certa continuitate non restaurantur nisi pauca ex eis et in paucis habitudinibus, et in etate pueritiae, sicut ossa et rami venarum parui. . . .

something else, as is apparent in bone fractures and the break of the continuity of the nerves[21] of members by their own exchange or from the union of other members. This aid has been given to the skin because it is a covering which protects the entire body and therefore suffers more from external causes and dries out. Hence it requires an assistance not provided by the fat in fat people even if in other respects it resists external harm. The third assistance offered by this membranous muscle which both sense and reason indicate is the shaking or shuddering motion of the entire body and the cooperation of some of its parts, for this panniculus preserves the nature of a muscle and since it surrounds the entire body it moves at the soul's bidding, thus shaking the body and assisting in this action with the other parts. You will also notice in this member very many veins for the nourishment both of itself and of the skin. Therefore you will observe that flesh from it grows in wounds. These are the services which as I have observed both by reason and experiment are offered by this panniculus. If anyone after me finds something better to add to this description I beg of him to do so in order that the truth may shine forth more from day to day, and may he receive the reward of his merits from posterity. Therefore this is the muscle which the most learned Avicenna, faithful interpreter of Galen, called the fleshy panniculus in *Canon* III, Fen thirteen, Chap. 1.[22] If all modern anatomists have not seen it or have not recognized it in their reading and disputations over distinctions let them blame the laziness of their intellects as well as of their hand since in the investigation of truth one must not spare labors. In a matter of doubt I never rest in sense or discourse until I find the truth. Do I not have you as a truthful witness, Gerolamo Ferro, most learned in both Latin and Greek, I who am joined to you by relationship? Even when I was ill I spared no labor, vigil, or inconvenience in the most willing pursuit of these studies. There are very many most learned physicians of my acquaintance who come [to my dissections] for the sake of seeing something; others also whom I recognize as studious men I have often invited, lest I perish as I walk along a perilous path. Among them are my friends Victor Trincavella [o], Donato de Mutis, Francesco Vittorio[23] (may God

[21] Note the use of the phrase "continuitatis solutio" by the sixteenth-century anatomists; it is a translation of Galen's συνεχείας λύσιν as it appears, e. g., in *De Locis Affectis*, II (Vol. VIII K., 80); earlier writers also use the phrase, as Francis of Piedmont, *Complementum Mesuae* (Venice, 1502), fol. 141. "By their own exchange" translates the expression *a proprio cambio*, from "cambium," a fancied nutritive juice originating in the blood which repaired the losses of every organ and produced their increase: R. Dunglison, *Dictionary of Medical Science* (Philadelphia, 1860), p. 159.

[22] Avicenna, *Canon*, III, Fen 13, Chap. 1 (translated by Gerard of Cremona, Venice, Giunta, 1595), p. 691 (Venice, 1505), fol. 211. Berengario da Carpi, *Commentaria*, etc., fol. L, cites Avicenna thus: cutis et panniculus qui est post ipsam et est carnosus et lacertosus positus in tunica quae est infima tunicarum lacerti uentris cum panniculo subtili qui est secundum uirtutem siphac; est ex summa mirach. When Massa later claims to have discovered this fleshy panniculus he is quick to add "in our time." Lauth, *op. cit.*, p. 332, takes Massa's boast too seriously. The brief treatise entitled *Anonymi Isagoge Anatomica* in which Lauth finds the first mention of the panniculus (p. 13, Caput VI. Panniculus carnosus pinguedini subjectus, totum corpus incingit.) can scarcely be taken as evidence in this dispute since, although its Greek version is not found in Aristotle, the entire text (made up of excerpts from Aristotle's works) cannot be accurately dated; see the edition of both Greek and Latin forms of the treatise by Daniel Wilhelm Triller and Johannes Stephanus Bernard, Leyden, 1744. See for a discussion of this work Wilhelm Schmidt's Berlin dissertation of 1905 summarized in *Mitt. z. Gesch. d. Med. u. Naturwiss.* **5** (1906): p. 382; it originates some time after 1595 since it utilizes the *Historia anatomica humani corporis* of André du Laurent (1595). Berengario (above) declares that modern writers misunderstood Avicenna here and mistook the cords of the oblique muscles above the long muscles for a panniculus since Avicenna had used the word "fleshy" to describe it. This panniculus actually occurs as a thin subcutaneous muscle which moves the skin in the forehead, neck, and scrotum.

[23] Like Alessandro Benedetti before him in the *Anatomice* (1497) Massa mentions his friends as witnesses of his dissections, probably following, as Benedetti does, the words of Galen (who in his turn follows Plato) in *Methodus Medendi*, VII 1 (Vol. X K., 456), translated by Charles Singer in his translation of Mundinus, p. 59: "a work in any science or art is published for three reasons: first, for the satisfying of friends. . . . " For Gerolamo Ferro see note 2; for the other men mentioned here and elsewhere see Cosenza, *op. cit.* in note 2, although his work is not complete and some are not listed; they were perhaps not of sufficient importance to have made much impression on the history of their time. Vittore Trincavello, born at Venice in 1491 or 1496 and died there on August 21, 1553 or in 1568, was a philosopher, physician, and an Aristotelian who studied at Padua and Bologna. He was professor of medicine at Padua in 1551 and still there in 1557. He was a friend of Cardinal Pietro Bembo, Gasparino Contarino, and Agostino Ricci; he wrote *Quaestio de Reactione Iuxta Doctrinam Aristotelis et Averrois* and medical works. He became a member of the Venetian medical college July 18, 1523. I can find no information on Donato de Mutis or Francesco Vittorio in Cosenza or elsewhere. On Jacopo Berengario da Carpi (*circa* 1460–1528 or thereafter) see Vittorio Putti, *Berengario da Carpi: Saggio biografico e bibliografico seguito dalla traduzione del "De fractura calvae sive cranei."* Bologna: L. Cappelli, 1937, and the long introduction on his life and work in my translation of his *Isagogae Breves: A Short Introduction to Anatomy* University of Chicago Press, 1959, with a bibliography (pp. 217–221). He wrote several important works in anatomy and, together with Zerbi, is the most important and voluminous writer on the subject between the time of Mundinus and that of Vesalius. His large work entitled *Carpi Commentaria cum amplissimis additionibus super anatomia Mundini una cum textu eiusdem in pristinum et verum nitorem redacto* (Bologna, Hieronymo de Benedictis, 1521) still remains to be properly assessed for its contributions to knowledge. See note 14 for his discussion of the fleshy panniculus, which is more detailed and professional than Massa wishes to acknowledge.

Gabriele Zerbi, whose life and work are treated elsewhere in the present work on pre-Vesalian anatomy, was born *circa* 1445, according to Ladislao Münster, and died in 1505. He is listed among the lecturers in medicine and logic at the University of Bologna from 1475 to 1483. He practiced medicine at Verona and wrote *Gerontocomia, scilicet de senum cura atque victu* (Rome, 1489),

have mercy on his soul!) and those who are named in memory, to whom I first showed the fleshy panniculus or membranous muscle or the envelope of the entire body upon some cadaver in the hospital of Sts. Peter and Paul. All of these men are witnesses to the fact that I was the discoverer of this panniculus in our time nor will you see anyone else after Avicenna who saw or recognized this panniculus, since Mundinus did not show it in his writings, although he said there is a fleshy panniculus below the fat, nor did he point out its origin nor size. In this also Jacopo of Carpi was deceived, a man learned in other respects, who claimed he had dissected so many men, and very many others such as Zerbi and certain moderns who recognize this member in name only; it is certainly strange since the member in question is not difficult to find. But let all who write on the dissection of the human body as well as those who actually dissect it spare me since they have placed their trust in the statements and questions that run through the schools and have lost their sense and with the fat itself they have removed this panniculus. Therefore one must turn his attention to this part and first carefully lift the fat with a razor by skinning it from the panniculus itself, since the latter is distinct from the fat; when the fat is lifted off the panniculus immediately presents itself. And therefore Galen the eyewitness said in the second book of *On the Use of the Parts of the Human Body*[24] that when one wishes to become an inspector of the work of nature he should not trust the books of anatomy but his own eyes, or he should come to me or join one of my associates or exercise diligently by himself in anatomical dissections. Very many other physicians and most learned men of another rank have watched me as I dissected. They are worthy witnesses of this pursuit and of many other facts which I have discovered and shall discuss in their place. Among these men were Maffeo dei Maffei, Francesco Marino, Benedetto Arino, and other true physicians whom I have wished to name as an honor to them. Nor are there lacking most weighty witnesses of another rank, such as Alessandro Businelli, secretary of a most serene master, and Vincenzo Ricci,[25] a clerk and doctor of civil and canon law. Laying these matters aside, after you have seen the fleshy panniculus or muscle it makes no difference what you call it since it is sufficient to have found the member denominated by the ancients but which they did not recognize; we need not dispute whether this member is the skin called derma and epiderma by Galen since it is clear that the member is not the same. Those items occur in a layer and constitute what is now called the first skin; this panniculus, as you have seen, is beneath it. Therefore cut the panniculus in the middle from the cartilage called pomegranate, or, if you like, in another direction since immediately beneath it is the abdomen, or mirach. Nevertheless, preserve this panniculus so that you can see its process from the back to other parts of the body and recognize the broad muscles which move the arms downward as they are distinguished from the panniculus.

Chapter IV

On the Method of Dissecting the Abdominal Parts, or the Mirach

When you have cut but not lifted off the fleshy panniculus or muscular membrane, or the covering of the entire body (as it pleases some to whom I have shown it to call this member), the mirach is immediately visible. This is composed of eight muscles; two of them are longitudinal, four are transversal, and two lateral. But before I come to the separation of these muscles, which coalesce into one, I wish to remind you that in the dissection of the human body those features are to be noted which Galen was wont to list. These are the substance of the member, its number, positions, shape, size or quantity, continuation,[26] and operations. Concerning complexion and other features I say nothing since they are distinguished rather by discussion than by sense-perception. These features as listed you should note not only here [i.e., the abdomen] but in all other parts of the body in order to benefit by these dissections. Look first then at the two longitudinal muscles. Cut their substance with a razor; it is sinewy and white and covers the longitudinal muscles. The white substance is that of a tendon or muscle-cord of the transversal muscles; these cords cover the lon-

the first work on geriatrics; *Anatomiae corporis humani et singulorum illius membrorum liber* (Venice, 1502; sec. ed., 1533, a much more readable form), the most considerable work on anatomy since that of Mundinus and comparable only with Berengario da Carpi's commentary on Mundinus (1521) for thoroughness and detail before Vesalius's *Fabrica* (1543); *De cautelis medicorum liber* (Venice, and various editions later at Pavia and Lyon), on legal medicine; *Anatomia matricis. De anatomia et generatione embrionis* (Marburg, 1537).

[24] Massa uses the fairly uncommon word *oculatus*, "conspicuous, visible," hence of one who uses his eyes for observation. See Galen, *De Usu Partium*, II, 3 (Vol. III K., 98): Quicunque igitur vult operum naturae esse contemplator, non oportet eum anatomicis libris crederem sed *propriis oculis*, aut ad nos accedere, aut cum aliquo eorum qui nobiscum versati sunt, aut ipsum per se diligenter exerceri, anatomis obeundis. Quamdiu autem legerit solum, omnibus me prioribus anatomicis credet eo magis, quo etiam plures sunt (italics mine).

[25] Concerning these friends of Massa I have been able to find only the following information about Vincenzo Ricci: Paulus Manutius borrowed some ancient copies of Cicero's *Ad Familiares* from Ricci while engaged in bringing out his edition of Cicero; see Cosenza, *op. cit*.

[26] Continuation: the Latin is *continuatio*; Mundinus has *continuitas*. This feature is called *colligantia* by other sixteenth-century anatomists including Massa himself later in his text.

gitudinal muscles and terminate here in the middle of the body cavity. I should not say that all the cords, except that of the diaphragm muscle, terminate at the bones since there are many cords which do not terminate there, such as the cords of the two muscles in the anterior part of the gullet which are in the middle part of the muscle and whose fleshy parts stretch toward the bones. Not everyone has recognized these muscles, which I shall discuss when I take up the glottis in the dissection of the neck. There I shall describe their uses. Begin this incision therefore from that frequently mentioned cartilage [processus xiphoideus] above the stomach and proceed in your dissection down the length of the body to the pubic bone, carefully removing the skin. You will see two descending muscles [rectus abdominis] which originate from the second of the true or intact ribs which lie immediately above the false ribs and terminate at the pubic bone by means of a little cord. I say by means of a cord because that final part of the muscle is a cord. Although small it is sensitive; I have often seen a convulsion or spasm and feeling of pain follow its puncture. Examine the substance of these muscles; it has filaments or threads which proceed lengthwise. You will see in this substance of the muscles two and sometimes three tendinous or cordlike transversal divisions; the purpose of these tendinous divisions is to prevent muscle-relaxation from accidental or natural extension since they are fleshy and soft. Thus with these tendinosities the muscles are strengthened; but the ancients wrote nothing about them.[27] You will note, however, that these muscles do not lack a panniculus to cover them, as some have said they lack, induced by the fact that it was sufficient for nature to have given the cords of the other muscles to protect them. But this is contrary to sense since they have a very thin panniculus that covers them; you can see it if you look carefully. Lay these longitudinal muscles aside after you have thus stripped them. You can cut them in the middle or in the upper part near the cartilage called the pomegranate. But do not lift them up entirely so that you can see the rest of what lies under them and shares sensation with them. When you have done this begin to skin the transversal muscles on each side or on one alone, separating them from the others, for on each side there are two intersecting each other in the form of an ancient letter *X* [external and internal abdominal oblique] or in the form of a Greek chi. In these muscles there are fibers which proceed transversely according to the descent and ascent of the muscles in their flesh as well as in their tendons; the latter are not round but broad like a membrane. The upper muscle called descending begins or takes its origin from the ninth rib, beginning the enumeration from the first of the false ribs below and terminates by means of a broad tendon which covers the muscles called longitudinal. You will keep the tendons of the transverse ascending muscles since they pass under the longitudinal muscles. You will thus see that the tendon of the ascending muscle is divided so that it embraces the longitudinal muscle on one side as well as on the other and covers it as though it were a garment. But to see this you must carefully separate little by little the tendons themselves from the tendon of the lateral muscles. The transverse muscles called ascending take their origin from the hips. Begin there to strip them with a razor for they ascend to the heads of the false ribs and terminate in a broad tendon which proceeds transversely in which you will also see fibers or filaments proceeding transversely from right to left in the fleshy part as well as in the tendon. The ascent and descent of the filaments in the two muscles make an ancient letter *X* or something like it on each side, right and left. Keep all the muscles you have thus stripped where they terminate by means of tendons since you will see a fleshy and tendinous mass and a panniculus that covers the muscles. For all muscles are covered by their own panniculus; I shall say nothing of its origin since a panniculus is something distinct from bone, ligament, nerve, and flesh. Also examine the termination of the aforesaid muscles in the fleshy part up to the substance of the longitudinal muscles in the extremity of whose flesh appear their tendons. When you have stripped these you can lay them aside and remove them, if you wish. After the transverse muscles appear two lateral muscles [transverse abdominal], one on the right, one on the left, which originate in the back from the corresponding parts of the hypochondria, ilia, hips, and false ribs. But the fleshy part of the aforesaid muscles does not proceed far for it is small in respect to the tendinous part. These muscles have their filaments passing laterally toward the umbilicus in both the fleshy and tendinous parts. These tendons do not terminate at the bones but are united among themselves and proceeding under the tendons of the transverse descending muscles they are joined to the siphac panniculus or peritoneum from the pomegranate [processus xiphoideus] to the pubic bone. I must not, however, omit the fact that under the longitudinal muscles in the lower part toward the pubic bone are two notable veins [inferior epigastric] which nourish these muscles. They arise from the vena cava and pass to the muscles by perforating the siphac [peritoneum] and the tendons of the other muscles. These tendons of the lateral muscles you must

[27] Lauth, *op. cit.*, p. 328: Les muscles pyramidaux du bas-ventre nommées succenturiaux par Sylvius, sont designés par Massa, qui décrit aussi avec précision les enervations des muscles droits, que Berenger (*op. cit.*, fol. LXXXV b) avoit prises pour les tendons de trois muscles particuliers . . . These are the pyramidales abdominis musculi or musculus Fallopii or succenturiati or auxiliarii.

carefully strip from the substance of the peritoneum or siphac since they strongly adhere to that panniculus itself. Mundinus did not recognize them.[28] You will also note that the didymi come out of the substance of the peritoneum to the scrotum. They perforate the tendons of the muscles because in the descent of the intestines or of the omentum (zirbus) the tendon is often cut. Therefore let those who try to cure these members open their eyes lest by a violent intromission the tendon should be lacerated because terrible consequences such as pains, spasm, and even death, proceed from the laceration. Let those also be cautious who cure by incision or inner ligature, or laqueation, as they call it, lest they make these mistakes since many are deceived to the very great danger of their patients. You must take note of all I have said above nor expect to find in this little book the possible ills which can occur in the members of the body except for a few of which incidental mention can reasonably be made. God willing, I have decided to write about them in another book I have already begun.[29] But what I ask you to note here is the tendons of the aforesaid muscles of the abdomen which do not move or lift heavy bones. One of their uses is to serve the expulsion of superfluities such as the feces and urine, as Galen learnedly said in his book *On Hygiene* I (*De Sanitate Tuenda*) and in his little book *On the Motion of the Muscles*,[30] with the assistance of the diaphragm and other members. Hence one must not say that the tendons are given to the muscles only to move or lift the bones but one should rather more reasonably say that tendons were given to those muscles which work the most in movement and gestation and so that they might not be loose, which would be the case if they were fleshy alone. This was done so that they might be strong to bear the burden of the intestines and of the foetus in women. At the end of the longitudinal muscles are two small muscles which proceed toward the groin; hence preserve these until we come to the anatomy of the penis.

Chapter V

On the Anatomy of the Peritoneum, or Siphach, and Omentum, and of the Umbilical Vein

The peritoneum, or siphach, is a panniculus or membrane which is immediately visible when you move aside the eight muscles of the stomach. It is slender but hard in order to sustain the burden of the intestines and of the uterus in woman. It is, however, more compact around the groin where frequently rents or tears [herniae] are apt to occur whence the omentum sometimes and the intestines often protrude like an abscess. A tumor, however, is not an abscess since there are not in it three kinds of ills which make abscesses, nor does it follow that if anyone says it is a tumor therefore it is an abscess, since tumors are often accustomed to result from an abscess. Let those be cautious then who cut the groin with steel lest they be deceived; let them learn the difference between a tumor, an abscess, and a rupture of the peritoneum. Note that this panniculus when it passes toward the groin and the pouch of the testicles is more firm and thick in that direction. Note also that in this panniculus are contained all the natural members such as the liver, spleen, stomach or ventriculus, intestines, kidneys, testicles and all the rest, including the spermatic vessels, the bladder, and their ducts which carry off the urine. All these members are embraced or contained by this panniculus as though by a sack. It adheres to the diaphragm and to the muscles of the cavity, as has been said, as well as to the back and the remaining parts, as you will see when you strip it. When it descends to the scrotum or pouch it contains the testicles and in woman it contains the uterus. You can strip this panniculus, if you wish, with your hands and fingernails since it is a single unit which contains all the members I have mentioned. I have often stripped it and drawn it forth with the aforesaid members and thus by means of an incision I have inspected the remaining members contained within it. But before you approach the dissection of this panniculus preserve intact the umbilical vein [ligamentum teres] which is composed of one vein passing to the hollow of the liver and of two arteries [common iliac] which pass through the groin and area of the hip bones toward the back to the aorta artery. This artery passes to the heart through the spine of the back. There in the hip bones it branches into many parts and to the groin as far as the umbilicus. The umbilicus is composed of another vein called the urachial [urachus], which passes to the bladder. Since you can have an understanding of its composition, substance, position, and connection (*colligantiam*) after you have stripped it, lay aside this umbilical vein and look at the rest of the parts. You cannot see the process of the aforesaid veins which compose the

---

[28] Massa is careful to record the oversights of Mundinus, although the latter's work is so slight in comparison with Massa's that the earlier anatomist may well have failed to mention features of the body with which he was familiar. The argument from silence, however, is of slight value in the history of anatomy, especially in the sixteenth century with its greatly increased activity in dissection. It is true that Massa is the most energetic of these anatomists in claiming priority for his discoveries, as he regarded them.

[29] This book, like the one mentioned at note 10, was never written.

[30] Galen, *De Sanitate Tuenda*, I, 11 (Vol. VII., 65) deals with the use of the epigastric muscles and the diaphragm in expelling the feces. The passage is translated by Robert Montraville Green in *Galen's Hygiene*; Charles C. Thomas, Springfield, Ill., 1951, p. 39. *On the Motion of the Muscles*: see Galen, *De Motu Musculorum* (Vol. IV K., 455–456).

umbilical vein except after cutting the peritoneum because this panniculus prevents you from seeing that process of the vein that passes to the hollow of the liver. Through this vein the infant in the uterus sucks the purer blood from the secundina [placenta] for its nourishment. In adults, however, you will recognize the hollow of this vein not without difficulty since it is rendered ineffective for use [*frustrata ab operatione*].[31] Therefore you must place a rod within it and thus you will see its passage into the substance of the hollow of the liver in which it is fixed and branches to all parts of the liver; but beyond the liver it is one vein only as far as the umbilical skin. But when it comes out of the skin and passes to the secundina it branches into two parts and these in turn into many branches so that each of them can more easily bring blood from the secundina to nourish the foetus. Pay no attention to Avicenna and Mundinus[32] when they say that the vein just mentioned passes to the hump of the liver since this is contrary to sense. But whatever may be the error of the transcribers of the manuscripts or of the author let those conjecture who recognize sense and reason. The use of the two arteries which compose the umbilical vein is by means of these arteries passing to the great artery as far as the heart to cool the spirit since infants in their mother's uterus do not breathe through the trachea and mouth, etc. The function of the urachial vein which passes through the middle part to the bladder is to send forth the urinal matter. You will see this urine in the newly born if you take the sumen and press it with your hand toward the umbilicus, for this is what midwives do before they bind and cut off the umbilical vein from the secundina. Therefore if you wish to know this you must see it in the newly born. It will be pleasant to speculate at this point whether fecal matter is found in the intestines of those newly born; it seems that this is true because newborn babies before they are fed sometimes void a certain matter, saffron in color occasionally verging on green. Someone might ask why feces are found in the intestines of the newly born. You must answer that it is true; in the intestines is found a bilious material like veal, as though a mixture of yellow bile and phlegm, or, as they say, from bile and rheum. Sometimes it is also green from an admixture of some humor which has been well heated and parched. Thus you can say that in the intestines there is no fecal matter from the superfluities of the stomach which in those who eat or take food through the mouth receive form in the intestines after the food has been made into juice by the meseraic veins. For in the flow proceeding from a stoppage of the meseraics the feces appear or descend in liquid, not solid, form, but in children newly born there is some matter which is transmitted from the gall bladder to the intestines by means of a duct which passes to the latter. There it is mixed with some phlegmatic material, the superfluities of the nourishment of the intestines. This material so mixed is expelled after the child has been born by the expulsive force excited by the coldness of the air acting upon the inner warmth of the uterus. Thus these superfluities are not truly fecal, that is, from the stomach, but of those members aforementioned. Therefore infants do not void in the uterus since they are not nourished by food taken through the mouth and sent down to the stomach but are nourished by that aforesaid umbilical vein which brings blood to the liver. From this blood nothing is cut off except urinal matter which is filtered from the blood. Thus infants in the womb urinate involuntarily through the urachus. From this we may also see and understand that the stomach is nourished with prime blood through veins since as has been said infants in the womb are nourished by the umbilical vein which carries blood nor is there found in the stomach any material which could nourish it. Therefore the stomach is nourished with blood from the veins which are scattered through it. If in the inspection of the umbilicus you have separated some part of it or some vein you can bind it with thread so that after the rupture of the peritoneum you can see the course of the remaining parts. In order to see the omentum and the other members contained in the peritoneum itself cut the peritoneum in that part where you will least injure the other members contained in it and you will immediately see the omentum as it descends from the stomach in such a way that you might believe it originates from the stomach. Its substance is sinewy or like a nerve; it makes no difference which term you employ. In it appear many veins scattered like the branches of a tree. It has a notable quantity of fat covering it because the omentum descends to the groin and covers all the intestines to the groin and

[31] That is, when the umbilical cord is tied off. Compare Berengario da Carpi, *A Short Introduction to Anatomy (Isagogae Breves)*, translated by L. R. Lind (University of Chicago Press, 1959), p. 42, and Mundinus, *Anathomia*, translated by Charles Singer (*The Fasciculo di Medicina*, Venice 1493, Part I, Monumenta Medica, Vol. II, R. Lier and Co., Florence, 1925), p. 61: "Nevertheless, this vein is utterly empty of blood, since, *after birth it is robbed of all proper office* and therefore dwindleth by little and little . . ." (italics mine).

[32] *Ibid.*, p. 61: "On that account within the umbilicus a vein may be traced which doth travel by the *zirbus* of the liver unto the vena chilis." Singer's arbitrarily archaic English, chosen for no satisfactory reason that I can discover, is not required for a translation of Mundinus's entirely normal medieval Latin. The numerous misprints scattered throughout this work should be removed in the re-issue that is urgently needed since it has become hard to obtain. I add the Latin text of Mundinus in the edition of Berengario da Carpi: Et propterea in intrinsecis apparet uena quaedam, quae continuatur cum ipso, et pertransit per gibbum hepatis in chilim; ista tamen uena priuata est sanguine, quia frustrata est post partum a propria operatione.

pubic bone. There at its bottom or in the extremity of the body cavity it is attached toward the back and returns again to the ventriculus or stomach whence I said it descended. Thus it makes a kind of sinus in the manner of a sack. This ample hollow is the same size as the zirbus, or omentum, itself, and from its composition in this manner they say it is composed of two membranous tunics. But only one of them turns back or is folded in two, whence it can be said to be composed of two [tunics]. I have often seen it attached to the colon intestine and to the peritoneum in the right hand region near the liver. The function or use of the omentum is to cherish the intestines and to maintain their heat for a better concoction and expulsion of superfluities. When you have examined these items you can see the remaining parts of the umbilicus which you bound with a thread, separating them with a razor or some other instrument, and all the aforementioned veins which compose the umbilical vein. Place a rod within these veins throughout their course or cavity. One of these is the course of the vein to the hollow of the liver which many moderns have not seen. Place a rod into the urachium which passes to the bladder; finally look at the course of the two [umbilical] arteries from the umbilicus toward the back to the aorta artery. But all these veins are seen far better in the newly born since while they are functioning properly they are more easily visible. In adults, however, since they lack their proper function we indicate their course and cavities not without difficulty. Yet I have always shown their course to other members and cavities.

Chapter VI

On Some Operations Which Take Place in the Lower Body Cavity With Reference to Manual Operations

Since modern anatomists have written about many matters not necessary to a knowledge of dissection I too have spoken incidentally about many things outside the ordinary routine of dissection. I am therefore happy to discuss here also certain manual operations which take place in these members both because they are necessary to the physician as well as because they are not perfectly understood by all physicians, since they are accomplished with some danger to human life. Therefore lest I seem to be at fault in a matter necessary and useful, although I believe it is of very little relevance to this part of my introduction to anatomy yet I thought it would be pleasant to teach younger medical men these items since they have also thought this an excellent decision. In this body cavity or in its muscles or in other parts of it are accustomed to occur ills which require manual operation, such as abscesses, either warm or cold, which must be incised, suppurated, or removed, such as knots and melancholic abscesses. Care must be taken to avoid a deep incision lest muscle tendons be cut since, on the testimony of Hippocrates,[33] cords or tendons and nerves are apt to cause spasms or revulsion due to their puncture or imperfect incision. Second, there is no question but that one must sew up in the lower body cavity a break in the continuous surface which proceeds to the intestines, that is, with incision of the peritoneum. There are other operations which cannot be performed without the greatest danger, such as the extraction of water from those who suffer from dropsy, castrations, extraction of stone from the bladder, and so on. But since I have decided not to write in this book on methods of treatment or diagnosis of illnesses I shall not set all of them down here but try to describe a few necessary ones accurately handed down by the ancients, as you can see by their words. Let me say first for the sake of example that a wound is made in the body cavity when the peritoneum is incised and that from this wound the intestines or the omentum either protrude or do not protrude. But if they do protrude the intestines must be thrust back as soon as possible. If they swell you must boil camomile in wine and after bathing them in this fomentation and warming them they should be returned to their place. But if they swell too much with flatulence or some other cause so that you cannot put them back either by pressure of hand or emollients or other means of resolving them, it is necessary to broaden the wound in that area which through your knowledge of anatomy you recognize as more fleshy and to avoid incision in the tendinous part, cutting with a blunt edge in order not to perforate the intestines. The best operators place their fingers or a thin plate of silver or brass down toward the edge of their instrument and thus proceed safely in making the incision until they can put back the intestines. Instruments adapted for performing this operation are in the form of a sickle with a sharp rod at the end with which you can safely cut, introducing the rod into the wound and perforating the abdomen with the sharp edge of the sickle, drawing it toward the wound. Thus when you have widened the wound put back the intestines. When you have done so sew up the wound in the manner described below. If any intestine should be cut it is useful to know which one, slender or thick. If it is a slender one and the wound is of some size, Hippocrates said this is fatal. But if the wound is not very

---

[33] The translation of the works by E. Littré gives no helpful reference under any heading, but see *Coan Prognosis* p. 493: "If the small intestine be severed, it will not reunite. p. 494. If a nerve is severed . . . the parts will not reunite. p. 496. A convulsion following a wound is bad" (translated by John Chadwick and W. N. Mann, *The Medical Works of Hippocrates* [Springfield, Ill., 1950], p. 263). Galen, *Ars Medica* 31 (Vol. I K., 388) says: Nervi vero et tendonis punctura . . . prompte accersit convulsionem. See Aristotle, *De Historia Animalium,* III, 5, 515 b 15–20.

large but small this too is fatal, although sometimes some patients have escaped and are witnesses to the fact, since I was present and carried on the treatment; these patients had to have their wounded intestine sewed up with the best skill possible. Certain ancients have shown how to do this with the heads of large ants,[34] although these creatures are hard to find, as they are at present, when someone by chance is wounded. It is hard to imagine how they might be applied in such a way that the parts of the intestine could be joined through biting if the rest of the ant's body has been cut away. Therefore I use another method, the invention of modern operators, the best of all methods we have in our time. I take a kid's membrane or sheep's skin made very thin for writers. I cut it into threads and use a needle with a long eye. After replacement of the intestines I moisten the thread made from the membrane and sew up the intestine gently in the way in which furriers sew a pelt. This is a wonderful secret because that membrane becomes sticky and glues the intestine and unites its parts without laceration. In the same way you can sew up the larger intestines if you find they are damaged. But suppose the omentum also protrudes outside the wound, either with or without the intestines. You must be as quick as possible to thrust it back if no alteration has appeared in it. But if some blackness or lividity has developed in it then bind the altered portion with silk or linen thread and cut it off below the ligature with a razor, as Hippocrates taught, so that the thread may remain attached to the sound part. Then cauterize the incision with a hot iron and push back the rest of the intestine so that the thread remains outside the wound in order that when the scab falls off you can extract it from the wound with the thread. You should do the same thing if any part of the mesentery has protruded beyond the wound and has altered in color, as I have often done. For I have cut off and bound the altered part, cauterized it with fire and then sewed up the wound not in the way other wounds are sewed up but in the way an incised peritoneum is sewed up. Since it is a spermatic member it is not healed like flesh, as Galen said in his *Method of Healing* VI,[35] for the peritoneum coalesces with the abdomen or mirach and must come together with it. Therefore he describes different methods of sewing it up; thus I do not intend to list them or others described by other most skilled and learned physicians such as Avicenna, Paulus [of Aegina], Albucasis and the rest. I shall try to show briefly what must be done. First, you must have an assistant who will firmly and carefully press the lips of the wound together with his hands and gather together all the parts of the abdomen and the peritoneum which have been cut. The physician should have a middle-sized needle and begin to sew the wound, perforating first the mirach or abdomen together with the peritoneum beginning from the cutaneous part outside. But on the other lip of the wound the peritoneum should be pressed down to penetrate into the mirach alone up to the other side, perforating the skin, and should be firmly brought together so that the lips of the wound coalesce perfectly. Perforated in this manner the three parts of the wound, that is, in one part the mirach and siphac or abdomen and peritoneum and on the other side or on the other lip of the wound the mirach alone since the siphac has been released, and the method of releasing it has been described. Because it is tendinous or sinewy it cannot be united with something else sinewy but must be joined to something soft. Let it proceed into the side where the peritoneum has been released and let it receive the abdomen with the peritoneum and penetrate with the needle. On the other lip of the wound let the peritoneum be released, penetrating with the needle in the abdomen alone. Then pinch it so that the lips coalesce as you did in the other perforation and keep on sewing until the entire wound has been sewn up, that is, by releasing the peritoneum on one side and by perforating the abdomen with the peritoneum on the other side, leaving a space of one finger between each needle-thrust, as Albucasis advised in the second book, eighty-seventh chapter.[36] This method of sewing has been tried by other men, better and more expert than I, but I do not see how a wound can be well healed if the peritoneum at any one point touches another part of itself. It remains to take up the method of extracting water from dropsy paitents, an operation which I have been often called upon to perform. I can assert that this can be done not without danger for different reasons. Among those, the first concerns the situation when the water is retained between the zirbus or omentum and the peritoneum or abdomen. In such an incision whether one wishes or not one must cut the cords of the muscles not along the course of their fibers, as they best know who recognize this part of the body cavity by sense-perception and not from books of anatomy alone in which Galen said[37] (in the passage cited above) one

---

[34] See Mundinus, Singer's translation, *op. cit.*, p. 64.

[35] Galen, *Methodus Medendi*, VI (Vol. X K., 416): Quoniam coire coalescereque cum peritonaeo oportet abdomen. For the incision of the intestines see Hippocrates, *Aphorisms* (ed. E. Littré, Vol. 5, Paris, 1844), 569, no. 24: Si quelque partie des intestins grêles est coupée, elle ne se réunit pas.

[36] Albucasis II, 87: see Albucasis, *Chirurgia*, in Pietro d'Argelata, *Chirurgia*, etc. (Venice, Giunta, 1531), fol. 145 r, v, who describes this method of suturing the wound as well as the more amazing use of ants for bringing the edges together: et pone inter omnem suturam quantitatem grossitudinis digiti minoris; hanc autem suturam sciunt homines.

[37] See note 24 for the reference to Galen, *De Usu Partium*, II, 3 (Vol. III K., 98).

should place very little faith. From the puncture of the cords or their partial incision a spasm is often accustomed to result, according to the testimony of Hippocrates and experience. The second reason, since in the abstraction of water the patient's strength is greatly weakened, is the debilitation of the liver which prepares the blood insufficiently; few cool spirits are generated. The third danger is perforation of the intestines or any other member which lies in this body cavity, such as the vein and artery which pass to the umbilicus and others. But the method of incision for dropsy, although accomplished not without danger, is as follows: let the patient stand straight on his feet with a prop behind his back and let a servant press the body cavity with his hands and squeeze the water off into the lower part of the body. Then let the operator with a sharp knife cut the skin for three fingers under the umbilicus and from the side for four fingers, drawing it lower with the hands where the cut skin should be removed from the underlying parts to the space of one finger-knuckle so that you can see the substance of the underlying muscles and make an incision along the course of their fibers up to the place where the water lies. Make this incision carefully in the lower place where the skin has been stripped so as not to cut the intestines. When the foramen has thus been made up to the location of the water (which can be recognized by its outward flow) place a silver or brass or lead cannula in the foramen and release the water according to the tolerance of the patient's courage or, better still, evacuate the water a little at a time and many times, as carefully as possible. When all the water has been removed allow the skin to ascend since it will close the wound in its ascent and when you wish extract the water on the next day. Let the skin extend below by drawing it down, put in the cannula and again extract the water as much as you like according to the tolerance of the patient's courage; you can thus continue the process of extraction until all the water has been removed. Note, however, that this is not true care of dropsy since it is not done contrary to the cause. The method by contraries is to reduce the liver or any other member, if it is the cause of the illness, to its proper temperature and thus, however it may be evacuated, to alleviate its pain for some time; it is not healed however, by this aid alone. You will also note that Galen advises the physician not to operate unless his courage is strong.[38]

[38] Galen, *In Hippocrat. De Acutorum Morburum Victu*, IV (Vol. XV K., 891–892) has a brief discussion of dropsy, and in *Hippocratis Epidem. VI et Galeni In Illum Commentarius*, I (Vol. XVII K., 144–152) he gives some excellent advice on the physician's manner of approach to his patient, his dress, speech, and grooming. He advises the physician to be neither humble nor arrogant.

Chapter VII

On the Anatomy of the Intestines, Beginning with the Rectum (Straight) Intestine

I decided to interpose at this point in my writings some advice, if not necessary at least useful, in regard to certain manual operations because in our time very few are found who operate with knowledge of what they are doing. These remarks are sufficient, and I proceed to my original plan of viewing the other organs. When you have lifted the omentum therefore by stripping it from the stomach, gently in order not to disturb the arrangement and location of the other members, you may be able to see its origin and its doubling from the stomach, which takes the form of a sack. If it is not cut off but turned back to the upper part you should inspect it to see the veins coming to it; this will be useful. When the omentum is lifted or doubled back you will see the intestines which fill the entire venter or body cavity from the stomach downward throughout the entire cavity to the anus. These are round and hollow with many involutions in every direction so that they may retain the food and the excrement for their required time and so that the chyle may be better attracted from the liver. Their substance is white and appropriate to themselves and unique, as Galen bore witness in his book *Method of Healing* X (*de Ingenio Sanitatis*).[39] You may call it tendinous or sinewy; it makes no difference. It is composed of a double tunic with all kinds of filaments or fibers, that is long, transverse, and broad, not all of them in the same measure in all intestines since in some the lateral and transversal filaments are more apparent, as in the colon. In some intestines longer and broader fibers appear as in the rectum intestine. But there are many broad fibers in all the intestines; this is to increase retention and expulsion by pressure through the assistance of the fibers in the lower

[39] *De Ingenio Sanitatis* X: This treatise is better known as the *Methodus Medendi* or *Megatechne* (*Therapeutike Methodos*) in the titles of certain manuscripts of translations of Galen listed by H. Diels, "Die Handschriften der antiken Ärtze. I. Hippokrates und Galenos," *Abhandlungen der königl. Preuss. Akademie der Wissenschaften* (Berlin, 1905), p. 92; see also Ignatz Schwarz, *Die medizinischen Handschriften der Kgl. Universitätsbibliothek in Würzburg* (Wüburg, 1907), p. 26. The medieval titles of some of Galen's writings differ from their later titles and present a certain confusion. *Methodus Medendi* is analyzed from the surgeon's point of view in E. Gurlt, *Geschichte der Chirurgie*, etc. **1** (Berlin, 1898): pp. 451–458. I can find nothing in it which resembles Massa's very specific description of the substance of the intestines nor in the text of *De Ingenio Sanitatis* in its earliest printed form, the Opera Omnia of Galen edited by Diomedes Bonardus (Venice, 1490). See, however, *De Usu Partium*, IV, 17 (Vol. III K., 324): Iam vero intestinorum substantia haud multum a ventriculo est diversa . . . ; *De Anatomicis Administrationibus*, VI, 9 (Vol. II K., 572): Iam vero intestinorum natura, quatenus intestina sunt, omnium eadem habetur. . . .

intestine. For Galen said in *On the Use of the Parts VI*[40] that retention is carried on by means of all the fibers; whether this is due to the attraction made by the long filaments I shall not say. You can understand this for yourself, you can also inspect the tunics which compose the intestines since you will see clearly that there are two in those intestines which are called the thicker ones. But in the slender (gracile) intestines the tunics are seen with difficulty because they are thin. You will also see many veins which they call meseraics [mesenteric] in all the intestines. In some intestines there are more of these, in others fewer according to the difference of the intestines. These veins draw the chylous material prepared by the stomach for nourishment and conduct it to the liver. Inspect the diversity and location of all the intestines since they are six in number. You will see their shape when you lift up part of each; they are chiefly round. Begin then in the lower part, that is, from the rectum intestine, which you must bind with a thread at its extremity toward the colon in two places so that the stinking excrement contained in it may not issue forth. Then cut between the ligatures so that you can strip it. But before you do so, separate it from the mesentery. Carefully note the veins called meseraics as they proceed through the substance of the intestines to the mesentery. These veins terminate at the portal vein which passes toward the hollow of the liver. In the mesentery itself you will see eight more prominent branches. If you gently strip the surface of the mesentery itself, note that in some people who have large veins or veins full of much blood you will see more than eight branches. You can inspect all these veins, if you like, before the intestines are moved aside, not, however, without difficulty on account of their weight. But in order to see the progress of these veins more clearly place a rod in each of the branches and thus you will be certain about all of them. When you have seen them begin to lift the intestines gently from the mesentery. First in order will be the rectum intestine which begins from the anus and rises straight for the space of a palm and sometimes more according to its larger size in relation to the bulk of the different intestines. You must then release this intestine here in order to see the anatomy of the muscles of the rectum intestine. The substance of the rectum is thick, with not a little fat clinging to it. It is composed of a double tunic and has a few small meseraic veins. This substance of the intestine is almost fleshy but since, as I said, it has a substance appropriate to it this is not fleshy. Its shape is round and it has broad, long, and slanting filaments; these are few. In the outer tunic there are long filaments; the rest you will see in the inner tunic. This intestine has two muscles at the end toward the anus and hemorrhoidal veins, of which I shall speak in their place. You will note, however, that although the intestines differ in name and a certain characteristic they are nevertheless continuous from the gullet or meri as far as the anus.

[40] Galen, *De Usu Partium*, VI, 8 (Vol. III K., 438–439), on the multiple fibers of the heart: . . . ob eamque causam multijugae in quoque horum [ventriculus, matrix, vesica] insunt fibrae, ut ostendebamus: rectae quidem, ut quum contrahunt sese, trahatur aliquid; transversae autem, ut excernatur; at quum omnes simul iis, quae ipsis insunt, undique adstringuntur, detineatur.

Chapter VIII

On the Anatomy of the Colon Intestine

When you have seen these members begin to move the intestines aside where you tied and separated them by incision from the rectum because the latter tends somewhat to the left; at its end begins the colon. This intestine is not of so thick a substance as that of the rectum but is thick nevertheless. Examine its course as it passes through the left region toward the ilia and braces itself near the kidney. In its course it is bent back in various ways. Then it passes near the spleen tending toward the right, soon proceeds under the liver and rides above the region of the stomach. In this region is the black bile cyst [gall bladder] which is located in a third, or a lobe, of the liver. It touches the colon itself. In this region you will see the intestine change color to darkish with a lemon tinge. After it rides over the stomach it descends toward the right hip and is joined to the sack intestine or caecum or monoculus; it makes no difference what you call it. This intestine has stamens or visible broad fibers; the rest of the fibers, that is, the slanting and the long, are not so visible. The substance of this intestine in those who are emaciated is solid and, as I said about the rectum, it is almost fleshy with a little fat. It has larger and many more meseraic veins. It also has cells in which the feces take shape. It has various involutions by means of which it touches different members and occupies different places. Therefore its illnesses are diverse and difficult for often because of contact with the stomach the bile overheats the cyst or is agitated in some other manner and causes nausea and upset of the stomach, vomiting and lipotomy and syncope (fainting) if the harm reaches the mouth of the ventriculus or stomach. Pain is also created by the retention of the excrement because of the multiple involutions of the intestine; this illness is called colic. From its passage close to the left kidney which is nearby physicians often cannot distinguish this illness from kidney pain. But since I decided not to treat illnesses nor their causes it will be sufficient to know in addition to those facts that the size of this intestine is not uniform in all people, for sometimes it is found to measure ten feet and often less than that.

Chapter IX

On the Anatomy of the Sack or Monoculus

When you have separated the colon from the mesentery itself and have examined the meseraic veins which come to the colon and its termination under the right kidney near the hip, examine also the sack intestine or the monoculus [cecum]. This intestine forms one path from two heads of the intestines; through this path from one region of the intestine the feces descend from the ileum intestine and pass out from the other region and enter the colon. In this way there appears to be only one orifice. You can recognize that orifice by the greater thickness of the intestine in that place [ileocecal valve]. If you examine it closely you will see one orifice composed of two extremities and two orifices. The location of this intestine is in the right side near the hip. Its substance is thick and composed of two tunics like that of the colon. There are all kinds of fibers in this intestine, but the broad ones are more visible than the others. There are not many meseraic veins in this intestine. Its shape is not extended but hangs like a pouch in the heads of two intestines, that is, the colon and ileon (ileum). In it are retained the feces so that by its force the complete nutriment may be taken up from the food by the meseraic veins. This intestine fills the right cavity between the kidney and the hip but its size is not the same in all people; it is small in some but larger in others. Sometimes it is of middling size and often quite small. Frequently also its division is almost undiscoverable since it forms part, as it were, of the colon and forms a larger cell of the latter. I have dissected many people in whom I did not find this large intestine; in fact, it appears to be of the same substance with the colon. But in these [cadavers] I found a certain substance hanging down about one finger length and about the same thickness as the pen with which I write. Since this addition [the appendix] is not found in those who have a visible and ample sack intestine, I have often thought it was the sack intestine which had been rendered inoperative at the time of infancy,[41] for instance, through a daily flux which did not allow the feces to delay in the aforementioned intestine but passing through continually rendered the intestine empty of food. Thus it remained without function and when its single orifice was constricted could not receive the feces thereafter and lay abandoned. This is my opinion about the matter; if anyone finds a better explanation I urge him to produce it in writing not only about this member but about any other so that his thoughts may without rivalry be profitable to posterity.

[41] See note 31 on the umbilical cord.

Chapter X

On the Anatomy of the Ileon [Ileum]

You have seen the thicker intestines of those mentioned, according to their three differentiae, which you have carefully stripped from the mesentery. Begin also to strip the gracile (or thin) [small] intestines. You will see these also according to three differentiae. The first of these intestines is the ileon. It begins from the foramen or mouth of the monoculus and passes with many revolutions toward the left region of the venter (or body cavity) which it fills in its passage and the places called the ilia from the bone which is in the hips. For it is the longest of the intestines as we see. It is twisted about everywhere from the hips through the body cavity around the right and especially the left region. You will see it proceed especially to the left toward the stomach and there it is joined to the jejunum intestine. This ileon is composed of two very thin tunics and is not divided into cells like the colon but is round. In it appear very many meseraic veins and it has stamens or fibers of all kinds, although the broad ones are more visible. Since it is thin and altered by its extension or whatever other cause it is subject to terrible accidents and pain which destroy most men to whom they occur. Its size is not the same in every person. I have found it twenty feet long but very often less.

Chapter XI

On the Anatomy of the Jejunum

Toward the stomach in that same left-hand region begins the jejunum intestine which in substance and stamens and color is similar to the ileum. But if you inspect it carefully you will find its color tends toward a certain lemon shade from the propinquity of the passage of the yellow bile through the bile duct of the gall bladder which leads the yellow bile to the duodenum intestine. In this intestine there are many and larger meseraic veins than in any other of the intestines in respect to its size. This is not as great as the size of the ileum; in fact it is no longer than five feet. It begins in that region where the ileum terminates and passes to the right toward the liver. This region seems more attached to the mesentery; therefore you must proceed carefully in stripping it in order not to disturb the arrangement of the members or any of the vessels therein. In that region where the ileum terminates the jejunum begins to appear empty. I say this because near the duodenum the jejunum actually is empty, that is, of feces. But toward the ileum it does contain something. In this region it is twisted about but in its ascent it is straight as it proceeds toward the duodenum; thus according to their direction you will see both the part which is twisted and the part which is straight. Galen gave the reasons why it is empty in *On the Use of the Parts* IV,[42] that is, because of the great number of veins which carry the chyle and the biting quality of the bile as it passes and stimulates expulsion. This intestine is the fifth in order. When you have stripped it you must bind the beginning of the duodenum intestine with a thread in two places under the bile duct which descends to that intestine from the liver to keep the rotten excrement from coming forth. Then cut it between the ligatures, releasing the duodenum intestine in its place so that you can, if you wish, be able to see the anatomy and the course of the eight, and sometimes more, branches of veins in the mesentery. These veins pass to many members and to the stomach. You will also see the duct from the bile cyst which descends to it; but if you wish to know the size of all the intestines you can measure it by the intestine which has been cut off. You will then discover that their size is not the same in all because some are longer, others are shorter. You can also see their diversity of substance when you have washed them and cut them lengthwise, although you will see it far better after boiling them. Then you can more easily separate their tunics and stamens or fibers, all very clear, and you will see their course, as I have often done.

Chapter XII

On the Anatomy of the Duodenum

This, the last of the intestines, is called the duodenum because its size is twelve fingers. I mean it is last if you begin to number from the rectum intestine because if anyone began to number from this duodenum it would be the first. Since you can take your choice about this, it makes no difference whether you begin to number from the latter or the former intestine. Look at the beginning of this intestine at the end of the jejunum in the right hand region. It tends straight to the portanarius or pylorus of the stomach; take your choice of terms since (as Galen said)[43] there is greater advantage in the recognition of the thing itself than in a discussion of its different names. Thus he who has gathered the errors of physicians from other sources into one collection inveighs quite fruitlessly against those who wished to call the duodenum the pylorus (if he has gathered true facts) since they clearly indicate to all by its size and substance that which I call duodenum and you call pylorus. Others call it duodenum because it terminates at the portanarius, which the Greeks and more recent physicians name pylorus, when they indicate its size and substance and the bile duct which passes to it. Nor do these omit the orifice of the stomach through which whatever is digested or not digested in the stomach passes to the intestines. Hence, while he has himself missed the mark in the process of recognizing as many as possible of the errors of the ancients, when he accuses them of a mistake in this matter (since there is none which they have made) he shows that it is rather he himself who has gone wrong. You will see this foramen in the bottom of the stomach which they call portanarius or pylorus. Through it (as I said) the material in the stomach passes to the intestines both for nourishment and for purgation, for it is not unsuitable for one member to have many uses. To this intestine you will see that the [common bile] duct of the bile cyst passes and is fixed in it between the tunics of the intestine. Therefore proceed cautiously in order not to break the bile duct. This intestine has all kinds of stamens but the broad are more visible. It has the fewest meseraic veins of any of the intestines already discussed. You have seen all the intestines and even if they have been allotted so many names they have nonetheless one feature in common, the fact that they are continuous, proceeding from the rectum to the stomach and attached to the back by means of the mesentery. Together with the mesentery, the stomach, and the bile duct you must also release the duodenum intestine so that you can better see the other members to be described.

Chapter XIII

On the Anatomy of the Mesentery and the Veins Contained in It

From the back proceeds the part called the mesentery. You must strip it from the peritoneum which covers the back. Its substance is diverse, that is, sinewy and glandular with a notable quantity of fat. There are also almost innumerable veins in the mesentery, all of which proceed from the portal vein that descends from the hollow of the liver and branches throughout the mesentery in many directions as well as to the intestines and other members, such as the stomach, kidney, and spleen. In the mesentery are also arteries which pass to the intestines. Its substance seems to vary according to location since in the upper region where it is attached to the duodenum and jejunum its membranes (or tunics) are not doubled nor are they so thick; the substance is more glandular than that of the remaining lower part. This lower part, to which the

---

[42] Galen, *De Usu Partium*, IV, 3 (Vol. III K., 348): Primum omnium intestinorum jejunum alimentum in ventriculo in chylum mutatum atque concoctum accipit. . . .

[43] See note 12.

remaining intestines are attached, has a thicker double panniculus, in which less glandular flesh is found. The substance of the mesentery is curled (or wrinkled) through the length and winding of the intestines and proceeds along their course and extent. But before you cut off the mesentery you can see the portal vein which proceeds to the mesentery from the liver (if you did not see its anatomy before that of the intestines). To investigate it you must strip the surface of the mesentery itself near the back where the jejunum intestine is located. You will find the vein passes from the hollow of the liver through the mesentery itself. This vein is composed of a single tunic and branches to other directions. See also the eight branches often mentioned (as I said, there are even more than eight of them) which proceed from the [portal] vein. Mundinus[44] said that two of these branches are not apparent because of their thinness, but this must be understood as applying only to some people whose veins are large. I have very often seen all these veins and even more than eight. It will not be unsuitable to trace their progress here, when you have stripped the panniculus of the mesentery on its surface from one side or the other if you wish, not with a cutting instrument but with one that is blunt-edged. You will see the first vein, one of the small ones, which passes to the duodenum intestine [pancreatico-duodenal]. The second passes across to the stomach near the foramen called the portanarius or pylorus. The third, which is more visible and is found in almost all people, except when bloodless, passes to the right hand of the ventriculus or the external part of the stomach [rt. gastro-epiploic]. The fourth passes across to the spleen [splenic] and is branched in its progress to the stomach [short gastrics] in the lower region. Of these branches, some pass again to the spleen and turn back in the upper region of the stomach and to the mouth of the stomach and are thus branched into many parts; examine these as well as one part of the vein which later passes to the omentum. The fifth [inferior mesenteric], descending through the mesentery, passes to the rectum intestine. The sixth branch [superior mesenteric] is itself branched through the substance of the mesentery into many thin parts which pass here and there, that is, to the stomach and to the omentum. The seventh branch passes to the colon. The eighth passes to the jejunum and is branched also to the ileum. But since as I said sometimes even more branches are clearly visible it will be your duty to investigate carefully the progress of all the branches. The purpose or use of these veins is to draw nourishment from the intestines to the liver and to nourish it and to carry nourishment to those members at which they terminate, that is, to the stomach, spleen, etc. Those veins which pass to the intestines do the same thing, nor will you err if you say so, since if there is a principal single virtue for one member it is not unfitting that other virtues which are not principal should follow from a member of the second order. The question whether *melancholia mirachialis*[45] is found in these veins or elsewhere has no bearing on this matter. But since there is more sensation here than in another place I say it [*melancholia mirachialis*] is found here for the most part and hence it is that pain in back and loins follows upon this illness. When you have seen all the branches of the portal vein and their progress you can cut off the mesentery in the place where you bound up the duodenum intestine and, descending, strip it from the peritoneum which mediates between the back and the mesentery. You must not, however, move the duodenum intestine but save it for the anatomy of the stomach and of the bile cyst.

Chapter XIIII

On the Anatomy of the Ventriculus or Stomach and of the Bile Cyst

So that you may more readily examine the size of the stomach and its location inflate it with a bellows or some other instrument or means through the duodenum intestine which you bound off and you will see the stomach or ventriculus located in the middle under the diaphragm, tending, however, more toward the left side on account of the liver and more elevated in that place. On the right side it tends to go lower. In order to see its location more readily cut some of the false ribs, as modern anatomists, who are few in number, are accustomed to do in order to point out this member. They do not, however, cut the ribs off entirely but only as much as they need in order to bend them toward the back. This done, lift up the cadaver to a sitting position; thus you will more easily see the location of the stomach and its substance, white and appropriate to the intestines (as I said). I cannot really call it sinewy, if that is called sinewy (*nerveum*) which is composed of nerves; even if this were true nature would not have sent nerves to it for sensation since it does not abound in superfluities. What I say about the stomach I wish you to understand is true also about all members to which nature sends nerves. Therefore its substance is white, or tending toward white, round and oblong, somewhat curved. You will see in it many ramifications of

[44] Berengario da Carpi, *Commentaria*, etc., fol. CXL: Egredientes uero extra hepar sunt octo, duae paruae, in quibus non labores ad discernendum, quia sufficit tibi si sequentes discernes. Singer, *op. cit.* p. 71, translates: "two of which are small and need not be considered."

[45] *Melancholia mirachialis* is defined by Singer, *op. cit.*, p. 102, note 29: "By *melancholia* is here meant that condition due to an excess of the primary humour *black bile*, the supposed secretion of the spleen, in contrast to the yellow bile that is the secretion of the liver."

veins which bring nourishment to the stomach. On its right side is located the liver, descending from above; it passes toward the left as far as the false ribs. On the left side of the stomach itself under the false ribs the spleen is propped against the stomach, but around the middle of the stomach, going lower, is the origin of the omentum or zirbus, whose anatomy I explained above. In the bottom of the stomach on the right side is the orifice called portanarius or pylorus at which the duodenum intestine terminates and through which the stomach was inflated. In the upper region toward the left is the mouth of the stomach to which the esophagus is continuous. The latter perforates the diaphragm and proceeds upward to the mouth proper; it makes no difference what you call it, meri or gullet. In the posterior region the esophagus is attached to the thirteenth vertebra of the chest. In this region there is a certain fleshy substance upon which the mouth of the stomach rests lest it touch the hard surface of the vertebrae without a buffer between. This flesh the Greeks call pancreon [pancreas]; the Latins translate the word "all flesh." The substance of the stomach is made of two tunics like the substance of the intestines, but that of the stomach is thick and has somewhat the color of flesh on the outside. Thus they call it the outer fleshy part, but the inner part, since it is whiter on the outside, they call sinewy; it is thinner, harder, and somewhat corrugated and rough, but since it is covered on the outside with a certain thin panniculus proceeding out of the peritoneum from the back you will notice it. It also has a panniculus which covers the inner tunic; therefore there are some who say the stomach is composed of four tunics. It has two kinds of stamens or filaments in the inner tunic, that is, in the long inner part. In the other exterior part of the same tunic the fibers are oblique. In the external tunic which they call fleshy the filaments are broad. Thus attraction is accomplished by the former fibers and retention by the oblique fibers. Expulsion is brought about by the broad fibers with the assistance of the others mentioned. Also look at the veins, if you wish, which pass from the branches of the portal vein. You will also see the branches of the aorta artery [left gastric] that comes to it [the stomach] from the great artery itself after it has passed across the diaphragm. But you will see the reversive nerves [nervi vagi] coming to the mouth [of the stomach] when you note the progress of the meri or gullet to the stomach. The stomach's use is to serve the liver for nourishment of the entire body. Nor is the stomach first nourished from itself, as some have believed, although it receives some alleviation from the food it receives, because if it were nourished entirely by the food nature would not have given it so many veins as you have seen. And when the duodenum intestine to which the [common] bile duct passes has been laid aside this duct is located in the third lobe of the liver [cystic duct]; therefore inspect, since you can best do so here, the other duct [hepatic] which is continuous with the bile duct for it passes across in the substance of the liver; you will see it in the elevation or stripping of the bile cyst [gall bladder] from the liver. You will note here also that sometimes there appears a third duct which passes to the portanarius or pylorus. Those people are rare, however, in whom this third duct is found; it passes to the stomach. If it is large these people are unhappy since much yellow bile is brought to the stomach itself and stimulates the man immediately to vomit. You can carefully strip and move the bile cyst from the substance of the liver and see its tunic, sinewy, thin and quite hard. It is single, covered by the panniculus which covers the liver. Its shape resembles that of a small pouch; it has two visible ducts, one in the liver [hepatic duct] through which the yellow bile passes to the cyst. From this branches a second which passes to the duodenum [common bile duct]. In order that you may show this duct in the stripping of the bile cyst from the liver note how from this one which proceeds from the liver you will see the second duct which passes for a distance of four fingers to the duodenum. But sometimes this duct is found to be longer. If you find the third duct, as I have sometimes found it, you will note its progress to the stomach. The bile cyst has long and oblique fibers, or filaments, in its inner part and broad fibers in its exterior part. To the bile duct come an artery [cystic], a vein, and a nerve, but because of their thinness they cannot be observed in all people. I have sometimes found them but not all in the same person. Once in flesh which had been wasted away in the hospital of Sts. Peter and Paul I saw a nerve passing to the bile cyst. I also saw in the convent of Sts. John and Paul a clear artery and also a vein, but I could not find them in many bodies. Therefore, let those who desire to know the perfect anatomy of man realize that it can by no means be attained except through many, and as many as possible, dissections of different bodies varying in age, sex, complexion, amount of flesh, and in emaciation, since in one body certain members are better examined than in another.

Chapter XV

On the Anatomy of the Spleen

When the stomach was inflated, the false ribs bent a little toward the back on both right and left, and the cadaver arranged in a sitting position, you saw the location of the liver and of the spleen. Therefore, in order to inspect all the parts of these members, deflate the stomach somewhat but not entirely, that is, by relaxing the ligature of the duodenum. First look at the lien or spleen, which is placed on the left-hand side of the stomach toward the back under the hypochondrium. Its substance is soft, thin, and spongy, blacker than the liver; there are no filaments in it. Its shape is somewhat convex toward the stomach against which it is propped, but toward the ribs it is rounded and embraced by a thin panniculus originating from the peritoneum. It is attached to the back with a visible ligament and is braced there in the upper region against, or touching, the pancreus (or pancreas); you will see its shape is somewhat quadrate. Its size in healthy people is not large. I have seen it no more than a half-foot long in many [bodies], but in those suffering from illness of the lien or spleen I have seen it very large, descending lower and [spreading] on this side and on that, swollen and hard. Do not be surprised therefore if you find it sometimes oblong in some bodies or of another shape. You will sometimes see some fat clinging around it. The spleen is also attached to some of the false ribs by slender ligaments. It communicates with the liver through the portal vein which I said passes to the spleen. It communicates with the heart through a branch of the descending artery for after that descending artery passes into the lower body cavity it branches out to various members and thus sends a branch to the spleen [splenic] which you must look at. The reversive nerves [nervi vagi] also come to the spleen; they give sensation to it, as will be said when I write about the inspection of the reversive nerves.

Chapter XVI

On the Anatomy of the Liver

In the right side is located the liver, lying above the stomach and under the diaphragm. But since at the time of death the spiritual members labor greatly, they are contracted for the protection of the heart and the union of the spirits in what you might call spasms. Therefore, in such a great contraction they draw to themselves the members of the lower cavity, such as the stomach, spleen, liver and the rest. Hence, in order to see them in the best possible fashion these organs should be brought back to their natural location. The body must be elevated a little, as I said above, and bent into a sitting position. When the body is so arranged inspect the substance of the liver. It is reddish, since it is coagulated blood, and is covered with a very thin membrane arising from the peritoneum. This substance is similar to flesh, but there are no fibers in it. The liver is concave in the lower region where it touches the stomach and humped (or gibbous) in the upper region where it is attached to the diaphragm by a very strong ligament [coronary], with the peritoneum as a mediary. The liver is also attached to the back where it touches the pancreon (pancreas). Lying thus above the stomach on the right side it proceeds from the upper region, verging toward the left, and descending under the false ribs it embraces the stomach. In this lower region the liver is divided into five parts which they call lobes or fibers. But these divisions are not found in all bodies. I have often seen one division only in the middle where the umbilical vein is fixed. Sometimes I have seen two and also all of the divisions. The size of the liver is wellknown. Inside it, through all of its parts pass veins here and there so that their composition is similar to a net. Of these veins there are two terminations, one in the hollow of the liver where the portal vein proceeds. This vein passes through the mesentery to the intestines and other members of which I spoke above. This vein is divided in the substance of the liver into five veins through the five lobes. Passing through these the veins are later branched into many branches. The second termination is in the humped portion of the liver where the chilis vein [vena cava] passes forth; I shall soon describe its anatomy. You have also seen the bile cyst hanging in the cavity of the liver itself, receiving the yellow bile from it through a duct which is visible and located in the middle or in the third lobe of the liver, as you saw in its anatomy. The reversive nerves also which give sensation with their panniculus pass across to the liver itself. Similarly the branches of the [hepatic] artery proceed to it, giving life; therein two of the branches are clearly visible.

Chapter XVII

On the Anatomy of the Chilis Vein or Hollow Descending Vein and of the Aorta Artery Which Descends at the Same Time from the Diaphragm and On the Emulgent Veins and Arteries

You must move aside the substance of the liver and of the spleen, but in order to see the anatomy of the chilis vein [vena cava] you must release or let down a bit of the liver in its humped part, from which the chilis vein comes forth. Put the stomach back into its place for the anatomy of the reversive nerves and other nerves. When the liver and spleen are cut off look at the ample vein which comes out of the hump of the liver. The Greeks call it chilis, the Latins hollow. It ascends, perforating the diaphragm as far as the right ventricle of the heart and is fixed there in the place they call the auricle of the heart as if it arose from the heart itself. It is composed of one thin tunic alone, wherein are all kinds of fibers. Near the heart this vein is thicker and harder since the blood in it is warmer and thinner and therefore requires a harder and thicker wall so that it may be better contained. I said above that this vein branches out in the substance of the liver to many parts of its humped portion; but you will notice that it is branched out toward the hollow of the liver into the smallest parts. These very small branches are united in fact with the branches of the portal vein. You will see this better if you macerate the liver for several days. Place it in a cooking pot with water and boil the flesh thoroughly;

thus you can more easily separate the flesh from the veins and see the substance of the veins which are entwined among each other like a net. But since there are many differences of opinion among physicians and philosophers about the origin of the veins and likewise of the arteries and nerves, as to whether the veins originate from the liver, the arteries from the heart, and the nerves from the brain, you must know to begin with that the glorious Lord has given to the sperm its due virtue, as all philosophers and physicians affirm. By means of this virtue are prepared for all principal parts at the same time the elements necessary for the conservation of individual complete life, as Aristotle[46] declared and his faithful interpreter, Averroes, saying that in its action it gives form and also all elements necessary for form. Nor would a subject be made of such and such form if the necessary elements were lacking. Therefore, it is better to say that from the virtue of the sperm are first generated those three principal members at the same time with their necessary instruments, which are arteries, veins, nerves and other instruments if there are any, nor does one depend from the other, although the virtue of one member, that is, of the liver is to nourish and cause growth, for nourishment and growth are received in all parts, and even in the bones. The same reason or cause does not operate in these parts since there is one cause for origin and another for nourishment and growth. Therefore you may say that all the members which have been mentioned have been created by the aforesaid virtue and the sperm for the introduction of form and the conservation of the individual. One can be persuaded of this fact because the chilis vein [vena cava] is dissimilar from the liver and from the diaphragm below. It does not seem to take the blood first from the heart but from the liver which is closer by, which shows that the chilis vein does not arise from the heart. There is another argument, that is, that the substance of both veins and arteries is quite different from the substance of the heart and of the liver, as I said; although the substance of the nerves resembles that of the brain in color it is not the same substance, for the substance of nerves is hard and that of the brain is soft. I am not disturbed in this view by that softness which nerves have when near the brain since the nerves which arise from the spine of the back are also hard near the spinal medulla. Thus the substance appears to be more similar to the covering of the medulla than to a soft substance; but not on account of this would I say that the nerves are of this or that substance but of their own, and different from the substance of the other members. But what is to be said of the reversive nerves that descend to the mouth of the stomach? For they are softer in that place. Thus on this basis someone could say they originate in that place, if they arise from the place where they are softer. But what would these people say about the substance of the bones which is not from any of these sources of origin? Therefore, let it be said that nature made these spermatic members from the first substance of the sperm, or sperms, and of the menstrual blood together with the principal members from the diverse substance of the sperms themselves or from different parts nor did one arise from the other but diverse virtues were manifested in these parts and through their instruments were transmitted to other parts of the body. Hence, nature, not deficient in necessary elements, generated with the principal members also those instruments through which her powers might operate even before form was introduced, whether vegetative or sensitive. For if the heart was first needed to give heat and the liver to nourish and the brain to give animation [to the human body], as I shall say, there are first the necessary instruments. Nor is it any objection to this view to say that Aristotle[47] declared the heart was the first member to live and the last to die, since this must be understood of the operations of this member and not of time or of that which proceeds from the member. For nature always operates according to the need of the subject, as is apparent from bones, which are generated after the generation of the principal members. All these matters [or substances] are different, as I said about the stomach and the intestines, which are called sinewy, whose substance is by itself (*per se*) from the sperm. But let the philosophers and physicians beware lest they con-

[46] Aristotle, *De Anima*, II, 3. 414 a 30 lists the psychic powers as the nutritive, appetitive, sensory, locomotive, and the power of thought. In *De Generatione Animalium*, II, 1, 735 a 5–10 he argues that semen both has soul and is soul, potentially. Both of these ideas are combined in Massa's thinking here and in his later discussion of elements, form, and cause. Galen, *On the Natural Faculties*, I, 6, 7 (Vol. II K., 11–18) treats the causes for origin, nourishment, and growth which create the various parts of the body by their special alterative functions. Aristotle, *ibid.*, 2, 736 a speaks of the nature of semen as composed of spirit (pneuma or warm air) and water.

[47] Aristotle, *De Partibus Animalium*, III, 4, 666 a 10: "For the heart is the first of all the parts to be formed . . . " (translated by William Ogle in the Oxford Aristotle, edited by W. D. Ross); *De Generatione Animalium*, II, 6, 741 b 15–20: "And what comes into being first is the first principle; this is the heart in the sanguinea and its analogue in the rest, as has been often said already. This is plain not only to the senses (that it is first to come into being), but also in view of its end; for life fails in the heart last of all, and it happens in all cases that what comes into being last fails first, and the first last, Nature running a double course, so to say, and turning back to the point from whence she started" (translated by Arthur Platt in the Oxford Aristotle, Vol. 5; but see the translator's reservations based on *De Iuuentute*, 468 b 15 and *De Respiratione*, 479 a 5 and the fact that the vertebral column begins to develop earlier, according to microscopic examination of the embryo). Pliny's account follows Aristotle: *Natural History*, XI, 59. 181, p. 545, Loeb Classical Library: hoc [the heart] primum nascentibus formari in utero tradunt, deinde cerebrum, sicut tardissime oculos, sed hos primum emori, cor novissime. See p. 214 for another reference by Massa to the heart as the organ which dies last.

tradict these statements with the products of their imaginations since when they give the matter consideration without jealousy they will realize how prudent nature is; she never generates anything without the parts necessary for its conservation. But, leaving these disputants, let us come to the knowledge of this vein that is gained by sense-perception. This chilis, or hollow, vein when it first issues from the liver sends out two branches or is divided into two veins; these branches come from one trunk that proceeds from the hump of the liver. One of these two veins perforates the diaphragm and ascends to the heart, as I said. I shall discuss its anatomy in its place when I describe the anatomy of the ascending chilis vein [vena cava]. The other branch or other part of the vein, which is commonly called the descending chilis vein, passes below through the back on the right together with the aorta artery descending from the left ventricle of the heart. Here you will note that the vein is above the artery in such a way that the artery adheres more closely to the back. Both vein and artery first send many branches to the panniculus which covers the kidneys and thus both proceed together, that is, the vein above and the artery below, through the back. As they proceed the vein sends out two branches; the artery likewise sends out two branches. The first branch of the vein and also of the artery passes from the right side of the vein to the right kidney and enters its substance; it is called the right emulgent [renal] vein. The artery proceeds in a similar fashion. Later, a little lower down, from the left side of the vein there proceeds another branch; it passes to the left kidney and enters its substance, just as with the right kidney. The artery thus sends out a branch and is called the left emulgent [renal] vein. These emulgent veins are notably ample, but the arteries are thin and you will see that the right emulgent vein is shorter than the left, as you will also see in the arteries. In order to view its progress and entrance more readily place a rod or stylus in any one of them, penetrating through the vein to the substance of the flesh of the kidneys, for the branch of the vein and similarly of the artery, passing to the left kidney, proceeds above the spine of the back. This is because there are similarly a vein and artery in the right side, from which the branches proceed. After the vein and artery have sent out the four mentioned branches which are the two emulgent [renal] veins and the two arteries, passing through the back both vein and artery descend directly, sending more branches to the different members, for example, to the back [lumbar]. These branches also enter somewhat between the vertebrae and through the [intervertebral] foramina from which the nerves proceed to the spinal medulla and also to the rectum intestine. From these branches the [superior] hemorrhoidal veins are composed. Some branches pass to the bladder [vesicle] and the penis [int. pudendal] and, in women, to the matrix [uterine]. But those [vessels] which branch out in the back pass not only to the vertebrae but to the muscles [obturator, iliolumbar, lateral sacral, gluteals] found therein, both for nourishment and the maintenance of life. These branches also pass to the muscles of the abdomen (*ventris*), its membranes [inferior epigastric], and to its fat, making those two veins ascending to the breasts which are branched to other parts as far as the skin. But the large vein and artery after passing across the vertebrae of the kidneys are bifurcated into two notable veins near the os sacrum [common iliacs]. These veins [external iliac] pass through the groin to the hips or coxae, one on the right, the other on the left. Inspect first the branches of the artery [internal iliac] since you will see that they proceed toward the bladder. These branches in their later progress pass as far as the umbilicus, composing with others the umbilical vein, as it was called above in the anatomy of the umbilicus. But the branches of both the vein and the artery which pass through the groin to the coxae, before they descend to the coxae send out many branches to the os sacrum and the neighboring muscles; proceeding thus through ramifications they pass to the pubis and other parts of it, such as the scrotum, and to the vulva in women. Similarly the artery branches to many parts, and since sometimes all of these ramifications are not apparent, either because the veins are bloodless or because they are thin, therefore in order to gain knowledge of them in some fashion, cut both the vein and the large artery; in their substance you will see foramina which pass to those parts. Into these foramina you can insert very thin styli and thus you will be able to trace them through the entire substance of both vein and artery. By this method you will see their progress when the branches cannot be found by stripping; it is helpful to use both in the back as well as in the coxae and in other parts of the body in order to learn the origin and progress of the branches of both veins and arteries. Nevertheless, I shall not overlook one item; many more branches of the chilis vein [vena cava] proceed from this vein to the muscles and other parts of the body than branches of the aorta artery to those members. After you have seen the progress of the vein and artery through the back to the aforesaid parts you should see the remaining parts of the vein and of the artery, which, divided into two parts, pass from the os sacrum to the coxae and to the legs. But since I have mentioned the progress of the branches which pass to the rectum intestine, the penis, and to other parts, in order not to interrupt the order of the dissection remember to look at these veins and arteries in the dissection of those parts, that is, of the bladder, the rectum intestine, the penis, the vulva, the matrix (uterus) and of other members to which these veins and arteries pass. Therefore you must follow their remaining progress to the coxae, as I said, and to the legs. It is as follows. The branches which descend to the os sacrum toward the coxae, after having sent the aforesaid branches to their members, descend through the interior of the coxae themselves or through the pars domestica [inner region], and in their progress send some branches to the muscles. The most evident of these branches is one which passes to the muscle which is above the hip joint and the scia. Hence it is that bloodletting of the sciatic vein is beneficial for sciatic pain. The progress of the vein to this joint was hidden from Avicenna[48] and others, as is apparent from his words in his chapter on bloodletting. After this vein has sent these branches to this joint and also to other muscles a large part of it and of the principal vein and [femoral] artery as well descends toward the posterior joint of the knee which they call the ham. There it is divided or branched into three branches. The larger of these descends through the inner part toward the ankle bone and as far as the big toe. In its descent it sends out many branches to the muscles. This is called the saphena vein [small saphenous]. Another branch slants under the knee and passes to the outer part of the leg [posterior tibial] descending toward the clavicle [ankle] and is apparent there. It proceeds to the foot toward the smallest toe and is there branched again; this is called the sciatica. The third branch proceeds between the two aforementioned branches in the knee joint and to the exterior muscles of the leg. It also sends out some little branches to the inner muscles which pass to the inner part of the ankle. Going deeply therein, they proceed to the sole of the foot. From what has been said you have the progress both of the descending chilis vein [vena cava] as well as of the aorta artery because they are companions of each other. This is not always true, since, although the artery is always found with a vein, the same is not the case with the vein; you will often find a vein without an artery. This is common to all parts of the body. But since it is a matter of desire to know the progress of the veins which are phlebotomized I shall not hesitate to speak of them in the anatomy of the feet; that will be the place to see their progress for the sake of a clearer instruction about them. Remember, as I said, that the substance of the vein is composed of one tunic only. This has all kinds of filaments. But the artery is composed of a double tunic and is harder. Certain people on the basis of the companionship of the artery to the saphena vein attribute the reason for its greater weakness in bloodletting, or in the opening of the saphena than in the opening of other veins, to the breakdown of the spirits sweating back from this artery to the flow of blood from it; but it is better to say that its greater weakness is to be explained through other causes. The first of these is the fact that through bloodletting, or the opening of the saphena or the sciatic vein, whichever it may be, the blood is drawn first from the body of the liver, since this vein proceeds from the liver to the legs. Thus when the liver is emptied the blood enters, in order to prevent a vacuum, from the right ventricle of the heart to the liver in place of the blood which has been drawn out. Therefore the left ventricle imparts something to the right, and an exhaustion of spirits is caused through the fact that spirituous blood has been transmitted from the heart to the liver. If you wish to add an excellent argument based on anatomy this will be the true and sensible cause of the weakening [of the liver's] faculty: because certain branches of the artery proceed to the vein and enter into it, branches well known to good, sharp-eyed anatomists, thus in the emission of blood from the vein the spirit enters the vein, hence creating a certain inanition or exhaustion of spirits in the artery, and a greater weakness follows. But the preceding reason [as given in the previous two sentences] is the more true, and this is enough discussion of the subject.

Chapter XVIII

On the Dissection of the Kidneys and the Urinary Vessels up to the Bladder

From the progress of the chilis vein [vena cava] and the aorta artery you have seen two veins and two arteries passing to the kidneys; they call these the emulgent [renal] veins and arteries. Both of them are fixed in the substance of the kidneys in order to cleanse the blood and the spirits of their aquosity. Therefore these emulgent arteries are large in proportion to the size of the kidneys since those arteries which occur in a member for the sake of life are not so ample in a small member unless there is a duct to carry off the spirits. From this proceed other branches as is apparent in large members. The substance of the kidneys is fleshy, red, hard, and dense, of oblong shape in the middle of which (where there is a certain small cavity) the aforesaid veins and arteries enter. These make by their progress inside the kidney a certain whitish substance the color of the nerves. This substance is composed of a panniculus of the emulgent vein which is scattered through the kidney. This pannicular substance seems to have many openings in a dead body; they are not, however, perforated. Through these openings, which they call the colatorium or sieve of the urine, is strained off the watery material sent down from the liver to the passages of the urine which proceed to the bladder. But perhaps these passages are open in a living creature; in the dead body I could not penetrate them with a rod. The kidneys likewise have a pannicular covering which

---

[48] Avicenna, *Canon*, I, Fen 4, Doctrina 5, pp. 218–224 (translated by Gerard of Cremona; Venice, Giunta, 1595). He mentions the sciatic vein but does not describe its bloodletting.

proceeds from the peritoneum; by means of this covering the kidneys are attached to the back above the loins. Around the kidneys you will see a notable fat and very small veins in their substance for their nourishment. You will also see some branches of the reversive nerves [nervi vagi] beyond the nerves of the spinal medulla. The latter nerves also pass to the other members just mentioned. Hence in kidney pain the mouth of the stomach also suffers. Furthermore, the right kidney is higher than the left and less elongated by the diaphragm; it is almost immediately attached behind the stomach near the liver to the spine of the back above the right loin, as I said. But the left kidney is lower under the spleen, also attached to the back above the left loin. In order to see the substance of the kidneys and their inner parts, carefully lift the kidney as you strip it without breaking any of its vessels. The first thing which will be fine to see and to know is whether the inner foramina of the sieve are perforated or not. Therefore in order to see this place a reed through the emulgent vein and inflate the kidney; you will see it swell. The spirit or wind [of your breath], however, does not come to the urinary ducts or pass out of them. When you have seen this, cut the kidney itself in its round part lengthwise, and you will see a carneous substance in which there are white cavities. But toward the cavity where are located the urinary ducts you will see some caruncles which have the shape of a nipple of the breasts [pyramids]. They are softer than the substance of the kidneys. They say that by means of a nipple of this sort the aquosity sent out from the emulgents is sucked and sent through the nipple's extremity to the urinary duct [ureter]. In order to see more clearly the progress of the emulgent vein and of the urinary duct place a rod in one or other of the passageways or in the veins themselves and their orifices which pass to the urinary ducts. You will see some pathways to the urinary ducts but not in the substance of the sieve since the urine sweats back through the sieve. Anatomists say they have often found the location of the kidneys reversed, that is, the left kidney above, the right kidney below; I have seen this situation twice up to the present day. And since stones are generated in different places in the human body, a fact which is evident to the senses, I do not wish to discuss but merely to state what I have seen. I have often found both kidneys full of stones; sometimes they were red but most of them were of a lemon color shading toward ash gray. Thus those who distinguish the place of generation of stones by their color are mistaken for also in the bladder sometimes a red stone was found while I was present by the very expert dissector Peter Martyr, who labors to cure by incision those who suffer from stone and cures many of them. It is better to say that the color of the stones is due to the admixture of material and its digestion. I have also found stones in the urinary ducts. These were large; among others I once found that a stone had broken or lacerated a urinary duct for four fingers below a kidney. In the kidney itself were almost innumerable stones and larger ones the size of a bean. In this body there were also many stones in the bladder. But let us return to view the anatomy of the urinary ducts. Look then in the lower part of the cavity of the kidneys [pelvis] since you will see a vein or a urinary duct descending from almost the middle of the substance of the kidney through the back to the bladder. This passageway has a covering panniculus originating from the peritoneum which you will carefully strip from the duct in order to see the duct itself more readily. It descends through the back toward the hipjoints and passes to the bladder. When it reaches the bladder it branches through the latter toward the bladder neck. However, I have often seen larger branches; this ramification is not separate from the substance of the bladder. In order to see these branchings, gently move aside the panniculus which covers the bladder; when it is moved look at the numerous little veins in its substance which nourish the bladder. This process of ducts in the substance of the bladder toward its neck is one of the marvels of nature, as Galen said in *On the Use of the Parts* VI,[49] when they enter between tunic and tunic. The substance of these ducts is similar to the substance of the emulgent vein. There are those who say the urinary ducts arise from that substance of the emulgent vein; this is later disseminated through the kidney, united again, and proceeds to the bladder. I consider this a more cogent argument because it is hard to believe how the urinary ducts can be of the same substance as that of the bladder when the ducts are made of one single tunic while the bladder is made of two tunics, as you will see in its dissection. The tunic of the urinary duct, however, is hard in order to resist the sharpness of the urine. Certain passages or ducts often pass from the kidneys themselves; they come from the emulgent artery together with the urinary duct. But you will not see them in all bodies although many have believed that they are found in every body, for I saw only one in 1532 in a cadaver which I dissected. Put back the kidneys and the urinary ducts in order to see the parts which proceed to the seminal vessels and to the bladder in their anatomy.

Chapter XIX

On the Anatomy of the Seminal Vessels

You must put back into their place the kidneys and the urinary ducts and the other parts which com-

[49] Galen, *De Usu Partium*, VI (actually V, 16) (Vol. III K., 405): Atque haec omnia plena sunt excellentis naturae artificii. On the ureter see also *De Anatomicis Administrationibus*, VI (Vol. II K., 581).

municate with them, such as the emulgent veins and the aorta artery, since from these members the spermatic or seminal veins draw their origin. But since the spermatic vessel is covered with a panniculus just as is the urinary duct, gently strip this panniculus and you will see the seminal vessel tending toward a reddish color. This is an accidental difference which is useful in distinguishing the urinary duct from the spermatic vessel. Begin from the left side and you will see its composition. It begins from the emulgent vein and also from the emulgent artery through a vein and an artery which descend from the emulgents, making the spermatic vessel. You will also see that sometimes the aorta artery sends a branch to this seminal duct. It is then composed of a branch of the emulgent vein and of the emulgent artery and from a branch of the aorta artery. These branches coming together in their progress compose the spermatic vessel, as I said, sometimes from two and sometimes from three [branches]. But on the right side the spermatic vessel arises from the great emulgent vein and from the great aorta artery, sometimes above the kidney and sometimes below the kidney, and not from the emulgents as on the left side. From these branches of the artery and of the vein the spermatic vessel is formed as on the left side. This process of branches sometimes appears also in the right vessel. Its substance is well known from what has been said; that is, it is composed of a vein and of an artery, from which one duct is made. You will, however, note that sometimes these branches of both vein and artery do not unite in their progress; in fact, they pass separately as far as the testicle. Thus you will see two vessels descending from one side, that is, a vein and an artery. Nevertheless you will see at the most one duct from two or three branches, as I have said. It descends through the back above the loin and the hip bone, proceeding toward the pubic bone, through the didymus to the testicle in man. I say this because in women the vessels pass to the testicles and to the matrix or uterus, which are within the lower body cavity. In males when these vessels are near the pubic bone they pass to the testicles through a foramen of the didymus, which is composed from the peritoneum, as I said. Up as far as this part the vessel is called seminal descending, or preparatory, since it prepares the material for the sperm and since it is composed of vein and artery. Therefore its substance is harder than a vein, but near the testicle and around it the vessel is softer. Thus this vessel and the right one also after it has reached the substance of the testicle are wrapped around with many involutions as far as the fundus. In this part there is a certain carneous substance which the Greeks call the epididymus. But the moderns call epididymus the veins upon which the spermatic vessel is based around the substance of the testis, and because of this they say that material is sent to the substance of the testis so that it acquires a whitish color there. Nevertheless the ducts (*viae*) are not known. In this preparatory vessel is sometimes found a whitish material. This, however, does not contradict those [who say that material is sent to the substance of the testis] since one can also believe that in the vessel near to the testis there is a power or virtue which changes the blood into sperm, as in the testis. In the dissection of the largest part of the substance of the testis of a young animal which is able to generate offspring there appears in it a certain white humidity, quite similar to the color of sperm although not so compact. This is better seen in a bull when he is ready for copulation. Let it be said therefore that the material of the semen is brought down from the preparatory vessel to the testis and by means of those involutions and resudation. It is retained there and made white. After the whitening process it is sent to another part of the seminal vessel which they call the vas deferens. In the process the semen becomes denser, beginning in the testicle itself after the aforementioned involutions where there is that substance called epididymus. This substance ascends from the testicle to the pubic bone where there is the preparatory vessel and together with the latter through the same didymus it ascends above the pubic bone into the lower body cavity and is fixed there near the neck of the bladder. There it is united with the foramen [urethra] which passes through the penis that carries the urine and the semen outside the body. Thus from three foramina or passage ways (*meatus*) is created one duct which carries the urine and the sperm, or semen, to the canal or duct which is in the penis, that is, from two seminal vessels and from the duct of the neck of the bladder. In this part where this duct is composed you will find a glandular flesh [prostate gland] upon which rests the neck of the bladder and the extremities of the aforementioned vessels in order to protect it [the duct] from the sharpness of the urine. When these pieces of flesh are consumed there occurs a burning in the urine, a very bad illness. As to this part evident to the senses there is speculation upon its origin because since the spermatic or seminal vessel arises from the chilis vein [vena cava] and from the aorta artery, as we have seen, nevertheless from the phlebotomy of the veins around the neck, according to Hippocrates, sterility follows. This does not seem reasonable if no vein descends from the brain to compose the spermatic vessels. Furthermore it is also remarkable that we see the child which is born very often resemble its parents in all of its members, that is, in legs and arms, head and eyes, and all parts, because if the material first prepared by the principal parts and all consequent ones did not fall downward this resemblance not only in shape and size but also in color would not be created. For the investigation of this truth anatomists are not unaware that the processes of both veins and arteries communicate with the entire body and that the spirit and the blood which are generated in the heart and in the liver and the powers of the soul proceed to all the members by means of these processes. Through the arteries and veins the spirit and the blood have a flux and reflux, especially from the heart first to the liver. That which ascends through the jugular veins and arteries to the brain not only flows to the brain itself but flows back to the lower members. Hence it is that from such a reversion of part of the spirit and of the blood returning after it has taken up virtue (or power) through its alteration from the heart and liver and also from the brain which is accomplished by nature when operating correctly the material is sent to the seminal vessels for the conservation of the species. It is superfluous material which receives its power from one or other of the principal members; by this power it very often generates a child similar to its parents. But this is more true of the spirit, since the informative power proceeds from it, through the alteration made by the members and not only by the members but by their neighboring parts. For power is not alone in the member but also in the parts near to it, as is said about the meseraics, which have the power of making blood from the liver. Thus also the neighboring parts prepare material for the sperm by means of their power. This material is directed through the branches of both artery and vein to the seminal vessels, nor is this contrary to sense perception. But that which can convince you more than this of the truth of this observation is the elective evacuation which is daily carried out by the use of drugs, turning back the particular humors from a particular member through a drug of this kind to the passageways of the fecal superfluities, as, for example, the evacuation of the material of the phlegm or pituita in gout. If we recognize this by the test of sense perception, how much more are we forced to believe that nature does this for the conservation of the species. But not only in reverting to the members does the blood return through the larger passageways but also through the smallest parts of the veins and arteries. Therefore the big artery and vein have many branches passing to the spinal medulla through which also the altered blood and spirit return according to their need and pass to the mentioned vessels. Hence from what has been said emerges a solution to Hippocrates' problem: how does it happen that when the veins of the neck are cut sterility results in man?[50] Because an excessive flow of blood creates an alteration of the virtue (or power) of the brain so that it cannot prepare the material suitable for generation when it descends to the spermatic vessels. The sign of this debility is the fact that men who have had the veins of their neck cut in bloodletting become somnolent and cold; thus their semen is rendered infecund since it does not have all the required alterations of all the principal members, for not just any [element] acts upon any other [at random] but a determined element upon a determined one with fixed conditions.

Chapter XX

On the Dissection of the Bladder, Didymi, Testicles, and Their Pouch [*Scrotum*]

Before the spermatic vessels or seminal delatories are moved aside inspect the anatomy of the bladder by first inflating it with a syringe and you will see its size and substance. This is white and appropriate to the bladder, as the substance of the intestines is appropriate to them. In the bladder appear many veins which nourish it. It is composed of a double tunic, as is very clearly apparent if anyone carefully inspects it and separates one tunic from the other when the panniculus that proceeds from the peritoneum is first removed. This panniculus covers the bladder. If you wish to divide the tunics more distinctly or to separate them allow the bladder when stripped from the panniculus to wither for a few days but not to the point where it smells. When the bladder is thus shriveled then strip it and you will most easily separate tunic from tunic since there are two. In them you will see all kinds of filaments; the long ones are fewer in number. You will also see the entrance of the urinary ducts in the bladder. These ducts are united with the substance of the bladder; in that part where they enter the substance of the bladder is thicker in proportion to the progress and the size of the ducts. The bladder descends and passes toward the anus and rests with its neck on the rectum intestine in men; but in women it rests above the neck of the uterus and within the pubic bone. When the bladder is full, however, it ascends toward the sumen and, as I said above, the two ducts descend to it, leading the urine into it from the kidneys. The entrance of this urine to the bladder itself is marvelous since there are no passageways apparent by which the urine may enter. Thus the bladder is a member in the extremity of whose neck there are muscles or a muscle with a certain width which squeezes the neck in order to prevent the urine from issuing forth before its time or allows it to issue without delay; this is done in the lower part. This muscle also assists in the emission of the urine through the upper part where the urinary ducts come by applying pressure in that part so that the urine which passes through the ducts to the bladder may

---

[50] Hippocrates, *On Airs, Waters, Places* 22, who writes on varicose veins among the Scythians: "They treat themselves by their own remedy which is to cut the vein which runs behind each ear . . . vessels behind the ears which, if cut, cause impotence and it seems to me that these are the vessels they divide" (translated by John Chadwick and W. N. Mann, *The Medical Works of Hippocrates* [Oxford, Blackwell, 1950], p. 108). The passage is also quoted in my translation of Berengario de Carpi, *A Short Introduction to Anatomy* (University of Chicago Press, 1959), p. 191.

descend entirely and with more ease to the bladder because the neck of the bladder is thicker than its body and has fibers across its width since the muscle passes across its width. Therefore they say the muscle is fleshy and it proceeds tortuously, with a form resembling that almost of the ancient letter *S*. But since females have a small neck in their bladders the muscle is not so tortuous. Hence it is that when they wish to pass water they make an attempt to do so and curve inward since the muscle is small, for the neck of the bladder is small and hence cannot squeeze out the urine; but when women curve themselves they press the bladder between the uterus and the intestines, while the muscles of the stomach assist. Thus women make water.[51] But men, since their muscle is larger, and also because they can squeeze better through the bending of the neck of the bladder, do not curve themselves to make water, as women do, although they curve slightly in doing so because the neck of the bladder rests upon two glandular carunculae, one on the right and the other on the left. These glandular pieces of flesh humidify the neck of the bladder so that it can resist the sharpness of the urine. Through these carunculae also pass the seminal vessels, which carry the sperm to the duct of the penis. There you will note three orifices coming into one, that is, two for the seminal vessels and the third to the neck of the bladder. These three in one send [their contents] to one common duct which passes under the penis and through which the urine and sperm pass outward. But these orifices can scarcely be seen unless the pubic bone is cut away. You must move the latter away toward the hip joints in the groin from one side as well as from the other, cutting this bone with a strong instrument. You can also, if you wish, cut it in the middle. When it is cut off examine the pubic bone. It has a cartilaginous ligament in its middle by which it is inseparably united; this ligament is separated by a final boiling or by its putrefaction in cemeteries. But before you move aside the pubic bone look at the two didymi from the substance of the peritoneum, passing in the groin above the pubic bone on the right and left. Since these members have various names do not be deceived by them. You should know that the didymi are certain parts of our body which pass across the groin above the pubic bone. They are parts of the panniculus of the peritoneum that pass across to the pouch of the testicles and cover them. This covering of the testicles is called oscheum [scrotum], but the upper part from the groin to the testicles is called didymus. These parts not only cover but sustain the testicles. These parts do not exist in women since the latter do not have testicles on the outside of their bodies. Also examine there in the groin two small muscles [m. cremaster] which proceed to the penis, one on the right, the other on the left. They lift the testes a little when the sperm is emitted; these muscles pass along the line of progress of the didymi, clinging to the latter, toward the testes and terminate in a tendon which adheres to the penis and to the didymus. In these didymi you will find the spermatic vessels, both preparatory and delatory. One of them descends to the testis, the other ascends toward the bladder to the duct of the penis, as I said above. When the oscheum is moved away look at the testicles. They have a glandular substance and an oval shape, nor are they large in men. They differ in size; some men have large ones, some have small ones, about the size of a dove's egg. They are two in number in all men and are situated between the coxae hanging in their pouch from the groin. In a cadaver which I dissected I saw one testicle only, the right one, and to it there descended a wonderfully ample spermatic vessel. But on the left side there was no testicle, nor did any vessel descend to this place from the emulgent [renal vein] because there was no need for one; nature had created this man without a testicle. Some say it is possible for a person to be born with three testicles, but this is a freak;[52] just as being born with only one testicle is monstrous so it is with three, from the large amount of matter involved. Furthermore, in the exterior substance of the testicle you will notice involutions of the spermatic vessels passing here and there through the testicle in the manner of dilated veins. These involutions come from the descending vessel, or the preparatory; they proceed in the lower part of the testicle; the lower involutions are covered with a certain caruncula. The latter is not as white as the substance of the testicle; it lies at the end of the testis. They call it epididymus. You can strip it to see its substance. Also cut the testicle lengthwise, and you will see its inner, whitish, glandular substance with a certain whitish humidity in its porosities. This humidity is said to come from the blood; it penetrates to the testicle itself from the preparatory vessel and is later transmitted to the delatory vessel [vas deferens] after it has become whitish; this, however, is difficult to recognize in men. When you have examined these items, look also

---

[51] Thomas Lauth, *Histoire de l'Anatomie* (Strasbourg, Levrault, 1815), p. 338, takes exception to Massa on this point: "Massa auroit pu se dispenser d'expliquer péniblement un fait dont il lui eût été aisé de reconnoître l'inexactitude: en effet, les femmes urinent très-facilement debout, et cela arrive fort souvent aux femmes du peuple dans les rues." The question is first raised by Aristotle, *Problems*, III, 20 (*Opera*, Paris, 1629), II, 725–726. The glandular carunculae of which Massa speaks later are the prostate gland: Gernot Rath, "Pre-Vesalian Anatomy in the Light of Modern Research," *Bulletin of the History of Medicine* **35** (1961): p. 145, says: "his most important contribution is his description of the prostate." See also chap. XIX, fol. 34 r for the first mention of this gland.

[52] Francesco Filelfo, the Renaissance Italian Humanist, boasted that he possessed three testicles as the source of his prodigious energy, both scholarly and amatory: John Addington Symonds, *The Renaissance in Italy* **1** (New York, Modern Library, 1935), p. 457. Demetrius Chalcondylas, the famous Greek teacher and text editor of the time, was also said to be similarly equipped.

at the final cutaneous part, which contains the testicles, the oscheum, spermatic vessels, and the rest; this is on the outside and is called the pouch of the testicles or the scortum [scrotum]. In this skin are scattered very many veins from those veins which come from the chilis vein [vena cava] after it has passed to the os sacrum. You will recognize from this place and also from many other places in which in the outer skin veins are sustained that the use of this fleshy panniculus or membranous muscle or involucrum of the entire body, no matter what you call it, is not to sustain veins primarily to nourish the external skin but to protect and move it, as I said.[53] This is not the place to discuss whether this panniculus is single or double. Look then at the skin which makes the pouch. It descends and can be dilated to an incredible extent, as can be seen in those who suffer from hernia. Sometimes this [hernia of the scrotum] is dilated in its descent as far as the knees, filled with intestines or humors and sometimes with a fleshiness. You will also notice that both to the bladder as well as to other members contained in the scrotum there come little branches of a nerve proceeding from the caudal bone; similarly there come arteries in addition to veins; these are for life, sensation, and nourishment. Furthermore, this pouch has a medial panniculus (*mediastinum*) from the substance of the oscheum which divides the right testicle from the left. Thus the pouch has two sinuses so that sometimes it is extended from one side only through the descent of humors or inner members or flatulence or through the repletion of its own veins which creates a varicose disposition. When you have noted these items you can move aside the kidneys with the urinary ducts and the seminal vessels in order to see the loins upon which these members were lying. These loins are on each side of the spine. Two muscles, one large and the other small [psoas major and minor], begin under the diaphragm and adhere to the spine. The larger adheres more closely. These muscles proceed toward the groin and terminate in tendons, the smaller muscle in a smaller oblong tendon and the larger muscle in a larger tendon. The latter passes to the hip joint at the head of the adiutorium [femur] of the coxa near its elevation together with other [bones] and where these terminate begins a livid muscle which proceeds toward the exterior part of the coxa in its descent. When you have seen the loins, you can see as you dissect them the parts which compose the penis.

Chapter XXI

On the Dissection of the Penis

In the elevation of the pubic bone you stripped the penis also from that bone since it adheres to the latter by means of a strong ligament [suspensory ligament of the penis] as far as the hollow of this bone. Returning now to the penis, examine its substance. It is composed of four muscles, the duct [urethra] which conducts the urine and the sperm outside the body, veins, arteries, sinew, and glandular flesh in its head. It is covered with skin. When the skin is stripped look first at the muscles, which, as I said, are four in number. Two of these [corpora cavernosa] with their fleshiness arise in the lower part toward the rectum intestine; in fact, to some extent they embrace the intestine itself. These muscles proceed toward a gland, one on the right, the other on the left, and laterally. Between these muscles in the lower region is a duct through which we void urine and semen since three ducts are combined in this one, that is, two spermatic vessels and the head of the bladder, as I said above. The single duct of the penis, however, begins at the end of the rectum intestine where the penis begins, and proceeds to its extremity. In this region is the flesh called glandular [glans penis] for the protection of the heads of the muscles. It is called the glans of the penis. There are also two other muscles [corpus spongiosum] which have a fleshy part above beginning from the glandule and proceeding toward the pubis and the anus, sending the urinary duct outward in the lower part. They terminate in a tendinous substance near the anus. The substance of these muscles is thin and soft and is largely of tendinous nature. These muscles are difficult to separate. The nerve and skin which cover the penis are thin and doubled in the head when the penis is depressed. By this depression the penis creates the prepuce from the skin which covers the glandule. At the end of the glandule is a foramen from which the urine and semen issue forth. Under the foramen is a certain terminus of the skin of the prepuce which the common people call the little thread [frenulum] (*filellum*). Veins, arteries, and even the semination of the nerve you will see both in the substance of the penis as well as in its skin. These veins come from the ramifications of the chilis vein [vena cava] after it has passed to the os sacrum. Similarly arteries come from the aorta artery; these veins and arteries are large in proportion to the size of the penis. A prominent nerve which gives sensation passes to this member. Note that both Hebrews and Mohammedans amputate the prepuce by circumcision. Before you separate the penis from the entire body look at the two little muscles [cremaster] which proceed from the end of the longitudinal muscles of the stomach. Terminating in a tendon passing through the pubic bone to the penis and the didymi they assist in the elevation of the penis and the testicles beyond the proper muscles which compose the penis and elevate the testicles in the emission of sperm. This member, that is, the penis, serves for generation, as both philosophers and physicians unanimously declare. Its size is not equal in all men since some have shorter and others have a longer penis; this is true also of its thickness, its

[53] See notes 14, 18, 19.

total bulk, either too large or too small or mediocre. Those who write about the subject enumerate and explain many aids in generation and many defects in conception which are ascribed to the penis.

Chapter XXII

On the Dissection of the Rectum Intestine and its Muscles and of the Hemorrhoidal Veins

If you wish, when you have seen the items mentioned above, you can move aside the pubic bone, if this has not already been done, and the bladder as well with the testicles and vessels both urinary and seminal and the oscheum in order more comfortably to see the anatomy of the rectum intestine. The first thing to look at in this intestine is the heads of the hemorrhoidal veins which are sometimes swollen and prominent. There are five of them, corresponding to five veins which you will see when you strip the anus, or sphincter, during the inspection of the muscles which go around the intestine. In regard to this inspection you must not believe those who say that it is necessary to move the rectum intestine aside with a quantity of the flesh of the nates or buttocks since such a method of viewing is quite confusing. If anyone wishes to see the muscles of the rectum intestine let him suspend the cadaver by one foot so that its head hangs down toward the ground and begin to strip the skin around the anus with a razor. There will first appear the five heads of the hemorrhoidal veins which proceed through the intestine. But note that these veins are not equal in all people since in some they are small. As I mentioned above, they come from the descending chilis vein [vena cava] after it has crossed the os sacrum to nourish the muscles of the aforementioned intestine. They also come from a branch of the portal vein which passes to the rectum intestine. After these veins look first at a notable muscle which goes around the extremity of the intestine [external sphincter]. It is quite thick and fleshy. Its thickness does not exceed two fingers. It proceeds toward the penis and is attached to the extremity of the latter above its duct. This muscle squeezes the orifice of the anus and hence it is that when the anus is squeezed there results a certain movement of the penis. This muscle assists also in the final emission of the urine and of the semen. After you have stripped this muscle you will see another slender muscle which goes around the rectum intestine for four fingers and even more; it ascends through the intestine itself [internal sphincter]. In some people it is so thin that it can scarcely be seen. It terminates in the lower region at the heads of the muscles of the penis which, as I have said, embrace somewhat the rectum intestine. Thus through the contraction of both muscles there follows a certain motion of the penis. They say the use of these muscles is to cleanse the intestine of its residue of feces by certain inversions. But since their function is to squeeze as well as to cleanse as is apparent from a sense-perception of both muscles, it is better to say that both concur in cleansing and in constriction. I do not wish to be reprehended, however, for failure to agree with the statements of authority by certain ancients and moderns since sense-perception which prevails over reasoning dictates the same conclusion; for constriction is carried on by the muscles that squeeze as well as by the muscles which suspend the intestine. These proceed lengthwise of the intestine. Through this contraction the intestine ascends and something of a closure is brought about but the final stricture of the foramen is not made unless the muscle of the sphincter or of the anus has the ultimate strength to squeeze the final part of the intestine. When you have seen the muscles that go around the intestine, look also at the two muscles [levator ani] I have mentioned which proceed lengthwise of the intestine with their origin toward the back both on the right and left between the hip joints. Their function is to contract the intestine above after the emission of the feces and also to relax it when the feces are being emitted at the bidding of the will or through the stimulation of the fecal matter. Through the softening of these muscles the intestine descends so that it protrudes beyond the anus. This illness is cured by drying agents; it very often occurs in children. Look also at the size of the intestine and its direct progress toward the left side where the colon intestine begins. Its substance, as I have told you, is made of two tunics which have all kinds of filaments or fibers, long in the exterior, broad and transverse (the latter are few) in the interior tunic. You will notice also that since abscesses often occur in the anus or in the head of the rectum intestine they require opening by incision when they suppurate. Therefore beware of cutting transversely lest you cut the muscle that goes around the intestine or injure it. In such incision for opening the abscess you must cut circularly around the anus; they call this a lunar incision. Take care also in the case of fistulae of the anus which penetrate the intestine not to cut four fingers above the intestine since in such an incision you will cut both muscles that go around the intestine. Hence will result involuntary excretion of the feces. I have seen many people who after incision of the rectum did not lose the function of these muscles but retained their feces voluntarily with perfect completeness because they were incised at the consolidation [aponeurosis] of the muscle or at the best union of the parts.

Chapter XXIII

On the Dissection of the Uterus, or Matrix, or Vulva

The Wonderful Lord in His workings has varied the members of generation in sex, for those in males were given on the outside of the body but in females these members are inside of the body and hidden. In males they are compact, hard, united, in females lax, soft, and hollow, the former to send forth semen in better fashion, the latter so that they might better promote conception from the semen. Concerning the members of generation in men I have spoken above. Lest the order of sequence of dissection of the remaining parts in the lower body cavity should be interrupted I have postponed the dissection of the matrix to this place, which is the sequence required in woman. After you have seen all the members as far as the elevation of the intestines, lift them out with the mesentery and put the rectum intestine back into its place. First turn your attention to the kidneys since you will find the process of the spermatic vessels which have their origin on the left, that is, on the left side from the emulgent [renal] vein and from the emulgent [renal] artery. This process passes through the loin to the left testicle [ovary] of the matrix. This testicle is not outside as in men but inside, attached to the matrix toward the back. You will note, however, that often this left-hand vessel does not originate from the emulgent vein but from the trunk of the great chilis vein [vena cava] and also from the trunk of the aorta artery, just as the right-hand vessel does, as was said in the anatomy of the seminal vessels in males. Similarly you will see a seminal vessel on the right side, arising not however from the emulgents of both vein and aorta artery but from the trunk or from the large chilis vein and from the aorta artery coming into one vessel. This vessel descends through the back. The vein and artery have their origin for the most part above the kidney, but sometimes both ducts, both that which originates from the vein as well as that from the artery, proceed separately as far as the testicle, which, as has been stated, is above in males. This vessel [in women] terminates at the right inner testicle, as I declared, toward the back. These vessels are covered with a panniculus originating from the peritoneum. The matrix therefore remains above the right intestine and above the matrix lies the bladder. All these members are located in that [pelvic] cavity corresponding to the os sacrum in the back and in the anterior part of the pubic bone. You have seen the latter at the end of the lower body cavity (*venter*); at its sides are the inner parts of the hip joints where the matrix is attached by means of ligaments. Thus you will see that the matrix is placed between the rectum intestine and the bladder, both along the part of the neck of the matrix which they call vulva [vagina] as well as along the body of the matrix which they call uterus and also matrix. Do not be deceived by the writings of certain ornate authors in which you will find small grain and much chaff. Note that the name matrix is received from the Greeks for the name *mētra* is Greek for any kind of vessel. Thus there is no need to believe that the word matrix derives from the Greek word *mētēr*, which means mother or genitrix in Latin. But the word for vessel comes from the word *mētra*, as matrix does, resembling a vessel which preserves the foetus, and thus the Latins call a body or substance in which the foetus is contained sometimes uterus, sometimes loci [places], and sometimes vulva, in their confusion. But those who were more choice in their Latin called a woman a matrix in whom all her parts are at the peak of their strength for the generation of a foetus. Later, from this denomination they called matrix in common speech the member of generation in women. This is not unreasonable when one considers Greek and Latin etymology; many who speak Latin properly call uterus (*loci*) the female member of generation, but the lowest part of the matrix, in which copulation takes place, they name not without ambiguity vulva or volva from the word for turning inward (*involvendo*) since the foetus turns inward. However, when the foetus is not turned inward from the ultimate part of the lower member of generation, they do not correctly call this ultimate part vulva. It makes little difference whether the terms are derived from this word or another. I have made these statements for the sake of those who increase their learning in the meanings of words. In order to allow no question to remain in the matter of names, which was so displeasing to both Aristotle and to Galen, I shall describe the parts of the member of generation in women, a declaration with which Galen would not find fault since I am teaching here the inquiry into the members and parts of the body by means of dissection. Therefore let it be said that the member of generation in woman is a receptacle placed above the rectum intestine and attached to the hip joints and to the back. It descends along the course of the intestine and terminates lower down near the extremity of the intestine which is called the anus. In this extreme part of the receptacle there is a foramen which the penis enters in copulation. I have decided to call this entire substance the matrix lest a difficulty in the recognition of both the member and its parts arise from the diversity of names for them. Therefore you will see that the shape of the matrix in its upper part is circular; in the middle of its highest point is a sort of depression. The matrix is round, oblong, and in its lower part toward the coxae it has a neck in the form of a gourd or a gourdlike vessel. If for a better inspection of it you inflate the matrix with a bellows it will be not unpleasant but difficult since when the matrix is chilled it is not so easily extended. Thus during such an inflation you should warm it with frequent hot applications, such as sponges dipped in hot water. Place the bellows at the orifice below the aforementioned neck of the matrix where copulation takes place; the end of this neck they call the vulva. Proceed thus as far as the other foramen which is above. This foramen they call the mouth of the matrix [cervix]. Insert a tube into this mouth and inflate the body of the matrix; thus its substance will appear round and oblong in the non-pregnant woman or one who has not yet conceived, as in a small virgin. It is not much larger than the empty bladder, or a hen's egg; but its size varies according to the size of the woman's body, as is true with other organs of the body. Note, however, that the tube of the bellows should be sharp at the end. Childbirth or coitus will also dilate the mouth of the matrix. But you will not always find it dilated; in fact, it is usually very narrow, to such an extent that at the time of pregnancy the point of a needle cannot enter it.[54] At such a time the body of the matrix swells into a very large round swelling which ascends above the intestines toward the stomach, passing across the terminus of the umbilicus; this happens especially at the time near childbirth. When you have seen the body of the matrix, which extends from its mouth upwards toward the upper part, also inspect the shape of the neck of the matrix and of the vulva. For this inspection you must remove the pubic bone from the sides or in the middle, as was said above on the anatomy of the bladder, cutting the bone with a strong instrument. In order not to injure the neck of the matrix while you are removing the bone you must carefully strip it away from the bone. This neck differs in size according to different women. In virgins it is short, since its wrinkles are not relaxed from its ligaments and veins. These veins and ligaments are destroyed during the first coition, or deflowering. Therefore those people are in very great error who have postulated a hymen membrane as an obstruction to the mouth of the matrix, thinking it is broken in copulation and the matrix is opened. Look then at the neck of the matrix, round, oblong, and hollow like a reed. Its size is in proportion to the penis of a man. Inside it are many wrinkles which are attached to some very slender ligaments and veins. In the extreme outer region is that part called the vulva, of oblong shape, beginning near the orifice of the rectum intestine and extending up to the pubic bone. Within the upper region near the pubic bone is the orifice of the neck of the bladder or the duct [urethra] through which women pass water. In that place are two little skins which cover the mouth of the neck of the bladder or duct almost as far as the duct of the neck of the matrix; these skins [labia minora] are called the prepuce of the matrix. When they grow they descend toward the neck of the matrix inside the vulva so that sometimes they impede the entrance of the penis. I have seen many women who had these skins very large. The use of this prepuce of little skins is to protect the neck of the bladder from injury from the cold air and also in some fashion to guard the matrix itself. I have thought it superfluous and unnecessary to describe the size of the vulva since it is a member well known to the experience of all by daily sight, knowledge, and handling. The neck is broadened and relaxed by much copulation; this is true also of those who are pregnant. In them it is far more stretched, and not only the neck but also the body of the matrix. And since we proceeded from the seminal vessels to examine the matrix, I wish you to know that the vessels themselves and the testes are covered by the panniculus of the peritoneum. To this there pass branches of the reversive nerves; they give sensation. From their retention of the sperm women often fall into loss of reason and other very severe illnesses, such as weakness of heart and liver. This is on account of the connection of the seminal vessels with the heart and liver, as was said in the anatomy of these vessels. They proceed from the upper parts to the testicles of the matrix which are united to the body of the matrix toward the back by means of the aforesaid seminal vessels. The testes themselves are not large as in males but small and comparable in size to an unusually small dove's egg. Their substance is softer than that of a man's testicle. They have epididyma [epididymi: fimbria of the uterine tubes] which envelope them; all those I have seen were of larger size than that of the substance of the testicles and of a different color from the latter, since the color of the didymi tends toward bluish and is harder than the testicle. The testicle is wrapped around by the seminal vessel like a varicose vein, as was said to be the situation in men also. This vessel is double, as in man, that is, consisting of preparatory and delatory vessels. These testicles, as was said, are covered by a panniculus originating from the peritoneum. They are at a distance since they are located in the extremities of the body of the matrix and are not contained in one pouch, or skin, as the testicles of men, which are contained in the scrotum. Each of them has its own small muscle to move it and its vessels, which are double, as I said. The delatory vessel is small, or short, since the testicles are not far away from the matrix. These vessels which carry the sperm to the matrix are called the horns of the matrix [uterine tubes] according to Galen, *Commentary on the Aphorisms* [*of Hippocrates*] V, 48,[55] although in his

[54] Galen, *De Locis Affectis*, VI, 6 (Vol. VIII K., 446) has an eloquent comment on this amazing phenomenon: Quid enim natura potest esse admirabilius, quam exacte, ut ne specilli quidem acumen admittat, perfecto autem foetu, usque adeo quod novem perpetuis mensibus claudatur os uteri adeo distendatur, ut totum animal per ipsum egrediatur? See also *De Uteri Dissectione* (Vol. XI K., 897–898) and *De Anatomicis Administrationibus*, XII, 3 (translated by W. L. H. Duckworth *et alii* [Cambridge University Press, 1962], p. 116).

[55] Galen, *Commentary on the Aphorisms of Hippocrates*, V, 48 (Vol. XVII B, K., 841): . . . semen ex propriis testiculis per utraque cornua profusum. In *De Dissectione Uteri* (Vol. II K., 890–891) Galen, who uses the expression *mastoides*, discusses other views and terms for the "horns" of the uterus (as Diocles called them), although Herophilus likened them to a half-circle and Eudemus called them curls or tufts (*cirros*) and Praxagoras and Philotimus used the word sinuses. Galen then abandons the subject with the assertion that these physicians could not explain the use of these horns. They are the Fallopian tubes, according to Thomas Lauth, *op. cit.*, p. 341. See also note 54 above.

book *On the Dissection of the Matrix* he wrote otherwise. They are short or small because they do not send sperm into another person as men do, for women, according to Galen, have their own sperm in addition to the menstrual blood, which makes for conception because sperm is generated from the blood which comes to the testes through the preparatory vessels and is transmitted to the matrix, from which as they say cotyledons are created. Since these do not fall under the perception of the senses nor can be seen except in the pregnant woman and also because the facts are otherwise I shall speak of them later. When you have seen the testicles and the spermatic vessels both preparatory and delatory which are called the horns of the matrix since they are extended in the upper region as if they were horns, look also from the sides of the matrix toward the upper region since there are two white substances, round like the vessels [round ligament]. They are veiled by the panniculus that covers the matrix near the delatory vessels. These substances extend to the groin and are stretched lower down. They gave a basis or handle (*ansam*) for doubt to the moderns (*neotericis*) as to whether these or the spermatic vessels were the horns of the matrix since both substances somehow produce a horned matrix. If Galen is to be believed, I cannot say anything more than that the horns of the matrix are those to which the delatory vessels come, that is, those which carry down the sperm to the matrix. So that there may be no further question about this matter listen to Galen as he explains *Aphorisms* V, 48,[56] where Hippocrates says the male foetus is on the right, the female foetus more on the left. In his explanation he says that the female contributes something to the warmth of conception. The female semen is poured indeed from her own testicles through each little horn, one part into the sinus on the right, the other part into the left one. These are the true and not faulty words of Galen from the translation of the very learned Leonicenus. It is consistent with the Laurentian translation.[57] Do not be disturbed by the statements of Galen in his book *On the Dissection of the Matrix*, where he seems to place the horns of the matrix separate from the seminal vessels themselves, since these parts, with their operation located [here], can be named as freely as you wish. These substances which come to the groin are ligaments; they pass to the groin and bring the sperm thither; thus you will not be at odds with the ancients in any necessary matter. Furthermore, the matrix is attached to the hipbones on the right and left by very strong broad ligaments under the places where the testes are. It is also attached to the os sacrum and to certain vertebrae of the back, to the pubic bone, the neck of the bladder, and to the rectum intestine with fat in between by means of very slender and soft [cardinal] ligaments. These ligaments of the matrix are not soft here alone but in all places where it is attached. Thus it is that the matrix can be extended without hindrance. The neck of the matrix adheres also to the inner muscles of the coxae which originate from the hip joints and is attached with ligaments. Therefore in the elevation of the matrix you must strip its neck from these muscles. This neck terminates in the upper region at the mouth of the matrix, which is a foramen like that of a tench's mouth located at the farthest extremity of the body of the matrix. In that place the foramen has thick parts lying around it [cervix]; they are hard and round like some animal's head. This orifice is extended and opened at childbirth by the will of the blessed God to such a size that I have quite often easily placed both my hands in the body of the matrix in order to extract a dead foetus from it. This mouth of the matrix at the time of conception can also be squeezed so tightly that the point of a needle cannot enter it. If after you elevate the matrix you invert it you will make it appear like a male member since the body of the matrix will contain testicles like a pouch and the neck of it will resemble a penis. Hence the matrix is said to be a male member inverted and diminished, made for the needs of the species; its origin was its small heat due to which it could not extend itself outside the body. Although there are difficulties concerning the double matrix you will notice that the matrix is composed of a double tunic. These tunics are separable, especially in the pregnant woman. I have separated them in a dead woman pregnant eight months who died of a lasting illness. She had two male foetuses, one on the right, the other on the left, face to face. The two backs of the infants were turned toward the walls of the matrix, that is, to right and left, and the infants were almost in a sitting position above their secundina [chorion] membranes, for each of them had one. These members adhered to the walls of the matrix, one attached on the right, the other on the left; I separated them with some difficulty. The infants lived only for a short time. Therefore let those keep silent who say that the generation of male and female children is due to their

[56] The reference is more likely to *Galeni Comment. III in Hippocratis Lib. I Epidemiorum*, III, 31 (Vol. XVII K., 445) or to *De Usu Partium*, XIV (Vol. IV K., 153) where the statement is made that the male foetus is on the right, the female on the left. This idea goes as far back as the pre-Socratic philosopher Parmenides; see the note on the subject in my translation of Berengario da Carpi, *A Short Introduction to Anatomy* (Isagogae Breves) (University of Chicago Press, 1959), pp. 187–188.

[57] Niccolò Leoniceno (1428–1524), a distinguished philologist and translator of Galen, came from Lonigo near Vicenza (see George Sarton, *Appreciation of Ancient and Medieval Science During the Renaissance: 1450–1600* [New York, A. S. Barnes, 1961], pp. 25–26) and translated Galen's *Commentaries on the Aphorisms of Hippocrates* in 1509 (J. Mazochius, Ferrara): see Richard J. Durling, "A Chronological Census of Renaissance Editions and Translations of Galen," *Jour. Warburg and Courtauld Institutes* **24** (London, 1961): p. 251; see p. 250 for Laurentius Laurentianus, who published his translation of the same work of Hippocrates in 1494 (Florence, A. di Miscomini).

position in either the right or left side of the matrix since experience bears a contrary testimony.[58] Let them say that the male is generated by reason of the strength of the sperms and of the woman's member of procreation while the female is generated by reason of the defect both of semen as well as by reason of a chilled matrix or one otherwise disposed. For I know the woman's member is the place necessary for generation just as a father is needed also but not in the same manner neither in regard to the sex nor in regard to the member involved since what is made or generated in the place [matrix] is generated by necessity. I also know that to the left testicle of the woman as well as of the man a watery blood is sent from the emulgents. From this a sperm is created which is not very strong. But when active sperm is sent from the male it can fall both into the right as well as the left side; if the sperm is strong a male will be generated, if it is not, a female results both in one side as well as the other. But the place of conception has an accidental effect, so that if an infant is generated on the left side it will be an effeminate man; if a female is generated on the right she will be a virago. Thus it happens also with the sperm of a weak man or of someone affected in some manner if it is sent into the matrix of a strong robust woman, and the contrary results when a strong man copulates with a woman who has a weak matrix. It can be said that different characteristics can occur in the new-born but not, however, by reason of its place in the matrix. But I return to the point whence I digressed. Examine the matrix composed of two tunics, having laid aside the covering panniculus. The first of these tunics somewhat resembles the substance of flesh and is on the outside. The inner tunic is whiter and covered with a thin panniculus; therefore some have said that the inner tunic is double. In this inner tunic you will see many heads of veins; these are round, as if they were pustulent and livid, and cover the entire inner substance. This is better seen in dead bodies on account of the suffocation of the matrix due to retention of the menses, as I saw [in a body] in the year 1526. These heads of veins are from the chilis vein [vena cava] which descends through the back to the coxae and in its course sends branches to the rectum intestine, to the bladder, and to the matrix. After it has in similar fashion passed through the region of the kidneys many branches from the descending artery pass to the matrix itself, nor, as many have believed, are all these heads of veins from the seminal vessels themselves which enter the substance of the matrix. For the most part they are from the branches of the aforesaid chilis vein. These branches of both vein and artery are fixed in the body of the matrix near the testicles. Look at these also since very many heads of veins terminate at the mouth of the matrix. When these swell some people call them hemorrhoids of the matrix. Note, however, that in the neck of the matrix there is not a multitude of veins and arteries but only enough for its nourishment and to give life. These veins are from the branches of the chilis vein. This is due to the fact that the body of the matrix requires blood not only for its nourishment but for the nourishment of the foetus. When the woman is not pregnant nature expels this blood by means of the menstrual flow as something superfluous to nourishment. You will notice also that nerves come to the matrix from the vertebrae; you will be able to inspect their passage. Furthermore, you will notice the neck of the matrix, which is not composed of many tunics but from a substance quite hard and mingled with flesh. It is harder in the anterior part where it adheres to the pubic bone. This neck is entirely muscular since women sometimes draw it and the vulva upward with a small voluntary motion, as I have often seen. Since there is in the upper part of the body of the matrix where its fundus is situated a certain bend or small depression some people, deceived by it, have believed the matrix is double. This is not true. The actual duplicity of the matrix arises from the difference of its two tunics. Many have also been deceived in their belief that the cotyledons are the heads of the veins which appear in the inner tunic of the body of the matrix. This tunic can be called the inner matrix, if you wish. But note that the cotyledons [placenta] are certain ligaments which bind the secundina [chorion] to the substance of the matrix. These ligaments proceed from the heads of the veins to the substance of the secundina and bind it to the matrix. The cotyledons are found only in a pregnant woman; I have seen them many times as they bind the secundina, and I have separated the latter from the substance of the matrix. In the anatomy of the non-pregnant matrix, therefore, one must not seek the cotyledons since they are not present unless the cadaver is that of a woman who died immediately after giving birth; they are found indeed in such cadavers. They are deceived also who believe there are many cells in the matrix, since actually there is in it only one cavity without any division, as sense perception indicates, unless perhaps by the phrase seven cells they mean the different positions, that is up, down, in front, behind, right, left, and the middle position. To call these positions by the name of cells is quite inconsistent. Therefore from these remarks you can see the diligence of certain anatomists and gain knowledge of the substance, size, number, shape, location, parts, and connection (*colligantiam*) of the matrix. But if you notice anything mentioned

[58] Massa finally explodes this fallacious theory of sex-determination although his own explanation also leaves something to be desired. For a brief account of the theory see L. C. MacKinney, "Sex Determination: A Scientific Superstition," *Medicine Illustrated* **3** (1949), No. 1: pp. 8–10. Aristotle, *De Generatione Animalium*, IV, 1, 763 b 20–765 b, discusses the idea and cites Anaxagoras, Empedocles, and Democritus among those who held it.

by me as in the lower venter (or body cavity) which has been placed in a different region by ancients and moderns, this should not be ascribed to me but to sense and reason since I have seen these things in a dissection which I have corroborated by reason in addition. These words therefore I intend as a sufficient introduction for young men on the method of dissecting the lower body cavity.

Chapter XXIIII

On the Method of Dissecting the Middle Body Cavity In Which You Will First See the Anatomy of the Breasts

Approach the incision of the middle venter, having disposed of the anatomy of the lower venter. But since it is useful to be acquainted also with the external parts, note these first in the lower venter beginning from its lower region where the venter terminates. This part is called the lower fork of the chest, where the scutal cartilage, or pomegranate [processus xiphoideus], is. On its right and left are the false ribs which begin from the vertebrae of the back and ascend with the true ribs which are also called the side. Hence arise pains in the side which trouble the ribs in man. The anterior part from the lower fork to the upper, where there is the gullet, or iugulum, is called thorax, chest, and cassum, from its emptiness, since although it is not a vacuum it is nevertheless spacious. In this anterior part are the breasts, one on the right, the other on the left. These are round. In the middle of each there is a part which protrudes somewhat. It is round. They call this the papilla, or teat. The breasts in women are larger and more prominent than they are in men. The parts under the arms where hairs originate are ascelares [axillares, armpits] or the parts under the wing; the metaphor is drawn from birds. They call the upper part toward the gullet the upper fork. From the upper fork going higher is the gullet and the lower jaw, with its parts. At the top of the back the posterior part corresponding to the upper fork, gullet [neck] and lower jaw they call the shoulder blades and neck. Now, lest I seem to waste time on what is well known, proceed to the dissection of the parts. Begin by cutting with a razor the skin from the scutal cartilage, drawing it toward the gullet as far as the upper fork where the gullet begins, stripping the skin carefully. When it is stripped the first thing you will see is fat (in those bodies which have fat) since, as I said, fat is not found in emaciated bodies for fat is not a member but a superfluity of the blood. After the fat you will see a fleshy panniculus [fascia] or membranous muscle, and in this part under the armpits you will see parts which are more fleshy than in other places. This panniculus proceeds to a cavity which is under the armpits among the muscles near the ribs and to the adiutory bone of the arm [humerus] through certain tendinous parts among the inner muscles of the arm which terminate at the adiutory bone and are united to it with many processes here and there. Look at these tendinous parts. Whether they are the tendons of another muscle than the aforesaid membranous muscle although they are multiplied [cannot be determined], since a number of tendons does not merely imply a number of muscles, as is clearly evident from the tendons which move the fingers. Many of these come from one muscle. Carefully lift the fleshy panniculus as far as the fork so that if it is possible you can find the membranous muscle which ascends from the end of the chest to the lower jaw and to the lip, which you will not easily find. When the fleshy panniculus is lifted look under the lower fork at the scutal cartilage [processus xiphoideus], which they call the pomegranate, located in the middle at the end of the seven complete ribs which make the fork. This cartilage has three angular parts, as if it were the flower of a pomegranate; two of these are in front, but the third is inside toward the stomach. These three parts compose the cartilage itself; it has a visible cavity which is empty. This emptiness ascends above between the diaphragm and the ribs to the bone cage (*clibanus*, oven) of the chest within the mediastinum. In that place there is a very narrow foramen, and therefore it is not always found except with great diligence in a few bodies. I was the first to discover this foramen and its use.[59] Beyond the fact that this cartilage protects the area of the mouth of the stomach from external injury by its width, it is useful by reason of its hollowness since through it evaporate certain vaporous materials from the mediastinum and the parts near the chest to the lower cavity as to a more ignoble place. Nevertheless, as I said, this cavity of the cartilage is not always recognized and thus it must be investigated with accurate diligence. When you have seen the cartilage look at both breasts. The first thing you ought to notice in them is a prominent round part in their skin called the papilla [nipple] since in this area the skin is thinner and mixed with some small amounts of glandular flesh about the size of a Venetian gold piece. Sometimes its color in a living woman is reddish tending toward a sort of purple. In the dead body it is, however, always black. Its thinness was created so that the child might suck the woman's milk more easily. But in the male this part is not so prominent nor so thin. When you have looked at the papillae look also at the substance of the breasts, or ubera, which remains under the large muscle [pectoralis major] that begins with its fleshiness in the middle of the chest on one side and on the other from the lower fork to the upper and terminates through a

[59] The foramen of the scutal cartilage (the ensiform or xiphoid process) is claimed as his discovery by Massa.

tendon at the anterior part of the adiutory bone of the arm [humerus] toward the armpit. In this muscle you will note two courses [incessus] of fibers. One of these descends from the upper fork toward the arm; the other ascends from the lower fork and the area around it toward the armpit. This muscle performs the function of two muscles although their fleshy substance is one. Note, however, that the tendinous part which terminates at the adiutory bone is doubled by being bent somewhat. It is continuous and united, nevertheless, down the length of the tendon since this muscle has two movements, one in front to move the arm as it ascends, the other as it descends, but with a small movement. Above these muscles is a sort of glandular flesh, whitish, rather thin in a woman, round and projecting so that in it the blood may be converted into milk. In this substance there are many veins and arteries; among these veins this glandular flesh is contained, white and insensitive, since although nerves come to the surrounding parts of the breast, they do not reach the glandular part itself. These veins and arteries are not so noticeable in men, and this fleshy panniculus is mingled with the substance of the breast. But before you proceed further look at the two veins [inferior epigastric] which ascend from the matrix through the fat of the breasts which you saved in the anatomy of the lower body cavity. These veins carry the blood from the matrix to the breasts and are scattered among them. This takes place at the time when women nurse their infants. Hence the doubt is solved as to why many women while they nurse children have no menstrual flow for some months and some of them lack menstruation for the entire time of their lactation. This happens because blood is carried through these veins to the breasts and it is not a superfluity. You see also why it is that the application of cupping-glasses under the breasts at the time of the greatest flow and most excessive flow of the menstrual blood checks it. See also the process of other veins which come from the armpits and from the upper fork to the breasts. These enter the substance of the breasts, bringing also blood for the creation of milk. From what has been said you see the error of that very great philosopher[60] who said the breasts are composed of the muscles of the chest with veins, arteries, and glandular flesh when the fact is that no muscle is mingled with them. He was deceived by the muscle which is under the breast. Lest such an error deceive you I wished to describe that muscle before inspecting the breast. You will see that the breasts vary in size according to the difference in women. The breasts are larger in pregnant women and in those who are nursing infants since they swell with the blood sent to them through the veins. The breasts, furthermore, protect the heart, both conserving its heat and tempering it by means of their complexion. With their weight they defend the heart from external forces which might penetrate and alter it. Also cut the substance of the breasts so that you may better see it together with its veins and arteries, for nerves do not come to the glandular flesh but to the region of the skin and especially to the papilla (or nipple) and it is in this region particularly that the breasts have sensation.

Chapter XXV

On the Anatomy of Certain Muscles of the Chest

Looking from above you see that the breasts lie above two muscles [pectoralis major] which cover the entire anterior part of the chest, one on the right, the other on the left. They are of triangular shape. Their base lies toward the thorax in the middle of the chest and the upper end of the triangle points toward the adiutorium of the arm [humerus]. Of the two angles of the base one points to the upper fork, the other to the lower fork. These muscles have a diversity of fibers, as I said, on account of their diversity of movement. When they are lifted up you will see two other smaller muscles [pectoralis minor] that move the arms toward the sides of the chest. You can inspect this movement as it draws the muscle. These muscles lie under the first large ones, above which you saw the substance of the breasts. You can also see the muscles which draw the arm downward, those which proceed to the shoulder blades, those which pass under the shoulder blades [serratus anterior], two which proceed to the neck, and two broad muscles which are continuous with the spine of the back; these are called deltoid muscles by the Greeks since they are similar to the capital letter delta. These muscles are also called cervical muscles and are of almost the shape of a certain winged bird. Look at them after they are stripped; they have four angles. But since I have decided not to describe here the anatomy of all the muscles I shall pass on to view the remaining parts of the chest, with the promise that I shall write about the anatomy of the muscles to the extent of my ability in another treatise, the great Lord willing. I shall there discuss the anatomy of certain muscles unknown to the moderns, such as that one which proceeds from the chest to the jaw [sternocleidomastoid]; you will see there many things passed over by the ancients, such as also those muscles which move the glottis [sternohyoid, sternothyroid, etc.] or the head of the bronchus in the formation of the voice and the inspiration of air.

[60] One thinks immediately of Aristotle, but he does not mention muscles in his reference to the breasts or in his detailed description of them in *De Partibus Animalium*, IV, 10, 688 a, b. Neither does Galen in his account of the reasons why the breasts adhere to the chest in *De Usu Partium*, VII, 22 (Vol. III K., 602–608). I find nothing in Hippocrates that might identify him as the philospher to whom Massa refers.

These have a tendon in the middle. There are also two muscles [genioglossus, hyoglossus] which move the tongue outward, and other parts not known to the ancient philosophers and physicians. Let it not be ascribed to my arrogance if I point out parts of the body not seen by Aristotle and Galen and many others, since it is given to each person to seek better things to the extent of his ability. For questions and difficulties not solved by Averroes as well as other wise men have made me a sedulous investigator of the parts of the human body and of other things. Therefore, once more I beg my brothers who will after me commit themselves to anatomy to discover the members and their uses and to write down their method of investigation without jealousy so that they may share their usefulness with the world in the future. But let these words suffice. You can see for yourselves the rest of the muscles which move the arms, the neck, and other parts. These muscles proceed from the chest both in front and on the sides and back. They are easily seen and do not require much investigation. Proceeding in this manner, first begin to lift from the area of the origins of the muscle where they adhere by means of their fleshy portion and proceed to strip as far as the end of their tendons those among them which have a tendon, since not all muscles have one. You will recognize the muscles by the panniculus which covers them since it both distinguishes and covers. In this work use care in stripping with different instruments according to need, for sometimes sharp or cutting instruments are best and sometimes blunt instruments are better so that you may not cut the panniculus or any other part of the muscle nor create confusion. So proceed to strip as far as the tendon in those muscles which have one, wherever it is found. You should not cut the tendon from the bone if it terminates at the bone before you experience the motion of the muscle. If you wish to do this place the muscle again in its place from which you stripped it and draw the tendon through the fleshy part toward the origin of the flesh: thus you will see how the muscle moves. In each muscle that has a tendon the fleshy part draws the tendon to itself since the nerves from which motion originates come first to the fleshy part; then they proceed through the entire muscle. But muscles which have no tendon are contracted to their origin by means of fibers, or filaments. I wish these words to be sufficient concerning the muscles since if I wrote in detail about everything my discussion would exceed the limits of an introduction. Therefore when you have lifted the muscles which cover the ribs of the chest there appear the ribs which are covered on the outside by their own thin panniculus adhering to them since each bone is covered by its own panniculus. They are covered on the inside by the pleura panniculus or membrane. Between these ribs are [intercostal] muscles which dilate and contract them; you will recognize these muscles from their diversity of stamens, that is, transverse, of which there are many, and broad, of which there are fewer. These things you will see, since they are evident to the senses, although difficulties about their number are not lacking among philosophers and physicians. On this account you will note that between rib and rib there is flesh which has filaments or fibers that proceed transversely and also laterally. Their courses are four in number, as is apparent to the senses, and therefore some have excellently said that there are four intercostal muscles. But those who consider that there are only two uses for these muscles (although they have many filaments,) and who inspect their different courses nonetheless declare that there are two muscles, that is, the superior-external, whose filaments proceed transversely, which they say are dilating muscles; and the inferior-inner, which they say are contracting muscles; these have broad filaments. These muscles have very thin tendinous parts which adhere to the panniculus which covers the ribs on the outside. I have often pointed out four distinct muscles between the ribs, not, however, without difficulty. Thus even if the matter is not of much importance to physicians and philosophers yet when the fibers are different and distinct, of necessity one must say that the members are distinct, although one contradicts the chief anatomist [Galen]. But while you note the intercostal muscles, also inspect the [intercostal] nerves which are between the ribs with the muscles themselves and proceed lengthwise of the ribs. Look also at the diaphragm or transverse septum. It is a muscle which separates the natural members from the vital members and moves the chest by dilating it. While it is moved it presses the chest. While it is quiet between movement and movement (for Aristotle said there is a quiet period midway between two movements)[61] this quiet period constricts the chest and motion dilates it, drawing every tendinous part toward the ribs. This part is drawn from the fleshy part which adheres to the ribs. For according to Galen's teaching the diaphragm is a muscle with a tendinous part in the middle and a fleshy part around it. This formation is necessary since if the fleshy part were in the middle and the tendinous part passed to the ribs it would not dilate but would firmly constrict the chest. Whether this motion is voluntary or natural (involuntary) you know from the teaching of Galen, *On the Motion of the Muscles II*,[62] where he says he recognized that all

---

[61] Aristotle, *Physica*, V, 6, 299 b: "A motion has for its contrary in the strict sense of the term another motion, but it also has for an opposite a state of rest (for rest is the privation of motion and the privation of anything may be called its contrary) and motion of one kind has for its opposite rest of that kind, e. g., local motion has local rest" (translated by R. P. Hardie and R. K. Gaye in the Oxford Translation of Aristotle edited by W. D. Ross, Vol. 2).

[62] Galen, *De Motu Musculorum*, II, 8 (Vol. IV K., 455): Quae igitur sunt huius actionis [excretionis] instrumenta? Plura quidem particularia, sed duplicia genere; eorum enim alia animae, alia

the muscles have in addition to a voluntary motion a natural motion also. Thus we must say that the diaphragm, or transverse septum, has both a natural and an animal motion which we recognize in the evacuation of the excrement from the bowels and also in respiration at the time of necessity. The shape of the diaphragm is round like the periphery of the lower chest where it is situated, beginning from the back from the first of the false ribs and proceeding as far as the scutal cartilage [processus xiphoideus] through the extremities of all the false ribs. From what has been said you will recognize their remaining features; hence enough concerning the muscles of the chest.

Chapter XXVI

On the Anatomy of the Bones of the Chest and Ribs

When the middle body cavity has been thus denuded of its muscles observe the king's citadel constructed by its bones in order that the heart, as the chief member in the middle and located in the safest place in the body, might be protected like a king. This chest or place of the heart is composed of twenty-four ribs. Twelve of these are on the right side and the other twelve on the left side. The substance of the ribs is diverse, for they begin from the spine or from the vertebrae of the back where they are attached and proceed in an arched position on right and left, beginning in front from the upper fork and from the top of the back and behind where the vertebrae of the neck become continuous with the first vertebra of the chest and toward the anterior part. There they change their substance for when they proceed to the anterior part of the chest they are cartilaginous. The first seven of the ribs, beginning above, are complete; that is, they have a perfect arched shape both on one side and on the other. They are joined to the bone of the thorax which is in the middle of the chest by means of a cartilaginous portion. This bone of the thorax is not a single continuity but a single contiguity, since it is composed of seven bones which are united with very strong ligaments so that you cannot separate them with a sharp knife. They are separable, however, by means of a thorough boiling. These seven bones correspond to the seven whole ribs on one side as well as on the other. Below these ribs there are ten false ribs, or ribs called untrue, five on the right and five on the left. Some of these with the whole (or entire) ribs compose the lower fork, where the scutal cartilage exists. These also according to their parts are diverse in substance since they proceed from the spine through the bony part and terminate toward the anterior part in a cartilaginous substance. The last of these is the smallest and is almost completely cartilaginous, but in the upper part toward the gullet above the first of the whole ribs there is the bone of the upper fork, which is not separated from the bone of the thorax but is part of this bone. Above the bone of the upper fork is the part called iugulum or neck. This bone is in fact composed of three bones, one of which is in the middle and is the upper part of the bone of the thorax [manubrium or sternum]. It has a cavity, or curved shape, toward the upper part. There are two other bones [clavicles], one of which on the right points toward the adiutorium of the arm [humerus] and rests upon the adiutorium itself. The other points toward the left. These bones check the motion of the arm lest by attraction from its muscles toward the anterior part the adiutorium should be drawn forward more than it should be and thus it is made more firm in its other operations also.

naturae sunt; atque animae quidem instrumenta motu voluntario semper agunt, instrumenta vero naturae sine motu voluntario. Diaphragma quidem et omnes musculi epigastrii instrumenta animae sunt, intestinorum autem omnium structura una cum ventriculo naturae.

Chapter XXVII

On the Anatomy of the Pleura, the Mediastinus, the Capsule of the Heart, and of the Diaphragm

Cut all the ribs transversely in the cartilaginous part both on right and left in the place where they are joined to the bone of the thorax. Do this with a sharp knife, carefully stripping the ribs themselves from the panniculus of the pleura. You can, however, leave some ribs so that you may later see the intercostal muscles which together with the ribs touch the pleura itself in the interior region on one side as well as on the other. You will strip these ribs almost as far as the spine. When they are stripped move them aside, cutting also in the lower part, since there will appear the pleura. This is a panniculus which covers the ribs in the inner part of the chest and contains the spiritual members. These are the heart, lung, and the remaining parts, as you will see. The substance of this panniculus is hard, white, and thin. As I stated, it covers all the inner parts of the chest, anterior, posterior, lateral, superior, and inferior such as the diaphragm. It is a double panniculus with one on the right and the other on the left. Since it is in the middle of the chest, when it is flexed on one side, the other panniculus is similarly flexed on the other side. From this flexion and conjunction of the two panniculi there is created the mediastinus [mediastinum], dividing the chest into two parts, that is, one on the right and one on the left. Thus you will see that the pleuritic panniculus, which is double with a right and a left, creates the mediastinus panniculus. This panniculus is composed of two pleuritic panniculi. It is white and thicker than the pleuritic since it is composed of two panniculi and also because it is not extended above the bones and the other members on all sides like the pleura. This mediastinus panniculus is separable according to its parts. But in order that you may see it, cut the pleura on one side as well as on the other from the upper part near the bone of the thorax and allow it to descend lower down; this panniculus will appear together with the lung and the rest of the members. With its double substance it covers the capsule of the heart and separates the lung into a right and a left part; you will see in the right side three fibers, or lobes, of which one is small. This one is not accepted by some as a part of the lung which serves the heart but as a covering for a vein, and therefore it has been said the fibers of the lung are equal in number, as appears from Galen, *On Anatomical Procedures*, VII.[63] On the left side of the lung there are only two because the mediastinus mediates between the lobes of the lung and separates them into two divisions. This division was made in the chest so that if one part was injured the other would remain undamaged. But you will note that both to the pleura and to the mediastinus come veins, arteries, and nerves from the reversives. You must leave these parts untouched until you dissect the rest. Before you come to these parts note that this panniculus, called the pleura, covers the chest from the upper fork inside as far as the diaphragm from which you must strip it so that the parts of the diaphragm may be better discerned. It is this panniculus in which abscesses very often occur, but usually warm ones, in one side as well as in the other and also in the mediastinum. These abscesses are very alarming; cold abscesses, although not so alarming, are nonetheless a bad illness since not only is matter in such an abscess absorbed into the substance of the pleura but also into the substance of the intercostal muscles and often ages there so that pain becomes heavy and continual coughing troubles the patient, as I saw in the magnificent Lord Bartolomeo de' Panzati,[64] a very famous Florentine noble, who died of an old pleuritic abscess. He was troubled for many years by severe pain and a continual accompanying small cough. When both pain and cough increased a fever supervened. Whether the intensification of both pain and cough resulted from new matter in the abscess or from matter already there which began to putrefy is a question which cannot be settled here. But the physicians who attended him, paying no attention to the increase of his pain and coughing, insisted that the fever had another source since it was putrid and to this fever as to the more important illness and the more mistrusted they directed their remedies alone in their meddling way. He died on the seventh day, unless I am mistaken. When at the entreaties of his son, who wished to know the cause of his father's death, his body was opened in this region, those who dissected him in my presence found in the side where he felt pain in the horizon of the diaphragm a large abscess, hard, and partly suppurating, partly sclerotic, or very hard. In the cavity of the abscess I found two hard stones of reddish color, half an ounce in weight. The hardness of the abscess existed not alone in the substance of the pleura but in the intercostal muscles of the false ribs; it also included part of the diaphragm. I also saw many others who died of pleuritic abscesses; these were large and sometimes in one side, sometimes in the other, in the posterior and in the anterior regions, and in the mediastinum, which was worse. Though these remarks are not without their usefulness, let them suffice. I come now to the dissection of the other parts. When you have seen the panniculi both of the pleura and of the mediastinum you can strip the pleura in the lower part of the chest from the substance of the diaphragm and examine it before you move aside the bone of the thorax which you had left alone, attached in the upper part to the upper fork and in the lower part to the extremities of the lowest false ribs. In this way we can better inspect the spiritual members and their locations. After the pleura is stripped the first thing you will see, as I declared, will be the diaphragm, or transverse septum; this is a flat, round muscle with a tendon in the middle, as I said, and a fleshy part in its extremities. This fleshy part originates from the extremities of the false ribs, beginning in the back, where you will see less of it. The fleshy part here arises from the first vertebra of the false ribs, that is, the twelfth vertebra, and terminates with its fleshiness surrounding all the extremities of the false ribs as far as the cartilage called the pomegranate [processus xiphoideus]. The filaments of the diaphragm proceed from the circumference toward the center, where there is a foramen through which the meri [esophagus or gullet] passes and makes the mouth of the stomach. But on the other side toward the lower body cavity the diaphragm is covered by the panniculus of the peritoneum. Look at the diaphragm on one side and the other since they are of the same substance and have the same processes of fibers in one part as well as the other, and you will see that it divides the natural members from the spiritual or the lower body cavity from the middle cavity. Nor do I wish you to lift an eyebrow if I have placed the anatomy of the muscle of the diaphragm here amid the dissection of the panniculi, for this would imply a frivolous criticism; by dividing cavity from cavity or member from members the diaphragm with its thickness protects the spiritual members from the evil and fetid vapors of the intestines. If it is a muscle it performs nevertheless in this case the function of a panniculus rather than of a muscle; therefore it is not without reason that I have spoken of it here. Look, I beg of you, rather at its use and abandon frivolous speculations. The diaphragm has also another foramen near the back through which pass the chilis vein [vena

[63] Galen, *De Anatomicis Administrationibus*, VII, 1–4 (Vol. II K., 589–596) describes the lungs and their vessels and panniculi.

[64] I have been able to find no information on Bartolomeo de' Panzati of Florence.

cava] ascending and the aorta artery descending. In this part you will note that both vein and artery send forth on right and left two [phrenic] branches to vivify and nourish the diaphragm. See also the branches of the reversive nerves which come to it. These nerves cross to the mouth of the stomach through the gullet. Look in addition at the transit of the nerves which come to the diaphragm from the vertebrae. Also see the rest of the members since I have decided not to describe everything I have seen but only certain items worthy of note. For example, since the lung is attached to the diaphragm by certain ligaments it is through them that the motion of the lung succeeds the motion of the diaphragm. The principal use of the diaphragm, as Galen bears witness,[65] is to move the chest for breathing; he justly praises himself as the first discoverer of this operation. The diaphragm has other uses, such as the expulsion of the feces and of flatulence and of the foetus in childbirth. In women it serves for the retention of the breath and for other purposes. When you have seen the diaphragm and other parts which you have released into their place as far as you were able, return to strip the pleura from one side of the bone of the thorax. You will see between the right and left panniculi which form the mediastinum a notable cavity in proportion to the contents of the bone of the thorax. This cavity [anterior mediastinum] exists before [the panniculi] unite and cover the capsule [pericardium] of the heart. When you have noted this cavity, continue to strip the panniculus itself from the capsule of the heart as far as the back and you will see that in the back these panniculi are united as far as the spine nor is there any cavity therein. But when the capsule is stripped examine it; it is composed of a pannicular substance, thick and quite hard, white, and dense in comparison with other panniculi. To this capsule come veins and arteries which proceed from the base of the heart itself. Its size is well known, for it is quite ample so that the heart may dilate within it to the ultimate dilation. It is full of clear water [pericardial (serous) fluid]. See also how this capsule is partly covered by the lobes of the lung. Therefore, in the process of stripping it you can proceed very well when you have lifted away those lobes. There are some people who deny that this water is found in the capsule of a living animal. They say it contains a multitude of spirits because it does not possess reason or sensation. Yet I have always found water in the capsule not only in a noticeable but a large quantity, nor can I believe that it is generated while the animal is dying from the dissolution of the aforementioned spirits which certain people say are found in the capsule, since this water is quite thick in reference to its substance and spirits are like vapors. It is rather from the vapor or spirits made heavy by the cold resulting from the deprivation of heat in the dead body than from dissolution of the spirits, as they say, that a very small quantity of water is generated, the thickened substance of the spirit. However, when this quantity is large it means the water is not generated after death or while the animal is dying but that it exists in the living body on account of its many uses, mentioned by Galen as follows: the heart is first bathed by this water lest the heart be dried out through its continual motion; hence that languishing ape, about whom Galen writes,[66] was found without water in the capsule of its heart. Add also the fact that from its actual humidity the heart is moistened, since it is the warmest of the members on account of its spirit, and because this member needs continual refrigeration and removal of secretions by other members, such as by the brain and the lung, so that the sharpness of its heat may be blunted, the sharpness created from its association with dryness. Dryness is the creator of heat and humidity is what blunts or checks dryness. In order that the heat of the heart may be somewhat cooled by its humidity, nature usefully for this reason employs water in the capsule of the living creature. The substance of this water is made by straining the blood which comes from the liver to the auricle of the heart by means of a continual resudation (or sweating back). The shape of the capsule of the heart conforms to the shape of the heart itself, that is, it has a wide upper part but a narrow lower part in the manner of a pine cone. It is very sensitive on account of the nerves which you see coming to it. You would not for this reason call it nervous, that is, composed of nerves, for the reasons mentioned[67] before in the discussion of the lower body cavity. The capsule is pannicular, of its own substance, and created, beyond

[65] Galen, *De Anatomicis Administrationibus*, V, 5 (Vol. II K., 503) describes the function of the diaphragm in respiration. See *Ibid.*, 657: Verum praeceptores nostri non recte censuerunt, solum septem movere thoracem in respirationibus, attollens quidem, quum tenditur, collabi autem in sese . . . ; compare *De Usu Partium*, IV, 14 (Vol. III K., 314); *De Motu Musculorum*, II, 5 (Vol. IV K., 442–444) on breathing as a function of the diaphragm and the muscles of the thorax: Caeterum (ut ego arbitror) nemo adhuc causam dixit, sed dubitatione sola (cuius iam mentionem feci) adscripta causam se putaverunt invenisse. While recording the doubts of his predecessors concerning the function of the diaphragm Galen does not specifically claim that he himself has discovered it. See also *De Causis Respirationibus* (Vol. IV K., 465–469), to which Galen refers his readers in *De Anatomicis Administrationibus*, cited above. Plato called the diaphragm the transverse septum, Aristotle called it the hypozoma or diazoma; see the long note on the subject in Charles Singer's translation of Mundinus, p. 105.

[66] Galen, *De Locis Affectis*, V, 2 (Vol. VIII K., 303) gives the famous account of his ape and the capsule of its heart.

[67] That is, the reasons given on p. 192, "I cannot really call it sinewy, if that is called sinewy (*nerveum*) which is composed of nerves. . . . " The word *neuron* in Greek can mean both nerve and tendon or sinew; see Aristotle, *De Partibus Animalium*, III, 4, 666 b 10 on the confusion of sinews, tendons, ligaments, and nerves; Galen, *On Anatomical Procedures*, Singer's translation, pp. xix, 243, note 57; and Alessandro Benedetti, *Anatomice*, V, 14, where the confusion is especially prominent. Jacopo of Forlì, *Expositio*, on Galen's *Ars Medicinalis* (Venice, Giunta, 1547), fol. 111 v, comments on the matter.

the uses already described, for the protection of the heart lest it be injured either by external members or vapors. The capsule is created simultaneously with the heart. I intend this to be sufficient concerning the anatomy of the members described. Put back all the members therefore in the best way you can into their place so that you may be able to see the connection of the other members contained in the chest, such as the lung, windpipe, gullet, reversive nerves, and the rest.

Chapter XXVIII

On the Anatomy of the Heart

Cut the capsule of the heart, or the royal chamber, so that you can see the king himself, who is the heart, for it is justly called by everyone the chiefest (*principalissimum*) of the members, first to live, last to die,[68] lying in the capsule as in the safest of chambers and in the middle of the chest between the lobes of the lung. The shape of the heart is that of a pine cone somewhat flattened or depressed, with the broad upper part inclined toward the right side of the chest covered by the substance of the lung beyond the capsule. But the lower part of the heart is pointed and inclines toward the left-hand lower side of the chest. This cuspis, or point, of the heart is not covered by the lung since there are only two lobes on the left side. These are smaller than the two large ones on the right side. Hence as a result the motion of the heart is felt more strongly on the left side under the breast since here it is not covered by the lung. The substance of the heart is fleshy, not like the flesh of the muscles but hard and dense like much of the flesh of the kidneys. But it differs in substance since the heart seems to be composed of quite firm fleshy fibers. The substance of the heart is covered with a thin membrane upon which you will see a notable quantity of fat. This fat, in the hottest of the members, is not, as some people have said, generated by the cold since there is no cold in the heart, but is created from the thicker part of the superfluous blood united through its viscosity with some of its thin parts, but not all, broken down by the agency of heat and converted into white substance. I know it is a common belief that what is congealed by cold is dissolved by heat. The converse is not true, however, for salt is congealed or hardened by heat with some of its thin parts broken down by the heat of the sun itself. Salt is not, however, dissolved by cold; it is well dissolved by humidity. But fat is congealed by the complexional heat of the animal; it is not, however, dissolved by that heat, although it is dissolved by the heat of the sun or of fire. In the material of fat when the moister parts which rarefy its substance are broken down and remain both viscous and conglutinated they make fat from their own substance, since it is not made from all material by the agency of heat, as I said. This fat is located by nature upon the membrane mentioned above for the protection of the heart and for the conservation of its heat as well as to resist the drying out of the heart's substance. I am not unaware of the fact that combustive heat acting upon a humid substance turns that substance black. This is not true, however, of a moderate heat, as is apparent from wax which turns white in the heat of the sun if it is often moistened or sprinkled with water so that it is not burned by the sun. But enough of this. It is said that those who have the broad upper part of the heart inclining toward the left use their left hand instead of the right; but those who have their heart straight in the middle, not inclining toward right or left, are ambidextrous, that is, they make use of both right and left hands in the same manner at will. Look at the heart lying in the middle and inclining with its broad part toward the right. In this part there are two partly membranous substances which they call the auricles of the heart, one on the right and one on the left. The right auricle is more ample than the left; the left one is far smaller. Furthermore, the left auricle is harder and more dense than the right. This is due to the fact that the right auricle contains blood which is thicker and not very sharp. But the left auricle contains spirits which are thinner and warmer than blood. The auricles have been given to the heart by nature in order to serve the heart's ventricles, that is, the right and left, since the auricles are united at the extremities of the latter. They are called the wrinkled auricles because they can be dilated and constricted according to opportunity. They have another use: the right auricle holds back the blood which is prepared when by a continuous transmission from the right ventricle the blood elaborated by it is sent across to the left ventricle lest the latter should lack some of the material prepared by the right ventricle and also that the blood pressed out there in the constriction of the heart should be held back. Thus this auricle saves the material sent from the liver, or attracted from the heart. I do not wish to enter upon this difficulty; it is enough to say that the material produced by some preparation is kept there and when the ostiola [atrioventricular valve] are opened it enters the heart. You will see these ostiola around the extremity of the ventricle; they have thin ligaments [chordae tendineae] which draw them within the substance of the heart. The function of the left auricle is to preserve there the spirit transmitted to it and elaborated and perfected in the left ventricle of the heart and to reduce somewhat the spirit's sharpness by a certain delay of it in the left auricle. Thus by means of a better temperament better operations result from the spirit transmitted through the arteries. Hence ostiola are given to the left ventricle of the heart within its substance. These are opened within the heart's substance in order that they may yield to the spirit as it enters the inner parts

---

[68] See note 47.

of the ventricle. They also have suspensories or ligaments within the substance of the heart. But to the right ventricle are given ostiola which also open inward so that they may yield to the blood which enters the right ventricle. In that part also of the upper heart which is broad, as I said, where you saw the auricles you will see two large vessels united to the auricles themselves. One of these on the right is called the chilis vein [vena cava] ascending. It proceeds to the heart from the gibbous part of the liver and perforates the diaphragm. The other vessel, on the left side, is the aorta artery which first carries spirit to the other members from the left ventricle of the heart; it later branches out in its descent and ascent together with the aforesaid chilis vein [vena cava]. The function of the vein which ascends to the right ventricle of the heart is to carry the blood received from the liver to that same right ventricle so that it may be converted into spirit. The function of the artery is to carry spirit from the heart to all the members so that they may live and also, changed through some process by the members themselves, to become animal and natural, as the physicians say, owing to some alteration by the proper member. You will also note in the heart that the chilis vein [vena cava] before it reaches the right ventricle sends a notable branch [coronary sinus] to the upper and exterior part of the heart. This is ramified into many branches throughout the entire exterior substance of the heart in order to nourish it. Thus from this fact you will understand that if the blood were generated in the heart nature would not send a vein from the liver to nourish it just as the artery sends two little branches [coronary arteries] to the exterior substance of the heart. You will also note the roots of the heart which are in the upper part where it adheres to the back near the auricles. These are sinewy or ligamental and hard in order to withstand the continuous motion of the heart. When you have observed these items cut the heart in its right hand part beginning from the point (*cuspis*) lengthwise toward the upper part but not as far as the substance of the auricle. The incision should not be deep lest the intermediate wall be injured. When the incision is made there will appear a notable cavity which is called the right ventricle in which, as has been said, the blood is prepared so that it may be converted into vital spirit and so that part of the blood may be sent to the lung to nourish the latter through the arterial vein [pulmonary artery]. Therefore at the top of this ventricle where the auricle is you will see two orifices. The large one of these is the orifice of the chilis vein which carries the blood from the liver, as I said. In this extremity there are three ostiola [tricuspid valve] already mentioned, of a white pellicular substance, semicircular in shape, which close the cavity of the heart and are opened within the substance of the heart so that in their opening they are based upon the wall of the heart. They have slender ligaments which pass inward in the substance of the heart and which restrain the ostiola. Another orifice [pulmonary artery] passes into the substance of the lung, carrying nourishment to the latter from the blood prepared in the right ventricle of the heart. Through this vein the lung also sends air to cool the heart. This is accomplished in the following fashion: when the ostiola are opened to receive nourishment there passes out the spirit or air which was detained by the lung, which is cold in comparison to the heat of the heart. This vein is called the arterial vein since it also brings blood to the lung. It is composed of two tunics in order to be strong in resistance to the continual motion of the lung and the sharpness or heat of the blood. In this orifice of the arterial vein you will also see three ostiola [semilunar valve] of the same shape and substance but smaller than those in the ventricle of the heart since this vein is far smaller. These ostiola open within the substance of the lung. Note also that the substance of the heart in the right side is not so dense and thick as the substance in the left side. This is due to the fact that the blood is thicker and less active than the spirit and therefore the walls of the left ventricle are thicker and harder. In order to see the left ventricle itself cut the substance of the heart in the left part as you did in the right part beginning from the cusp and ascending toward the base. In the left part or side you can cut more deeply since (as I said) the walls are thicker because of the reasons mentioned above, in order to contain the subtle spirit and to preserve it. When you have made an incision you will see the left cavity, smaller than that in the right side. They call this the cavity of the ventricle; it is smaller since the quantity or bulk of the spirit is far less than the quantity of the blood. Look also at the extremity of this ventricle since there are therein two vessels such as you saw in the right ventricle. They have their orifices, one of them large but not as large as that of the chilis vein [vena cava]. This large orifice is of the aorta artery which, as I said, conducts the spirit into all parts of the body. In this orifice are ostiola [mitral valve], already described as to their shape and substance, but situated a little lower, that is to say in the cavity of the ventricle with its ligaments, as you saw in the orifice of the chilis vein [vena cava]. But note at this point in my description that of the three aforementioned ostiola I have always found one which is not entirely continuous with the wall of the heart. It has a separation through which the spirit ascends to the aorta artery so that it may pass throughout the entire body. Many of the ancients did not recognize this separation, but I have pointed it out.[69] These ostiola

[69] It is difficult to ascertain the exact truth of such an assertion by any early sixteenth-century anatomist, and the attribution of "discoveries" of certain features of the body is a perilous business. On some attributions see Bartolomeo Corte, *Notizie istoriche intorno a'medici scrittori milanesi e a' principali ritrovamenti fatti in medicina dagl' Italiani*, etc. (Milan, G. P. Malatesta, 1718), p.

open within the substance of the heart toward the cusp just as in the right [ventricle] in order that they may yield to the spirit as it enters the inner part of the heart. Look also at the beginning of the artery since I have often found there three other ostiola [semilunar valves] which have no ligaments; thus the spirit may issue freely to all the parts of the body through the artery and [the ostiola] open within the artery. Another orifice of the vessel is the orifice of the venal artery [pulmonary vein]. This artery passes through the substance of the lung and conducts the air which cools the heart when the ostiola are opened to receive nourishment; as they are opened the air passes out from the arterial vein and cools the heart of its heat. This artery also conducts the blood, or the thicker part of the thin blood not converted into spirit, for the purpose of nourishing the lung. In this orifice of the venal artery there are only two ostiola, of pellicular or skinlike substance and semicircular shape. These are smaller than all the others and open outward and inward. They do not close perfectly nor do they have suspensories since in this part there is not such a great impulse of blood nor so great a quantity of it and also because the latter requires a greater refrigeration. Hence the foramen in this part is more free and the ostiola unattached. They do not close the foramen perfectly and it is thus free to receive nourishment and refrigeration. You will also note that the venal artery consists of one tunic only since for the greater part it serves the heart, carrying air with which to cool it. Part of the blood then passes to the lung for its nourishment, for on the left side there is a smaller quantity of lung not only in respect to size but also number [of lobes]. Difficulties exist among the most learned of the ancients as to whether a third ventricle is found in the heart; this ventricle does not always appear in the dissection of the dead. Therefore you must note that between these ventricles there is a rather thick wall [interventricular septum] at the top of which toward the base of the heart or toward the broad upper part there is a certain small cavity which I found and pointed out in 1534 in a certain person who had a large heart of such great size as I have never seen in another man. I have seen this third cavity or ventricle also in others, but not in all [the bodies I have examined]. Therefore you will know from these words that there exists a third ventricle not only in large animals, as Galen said,[70] but also in men. From this you will also recognize the error of those who say the third ventricle is in the middle wall since this wall is dense, hard, and without a cavity. Note also that in the ventricles of the heart there are certain fibers or certain fleshy substances which intersect each other, proceeding in different courses. They have the shape of the small muscles, round and oblong. If anyone inspects these parts carefully he will see that they compose the bulk of the heart for they proceed almost to the ultimate surface of the heart although they are united. It is true, however, that in the surface of the heart's exterior there appears no further process of these fibers, but the substance is dense and hard, especially at the point of the cusp. In this substance of the heart you will also note different courses of filaments which cause the different motions of the heart. The use of these fibers as described is to prevent the heart from dilating beyond a proper limit. Note also that in large animals such as oxen and old stags there is found in the upper region of the heart a certain hard cartilaginous part, a substance I have not found in men. But since those who write about the heart say that they have seen many things, therefore since I too have often seen wonders it will not be displeasing to note some of them here. In 1533 in the monastery of Sts. John and Paul I examined the heart of a certain dead man I had dissected. He had died of a head wound. I knew him while he was alive and well and saw him while he was suffering from his wound and under the care of a certain surgeon. I judged that he would die from the wound since it involved an incision of the panniculi. I asked the surgeon whether he would be willing to inform me when the patient died so that I could dissect him, since I possessed a license from the presiding authorities to carry out dissections.[71] The surgeon did so, for he was a stranger passing through town on a pilgrimage. During the dissection I saw many things worthy of admiration. The first of these was as follows: in the right ventricle of the heart toward the ostiola in one of those fibers was a notable

51, on Eustachius, Falloppia, Casserius, Achillini, Massa, Aranzio; p. 52, on Aranzio and Berengario da Carpi. See also Luigi Belloni, *Giovanni Battista Morgagni: Gli inventori anatomici italiani del xvi secolo nel carteggio col medico milanese Bartolomeo Corte*; XV Convegno Nazionale della Società Italiana di Anatomia (Milan, Industrie Grafiche Italiane Stuchi, 1953, 23 pages).

[70] On the difficulties presented by the third ventricle of the heart, whose existence Massa persists in defending down to 1534, see Arthur Platt, "Aristotle on the Heart," in Charles Singer, *Studies in the History and Method of Science* **2** (Oxford, 1921): pp. 521–532. Massa erroneously attributes to Galen the belief held by Aristotle on the three ventricles. Galen himself says in *De Anatomicis Administrationibus*, VII, 10 (Vol. II K., 621), Singer's translation, p. 187: "What wonder that Aristotle, among his many anatomical errors, thinks that the heart in large animals has three cavities? It is not surprising that, lacking anatomical experience, he failed to find the parts, and he deserves to be forgiven." See also *De Usu Partium*, VI, 9 (Vol. III K., 442) and *Galen on Anatomical Procedures–The Later Books,* translated by W. L. H. Duckworth, M. C. Lyons, and B. Towers (Cambridge University Press, 1962), Book XIII, 9, p. 173. Singer in his translation of Mundinus (*The Fasciculo di Medicina*, etc., Florence, Lier, 1925), p. 106, note 85, points out that Mundinus copied Avicenna in maintaining the false notion of the three ventricles in the heart. Lauth, *op. cit.*, p. 336, refers to Massa's description of the papillary muscle of the heart's ventricles.

[71] Evidently all reputable physicians at Venice in the sixteenth century held a license to make post mortem examinations; in addition, public dissections were required by law since the early fourteenth century in that enlightened city.

abscess which ascended as far as one of the ostiola. This ostiolum was also swollen with a notable tumor and was abscessed. In this man I also saw a left auricle of the heart which was very small, completely ulcerated on its exterior with polluted blood or pus visible. Also in the substance of the lung in the left side there was a large concave putrid ulcer much larger than half a hen's eggshell with its usual depth. In this cavity was much bad, putrid blood. This man, however, while I knew him in good health before he was wounded in the head never complained of any pain nor did he cough. After he was wounded he remained in his bed. I then found in his head two abscesses, one in the substance of the brain near the wound, the other in the posterior cerebellum, for it was a wound in the right side near the sagittal commissure with incision of the panniculi. These abscesses were open; they suppurated with abundant pus and from them he died, in a state of madness before death with paralysis of the side opposite to the wound. Hence there occurs a speculation not entirely without merit as to how it happens that those who are wounded in the head usually before death fall into a paralysis of the side opposite the wound and universally down the entire half of the body. But since it is not my intention here to give the causes of illnesses but to recognize the parts of the body through dissection you can consult my treatise on wounds of the head[72] on this matter. Hence enough has been said for those who are intelligent to realize that the heart also can suffer not only a bad complexion, destruction of its continuous function, and a bad, disjointed composition[73] but also abscess. In order that you may not believe I saw this in a dream you should know that many very learned physicians and philosophers were present whose reliable testimony I am happy to possess. Although I have mentioned them earlier it will not be out of place to name them also here. They are Vittorio Trincavella, Maffeo de' Maffei, Francesco Marino, and others. But did you not also see these things, most learned Gerolamo Marcello, and Gerolamo Ferro and our doctor Ricci the scribe, learned in both civil and canon law? Other very learned noblemen saw these things. But in order not to draw matters out into boredom because of the excessive length of my discussion I have decided to omit many items which I saw. One of them I shall not pass over without mention, for you should know (emending *sis* to *scis*) that I dissected a freak of nature born at Fossa Clodia.[74] It seemed to be two males joined in such a way that the joining began from the lowest part of the thorax and extended to the umbilicus. To these bodies thus joined there were two heads each propped up with its own neck, two chests, four arms and hands, but there appeared to be only one lower venter for there was one umbilical vein. There were also four feet and legs; these were external parts and a form or measure of them: i.e., they were rudimentary. When I cut the freak internally I found in the first chest two lungs in their proper places. Both of these lungs, however, covered only one capsule of the heart. In this capsule there were two separate hearts, touching each other and depressed or flat at the point where they touched. In other respects the hearts had their natural form for there were auricles, veins, arteries for each heart. These hearts were situated in that farthest part of the chest where the bodies were united. One diaphragm also separated the natural members from the spiritual ones with its sinewy part perforated in the middle for the meri (or lower part of the gullet) and with other veins. Under this diaphragm there was only one ventriculus [stomach] covered with one liver which was divided by a panniculus with an imperfect division into two parts. The intestines were also imperfect since they were not natural in bulk, shape, or location nor did they extend naturally for they did not stretch to the anus through that cavity where the bladder lies between the pubic bone and os sacrum. In fact, there was no foramen in the place of the anus. The virile members were also most imperfect for the testicles did not hang down nor could they be found. I found also one spleen and four kidneys, which were not in their proper places. One of them I found under the left shoulder blade between the ribs and the shoulder blade. They said this freak lived four days after it was born.[75] I ask my readers not to be vexed because

---

[72] No such treatise exists among Massa's writings. There is a description of bones of the head in his *Epistolae Medicinales* (Venice, F. Bindoni and Maffeo Pasini, 1550), Epistle V, folios 54 r-55 v, addressed to Antonio Fracanzano, of Padua, Jan, 20, 1543. These thirty-four letters, dating chiefly from the 1540's, deal largely with diet, fever, stomach troubles, intestinal ailments, the "morbus Gallicus," and ulcers; the last two letters discuss dialectics, the creation of the world, and the immortality of the soul.

[73] The distinction between the terms complexion and composition is discussed by Singer, translation of Mundinus, pp. 101–102: "The *complexion* is the particular manner in which the four elements or four humours are mingled together and is to be distinguished from composition which refers to the coarser structural nature." Avicenna's definition of complexion is as follows (*Canon*, I, translated by Gerard of Cremona, Venice, Giunta, 1595, fol. 11): Complexio est qualitas, quae ex actione ad inuicem et passione contrariarum qualitatum in elementis inuentarum, quorum partes ad tantam paruitatem redactae sunt ut cuiusque earum plurimum contingat plurimum alterius prouenit, quum enim ad inuicem agunt et patiuntur suis virtutibus, accidit in earum summa qualitas in toto earum similis, quae est complexio.

[74] Fossa Clodia is first mentioned in Pliny, *Natural History*, III. 16. 121 (Loeb Classical Library, translated by H. Rackham, p. 88): "A part of these streams also forms the neighboring harbour of Brondolo, as likewise that of Chioggia is formed by the Brenta and Brentella and the Clodian canal (fossa Clodia)." the *Enciclopedia Italiana* identifies Fossa Clodia with modern Chioggia, a few miles south of Venice.

[75] Massa's interest in freaks and anomalies is shared by other sixteenth-century anatomists and scientists from Achillini to Aldrovandi; see, for example, the pictures of freak chickens

I have brought in these facts incidentally. This is certainly something at which to marvel; therefore since there were two hearts in one capsule I wished to describe it. These hearts touched each other and lived for four days. Now, dismissing the anatomy of the heart, I urge you to turn both to the dissection and to the viewing of the rest of the parts of the body.

Chapter XXIX

On the Dissection of the Lung

Put back the heart into its place. After you have seen everything which I have mentioned inspect the substance of the lung. It is composed of thin, light flesh, reddish in color tending toward white, and is covered with a very thin panniculus by which it has sensation through the reversive nerves [n. recurrentes] that come to the panniculus. The lung is divided into two parts by the mediastinum to prevent any harm to one part if the other is injured. The entire substance of the lung is made up of five fibers or lobes; three of these are on the right side of the chest and two on the left side. Those on the right side are attached by some ligaments to the upper and lower ribs as well as to the diaphragm. Of these three lobes the two lateral ones are longer than the one in the middle. Under this lobe passes the chilis vein [vena cava] in its ascent to the heart. There are some people who do not include this lobe among the parts of the lung since it is a sort of cushion for the chilis vein. Hence you will see that this lobe is hollow at the point where the vein lies upon it in order to provide a better resting place. The two longer lobes extend rather toward the back than toward the anterior part of the chest. You will also see that the lobes on the left side are attached to the upper ribs but not to the diaphragm and the lower ribs because they are smaller in this region and do not reach the terminus of the diaphragm in the anterior part of the chest. As is apparent to the senses the substance of the human lung fills the entire upper cavity of the chest and adheres to the back and to the ribs by means of ligaments which are not weak. Thus the lung is separated from the ribs not without difficulty, or from the pleura in the upper region, as you will see in many cadavers. I say this because sometimes (although rarely) I have found a lung which was not attached to the ribs by such ligaments as those mentioned; therefore some people have concluded that these ligaments are not necessary. But considering the nobility of the heart I say that nature has done its work very well in preventing pressure upon it from the great weight of the lung by giving the latter these ligaments or suspensories. They are thin and do not hinder motion since they are yielding and can be extended. I find them almost always adhering to the lung with the pleura and I have separated the upper parts of the lung from the pleura not without difficulty especially in the regions at the sides. The function of the lung is to cool the heart with its intake of air drawn through the rough artery [trachea]. This air is not a pure element but an airy body and thus can nourish, although Galen, that great philosopher, thought he understood it in another sense, for air prepared in the lung itself nourishes or restores the spirit. You can experience this attraction of the air if after you have skinned the gullet and perforated the trachea you move the lung by blowing in some part of it with a bellows for you will see how the lung is moved by the inhalation of air. There are in the substance of the lung textures of the veins mentioned in the anatomy of the heart; that is, on the left side is the pulsating venal artery [pulmonary vein] which proceeds from the heart and is spread through the substance of the lung. This artery is composed of a single tunic, as I said, because it chiefly supplies air to the heart. On the right side of the lung is the quiet arterial vein [pulmonary artery], non-pulsating, which is composed of a double tunic because very hot thin blood in large amounts moves continually through this vein. Thus the vein can resist pressure upon it. Examine also in the lung in addition to the ramifications of the veins and arteries I have mentioned, which are very numerous, the thinness and porosity of its substance in which catarrhs and other humidities are absorbed. These are foamy with the admixture of air and are spat forth while still foamy. The foaminess exists not only in these humidities but in the blood which is expectorated when the veins of the lung are ruptured. Thus the lung is made thin and foamy by the mixture of air with blood, by the continual motion of the lung, and by its heat. You will not see the blood so foamy in other veins of the chest as it is in the lung. Therefore the blood of the lung is distinguished by physicians from other blood through its foaminess. The trachea artery also enters the substance of the lung, separated for the most part in its descent from the substance of the lung by its cartilaginous and annular nature and its attachment with very hard membranes. But beyond the substance of the lung as the trachea ascends toward the tongue it is cartilaginous only in part. The rest of it is membranous in the posterior region; more than half of the trachea in the anterior region is cartilaginous. It is composed of this material in order that the meri [gullet] may not be injured as it descends to the stomach. The trachea artery is divided near the lung as it enters the substance of the latter both on the right and left sides. These branches after being distributed to all parts of the lung by means of many ramifications even to the thinnest and tiniest parts

---

reproduced in *Aldrovandi on Chickens: The Ornithology of Ulisse Aldrovandi* (1600), Vol. II, Book XIV, translated from the Latin with introduction, contents, and notes by L. R. Lind (University of Oklahoma Press, Norman, Oklahoma, 1963).

form together with the two veins already mentioned, that is, the venal artery and the arterial vein, a substance in the shape of a net, as I said when I described the liver. Through this artery, which is also called rough, the lung draws air or an airy body by inhaling it toward itself by dilation; when the lung is constricted bad vapors are expelled. The lung also expels its superfluities by coughing. If you wish to see the inner veins of the lung you can do so by stripping the flesh from them, as I said concerning the liver, that is, by first allowing the lung to wither somewhat for a time and then stripping the flesh from the veins. From what has been said you have seen three kinds of veins which enter the substance of the lung and its composition, that is, the venal artery, the arterial vein, and the trachea artery; these make a texture filled with soft flesh and covered by a panniculus which originates from the pleura. By means of this panniculus the lung has sensation with the mediation of nerves. What is called the lung, that is, its substance, if you inspect it properly, lies in the middle of the chest verging more toward the left side so that it may adapt itself to the location of the heart. For the heart seems to slant more toward the left with its point. Thus it is not unreasonable that the lung should also slant toward the left. In that direction it has also smaller and fewer lobes in order not to impede the motion of the heart either by their size or number. The lung is also attached not only by ligaments or panniculi to the pleura and diaphragm but to the back. Thus attached, the lung moves according to the motion of the diaphragm and of the intercostal muscles for the greater part. From these facts you may know that the lung serves the heart, drawing in air to cool the latter and preparing the air for mixture with the spirits, for spirit is made from blood and air prepared by the lung. Hence they say that spirit is created from thin airy blood. The lung is also a sort of cushion for the heart; I say sort of because the substance of the lung does not touch the heart since the latter is covered by its capsule. The lung is also necessary for the formation of the voice and for strength; it is useful for containing good air in itself and keeping bad air out while the good is retained when anyone smells bad odors. You should not neglect the connection of the gullet which proceeds from the outermost part of the mouth and descends as far as the diaphragm, touching the trachea under which there are the gullet and the chilis vein [vena cava] descending and the aorta artery. All these members pass across through the spine of the back; in the upper part of the latter above the heart you will see a certain glandular flesh above which the chilis vein [vena cava] is braced as upon a cushion. This flesh is called morus [thymus gland]; its function is to prevent the members which serve the noblest member from being injured by the hardness of the dorsal spine.

Chapter XXX

On the Anatomy of Certain Parts of the Neck, of the Lower Jaw, and of the Teeth

You cannot make a perfect dissection of the trachea artery and of the gullet if you move the lung and heart aside. For this reason put these organs back in their place. Even so you will not be able to see the trachea and gullet unless you first move them aside from certain parts of the throat or gula. Next begin to strip the gula or throat, which is above the upper fork of the chest as far as the jaw. It exists for the sake of the lung, as Galen learnedly said in *On the Use of the Parts VIII*.[76] When the throat is stripped examine the substance of the fleshy panniculus. When you have carefully stripped the latter you will see two veins [internal jugular] ascending, one on the right and the other on the left; they proceed from the chilis vein [vena cava] ascending and are accompanied by two arteries [common carotid] that come from the aorta artery ascending. These veins [external jugular] pass across above the two longitudinal muscles [sternocleidomastoid] in the throat or gula. Both veins and arteries [external carotid] branch toward the ears in anterior and posterior parts as far as the most minute branches which they call capillaries. These nourish and vivify the temporal muscles and the remaining upper parts of the head. These veins which thus ascend through the throat are called the guidez manifest [external jugular] because they are visible above the muscles to distinguish them from the two veins called guidez hidden [internal jugular], which are under the muscles; but some call them the apoplectic arteries. Look also above these longitudinal muscles at certain branches of the reversive nerves [vagus]; they come from the sixth pair if you begin to number from the optic nerves. But if you begin the count from the first pair of nerves [olfactory] which descends to the nostrils those branches will be the seventh pair, mingled also with other joinings of nerves. They pass to the tongue and to the muscles of the glottis artery. These are called the nerves of the voice since when they are broken or otherwise hindered the living creature loses its voice. These nerves also pass under the ear. Thus those who handle illnesses in the places mentioned, such as abscesses and ulcers, should take care not to injure these nerves with cutting instruments or biting unguents or caustics since loss of

---

[76] Galen, *De Usu Partium*, VIII: The closest approach to such a statement is chapter 1 (Vol. III K., 610–611): Reliquum igitur adhuc est asperarum arteriarum genus, collo ac pulmoni commune . . . Ea igitur ratione pharynx pulmoni est affinis, tantosque animalibus praestat usus, eiusque gratia collum extitit. Another passage worthy of notice in this connection is *ibid.*, VII (Vol. III K., 548): Unde perspicuum fit, asperas arterias sic a natura fuisse institutas, ut pulmonem in maximam distentionem agerent. . . .

voice will result from the action or power of such instruments or materials. Galen also advises us about these operations. He disapproves of the ignorance and the lack of skill in anatomy exhibited by the man who, wishing to root out scrofula in someone's neck,[77] lacerated the nerve and left his patient with half his voice because the nerve was lacerated only on one side. If the nerve had been lacerated on the other side the patient would have been made entirely voiceless. You can learn this from what Galen said if you take a brute animal and strip the skin on its neck under the ears so that the nerve becomes visible. Then tie the nerve itself with a thread. When this nerve is tied on one side of the neck the animal retains only half its voice, but if you bind the nerve on the other side of the neck the animal will become completely voiceless. If you loosen the thread on one side he becomes semi-voiceless. If you loosen the thread on the other side the entire voice will return. This is easy to do for those who are skilled in dissection. But if you wish to see the remaining process of the reversive nerves [vagus] begin to remove the skin in this region of the neck behind the ear and, laying aside the other parts, continue to descend since you will best see them as they proceed to the cane of the lung, both descending to the meri [gullet] and ascending through it to the mouth of the stomach and even as far as the intestines up to the os sacrum, where you will see many reversions of the meri. In this part of the stomach you will note that these nerves are softer. But this process of the nerves is not revealed in dissections which are performed to show young men the introductory method of cutting cadavers. One must have a single particular subject and examine in it, as I have often done, some organ as long as I wish in order to acquire a full knowledge of it. When you have noted these items examine in the throat two longitudinal muscles [sternocleidomastoid] which draw the head forward and downward. One of these is at the right, the other at the left. They are broad and proceed from the ears, descending to the lower region as far as the fork of the chest. As has been said, above these muscles pass the veins, arteries, and nerves that have been mentioned. You must carefully lift up these muscles in order not to injure the aforesaid veins and nerves since under the muscles you will see the process of two veins, the guidez hidden [internal jugular], together with the apoplectic arteries, or carotids. All those veins and arteries pass to the substance of the brain, as you will see when I describe the anatomy of the chilis vein ascending and of the aorta artery. But before you come to inspect these veins you will notice two muscles [omo-hyoid] placed immediately below the longitudinal muscles. You will see that they originate from the head of the trachea artery which they call the head of the bronchus, that is, in the external region under the tongue. They descend to the juncture where the adiutory bone of the arm [humerus] is braced on the bone of the joint of the spatula [scapula]. Both in the upper region near the head of the bronchus as well as in the lower region near the juncture of the spatula these muscles have a fleshy part. In the middle they have a sinewy part, that is, between the two fleshy heads in such a way that the sinew is in the middle. These muscles are round. I have found no description either of their substance or of their function among the ancient anatomists.[78] Thus I thought it would not be displeasing if I said something incidentally both about their shape and function. They are round, oblong, fleshy at their heads and sinewy in the middle. Nature who does not err made them in order that the head of the bronchus could be dilated by means of the attraction of the sinew from the head of the muscle which terminates at the juncture of the spatula through the fleshy part when we speak. Then that head is constricted when another head of the muscle near the head of the bronchus draws the same sinew tight as well as in the necessity of inhaling a little air and in the lack of a greater retention. These two operations proceeding from one member are in accordance with nature, which does not abound in superfluities nor is deficient in what is necessary,[79] using two operations for one member as in this instance for only one pair of muscles. But to say that the sinew here in the middle was given in order that it might yield to the supervening muscle is not characteristic of that man of talent and goodness[80] if he judged its prior function. Here you see that not only the transverse septum has a sinew in the middle but also these and

---

[77] This incident occurs in Galen's discussion of the recurrent laryngeal nerves which he (53) claims he discovered: *De Locis Affectis*, I, 6 (Vol. VIII K., 48–55; incision for scrofula, 55): Quum ergo ex cervice quidam strumas in profundo sitas excideret, ac deinceps ne secaret aliquod vas, membranas non specillo caederet, sed unguibus evelleret, imprudens ob ignorantiam simul recurrentes nervos distraxit; atque hoc pacto puerulum liberavit a strumis, sed mutum reddidit.

[78] Thomas Lauth, *Histoire de l'Anatomie* (Strasbourg, Levrault, 1815), p. 328: "il a découvert les muscles génioglosses." See the description of the extrinsic muscles of the tongue by Vesalius, *Epitome* (translated by L. R. Lind, New York, Macmillan, 1949), p. 16, and the careful analysis of them by Dr. C. W. Asling, note 19, p. 34. The ninth muscle of Vesalius is the genioglossus.

[79] "Nature, which does not abound in superfluities nor is deficient in what is necessary" is a favorite theme in Aristotle, together with the also much-quoted "Nature does nothing in vain"; the sixteenth-century scientists quote these phrases frequently. See Aristotle, *De Partibus Anamilium*, II, 13, 658 a 5; III, 1, 661 b 24; IV, 13, 695 b 18; *De Incessu Animalium*, II, 704 b 15; VIII, 708 a 10; XII, 711 a 17; *De Generatione Animalium*, II, 4, 739 b 20; 741 b 5; 6, 744 a 35; *De Anima*, III, 9, 432 b 22; 12, 434 a 32; *De Caelo*, I, 5, 271 a 35; II, 11, 291 b 14; *De Respiratione*, 476 a 13; *Politica*, I, 2, 8.

[80] Galen, *De Usu Partium*, VII, 19 (Vol. III K., 591–594, on the muscles of the tongue; *De Motu Musculorum* I (Vol. IV K., 368–369), on the muscles, tendons, and ligaments in general.

other muscles likewise, as I shall say at greater length when I write a book on all the muscles; for if I were to describe all the parts of the body here confusion would result for those who are beginners in anatomy and my discourse would also become tedious owing to excessive prolixity and a confusion in the order of my original plan of procedure. Therefore I shall return to the description of the remaining parts of the guidez veins and of the apoplectic [external and internal jugular] arteries. These guidez veins and the arteries called hidden are much ampler than the guidez veins and arteries called manifest because the veins carry blood to nourish the bulk of the brain and the arteries carry vital spirits since they are more than full as they vivify the brain. This spirit is altered by the faculty and complexion of the brain; it is called animal spirit by physicians and carries out animal operations, i. e., the function of animation. But in order to see these hidden veins and arteries better you must lift the upper fork of the chest carefully both on one side and the other in order not to rupture any of the manifest veins since, as I have said, the manifest guidez veins and arteries ascend above the fork and the muscles. When you have seen these veins and arteries you can if you wish inspect the muscles of the neck which move from right and left, inclining or drawing the head, and in the neck those muscles which draw the head backward and turn it around, for I have spoken about the anterior muscles which are in the region of the throat above which pass the veins and arteries mentioned. In order that you may not be confused by the word neck, I mean by that word the entire material from the shoulders to the head under the basilar bone. Its posterior part is called cervix by the Latins and the anterior part is called gula or throat. In the latter's anterior part you will sometimes see some glandular substances [lymph nodes], which are not natural;[81] they lie above the longitudinal muscles. Therefore [because they are not natural] these substances are not found in all people, but natural glandular pieces of flesh which they call glands [thyroid] or tonsils are found under the aforesaid longitudinal muscles toward the tongue. These glands are to be seen inside rather than outside near the gullet; above them ascend the hidden veins and arteries and they lie between the manifest and hidden arteries and veins. These tonsils or glands receive the humidity from the brain in order to humidify the nearby members in time of need, such as thirst. There are two glands, one on the right, the other on the left near the gullet. It makes no difference whether you use the word meri or esophagus or any other for that part of the body through which the food descends to the stomach or ventriculus, as long as you recognize the member by such a term. But I am not unaware of the great difficulties of the grammarians who have laid chief emphasis upon words for the recognition of things, not realizing that true recognition arises from the knowledge of both material and formal parts as far as the elements, according to the learned teaching of Aristotle, whom Galen always imitated to the extent of his ability, ever confuting those who dispute about names. Thus dismissing grammatical matters, lift up the aforesaid fork after saving the parts mentioned and look at the remaining part of the trachea artery which is attached to the mouth and tongue. Begin to lay bare the lower jaw, first moving aside the skin off from the muscles and tendons that come to the jaw. You will notice that the hairs of the beard in men proceed as far as the membranous muscle with their roots, or, to speak more correctly, the material of the hairs apparently begins to receive the form of hair at that point. Since, as I have often said, it is impossible to inspect all the members in one body, it will suffice here if you diligently proceed to strip the muscles which open the lower jaw. They originate in the occiput from the bone of the head behind the ear, one on the right, the other on the left. This bone they call the sagittal or acular [mastoid process], because it is sharp (*acutum*). These muscles [digastric] which come to the jaw have a sinewy part. This sinew is joined to the jaw, moves the latter, and opens the mouth. But since the jaw is bony it is heavy and tends to move downward. Prudent nature gave to this sinew at its end another fleshy part so that by contracting the sinew in the opposite direction it might hold up the jaw lest the mouth be opened farther than is proper. You also see here that the muscles which open the mouth have a sinew in the middle and a fleshy part at their ends. This flesh proceeds from one side from the sharp bone behind the ear and on the other side from the end of the maxilla or lower jaw. Proceed then to strip these muscles carefully up to the place where they [reading *uniuntur* for *unitur* of the text] are united with the jaw and look at the two temporal muscles which lift the lower jaw and close it. In the inner part of the jaw you will strip also the muscles of mastication and those which proceed within the jaw from the palate to the jaw in order to see more comfortably in dissection the members which are under the muscles, or covered by the latter. But before you move aside the lower jaw look at its joint [temporomandibular] marvelously constructed by nature which binds the jaw to the upper jaws with very strong ligaments. Look at both lower and upper jaws. They are covered with a certain simple flesh called the gums. I say simple because it is not muscular for the sake of movement but as a covering for the bone. It has little sensation since certain very

---

[81] Galen, *De Usu Partium*, VII, 17 (Vol. III K., 589): Quin et glandulae ipsae, quae laryngi adjacent, idem ipsum indicant; quas semper, quam alias glandulas, reperias *laxiores* ac fungosiores. At inter omnes propemodum anatomicos convenit, eas in eum usum a natura fuisse factas, ut partes omnes, quae tum ad laryngem, tum ad pharyngem attinent, humore perfunderent.

slender nerves are mingled with it. These gums also partly cover the teeth in their lower part. You will also note in both lower and upper jaw the substance, number, and difference of the teeth. Their substance is hard, white, with some sensation similar to a bone. Their perfect number is thirty-two, that is, sixteen in the lower jaw and the same number in the upper jaw. Nevertheless you should know that usually young people have twenty-eight teeth since the last four generally grow when the individual is more advanced in age: that is, they grow for some people in youth, for some at a later age, and for some even in old age. I have seen a decrepit old man who had two teeth which grew after the age of seventy. Since there are different names for the teeth the wise ancients and moderns are not in agreement about them. Hence for the present I shall point out only two differences in the teeth, of which one concerns those in front. There are twelve, six in the lower and six in the upper jaw. These twelve teeth both cut and separate the food. You may call the rest of the teeth on right and left sides both in the lower and upper jaws either molars or maxillars, it makes no difference. Their number is uncertain, as I said, because the last four come in at an advanced age. Note also that the teeth differ in their roots since some have many of them. I have seen two, three, and even four roots in the molars, but the front teeth have only one root and sometimes two but these are united and almost touch each other. The molars have more roots because their work is heavier and more continuous in chewing rather than cutting. Hence the teeth in front do not have so many roots. All the roots of the teeth are fixed in foramina in the jaws. At the extremity of these foramina there is a certain white substance which proceeds from a nerve that passes through the cavity of the jaw bone. This sinewy substance is mingled with the substance of the teeth and not only binds them together but gives sensation to the teeth, for sense and experience bear witness that the teeth possess feeling. Whether or not their substance is bony or something else which differs from bone, I am convinced that the bones are the support and base of the entire body. The flesh and the other parts are distributed around the bones as well as around each member like a wall, as Galen said, *On the Use of the Parts XVI*,[82] beyond their primary purpose. The teeth do not offer a support of this kind but are rather members which serve the stomach and also the voice. Therefore I cannot say that the teeth are bones but another substance similar to bone which possesses sensation from the admixture of parts of the nerves which pass through the jaw. Thus the teeth communicate with the brain. When a tooth is perforated an alteration ensues on account of the humidity which enters and from the outside air. A very sharp pain follows. I swear to God that I have seen a man who had an ulcer in his leg where the bone was visible. In this bone there was painful sensation to such a degree that he did not permit it to be touched by a rough instrument; the bone was denuded of its panniculus. Nevertheless I perforated the bone; inside there was the same feeling of pain. I wished to make this explanation so that anatomists might discover whether or not some scattering of nerves penetrates to the substance of the bone.

Chapter XXXI

On the Anatomy of the Trachea Artery in Its Upper Part, and of the Meri or Esophagus, Which Terminates at the Stomach

In the dissection of the lung you saw the trachea artery divided into two parts and entering the substance of the lung both on right and left and scattered throughout all the parts of the lung. This artery is a sort of suspensory of the lung. It is hollow, covered on the outside with a very thin panniculus which is thicker and harder in order to resist catarrhs and other operations of the cane of the throat, in whose cavity the external air is received for the refrigeration of the heart. Therefore examine that cavity as far as the substance of the tongue since it is made up of many parts both cartilaginous and membranous. It is harder in the upper part; here is the head of the aforesaid cane of the throat or of the rough artery, which they call the head of the bronchus. This rough artery is composed of cartilage which makes many rings that are incomplete where the cartilage touches the gullet; in the parts thus incomplete the substance is membranous, for a hard substance would injure the gullet. These incomplete cartilaginous rings are continued by means of ascending membranes as far as the head of the bronchus [larynx]. The latter is entirely cartilaginous, covered with membrane, and composed of four parts, including the muscles and ligaments which cover the head itself. The first part is cartilaginous substance, the second is the panniculus which covers that substance, the third is the ligaments, and the fourth is the muscles that move the head; these are inside and outside: God willing, I shall write about them in the chapter on the anatomy of the muscles.[83] Inspect the cartilaginous substance which composes the head of the bronchus. There are three distinct cartilages bound together so that this part of the body can be dilated when necessary and also contracted with its yielding components. The first of these components is in front and is called the scutal cartilage [cartilago thyreoidea of the larynx]

---

[82] Galen, *De Usu Partium*, XVI, 2 (vol. IV K., 268): alibi etiam passim velut propugnaculum quoddam ac murus. . . .

[83] Massa here reduces his earlier promises (see note 10) from a *book* on the muscles to a *chapter*, if we can draw this conclusion from his usual method of citation in this book.

since it is shaped like a shield and is larger than the other components. It is not continuous in substance circularly to the gullet. The second component [cricoid] of the head of the bronchus is also cartilaginous; it lies toward the gullet or esophagus and begins with a perfect circle in the upper part of the artery. As it ascends it does not preserve the complete circle but is united with the scutal part by some strong ligaments at its very end toward the gullet. They say this part has no name by which it is called but we shall call it second, from its order; thus from these parts you have the composition of the roundness and emptiness of the passage. They call the third cartilaginous component the cymbalaris or larynx; it is smaller than all the others and bipartite. It lies in the upper region where the gullet is continuous with the head of the bronchus. This cymbalaris [arytenoid] cartilage is attached with strong ligaments. While it is in its own place it opens; you will see the opening in that region. When it moves inward like a double door the artery or cane of the lung is closed, as at the time of vomiting so that nothing may enter the lung. Just as the uvula (lingula) covers the opening so that nothing may descend to the lung when a person swallows, so these parts close at the time of vomiting. There are also those who call this part the larynx; it makes no difference so long as you recognize its substance, position, and operations. These are the parts which compose the head of the trachea artery. There are many muscles, as I said, which move it. At the upper head of this artery there is a substance which has the shape of a small tongue; therefore they call it by the latter name [lingula]. This epiglottis covers the cane of the lung so that no food may enter, as I said, nor anything harmful. Its substance is not solely cartilaginous but also membranous; thus it is midway between cartilage and membrane, or something composed of both, so that it may bend easily when opening the head of the rough artery, or head of the bronchus. It is continuous not only with the head of the cane but also with the tongue; it moves also with the movement of the tongue and of the gullet. Concerning the name of this part or member there are difficulties in our time, for some call this lingula the epiglottis, which is the same thing as "before the glottis" for it is in front as a covering of the glottis. Others have believed that the glottis itself is the epiglottis because they do not correspond in the composition of the upper part. The Greeks call it glottis, or fistula, or part of a pipe, rather than the covering of a pipe. You can assign another derivation also to the epiglottis from the fact that the word is composed of epi, which means in front, and glotta, which is tongue. Thus the word signifies in front of the tongue, that is, placed in the extreme inner part, which is the sensible truth. You can at your pleasure select whichever you like of the two meanings since both among the Greeks as well as among the Latins these two letters are often confused in pronunciation and one put in place of the other, as here glotta for glossa. Not to delay you with names, I consider what I have said to be sufficient on the subject since from what you have seen I think the cane which brings air down to the lung, together with its parts, has been fully described. But before you come to lift it up you will notice the branches of the reversive nerves [vagus] which come to the cane; they are notable. Put back the trachea or cane and move aside some parts of the lung in which you have seen the process of the veins and arteries in order more comfortably to see the meri or esophagus, which is the path whereby the food is brought to the ventriculus or stomach. It is composed of two tunics, as I said when I discussed the anatomy of the stomach. These have all kinds of stamens; the long ones are in the inner tunic for attraction from the inner part. The stamens which attract from the outer part of the same tunic are transversal and fewer in number. The inner tunic is similar to the inner tunic of the stomach since part of it tends toward white in color. But the outer tunic has broad stamens and tends toward a flesh color. Hence it is called fleshy not because it really is, since its substance is distinct from flesh, but because it resembles flesh. Let not the spectators wonder at the fact that nature has given all these fibers to this member which seems to serve only for attraction. By means of them, the transversal, which are few, and the broad fibers, nature not only effects a better and more comfortable attraction, creating retention through the transversal fibers in vomiting and swallowing and expulsion by means of the broad fibers as well as in swallowing that which is thin and liquid by uniting the parts of the esophagus. The esophagus also checks these operations if its nature is ever irritated. It moves above in the case of vomiting, contracting the gullet by means of these lateral fibers, or when a person is bent over in some fashion, and operates in accordance with the particular occupational need, as with men who tan hides or women who wash clothes and are thus bent over at their work or with many others who work bent over almost to the ground. Examine this part of the body beginning from the mouth of the stomach, which is surrounded by the diaphragm, and proceeding up to the mouth at the upper end of the body, for the inner tunic is continuous between these points to the lips and is united almost inseparably with the palate by longitudinal fibers. However, by means of the external tunic the esophagus is almost inseparably continuous with the head of the bronchus. Notice as well how not only the head of the bronchus is continuous with the esophagus by means of this tunic but also by means of the inner tunic, as will be clearly apparent if you inspect the cymbalaris or cartilaginous larynx. This continuity exists both around as well as toward the rough artery and toward that bone [hyoid] where there is the part of the body they call the pharynx.

Modern students of anatomy labor at the recognition of the pharynx, blaming first one and then another of the ancients because they did not call every member by its own special name; these students are heedless of the fact that different languages in the course of time change such names. It should suffice for these moderns to indicate the parts of the body and their positions, substance, and operations by whatever names there are for them at the present time. I do not wish to be reprehended by odious people for having occasionally coined some name or other in order to make my teaching more explicit since I confess I have done this according to the precepts of Galen. Continue therefore to examine the substance of the meri [gullet or esophagus] and its progress since you will see two scatterings of the reversive nerves on one side which as they descend and ascend turn back in many places to the mouth of the stomach where they terminate in that duplication. They are very soft and from the fact of their softness you may argue against those who say that because the nerves near the brain are softer therefore they originate from it. For here also in the gullet these nerves are soft. They descend from the sixth pair of nerves according to the common enumeration, as I shall say in the chapter on their anatomy. On account of these nerves the mouth of the stomach is very sensitive together with the esophagus since there is a multitude of them scattered about the mouth of the stomach. You may also see how close the chilis vein [vena cava] and the aorta artery are to the gullet as they pass through the dorsal spine as far as the diaphragm near the thirteenth vertebra; here the esophagus or meri terminates at the mouth of the stomach. You will notice also in the upper region where the meri touches the trachea artery that the latter is not cartilaginous but membranous in that area. See also how the meri is based in the upper region upon the glandular flesh called the morus [thymus gland]. When you have seen these parts of the body you can gently strip the upper part of the meri [esophagus] with a razor as far as the head of the bronchus. Do not, however, cut it off entirely before you have inspected the anatomy of the tongue.

Chapter XXXII

On the Dissection of the Tongue

The difficulties concerning the dissection of the tongue are not small both as to the number of its parts and their substance as well as their operations. All these difficulties would be removed if sometime writers would listen to Galen's advice and trust the anatomy books less and if with their own hands they would touch the bodies of men and write down what they see and feel. For this reason I shall not cease to desire that the disputants in these matters should trust themselves entirely to their senses and put aside sophistical reasoning, as the ancient wise men were accustomed to do; they did not disdain to dissect the bodies of various living creatures in their search for the truth. Therefore, since we must discuss the tongue and its parts I should like to advise you at the outset that the tongue is a certain soft fleshy substance. But since there is no flesh without muscle except the glandular material which serves as a cushion for certain members or as a covering, such as the gums, we should more correctly speak of the tongue as muscular flesh but not distinguishable, just as the muscles of the lips which can scarcely be distinguished as to their flesh and muscle in their dissection. The motion and substance of the lips, however, are apparent; in the substance there is an evident mixture of fibers, by which most of the movements of the lips are carried out. It is possible to see this mixture also in the upper part of the tongue's substance when the other manifest muscles are laid aside. This upper part has various fibers which are not small; you can recognize them by their different processes. These are deeply embedded in the substance of the tongue, and I say that from them proceed the varied movements of the tongue since apart from the motions caused by the apparent muscles—downward, upward, sideways, contraction, and extension outwards—the tongue has other movements not caused by those muscles. While these are voluntary movements there remain those which are made by a muscle or [intrinsic] muscles which are mingled in the substance of the tongue. These muscles are inseparable; you can see this in the substance of the tongue, which has mingled fibers quite similar to those in the lips. Because of this fact you can be certain that in this part of the tongue there are muscles so mingled with each other that they cannot be separated or their component parts distinctly seen. But, as Galen said, they are recognized by their operations since voluntary movements are not made except by muscles.[84] In addition to other movements, the substance of the tongue has one by which, when it is flat, the tongue is voluntarily rounded and dilated to a greater amplitude not only in the part corresponding to the muscles but at the tip of the tongue. The tongue is also contracted in its substance. There are other small movements which do not proceed from these manifest muscles revealed by dissection, as is most apparent to people of sense who recognize both animal and natural operations and can distinguish between them, a fact which each individual can experience for himself. Thus we must say that the tongue is composed of muscles intermingled among themselves as are those of the lips but that it has a substance softer and more mixed, covered by a membrane common to the entire mouth and palate as far as the head of the bronchus and the meri or gullet. The gullet proceeds from this point, as has

[84] Galen, *De Motu Musculorum* (Vol. IV K., 367): Instrumenta motus voluntarii musculi sunt.

been said; it has longitudinal filaments. Scattered through the substance of the tongue there are notable and larger nerves than are found in any other member of the body in comparison with its size.[85] These are both nerves of motion and of sensation; they come to the tongue from the brain. The nerves of motion come from the eighth pair, beginning the enumeration of the first pair from the nerves which give the sense of smell to the nostrils. These motor nerves [hypoglossal] are deeper in the tongue itself than those which give sensation [trigeminal] since the latter are scattered more on the surface and proceed from the fourth pair, beginning the enumeration from the nerves which come to the nostrils, as stated above. If this manner of enumeration is not suitable to you, you can begin the enumeration from the optic nerves; in this case the motor nerves will begin from the seventh pair and the nerves of sensation from the third pair. There are also veins which come to nourish the tongue and arteries to maintain its life. These veins are visible under the tongue, where there are muscles that move the tongue outward. These muscles, which I was the first to recognize,[86] have this function. The veins are very frequently bled in many illnesses such as angina, quinsy, headache, and abscesses. These veins, both large and small, come from the guidez veins [jugulars]: the large ones from the hidden, the small ones from the manifest, guidez veins. Similarly, the arteries of the tongue come forth from the arteries which accompany veins. Furthermore, the tongue has its own bone, which you will see after you lift up the tongue; the bone [hyoid] is located near the scutal cartilage of the head of the bronchus and is attached to the latter. At this bone terminate two lateral muscles [hypoglossus] of the tongue; other muscles are continuous with the bone but terminate elsewhere, as you will see when you strip them. Not only do the muscles of the tongue terminate at the bone of the tongue but the muscles of the head of the bronchus terminate there also; of these I shall speak in my little book on the muscles. The bone has a semicircular shape with some sharpness or angularity. It is composed of three bones bound together by strong ligaments in order to support the tongue and comfortably serve all its movements, both dilation and constriction according to need. Look also at the muscles under the tongue at left and right of the pieces of glandular flesh [submandibular gland]. In that place and near these there are some loose places which together with the glandular flesh retain or (as they say) generate the salival humidity; it is more correct to say that they retain it. These places are called the salival fonts. Note that these pieces of glandular flesh are separate from those at the root of the tongue where the ascending veins and arteries are based. When you have examined these, begin to strip the manifest muscles which move the tongue. There are nine of them, as you will see, nor do I wish to be accused if I say things not said by other anatomists since the senses themselves always defend and protect me. Look first at the lower region after you have stripped the skin under the chin toward the gullet as far as the lower jaw. Two muscles will appear which, if correctly observed, will be seen in the middle under the substance of the tongue stretching lengthwise between its two extremities, taking as one extreme the anterior region toward the teeth and as the other extreme the inner region where the tongue terminates near the esophagus. These two muscles [digastric] are attached to the lower jaw in the region in front and toward the inside and at the side. They proceed toward the trachea and esophagus where appears the tendinous part of the aforesaid muscles. They terminate near the basilar bone behind the ears, near that part they call the sagittal bones. This tendinous part is united with a certain fleshy part that terminates in the sagittal bone; this flesh has the shape of a round muscle. Nature has carefully formed these muscles in such a way that when one of them draws the tongue inward it may be a single unit and make the necessary movements of extension and contraction. Hence, in order that a single muscle should not draw the tongue inward in a sluggish fashion nature gave those parts which draw the tongue back after it has been extended. Hence it is not without labor that we hold the tongue extended for a long time because that flesh operates in a contrary way by retracting the tendon toward itself and is, as I said, attached to the sagittal bones. This movement of retraction is made first by the muscle alone with the assistance of those muscles I have mentioned, that is, the transversal and lateral, which in their withdrawal are contracted to themselves. The muscles mingled in the substance of the tongue operate in the same way, contracting the tongue toward themselves, as may be seen in the contraction of the lips. From what I have said you can understand how the tongue is moved outwards, that is, by these two muscles. When these contract the fleshy part toward themselves, the chin, and the front teeth, they draw the tongue outward of necessity while the other muscles assist until they are relaxed in the thrust outwards. Therefore let all questions concerning the outward motion of the tongue cease among the most eminent wise men because I have discovered the function of the aforesaid muscles. When you have seen and diligently examined them, proceed to strip the two muscles [genioglossus] at the sides of the tongue, one on the right, the other at the left, which are united in the middle. Their function is to move the tongue this way and that since they have a fleshy part united in the middle of the tongue and

---

[85] Berengario da Carpi emphasizes the veins, not nerves, of the tongue: *A Short Introduction to the Anatomy (Isagogae Breves)*, translated by L. R. Lind (University of Chicago Press, 1959), p. 122: "Therefore, like the penis, it has more and larger pulsating and quiet veins than any other member equal to it in size."

[86] See note 78.

another part which extends to the lower jaw in front. After these muscles you will find another called solitary [mylohyoid] because it is alone without a companion. It is not like the rest of the muscles, which come in twos. This muscle draws the tongue inward in the manner described, with the assistance of those muscles near the sagittal bones and by means of that fleshy part. The solitary muscle has a certain sinewy part by which it is united to the anterior region of the lower jaw. It makes no difference if you refuse to call this part a tendon, lest you contradict Galen, for Galen himself called the sinewy substances in muscles by the name of the tendons in the passage referred to above.[87] This muscle is united to the tongue and proceeds as far as the bone of the tongue and adheres to it. There are also two other muscles [hyoglossus] in the middle of the tongue joined in such a way that they seem almost united; nevertheless, they are separable and proceed the length of the tongue to the region of the head of the cane of the lung where the head of the bronchus is located, which is the same thing as the head of the lung's cane. These muscles draw the tongue around on right and left. Then there are two other superior muscles which move transversely toward the roots of the tongue; these terminate at the sides of the bone of the tongue. Notice that the two muscles proceeding the length of the tongue touch each other; if they were united in the middle where you see a sign of their division they could not be separated without some laceration of the parts since they would be very strongly united. But their division is more apparent where there are two muscles [styloglossus] that go deeply into the substance of the tongue and terminate at the tongue's root at the end of its bone. These muscles move the tongue also downward by contraction. However, since what I have said does not conform to what is said on the subject by ancient writers and will thus be ambiguous to my readers, I beg them to lay aside arguments and discuss these difficulties with a knife that cuts by dissecting various human tongues. For, as Galen said in his book *On Anatomical Procedures*,[88] the tongue of a brute animal differs from a human tongue. You can confirm this fact by its operations since a tongue was given to humans not only for taste and for turning one's food from side to side but also for a clear and distinct speech. In order to avoid prolixity I shall say nothing about the other muscles which come to this region of the body but do not form part of the composition of the tongue since you can see them for yourself and enumerate them.

Chapter XXXIII

On the Anatomy of the Palate, the Uvula, and Parts of the Upper Jaw, and of the Lips

When you have examined under dissection and understood the parts of the tongue and their operations, look also at the upper part of the mouth which they call the palate or the sky (*celum*) of the mouth. This part is within the teeth, above the tongue. At its extremity there is a certain particle which they call uvea and columnella (little column). It hangs above the extremity of the tongue where the epiglottis or lingula is. The upper part called the palate is round or convex like the upper part of an oven; it is composed of bone covered with flesh continuous with the upper jaw; the fibers of nerves are scattered through this mingled bone and flesh, giving to this part of the body as well as to the tongue some sense of taste. Anyone can discover this fact for himself by placing something salty or bitter on the place where the mingled flesh is covered by the previously mentioned panniculus which is part of the esophagus. You will, however, notice that part of the columnella or uvea is more fleshy than the palate and its flesh less mingled [with bone]. Hence it is softer and thinner and thus swells with humidity received from the brain. The palate and the uvula assist speech, but the palate not only aids speech but does other things such as tempering the air for cooling the heart and preparing the food, since the latter is noticeably altered in the mouth. The palate also assists somewhat by attracting both food and drink with its fibers, although this attraction is very small. The uvula exists in humans alone so that they may be able to articulate words more readily and to prevent the passage of air without an intervening space before it reaches the inner region. You can lift up the fleshy parts I have mentioned and see the veins and arteries which come to them. When you lift them up you will see the substance of the gums both inside and outside. When the gums are lifted the bones of the palate and of the upper jaw will appear, where the teeth are fixed in their sockets as I have shown in the passage where I spoke about the difference of the teeth. The bones of the palate are distinct from the basilar bone; thus you will see them before you dissect the upper body cavity. But before you proceed further in the elevation of the parts look at the lips of the mouth, which

---

[87] Galen, *De Motu Musculorum*, I (Vol. IV K., 368–369). Massa could not have known and used the later books of Galen's *De Anatomicis Administrationibus*, IX, 7–XV, 8, which have been recently translated from the Arabic text by W. L. H. Duckworth, M. C. Lyons, and B. Towers (Cambridge University Press, 1962) in which Book X, chaps, 6, 7, and 8 present a detailed analysis of the tongue, its extrinsic muscles, the salivary glands, the arteries, and the nerves, although Galen does not explain the functions of the muscles. Massa's translation of the life of Avicenna was made from an Italian translation prepared for him by a friend. In *De Usu Partium*, VII, 19 (Vol. III K., 591) Galen points out the origin of most of the muscles of the tongue from the hyoid bone.

[88] Galen, *De Anatomicis Administrationibus*, II, 3 (vol. II K., 291), Singer's translation, p. 36: "Yet they dissect the heart or tongue of an ox without realizing that these are utterly unlike those of a human being."

are composed of muscular flesh or of muscles so intertwined among themselves that they cannot be separated by the use of instruments. Those muscles, however, which come to the lips but do not form part of the latter, such as the muscles common to the cheeks and those which come from the lower jaw, can be separated. It will be useful to separate them if you wish to understand some movements of the lips. You will not see the complete termination of the muscles I have mentioned since they are embedded in the lips. The lips are, as is well known, covered with skin on the outside, but on the inside they are covered with the panniculus common to the tongue and the mouth which is part of the esophagus, etc. The voluntary movement of the lips is evident to all. Their embedded muscles, however, are not separable; a certain movement is caused by these, as I said in discussing the tongue. The well-known operations of the lips are to assist speech and to close the mouth; it would be superfluous to speak of their remaining operations. Also examine the chin, lest you overlook any well-known part. The chin sometimes has a sort of depression at its end which enhances the beauty of the face with its other parts. This [dimple] appears both at the right and left [parts] which they call the cheeks (*genae*) and the beard since in this region hairs grow in men and make them bearded. But before you approach the upper body cavity examine the parts which you had to leave in their place in order that the veins, arteries, nerves, and other parts near them and with them could be better recognized in dissection.

Chapter XXXIIII

On the Anatomy of Certain Parts of the Chilis Vein [Vena Cava] and the Aorta Artery Ascending

When I wrote earlier on the anatomy of the liver I said that during the removal of the substance of the liver you should put back a portion of its gibbous part where the chilis vein passes. This vein is divided into two parts, one of which descends through the dorsal spine to the hips; I spoke of this one above. The other part perforates the diaphragm and ascends to the right ventricle [atrium] of the heart; this is called the chilis vein ascending. Thus return to that portion of the liver which you left untouched and you will see this vein proceeding as described from the liver. When the chilis vein approaches the diaphragm it sends out two manifest branches to nourish that organ and other small branches to the farthest vertebrae of the false ribs. After sending forth these branches it proceeds through the spine and enters the cavity of the chest, sending in its ascent many branches to the vertebrae of the chest and to the mediastinum and other places. It then ascends as far as the right auricle of the heart. This vein is large, but when it enters the heart it sends a notable branch to the heart's exterior substance [coronary sinus]. This branch is in turn divided into various branches throughout the entire exterior substance of the heart to provide it with nourishment. The larger branch, that is, the larger chilis vein, after it is joined with the right ventricle of the heart proceeds through the auricle of the latter and ascends above the spine [superior vena cava]. When it is near the ninth vertebra it sends out two branches that pass below through the spine [brachiocephalic], one branch to the right side and the other branch to the left side. From these branches proceed other branches [internal thoracic] to nourish nine and sometimes ten lower ribs of the chest both on right and left, although some people have said eight ribs only.[89] The truth based on sense perception, however, as I have always seen, is that there are nine and sometimes ten ribs which are nourished by this vein. The latter also nourishes the rib muscles or the muscles which come to the ribs and other parts by means of branches proceeding from itself both on right and left. Note that the branch which passes to the left side descends in its progress and, riding above the spine, scatters ramifications just as it does on the right side to the farthest region. After the chilis vein has sent these two branches above the heart, one on the right and one on the left, to nourish the ribs, it ascends in its progress and is based upon the glandular flesh they call the morus [thymus gland], as I have said. Then it proceeds with the aorta artery, which comes from the left ventricle of the heart, and is branched toward the arms [subclavian]. In this progress it sends two branches, one on the right and the other on the left, to nourish three ribs and sometimes two remaining superior ribs, on right and left and the parts in these areas. These branches to the upper ribs are smaller than those which descend to the lower ribs. These words do not conform to Galen's statements in the tenth commentary on the second book of [Hippocrates'] *Regimen of Food in Acute Diseases*,[90] but are quite consistent with sense perception, as you will see if you carefully inspect the progress of the veins. But before you proceed to strip further, look at the aorta artery on the left side of the heart from its left ear which is continuous with the left ventricle. This artery early in its progress is divided in two a little distance above the heart. One of these divisions descends and, perforating the diaphragm, it proceeds together with the chilis vein descending, as was said

[89] Galen, *Hippocratis De Acutorum Morborum Victu Liber et Galeni Commentarius*, II, 10 (Vol. XV K., 527–535): Propagines . . ., exilium autem ad inferiorem thoracis ipsius partem octo costis constantem *perrepunt*. Massa is speaking of the two azygous veins sent out by the vena cava superior toward the arms and from which proceed the superior intercostal veins: compare Lauth, *op. cit.*, p. 330.

[90] See note 89.

in reference to the anatomy of the chilis vein in its descent through the back to the legs and the other places mentioned. The other division of the aorta artery ascends and in many places accompanies the ascending chilis vein. I say this because a vein does not always have an artery as its companion, although it usually does. Later this vein, unaccompanied by an artery, sends out branches which ascend to the upper ribs; in its progress toward the arms it is divided in two, on right and left, and sends out branches to nourish the exterior muscles of the chest. These branches pass under the armpits toward those parts under the latter; the branches, since there are many, pass also to the upper regions, as you will see. But after this ramification of the armpit veins, as they are called, the chilis vein together with an artery proceeds to the inner part of the arm; I refer to the upper vein and the lower artery. This vein [basilic] often appears more clearly at the elbow joint and is opened for bloodletting in many illnesses, such as fever, abscess of the inner organs such as the liver, spleen, stomach, chest and others, inflammation of the lung and of the pleura in the inner panniculus of the chest. Concerning the opening of this vein and emission of blood in case of such abscess there is much discussion among the physicians of our day; this discussion would be resolved, first, if a correct distinction as to the times of the abscesses were made and if the physicians were acquainted with the anatomy of these veins by sense perception. With God's assistance I shall write a special treatise[91] and set down a method for investigating them to be used by those who do not know or recognize the progress of these veins as it begins from the liver and extends through the entire body. If I should set down this information here concerning the progress of the chilis vein my introduction to anatomy would become too large and tedious for readers. It is sufficient here to describe in one cadaver the progress of certain veins which are more visible and their venesection in the relief of certain illnesses. I cannot, however, refrain from insisting that young men should understand the progress of the chilis vein [vena cava] in the venesection of abscesses of the inner organs, both spiritual and natural, since almost all doubts created by those who write on the subject would then cease. First of all, they should know that the first evacuation takes place from the opening of the vein called basilic [vena basilica] and immediately from the right ventricle of the heart for the greater part since this vein is the large trunk and more adapted to the purpose because it has less bending and greater straightness and also because some small branches are closer to it although they are curved and proceed in a tortuous fashion, such as two branches which extend to both lower and upper ribs for the purpose of their nourishment. Blood does not issue from those branches so swiftly or freely as it does from the large trunk that proceeds from the ventricle of the heart. This ventricle is always full of blood and is of a notable capacity. Since it is the source of heat the blood is made thin within this ventricle by means of that heat and is thus rendered more swift of flow and more adaptable for

[91] Massa's promise to write a special treatise on venesection was duly carried out in his *Examen de Venae Sectione et Sanguinis Missione in Febribus ex Humorum Putredine Ortis, ac in Aliis Praeter Naturam Affectibus*. 3 p. l., 51 pp. sm. quarto, (Venice, J. Ziletti, 1568). The book was dedicated by Massa to his nephew, the lawyer Nicolaus Crassus, in Massa's eightieth year; the publisher's letter to Lactantius Ferrus, a Bolognese physician, reads in part (p. 50): Hactenus de Venae Sectione . . . quae tibi impraesentiarum satis esse uolui, ut pote a Nicolao Massa tui amantissimo, in ultimo senio constituto . . . Massa died at the age of eighty-four in 1569. Pages 33–51 contain *Canones Observandi*, helpful hints on the technique of bloodletting, and at p. 29 occurs a passage of biographical interest: . . . uidique alios etiam doctissimos medicos ita operantes inter quos, dum essem puer, et frater meus nomine Dominicus aetatis annorum octo laboraret pleuritico morbo, Medicus ille prudens iussit venam tundi, ac sanguinem educi, ex qua salutem consequutus est. Nostris quoque temporibus Clarissimus Medicus Victor Trincavella, anima cuius requiescat in sempiterna pace, dum correptus febre fuisset anno aetatis suae septuagesimo, permisit uenam sibi secari, ac sanguinem mitti, quod remedium faeliciter ferens, sanitatem comparauit. On p. 18 he uses the expression *euacuetur ad virtutis tolerantiam* and again at pp. 27 and 41 just as he does in the present treatise; it is clearly a technical expression. He does not, however, employ the terms *laxative* and *eradicative* in the *Examen*.

A long and informative article on bloodletting, with a bibliography beginning in 1525 and extending to 1876, is that under the title "Saignée" in *Dictionnaire Encyclopédique des Sciences Médicales* (Paris, 1878), pp. 145–179. Unfortunately, it does not treat the early history of the subject in sufficient detail or employ any of the early technical terms. Galen in three treatises (Vol. XI K., 147–316) discusses the subject but without the use of such terms as *laxative* and *eradicative*. Stephen Blankaart in his *Lexicon Medicum* II (new edition by C. G. Kühn, Leipzig, 1832) under "venae sectio" gives the following types of bloodletting, according to purpose: *depletoria, revulsoria, derivatoria. Laxative* probably comes under the general heading *depletoria* while *eradicative* is presumably a variety of *derivatoria. Derivatio* is thus defined by Blankaart: actio, qua humor, pus, urina, sanguis, etc., qui in partem influxit et nondum impactus est, artis ope per locum vicinum educitur, vel ex loco uno in alium allicitur, cuius resistentia aliquo modo tollitur; uti fit per venaesectionem, cucurbitulas, epispastica, frictiones, etc. There is a short account of the practice of bloodletting in George Sarton, *Introduction to the History of Science*, II, 1 (1931), 76–79; at II, ii, 1088 he cites Arthur Morgenstern, *Das Aderlassgedicht des Johannes von Aquila*, etc. (Leipzig Dissertation, 1917), which contains a list of writings on bloodletting down to 1917. Since around 1930 the *Index Medicus* has ceased to list the heading "Bloodletting," and it is perhaps time for someone to write a thoroughgoing history of the subject. Some of the old treatises are given by Guilielmus Godofredus de Ploucquet, *Literatura Medica Digesta sive Repertorium Medicinae Practicae, Chirurgiae atque Rei Obstetriciae* (Tübingen, J. G. Cotta, 1809, 4 vols, and supplement), *s.v.* venaesectione. See also the discussion of revulsive bleeding and the controversy between Brissot and Denis by J. B. de C. M. Saunders and C. D. O'Malley, "The Bloodletting Letter of 1539," in *Studies and Essays in the History of Science and Learning Offered to George Sarton,* ed. M. F. Ashley Montagu (New York, Henry Schuman, 1946), pp. 6–19.

evacuation. Nature has no intention of detaining the blood within the ventricle; it flows therefrom to nourish those parts of the body toward which the veins are led off. Induced by this sense perception I am accustomed often to say that phlebotomy of the basilica vein first empties the heart, not the liver as some people think. When the heart is emptied of blood, then the liver is emptied, since the blood flows from the liver through the chilis vein [vena cava] to the heart. Hence it is that in pestilential fevers, in which a deposit of putrid and poisonous humor is found in the phlebotomized liver, patients often die after the fourth day, at which time the putrid matter has already acquired a poisonous quality. For when the right ventricle of the heart is emptied the liver supplies blood to the heart, which, mingled as it is with poison, proves fatal. These statements do not contradict what I have said in my treatise on pestilential fever,[92] where I described the cause of death of phlebotomized patients after the fourth day. They usually die since nature at that time begins to expel the putrid matter; when the vein is opened for bloodletting this matter is drawn from the circumferential area to the center and hence to the heart, whence death results. Returning to this subject therefore, I insist that when the poisonous matter is found in the liver it is more convenient to cut the saphena vein, for it empties more immediately from the liver and does not pass to any other nobler organ of the body, which would weaken the latter. This point is not part of my present purpose, but I can briefly discuss here laxative [minoratiua] bloodletting according to the tolerance of its power and not eradicative bloodletting. I shall, however, touch upon the cause of debility when I write about the anatomy of the legs; what I have said incidentally should be sufficient for the moment. Let me return to inspect the progress of the basilic vein [vena basilica]. In order to view it more readily you may tie off the armpit vein in the upper region before it descends to the arm. You must first, however, gently lead the blood toward the arm for in this way the branches will appear more clearly since they will be full of blood. You can repeat this procedure in the inspection of the arteries, for after the basilic vein passes the curvature of the arm in the inner region it passes toward the inner joint of the arm. In its passage it sends forth many little branches to the muscles of the arm for their nourishment. Among these branches there is a notable one which passes through the lower region and, before it reaches the arm joint, it ascends in that region and passes from the inner to the outer part toward the ring finger and the little finger; it is ramified abundantly in the upper region and is bloodlet between these fingers; a short distance above it is called the salvatella, which takes the place of the basilic vein, since it is a branch of the latter, although there are many questions of dispute on this matter. This, however, is the truth of sense perception. We may also open other branches of this vein in the hand if we wish to do so; they are not opened, of course, except when occasion demands and when it is necessary to open a larger and more suitable branch. When you have examined the basilic vein and its general course begin to lay bare the flesh toward the gullet near the fork of the chest where the neck begins. Here there is a large vein and an accompanying artery from which other veins branch out. You will see that from its trunk two other branches ascend through the neck on both right and left. Each of those branches in turn is divided into two branches in its ascent. One of them is larger than the other and is called the hidden guidez vein [internal jugular] since it is hidden by a longitudinal muscle and passes to the upper parts of the head. But the other smaller branch is called the manifest guidez vein [external jugular] because it passes above the longitudinal muscle and is visible in the neck. Ascending immediately from the fork it is divided into two branches; one of these proceeds around the fork to nourish those parts. The other branch ascends and is united with some branches of the aforementioned vein; but before it does so it sends forth one notable branch toward the joint of the shoulder blade near the adiutorium of the arm [humerus]; this descends toward the inner part near the elbow joint in its upper region and is more evident there, where it is phlebotomized. This vein proceeds lengthwise along the arm toward the thumb and along its entire progress it is called the cephalic vein; it is phlebotomized in illnesses of the head at the inner joint of the elbow, as I said, and also near the thumb where it is more visible. Note, however, that this vein in its descent transmits a branch which is united near the joint with a branch of the basilic vein a short distance under the joint toward the hand. Sometimes the former branch is united directly with the basilic vein, in which case it is called the common vein. It is phlebotomized when there is need for evacuation or diversion of blood both from the lower and upper organs of the body. The greater share of those who perform venesection in our time are deceived by accepting some branch of the basilic vein which has no connection with the cephalic vein in place of the common vein. Hence it is necessary to point out these veins and their progress and connection to those who phlebotomize since there are other branches which pass into one another. You will also note that there is one of the branches of the basilic vein which ascends from the arm joint toward the joint of the hand and to the thumb; this branch is close to the cephalic vein. Therefore some dissectors

[92] Nicolaus Massa, *Liber de Febre Pestilentiali ac de Pestichiis, Morbillis, Variolis et Apostematibus Pestilentialibus*, etc. (Venice, apud Franc. Bindomen et Maphaeum Pasinum, first edition, 1540; other editions: Venice, Arrivabeni, 1555, 1556).

have supposed that the cephalic vein does not proceed directly from the neck to the thumb. These veins, however, both the cephalic and the branch of the basilic, send forth visible branches and many branchlets here and there through the arm and to the hand, both to the palm as well as to upper regions of the hand. Nor do I wish to omit the fact that the cephalic vein near the elbow joint sends a notable branch to the exterior part of the arm which descends toward the middle finger or between the middle and ring fingers near the salvatella vein. Many dissectors have also been mistaken in believing that the salvatella is a branch of the cephalic and not of the basilic vein, just as they have been mistaken about the cephalic vein, as I said above. This error is the result of slight acquaintance with dissections of the body. After you have seen the veins which come to the arms and hands, reexamine the progress of the other branch of the vein which ascends. After the vein has sent a branch to the arm in order to form the cephalic it sends another branch up through the neck above the longitudinal muscle toward the ear. An artery ascends with this vein. These veins and arteries of the branch pass above the ear; but in its passage the vein branches out to the muscles of the lower and upper jaws and under the chin as far as the tongue. In its further progress toward the ear this vein sends branchlets to the cheeks and toward the eyes and to the posterior region of the neck toward the occiput, ascending both from the anterior part as well as behind the ear. These branchlets are scattered through the top of the head to its smallest parts, the vein for the sake of nourishment and the artery for the sake of life. Thus you have seen the progress of some part of the manifest guidez vein [jugular], which almost everywhere has its accompanying artery of the aorta ascending. This artery appears clearly in the neck and also temples of the living person and at times visibly pulsates there. Furthermore I do not wish you to be ignorant of the fact, as you have seen in the progress of the veins of the arm, that an artery does not always accompany a vein; sometimes the vein proceeds without one. In its relation to the basilic vein the artery abandons the vein near the curvature of the arm and proceeds toward the cephalic vein, associating itself with the latter for a certain distance. The artery, however, always proceeds under the vein in the external members of the body and is thus guarded as a favorite of nature. But in the case of the artery mentioned above, it leaves the cephalic vein in its progress and proceeds alone through the inner region toward the thumb near the tendons which close the four fingers of the hand. This [radial] artery pulsates visibly up as far as the elbow joint and even above it to the adiutorium [humerus], just as you have seen that artery pulsate which beyond the elbow joint is formed of two branches of an artery coming together. This artery is touched by physicians when they wish to know the disposition of the heart and its vital quality; it also branches out to other parts of the hand. Lest I spend too much time on them you may inspect the remaining ramifications of both artery and vein for yourself; I have wished to point out here only certain larger branches, more necessary for the physician, which proceed to the arms and to other parts. But since, as I said, the guidez vein [jugular] is two-fold, that is, manifest and hidden, and we have seen some branches of the manifest vein and of its artery, in order that you may be introduced to all parts which physicians desire to know, examine also the hidden guidez vein. This is larger than the manifest guidez and ascends near the meri (gullet) and the head of the bronchus. You must strip it carefully with an instrument that does not cut deeply in order not to injure it. You will see that it ascends under the longitudinal muscle together with another branch of the ascending aorta artery which is called the carotid and many other names by physicians. Here, however, it seems sufficient to me to call both veins and arteries by a single name, since the location and origin are more suitable as a means of tracing their progress than mere names. I shall, God willing, write in the future about these and all other names of the parts of the body so that each person may have access both to their dissection and designation.[93] Proceed therefore to examine the hidden guidez vein [jugular vein] and its artery by stripping it. They ascend near the head of the rough artery and the gullet, sending out branchlets to the muscles of the head of the bronchus and to the vertebrae of the neck. They also pass to other parts of the neck and are visible behind the ears near the beginning of the nape of the neck. As they ascend they are spread out through the skin of the head and the pericranium. From both sides they cross the place of the commissure called lauda [lambda] and are fixed in the bone between the sagittal and false commissures where there are visible foramina. But it is difficult to see whether these veins enter from the outside inward or pass from the inside outward since their progress appears both external and internal from the guidez vein. When these two branches are within the cranium they have ramifications that proceed between the membranes of the dura mater both toward the anterior and posterior regions. They enter the passage called the torcular and, as they proceed, are united in the torcular in the lower part of the commissure corresponding to the bone called lambda. In order to see the other remaining branches of the artery and of the vein continue to strip around the parts of the neck since other branches will then

---

[93] Nothing is known of a work by Massa on anatomical nomenclature. He seems to have been attracted in a number of directions by his scientific enthusiasms and with this airy promise he clearly admits he knew the importance of proper terminology, which is emphasized by both Alessandro Benedetti in his *Anatomice* (1494) and Berengario da Carpi, *Commentaria . . . super Anatomia Mundini* (1521).

become visible, that is, two veins and two arteries extending toward the basilar bone near the foramen of the nape of the neck. In that place they spread out to the sides toward the parts under the basilar bone like the veins and arteries and enter by certain foramina in the basilar bone near the lambda commissure in the lower part near the basilar bone; through these foramina the veins and arteries proceed to the dura mater and the pia mater. There are also other branches which enter through the foramen of the basilar bone; two of these enter the passage called the torcular, concerning which I shall speak in discussing the anatomy of the brain. Two other branches of the artery proceed toward the lacuna where there are foramina through which the arteries pass to the brain. These foramina [carotid canal] are near those of the optic nerves. The arteries then pass inward through the foramina of the cranium just mentioned. They compose the texture of the marvelous net within the membrane called the dura mater near the cranium, but the other veins which come to the brain proceed through the substance of the pia mater above the substance of the brain. These veins are numerous and large; very many arteries are scattered among them and the progress of both veins and arteries form a mass in the shape of a net throughout the entire circuit of the brain. This mass of veins and arteries is united with the pia mater. It does not enter the substance of the brain, since veins are not found in that substance, but only in the superficial portion in the membrane of the pia mater. You will also note that very few veins or arteries pass to the dura mater in contrast with those which pass to the pia mater, as is evident to sense perception, since the dura mater, unlike the pia mater, is almost bloodless, for, as I said, very few veins and arteries pass to it and are not found in the superficial portion but lie between the two membranes that constitute the dura mater. Notice also that these two arteries which pass toward the colatorium of the nostrils branch out as they ascend to the mass of the marvelous net and proceed with some veinlets and arteries to form two vermiform substances, which Mundinus called worms. They are not, however, the worms of which Galen spoke, since you will see those vermiform substances in the posterior cerebellum; they are white. But the mass of worms formed by the weaving together of the aforementioned arteries with some veinlets [choroid plexus] lies in the empty spaces which they call the anterior ventricles. Whether or not you count these spaces as one or two, it is reasonable to believe that in this mass of arteries together with the marvelous net take place all the operations of the mind since the spirits which proceed from the heart exist therein so that they may be prepared by the power of the brain and become animal spirits. Therefore since in this region there appears a larger quantity of spirits it is reasonable also to believe that the powers of the mind which proceed from the spirits also exist there. Whatever others may have said about the ventricles of the brain those who have vigor of intellect should recognize that in what is called the middle ventricle there appears no artery or vein or substance of blood but only a watery excrement and that this ventricle is the receptacle of that excrement and a passage for its purgation, as you will see in the dissection of the ventricles. What I have said here as incidental should be sufficient by way of demonstrating the fact that you cannot point out completely the progress of both veins and arteries in the ordinary method of introducing young men to anatomy. Thus I wish to say nothing further since an inspection of the progress of the veins requires an inspection of the veins alone, laying aside all other parts; in fact, it is necessary to lacerate and cut many organs of the body according to the individual necessities of the progress of the veins. Thus much, therefore, will be said as a sufficient introduction to the anatomy of the veins, arteries, and of the remaining parts of the middle body cavity.

Chapter XXXV

On the Method of Dissecting the Upper Body Cavity, That Is, the Head, Beginning from the Bone of the Palate, and First Concerning the Dissection of the Skin and of the Temporal Muscles

When you have completed the dissection of the middle body cavity proceed to the upper cavity. You will first notice the external parts which you will see are not without their usefulness since the external parts communicate with the inner parts. You will note the roundness of the head with a certain depression in it. Then inspect the anterior portion in which are located the eyes and nose together with the forehead and the other parts of the face. The forehead extends from the hair line down to the eyebrows; the part above the forehead is called the sinciput. Look also at the lateral regions where the ears are situated; these regions are called the temples, but the posterior portion above the neck is called the occiput. This highest part is at the top of the head, which they call the vertex; it is completely covered with hair. You must shave the hair away with a razor in order more readily to inspect the skin and the other parts of the head. When you have observed these parts and shaved the hair the skin will be immediately visible. The skin of the head is quite thick, not fleshy, as some have supposed, but pure and formed of those three skins mentioned in the description of the anatomy of the lower body cavity [Chapters II and III]. In this skin you will see the roots of the hair which are fixed within the fleshy panniculus; at any rate, the material of the hair seems to receive its form in that panniculus. These roots are notable; but in order to examine all the parts carefully, cut the skin in the shape of a cross, beginning at the extreme end of the forehead toward the outer angles of the eyes and proceeding transversely toward the top of the head and ending behind the ear. For instance, if you begin to cut on the right side of the forehead you will proceed as far as behind the left ear to the place called the occiput. When you have made the first incision return to cutting along a transverse line beginning from the left side of the forehead above the angle of the eye and proceed transversely up to a point behind the right ear. This method of dissection is employed to prevent the muscle of the forehead from being cut so that you can examine it if you wish. When you have made this cruciform section carefully strip the outer skin since immediately beneath it is a fleshy panniculus. Note that because this panniculus in this region of the body adheres firmly to the skin since there is almost no fat in it at this point many have been deceived by this fact and have said that the skin of the head is fleshy. For the skin exists only as a covering for the body and therefore it is an incorrect opinion among those who insist they have seen fleshy skin in some part not only of the head but of other regions of the body since they have been deceived by the fleshy panniculus thus far unrecognized by physicians and philosophers. Seeing the growth of the flesh they have supposed that in that place there was skin alone. But the growth of the flesh was not due to the fleshiness of the skin but to the fleshy panniculus beneath it. This is not a pure membrane like other panniculi but created by nature from both flesh and membrane in order that separate parts could coalesce by means of that which is fleshy. For if the skin is a spermatic member,[94] as those who hold an incorrect opinion about it affirm, it could not always be regenerated. But since prudent nature does not lack what is necessary to her functioning let those people realize how mistaken they are because the fleshy panniculus is that one through which the flesh comes forth when the skin is cut. Even if I have already spoken of this panniculus it is not unsuitable here also to say something toward a clearer doctrine about the same member, as I have done in many other passages of this introduction to anatomy about other members of the body. I shall do so here in order to explain these matters better by a repeated description. Therefore when the skin has been stripped off strip the fleshy panniculus also; it is especially thin in adults. Proceed carefully in order not to cut any veins, particularly those of the forehead which are usually phlebotomized in many illnesses of the head which proceed from the guidez vein [jugular]. When you have taken note of the veins and arteries examine the panniculus called the pericranium [galea aponeurotica] or almocatim which covers the cranium and the two temporal muscles at the sides of the head. Begin to strip these muscles first and look at their substance, shape, and terminations. They end at one side in the upper region at the small commissures above the false ones; you will see these better when you dissect the cranium. On the other side, they end near the lower jaw [coronoid process] with a quite ample cord which elevates the jaw itself. The shape of the latter muscle is flat, round in the upper region, and it ends below in an acute angle toward the cord mentioned near the jaw. To this temporal muscle come the nerves from the fourth and fifth pair [trigeminal], according to the aforesaid enumeration, beginning from the nerves which descend to the nostrils so that the muscle may better serve the motions of the jaw. Because of the large number of nerves and their closeness to each other blows struck in this region are deadly. Note that nature sometimes doubles these temporal muscles as a freak, as I saw in 1532. When you have examined those muscles look also at the skin in the anterior part of the forehead which you have stripped. You will see the fibers of the muscles [frontalis] which move the eyebrows and the skin of the forehead. These fibers proceed below toward the nose and eyes. When you have seen these parts also strip the panniculus called the pericranium which adheres to the bone of the cranium with strong ligaments that come not only from the commissures but also from the substance of the cranial bone and is so attached that it can be called continuous rather than contiguous, as you may also see in the other panniculi which cover the bones. This panniculus is hard but not thick and joined by means of the sawlike commissures as well as by those which are not sawlike to the panniculus of the dura mater in order to support the latter. Not only is the pericranium united to the cranium by such ligaments of the commissures but the pericranium supports the panniculus of the dura mater. This panniculus is joined immediately in the convex part of the cranium not only with these but other ligaments coming from the other parts of the bone. You will see this panniculus when you lift up the cranium.

Chapter XXXVI

On the Anatomy of the Cranial Bone and Its Parts

When you have seen the substance of the panniculus of the pericranium and stripped it from the cranium, white, hard, and thin as I have described it, look at the substance of the cranial bone. It is white tending toward some redness and of a round shape

---

[94] See note 20 and Aristotle, *De Generatione Animalium*, IV, 5, 774 a 20 (Vol. 5, Oxford Aristotle, translated by Arthur Platt and his note: "By 'spermatic' A. means that both sexes produce large quantities of residual matter which is worked up into semen by the male and remains in a sanguineous condition in the females." "Spermatic" animals were also supposed to be able to superfoetate.

somewhat depressed and divided into many parts by the commissures [sutures]. You will see the first of the latter in the anterior part of the head above the forehead, beginning from the farthest cavity of the nose under the eyebrow and measuring a hand's breadth from the end of the palm toward the fociles as far as the end of the middle finger. You can also find the place of the aforementioned commissure by taking a length of thread as long as the space from one anterior part of the ears to the other and then doubling the thread or taking half of it, placing one end of it at the end of the nose and drawing the thread upward. Where the string ends is the location of the anterior commissure which is called the coronal commissure. It is saw-toothed; the teeth issue alternately from both sides of the commissure to form it, passing in an arc from one temple to the other. The second commissure is called the sagittal; it is also serrated and passes directly through the top of the head, beginning from the coronal commissure toward the sinciput and ending at the other commissure at the occiput, which is called the lambdoid since it resembles the Greek letter lambda. The sagittal commissure is straight, at the top of the head, serrated, and divides the top of the head into right and left parts. The third serrate commissure is called lambdoid and is in the posterior part of the cranium, proceeding as far as the basilar bone. Sometimes a fourth serrate commissure is found in some skulls which divides the frontal bone and proceeds from the coronal commissure as far as the nose bone; this is not found in all skulls, as you will clearly see in the heads of dead people in cemeteries if you wish to examine them, although some heads are found in which you will see commissures of this sort. Some have said this commissure is found only in the heads of women,[95] but I have seen it also in the head of a certain young man, passing through the forehead as far as the nose. After you have examined these commissures look at the lateral ones above the ears which they call false. These are visible and continuous but not serrate. They ascend like scales and from this similarity are called squamous. Above these are two other commissures, one on the right and another on the left, which are almost imperceptible and are thus frequently not seen except in cemeteries; you will not always find these commissures clearly above the petrous bones or above the false commissures. Above these commissures, that is, between the last described and the sagittal, there are two foramina [parietal] through which the veins pass to the brain, one on the right and another on the left. There are other foramina but not so apparent, through all of which the veins pass to nourish the bone and to the substance of the dura mater. Sometimes other commissures, about which I shall say nothing, are visible; however, it is sufficient to have enumerated those which are visible. They are of the greatest usefulness and created by nature, which does nothing in vain. Many purposes are served by the commissures of the head. By means of them connections and divisions are established between the external and internal members so that if one part is injured the other parts remain uninjured. Another considerable use is that through the porosities of the commissures many evil vapors and superfluities are expelled. Similarly when the parts are dilated in equal fashion if there is need of some external assistance of air or something else both heating or cooling, drying or humidifying, it can reach the substance of the brain more easily through these commissures. But in order to inspect carefully the remaining parts of the skull turn to its inner region. You must elevate the upper part by cutting it with a small saw in a circle around the head above the ears, beginning from the anterior portion in the bone of the forehead and proceeding through the temporal bones toward the posterior part and returning to the forehead, where you began to cut with the saw. In order to prevent any laceration of the membranes contained within the skull lift its upper part and separate it like a bowl. When you lift it up you should note that the dura mater is attached everywhere in the convex part of the skull by very strong ligaments beyond the commissures, as you saw in the pericranium. You must lift the skull itself with the aid of some blunt stripping instrument in order not to lacerate any part of the brain since it is a soft substance; this instrument should be thin and somewhat curved. It will be rather difficult to lift the skull on account of the aforementioned ligaments. There are some readers who have denied the existence of this attachment of the cranium with the dura mater beyond the commissure, although I have shown it to exist, because they have never read of it anywhere, especially in Galen. If these people, whether young or old, were rightly to consider the words of Galen in the ninth book of his *Anatomical Procedures*[96] and in other passages they would know that Galen, while he marveled at the name meninx used by the ancients in

[95] Berengario da Carpi, *Commentaria*, etc. (1521), fol. CCCCXVIII cites "Arist. primo et tertio de hist. c. vii" on this odd notion and refutes it, as Massa does. See Aristotle, *De Historia Animalium*, I, 7, 491 b: "The skull has sutures; one, of circular form, in the case of women"; *ibid.*, III, 7, 516 a: "and in the human species the suture is circular in the female"; *De Partibus Animalium*, II, 7, 653 b: "Man, again, has more sutures in his skull than any other animal, and the male more than the female" (translated by D'Arcy Wentworth Thompson in the Oxford Translation of Aristotle, edited by W. D. Ross; the translator points out in a footnote that the sutures on the vertex do become more or less effaced in pregnant women, giving rise to "puerperal osteophyte," citing Rokitansky, *Path. Anat.* iii. 208, Sydenham Society Translation.) Berengario goes on to quote Pietro d'Abano, *Conciliator*: Conciliator tamen X. prob. XLVIII dicit se vidisse mulieres habuisse plures commissuras viris, etc. Galen, *De Ossibus* (Vol. II K., 739–746) does not make such distinctions between male and female sutures.

[96] Galen, *On Anatomical Procedures*, IX, 2 (Vol. II K., 716–717), Singer's translation, p. 230.

connection with the panniculus of the brain, said that this thin membrane girdled the substance of the adjoining brain beneath and that the thick membrane girdled the cranium, the brain ascending and descending within the intermediary space between these coverings when it was dilated and contracted. But of this matter the very learned Arab (Rhazes) has spoken more clearly in the first book of his treatise to Almansor.[97] However, since I had decided to point out the truth of sense perception, dismissing the authorities you will see this union if you make use of incisions in the basilar bone to which the dura mater is joined so that you can scarcely separate it with a scalpel unless the animal or cadaver is still warm or unless the parts of the creature have already begun to relax from humidity through the means of external heat. But the witness of sense perception is the roughness itself in the convex part of the bone; if you inspect the bone of the cranium in its hollow part it is not smooth and dry as in the gibbous upper part but is uneven with hollow places in its surface where the dura mater is attached along the progress of the veins which in the dura mater are beyond the commissures. The reason given will not hold, namely, that crabs in the new moon are half empty[98] since he (Galen) well concludes with Hippocrates and other wise men that at the time of the new moon the humidities of our body are restricted and at the full moon are extended and scattered and thus soft members, such as the brain, are more swollen. It was not because of this situation that Hippocrates and others said the dura mater was not joined anywhere to the cranium, because if sometimes in wounded people the dura mater is seen to be separated from the cranium this is on account of a blow which has caused it to break loose or because the ligaments have been torn or because some humidity has relaxed them; it makes no difference whether the humidity is bloody or not. It is sufficient to know that in some people this panniculus of the dura mater is found to be separated from the substance of the bone. This is not according to nature nor would it be reasonable, because the dura mater, if it were movable and separate, would have to touch the cranium at all points so uneven as it is, since nowhere in a living creature has nature made rough bones without a covering which adheres to them, as is well known to practiced anatomists. However, if you wish to ascertain the truth of this statement cut the lower bone, that is, the basilar, with a saw or some other instrument; after the substance of the brain is removed together with the pia mater, you will see the dura mater joined everywhere to the cranium beyond the commissures. But allow those who hold a contrary view to abide in their knowledge of words and ornate speech in which they have placed their highest glory. You shall, however, be zealous in the pursuit of truth and return to view the remaining parts of the skull. In these you will notice one substance of bone, white, hard, dense, made up of three parts. One of these is above, contiguous to the pericranium; this is dense and hard but not equal in thickness in all people since it differs according to age, complexion, and sex. This bone, or rather the first part of the bone which they call the first table is of the thickness of a piece of undried oxhide. The substance of the second part of this bone is spongy and red [diploë]; they call this the second table. It is not dense and hard but thin (as I said), reddish and not soft, but because of many porosities in which blood appears or the material of nourishment from which the qualities of blood have not been removed it is less hard. This second table ought rather to be called the spongy middle part of the cranium than the separated substance of the bone, for the brain (since it is a principal member) is not the marrow, as certain ignorant people have believed, having misunderstood the words of the ancients. Therefore, the second table is the spongy substance of the cranium, where the nourishment of the bone is digested; this part is also called medullaris by some. After this second table there occurs the last part of the substance of the cranial bone, called the third or glassy table because of its thinness, since it is the thinnest part of the skull. This is united or attached to the dura mater. If you inspect its hollow portion you will see that this part is rough, uneven, and contains some cavities in which the dura mater is more closely attached. The veins proceed better through the dura mater itself since they pass through these cavities where the dura mater is more swollen. Thus you have here the number of the parts and the diversity of substance of the cranial bone. But since there are other parts which compose the cranium in order to complete it note first the anterior portion from the coronal commissure downward toward the eyebrow, which is called the sinciput. Above the commissure that part of the bone which is divided into right and left by the sagittal commissure ought also to be inspected. There are divisions or commissures which proceed to the ears on both sides. The larger and more visible of these is the one called false. Above this commissure is another not always visible in those recently dead but very often apparent in bodies found in cemeteries. This inspection is useful in manual operations of the head by cutting or by striking when on account of an inner damage it is necessary to cut the bone with a trepanning tool or grinding instru-

[97] Muhammad Ibn Zakarīyā, Abū Bakr, Al Rāzī (Rhazes), *Liber Rasis ad Almansorem* (Venice, Jacopo Pentio de Leuco, 1508), I, chap. 7: Grossus [panniculus] autem cranio adheret, etiam in quibusdam locis cerebro fit multum vicinus.

[98] This idea may proceed from Pliny, *Natural History*, 2. 41. 109, p. 251, Loeb Classical Library, translated by H. Rackham: "Indeed persistent research has discovered that the influence of the moon causes the shells of oysters, cockles and all shell-fish to grow larger and again smaller in bulk, and moreover that the phases of the moon [i. e., the number of days from the new moon] affect the tissues of the shrewmouse, and that the smallest animal, the ant, is sensitive to the influence of the planet and at the time of the new moon is always slack."

ments or other means; in order to remove that ill or damage we operate upon the commissures with resulting injury. The lateral parts have different names; those near the sagittal commissure are called the bones of the bregma. Under these are the false commissures caused by the petrous bones. These bones are called petrous because they are very hard. They are also called squamous because they ascend above the other parts of the cranium like scales. Some people take all these parts together and call them temporal, but there is not much force in names. Look also at the posterior part of the bone beyond the lambdoid bone; it is also called stern or poop of the head; it is very hard and dense in comparison with the other parts for the posterior part of the cranium [occipital] is harder, thicker, and more dense than the anterior part with the exception of the eyebrows, where the eyesockets are. This part, although thicker, is not harder. I shall not pass over one item that is worth noticing and quite useful to physicians: this is the fact that the bone around the coronal commissure is thinner in the posterior part. In the lateral regions the bones are also thinner than they are in other parts of the cranium. Look also at the foramina in the part of the cranium near the sagittal commissure through which the veins pass to the brain. Examine also other foramina through the substance of the bone; these are not so large and through them the nerves pass from the dura mater and pass out to the parts of the head and of the face, to the tongue and to other members. There are also veins which pass from the dura mater through the cranium outwards. Therefore lay aside the remaining parts of the cranium, such as the basilar bone and the sagittal bones behind the ears, for you will see these parts after you have elevated the substance of the brain, lest you should believe the head is composed of only six or seven bones since there are many more if you count all the parts of the basilar bone. Thus after you have noted everything worthy of notice and have seen the places and divisions between them in the manual operations both in moving and dilating the bone itself you must prepare your eyes and mind to see the remaining parts.

Chapter XXXVII

On the Dissection of the Membranes of the Dura and Pia Mater and Their Contents

Immediately under the cranial bone which you have lifted up is the membrane called the dura mater. It is white, hard, quite thick, and resembles the substance of the nerves. It is composed of two tunics; between them are many nourishing veins but not large, stretching out in all directions. The larger veins extend down the longitudinal aspect of the dura mater; these are not apparent as they are in the pia mater because they are inclosed between the two tunics mentioned above. Because of this concealment the dura mater appears almost bloodless. Not many arteries pass to the dura mater to give it life. The dura mater covers the substance of the brain which is veiled by the membrane of the pia mater. The dura mater is divided into two parts in the anterior portion, on right and left, beginning from the anterior portion and proceeding to the posterior and the cerebellum along the course of the sagittal commissure. This division is deep. By means of it the dura mater is doubled and united with a perfect union. Where it is thus united it is thin and fixed or joined in the anterior portion with the substance of the brain and on the other side, that is, toward the cerebellum in the form of an arch so that in this division it is joined only with two heads. This part is the middle of the dura mater [falx cerebri], which divides the anterior brain into right and left and hence into two portions. Under this division is the part of the brain between the right and left ventricles, which is called the diaphragm of the brain [septum pellucidum]. It is transparent and gleams like a jewel. You will not see this transparent gleam perfectly until you open the ventricles and gently spread apart that substance which is between them and distinguishes the anterior ventricles in order to keep them from being broken in their lower part. Thus you shall see it shining and transparent, as I shall say more clearly in due course. Galen, the greatest of anatomists, discussed this diaphragm in the ninth book of his *Anatomical Procedures*.[99] The dura mater, proceeding toward the posterior brain, which is called the cerebellum, is doubled (or folded) also laterally or along its transverse dimension, dividing the anterior brain from the posterior cerebellum [tentorium cerebelli]. However, in this division parts of the dura mater are not united as they are down its length in the anterior part of the brain but are separate and apart so that each part is covered by its own portion of the dura mater. One of these parts divides the anterior portion of the brain into right and left; the other part divides the entire substance into anterior and posterior portions. But before you move the dura mater aside you should examine a certain passage composed of the substance of the dura mater. It proceeds from the anterior part of the brain through its top along the sagittal commissure; it makes no difference whether you say it proceeds from the posterior region to the anterior or not [superior sagittal sinus]. This passage is called by some people the torcular, but the ancients called torcular that place in the posterior part where two veins enter this cavity and where the posterior brain approaches the foramen of the spinal medulla. Galen and other wise men called this cavity a vein.[100] In this

---

[99] Galen, *On Anatomical Procedures*, IX, 3 (Vol. II K., 721, 726), Singer's translation, pp. 232, 234: the *septum lucidum*.

[100] Galen, *ibid.*, IX, 1 (Vol. II K., 712), Singer's translation, p. 228: the torcular of Herophilus: see Singer's note 173.

vein, cavity, or torcular as some call it there pass in the posterior part toward the spinal medulla the two veins mentioned above. They ascend through the cavity and in their progress their tunics grow thin; I have often seen the progress of these veins. In some people, however, the continuation or progress of these veins is not visible nor have I seen anything except the cavity itself. But when the veins appear in the cavity they are united for their substance is rarefied as they pass. In that part of their tunic which is joined together they hold the blood by means of their thinness. The blood seems to be outside the veins because of the thinness of the membrane which retains it. This is not true, of course, since after this apparent rarefaction and expansion of the blood detained by a very thin membrane you will immediately see the blood or the vein once more uniting its parts to form one substance similar to a vein in every respect. It ascends toward the forehead by the aforementioned passage. You can see this by sense perception, without any discussion, if you carefully lift up these veinlets. But when you come to this part in which the blood appears so rounded into a ball, inspect it carefully by lifting it up gently, with lighted candles in your hands, for you will see the membrane of the veins stretched out very thinly as if it were a spider web encircling this blood, which later is united in its progress and forms one substance similar to the vein which I said passed through the torcular or that cavity and sends little branches through the foramina of the dura mater to the pia mater. Hence it will be of some advantage to doubt whether this cavity formed from the substance of the dura mater can be called a vein or whether one should believe those who say that there is blood within it outside of the veins since you have already heard and seen that the veins pass within the passage and often pass through its entirety. Sometimes, however, owing to lack of blood or the thinness of the veins their final transit is not visible. But in order to see what I have thus far described cut the dura mater in its anterior portion at the top along the coronal commissure with an instrument shaped like a sickle in order not to break any of the parts contained within the cavity or torcular. You will assist the incision by placing in it a blunt stylus and then cutting. Proceed finally to the end of the passage both in the anterior and posterior portions until the entire cavity is opened to view. In the lower anterior portion toward the spinal cord you will see the progress of the two veins mentioned above [transverse sinus] coming from the guidez veins [jugulars]. These proceed through the torcular and are united, making one substance. This proceeds to the anterior part toward the forehead and, descending, is fixed in the depths of the fold of the dura mater, making a division of the anterior brain into two parts. Note that the branchlets of the vein which passes through the foramina of this large cavity to the pia mater are united with branchlets of the veins of the pia mater. There are some people, however, who believe these veins proceed in a contrary manner, that is, from the pia mater and through those foramina enter the passage or this cavity. But when you have seen below two veins coming into the torcular and forming one substance they will be sufficient for a knowledge of the truth since sense perception is more powerful here than dispute. When you have noted these items, in order to see the substance of the dura mater and the veinlets which proceed from the vein in the cavity through its foramina to the veinlets of the pia mater cut around the dura mater in its lower part when the cranial bone has been removed with a saw. Carefully lift up the dura mater thus incised, turning it toward the top of the head both on right and left. As you lift it and approach the torcular or cavity you will see the progress and conjunction of the aforementioned veinlets which I said proceeded from the torcular. After this inspection look at the dura mater with its two tunics. Between these you will see the progress of the veins as they pass across through the dura mater. Thus you will recognize the substance, parts, uses, and connection of the dura mater, since it descends through the spinal cord to the vertebrae, as you will see later. Note also that the dura mater is joined here and there to the pia mater by the mentioned veinlets. After you have examined everything in the dura mater examine the diaphragm which lies in its place beneath the dura mater. Look also at the membrane of the pia mater and the space between it and the dura mater of which Galen spoke in *Anatomical Procedures*, IX.[101] You will thus recognize that the membrane of the pia mater adheres to the substance of the brain and that the membrane of the dura mater also adheres to the substance of the cranial bone. The membrane of the pia mater is thin and almost like a spider web, tough, however, and strong in comparison to its thinness. Through this membrane pass many notable veins full of blood which you will see throughout their entire progress. Branches of an artery also pass through this membrane, but not so numerous or large. From this passage of veins and arteries is formed a netlike texture. Keep these veins thus complicated with one another in order to see the connection they have with the veins passing through the anterior ventricles of the brain; these are at the bottom of the ventricles and resemble red worms [choroid plexus]. Of these I shall speak later since anatomists have called these veins by the name of worms. They are not the worms, however, of which Galen spoke, since their substance is white and they are located in the cerebellum quite near the basilar

[101] Galen, *ibid.*, IX, 2 (Vol. II K., 716), Singer's translation, p. 230.

bone. Thus you have seen the substance of the pia mater together with its veins and arteries.[102] Its function is to sustain these veins and arteries, to clothe the substance of the brain, and to divide many parts of the brain more immediately, for if you observe correctly you will see that the pia mater not only covers the exterior substance of the brain but also the inner grooves and walls of the ventricles in which there is the mass of veins and arteries composing the substance which resembles worms.

Chapter XXXVIII

On the Method of Cutting the Substance of the Brain So That You May See All the Ventricles and Other Parts

According to the judgment of all the experts the white, soft substance of the brain is larger in man in comparison to his size than it is in any other animal, and therefore man excels other animals in knowledge. This substance has been wrongly called marrow by many people since the brain is in itself a principal and noble member in which the mind is located. Therefore nature has striven to protect it by all means lest the operations of the mind should be injured. Look then at the brain's substance, full of grooves [sulci] (sinous depressions), cutting it in the upper part with a broad thin knife, gently lifting it piece by piece. If you wish you can, according to Galen's words, cut it lengthwise on right and left as far as the cavity of the ventricles. The first thing you will see is that broad cavity called the oven of the brain or psalidoeides, a Greek word which seems to mean "shaped like an arch" [corpus callosum], for the cavity of an oven is broad.[103] When you have found the place of this cavity use a wooden stylus, or many of different sizes, to lift up its parts more readily in order to inspect it, for the moderns have not noticed it. This cavity of the oven has within it a convex upper part and on the outside it is humped. If you look carefully you will see that its substance also differs somewhat in color from the other neighboring parts. The use of this oven is to cover the anterior ventricles. It is supported by the extremities of the diaphragm. When you have seen this gradually remove the substance of the brain which forms an oven and the anterior ventricles will become visible; you can call it only one ventricle if you take the two parts as one ventricle, for there are two cavities, one at right and left. These begin toward the forehead and proceed in a circuit, winding almost as far as the end of the anterior brain. In these cavities there are certain elevated and more whitish substances called nates or anchae by modern anatomists, for they are humped, whitish, and elevated like human buttocks [caudate nuclei] in the right and left cavities, or in the ventricle.[104] Above these anchae pass the nerves which proceed to the ears from the fifth pair of nerves, according to the common enumeration. If you begin to number from the nerves which proceed to the nostrils then these nerves above the anchae proceed from the sixth pair. They should be inspected, for their transit is very beautiful. You ought to know, however, that the buttocks of which Galen spoke and which open and close by the motion of contraction and constriction of the worms in the posterior brain they call the cerebellum you will see when you inspect the middle ventricle. When you have inspected the former items you must dilate the walls of the ventricles in order to see the other parts more comfortably with the wooden stylus I have mentioned, whether of boxwood or any other kind makes no difference so long as it is clean. Use this lightly along the course of the walls in order not to break the membrane of the pia mater which also adheres to the walls of the ventricles. You will see in the lower part or bottom of the ventricles the two substances already called reddish and wormlike woven from veins and arteries coming from the pia mater. This mass of veins and arteries passes through the entire extent of the anterior ventricles around the anchae (or buttocks); it is called worms by Mundinus and other moderns, as I said above. When you have seen this remember what I said in the dissection of the membranes known as dura and pia mater, that in the division of the brain made by the dura mater into right-hand and left-hand parts under the dividing membrane there is that part of the brain between right and left ventricles called diaphragm [septum pellucidum] by Galen, shining and transparent.[105] In order to see it you must gradually lift the substance itself with care since it is soft in order not to break it; the warmth of the spirits in the arteries you saw in the ventricles which furnish spirits has created this shining transparency like a wall between the right and left ventricles. Examine also in these cavities the watery superfluity [cerebrospinal fluid] deposited there by other parts of the brain which is purged through a certain foramen which comes from the ends of these ventricles. Thus proceed to separate their walls with the stylus I mentioned so that you may recognize everything. But what is more worthy of note is the fact that I have always found these cavities full, or half-full, of the aqueous substance just mentioned. It is the rational superfluity of the brain itself since it is a cold and humid member and is

[102] Galen, *ibid.*, IX, 3 (Vol. II K., 719), Singer's translation, pp. 231–232: the choroid plexuses.

[103] Galen, *ibid.*, IX, 4 (Vol. II K., 724–726), Singer's translation, pp. 233–234: the fornix.

[104] Galen, *ibid.*, IX, 5 (Vol. II K., 729–730), Singer's translation, pp. 236–237: the buttocks of the brain and the vermiform process.

[105] See note 99.

nourished by a nutriment similar to itself. Hence you will say, and reasonably too, that when nature gave each member proper places for the reception of superfluities and passage ways for expelling them, to such a member she also gave cavities in which to receive these superfluities and for their purgation placed a passage in that member's lower region by which the superfluities could be drawn away. Thus it is more consistent with reason to place the power and the operations of the soul where there are members that retain the instruments of the soul. These members are the mass of arteries which contain the spirit. This spirit is the instrument of the soul, according to Aristotle,[106] and these spirits exist both in the marvelous net formed from those two arteries ascending to the dura mater as well as in those arteries which come to have, in combination with those aforementioned veins, the form of worms. When I speak thus I do not contradict the words of those ancients who spoke wisely, for nature does not abound in what is superfluous nor does she lack what is necessary for her proper functioning since she can impart many uses to a given portion of the body. What she always does she has also done in these cavities, which have the purpose of draining the watery material from the brain to the other members. There are also places in which there are the instruments of the mind and their powers; these are the spirits contained in the arteries. Parts of these arteries pass through these ventricles as a necessary place for containing the arteries. In the ventricles are the spirits which operate by the power of the soul; the instruments or organs of power are the spirits detained in the arteries, nor are the spirits found outside the arteries. For the soul does not operate without a corporeal instrument, and therefore those who postulate powers of the soul without a corporeal subject do not think correctly, according to my opinion at any rate, since power is given to the external members by a nerve, which is a motivating and sensitive (agent), because the powers of the mind are sent to the nerves. These powers are principally in the spirits as in a prime organ and are therefore nobler since they exist in the first and proper instrument. Concerning the powers of the theoretical soul it is necessary to speak otherwise from the context of Aristotle's books on the soul.[107] Having completed the inspection of the parts described above, continue to strip and dilate with the instrument indicated and to look at the place where the optic nerves begin around their crossing [optic chiasma]. In that place is a foramen or aperture through which the aforesaid superfluities are purged to the cavity they call a lacuna [sella turcica], outside the substance of the brain, to the basilar bone, and to the palate in their descent. Proceed to elevate the substance of the brain and to separate the parts with your wooden stylus since at the end of the ventricles toward their posterior part you will find another cavity, small and oblong, situated above the aforesaid lacuna so that it may send the superfluity to the latter. They call this cavity the middle [third] ventricle. Near there you will find a pine-cone-shaped object [pineal body] according to true sense perception, harder and proceeding with the substance of the brain but different in color. You will recognize this pine cone[108] by the combination of veins which pass through it to the brain; that is, this mass of veins is based upon the pine cone in their transit where it is located in the middle ventricle. The harder part of the pine cone comes to a sharp point. The middle ventricle seems to be a passageway to the other parts of the brain, for the walls of its cavity [thalamus] are humped and almost touch each other in a dead body. Galen called these parts the buttocks, as I said above. Whether or not the middle ventricle is bipartite, as some people have said, is difficult to discern by sense perception. Those who say this cavity is bipartite present this reason: that it proceeds from the anterior ventricles, which are two in number. Note, however, that in this middle ventricle there is found no mass of veins and arteries and therefore there is in it no power of the soul. All the powers of mind are in the anterior part of the brain where the spirits exist in its vessels. It is consistent with reason to assume and to say that this middle

---

[106] Aristotle, *De Generatione Animalium*, V, 7, 789 b: "so it is reasonable that Nature should perform most of her operations using breath as an instrument . . . " (translated by Arthur Platt, Oxford Aristotle, Vol. 5, edited by W. D. Ross); *De Spiritu*, 4, 482 b: "Now clearly the respiration has its motive principle from the inward parts, whether we ought to call this principle a power of the soul, the soul, or some other combination of bodies which through their agency causes this attraction . . . " (translated by J. J. Dobson, Oxford Aristotle, Vol. 3).

Galen, *Hippocratis Epidem. VI et Galeni in Illum Commentarius V. Sectio V* (Vol. XVII B. K., 248): Illud praeterea persuasum habeo spiritum in eius [cerebri] cavis positum primarium esse animae instrumentum, quem ut animae substantiam esse dicerem, mihi temere inclinabat animus; *idem, Quod Animi Mores Corporis Temperamenta Sequantur*, 4 (Vol. IV K., 783): In hoc substantiae genere Stoicorum quoque opinio comprehenditur; animam siquidem spiritum aliquem esse volunt, quemadmodum et naturam, verum humidiorem ac frigidiorem naturae spiritum, sicciorem ac calidiorem animae. Quamobrem hic spiritus peculiaris quaedam animae materies est.

[107] That is, Aristotle in *De Anima* does not postulate organic means such as nerves and respiration (or spirit) as the organs of sense, according to the manner of Massa. Aristotle speaks in terms of soul as "the first grade of actuality having life potentially in it" (412 a 30). Compare *De Generatione Animalium*, II, 4, 738 b 25: "the soul is the reality of a particular body." In other words, the neurological and physiological apparatus for sense perception are not dealt with in Aristotle's account of the theoretical psychic powers; he is more interested in physics and metaphysics than in biology in his treatise on the soul.

[108] Galen, *On Anatomical Procedures,* IX, 4–5 (Vol. II K., 728–729), Singer's translation, p. 236: the pineal gland. See Lauth, *op. cit.*, p. 333.

ventricle is the passageway which leads off the superfluities just as the other cavities of this very humid member do. In order not to interrupt the sequence of dissection proceed to dilate this passage toward the posterior part of the head. You will see a passage descending to a cavity toward the space which is in the middle of the anterior brain and the posterior, or cerebellum. Many people have believed that this space is the posterior ventricle. This is not reasonable for if you carefully inspect the region you will find another passage also proceeding to the posterior cerebellum. Although this passage [cerebral aqueduct] is small and not to be found except by a very skilled anatomist yet it exists in all bodies and leads to the last (or rearmost) [fourth] ventricle. The ancients have spoken in various terms of this ventricle; but that place below and between each part of the brain is the place by which the middle as well as the posterior ventricle is purged. Under this place you will find veins adhering to the pine cone and proceeding to the substance of the pia mater. From what you have seen you can realize that these ventricles exist rather for the collection and expulsion of superfluities than for the various operations of the mind which must be accomplished by different parts, since in the last ventricle which proceeds in the posterior cerebellum no vein or artery is found. It is a pure cavity through which the superfluities of the brain are drained off to another cavity aforementioned. Thus you see from what has been said that these ventricles, that is, the anterior ones, the middle, and the posterior, have no primary function in the operations of the soul since there are no spirits in them as in an appropriate receptacle, as some have believed there are. But there are spirits in the arteries of the anterior ventricles; these spirits are the powers of the soul acting through inner operations which according to the different degrees of spirit bring about different operations of the mind. Thus these people have erred in their opinion since they do not know that, just as the proper place of the blood is in the veins so the proper place of the spirits is in the arteries, for it is not possible to find spirits outside the arteries except in the left ventricle of the heart. Similarly one can find blood outside the veins only in the right ventricle of the heart. Add also that the soul uses its power in the organ. When you have seen the ventricles, and especially the anterior ones, proceed to lift the substance of the posterior brain, which they call the cerebellum. This substance is placed under the substance of the anterior brain since part of the anterior of the brain which you have seen covers or is above the substance of the posterior cerebellum. The anterior of the brain which you have seen covers or is above the substance of the posterior cerebellum. The substance of the cerebellum is harder and thicker than that of the anterior of the brain (or cerebrum) and darker in color. It has a small cavity within it called the posterior ventricle. Thus the cerebellum is surrounded by two membranes, the dura and the pia mater, just as is the anterior brain, and has a round shape since it tends to assume a point toward the foramen of the spinal cord and, as I said, is divided from the anterior brain by the two aforementioned membranes. Cut these membranes in order to see the substance of the brain, but before you move aside the dura mater look at two veins which pass through it, of which I have spoken before, because they pass to the torcular since in their transit they are branched out through the dura mater and are especially visible in the place corresponding to the lambdoid commissure [transverse sinus]. But I have spoken of them above. Since you have opened the torcular or that cavity, continue, after examining it, to elevate the membranes which enclose the cerebellum in order to move aside its substance more comfortably. It differs in color and hardness from the substance of the anterior brain. Descending toward the basilar bone you will find two substances, white and wrinkled and somewhat reddish, different from the substance of the cerebellum and resembling worms found in tombs. They are located near the buttocks and the spinal cord. The anterior one is also superior, while the posterior substance is near the spinal cord. Galen compared these substances to worms that grow in wood in *Anatomical Procedures*, IX.[109] These substances tend toward the anterior brain near the buttocks which you saw when you looked at the veins adhering to the pine cone. Under these veins is the posterior ventricle with a notable cavity. Look at it, for many moderns have not seen it. These worms in their dilation, cause, as they say, the retention of the spirits and when they are contracted the spirits pass forth or proceed freely. When you have seen these parts continue to lift the remaining parts of the cerebellum. You will see a certain white substance under the posterior ventricle; this is continuous with the anterior ventricle. Therefore on account of this continuation they say that the marrow proceeds from the anterior and not from the posterior brain. You will see that this substance passes to the foramen of the marrow and unites with it or composes it. Not without pleasure you may speculate and with the aid of your senses see whether the substance of the marrow proceeds from the posterior or from the anterior substance of the brain. You will agree that both by virtue of its power and its visible progress that the marrow arises from both anterior and posterior parts of the brain. Thus from what has been said you know the substance, bulk, parts, shape, and the remaining attributes of the brain's substance, such as its operations and uses. These parts, which do not agree with the views of many who write on anatomy, require a true judge, not authorities or apparent reasons, for the true judge is sense percep-

[109] Galen, *On Anatomical Procedures*, IX, 5 (Vol. II K., 730), Singer's translation, p. 236: the vermiform process.

tion itself. Proper truth is anatomical science applied to these parts of the human body.

Chapter XXXIX

On the Method of Procedure in Viewing the Progress of the Nerves from the Brain, On the Location of the Marvelous Net and On the Parts of the Basilar Bone

We have now seen many things which you will note have not been described in this manner by all the ancient writers on anatomy. In the process we have seen many things worthy of admiration which require the good offices of the senses and of reason since they are not found in the same manner as some people have supposed in respect to many details. First you will see not without admiration a pair of nerves descending to the nostrils [olfactory tracts]. Their progress is quite apparent. They are first visible near the optic nerves and pass toward two little eminences shaped like the nipples of the breasts [olfactory bulbs]; from this similarity anatomists call them the mamillary carunculae. They appear when you gradually begin to lift with your hands the rest of the brain from the anterior part. You should do this gently in order not to injure the two nerves which descend to the nostrils so that you can see each of them clearly since they proceed from a place near the optic nerves and come down to these carunculae. Afterwards they [olfactory nerves] descend through two passageways to the nostrils and are fixed in the walls of the latter. The sensation of smell is accomplished by means of these nerves, nor is this statement inconsistent with the truth although Galen, *On the Use of the Parts* IX[110] was unwilling to call these substances nerves, for if anyone correctly recognizes the instruments of sense-perception I do not know by what other name he may call these substances than nerves. Our question here, however, does not concern names. Let us see whether these substances are similar to the substance of other nerves of sense. Indeed they are in shape, color, and other features since their substance is similar to that of the brain in color and softness, as the optic nerves are also similar as well as those nerves which ascend to the ears above the anchae or buttocks of the brain in the anterior ventricle. This similarity exists in shape, round, extended, and oblong. As to their operations these nerves are the instruments of the sense of smell as the other nerves are of vision, taste, and hearing. Just as Aristotle said in *De Anima* II,[111] taste is a sort of touching, and therefore these nerves are harder because taste is not achieved without contact, which is more material. Since smell is not very material but is accomplished by means of a smoky evaporation it does not require a strong instrument by any means and therefore nature made those nerves soft so that they might take up weak vapors and could be changed by the latter. But the eyes have stronger nerves since sight is a power more necessary for distinguishing between different things. Thus similarly the nerves of vision are greatly fatigued and the nerves of hearing suffer from the blows of beaten air and its concussion as these come to the ears. It is better, for the reasons given, to call these substances the nerves which come to the nostrils as sense perception demonstrates. Add further that these nerves pass through two foramina in the basilar bone to the nostrils. They form the first pair of nerves which descend to the nostrils, as you have seen, and are of soft substance and the shape mentioned. Continue to lift the substance of the brain and there will appear the second pair of nerves of sensation. These are much larger than those previously described and are intertwined in the shape of a cross. They call these the optic nerves since they descend to the eyes and give them the sense of vision. These nerves are also soft but whether their crossing is made from their transverse passage or whether they are joined by contact it is difficult to discern in man as well as in an ox.[112] These nerves seem to be perforated in large animals, which reason also persuades us to believe is the case in all other nerves so that their power may be carried down through them. Near these nerves last mentioned you will see a third pair of nerves [oculomotor], more thin and much harder since they descend to the muscles of the eyes through other foramina to create motion for the eyes. Continue to lift and there will appear the fourth pair

---

[110] Galen, *De Usu Partium*, IX: the reference seems rather to be Book VIII (Vol. III K., 647), where Galen localizes the sensorium of smell in the anterior ventricles of the brain, "qui etiam vaporosum quendam spiritum continent," but does not speak of nerves or mammillary carunculae. Book IX, 6 (Vol. III K., 640) may, however, be the passage Massa has in mind; after speaking of the four sense instruments in the head (eyes, ears, nostrils, tongue) he discusses the pathways (viae) of the senses, some of which are the nerves. In the anonymously translated *De Iuuamentis Membrorum* (a Medieval form of *De Usu Partium*), X, 5 (edited by Diomides Bonardus, Venice, 1490, in the first printed *Omnia Opera* of Galen) Galen says: (fol. 31 r) Modo autem volumus narrare dispositionem alterius rami nervorum orientium ex cerebro, et est par quattuor parium nervorum quoniam anothomici non computant duos ramos qui ueniunt ad nares ex numero partium nervorum, quoniam isti duo rami non sunt nerui sicut alii nerui. . . .

[111] Aristotle, *De Anima*, II, 421 a 15; 422 a 10; *De Partibus Animalium*, II, 17, 660 a: "touch, of which sense taste is but a variety" (translated by W. Ogle, Oxford Aristotle, Vol. 5).

[112] Alessandro Achillini, *Annotationes Anatomicae (Anatomical Notes)*, translated above, p. 59, ff. cites Galen, Rhazes, Avicenna, and Averroes, who say these nerves are joined by contact: *Liber Rasis ad Almansorem* (Venice, Jacopo Pentio de Leuco, 1508), fol. 3 v, chap. iiii; Avicenna, *Canon*, translated by Gerard of Cremona (Venice, Giunta, 1595), I, 60, col. 2; Galen, *De Usu Partium*, X, 12 (Vol. III K., 813); *De Nervorum Dissectione Liber* (Vol. II K., 833), where Galen says: "some call them passageways, not nerves."

of nerves [trigeminal] according to this enumeration. These descend through foramina from the sides of the bone to the muscles of the face and have branches which pass to the fifth and sixth pairs of nerves according to this enumeration and are united with them. Continue to lift the brain's substance gradually and you will see the fifth pair of nerves [trigeminal] which descend through foramina of the basilar bone to different parts of the palate and mouth. After these look at the sixth pair [facial, auditory] which proceeds through the substance of the anterior brain above the anchae or buttocks which are in the anterior ventricles and perforates the bone toward the ears where there are more prominent bones in which are found the hammers and the drum which create the harmony of hearing for the ears or for the ear drum. Then look at the seventh pair of nerves [vagus, glossopharyngeal, accessory]. These are harder and thicker than the other nerves of motion and pass through foramina of the palate to the tongue and are scattered throughout the latter. When you lift the substance of the brain in the last place you will see the eighth pair of nerves [hypoglossal] which descends to the muscles if the tongue and of the head of the cane of the lungs. Thus from what has been said you have seen eight pairs of nerves in the brain, enumerating them as I have done, that is, beginning with the nerves which come to the nostrils. I do not wish you to think that I have been too brief in my description of the progress of these nerves to many other members and in some of their conjunctions for by such a method of procedure their pathways and connection might not be clearly visible. But in order not to protract my discussion unduly I have not chosen to explain their progress completely since something has been said here and there about their conjunctions and progress in the dissections of the bodily members to which they pass; you will be able to see these best in bodies which have been entirely stripped of flesh. This is especially true of the progress of the reversive nerves [nervi recurrentes] which according to my enumerations come from the seventh and from the fourth pairs, which are joined together and proceed to the muscles of the neck. But since these nerves are not demonstrable by this method of dissection they must be abandoned. Look into the cavity in the lower region behind the location of the optic nerves where foramina descend to the palate and you will see the contexture of veins and arteries in the manner of a net composed of their various forms of progress above the dura mater, which adheres to the basilar bone. This net [circle of Willis] is called the secundinus and differs from the marvelous net not only in its location but also in the veins which form it, for the marvelous net is composed of carotid arteries alone without veins. But since sometimes the veins are dissipated when the brain is lifted this net does not appear; this also happens because the veins are bloodless and are thus hidden from view. After you have seen this net remove it, having first noted that sleep is caused in these veins by an obstruction brought about in them by thick vapors ascending to the brain or by thickened or chilled blood. This is caused more easily in the blood which more quickly grows thick than in the spirits. The thickened blood near the heads of the nerves weakens their powers and hence cause sleep. The spirits, however, in the arteries are not so thickened at first but only in their progress, and thus visions are caused and many operations of the mind are brought about because the spirits acting by means of inner powers have a certain latitude in their qualities, also when a man is asleep. For if the spirits are weakened so that they go forth beyond their latitude at the time of sleep, sleepers would be without sense and motion as apoplectics are. From what I have said you will recognize the use of this net beyond its function in the nourishment of the brain. Having seen these items, examine the composition of the marvelous net, its location and substance composed from the arteries which proceed from the carotid arteries alone ascending to the brain, that is, without accompanying veins. When these arteries reach the foramina of the colatorium they pass through the foramina of the basilar bone to within that bone and are scattered there or branched out between the bone and the dura mater. But it is better to say that the arteries proceed between the membranes or tunics and when the lower part of the dura mater, which is thin above the basilar bone, is visible then the arteries that form the marvelous net are immediately above the bone and placed under the dura mater, making a very thin netlike shape and hence the marvelous net itself. You will see that it spreads a great distance both toward the ears as well as toward the posterior cerebellum beginning at the place where the optic nerves cross. In order to see the net cut a piece of bone under the crossing of those nerves near the carotid arteries through the foramina of the basilar bone which are near the foramina of the optic nerves and the foramen of the colatorium. Make a small incision here so that you can more comfortably strip away the dura mater, proceeding toward the spinal cord, and you will find the marvelous net between the dura mater or between the tunics of the dura mater, which clings to the basilar bone. The texture of the net is certainly admirable since it is composed of the very thinnest ramifications of the arteries. Anatomists should not be surprised if sometimes they do not see this net since occasionally the arteries are empty of red spirit and are almost imperceptible. With the use of candles and eyeglasses, if ordinary vision proves insufficient, you will see at any rate the progress of the very small arteries in the manner of the thinnest fibers. But when the arteries are full this net will appear red, ample, and visible. Sometimes it is not so visible, owing, as I said, to the absence of spirits, as often happens with death when the spirits retreat to the heart. Thus some men have dared to say that this marvelous net was a fiction created by Galen; perhaps the reason why they said this was the fact that nothing but drained-out bodies came under their hands for dissection or those which had small and very narrow arteries. But this is no justification for denying the validity of the senses by the use of reasoning, since according to the opinion of Aristotle this argues a weakness of intellect.[113] I confess that I have found this net many times and pointed it out to people standing nearby. Sometimes it was very large, so that no one could deny its existence, not even an idiot. But sometimes it was so slender that I could not move it from its place because of the reasons already stated. In order to see these nets more suitably it will be of the greatest assistance if in the stripping bare of the neck you tie off the guidez veins [jugulars], both manifest and hidden, and the carotid arteries with a thread before the blood and spirits are let off to the upper regions, as I said in the inspection of the veins in the arms. When you have stripped the entire dura mater from the basilar bone, a laborious task on account of its adherence which proves that it is also by nature united in the upper cranium, look at the basilar bone in which there are very many foramina both of nerves and of veins and arteries through which superfluities are drained to the palate. Among these foramina is a large one through which the substance of the brain passes to form the marrow [spinal cord]; the latter passes through foramina of the vertebrae, of which I shall speak later. Gather together then from what I have said all the differentiae of the aforementioned nerves, especially as to hardness and softness, since the nerves which are only for sensation, the anterior ones, are softer while those which are motor nerves and posterior are harder than the former. Some of the nerves are more flattened and not so round; such are the optic nerves and those which pass to the nostrils and ears by means of a perforated bone, for these also pass through foramina. Thus nerves ought also to be described according to this characteristic, although the reason for doing so is quite silly.

Chapter XXXX

On the Method of Dissecting the Parts of the Eyes

Before you come to the dissection of the parts of the eyes (having removed a portion of the frontal bone that lies above the eye itself) I advise you first to inspect something useful, that is, the veins which come to the forehead from the guidez [jugular veins], if they are visible. These veins do not always maintain the same position and progress but sometimes appear higher and sometimes lower. They are phlebotomized in many illnesses of the head, as I said above. Look also at the muscle of the forehead [frontalis], which is of one substance. Although it is not wide in the middle it is more elevated, broad, and extended at right and left toward the coronal commissure and appears to be two muscles. It is in fact only one and very thin with filaments down its length and descending lower. Therefore those who cut deep abscesses of the forehead or the skin without abscess for any reason should beware of cutting this muscle lengthwise since they may injure the muscle which lifts the eyebrow and the skin of the upper forehead. One should not look for a tendon in this muscle since it is slender and mingled with skin in order not to become greatly wearied in lifting the light skin. In the same way many other large muscles in other members of the body have no tendon, since they have no great task or the need for frequent motion as is the case here. But in small muscles which move a small part or one hard to move there are tendons usually at their ends; some muscles have a tendon in the middle. Since this description is not accepted by a certain person (Galen)[114] as you saw in the chapter on the tongue and as you will see in connection with the muscles of the eyes, I should like to find a man so learned and expert among humankind that he could without contradicting himself give a name to that non-carneous substance in the heads or in the middle of these muscles, which Galen simply said had no tendons. But returning to our dissection, after you have seen the muscle of the forehead look also at the eyebrow. It is the extreme part of the eyebrow where the hairs are fixed in the skin. This hairy part is called the eyelashes, at right and left, continuous with the eyelids which protect the eyes and like doors continuously close and open. The hairs are fixed at the ends of the eyelids to guard the eyes from external and very small things, such as dust. This part of the body is sufficiently well known. Its substance beneath the skin can be called membranous and somewhat hard; in man I cannot call it cartilaginous since cartilage is a certain substance less hard than bone but harder than a nerve. Therefore examine the eyelids, which are also softer than a nerve; they have muscles that are confused or not distinct, such as those in the lips, which move them; but whether there are two or more muscles I shall say when I write about the remaining muscles in a special work on them. The eyelids are continued in this manner: the upper lid is united with the pericranium, the lower lid is united with the membrane which covers the lower part of the face.

[113] Aristotle, *De Generatione Animalium*, III, 10, 760 b 30: "Such appears to be the truth about the generation of bees, judging from theory and from what are believed to be the facts about them; the facts, however, have not yet been sufficiently grasped; if ever they are, then credit must be given to observation (aisthesis, sense-perception) than to theories, and to theories only if what they affirm agrees with the observed facts" (translated by Arthur Platt, Oxford Aristotle, Vol. 5).

[114] See note 87.

This part is not moved except in the attempt to open the eyes. But examine it, since there are some very thin carneous fibers in it, confused and scarcely perceptible. Proceed to strip the eyelids. You will see in the eyesockets, in both corners, foramina [lachrymal] toward the nose; therein is a certain glandular caruncula which retains humidity. Note that when a person weeps the humidity or tears descend through these foramina to the nostrils and palate, if they overflow. When matter is often stopped up there and it suppurates then what are known as lachrymal fistulae result. Also, when collyrium is placed within the eyes, through this foramen it descends to the palate and nostrils and its taste, bitter or sweet, can be felt. There is another foramen above through which humidities descend from the brain. When you have seen these parts and noted their uses cut the forehead bone (frontal bone) with a saw at one end and the other of the eyesocket so that you can lift the bone and see the substance and location of the eye. Its substance is surrounded by muscles which move the eye. Above these, as you will see, there is fat which fosters the eye and its instruments. The shape of the eye is round, surrounded with seven muscles. In the posterior part of the eye you will see the optic nerve stretching toward the pupil, which is surrounded by one muscle. The remaining six muscles surround and move the eye. It must be noted here, however, that Galen, *On the Motion of the Muscles* I,[115] said that the muscles of the eyes have no tendon but terminate in a certain hard nervelike membrane. I beg you to inspect it since tendons are made from nerves and ligaments which contribute to the formation of muscles and this part of the eye muscles, nervelike and membranous, has all the attributes of tendons although it moves no bones. In the muscle which turns the eye around I should particularly name this part a tendon for in every respect its substance is clearly what can be called tendinous, since it has the operation, substance, and location of a tendon, without multiplying entities beyond necessity, especially since Galen himself said in *On Anatomical Procedures I*[116] that he called the nervosities (or thick parts) of the muscles by the name of the tendons. While these muscles pass to the eye they spread out their tendons about the eye in various directions. However many voluntary muscular movements of the eye there may be, each one can be experienced by itself and seen in the living person. You will be able to see this also in the process of stripping some muscles by noting their terminations and moving them aside. When they are moved aside you will see the substance of the eye, composed of four membranes which total seven (when, as you will see, three of them have been divided in turn) and of three humors. In order to see these parts, that is, both the membranes which are called tunics of the eyes as well as the humors, cut the eye in two anterior and posterior halves down the middle. Then you will see the first tunic called the conjunctiva [bulbar fascia] since it joins the eye with the eyelids on the inside and proceeds from the inner eyelid. Since the eyelids originate from a membrane of the cheeks and from the pericranium, they say simply that the conjunctiva arises from the head, that is, from the pericranium. The conjunctiva begins in that part of the eye where the terminus of the tunic called the cornea is located; I shall speak of the latter in a moment. The conjunctiva terminates in the posterior region at the optic nerve. Note that the conjunctiva does not cover the eye entirely in the anterior region since by its thickness it would impede vision; therefore it is a single entity which is not divided. The second tunic, however, is divided into anterior and posterior portions; it is transparent and hard and is called the cornea in the anterior part since it is quite similar to transparent horn, to which there corresponds in the posterior region a tunic called the sclerotic, not so transparent but thick and obscure, truly continuous with the cornea itself. However, one part differs from the other in substance and color. There are, nevertheless, some people who believe the sclerotic is joined with the conjunctiva and in their enumeration they say the first membrane of the eyes is the cornea. But this difference of opinion is of no importance since such matters are decided by sense perception. From what has been said, you have two tunics, of which one is divided into two, making a total of three. Note, however, that the cornea is smaller than the sclerotic. This tunic arises from the dura mater and is therefore hard. After this tunic comes the third, which surrounds the entire eye and is different in color, for all the tunics are not of the same color, as you can see in living people, since not all have the same color of eyes. For some have black, others blue eyes, and so on. Note also that this tunic is perforated for the passage of spirit and resembles a grape seed. The perforated part is the pupil and from this anterior portion it is called the uvea [iris], on account of the similarity just mentioned. The posterior part corresponding to the uvea is called the secundina [pigment layer]. It is continuous with the uvea itself. These parts do not differ much in substance or color, except that the secundina is thinner than the uvea. Thus by dividing this third tunic into two you have seen the five tunics of the eyes. This tunic arises from the pia mater. Behind this third tunic you will find the albugineous

[115] Galen, *De Motu Musculorum*, I, 2 (Vol. IV K., 373–382) discusses the nature of tendons in general. At 381–382 he says: Musculi igitur oculorum membraneis quidem, sed validis aponeurosibus in duram et nervosam tunicam adjacentem rhagoidi perveniunt.

[116] Galen, *On Anatomical Procedures*, I, 5 (Vol. II K., 245), Singer's translation, p. 14: "You may call the fibrous end (aponeurōsis) a 'muscle-tendon' (hymenōdē tenonta)." See also *ibid.*, I, 3 (Vol. II K., 233), Singer's translation, p. 8. Note that Massa is acquainted with William of Occam's "razor": "Entities must not be multiplied beyond necessity."

[aqueous] humor surrounding the entire space between this third tunic and the fourth, of which I shall speak immediately. This humor is very clear in the anterior part and thinner than in the posterior part. Because of this difference there are not lacking people who divide this humor into two, a matter of no importance. Behind this humor is the fourth tunic, which contains two humors, that is, the vitreous [body] and the crystalline [lens]. This tunic is thin, clear, and dense in its anterior part and called the aranea (spider-web) [zonular fibers] on account of its thinness, but in its posterior part it is called the retina since it is thin like a net and not so clear. This tunic, as with the others, is numbered as two when divided: thus you have seen seven terms for the tunics according to their different locations and substance. The aranea and retina contain two humors, as I said above, for the humor in the anterior part is called the crystalline on account of its exceeding clarity, but the humor in the posterior part is called vitreous since it resembles liquified glass. It is larger than the crystalline humor but softer than the latter. Therefore, by way of summary, you have seen from what has been told you that the eye is composed of four tunics, of which three are divided into six parts and thus form seven tunics. You have also seen three humors. But it is difficult to recognize in this method of dissection whether the dissemination of the optic nerve creates a fourth tunic. Enough has been said for you to understand the substance of the eyes, their shape, parts, and uses.

Chapter XXXXI

On the Dissection of the Ears, Nose, and Upper Jaws, Together With the Basilar Bone

Look also at the substance of the ears after stripping them although the ears are clearly visible, for their substance below the skin is cartilaginous with some slender carneous irruptions. In the lower part of the ear is a piece which is completely fleshy without cartilage. Therefore since men's ears are not moved by voluntary motion although we see that they are moved in many other animals and certain muscles pass to them, hence you should not pass them over without consideration. Examine the hollow of the ear, in whose depths there is a tortuous foramen [external auditory meatus] which extends to a perforated bone placed in the direction of the ear. This foramen passes within the bone, where there is a round cavity the size of a fingernail covered by a thin, hard membrane [tympanic]. They say that in this place is contained the stricken air for receiving the sounds of voices. Above and within this membrane there are two little bones like drum hammers [malleus]; hence they are called the hammers. They move and cling and are set in motion by the movement of the membrane; in imitation of the Greek they call this membrane a meninge. You will see its motion in its place since you will recognize that by the motion of the hammer bones sound, harmony, and hearing are caused. If you wish to find these hammers and the meninge note that at the sides of the cavity of the cranium above the basilar bone where the ears correspond on the exterior of the head there are two bony eminences, one on the right and the other on the left. You must cut these eminences carefully with a scalpel in order not to injure the inner parts within the cavity of the bone. When the bone is thus elevated these bones will appear lying above the meninge or eardrum which corresponds, as I said, to the foramen of the ear in the exterior bone. These little bones and their uses were not recognized by the ancients since if you say this inner harmony was not unknown to Aristotle when he wrote in *Problems*, Book XI,[117] that hearing is a sort of discourse you would only be praising his genius. Hence these hammer bones must be examined since after you have seen them you will not cease to wonder at the admirable wisdom of the Creator. Examine carefully the windings of the inner bone where the incision was made for you will see a nerve that passes through the substance of the bone to the eardrum. When you have seen these items, the substance of the ears and the parts which lie around them, and have been able to note them carefully not without admiration, then proceed to strip the skin of the nose. The nose begins from the cavity below the eyebrows and proceeds as far as the upper lip. It is composed of a twofold substance beyond skin and flesh, that is, a substance both bony and cartilaginous. The bony part begins under the bone of the eyebrow where there is a sawtooth commissure by which it is united and proceeds to terminate in cartilage. This cartilage begins in the middle of the longitude of the nose. Furthermore, the nose consists of two parts on right and left, which are actually divided. They are hollow as far as the basilar bone, where there are two foramina corresponding to the two cavities through which odors and the air may be drawn in and the superfluities from the brain be expelled. The bony part dividing the bone of the nose proceeds as far as the palate. Later, the cartilaginous substance divides the remaining portion. Note also that the substance of the nose bone is separated by a commissure also at the angles of the eyes in which there are foramina passing to the nostrils of which I spoke in discussing the anatomy of the eyes. In the stripping of the nose one must proceed carefully since

[117] Aristotle, *Problems*, XI, 898 b 29: "Why is it that of all the senses the hearing is most liable to be defective from birth? Is it because the sense of hearing and the voice may be held to arise from the same source?" (translated by E. S. Forster in the Oxford Aristotle, Vol. 7); see also *De Anima*, 426 a 27. Massa's position in the history of the discovery of the auditory ossicles is carefully examined by C. D. O'Malley and Edwin Clarke, "The Discovery of the Auditory Ossicles," *Bull. History of Medicine* **35** (1961): pp. 419–441.

there are very slender muscles which are difficult to recognize since they are almost inseparably united with the skin itself and extend toward the cheeks. These muscles move the nostrils, or the extreme lower portion of the nose. Furthermore, the inner part of the nose is covered with a thin membrane in which veins are scattered which cause a great flow of blood when they swell and burst. These veins proceed from the guidez [external and internal jugular] veins which ascend to the brain. These are the parts, substance, shape, connections, uses and so on of the nose. When you have seen them continue to strip the cheeks and you will see the two muscles just mentioned united with the skin itself, one on the right and the other on the left, covering the bones which are called the cheeks. They are somewhat round and under the eye sockets, not continuous with the latter but quite close to them with the lower bones of the eye. These muscles not only move the skin and the parts which form the cheeks but also lift the upper lip to which motion the forehead muscle gives assistance. Therefore there are some people who believe these muscles are parts of the forehead muscle spread out in this area. This is not true, for these muscles are separate from the forehead muscle, although this fact is well known only to an expert anatomist. It is not inconvenient if this muscle of the forehead does assist at these other movements since nature has after all done the same thing in other parts of the body, that is, caused one member to assist another beyond its own proper function. See also how these bones are covered by a membrane common to the entire face. The region in the lower part of the cheek bones and under the nose bone is called the upper jaw, corresponding to the lower jaw. It has teeth which are the equivalents of those in the lower jaw fixed in hollows of the same jawbone, of which I have spoken when I wrote about the lower jaw. This jaw is composed of two bones, as you can see by the visible commissure under the foramina of the nostrils. Look also at the inner part of this jaw where there are two muscles called those of mastication within the jawbone which move the lower jaw in chewing. But the exterior muscles which descend from the temples and are called temporal, as you have seen, move the lower jaw, contracting it upwards. When you have seen these muscles look once more at the lower part of the cranium, which is the base that supports the brain itself; therefore they call this the basilar bone. You must first note its substance as far as the palate; this is not continuous but composed of many hard parts and perforated by many foramina, as I said above. In the anterior region toward the nostrils there is a certain spongy part where the humidities descending from the brain grow thick and are expelled through the nostrils and the palate. In the posterior region there is a part more dense and hard. Since the bones which compose the basilar bone are enumerated in different manners I must enumerate them as follows; first, in the anterior region the upper bones of the eye sockets since they sustain the brain in that region. You must not count the petrous bones at the sides among the parts of the basilar bone, as Mundinus[118] does, since they do not form part of the base except in a very broad sense. But you must include those bones where the foramina descend to the nostrils behind the foramina of the optic nerves near the foramen of the colatorium which draws the superfluity to the palate, likewise the perforated bone with a broad foramen [magnum] through which the marrow passes since it is continuous with the lambdoid bone by means of its commissure. You will also count the two sagittal bones, often mentioned above, which are behind the ears and the two little oblong bones behind these stretching toward the foramen of the colatorium. But since these are very well known and can always be seen in cemeteries I intend these words to suffice since I have now discussed the three bodily cavities about which I set out to speak, that is, the lower, middle, and upper. In order to be introduced to the remaining parts of the body after seeing these three cavities begin to inspect them by means of dissection, and in some manner only, since all parts cannot be perfectly examined in a cadaver which has been thus lacerated.

Chapter XXXXII

On the Dissection of the Remaining Parts of the Body and First Concerning the Anatomy of the Arm and Hand

Beyond the three bodily cavities which I enumerated at the outset the extremities have also been enumerated. Therefore, in order to view the remaining parts which are included in an introduction to dissection according to their sequence, begin to strip the arms carefully since beneath the fat, as I have said elsewhere, if there is any fat, you will see a fleshy membrane or covering of the entire body. What I wish you to note first is the progress of the veins which come to the arm. That which I said about them in the chapter on the dissection of the chilis vein ascending [vena cava] it will not be displeasing to make clear by repetition here also in part for the benefit of young students and bloodletters. The first of these which I advise you to view once more is the basilic vein proceeding under the armpits to the lower and inner part of the arm and more visible in the curve of the elbow; this vein is also branched off to the exterior part of the arm. Also examine the cephalic vein which comes from the guidez vein ascending [jugular] to the head and sends out a branch also through the inner part of the upper arm and in many bodies proceeds directly up to the

[118] Mundinus, *Anathomia*, translated by Charles Singer, *op. cit.*, p. 94.

thumb. I emphasize this point because in some bodies it does not proceed directly and is more visible in the curve of the elbow and near the thumb. You will note that the cephalic vein near the curve of the arm joint sends out a branch which is united sometimes in its progress under the joint with a branch of the basilic vein and sometimes with the basilic vein itself. When it is thus joined it is called the common vein; bloodletters in our time are often deceived by taking a branch of sometimes the cephalic and sometimes of the basilic vein in the curve of the arm for the common vein. Therefore notice how the common vein is made of the two veins mentioned. Look also at its location, which is such that in many bodies you should not be led astray since very rarely is this conjunction established in the curve of the arm but three or four fingers below the curve, and even less. But I have spoken at greater length on these matters in the passage where I wrote about the chilis vein ascending [vena cava]. Thus if you wish a clearer description of the remaining progress of these veins, read that passage, for you will see the progress of both veins as far as the hand. One vein proceeds to the outer part of the hand. This is the basilic, which is commonly opened for bloodletting between the ring finger and the little finger under the name of salvatella. The other vein, which is the cephalic, as I said, proceeds toward the thumb and is phlebotomized there between the thumb and the index finger under its name of cephalic. There are other branchings to different muscles of the arm, which you will see as you choose; you will be able also to see the branchings in the hand. Examine the progress of the artery with the armpit vein together with the nerve under the artery itself as these three proceed together, the vein [brachial] above the artery [brachial] and the artery above the nerve [median]. The nerve is a notable one and proceeds to the hand and is branched out to all the fingers and not only to the fingers but to other parts of the arm, that is, to the muscles. Thus also you will see the artery send forth various branches just as the vein does, but the artery does not always send out branches when the vein is branched. This is enough on the subject. When you have seen the progress of the veins, artery, and nerve, examine also the muscles which cover the adiutory bone of the arm [humerus] which move it and also other parts and come from the upper regions, that is, from the anterior part of the chest as well as from its posterior part and from the shoulder blades which are different in size and shape and move with different motions. They either elevate or lower by curving or extending or turning around in one direction and another; I have spoken of some of these motions on the anatomy of the middle body cavity and you can see some movements by experimenting, as I said, and drawing the tendons through the fleshy part. I say that they have tendons since, as I said, tendons are not only attached to muscles which move the bones but to other muscles also. The proper and principal reason why tendons are attached to muscles is not because they must move the bones alone but because muscles have a constant and laborious task, as proven by the muscles of the stomach, which lift no bones and yet terminate in tendons. Therefore, in listing the uses and reasons for the parts of the bodily members one must remember, since it is better to say so, that tendons were given to muscles when these labor greatly in movement, or on account of the size of the member, or because of its hardness or frequency of movement or for whatever other reason, as I also said in the anatomy of the lower body cavity. But in order that you may be introduced to the dissection of the muscles of the arm note that of the muscles mentioned there is a descending one [biceps] which covers the joint of the adiutorium [humerus] and passes above the other muscles. It has a broad tendon which descends through almost the entire arm, by which the arm is turned around. You will also note the terminations of other tendons of the muscles which come to the elbow joint. Hence you will discover why wounds of the joints are dangerous and create spasm or convulsion. After you have seen the number and remaining features of these muscles look also at the others which cover the fociles. These are two bones [radius, ulna] which immediately behind the adiutorium [humerus] proceed to the hand. The muscles which cover the fociles are many and move not only the fociles but also the hands and fingers with different motions. You should not be deceived by the muscles which have many tendons that proceed to the fingers and move them, both closing and shutting, for the inner muscle which draws the hand downwards and closes the fingers is a single one and begins at the inner joint of the elbow and terminates in four tendons [musculus flexor digitorum superficialis] which multiply as they move the hand and fingers and close them, drawing the hand itself upwards. There are other muscles in the outer region which lift the hand and fingers and begin at the elbow joint in the upper region through the fleshy portion and terminate in four tendons that proceed to the fingers and lift them or open and lift the hand [flexor digitorum profundus]. Hence I urge you to examine these muscles on account of the variety of their enumeration by the learned men who write about them, since those who enumerate muscles by the number of their tendons count more muscles in this fashion. Those who enumerate muscles by their fleshy substance and origin count fewer muscles. Examine and count them so that hereafter you may know and declare without hesitation that one muscle moves many members or moves in a single motion the fleshy parts of many muscles which come to form one. What is most beautiful to see, not without giving the highest praise to the most blessed Lord, is the composition of the hand itself or of the instruments which move the hand. When you have carefully, I say carefully, stripped the hand, for this is a most difficult task, look first at the four tendons of the above mentioned muscle [musculus flexor digitorum superficialis] which pass through the inner part of the arm to the hand and are united and covered by a single membrane as far as the joint of the raseta [wrist] or brachial bone [carpus]. After they pass this joint they are divided along the progress of the four bones of the metacarpus corresponding to the four fingers, that is, index, middle, ring, and little fingers. These four tendons proceed to the second joint of the four fingers and are united there to the bones. But from the first joint proceed tendons [flexor digitorum profundus] that perforate these tendons which are fixed in the second joint by a ligament that covers them, and through those foramina they pass to the third joint of the fingers and move them; they are united with the extreme ends of the fingerbones. Thus you will see that the tendons which move the first joints move also the third joints. However, you will not be able to see these admirable perforations and conjunctions of the tendons unless you first cut their covering membranes which enclose these tendons throughout the entire hand and the progress of the fingers; this incision should be made lengthwise. This covering membrane [synovial tendon sheath] is one which Galen, *On Anatomical Procedures* I,[119] left free to be named at will as either a membrane, ligament, or membranous ligament or membranous tunic, all of them appropriate terms. It is a member distinct from a tendon or a nerve, performing the function of a ligament in part and in part that of a membrane which covers a bone, as you will perceive. When you have seen this look also at the disseminations of the nerves between the fingers which come from the large nerve already mentioned up to the extremities of the fingers and are there mingled with a certain caruncula. From this mingling proceeds a very noble and temperate sense of touch at the finger ends. Look also at the tendons which proceed to the fingers in the outer region to raise them and open the hand which come forth from two muscles [extensor digitorum] in the inner region of the arm. Of these tendons four which proceed from one muscle move the index, middle, ring, and little fingers, but the tendon which moves the ring finger is branched out to the little finger. This finger has another tendon which proceeds also from one muscle alone [extensor digiti minimi], as you have seen. But then it has another tendon coming from the larger focile arising from the middle of the latter; this tendon is tripartite. Look also at the movement of the thumb by the tendons which lift it and the muscles of the tendons which are in the outer part of the arm in the middle of the larger focile and lift and draw the thumb toward the index finger [abductor pollicis longus, extensor pollicis brevis]. But the tendon which lifts and extends the thumb from the index proceeds from another muscle which is in the middle of the lesser focile in the inner region [extensor pollicis longus]. The finger is closed by two small muscles which proceed from the palm of the hand. One of these proceeds along the finger [first dorsal interosseus], the other passes transversely above the raseta and carpus to the thumb [abductor pollicis brevis]. The finger is depressed or bent down by one muscle coming from the elbow joint. There are also very many other muscles which move the fingers; these (to be brief) I do not describe, for I shall write about them all, God willing, to the best of my abilities. However, under my direction examine not only the items discussed by me but others which can be seen.

[119] Galen, *On Anatomical Procedures*, I; see Singer's translation, p. 8.

Chapter XXXXIII

On the Dissection of the Legs and Feet

Before you strip the legs you may strip the parts of the back which were left aside and note their substance, size, number, connections, and other features. Then strip the entire skin from the legs and feet and begin to look under it since there comes after the fat, if any exists, a fleshy membrane or covering of the entire body. This membrane is continuous with that one first seen in the lower body cavity and covers the entire leg as far as its extremity. Having seen it, examine the progress of the chilis vein [vena cava] descending and of the aorta artery, although I wrote about them in discussing the anatomy of the chilis vein, for it is beneficial to young men to repeat such things from time to time. Look first at the vein [great saphenous] which in each leg proceeds through the inner region where the groin, or the emunctory of the liver, is located to the femora and extends toward the joint of the knee or to its posterior part. But before it reaches this joint the vein sends many branches to the muscle of the femur, both interior and exterior; among others, examine a notable branch which passes to the exterior muscle that covers the hip. This branch is ramified to many parts around the hip joint, but when the big vein is under the knee joint it is split in two or three. From these parts an inner branch descends through the inner region to the cavilla (clavicula *Massa*) or ankle and proceeds toward the big toe, but in its descent it sends many branches to the muscles of the leg; this vein is called saphena. Another branch [lesser saphenous] of the three branches passes through the calf of the leg, descending toward the exterior part as far as the ankle, and proceeds toward the little toe. In its progress in both leg and foot it sends little branches here and there. This vein is called sciatica and is phlebotomized in sciatic pain with reasonable relief since the chilis vein [vena cava] descending into the femur and sending a branch to the muscle of the hip joint is a more direct vein to this region than other veins. Hence bloodletting of this vein empties from the joint itself and is of benefit. This branch of the vein was not known to many of the ancients. After this vein look at the third branch in the middle of the branches just mentioned, that is, between the sciatic and saphena veins, which descends to the foot or to the heel and is there branched out under the sole of the foot. Do not fail to note first in the femora and in the groin the progress of the aorta artery descending and accompanying the vein itself as well as the progress of the nerve according to the following arrangement: the vein [femoral] above, the artery [femoral] under the vein, and the nerve under the artery. This accompaniment is not maintained throughout the entire progress of the vein since, as was said elsewhere, there is no artery without a vein, although the reverse is true since often a vein is without an artery. When you have seen these veins, arteries, and nerves look also at the muscles of the femora and of the legs, and first near the outer part of the femora a muscle proceeding from the hip with a broad tendon that descends to the leg in its exterior part under the knee joint [tensor fascia lata]. This muscle moves the entire leg in an oblique direction, and this tendon is large and broad like a membrane. See also in the inner region the muscles [hamstrings] which descend through the femora and terminate under the knee joint for a distance of four fingers in round, broad tendons and also two muscles [quadriceps] terminating in one anterior tendon which descends to the kneecap and is united with it by a tendon. This is large and after it has been attached to the kneecap it descends under the knee and adheres to the tibia bone and together with other ligaments creates an adherence which protects the joint of the kneecap. Look also at the other muscles, since I shall not enumerate all of them here, for this inspection is very useful for the treatment of wounds, sprains, and the pains of gout. When matter becomes lodged around these tendons, by its extent and nature it causes great suffering and in wounds and other ills of the nerves it creates spasm. Next, look at the muscles which descend through the fociles [tibia, fibula] and cover them, creating the calf and moving the foot. Among these you will notice muscles with four tendons [extensor digitorum longus] which cause the motion of the toes. As you have seen in the arm, the tendons preserve almost the same sequence as those which move the fingers. I say almost, because the tendons which move the toes maintain one order in all the toes, that is, they pass to the first, second, and third joints. The tendon of the big toe passes to the second and third. At this point occurs a pleasant speculation concerning the foramina of the tendons in the foot and the passage of the tendon through the foramen to another joint since the foot was made primarily for walking and not for grasping; but because walking cannot be done without a certain amount of grasping by the toes, thus nature in her excellent fashion has sent through the foramina of one tendon another tendon to another joint, just as in the hand. Examine in the foot the disseminations of the nerves among the toes, as you saw them in the hand. These nerves proceed from that large nerve which passes through the femur together with a vein and an artery. When these parts are raised, the bones remain with their cartilages, membranes, and ligaments, but you will note in the vertebrae of the dorsal spine (as I shall explain shortly on the anatomy of the bones) certain notable nerves descending to the muscles of the femur or hip, giving motion to the legs, for it is profitable to know the communication of the parts with other parts.

Chapter XXXXIIII

On the Anatomy of the Bones, Cartilages, Ligaments, and Membranes Covering the Bones, and On the Nerves Which Come from the Vertebrae or the Spinal Cord

It is very clearly apparent from the admonitions of Galen how great is the usefulness of a knowledge of the bones, since the bones are the foundation of the rest of the parts of the body and all the members rest upon them and are supported, as proceeding from a primary base. Thus if any one is ignorant of the structure of the bones it follows necessarily that he will be ignorant of very many other things along with them. How then can either a philosopher or a physician know the operations and parts of the body if he has not seen the substance, size, shape, location, number, and other features of the bones? Therefore let not anything necessary be lacking from my method of introduction to anatomy. After I have introduced young men to a recognition of other parts of the body I am eager to introduce them also to the knowledge of the bones because of its great usefulness, for, as Galen said,[120] this knowledge is of the greatest value to healers not only in the treatment of dislocations and fractures but in the recognition of many illnesses. Thus when you have lifted all the other members from the bones in the cadaver, these remain to be examined with certain other members or parts adhering to them, such as cartilage, ligament, and membrane. Begin from the head, because the bone of the head is called skull or cranium. As I said in the account of the cranium, it is composed of many bones joined by saw-toothed and non-saw-toothed

[120] Galen, *De Ossibus Ad Tirones* (prooemium) (Vol. II K., 732): Ossa singula per se qualia ipsa sint quamque inuicem constructionem habeant, nosse dico medicum oportere, si recte ipsorum fracturas et luxationes curaturus est. . . . Quare neque nosse morbos, neque recte sanare poterit [i.e., without a knowledge of the bones]. Charles Singer has translated this work of Galen in the *Proc. Royal Society of Medicine* **45** (1952): pp. 25–34.

commissures, as are the petrous bones, so called because of their hardness. But since according to my method of incision the cranium was cut by a saw in order to present the substance of the brain to view and the frontal bone to provide an inspection of the eyes and of the hammer bones above the eardrum, the bone was cut within the cranium also. It is sufficient to have seen these parts in the dissection of the cranium. But since a description of it will be pleasing it follows herewith. The head bone, called cranium and skull, is almost round in shape, a little flattened or depressed at the sides, made up of many bones, named thus: the bone of the anterior part from the eyebrows to the coronal commissure is called bregma in Greek but sinciput in Latin [frontal]. The bone or part of a bone opposite to this one in the posterior region is called lambda, so called from the lambdoid commissure which both separates and unites it with other bones. This posterior part is called the occiput. But between the coronal and the lambdoid commissures there is a third, saw-toothed and straight, called the sagittal, which divides the upper bone into right and left parts; these bones are called bregmata, vervalia [nervalia],[121] and the bones of the pair [parietal]. In these bones are foramina through which the veins pass, as I have said. Under these bones descending toward the ears you will see two non-saw-toothed commissures in the shape of a little fissure. These are often not visible because of their slenderness in those who have just died but are frequently found in cemeteries in heads which are completely dried out. These commissures are a little above the false commissures. From the latter and below there begin the two petrous bones with the foramen of the ears. The false commissures are not saw-toothed but ascend in the manner of a scale; these are also called the temple bones. From what I have said you have the upper part of the cranium, composed of eight bones and divided by seven commissures. Leaving aside those two small commissures above the petrous bones, some people say this part of the cranium is composed of only six bones, divided by five commissures. Behind this upper part of the skull is the basilar bone, harder than the other bones with the exception of the spongy part which is in the anterior region. It is called the basilar bone because it is situated below as the bottom and base of the entire head, containing the substance of the brain. This bone is perforated by many foramina through which the nerves descend from the brain and the veins and arteries ascend. There are also colatoria, or sieves, by means of which superfluities are drained from the brain. In addition there is also a foramen [magnum] much wider than the others through which the marrow passes to the foramina of the spinal vertebrae. In order to view these parts you will first know that the upper bones of the eye sockets and other bones, of which I spoke in the dissection of the cranium, that compose the basilar bone are enumerated among its parts, and thus you will see that it is composed of many bones. For the sake of brevity I urge you to read the passage to which I refer, for I spoke about the bones of the nose and its cartilage. These bones are separated from the basilar bone. I have spoken also about the bones which compose the cheeks and about the upper jaw and the cartilage of the ears. You can see here the bones of the palate, continuous with the upper jaw, two of them joined by a commissure; they are united with the bones of the colatorium. You will also note that one part of the upper jaw [maxilla] extends in front and is united with the bone of the palate in such a way that it must be integrated with that bone. You must read about the remaining parts where I told you to do so since by the present method of dissection all these parts have been seen and also removed to some extent. You have seen the lower jaw, with its teeth and other parts which were removed. Since not all the parts are available, having been cut off according to their sequence, these words should suffice about them. Turn your attention to the bones of the neck which support the head. They are continuous with other bones of the spine and are perforated to allow passage for the nuca [spinal cord]. You must not believe the nuca is the marrow of these bones, which they call the vertebrae and the spondyles, since, beginning here from the neck and going down to the bone of the tail, all the bones are spongy and contain the medullar humidity for their nourishment. But the substance of the nuca is from the substance of the anterior and posterior brain; therefore it is called the proxy (*vicaria*) of the brain since from it proceed the rest of the nerves which give motion, as has been noted in dissections. Examine the nuca in its location, cutting the vertebrae lengthwise, and you will discover that it is composed of two parts which touch each other in such a way that they can be said to be almost united but are in fact divided. It is white, a little harder than the substance of the brain, and covered by a double membrane, of which one is hard [dura mater] and adheres to the bone and the other is thin and adheres to the substance of the nuca itself [pia mater]. Some people say that the progress of the nuca in different bodies is different since sometimes it is larger and sometimes smaller through the foramina of the vertebrae. They also affirm that it rarely passes through the lumbar vertebrae but very frequently to the last vertebra of the ribs. But note that I have always found the nuca proceeding to the end of the tail bone, which is reasonable if the nerves receive power from the nuca itself, as from the proxy of the brain. When you have noted these items, return to the examination of the vertebrae of the neck from which I digressed. They are seven, round and of hard

[121] *vervalia* is an error for *nerualia* in Hippocrates; these are the nerualia, or ossa paris (os jugale): see J. Hyrtl, *Das Arabisches und Hebräisches in der Anatomie* (Vienna, 1871), p. 195.

substance. The first of the four is smaller than the others but broader and has two foramina at the sides from which issues the first pair of nerves of the nuca. The remaining six vertebrae of the neck are longer. It makes no difference whether you call them bones of the neck, true spondyles, or vertebrae, or untrue so long as you recognize their substance and remaining features. Thus the bones of the neck are seven, from which proceed eight pairs of nerves. Not all of these bones are perforated, however, but only the first bone which is smaller and broader and lies immediately under the basilar bone. The first pair of nerves issues from the two foramina in the bone just mentioned. The remaining nerves do not issue from the foramina since in the remaining six vertebrae there are no foramina; therefore they come forth from the intermediate spaces of the vertebrae. The second pair of nerves issue from the space between the first and second vertebrae, the third pair from the space between the second and third vertebrae, the fourth pair from the space between the third and fourth vertebrae, the fifth pair from the space between the fourth and fifth vertebrae, the sixth pair from the space between the fifth and sixth vertebrae, the seventh pair from the space between the sixth and seventh vertebrae, the eighth pair from the last interspace between the vertebrae of the neck and the first vertebra of the chest. The nerves that come from the vertebrae of the neck proceed to the different parts of the face, neck, and shoulder blades. Since they had to be dissected in such a manner as this, the muscles and other parts were removed; hence you can scarcely see their progress. Thus you must look at the remaining spondyles or vertebrae and their parts. But before you come to their inspection I wish you to note that all the vertebrae are bound together by a certain membrane which covers them and also by cartilaginous parts at their extremities [intervertebral disc]. This cartilage lies between vertebra and vertebra except in the two first vertebrae of the neck which are bound to each other by a certain eminence [odontoid process] of one and a cavity in the other. Similarly, in the first vertebra there are two cavities into which enter two eminences [condyles] of the lower bone of the head, that is, the basilar. These eminences are near the foramen where the nuca issues forth. The head is attached to the neck by these eminences and by some ligaments that bind it on the outside. The first vertebra, however, is not thus attached to the second since it has no ligament except that cavity into which the aforementioned eminence enters. When you have noted these parts look at those which remain for after these seven vertebrae of the neck you will see those that follow down to the os sacrum. These are eighteen, if you wish to enumerate them. Of these vertebrae there are twelve [thoracic] to which twenty-four ribs are attached, twelve on each side forming the cage of the chest. Of the twelve on one side seven are true ribs and five are false, incomplete. But since in the dissection of the chest these ribs were for the greater part broken and removed, return to the chapter on the dissection of the chest to see what was said about them. For both true and false ribs after they are extended from the back where they are attached to the vertebrae (two ribs to each of the vertebrae, one on the right, another on the left) proceed toward the anterior region and terminate in a cartilaginous [costal] substance. Note that the seven true ribs terminate with their cartilaginous substance at the bone [sternum] of the thorax in the middle of the chest. Although this appears to be one bone it is composed of seven so bound together that they appear one. At the end of the bone of the thorax toward the stomach is the scutal cartilage [processus xiphoideus] called the pomegranate to whose upper extremity there cling the two bones of the upper fork which were removed in the dissection of the chest. You will also note that the false ribs are less hard than the true ribs even near the vertebrae. But since I have said here that all the ribs terminate in a cartilaginous part you should know, according to the opinion of Galen, *On the Use of the Parts* VI,[122] that cartilage is a certain substance like bone, but softer, which you will find at the extremities of all the bones, large and small, according to the need of the member, as it is possible to see in the shoulder blades, for there the cartilage is large as well as at the extremities of large bones near the joints and in the vertebrae. This cartilaginous portion [articular cartilage] was reserved for many bones at their extremities in order to keep two hard surfaces from coming into contact and being broken by movement and to maintain something between the final hardness of the bone and the soft flesh, as is apparent in the shoulder blades and in members which require dilatation, such as the chest. For a large cartilage is given to the ribs so that they may be flexible in dilatation and so that movable members should be located in a place less subject to breakage. But when you have seen the ribs and the rest of the bones which contain the chest look at the twelve vertebrae corresponding to the twenty-four ribs, that is, twelve on the right and twelve on the left. From these vertebrae there pass forth the remaining nerves, giving motion to the rest of the parts of the body. But since by the present method of dissection you cannot observe their progress and also because they are sometimes diversified as they issue from the foramina of the vertebrae it will be sufficient here to know that the remaining nerves issue from the foramina of the vertebrae since the latter possess them. The order in which the nerves issue forth from

[122] Galen, *De Usu Partium*, VI (really VII, 3) (Vol. III K., 519): Est sane pars quaedam simplex in animalis corpore, (de qua ante etiam dictum est, quum de manu ageremus), quae aliarum quidem est durissima, solo cutem osse est mollior, cui nomen omnes propemodum medici imposuerunt cartilaginem.

the foramina is not always the same since you will sometimes see two nerves passing through one foramen. This anomaly is not entirely freakish for nature occasionally does this intentionally for the greater strengthening of the body or of some particular member. Note also that through those foramina from which the nerves issue there also issue the branches of both a vein and of an artery, passing to the nuca. After these twelve vertebrae there are six lumbar vertebrae down to the os sacrum. The latter is broad and made up of four bones united by a very strong ligament. The last bone of the spine is the tail bone [coccyx]; it is soft as if it were of cartilage and has one foramen in the middle from which one nerve only comes forth. According to what has been said, beginning from the head and going to the tail bone there are thirty vertebrae. The nerves issuing from these are thirty pairs plus one nerve alone which issues from the single foramen of the tail or caudal bone. After you have seen these bones you may cut all the vertebrae lengthwise and note first that those of the neck have a larger cavity than the others. Next examine the two bones of the shoulder blades [scapula]; they are broad and have a notable portion of cartilage at their extremities. These are also broad and extend toward the spine since the shoulder blades are located in the posterior region of the chest, proceeding above the ribs with their flat part. Another part which is not flat but tends to become sharp is the point where the adiutory bone [humerus] of the arm is located. This part is hard and has a cavity [glenoid fossa] at its extremity in which the adiutory bone is fixed. Hence because of this the shoulder blade alone could not support the adiutorium; thus there are two bones at right and left named from the upper fork [clavicle]. These bones terminate at the heads of the bones of the shoulder blades and thus at the same time hold the adiutory bone [humerus] in place. This bone is round, with a cavity in it in which there is a notable quantity of marrow. Not all the bones have a cavity with marrow in it, such as the ribs and vertebrae and the rest; you will note this difference of the bones in arms and legs. The adiutory bone [humerus] is attached to the spatular bone or shoulder blade in the upper region with ligaments which arise both within and outside of the joint. This upper part is round and thicker because the adiutorium [humerus] extends as far as the elbow joint. In the exterior part at the end there is a notable cavity in which the large bone of the fociles [ulna] of the lower part of the arm is fixed to the adiutorium [humerus] itself by its extreme portion formed like a hook so that it can sustain any sort of weight. You will notice, however, that the extreme lower end of the adiutorium [humerus] has two heads, or a bipartite extremity, in order to support the two fociles. Of these heads of the adiutorium one is longer. They are not round like the upper part but flat and somewhat depressed or sunken. Observe the great providence of nature and of God in the formation of these parts by which a foramen was made in the outer part of the adiutorium [humerus] into which the extreme portion of the larger focile [ulna] enters. It is curved like a hook. This focile terminates at the joint of the hand toward the little finger. The other focile, called the radius, is braced upon the other extremity of the adiutorium [humerus] on the inside and is strongly united with the thinner larger focile [ulna]. It also proceeds toward the thumb in the joint of the hand. You have thus seen the two fociles, the larger of which according to the different circular motions of the hand and arm sometimes appears in the inner region and sometimes in the exterior region. In a similar manner the smaller focile [radius] according to the circular motion of the arm sometimes appears on the exterior, sometimes on the inner side. These fociles are attached in the upper region with very strong ligaments to the adiutory bone [humerus] both within and outside of the joint beyond the entrance of the curved focile bone into that cavity of the adiutorium [humerus]. In the lower extremity of these fociles are located the first bones of the hand, called raseta [wrist]. But since I have said that bones are bound together with ligaments, you will note first that this is not always true. They are joined also by means of serrate and non-serrate commissures, as you have seen in the bones of the head, and sometimes by means of a covering membrane or by the entrance of an eminence [process] of a bone into the cavity of another bone, as you saw in the neck. You should also understand the substance of a ligament. That which is properly so called is white, similar to a nerve in hardness and softness, flexible, and it proceeds from the heads of the bones, where the parts are quite cartilaginous. Hence the most learned Avicenna said in the first chapter of the first Fen [of the *Canon*][123] that it is admirable of nature to have made from bone a ligament similar to a nerve without feeling, and the ligaments were for the conjunction of bones and many other parts. When you have noted these items examine the bones of the hand and first eight bones of the raseta [wrist], four of which terminate at the joint of the two fociles they call the brachial bone and are bound with strong ligaments. The heads of the fociles have a notable cavity in which the bones of the raseta are fixed. These heads are round in order that the hand may be better adapted for different movements in addition to other uses. The remaining four bones of the raseta are larger and are united with the smaller bones. They correspond to the four extremities of the bones

[123] Avicenna, *Canon*, I, First Fen, Doctrina 5, Chap. 1 (translated by Gerard of Cremona, Venice, 1595), p. 29: . . . et haec quidem [chorda sunt], quae ligamenta uocantur que etiam secundum visum et actum neruosa existunt et ab ossibus venientia ad musculum vadunt. . . . Post eas autem sunt ligamenta, quae praediximus, quae etiam sunt corpora neruo similia. . . . Ligamentorum nullum sensum habuit.

of the pecten or row of the hand which are attached to the extremities by ligaments and by their cavities. After the four bones of the pecten [metacarpus], which are oblong and not so united, are four bones of the first joint of the four fingers, that is, of the index, middle, ring, and little fingers; at the extremities of this joint are attached the bones of the second joint of the fingers. To these second bones are attached the third bones of the fingers. These are the last and attached to the second bones at whose extremities are the fingernails. There are also three bones of the thumb, but they do not proceed according to the order of the bones of the pecten [metacarpus] and of the raseta [wrist]. First the lower bone is attached at the side of one of the bones of the raseta or wrist where there continues the bone of the pecten [metacarpus] corresponding to the index finger. The remaining two bones follow. At the end of the thumb is a fingernail, as in the other fingers. But you will note that the nails are not attached to the ends of the fingers by a ligament or suture as the bones are attached to each other but grow out of the superfluity by unerring nature for the sake of their uses, that is, to guard the extremities of the fingers and to assist other operations of the hand. They grow or increase as the hairs do and thus it is necessary to cut them as we cut the hair. Let these words be sufficient concerning the bones of the arms and hands. Look also at the lower bones. The first to appear is the bone of the comb or pubis located in the extreme lower body cavity and united in the middle with a very firm ligament, for the pubic bone is part of the hip bones, as you can see in cemeteries. There are people who say this bone is separate from the hip bones, but there are no places where it is joined to the hip bones, or, to speak more correctly, distinguished from them. It is continuous everywhere without a ligament except in the anterior region in the middle where the penis lies. Here its substance appears to be cartilaginous. Hence I do not hesitate to state with many other people that the comb is part of the hip bones. But since in the demonstration of the bladder and of other members this pubic bone was cut, examine the remaining parts of the two hip bones, which are also called by other names and are attached to the os sacrum at the back. The lateral parts are broad and make an inner space where are located the bladder, straight intestine, and other parts. These bones as they proceed to the anterior region are called the pubic bone in front and the hips [ilium] at the sides. In the extremities of the hip bones are cavities [acetabulum] shaped like a box into which the extreme ends of the superior parts of the bones of the coxa or femur, which are round, enter, one on the right and one on the left, for the femora are proportioned to the bones of the adiutorium of the arm [humerus]. This joint is called by some people the vertebrum, by some the sciatic joint, whence comes the name for the sciatic pain in this joint. The hip bones are attached to the coxae bones with strong ligaments. You will note here that the coxae bones are round, oblong, and larger than the other bones of the body and proceed to the knee joints. The lower parts of each bone of the coxae is united to the two heads of the fociles of the tibia by strong ligaments and thus they make the knee joint. But in the anterior part of this joint there is a round bone which they call the patella. It is not cartilaginous but hard and like a shield which prevents the bones from sticking out in the bend of this joint, which is also called the ham of the knee joint. The patella bone is attached by two tendons of large muscles which descend through the coxa and terminate with one tendon which after it is united with the rotula descends under the knee and is united with the tibia, as you saw in the anatomy of the legs. Examine also the two bones of the tibia, which are called fociles. One is larger and located in the inner part of the leg. It is called the large focile, or the tibia, it makes no difference which. Above this large focile is based the femur bone mentioned above upon its round part. The lower part of this focile, which is hollow, terminates at the ankle bone, or cahab (chaib), and lies above its roundness. This ankle bone [talus] lies above the bone of the heel or calcaneus, which is large. These bones are attached by strong ligaments. In the anterior part toward the toes and near the raseta lower down is the navicular bone or scafoides, so called because of its cavity, above which the ankle bone lies in the anterior region. After these look also at the smaller focile [fibula] in the exterior and upper region. It is bound to the knee joint at the extremity of the coxa bone [femur] and in the lower part terminates at the ankle bone and is there attached with its ligaments so that it may be an auxiliary to the tibia or larger focile and thus be better accommodated both in running as well as in other motions. Next examine the remaining bones of the foot, which are continued by means of very strong ligaments: first, the four bones of the raseta, which are continuous with the navicular bone. To these bones of the raseta at the external side is attached a small bone, which they call the sesame bone. Thus the bones of the raseta of the foot proceed lengthwise down the foot and are four in number, to which are attached the five bones of the pecten or pubis [metatarsal] which proceed as far as the first toe joints and are attached to the five toes. Look also at the bones of the toes because the first toe, which is the pollex or great toe, has only two bones. The remaining four toes have three bones each, in whose extremities there are toenails. Note also that the bones of the toes are attached with strong ligaments, and the great toe has two joints. The other toes have three joints. From what has been said you will also note the difference between the feet and hands not only in shape but in number and location of bones since their operations and uses are different. These are the words, most learned Marcellus, which I set out to write as an introduction for young men to the dissection of the human body. I could point out very many other things which would be useful for both philosophers and physicians, but in such a method as the present one many members have necessarily been cut away in order not to destroy the sequence of procedure. Many things are also very difficult to find and are not discovered in all bodies; among these are the membranous muscles of the hand and of the gullet as far as the lip, and the progress of the nerves, veins, and arteries, which can scarcely be seen in this method of dissection for not only do they lack the other parts which were laid aside in dissection alone but they do not appear equally in all bodies nor preserve the same order. In order not to fatigue the minds of the young men by an excessively long description I have been as brief as possible. However, if in some part I have digressed farther than the scope of this introduction required I confess I have been forced to do so either by material which needed more detailed treatment or by various authorities contrary to sense perception. For I have not wished to reveal myself as a follower of others but of the truth of the senses, although I have read many writings of the ancients by which I have been instructed. I have also often repeated my words according to the needs of a given passage to attain a clearer doctrine. Therefore once more a suppliant do not cease to pour forth your prayers for me to Almighty God so that what I began under His protection may be completed also under His protection, and that my readers may beg the blessed Lord according to their paternal religion that He should judge me worthy of His aid to direct them to the knowledge of other parts of the body, as I have resolved. Farewell.

Andrés de Laguna
(1499–1560)

LIFE AND WORKS

THE most colorful and individualistic of the pre-Vesalian anatomists, the most varied and prolific in his production, and the most picturesque and realistic in his style, both Latin and Spanish, Andrés Fernández de Laguna, to give him his full name, was born at Segovia, Spain, in 1499; the exact day is not known.[1] Few documents survive to give us details about his early life; no baptismal record nor last will has as yet been discovered in the Spanish archives. We know that his father, Diego Fernández, who died in 1541, was a converted Jew and a prominent physician who attained noble rank; his mother, Catalina Belázquez or Velázquez, of noble birth, died in 1568. Andrés, one of five children, spent his boyhood in the house where he was born in the Calle del Sol, the parish of San Miguel. Among his teachers at Segovia were Juan Oteo and Sancho de Villavesano, who taught him classical languages, including the Greek which he was to display with so much vanity in his writings.

Autobiographical details concerning his boyhood and later years can be gleaned from his books, especially from his commentary on the *Materia Medica* of Dioscorides (a task performed by Dubler), and to a certain extent from the *Anatomica Methodus* (1535). In the former he records the frightful scarification performed on him by a brutal barber-surgeon in 1500 in an attempt to cure an attack of fever when he was a child of only fourteen months. Even when he was a grown man de Laguna trembled with sensitivity when he touched the welts this treatment had left upon his leg. At the age of twelve he was beset with the temptation to steal a purse in the bedroom of one of his father's patients (*Anatomica methodus,* fol. 57); the personal incident is vividly described in the course of a discussion of the effect of strong light upon vision, written twenty-five years later. In his boyhood he had also experimented with the deadly result of human urine (his own) when drunk by barnyard geese; they died in an instant, as he discovered by repeated trials. The Commentary on Dioscorides also gives us glimpses of his student days at Salamanca,[2] such as his experience with eating figs (presumed to make the blood impure and to encourage the growth of body lice, as well as to obstruct the liver and the spleen); his conclusion was that students, like ostriches, could digest even iron, since without doubt they had wolves in their stomachs. Such humorous remarks are frequent in de Laguna's writings and stamp him as a master of wit and irony. A final incident from his days at Salamanca describes his ignominious escape from a wandering bull (who he said must have been a Lutheran) in the streets on the evening of St. John's day, when he was saved from a goring by a lame holiday-maker who happened to be nearby.

Probably in 1530 de Laguna went to Paris to continue his studies, taking with him a servant, who also had a frightening experience with a cat whose eyes glowed on the hearth one night like those of a demon and who leaped out at the poor fellow and scratched his face. Among de Laguna's teachers at Paris were Jean Ruelle and famous men at the recently founded Collège de France, such as Pierre Danès and Jacques Toussaint, also the teacher of Henri Estienne. De Laguna became well grounded in medical botany and pharmacy at Paris, laying the foundations for his considerable writings in these fields. There de Laguna found many Spaniards, some of them professors, and other prominent Europeans including Erasmus, Ignatius Loyola, and Luis Vives.[3] There were also a number of men from Segovia itself, some of them at the Sorbonne and other colleges which had a reputation for conservatism in their teaching methods in contrast to the Collège de France. Here were to be found Guillaume Budé, the famous French humanist, Barthélemy Masson, and Johannes Gelidus, a native of Valencia who became a French citizen.

At Paris, de Laguna received his baccalaureate in medicine in 1532 and began his long series of publications. In 1533 Vesalius arrived at Paris and met de Laguna, whose influence upon the more famous anatomist is acknowledged by C. D. O'Malley in his recent biography of Vesalius;[4] he is said to have pos-

[1] The latest works on de Laguna, both of which give incomplete lists of his books and a considerable bibliography of writings about him, are César E. Dubler, *La 'Materia Medica' de Dioscórides—Transmisión Medieval y Renacentista, Vol. IV. Don Andrés de Laguna y su época* (Barcelona, Tipografía Emporium, 1955), longer and more dependable than Teófilo Hernando y Ortega, *Vida y labor médica del doctor Andrés Laguna* (Segovia, Instituto Diego de Colmenares, 1960; Estudios Segovianos, XII). Both books are poorly organized and full of printer's errors but assemble an abundance of material, including plates showing book title pages which have been useful in preparing a check-list of de Laguna's books; pictures of the man himself are also given in Hernando y Ortega, with one of the house in which he was born. Dubler prints the single autograph letter by de Laguna which has survived, dated July 7, 1554, and addressed to Don Francisco de Vargas, imperial ambassador at Venice.

[2] Hernando y Ortega's impression (*op. cit.*, p. 43) that de Laguna preserved excellent memories of Salamanca is not borne out by *Anatomica Methodus* (1535), the final letter to Rodrigo de Reinoso, in which he calls all the rest of the Salamantines, with the exception of a few learned men, nothing but lovers of abuse, babblers, and sophists.

[3] Gregorio Marañón, *Luis Vives (Un Español Fuera de España)* (Madrid, 1942), writes of this group of scholars and mentions de Laguna at p. 42.

[4] *Andreas Vesalius of Brussels–1514–1564* (University of California Press, Berkeley, 1964), p. 58. O'Malley mentions only Hernando y Ortega's book on de Laguna and does not seem to know the much more detailed and scholarly work of Dubler. He accepts 1511 as de Laguna's birth date instead of the actual

sibly even followed "in the pathway marked out by the Spanish student." The most prominent teachers of de Laguna were, of course, Jacobus Sylvius and Johannes Guinter of Andernach, who played also such a large role in the instruction of Vesalius.

De Laguna remained at Paris until at least the end of 1535, publishing there his first three books, all in that same year. This feat is an early indication of his prolific production, which was to continue until he had published more than thirty books. In 1536 he returned to Spain, where he was called to a professorship at Alcalá, according to certain biographers, although this is doubted by Hernando y Ortega (*op. cit.*, p. 20) on the basis of a careful search made by the Archbishop Alonso Muñoyerro in the archives of Alcalá and of Madrid; he discovered no evidence that de Laguna had taught there. De Laguna may have merely delivered a lecture or two in the city. At any rate, the dedication of his translation of Aristotle's *De Mundo* is dated November 1, 1536, at Alcalá; his translation of Lucian's works is also dated at Alcalá on October 21, 1538.

In 1539 he is said to have attained the doctorate of medicine from the university at Toledo although this statement is disputed on the grounds that the title of doctor appears in none of the books published by de Laguna between 1539 and 1545, in addition to the fact that it was most unlikely that he should have left the more famous university city of Alcalá, where he had published three books, to take a degree in Toledo, where he remained a very short time. Furthermore, the Council of the Inquisition had decreed at Valladolid in 1522 that the sons of converted Jews were barred from taking degrees at Salamanca, Valladolid, and Toledo. Hence, it is uncertain just where de Laguna obtained his doctorate. In 1539 he traveled to London, where he visited cock fights which he vividly describes in his commentary on Dioscorides. We have little more information than this concerning his stay in England.

FIG. 13. Andrés Laguna. Courtesy of National Library of Medicine.

Later in 1539 de Laguna visited the Low Countries, where at Middelburg in Zeeland he almost cracked his teeth on a large pearl in an oyster he was eating, proving to his scientific mind that pearls could be found in western as well as eastern waters. At Ghent he studied Galen intensively and collected books and manuscripts; one of the latter, in the possession of Adrian Corón, was the spurious *Prototypos* attributed to Galen. In 1540 Emperor Charles V and his court reached the Low Countries, giving de Laguna an opportunity to meet the great man. Soon after he was called to serve as physician in Metz in the dukedom of Lorraine.

date, 1499, which is clearly established by the anecdote from his childhood which he tells in *Anatomica Methodus*, printed in 1535, and in which he states that it occurred when he was twelve years old (fol. 57), hence in 1511. In a very brief communication entitled "Andrés de Laguna and His *Anatomica Methodus*" in *IX Congreso Internacional de Historia de las Ciencias*; Barcelona and Madrid, 1–7 Sept., 1959, Vol. I, published as No. 12 by UNESCO, p. 398, O'Malley gives, however, the dates 1499–1560 for de Laguna and speaks of the *Anatomica Methodus* as the first "modern" anatomical treatise published in Paris. He mentions de Laguna's description of the ileocaecal valve and suggests that the organization of de Laguna's book may have influenced the composition of the *Fabrica* since Vesalius was acquainted with the Spanish anatomist. De Laguna also denied that air entered the left ventricle of the heart before Realdo Colombo reached the same decision; thus de Laguna may have influenced Servetus.

The date 1532 for de Laguna's bachelor of medicine degree is drawn from the *Commentaires* of the faculty of medicine at Paris; but Hernando y Ortega (*op. cit.*, p. 17) refers to a manuscript of the same faculty cited by Luis de Matos, *Les Portugais à l'Université de Paris entre 1500 et 1550* (Coimbra, Portugal, 1950), p. 74, which states that in March, 1534, the Spaniard Andrés Laguna, before taking his examination in order to present himself for the baccalaureate in medicine after three years of study in Paris, furnished as witnesses Lupus Serranus and Ludovicus Gometius, evidently two fellow Spaniards. See now the recently issued *Commentaires de la Faculté de Médecine de l'Université de Paris* (1516–1560), ed. by Marie-Louise Concasty (Paris, Impr. Nationale, 1964), p. 231 (1533–1534), which corroborates de Matos's date, 1534.

At Metz de Laguna lived five years, from St. John's Day 1540 to the same day in 1545, according to documents in the municipal archives. Here he may have met Rabelais and certainly saw John Guinter of Andernach, with whom he shared the responsibility of fighting an epidemic of fever. Guinter had fled to Metz in order to avoid persecution as a Protestant in France. From de Laguna's experience with the plague at Metz sprang his writings on the subject published in 1542 and 1556.

At the end of 1542 de Laguna visited Cologne, where he remained three months; on the way, near Coblentz, he discovered a spring whose water tasted like some very delicate wine. In Cologne he lodged at the house of the rector of the university, Adolf Eicholtz. On February 11, 1543, de Laguna read a public lecture on the current condition of European affairs, later published in the form of an allegory in which Europe was the central character with the title "Europa . . . se discrucians." Its highly rhetorical and emotional staccato style seems to have made a deep impression on his audience; the theme dealt with the violence created by the wars of the European kings and the unity which could be imposed on the warring nations by Emperor Charles V and the Catholic Church. While in Cologne de Laguna translated into Latin from Italian a number of accounts of the deeds and origins of the Turks in a volume published in that city in 1543: *Rerum Prodigiosarum quae in urbe Constantinopolitana . . . acciderunt etc.* This work is only one of many indications that he did not cease to carry on his scholarly work in various fields in whatever city he found himself. Another phase of his activity was revealed in the polemic he directed against the Protestant physician, Janus Cornarius, in a book entitled *Castigationes Andreae a Lacunae . . . octo ultimorum librorum de Re Rustica, Constantini Caesaris, per Ianum Cornarium Physicum editam.* (Cologne, Lupus Iohannes Aquensis, 1543).

De Laguna returned to Metz perhaps at the end of February, 1543, fulfilling the promise he had made to the grateful citizens of that city who had not wished him to leave. In the course of an illness he became a virtual insomniac for two weeks until an old German woman with a waist graceful enough for a witch filled a pillowcase with henbane leaves, on which de Laguna swiftly went to sleep. In 1545, while still at Metz, he attended as the salaried medical officer of that city the Duke Francis of Lorraine, who had been rendered ill by witchcraft; two old people, man and wife, who lived in seclusion were accused, tortured, and destroyed for their supposed crime. Bataillon and others, however, suspect that this visit to the duke and the incident described did not take place but was merely added by de Laguna to his commentary in order to enhance its interest.[5] The story involves the use of herbs as a magical unguent.

In the summer of 1545 he made a short journey to East Prussia, where he saw a river whose banks contained marvelous springs (or pits) of bitumen; in his mention of Koenigsberg he referred also to the famous mathematician and astrologer, Johannes Regiomontanus. Returning on his way to Italy he passed through Nuremberg and Bavaria, stopping at Viterbo and arriving in Bologna on November 10, 1545. Here he was welcomed with all the honors his fame by this time had brought him and the degree of doctor of medicine was conferred upon him.[6]

In Italy de Laguna was to remain for nine years. Much of this time was spent at Rome, where Pope Paul III decorated him with various honors including the knighthood of the Golden Spur and status as Count Palatine. After a short trip to Germany to attend the family of Emperor Charles V he returned to Rome as physician of the Cardinal Don Francisco de Bobadilla y Mendoza. His stay in Italy was one of great activity, of many friendships and associations with important people, and of great literary and scientific production. He visited Venice in the course of his constant investigations in medical botany, finding it a valuable source of seeds and plants brought from the East. He also visited Falloppio at Padua. In 1548–1549 he spent some months in Genoa and the Ligurian Alps, always searching for medicinal plants and visiting botanical gardens, as the one at Pisa. Among the friends and patrons to whom he dedicated some of his books was Don Diego Hurtado de Mendoza, ambassador to Pope Paul III, who was more fond of the Spanish books *Amadis de Gaula* (a famous romance) and *La Celestina*; he was once regarded as the author of *Lazarillo de Tormes* (Antwerp, 1552), the most renowned of Spanish picaresque romances. In Venice de Laguna met Martin Stern, the cousin of Vesalius, who described the Spaniard in a letter to Andreas as "vir graecis latinisque literis iusta ac pari lege peritus."[7] He returned to Rome in 1549, where he continued his medical practice and his writing.

In 1554 he left Italy for Flanders, visiting Trento

[5] Hernando y Ortega, *op. cit.*, p. 32. See also Harry Friedenwald, *The Jews and Medicine. Essays* **2** (Baltimore, Johns Hopkins Press, 1944): chap. XXX: Andreas a Laguna, a Pioneer in His Views on Witchcraft, pp. 419–429.

[6] Vincenzo Busacchi, "Gli studenti spagnoli di medicina e di arti in Bologna dal 1504 al 1575," *Bulletin Hispanique* **58** (1956): Carta 86R. 1545 (December 10).

[7] Hernando y Ortega, *op. cit.*, p. 41. I find no mention of this letter in C. D. O'Malley's biography of Vesalius (Berkeley, California, 1964). It is reproduced by Hernando y Ortega, *op. cit.*, p. 43, but unfortunately he has misunderstood it and describes it as a letter by Vesalius to his cousin; the exact reverse is true.

and Augsburg on the way, where he met an innkeeper as large as the Colosseum at Rome, with five hundred keys at her belt, whose savory dishes made up for her steadfast chastity, although de Laguna found a large needle in one of them. From July, 1554, to September 25, 1555, we have few facts concerning his life and little about his travels. He met Rembert Dodoens, the famous botanist, in Antwerp. In 1555 he published his commentary on Dioscorides in that city and in 1557 his Spanish translation of the Catilinarian orations of Cicero, written while he was convalescing from an illness.

At the end of 1557 he returned to Segovia, where he spent the last years that remained to him. During this period he prepared his father's epitaph in the parish church of San Miguel; he had died in 1541. His mother's gravestone is also to be seen there today, together with his own. In 1869 de Laguna's remains were transferred to Madrid to rest in a national pantheon but were returned to Segovia when such a memorial building failed to be built. His death, probably from cancer of the rectum (he had suffered for some time from hemorrhoids), occurred during 1560, perhaps after February 5, when he left the reception committee which had traveled to the French border to meet Princess Isabel of Valois to return to Segovia.

I have tried to make as accurate a check list of de Laguna's writings as is possible although their great number (more than 30) together with their confusing similarity in certain instances has made this a difficult task. I have followed the BM and BN catalogs primarily since few of them are to be found in this country. Hernando y Ortega's and Dubler's works list certain copies in Spanish libraries and print facsimile title pages of some books which have been useful in the attempt to give a correct description. See also the bibliography in Palau y Dulcet, *Manual del librero Hispanoamericano* (2nd ed., Barcelona, 1954), pp. 336–337 (nos. 130019–61).

THE WRITINGS OF DE LAGUNA

1. Aristotelis Stagiritae de physiognomicis liber unus, per Andream à Lacuna Secobiensem etc.; apud Ludovicum Cyaneum, Paris, MDXXXV (dedicated to Louis Guillard, bishop of Chartres). BM lists another edition "ex officina P. Caluarini, Parisiis, 1541." Not listed by Palau.

2. Galeni de Urinis libri duo, antehac numquam in lucem emissi, Andrea a Lacuna Secobiensi interprete; apud Poncetum le Preux, sub lupo in via Iacobea, Paris, MDXXXVI. Dubler, *op. cit.* 113, lists a first edition "apud Ludovicum Cyaneum, 1535." Not in Palau.

3. Anatomica Methodus, seu de Sectione Humani Corporis Contemplatio, Andrea a Lacuna Secobiense authore . . . Parisiis apud Ludovicum Cyaneum MDXXXV; *idem*, Parisiis apud Iacobum Keruer MDXXXV. This book was published simultaneously by two publishers in Paris. Palau 130019.

4. Aristotelis de mundo seu de cosmographia liber unus ad Alexandrum Andrea a Lacuna Secoviensi interprete, nunc primum in lucem emissus. Alcalá de Henares, Juan Brocar, 1536, 1538; Coloniae, Ioan. Aquensis, 1543. Juan Catalina García, *Ensayo de una tipografía Complutense*; Madrid, 1889, no. 163, lists the earliest imprint as 1538; de Laguna's dedicatory letter to Charles V is dated Nov. 1, 1538. The volume also contains no. 5, below.

5. Tragopodagra Luciani, Andrea Lacuna Segobien. Interprete; Alcalá de Henares, Juan Brocar, and Segovia, 1538; Rome, 1551, 1552; Lisbon, 1560. (variant title: Tragoedia alia Luciani Occypus dicta Hipotesis). Palau 130020, 130021.

6. Ex Commentariis Geoponicis, sive de Re Rustica, olim Diuo Constantino Caesari adscriptis, octo ultimi libri . . . Andrea á Lacuna, Secobiensi Philiatro, interprete. Accedunt etiam eis quaedam castigationes in traslationem eorundem librorum, per Ianum Cornarium . . . Metz, 1541; Coloniae, prope D. Lupum Ioannes Aquensis excudebat. Anno MDXLIII. Palau 130023, 130024.

7. Compendium curationis praecautionisque morbi passim populariterque grassantis; hoc est vera et exquisita ratio noscendae, praecavendae, atque propulsandae febris pestilentialis; Argentorati, Per Vuendelinum Rihelium, 1542. Palau 130029. See A. C. Klebs, *Die ersten gedruckten Pestschriften*; Munich, 1926.

8. Rerum prodigiosarum quae in urbe Constantinopolitana et in aliis et finitimis acciderunt Anno a Christo nato MDXLII brevis atque succincta enarratio. De Prima Truculentissimorum Turcarum origine, deque eorum tyrannico bellandi ritu et gestis, brevis et compendiosa expositio. Coloniae, Joannes Ruremundus excudebat Anno MDXLIII. Antverpiae, 1544; Moguntiae, 1552. (variant title: Relatio ex Italia ad Germanos missa de ostentis quibusdam Constantinopoli Junio et Julio mensibus anni 1542 factis simulque Tractatus de Turcarum origine et successione deque moribus pace belloque illius gentis (translated from the Italian). An excerpt is printed in Torquati (Antonii), Prognosticon . . . De Euersione Europae, the first work in which volume is Joannes Placentinus, Catalogus omnium antistitium tungaborum. The title of the excerpt is: De origine rerum turcarum compendiosa quaedam perioche; De Turcarum cultu ac moribus enarratio quaedam breuiuscula . . . per Andream à Lacuna Secobiensem collecta. See Carl Göllner, *Turcica. Die Europäischen Turkendrucke des XVI. Jahrhunderts.* I Bd. București, 1961, no. 803 (1543 ed.) for a good description. Palau 130025, 130026 (Coloniae 1542), 130027, 130028.

9. Aristotelis . . . De natura stirpium liber unus

et alter . . . ex graecis latini facti . . . Andrea a Lacuna interprete; Iohannes Aquensis, Coloniae, 1543. Palau 130032.

10. Galeni Pergameni Summi Medicinae Parentis, De Philosophica historia liber unus . . . nunc vero ad fidem uetustissimorum codicum restitutus . . . Andrea à Lacuna, Secobiensi, Philiatro Interprete. Coloniae Iohannes Aquensis excudebat, Anno MDXLIII. Palau 130031.

11. Europa ΕΑΥΤΗΝ ΤΙΜΩΡΟΙΜΕΝΗ, hoc est misere se discrucians, suamque calamitatem deplorans . . . Andrea à Lacuna Secobiensi, Philiatro, Authore. Haec declamatio lugubris fuit recitata Coloniae in celebri Artium Gymnasio, coram maxima Principum hominumque doctissimorum corona . . . Anno 1543. Die Dominica XI. Cal. Febr. hora septima post meridiem. Coloniae Iohannes Aquensis excudebat, 1543. Palau 130035.

12. Aristotelis Philosophorum Principis, de virtutibus uere aureus . . . ex Graeco in sermonem Latinum per Andream à Lacuna Secobiensem, Medicum, summa fide atque diligentia conuersus scholiisque et exemplis multis locupletatus. Additae sunt ad calcem aliquot in Gryneum castigationes . . . Coloniae Ioannes Aquensis excudebat. Anno 1543. Dubler and Palau give the date 1544; see no. 24. Palau 130036.

13. Dubler, *op cit.*, p. 115, lists an omnibus volume published at Cologne by Johannes Aquensis in 1543 which contains items 6, 8, 9, 10, 11, 12, 14 and "Facetum exemplaris etc." Palau 130025.

14. Victus ratio, scholasticis pauperibus paratu facilis et salubris. Adjectus est quoque libellus de victus et exercitiorum ratione, maxime in senectute observanda, utilis plane et aureus . . . (Plutarchi . . . de tuenda bona valetudine praecepta. Erasmo interprete.) Cologne, Jaspar von Gennep, 1546; Paris, Jacobus Bogardus, 1547; Coloniae, Henricus Mameranus, 1550. Palau 130037. The first entry is by Jacques Dubois (first ed., Paris, 1540; Parisiis, Apud viduam Jacobi Gazelli, 1549). It is translated by C. D. O'Malley in *Jour. Hist. Med.* **17** (1962): pp. 141–151.

15. Quincti Tyberi Angelerii Epidemiologia sive Tractatus de Peste. secunda editione (contains Compendium praecautionis et curationis pestilentis morbi a Lacuna ad faciliorem intelligentiam). Matriti Ex Typographia Regia, 1598. See Cristóbal Pérez Pastor, *Biblioteca Madrileña* **1** (1891), no. 559: pp. 287–288. Like no. 14, this book is also an anthology of works on the same subject; only the Compendium is by de Laguna.

16. Epitomes Omnium Galeni Pergameni Operum . . . sectio prima per Andream Lacunam Secobiensem, Doctorem Medicum, atque ex sacro militum Sancti Petri apud urbem collegio, Auratum Equitem . . . Venetiis apud Hieronymum Scotum, 1548; Basileae, 1551; 1571; Lugduni, Guglielmus Rovillius, 1553; Argentorati, Lazarus Zetzner, 1604; Lugduni, I. Cassin and F. Plaignard, 1643. This book contains a letter to Vesalius from his cousin, Martin Stern. Palau 130042.

17. Vita Galeni Pergameni ex Galeno ipso et ex variis Authoribus per Andream Lacunam, Secobiensem, Doctorem Medicum, Militem Sancti Petri. Index locupletissimus omnium rerum et sententiarum . . . in Epitome Operum Galeni habentur. This work is contained in no. 16, Epitomes etc. Palau 130039, 130042.

18. Annotationes in Galeni interpretes . . . Andrea Lacuna Secobiensi . . . Venetiis apud Hieronymum Scotum 1548. This book, like no. 17, tends to be bound with no. 16. Palau 130040.

19. De ponderibus ac mensuris. 1548 (this short treatise is bound with various works of Galen and in the commentary on Dioscorides: see below).

20. Epitome Galeni Pergameni Operum . . . per Do. And. Lacunam Secobiensem . . . accesserunt eiusdem . . . Annotationes in Galeni Interpretes . . . item De ponderibus et mensuris medicinalibus utilis Commentarius, Item, Index rerum et verborum . . . Basileae per Thomam Guarinum, 1551; *ibid*, Isingrinius, 1551; Argentorati, Lazarus Zetztner, 1604; Lugduni, I. Cassin et F. Plaignard, 1643. Palau 130047–130051.

21. Epitome Omnium Rerum et Sententiarum quae Annotatu Dignae in Commentariis Galeni in Hippocratem extant, per Andream Lacunam Secobiensem, Medicum Iulij III. Pontif. Max. . . . cui accessere nonnulla Galeni Enantiomata per eundem Andream Lacunam . . . collecta. Lugduni, apud Guliel. Rouillium, sub scuto Veneto, 1551 and 1554. See no. 26 below. Palau 130043, with editions listed.

22. De Articulari Morbo Commentarius ad S.D.N. Iulium III. Pont. Max. authore Andrea Lacuna Segobiensi medico pontificio cui accessit Tragopodagra Luciani . . . per eundem Andream Lacunam in Latinam linguam conversa. Romae, Excusum apud Valerium et Aloysium Doricos fratres Brixienses, 1551. Palau 130044.

23. Methodus Cognoscendi Extirpandique Excrescentes in vesicae collo caruncul as, authore Andrea Lacuna Segoviensi medico Iulii III. Pont. Max. illustrissimique et Reverendis. D.D. Francisci a Mendozza. Card. Burgen.; Romae, apud Valerium et Aloysium Doricos, 1551; Alcalá de Henares, Juan Brocar, 1555. Palau 130052.

24. Georgii Gemisti Plethonis . . . Quatuor Virtutum explicatio, graece et latine, nunc primum edita, Adolpho Occone Physico Augustano interprete. De moribus philosophorum locus ex Platonis Theaeteto . . . eodem interprete. Adiunximus Aristotelis de Virtutibus et vitiis libellum . . . Andrea a Lacuna . . . ; Basileae, per Joannem Oporinum, 1552. (See no. 12: no. 24 is a later printing.)

25. Apologetica epistola in Janum Cornarium;

Lugduni, Gulielmus Rovillius, 1554; Coloniae, 1557. Palau 130056.

26. De contradictionibus quae apud Galenum sunt; Lugduni, Gulielmus Rovillius, 1554. See no. 21; this is probably a variant title.

27. Annotationes in Dioscoridem Anazarbeum, per Andream Lacunam Segobiensem . . .; Lugduni, apud Gulielmum Rovillium, 1554; at end of prologue: Romae 1553. Palau 130057.

28. Pedacio Dioscorides Anazarbeo, acerca de la materia medicinal . . . traduzido de lengua Griega, en la vulgar Castellana . . . por el Doctor Andrés Laguna; Antwerp, Juan Latio, 1555; Salamanca, Mathias Gast, 1563; 1566; 1570; 1584; Salamanca, Cornelio Bonardo, 1586; Valencia, Miguel Sorolla, 1626; 1636; Valencia, Claudio Macé, 1635; 1651; Valencia, Vicente Cabrera, 1677; Valencia, Heredero de Benito Macé, 1695; Madrid, Vol. I., Fernandez de Arrojo; Vol. II, Alonso Barbas, 1733.

29. Discurso breve sobre la cura y preservación de la pestilencia, hecho por el Doctor Andrés de Laguna Medico de Julio III. Pont. Max.; En Anvers en casa de Christoual Plantin, cerca de la Bolsa nueva, 1556; Salamanca, Mathias Gast, 1566. Palau 130060.

30. Quatro Elegantissimas y Gravissimas Orationes de M. T. Ciceron, contra Catilina, trasladas en lengua Española por el doctor Andrés de Laguna, Medico de Iulio III. Pontifice Maximo; En Anvers en casa de Christoual Plantin en el Vnicornio Dorado, 1557; Madrid, Francisco Martínez, 1632; Madrid, Manuel González, 1786; Madrid, Imprenta Real, 1796.

31. Liber de Parabilibus (Galen), translated into Spanish; Valencia, Macé, 1561. (Considered doubtful by Hernando y Ortega, *op. cit.*, p. 117.)

32. Hieronymi Cardani Medici Mediolanensis Contradicentium Medicorum libri duo . . . Addita praeterea eiusdem autoris de Sarza Parilia, de Cina radice eiusque usu, . . . Accesserunt praeterea Iacobi Peltarij contradictiones ex Lacuna desumptae, cum eiusdem Axiomatibus; Parisiis apud Iacobum Macaeum in monte D. Hylarii sub signo Pyramidis, 1565; Marpurgi, P. Egenolphi, 1607.

33. Galeni de Antidotis Epitome; Antwerp, Johannes Bellerus, 1587. The full title "[Galeni] De Antidotis Libri II, ab Andrea Lacuna in compendium redacti" is given by Everard, Giles, De herba panacea, quam alii tabacum . . . vocant . . . Antverpiae, Apud Joannem Bellerum, 1587. An earlier printing of this excerpt from the *Epitome* appears in: Cl. Galeni libellus De theriaca ad Pisonem, interprete et commentatore, Joanne Juvene, medico etc. Antverpiae, Apud Joannem Bellerum, 1575.

34. Discurso de como se ha de preservar y curar de Pestilencia. Dirigido a los iurados de la ciudad de Valencia; Valencia, Petrus Ioannes Assensius, 1600. A later printing of no. 29? See Palau 130061.

35. Magri, Domenico (Melitensis), Virtú del Kafé . . . con alcune osservationi di [A. de Laguna] per conservar la sanità nella vecchiaia . . . Seconda impressione con aggiunta, etc.; Roma, 1671. See for details Mueller, Wolf, *Bibliographie des Kaffee, des Kakao* etc. (Wien, Walter Krieg, 1960), p. 154, *s. v.* Naironi.

ANATOMICAL PROCEDURE, OR A SURVEY OF THE DISSECTION OF THE HUMAN BODY

By Andrés de Laguna, of Segovia, edited in the form of a compendium or even a handbook; you will find gathered here whatever is good and full of vigor which the most highly respected philosophers have written on the subject, purity of language being indeed observed but the niceties of words and, so to speak, their ornaments not too carefully sought after, for often where Latin terms could not suffice I have had recourse to Greek rather than to leave the matter itself without explanation. Buy it therefore, reader, and enjoy it. Paris, at Jacob Keruer's, 1535.

To the most noble and at the same time most polished of gentlemen, Don Diego de Rivera, most meritorious prelate of the church of Segovia, Andrés de Laguna, a friend of the art of medicine, greetings:

It is a cause for the deepest sorrow and sympathy, most distinguished father, that education has declined to such a state of misfortune where at the present time men who are unskilled, uncultured, and in fact not educated even in the slightest degree do not blush to approach with unwashed hands and, as the adage goes, without the dust of a struggle (such is their worthless nature) those forms of learning of which once, because of their sublime quality, even the most distinguished philosophers stood in awe. To such an extent does this intolerable evil increase day by day to the greatest destruction of education and learning that shoemakers, butchers, carpenters, and weavers have abandoned their servile duties for theories and loftier contemplation, as if they had set out upon some expedition. Although I keep silent concerning other professions I shall not forget my own; who does not grieve for the art of medicine, which in an earlier and happier era was practiced by Apollo, Aesculapius, Chiron, and Hippocrates, and has now fallen to such an evil pass that it is carried on as the hollowest mockery in the hands of cobblers, weavers, and at last public wine-sellers or any of the most common of craftsmen? Who, I say, does not grieve for the art of medicine, which once rendered men immortal or made them worthy of immortality and is now cast down to such a point of abject vileness that it is turned over to the tender mercies of the lowest of muleteers? For just as a garment which is used by many wears out much faster than if it is worn by one person alone so any noble discipline of learning, if usurped, as now unhappily has occurred, by a mass of ignorant people cannot fail to be ruined as swiftly as can be by a kind of contagious calamity. And the cause of this calamity and pestilence is the physicians and the beggarly professors. For who indeed are possessed by so great and insatiable lust for money, fame, and glory that not only do they urge mountebanks of any sort to join the medical profession but also promise, like a Thessalian magician, that these buffoons will learn the entire art of healing within six months, without any training or study? By such exhortation, by false declaration as to the ease of learning medicine, those physicians and professors incite their victims and attract them. In ancient times Pythagoras deemed no one worthy of his truly philosophical school unless he was trained in mathematics and could demonstrate that fact. But the medical teachers of the present era declare that not even grammar is required for the Apollinian art, which is superior to philosophy. It is the greatest of indignities that the art which lofty Apollo granted to Minerva alone and to her descendants has degenerated, with the deepest tragedy to those who have surrendered it, and has been yielded into the grasp of magicians and poisonous enchantresses worthy of punishment and execution. Now to our shame those whose hands have acquired rough callouses from constant use of auger and hammer do not fear to offer their judgment publicly concerning a patient's pulse rate. Others very carefully examine the urine of those who suffer illness, divining what is to come and constantly making reference to past instances; they themselves, moreover, are most skilled at tanning hides. If these people had come into the profession of medicine after being tested in philosophy or had been allowed to enter it at the bidding of the chief practitioners of the art we should not now be overwhelmed by a deluge of empirical performers nor should so many good fathers of family have been so foully murdered by such a cruel band of torturing scoundrels. But since to wish to recall these monsters to the light of learning is nothing other than (as they say) to beat a stone or to whitewash an Ethiopian, why, immortal God, do I waste time here and do not address to you, most worthy prelate, my entire discourse? Lo, here before you, most respected Bishop, is my commentary on anatomy, small indeed since it has been only three months a-borning but otherwise not to be despised since it has come into being not without considerable labor. But if I should have produced a work which is by no means comparable in its size to the dignity of the undertaking and have attempted something beyond my strength, I should certainly not beg your indulgence under this pretext nor appear to merit praise because I have tried to carry out a most handsome and noble project, for in matters of great importance (as Tibullus[1] clearly admits) it is sufficient to have wished to accomplish something. I have decided that I should write out these thoughts of mine for your Eminence not because I thought I should be making any worth-while contribution with them (this was the least of hopes allowed to my small talent) but because under the splendor and protection of your name they would come abroad under the very best of auspices. In addition to this reason, to whom, by the gods, should the fruit of my vigils more justly be dedicated than to you, most famous patron, who not only cherish the studies of all learned men with a certain rare enthusiasm but have honored my father with such an affectionate friendship that you have selected him to serve among the physicians who are closest to your person and have enriched him with ample and magnificent gifts? I pass over the singular constancy of your spirit, your recondite erudition, your extreme prudence in handling affairs; with all these qualities, you govern your bishopric at Segovia in such a manner that nothing remains to be desired in respect to complete piety, to such a degree, in fact, that you furnish us the example of that well-known phrase from Plato:[2] "The good leader creates a good following." The public business which you transact, although very arduous, does not hamper you so much that you cannot render faithful service also to private studies and you pursue them in such a way that, if anywhere, "the open doors of the Muses," as Zenodotus[3] says, are readily found in you. Thus, to the extent that men themselves are superior to other animals, so you excel among men in humanity, judgment, prudence, learning and, finally, in the greatest piety, which is the best quality of all. Indeed, you easily surpass Pomponius Atticus in faithfulness, Themistocles in prudence, Cato in equity, Domitian in gentleness and clemency, and finally Numa in piety and pure religion, as is agreed by all. Therefore, most worthy prelate, I have not the slightest fear lest I be branded as a flatterer in any way since I am telling you nothing new and, so far as all can understand, nothing which does not agree with the truth. However, because your noble spirit vigorously resists this commendation of your virtues, preferring that the testimony of its praise lie hidden in the hearts of men rather than fly upon their lips, I am convinced that I should refrain from offering it, imploring only that you accept in kindly fashion these writings of mine composed under great stress of other activities and employment so that they may represent no ordinary memorial of the love and duty I bear toward you as a most affectionate father. If you accede to my request, as I assuredly hope you will, I shall publish many more than these pages for the scholarly public, in gratitude for your noble indulgence. Farewell, most eminent among the heroes of the Church, and as a father does, bestow your love upon your son. At Paris, on the eighth day before the Calends of August, 1535.

[1] Tibullus, *Panegyricus Messallae* (III, 7 = IV, 1), 6–7.

[2] D. Fridericus Astius, *Lexicon Platonicum sive Vocum Platonicarum Index* (2 v., Leipzig, Weidmann, 1835–1836) does not list this saying.

[3] Zenodotus (3rd century B.C.), the first librarian at Alexandria in Egypt and an editor of Homer; he may have been the first to list the books of Homer's *Iliad* by Greek capital letters.

Preface to the Anatomy

There have been three sects of physicians: the empirical, methodical, and rational, as Galen[4] excellently points out. The empirics believed that they became trained only through contact with actual things and by experience, to the complete neglect of reasoning. The methodics, however, were so addicted to their most detailed methods that sometimes they took no account whatever of the results of clear demonstration or of the evidence of the senses. But the rationalists, as men who excelled all the rest of the physicians in dignity, obeyed their senses, it is true, but did not likewise acquiesce in them until they had consulted their reason more deeply and investigated it. In fact, it would be my opinion that the heresy of the sophists, as the most pernicious of all, should be included with these errors. For no heresy is more impudent than this, none more ready to jeer and deceive, none more alien to culture and, in brief, to a true philosophy. It alone subverts good training; it alone destroys tender talents. What more shall I say? It alone strives to remove the most ancient memorials of the philosophers from our midst and to destroy them utterly. However, although these most ignorant sophists turn most readily to their vicious purposes that of which they know they are especially deprived, a knowledge of medicine, there is no lack of learned men who incessantly seek out such education with an insatiable thirst for it. May they perish, perish miserably, those lazy scoundrels to whom learning is so hateful that they glare at it with a beastly gaze as if it were a foul poison. Let them depart hence; would that they might take themselves off to the unhappy isles or, if you prefer, be deported most wretchedly with themselves at the oars like the galley slaves they are.

With this preface, therefore, I should like to give only the following warning: if this book I wish to write about anatomists is destined to be useful to anyone, let him remove from his spirit all boastfulness, all impudence, all wanton petulance and jealousy, and abandon any foolish opinions he may have conceived before he turns to welcome the product of my vigils. As far as I am concerned, indeed, I

[4] Galen, *Definitiones Medicae* (XIX K. 353).

make bold to offer this promise in good faith: I shall write nothing in this commentary which I could not also prove by the authority of Hippocrates, Galen, Cornelius [Celsus], Plato, Aristotle, Pliny, or, finally Aphrodisiensis [Alexander of Aphrodisias]. Note, meanwhile, that if I do not convince you with my discourse, the credibility and authority of these illustrious men is by no means diminished. And now I gird myself for the task in hand.

Here begins the Anatomical Method of Andrés de Laguna, of Segovia, the Author

In every commentary there are two methods which are especially to be observed. One is called the method of nature by the dialecticians, the other the method of the discipline. Since first things come first and other things later in the arrangement of separate facts I use the order of nature itself, according to the relative value of the matters to be discussed. Compelled by utility and necessity, however, I employ the method of the discipline and have arranged my material in reverse order, placing that which is posterior before that which is prior. Porphyry[5] seems to have preserved this order of procedure in his introduction to [Aristotle's] *Categories*, for he had earlier followed the order of nature, promising to deal first with genus, then differentia, thirdly with species, always proceeding from prior categories. But while he considered separate terms singly he dealt first, of course, with species, which is posterior to differentia, and treated it with the most artful care and detail. Why do I make these points and what has genus to do with intestines? I have not drawn this comparison inaptly if indeed a double method is discoverable in the dissection of the human body, which the Greeks call anatomical theory; but one method follows nature and the other necessity. All who have written on anatomy have preserved what I call the method of necessity, beginning the dissection itself from the thicker intestines lest their smoky excrements which putrefy in short order interfere by means of their invincible heaviness of odor with our contemplation of the noble members. The second method which, as is clear from what has been said, is that of nature I shall employ in my investigation as far as is possible to the very end of the task I have undertaken so as to pursue each detail in a kind of sequence. This can be brought about if I simply follow the order in which food, however rough it may be, is gradually broken up and consumed. I shall therefore begin from the mouth, where the first alteration of the food is carried out. I shall end my investigation with the brain, from which the spirits themselves receive their ultimate and most complete transmutation. The animal spirit, indeed, does not proceed any farther than the brain and no other organ is discovered in which it could be further elaborated. But why should these matters delay me from girding myself for the business before us?

Since there are at most three elements by which man's life is sustained, food, drink, and spirit (breath), the mouth has certainly been most carefully fashioned by nature for their reception. Its construction is indeed most suitable both for what is necessary and best for its functioning, according to Plato. He calls necessary those elements which nurture the life of man and best those elements which contribute to man's education and training. Hence the construction of the mouth includes particularly two jaws, along with other parts. One of these jaws, the upper, is completely immobile since it is attached to the head. The other jaw which holds the lower region is employed in frequent motion. Both jaws are fitted with sixteen teeth at most. Hence there are thirty-two teeth in all, whose shape and function are widely different, for some of them are broad incisors (cutting teeth), some canine teeth, and others are molars, which are larger and more irregular than all the rest. The incisors occupy approximately the middle region of the jaws in order that with their sharp points they may first bite the food, then sheer it off, and cut it up; hence the Greek name for them—tomeis or cutters—is appropriate and comes from their action. The canines, which the Greeks call dog teeth, are located next to the incisors; they are more suited for breaking and separating the food than the incisors since they are much duller. Nature has very cleverly constructed them so that if any food resists the action of the incisors it can be readily ground up by the canine teeth. The third set of teeth, the molars, are located at the extreme ends of the jaws. These are called gomphoi or mulai by the Athenians but genuini by Cicero.[6] With their broad, strong surface these teeth gradually grind up the food cut by the incisors and broken up by the canines and reduce it to a certain even quality. There are eight incisors, four canines, and twenty molars; those which have the greater use have also been supplied in greater number by nature. One nerve is disseminated to each of the teeth from the third pair of nerves and is inserted at the base of each. As in many other instances, so also in this, nature has been provident for all parts of the body in order that they might not lack sensation but be endowed with the faculty of feeling to perceive external dangers and to repel them. Since by their nature the teeth are bony they would have no feeling or pain if nerves were not disseminated

[5] Porphyry, *Aristotelis Categoriae et Topica cum Porphyrii Isagoge ex recensione I. Bekkeri* (Berlin, G. Reimer, 1843) **1.**

[6] Cicero, *De Natura Deorum* 2. 54. 134.

to their base. In fact, they are completely without sensation; nerves therefore are designed by nature for the perception of their ills. The nerves themselves are affected by the mediation of the teeth, for when the tooth is chilled the inserted nerve is also chilled. This change of temperature in the nerve is straightway accompanied by a very severe pain. Hence we use the term inexactly when we speak of toothache since in fact the tooth feels no pain. The nerve itself when unduly chilled, heated, dried, or moistened beyond the just tolerance of its nature straightway feels pain through the tooth. The fact that the nerves, as I said, are inserted into the teeth themselves is very clearly proved by the extraction of those which I called canines a short time before. One of these teeth cannot be extracted with very great danger to the eyes for the latter lie above the canine teeth in the same region along the same diameter in such a way that there must be some necessary connection and feeling between the eyes and the canine teeth. Hence, if the nerve inserted at the root of the tooth is pulled out it is necessary to do some damage to the nerves of the eyes also. The tongue, like some stern doorkeeper, endowed with free motion and movable in any direction, is most carefully situated in the middle of the mouth. It seems to be composed of many arteries and veins and also of various offshoots of the nerves. Certain very hard nerves descend into its center to furnish it with free movement, but other, much softer nerves are scattered upon the upper surface of the tongue by which it must make contact with the food; it is reasonable to suppose that these nerves arise from the third pair of nerves. You should not therefore be surprised that, although the instruments or means of motion and sensation are different in the tongue, it sometimes happens that its motion is crippled while the sense of taste is completely unharmed. The reverse may also happen: the tongue still endowed with the faculty of motion may be deprived of the power to feel. The substance of the tongue is quite spongy so that it draws and attracts to itself whatever moisture there may be, like some old drunken woman. Not unworthy is the opinion of Plato, expressed in the *Timaeus*,[7] that certain small veins extend from the tongue to the heart, which, like most diligent escorts, carry pleasure and pain to that organ. Here are his words as they run in Latin: "For there are certain little veins which penetrate like messengers of the tongue as far as the heart and are scattered in the softer flesh of that organ. They contract and dry out when the terrestrial substance of the heart is moderately liquefied, so that if they are quite rough they are called astringent, if less rough they are called harsh."

Thus far Plato. Furthermore, it is a doubt of no small importance which delays my discourse whether taste is necessary to man's life or, while conducive to its maintenance, it is not necessary, in such a way that without it man can be nourished and continue to live, not, however, so well as if he were endowed with taste. To give you my opinion in a few words, I declare that taste is not at all necessary for the life of man but is nevertheless most conducive to it. There are many very cogent reasons to prove this statement. First, our human well-being and nourishment are the same as those of plants and yet the latter do not have the faculty of taste; therefore it is by no means necessary to us. Theophrastus presents an even greater proof of this proposition in the first book of his *History of Plants*.[8] A living creature carried in its mother's womb, very completely sealed off there and indeed more completely at that time than ever, is also nourished without the faculty of taste. Therefore we should not require taste as far as maintaining life is concerned. We also see many people whose tongues have been cut away at the roots; nonetheless they live their lives according to the desire of their spirit. Pliny, moreover, writes in his *Natural History*, Book VI,[9] about certain people in the East who are deprived of their tongues; however, in Book XI, 37,[10] he considers that a storehouse of flavors exists in the tongue. Hence, if anyone nourished on honey whose sense of taste is so debased that the sweetest honey seems to him the most bitter of flavors, although it weakens his nature and brings by no means any relief or cooling effect, such a person, to whom honey is neither sweet nor bitter, totally deprived of taste as he is, will be much more happily fed on honey. Aristotle also corroborates this statement both in other passages as well as especially in *De Anima* III,[11] last chapter, adding his view in this manner: "Therefore excess in those things which are perceived by touch not only destroys the sense-organ but the living creature itself since touch is the one sense that the living creature must possess. Yet the living creature is given the other senses also, not merely to exist but to exist happily, as I said earlier. For it possesses vision in order to see since it lives in air, water and in a transparent medium. It has the sense of taste in order

[7] Plato, *Timaeus* 65D. De Laguna's Latin does not make it quite clear that Plato says the earthy particles which enter by the small veins contract and dry up the latter and that those particles are the ones called astringent and harsh.

[8] Theophrastus, *History of Plants* I, makes a limited use of analogy between the parts and characteristics of plants and animals but does not make any specific reference to taste.

[9] Pliny, *Natural History* VI. 35. 188.

[10] *Ibid.*, XI. 37: Pliny, XI. 65. 174 says the tip of the tongue is the location of taste but with man taste is also situated in the palate. He says nothing of a storehouse of flavors.

[11] Aristotle, *De Anima* III. 13. 435b15–25. The later quotation from Alexander of Aphrodisias is not in his *Problemata*, at least in the 1520 text I have used (Basel, Cratander).

to distinguish between that which is pleasant and that which is unpleasant immediately in whatever food so that it may be moved to appetite by this sensation.'' Thus far Aristotle. Let me conclude therefore, if you like, that the sense of taste is very useful for leading a happy life but by no means necessary for the maintenance of life. The tongue itself, then, is the instrument or organ of taste. This organ is pleased by those foods which it recognizes as acceptable to the stomach, but it avoids and drives away those foods which are unpleasant to the stomach since the tunic which envelopes the tongue is also continuous or common with the stomach. Nevertheless it often happens that the tongue admits some harmful medicines which completely upset our body, not intentionally but because of some imposture or deception, as for instance medicines injurious to our powers and the substance of our nature which have been smeared with sugar or honey. The tongue does not resist their entrance, like a king's doorkeeper who, regarding them as servants and ministers of his royal highness, admits assassins and conspirators and at length the entire destruction and pestilence of the empire. Thus it happens that the tongue is very frequently deceived. The fact that it is divided into two parts is, I suppose, not unknown to anyone, so that if one part becomes dulled by paralysis or inflammation the remaining part may not be similarly afflicted. Under the tongue there is something called a frenulum (little checkrein); if by chance this becomes more loose than is convenient or reasonable in any person, as may be seen in women who are more talkative than any turtle-dove, you should not expect silence. But if this checkrein is tightly contracted so that the tongue can scarcely be moved, a man who has such a checkrein will speak very little. One must, however, note carefully that nature has attached such a checkrein only to the tongue and to the private parts, for in these organs especially she has desired that men should be modest. At the root of the tongue itself near the end of the palate there grow certain pieces of flesh which are called tonsils by the Romans and paristhmiae by the Greeks although they are called amygdalae (almonds) by the unlearned barbarians. These serve, particularly at the impulse of the will, to keep the saliva which distills continually from the brain through the colatorium [lamina cribrosa of the ethmoid bone] from passing down to the lungs through the rough artery, holding and restraining it since they are quite spongy and porous. Between the tonsils there is the columella [uvula] which is called grape by others because of its similarity. Nature has fashioned the uvula as a means of modulating the voice. Since indeed the material of the voice is nothing other than breath it could not form the voice if it were expelled roughly and freely (for all brute animals breathe like men) unless the columella were given deservedly to man alone of the animals for the production of the voice by adjusting the expiration of breath in various ways so as to compose man's speech. Thus if the columella is cut off a man immediately becomes mute. This part of the body among all the other parts is liable to dangers of many kinds especially because it is subject to various distillations of the brain and especially because the air first strikes the uvula as we breathe, whatever its quality may be. Therefore it was with considerable foresight that Alexander of Aphrodisias declared that this part of the body first tempered the breath to keep it from rushing unchanged to the lungs. All these parts of which I have been speaking are covered over by the palate as though it were the sky; the palate serves a great purpose in the preparation of the food. The outermost construction of the mouth is completed by two lips which serve both for the formation and production of the voice as well as for the special protection of the teeth, and for handling food in order to prevent it from slipping away and falling due to its excessive humidity. Nature created the lips for these necessary purposes. Furthermore, the lips are covered by the internal tunic which is continuous with the stomach, and for this reason they tremble exceedingly and seem to indicate imminent vomiting when the stomach is troubled by nausea. The food which provides nourishment for the body is first received by the two lips, which close upon it by a kind of mutual connivance to prevent that which has been taken into the mouth from jumping back out again. Later, when the food has been broken up and reduced to a certain consistency by the various kinds of teeth it is rolled around under the palate; for this service the tongue is particularly adapted. If the food requires some moisture to dilute its dryness or roughness the tonsils will supply it as quickly as possible so that the food may absorb the greatest amount of saliva. If there is something smoky and sooty in the food itself which would be injurious to the operations of the stomach because of its biting quality the colatorium is at hand in the middle of the palate to receive this injurious substance like an oven so that the vapor which results from such food may fly up at once to the region of the brain and straightway be congealed (for it is cold) and pass out through the nostrils by means of its own weight. When the raw food has thus undergone a preparation as described in the mouth it passes straightway through the esophagus to the stomach. At this point there is the root of the tongue which by means of the lengthwise fibers which its inner tunic has stretched out through it like straight hands draws and attracts the food prepared in the mouth toward itself but with its transverse filaments very greedily swallows and thrusts down that food. Nevertheless the gullet itself is by no means necessary for taking in food. Aristotle,

De Partibus Animalium III, 3,[12] seems to have spoken justly: "The stomach could have been joined immediately and without discontinuity to the mouth." However, because the rough artery was necessary for drawing the breath of life although the gullet was not needed for the same operation nature gave it to us as something useful but not exactly required. When Galen was searching for the proper shade of meaning for each term he was in the habit of calling it stomachos while the common and vulgar word for it was esophagus. It is formed of nerve, vein, artery and a double membrane; part of it seems to be inserted into flesh and part seems to be a muscle. This organ alone is endowed by nature with a very slight sensation and especially the lower part of it which is joined with the stomach, called by Galen in his customary way the mouth of the stomach, although sometimes he called the entire organ by that name, as in *De Locis Affectis*, V, 1 and 5,[13] where he writes as follows: "Therefore it is necessary to observe closely so that you may discern at the proper time the various conditions of this organ as they occur; from the actions of this part of the body it was at first considered to be the mouth of the stomach (which all the ancients used to call the heart but in our time is called the stomach)." The ancients called it the heart from the fact that it suffered pains like those of the heart (cardialgiae in Greek), deducing the name from the similarity of its pains ["heartburn"]. There is no wonder then (as Galen himself testifies) that it usually suffers very severe accidents such as syncope, convulsions, lethargy, epilepsy, melancholic afflictions, suffusions, and dizziness. The gullet or esophagus, whatever one wishes to call it, which I said is a kind of vehicle for the food, was not without cause inserted into the side of the stomach lying below it to keep the food, once it was thrust into the stomach, from regurgitating to the mouth when the head was bent to the side. There is no difference at all between the stomach and the gullet as far as their substance is concerned at least, that is, the substance of their muscles and tunics (I say nothing of their shape, which is entirely different). In fact, as Galen declares in *De Usu Partium* IV,[14] the gullet is the upper part of the stomach since its inner tunic, which is rather membranous, is equipped like the stomach itself with straight fibers from top to bottom and on the outer part which is rather fleshy with transverse filaments, called egkarsiae ines in Greek. Furthermore, one of the orifices of the stomach which the Greeks call pylorus, that is, doorkeeper, is not placed at the bottom but a little higher up so that any material which had not been thoroughly digested could not be prematurely carried down to the bowels. This orifice is much narrower and tighter than the orifice above, no doubt owing to the proportion they preserve in relation to whatever is strained through each. First, therefore, the stomach draws or attracts substances through the gullet by means, as I said, of its lengthwise fibers. Next, after the substance or food has been drawn to it the stomach embraces the food with another kind of fibers called oblique. The stomach retains the food until its function is completed; by that force which the Greeks called alloiotike, or alterative, the stomach changes the food into a kind of rather soft thick juice. Then, like an all-too-faithless cook, it takes the substance thus made gentle and thin by its art and greedily swallows it up, for the stomach is nourished partly by the chyle and partly by blood carried as a kind of remuneration from the liver. After the oblique fibers together with the muscle of the lower orifice (which we call the pylorus) have shown themselves to be quite strong in retention the transverse fibers are very firmly contracted and propel that humor called chyle by the Greeks as far as the intestines, which have very little power of attraction. Further on between the transverse septum and the intestines the stomach itself is surrounded by the spleen and the lobes of the liver, which are like hands grasping it. Thus the stomach may receive for stimulation of appetite some portion of the black bile by a sort of partnership from the spleen, that portion which the latter does not need for its own use. The nature of this kind of superfluity, called black bile (*cholē melainē*) by the Greeks, fortifies and strengthens with its acidity all the actions of the stomach which take place within its embrace. For since it is cold and acid by nature it clearly compresses the stomach and thus forces it by contracting itself to surround the food so closely that nothing which is not properly digested can descend to the intestines. Hence this excrement is a very useful friend to the functions of the stomach just as that other excrement called the yellow bile is harmful and a most pernicious enemy. Alexander Benedictus[15] was wandering in his mind when he observed that from the bile cyst a duct which carries the bile always empties into the lowest part of the stomach. For who, immortal God, does not know that owing to its biting quality the yellow bile does not allow even the most impure excrements to remain still anywhere? In fact, it almost expels the humors of the stomach before they have attained the proper and desired digestion. Hence it seems to me much more sensible to say that those people who have excellent health of body do not have

[12] Aristotle, *De Partibus Animalium*, III. 3. 664a25.
[13] Galen, *De Locis Affectis* V. 1 and 5 (VIII K. 332–333).
[14] Galen, *De Usu Partium* IV. 7 (III K. 281–282).
[15] Alexander Benedictus (Alessandro Benedetti), (1450?–1512 A.D.), *Anatomice* II, 15.

a bile duct leading to the stomach, although others whose disposition is unhappy have a bile duct inserted by some freakish act of nature into the stomach instead of the jejunum intestine where nature ought to have inserted it. Thus people who have such a constitution allotted them, although they are not bilious by disposition, nonetheless vomit up unmixed bile for days and perform their bodily functions so poorly that their digestion is quite faulty, they excrete a whitish substance, and frequently suffer from the most severe headaches, among other ills. But that the yellow bile duct is not inserted into the stomach, as I have said, is proven by experience and demonstrated most clearly by Galen, *De Usu Partium* V.[16] These, at any rate, are his words on the subject: "You will be amazed even more, I think, by nature's providence in not inserting some part of this duct into the stomach when the latter generates many superfluities." Later on he says: "For this reason the bile passes into the first intestines and irritates them with its biting quality, thus preventing the food from remaining in them. The bile would do the same thing for the stomach [if it passed into the latter] since it is endowed with keener sensation than the jejune intestine, that is, force the food downward before it has been properly digested." Galen adds the further remark: "For the stomach, bitten by the sharpness of this humor, bears it ill as though it resented the bile; stimulated by it, the stomach is forced to make a swift excretion of its contents." But perhaps you make the following objection: "How is it that almost all people suffer from bilious vomiting if no bile duct is implanted in the stomach?" This is a fair criticism; you must, however, examine the matter a bit more closely. Into the jejune intestine, which is situated very close to the stomach, there extends a fairly large yellow bile duct; hence that part of the intestine into which the duct is inserted is always observed to be empty since it is sometimes very keenly irritated by the sharpness of the humor and thus drives and propels it away. However, since the yellow bile is light and easily movable by nature, it often flies upward and completely upsets the operations of the stomach, disturbing and injuring it unless it is immediately voided by vomiting. I insist, nonetheless, as I have done above, that despite Galen's statement in *Temperaments II*[17] a bile duct does actually grow in the bottom of the stomach as a means of passage for the bile in some people, although very few, whose bodies have a faulty structure. Sometimes it is large, sometimes small, but the way it functions in such people you can observe in Galen's description. Sometimes they suffer from nausea; they are always attacked by dizzy spells and by suffusions or inflammations. Finally, to sum up the situation, the stomach is not injured so much by cold as by heat, for due to cold its powers are drawn and collected into the inner region and by heat they are broken up and relaxed in a wonderful fashion. The story is told of Aesop[18] that in order to establish his innocence before his lord and master when accused of eating some figs he took a hot drink and made the servants drink also. When they had done so they vomited up all the figs they had eaten surreptitiously. Thus Aesop was restored to his former position of trust by his master and the slaves were soundly beaten. Aesop could not have accomplished this demonstration if he had given them something cold to drink. Furthermore, the stomach appears to distribute and bestow the food to all the other members; hence all of them, noble and ignoble, seem to depend upon it for their sole source of sustenance. If its function is impaired, all other functions of the body are placed in the greatest danger. Titus Livy[19] relates the following fable: "All the other parts of the body once made a plot and conspired against the stomach. When this was discovered the stomach became angry with them and their unjust behavior toward itself, the father of the family, from whom depended the life of the entire body and against whom such hostile sedition had arisen. However, the members had not undertaken their conspiracy with impunity, since it resulted finally in calamity for all of them. For when the stomach denied its food to the rebellious parts they were quickly stricken with disease and languor." Thus the stomach is located in the middle of the entire body since it has many various connections with the other members, such as the sinewy membranes by which it is attached to the transverse septum. It is joined to the brain, the source and origin of the nerves, by many nerves. It is connected with the heart by arteries and with the liver by veins, as is shown by certain illnesses of those parts, for if the membranes which cover the brain undergo any rupture, the stomach, by reason of its connection with the brain, likewise suffers keenly and is seized by a sudden fit of vomiting. If the stomach is thus at-

[16] Galen, *De Usu Partium* V. 4. (III K. 354, 357).

[17] Galen, *Temperaments* II: *De Temperamentis* II. 6 (I K. 631–632) Galen says that for the most part there is a single duct inserted into the intestine which is between the pylorus and the jejunum; this duct some call an ecphysis of the stomach, as though it grew out of the latter.

[18] The life of Aesop in manuscripts G, W, and Bellunensis Lollianus 26 has been edited in a work of monumental scholarship by B. E. Perry, *Aesopica*, etc. (University of Illinois Press, Urbana, Ill., 1952); the story quoted by de Laguna appears on the first page of each recension, pp. 35, 81, and 111; the first two are in Greek, the third in Latin. Which text de Laguna may have used is difficult to say, since the *Life of Aesop* was first published by A. Westermann in 1845.

[19] Livy, *History of Rome*, II, 32.

tacked by an unconquerable desire to vomit, what vapors, what inflammations or suffusions, what dizzy spells in the brain will be observed? An irregular movement of the arteries will also be discerned in the pulse, now small, now large, now slow, now fast, and by an acceleration that is more than reasonable to such an extent that we conclude the patient has a fever. On account of this persistent and immoderate vomiting due to the connection I mentioned between stomach and liver such patients finally vomit pure blood because the stomach draws to itself completely unmixed blood through those same veins by which it transmits the chyle to the liver; there is no mention here of the meseraics which descend to the intestines but of some which are inserted into the stomach. I shall now show that small veins of this kind are necessary. After the stomach has broken up the food evenly with its denticulated hard flesh it does not cease its function but is nourished and restored by means of a rather mild thick juice [the chyle]. Later it sends the rest of the chyle to the intestines like a troublesome burden so that through those veins which are woven into the mesentery it may be carried up to the liver. Since this pathway is long and winding, to prevent frustration of the liver's operation because of long anticipation, the stomach selects from that substance which was ground up (before sending it downward) the part which is more thin and transmits it through the veins which I said were shorter, like messengers of the other veins, and are attached to both stomach and liver. This thinner substance goes to the lowest part of the liver. The rest of the food which cannot be sent through the veins is shut off by a lower valve and finally passes to the intestines which are close to the stomach.

The intestines, to sum up, when dissected are seen to be six in number. Although they are held together in one place nevertheless they can be easily distinguished from one another both by their shape and substance, for they are called by Galen the members of evacuation. They too have a considerable faculty for digestion as if whatever was not adequately prepared by the stomach should be completely ground up by the intestines. They are, however, deprived of an ability for attracting anything to themselves since they are equipped with a single kind of transverse fibers alone. The intestines of man are not the same as those of wolves, deer, and waterfowl, that is, stretched out straight in their length, in order to keep what they contain from flowing out quickly since that would be a serious impediment to outdoor activity. The intestines like the stomach receive a double nourishment, both with chyle and with blood: with chyle before the giving-forth (*anadosis*) and distribution of the elaborated juice, and with the blood by means of those veins through which the chyle is sent to the liver as well as through other veins which carry blood alone into the intestines. The first intestine which arises at the side of the stomach is called ecphysis or outgrowth by Galen; by more recent physicians, however, it is called duodenum because it is twelve fingers long. It has no coils or bends, which the Greeks call helikes; nature has made it in this fashion for if it had winding coils it would hinder the operations of the stomach and of the transverse septum. To this intestine there is attached immediately and continuously the jejunum, which Galen calls nēstis and the Latins sometimes hira; it receives its name from its continual state of emptiness. It retains nothing of that which is brought down from the stomach. There are two reasons for this. First, the impurities from the liver descend in large part to this intestine, which quickly seizes and destroys their juicy part. Second, it receives the most exuberant and fervent supply of bile from the liver through a large duct; the bile stings the tunics of this intestine with its incredible biting quality and irritates them toward a sudden excretion. The bilious humor acts more sharply when it flows pure and fresh from the liver and when it is not contaminated by the heavy superfluities of the feces. Thus it happens that the emptiness of the jejune intestine brought about for no purpose nevertheless may follow that which is brought about by necessity and produce such an effect. The yellow bile fulfills an important function in driving out the excrements of the bowels which are frequently impacted therein and block the pathway of excretion, bringing an inevitable calamity to the human body. Experience teaches us this as anyone who has even a moderate reason will know without a tedious demonstration. As soon as the yellow bile descends into this intestine it does not allow the chyle to remain there even for a moment but immediately drives it to lower regions. The ileon (ileum) intestine is joined to the jejunum. Because it is very thin and appears to consist of many coilings it is also subject to numerous and severe ailments. The one called iliac pain receives its name from this intestine and arises from the fecal humors which have blocked it since they are not easily cleansed away by any other biting humor in the cells and sinuses of the intestine itself. It often happens that men are seized with iliac illness accompanied by a certain flatulent exhalation that distends the tunics of this intestine and are so tormented by this illness that they expel their feces through the mouth and finally die a very wretched death. The cecus or monoculus intestine lies next to the ileum; it appears to be without any doubt the intestine which contains most feces. It is called cecus because it seems to have only one orifice through which it attracts and expels. But this intestine actually has two very small orifices placed close together so that many have concluded that it has only one orifice, betrayed by the shape of the intestine for it hangs like another stomach that is filled and at whose bottom there is no orifice. He who wishes to recognize accurately the care and industry of nature must present himself as a dissector of even the repulsive bodily members and most diligently examine their location, shape, number, and consistency. Once at Paris I attended an anatomy presided over by all the fellow practitioners of the medical art as well as the barbers to whom the actual task of dissection was committed. In order to avert from themselves the stink of the intestines and to leave the matter as one too well known to require explanation they declared without even looking at them that the cecus intestine had only one orifice. But I took up a scalpel, dissected this intestine and demonstrated to all with a small peg or stick that it had two small openings in line with each other, one through which it attracted and another through which it expelled the feces. For I had read in Mundinus, who is not so ignorant as he is barbarous, that such was the fact, as I discovered with my own eyes.[20] In earlier days I had dissected a little dog whose cecus had only one orifice, but you should not mix this instance with the former fact. Furthermore, there are many small veins scattered in this intestine and especially at its bottom. Hence, owing to the daily deposit of food in it, certain vaporous substances are found there by which through anastomosis the intestine feeds and sucks. The colon intestine is joined to the cecus; the more closely it approaches the stomach the more it surpasses the other intestines in number and thickness of its cells. This intestine winds around the left-hand region of the kidney; hence many physicians confuse kidney pain with colic and vice versa since each kind of pain seems to afflict the same part. The difference between these ills, however, has been explained in great detail by Galen in *De Locis Affectis*, I and VI.[21] In fact, as Galen teaches and our own experience tells us, in the affliction called colic the pain seems to move around very widely, troubling now this, now that part. Kidney pains, however, remain in the same place and trouble the same spot. If something in colic pains can be separated out by the bowels it is a certain flatulent substance with the consistency of cattle dung which often floats upon water. But this is enough of such matters. They will be discussed in greater detail when I come to an examination of affected parts of the body. This intestine stretches to the right side of the stomach and hence readily receives from the liver, which is located close to the stomach, some portion of the bile in order to expel the feces. Many veins descend into this intestine which carry back a large amount of milky substance to the hollow parts of the liver. Last of all in its position is the thick intestine which is called the rectum and the aluus (bowel). Because it is the storehouse and repository of the excrement it is of all the intestines most full of feces. This intestine does not wind about as the other intestines do nor coil and twist but slants upward to the ilia or groin and stretches to the right kidney. Then it declines somewhat to the stomach. If its shape were coiled and twisted it could not easily dislodge those excrements which are hardest to remove. Nature does not send even the slightest amount of bile through the bile duct into this intestine; she has decided that the bile which flows into the upper intestines will be of use also for this last intestine and that the muscles of the hypogastrium will easily drive out the superfluities collected in it. The body can be nourished through this intestine by injections sent up to the liver by means of enemas when the actions of the stomach proceed with imminent risk of death [owing to some malfunction]. The extreme end of this intestine is called hedra by the Greeks and anus or seat by the Latins. It is a transverse muscle which in a marvelous way inhibits untimely excretion of the feces since nature has endowed it with a very strong faculty of retention. But since (as I said earlier) the intestines not only contain the material which has been prepared by the stomach but also alter and change it they deserve to have the omentum, called epiploon by Galen, as a cushion and a support around them. Its fat, adipose substance appears to be interwoven with many arteries and veins. It stretches to the middle of the stomach, with one wing above and another below, covering almost all of the intestines and nourishing them with its vaporous heat. The barbarians [Arabs] do not blush to call this membrane the fatty zirbus, a name worthy of them. We, however, with Galen call it omentum or epiploon, as does Aristotle, *Historia Animalium*, I, 16, whose words run as follows: "The epiploon hangs from the middle of the stomach." This membrane assists greatly in concoction and digestion, that is, distribution of the food. Galen[22] states that some persons in whom this membrane has been ruptured have survived but have been thereafter able to perform what physicians call the natural functions only with the greatest difficulty. Contiguous to the omentum is another membrane which physicians call the peritoneum; Avicenna the Arab calls it siphac. Nature made this part for many uses. The first of these is its service as a covering for all the parts of the body below the diaphragm. It

[20] Mundinus: see Charles Singer, *The Fascicolo di Medicina*, etc. (Florence, R. Lier, 1925), p. 66 in his translation of Mundinus.

[21] Galen, *De Locis Affectis* I and VI (VIII K. 1–68; 377–452).

[22] Galen, *De Anatomicis Administrationibus* VI. 5 (II. K. 547, 556–557). Compare *De Usu Partium* IV K. 285–292, especially 286 about the gladiator whose injured omentum Galen excised almost completely. The patient recovered quickly but could not thereafter expose his unclothed midportion to the cold.

stretches lengthwise from the bones of the pubis to the transverse septum; in its width it stretches on both sides of the spinal vertebrae, called spondyles in Greek, and is attached to them at its ends. The second use of this membrane is to keep the intestines from slipping out of place when they are filled completely with nutriment sometimes humid, sometimes dry or flatulent. Its third use, and a very important one, is to expel and drive out the excrement, which is very often reluctant to depart. Galen has discussed the peritoneum in *De Usu Partium* IV at great length. Between this tunic, which is sinewy and quite hard, and that one we call the epiploon (or omentum) there often occurs a disease called dropsy by the Greeks and "water under the skin" by the Romans; not infrequently the disease occurs between the omentum and the intestines. A disease of this kind occurs also between the hypogastrium, of which I shall shortly speak, and the peritoneum. The hypogastrium, called abdomen by the Latins and mirach by the barbarians [Arabs], is interwoven with muscles, veins, nerves, arteries, and membranes; it serves as a sort of wall and shield for the stomach itself, extending around it on the outside. This part of the body has eight muscles by means of which it can attract, retain, concoct, expel, and perform many other natural functions. These muscles are not identical in form, since if they were the abdomen would be deprived of many important activities, but two of them especially with stamens or fibers that go up and down are extended from the sternum to the pubis in a straight line, very skillfully designed and especially suitable for traction. The substance of each of these muscles is not solid but composed of a mass of innumerable ligaments joined together. Nature has prudently put these muscles together in such a way that the affliction or injury of any part of them is not communicated to neighboring parts but is checked by membranes spreading like the sea water which holds in the injured part with its shores and limits. In order to avoid confusion I mean by sternum (*malum punicum*) that sword-shaped cartilage to which the false ribs extend, called rhoa or rhoia by the Greeks and the xiphoid cartilage by Galen. In addition to these two straight muscles there are two transverse muscles which the barbarians call latitudinal. They are not as long as the straight muscles but reach the latter after proceeding from the middle of the back and then degenerate into sinewy filaments by means of which they are attached and bound to the straight muscles. One of them originates from the left side of the back, the other from the right side. The task and purpose of these muscles as given to them by nature is to drive out the feces at once if they linger more than is proper in the bowels, with the assistance of the transverse septum, which the Greeks call diaphragm and phrenes. The transverse muscles would not be powerful enough by themselves to propel the smoky excrement downward rather than upward without the assistance of the diaphragm. Thus among other services which this septum furnishes for the vital function there is also this natural one: it expels as quickly as possible without any regurgitation to the stomach above whatever fetid and smoky excrements remain in the lower bowel just as though they had been squeezed between two hands until the wrists came together. Since it has seemed worth the effort to describe the straight and the transverse muscles, which are four in number, I should also say something about the oblique muscles which nature has designed especially for retention. Four of these appear in dissections; two are above, one from the right side of the abdomen, the other from the left side, extending as far as the transverse muscles with their ligaments and carried obliquely downward. Then there are two more oblique muscles which proceed in an opposite fashion from the chest, one from the region which stretches to

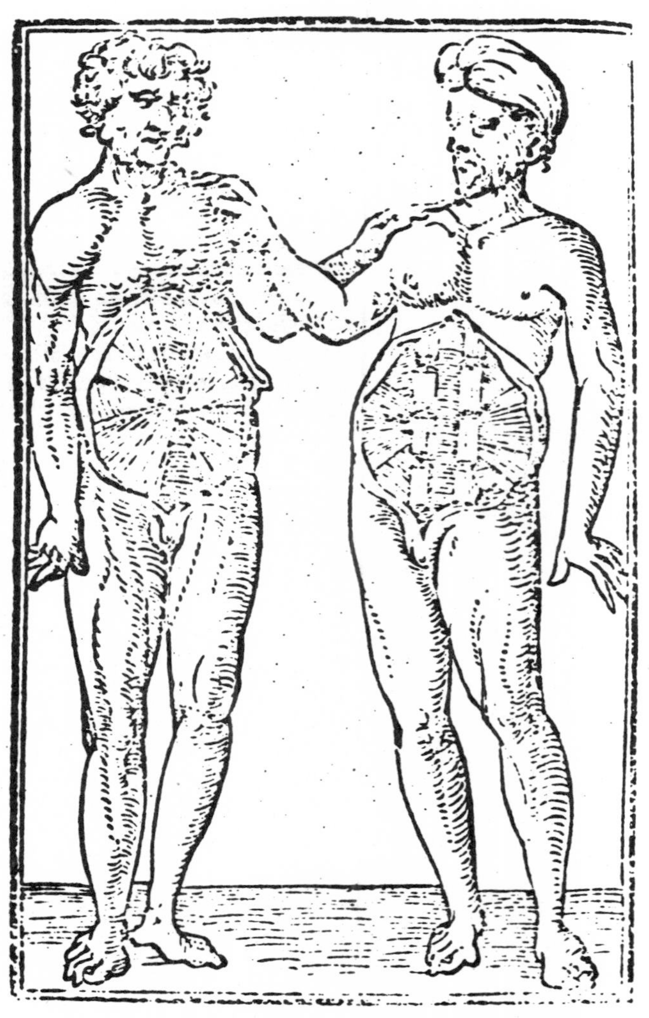

(*on left*: Illustration of the oblique muscles.) (*on right*: Illustration of the straight and transverse muscles.)

the right groin, the other from the region which stretches to the left groin, extending upward in a slanting direction with their filaments until they reach the descending oblique muscles described earlier and attain the region of the umbilicus. The person little skilled in anatomy should learn the arrangement and composition of these muscles from the following picture.

The eight muscles just illustrated are covered with a certain hard membrane which protects and defends them from untimely prolapse. Above this membrane there spreads as widely as possible a very abundant soft fat which originates from hardened blood. Upon this in turn there is placed as a covering and protection the very thick skin of the abdomen which has the consistency of drying flesh. You have here the description of the entire abdomen accurately set forth. It has no bone at all since sometimes it must subside as when it is without food; very often, however, it swells with a great amount of food or with the conception of a foetus, as in pregnant women. In these cases bones would be a great hindrance. I say nothing about the fact that we could not take a double refreshment at breakfast (or lunch) during Lenten fasts when it is forbidden to dine if our stomachs were surrounded by some ribs. Likewise in the coitus of male and female a mutual collision would take place if the stomach of each sex were bony; hence everyone would abstain from the act of generation as much as possible and the entire human race would perish very shortly. Therefore since the abdomen is free to swell outward there is no possibility for it to resist food itself if not too much is taken in at a time, so that the abdomen may attract, retain, and alter it with its muscles and expel the smoky excrements which overflow into the intestines. Hence from the intestines which are like large ships that carry a milky substance very many meseraic veins, like small boats or skiffs, bear away the purer juice and send it to the liver, similar to those tall ships which as soon as they have crossed the ocean come to Rouen with their cargoes on their way to Paris but transfer their cargoes at Rouen into small boats for the last stage of the journey up the Seine. Indeed, the intestines are rightly called ships since they carry the chyle and all the excrement through the entire region of the stomach as if through the Ocean Sea. Similarly I call the meseraic veins small boats which are scattered throughout the intestines. But what shall I liken to the Seine River? That small membrane called the mesentery which is attached to the back. Then it will not be absurd to call carriers what are commonly known as the humors, for whose sake all these parts of the body are constructed. And since no ship sails without a helmsman, exposed as it is to a thousand perils and shipwrecks, who, I ask, shall be that one who steers the ship, the kybernetes [steersman in Greek] who grasps the rudder? By Apollo, that one is Nature herself, who guides and disposes everything according to her will. But you ask, and properly too: "What performs the function of a rudder, since a helmsman cannot operate without a rudder?" I answer: "Those four first qualities of things, namely, heat, cold, moisture, and dryness." Is there not an analogy here with the guidance of ships, that is, if the helm is turned farther than it should be toward one direction the course of the ship is changed? Does not the same thing happen in our human body? If some one of these qualities becomes greatly increased does not the entire body suffer wretchedly? Thus the meseraic veins which I have compared to small boats receive from the intestines by means of anastomosis or the unlocking of vessels that which wanders about in them gently and unobtrusively, being neither biting (acid) nor heavy. The meseraics are scattered about in the intestines like the roots of plants which suck food and nourishment from the earth. In fact, the stomach and intestines perform a function analogous to that of the earth for man and other animals. Before the meseraic veins reach the substance of the liver in the upper region they send such a multitude of veins into one very capacious vein which is called portal from its resemblance to gateways, pylai in Greek. This is well done by nature to prevent so many veinlets from perforating the substance of the liver like a sieve. The meseraic veins are those which converge into this one substance which the Greeks call mesenterion; they are interwoven with each other. The mesentery is an adenous membrane adapted for providing a pathway and a support for all these veins. It clings to the back and to the transverse septum by means of very hard ligaments in order to keep the weight of the intestines which hangs from this membrane from being torn out of its position. The vein I have called portal winds itself into the lowest parts of the liver where the final process of blood-making takes place because they are filled with very many arteries and veins which possess a large capacity for attraction, retention, alteration, and expulsion. The substance of the liver is quite thick since it is nourished by the thickest blood. It is very rarely divided into five lobes, more frequently into four, most frequently into three lobes. It suffers dissolution. I have seen a liver which was divided into only two lobes or fibers. Plato[23] has clearly indicated that the part of the soul which desires and which feels pleasure and sorrow is located in this organ, for if I remember correctly these are his words on the subject in both Greek and Latin: "As for the liver, which I have just mentioned, the third part of the soul is based within it; the liver lies deservedly between the umbilicus and the transverse septum. It has almost no thought, mind, or reason at all but shares

---

[23] Plato, *Timaeus* 70D–71B, considerably condensed.

the senses of pleasure and sorrow and of the desires which pertain to those senses." The liver has also a faculty of divination, if indeed we can believe Plato,[24] who says in the *Timaeus*: "Such is the nature of the liver, designed for the purposes I have mentioned. It is made and located where it is for the sake of divination." Nature has placed this organ in charge of making blood for the body because it has those very strong faculties already mentioned for attraction, retention, alteration, and segregation. Anyone, however inexperienced in these matters, could gather this information from the facts of its red color and its consistency. For what else is the substance of the liver, to tell the truth, other than hardened blood? It is nothing other than this, by Jove, even if you constitute the sense of taste itself as the judge and arbiter of the matter. For if you should taste both blood and boiled liver you would notice little if any difference between them. Nor should it be surprising to anyone if among the other inner organs the liver should be nourished with thicker and more impure blood, and therefore its substance is very thick, especially since I said just now that it is the workshop of all the humors just like some kitchen which prepares nutriment for the entire body. And just as it is not unreasonable that the cooks should eat what is more impure and less good than that which is most palatable and, in short, are more mean in every respect than those they serve and to whom they minister so the liver, which performs the function of a cook, serves the other more principal members of the body and cannot thus distinguish the overflowing excrements from the blood itself but drinks them up also along with the blood. But since the blood is most abundant in the liver the latter obtains its color from what is poured into the liver not unlike those who are forced to grind grain in a mill daily and grow white with the flour they grind out. Everything, finally, attains the color of the humor which is predominant in it. Thus if the liver is tinged with red, the spleen with black, and the brain with white you should not be amazed since they acquire various colors in accordance with the varied nature of the humors. It will not be difficult to discern this fact in those persons who suffer from the king's disease, that is, jaundice, which the Greeks call ikterikos. Because the pale bile wanders throughout the entire body a certain ineradicable pallor seizes upon the entire surface of the body. This organ [the liver] is suspended with many veins and ligaments on the right-hand side below the pericardium. It suffers along with the brain by means of one very slender nerve, nor did it require a larger nerve than that by which it could determine friend from foe. It is humped on the outside and with its hollows or lower parts it embraces the stomach whose actions it assists or strengthens in many ways. It is attached also to the spine and to the transverse septum. Furthermore it is necessary for blood-making, which the Greeks call hematosis, and separates the benign portion from the feces, the pure from the impure, the thin from the thick. Nature has provided this organ for the care of this necessary excretion. Just as must freshly pressed from grapes when quickly placed in a cask is fermented by its inner heat and the dregs are thrust down into the lower region while some of the lees float upward, so also somewhat the same thing happens in blood-making. Of the contents of the liver, whatever is thick and muddy is sent at once to the spleen by means of a vessel designed for the purpose; whatever is foamy and capable of floating is immediately relegated, so they say, to the gall bladder, which is appended to the liver like a sort of pool adapted for receiving the thinner elements which are cleansed away from the liver. This vessel mentioned above separates that yellow humor by way of a duct which penetrates to the lowest parts of the liver; hence it is rightly called from its function the cyst which receives the bile. From it the bile is sent forth, in some people to the jejunum intestine alone, but in others, although, as I said, very rarely, and these most unfortunate, also to the bottom of the stomach. More frequently, however, the bile is sent down to the jejunum alone; this fact is made clear to us by the emptiness of that single intestine. But if the bile duct is in some manner stopped up and hindered in its action all the bile which otherwise is poured into the intestines is dispatched to wander throughout the body and causes that ill which the Greeks call icterus and the Romans call the royal sickness [jaundice]. We can discern that the obstruction of the duct is rather solid if the excrement of the bowels comes forth white, for it is then deprived of that humor of the bile which usually gives it a yellowish tinge, rendering the excrement whitish. Sometimes stone is created in the gall bladder. This at length seems to give rise to the bending sickness, as happened to a certain nobleman this year whose body I dissected; so that all may know his name, he was called Monsieur de la Ville de Saint Riquier. His gall bladder contained only two stones. In other respects he showed no signs of jaundice since all his bile was voided through the intestines. In the same month as the one in which I made this dissection or autopsy another noble soldier was dissected at Étampes by the very learned doctor Tagault.[25] The former's gall bladder was

[24] *Ibid.* 72B.

[25] Jean Tagault, elected dean of the school of medicine at Paris in 1534 and holder of that office during Vesalius's stay in Paris, was the teacher and friend of Sylvius. He came from Picardy, took his medical degree at Paris in 1522, and published a text on

occupied by two stones that were like very large acorns, although it had heretofore aroused not the slightest suspicion of jaundice among the doctors. Hence I believe that it is now abundantly clear how necessary the bile is to the functions of nature even if it is not useful for nutrition. Nor should we listen to that famous man[26] (whose name I withhold for the sake of his honor) who insists that some part of the body is nourished by this superfluity of the bile. For what, pray tell, is more absurd, what, immortal gods, more foreign to philosophy, than to say that what nature separates as refuse from the blood and as particularly alien to nutrition is actually nourishing to the body? Yet that man dares to controvert the opinions of Galen, that man who is the filthiest of all men in regard to the art of medicine. He accuses Galen because the latter said the other humors are residues from the blood and hence wisely denied them the term nutriment. O the outrageous boasting of the man, O his enormous audacity! Or did you not consider it sufficient to spread darkness and shadows upon the minds of young men when you brought your insane opinion to their attention, that you should also most violently enrage that great father of medicine because he professed the truth clearly and brilliantly? Furthermore, in order that we may understand your unheard-of reasoning, why, pray, O most excellent man (for you must now deal with me) should you say the bile cyst is nourished with the bile but deny that the gall bladder is nourished by a serous fluid? For each is a residue favorable to nutrition to both latter and former vessels. You have added to your discourse, if that is what it must be called and not delirium, the statement that the brain, because it is cold and humid, desires nourishment by means of the most phlegmatic blood, since all things are pleased by that which is exactly similar to them. To argue briefly with you, if you say the brain is nourished by phlegmatic blood because it is the coldest and most humid part of the body, will you not also admit that old age, the coldest and driest among the ages of man, must likewise be nourished by what is colder and drier? And if you declare that the warmer part of the body insofar as it is warm is nourished by a warmer fluid or juice, by the same token will you not admit that the warmer age of man desires to be nourished by warmer food? The same argument which applies to the parts of the body should also apply to its various ages or times of life. Yet old age, because it is colder, is pleased with warmer foods, but youth, because it is warmer, desires colder foods; hence the brain insofar as it is a colder part of the body desires warmer nourishment. But, you say, if the blood alone should nourish the members there would be no dissimilarity among the temperaments. For bone is harder than flesh because of no other reason except that it is nourished by the melancholy and filthy blood. These are your words and the rest of your argument, the mention of which creates disgust. Have you no other proof by which you may persuade us to believe in your imagining? This one is unworthy to give pause even to a little child. For to whom is it not crystal clear that the constitution which all particles of matter received in that first conformation of the entire body have been preserved forever, or changed only slightly? The brain, since it was formed from the more humid parts of the semen itself and from the menstrual blood, seems more humid than the other organs. But the bones, made of a harder substance, appear harder among all the bodily parts. Then as long as the single parts of our body are nourished they change their nourishment in proportion to their various temperaments and render similar elements to similar. Thus the brain, since it is by nature cold and humid (I realize that by comparison with the warmer parts, to put it positively, as long as the animal lives the brain ought to be warm) converts the purest blood (the less pure being sent to nourish the upper parts, as may be seen in the flow of blood from the nostrils and in the rupture of the pericranium) into a certain phlegmatic substance and renders similar to similar, which could not be done by languid warmth. But why do I labor so much to explain this when the process has already passed into a proverb among old people, lest I furnish hot things to the hot or cold to the cold, lest I seem to add fire to fire or snow to snow? O noble fellow, do not therefore, while zealous to propagate your heresy, carp at any part of the beliefs of those who have so amplified this art of medicine deservedly called Apollinian from whom you have stolen completely whatever you have written. The brain, then, is nourished by the purest and at once the reddest of blood; in fact, something which will seem even stranger to you who say that like must always be nourished by like, the spleen is nourished by blood far purer than that which nourishes the liver. I shall now show how this is done. Imagine first someone who is separating husks and chaff from wheat with a sieve; he turns the residue over to someone else for re-examination and further sifting. He could not on the first try separate the wheat from the useless parts so exactly and completely as to prevent some chaff

surgery, *De chirurgica institutione libri quinque*, (Paris, 1543). He died in 1545. See C. D. O'Malley, *Andreas Vesalius of Brussels 1514–1564* (University of California Press, Berkeley, 1964), p. 425, n. 27.

[26] Benedetti, *Anatomice*, I, 4, on the qualities of humors, does say that the two biles nourish the members, and since he is mentioned by de Laguna shortly in a derogatory manner Benedetti may be the unnamed adversary here; the alternative is some colleague at Paris, where de Laguna studied and where he published his *Anatomica Methodus*.

and foreign grain from remaining mixed in the wheat, nor could he thrust down that refuse so carefully that some of the purest grains did not slip downward with it. Thus he to whom the task of resifting the wheat is committed separates grains one by one with his fingers from the refuse. It will be a longer task but one not to be regretted, for he will gather purer grains than if he used a sieve. Let the liver serve therefore in this analogy for him who holds the sieve; the spleen thus corresponds to the second person who is charged with resifting the grain. For the liver cannot so cleanse the blood from other impurities that many do not remain in it, mixed with sooty residues. Nor is it reasonable to presume that the black bile should flow to the spleen so carefully separated that some pure blood should not flow down with it. Hence the spleen, since it does not obtain its nourishment from any other source than from this melancholic sediment which has been very carefully elaborated, separates in its veins the benign from the rotten refuse, the pure from the impure substance, with which it is later not unhappily nourished. That which is left after this elaboration or cleansing and which cannot be further elaborated is truly called the black bile. It is useful to the actions of nature for, carried to the stomach above, it constrains that organ and irritates it to appetite. But that part of it which is not useful to the stomach, or which has already performed its function, is carried off through the lower bowel. The substance of the spleen is soft, thin, and fistular, so that it is no wonder that it absorbs that thick humor like a sponge. The fact that the spleen was created solely for cleansing the blood may be deduced particularly from the fact that it is present in blooded animals alone and in none of the bloodless.[27] It is suspended by a thin ligament near the left region of the liver, not from the diaphragm but from the stomach. This suspension is the reason why it is frequently a hindrance to runners since the motion of the body causes it to slip and swing now to the right and now to the left. The bile sack in a similar fashion was planned by nature for purifying the blood although it is often missing from some men, and these quite healthy individuals, as Pliny records.[28] Now when the blood itself has been cleansed, that which had been prepared from that humor is carried through the meseraic veins like roots to the lowest parts of the liver, and from the opposite region of that organ which is called the gibbous portion a very large trunk of the veins is quite skillfully extended; this later spreads out into numerous branches. On this account even Aristotle himself truly reveals himself to be a mere man and simply wandering in mind when he held that the origin of all the veins was not the liver but the heart.[29] Nor is that Alexander Benedictus quite worthy of pardon. Although he boasts that he searches out all truths with method nonetheless that dissection of the human body[30] which he published is thoroughly *désorganisé*;[31] those who have taken up the art of medicine ought to avoid it as they would avoid all chasms and whirlpools. Thus the vena cava springing from the liver itself carries the blood, as I said, to the entire body, passing on both sides of the liver and moving up through the transverse septum to the heart and thence to nourish the upper parts of the body. It is as firmly thrust downward to the spine of the back above the large artery which arises from the left ventricle of the heart. Just as that part of the vein which moves upward brings food and sustenance to the upper regions, so that part which moves downward nourishes the lower regions. Moreover, on account of that necessity which I mentioned a certain serous humidity always accompanies the blood itself as a sort of vehicle of the blood. Nature very ingeniously joined the two kidneys to the vena cava to purge this serous humidity when it has performed its function as a vehicle. Since these kidneys are endowed with as great a power of attraction as any part of the body each of them takes over from its vein and cleanses whatever thinner matter is found in the blood. Furthermore, to prevent any quarrel from arising between the kidneys or any envy from being created should they encroach upon each other—for rarely do those who practice the same profession agree, especially if their houses face each other, as we may see particularly among physicians among whom (to say nothing of druggists and barbers) there exists such a great rivalry, jealousy, and lust, a very grievous fact, such as one could not easily find even among cobblers or other low artisans—in order that, as I say, the kidneys should not rend each other like gladiators with a similar rivalry, one was placed on the right side as the nobler region and slightly higher, while the other was placed on the left and lower down, so as to be of humbler position. However, at Paris I saw a man who had only one kidney. The veins which project from the vena cava to the kidneys are called the emulgent veins by

[27] The description of the spleen is similar to that in Benedetti, *Anatomice*, II, 12. See Aristotle, *De Partibus Animalium*, III. 4. 666a25; 6. 669b25; 7. 670a30, b30; *De Historia Animalium*, II. 15. 506a10–15; also Benedetti, *ibid.*, on the reference to runners; Vesalius, *Fabrica* 511a, attributes the statement to Erasistratus.

[28] Pliny, *Natural History*, XI. 74. 192.

[29] Aristotle, *De Historia Animalium*, III. 3. 513a20.

[30] The *Anatomice:* see my translation. There is no real evidence to support de Laguna's gratuitous insult and no reason why he should single Benedetti out for criticism among the pre-Vesalian anatomists.

[31] By the French word I attempt to reproduce the effect of the Greek word *amethodikotatos* used by de Laguna, who loves to display his knowledge of that language. In this respect he resembles Benedetti, whom he scorns.

physicians. Certain small net-shaped skins receive the heads or extremities of those veins in the lowest parts of the kidneys. In order to achieve small openings or anastomoses these little membranes (or skins) yield quite easily to the thin serous substance flowing through them but not in the same manner to the blood, since it is thick. When this serous humor penetrates the ventricle of each kidney it distills rapidly through one of the ducts which originates near the emulgent vein into the bladder below, if indeed there are two ducts or meatuses of this kind, one corresponding to the right kidney and the other to the left kidney. Both Greeks and Latins call these ducts by a common name, ureters. Stone tends frequently to be generated in the kidneys, that is, that stopped-up skin through which the urine customarily percolates from some thick hillock (*grumo*) hidden away in the ventricles (or sinuses) of the kidneys. This results particularly from the hardening effects of heat, nor can so great a cold dominate the region around the kidneys so as to congeal the blood but not to kill the individual. Very often also a large part of the blood drips down together with the urine when that membrane is relaxed and laid open. Since all things in our bodies correspond to the order of nature the membrane readily offers itself to the thinner humors, that is, the serous fluid and the bile. For just as in schools or monasteries there is a faithful doorkeeper who does no more than open the door to servants and ministers but detains them there according to usage and rule and allows no one to go out unless the latter makes a very firm effort to do so, thus also in the veins of our body it is easy to notice that the benign blood for nourishing the parts of the body is very vigorously detained as it were in a prison or workhouse but the other thinner humors, after they furnish no more use to the body, are sent off to the bladder. This membrane which is woven about the entire kidneys performs the function of such a doorkeeper. If it is not loosened and relaxed by a strong heat or corroded by a sharp humor or lacerated by the force of a thicker humor, the bile alone with the serous humidity will percolate through. Since the substance of the kidneys is hard, thick, and fibrous or sinewy it is properly nourished with pure blood. This blood sheds that sharp and serous humidity as soon as it reaches the region of the kidneys and finally is spread out through the substance of the kidneys. The kidneys are attacked, if any part of the body is attacked, by many and very severe diseases. All these diseases are lumped together under the term nephritis, from the part affected, although true nephritis is only an inflammation of the kidneys. Nature has made them so thick and dense in order that the blood may not be distilled (as I said) with the serous portion itself. Experience has proven and reason persuades us that the arteries are implanted in the kidneys in a series similar to that of the veins since it is possible to find in them not only spirit but a thin and vaporous blood which, however thin it may be, still requires a purgation of that serous humidity which it contains. The urine now secreted from the kidneys by means of the ducts I call the ureters flows completely into the bladder. These ducts are implanted in the stomach or neck of the bladder in a slanting manner by the great cleverness of nature. If some of the membranes which lie near the inner part of the bladder are dilated an entrance is easily made for the humors which descend to it. If these humors are regurgitated and move back upwards under some great force, these valves or membranes by coming together prevent even a little of the humors from passing back to the kidneys. Many, however, have doubted whether the urine distills in this way into the bladder; this doubt has surely not been justified by dissection since all the ducts become occluded owing to the cold air to which they are exposed so that these valves and foramina completely escape detection by the senses. In order to confirm by experience what I had conceived in my imagination, I placed a rather long probe through the ureters and dilated the closed meatus without much trouble so that no further occasion for doubt remained. Furthermore, the bladder itself consists of two very cold membranes; if these are injured they can rarely or never be consolidated or healed. It is large in proportion to the other vessels of the body so that it may be the receptacle of the most abundant humor. When this humor has attained such a great quantity that the bladder is rendered heavy and distended it is sent down by a duct or meatus to the penis as this is stretched forward. In that part of the neck which is nearest to the bladder there is found another muscle or pylorus such as there was in the stomach especially to check the untimely flow of the urine. But since in those recently born a muscle of this sort is humid and quite soft, hence languid and weak, the newborn are often troubled with stillicide. The bladder is nourished, as I said, by the purest blood in the same manner as the other remaining parts. However, I have already spoken much about the bladder. Now I must come to the testicles since they take the blood from the vena cava and the emulgent vein for making the prolific semen. If the testes were not enclosed, since men are warmer than women,[32] the testicles would grow warm with an excessive heat, and since they are the warmest members of the body, nature caused them to hang down so that their extraordinary heat might be

---

[32] *Cf.* Aristotle, *De Generatione Animalium*, I. 19. 726b30; II. 4. 738a10; *Problemata* IV. 880a10, *cf.* 879a25.

fanned by external cold. In women, however, since they are colder than men, nature placed the testes near the matrix. In males the testes, which are called didymi by some people, are each covered with tunics. The inner tunic, thin and white, does not have even a proper name. The external tunic, however, is called the scrotum; it is wrinkled, oblong, and marked with a distinct suture. This tunic can be stretched to such a width that the entire bulk of the intestines can be surrounded sometimes within the sinus of the scrotum when the intestines descend due to the rupture of the peritoneum. The Greeks have called this illness by the term enterocele although the Latins call it hernia or lapsus. Four vessels are inserted into the testes, two veins and the same number of arteries. The veins carry the substance of generation, that is, the semen. The arteries carry the vital spirit made into semen to be ejaculated with force into the matrix of the female. The vessels which come to the left testis take their origin from the emulgent veins that are carried to the left kidney. Those which reach the right testis derive beyond doubt from the vena cava and the dorsal artery. For this reason the right testicle must be warmer than the left and the right side of the matrix warmer than the left side, especially since the more impure and serous blood always distills into the left testicle. Hence it is that shepherds who are zealous about their flocks sometimes support the right testicle with a band or cord and sometimes the left testicle, in order to vary the offspring of the sheep. The fact that the right side, that is, the warmer part of the matrix, is more suitable for creating male progeny is fully vouched for by that oracular saying of Hippocrates:[33] "Male embryos occur more frequently on the right side and female embryos on the left side." This can clearly be seen in pregnant women, for if their right breast becomes weak they will suffer abortion of a male embryo since the breast has a very firm connection with the right side of the matrix. But if the left breast becomes weak, since it communicates with the left side of the matrix by means of a group of veins that side will eject an aborted female embryo. Furthermore, those veins which carry the substance of generation to the testicles do not enter their glandular substance, a fact not evident to anyone who has not dissected the testicles with a scalpel, but are scattered throughout their surface circumference in a varied twisting and turning. Within this plexus of veins that blood which has been poorly prepared and transmitted by anastomoses to the glands attains an uncommon whiteness. You should not be amazed if the very red blood can pass into a very white humor since even pharmacists can with great skill make an exceedingly white liquor from any amount of yellow honey by means of heat. But there he is with his objections, that wordy, babbling sophist! What are you saying, O Peripatetic? Is not heat the mother of blackness and cold the mother of whiteness? At any rate, I hold, O sophist, that heat is the mother of blackness if that heat exceeds the measure of nature. For if you subject bread or a pear or honey or sugar or anything else to fire and leave it there a rather long time each of these substances will quickly turn black. This statement is proven by the Ethiopians, who grow black under the strong and very ardent sunshine of their climate. But we dauntlessly refuse to believe that a moderate heat can accomplish the same effect. You should not be surprised, O sophist, if the heat of the testicles which is gentle and vaporous, tempered by the cold of the adenous tissues, generates not black but white humors. The coldness of the glands is also no small help in obtaining that whiteness which is evident to all in the breasts. The proof that the genital semen is produced from the entire body is greatly strengthened by the fact that the immoderate use of sex causes the entire body to grow languid. The man who has become practiced in anatomical dissections and examinations will scarcely hesitate to tell you how this comes about. For just as among populous and most distinguished cities there is a public treasury in which the common goods are stored and if anyone robs it he robs the entire people, so also you may regard that hollow vein which spreads out on either side to the back as a kind of repository and storehouse for the entire body. If any part of the body should filch blood away from this vena cava it will certainly filch it away from the entire body, namely, from the head, the heart, the liver and indeed from the other parts since all these, of course, seek their nourishment from no other source than the vena cava. If this becomes empty the other parts have no way by which they may obtain their livelihood. In immoderate sexual relations as much semen is emitted so very much blood disappears, for in sequence to that which is emptied a very copious supply of blood is always distilled from the vena cava to the testicles since two ounces of blood scarcely suffice to make a dram (one-eighth of an ounce) of semen. For the blood is reduced to a much smaller portion after it has steamed with warmth and has attained its final elaboration, not otherwise than if (as the philosophers say) you wished to make ten handfuls of fire from a single handful of air. The reason is that when the semen reaches the testicles in sexual relations men feel a certain voluptuous sense of lassitude around their loins and kidneys. This occurs when a supply of copious blood flows to the testicles from the vena cava in order not to frustrate their

[33] Hippocrates, *Aphorisms* V. 48 (Loeb Classical Library IV, p. 171).

action by a weakening lack of material. Men who are possessed by an insatiable lust feel also a great weakness of the brain since the testicles suck out and plunder the brain's purest substance by means of nerves. This fact has been revealed by a very lustful person whose skull was dissected after his death and found to contain a small and very dry brain. The eyes also suffer pain owing to excessive venery because of the connection of their nerves with the genitals. That some nerves are inserted into the testicles you can learn from epileptics, whose testicles are so compressed owing to the wretched convulsion of their entire bodies that they are finally forced to ejaculate semen beyond the control of their will. In fact, these nerves, according to Cornelius Celsus,[34] are entirely without sensation. But if venery is indulged in moderation it is quite helpful and none of these ills occur. For (as Paul of Aegina testifies in the course of the first book)[35] sexual relations dispose of superfluous humors, render the body agile, promote its growth, and make it more virile. As far as the mind is concerned, they dissolve hindrances to the power of understanding and soften those who are angry. Hence coitus, if anything can be, is a most suitable remedy for those who are melancholy. They likewise restore complete command of their senses to those who have gone insane. Venus is most helpful to those who suffer from too much phlegm. Others who vomit up their food have had their appetite restored through indulgence in sex. Others have been freed from continuous dreams of sex by indulging in it. This is what Paul of Aegina says, at any rate. In order to understand all these statements most happily, it is better to be Tarquin the physician than that very chaste Xenocrates,[36] who was enticed by neither beauty nor blandishments nor by the tickling allurements of a prostitute. Moreover, if the testicles are removed from men they lose their strength, vigor, beauty, voice, and beard, and finally their very talent and manners together with their private parts since all these qualities and possessions are completely created by that thin and steamy heat which is responsible for the semen. Hence castrates, who lack prolific semen, have no beards, for you know that the cold and humid nature with which eunuchs are endowed is quite hairless. A humor that is a kind of matter or material is beyond doubt required for creating hair and beard as well as a thin heat which drives the humor to the weaker parts of the body. This heat is not to be found in castrates. Therefore the humors in them which are destined for the chin, the armpits, and the pubis are closely concentrated and often fall down to their legs, whence it happens that most eunuchs are accustomed to suffer from gout. But what of that saying of Hippocrates:[37] "Eunuchs are not afflicted with gout." Shall I recant? Or shall I reject what the experience of my eyes teaches me daily? Or shall I, in this matter, consider the opinion of so great an authority as worth no more than a straw? Or shall I rather follow the interpretation of this passage by Galen and Oribasius,[38] that in the time of Hippocrates eunuchs were not susceptible to diseases of the joints on account of their temperate and moderate manner of life but that at the present time because of their great sloth and ease they are subject to such ills? By no means. What then? I must give my opinion as follows: Hippocrates, in that aphorism, has stated the same view as Aristotle's, set forth in *Problems XXXIX*. 10 [Book X. 37, Oxford Aristotle VII]. Aristotle's statement is as follows: "Why are castrates completely free from varicose veins or have fewer than other people?" and the rest that follows in the *Problem* quoted. In fact, my opinion is especially confirmed by a reading I have found in a certain manuscript; the Greek phrase when translated runs thus: "Eunuchs are not troubled by varicose veins." I am sure that no one is ignorant of the meaning of the word *varices*. It is used particularly of the veins in the temples and the legs which swell up with much blood. For this reason I see no objection to the statement that castrates do not suffer from gout and likewise do not have varicose veins, since gout results from thick humors which flow down into the joints of the leg and its other parts. Varices are the veins spread out in the surface of the legs which become swollen with very abundant blood. So I came to believe that Aristotle had taken this particular *Problem* from Hippocrates, whom he has followed entirely in this passage. Although the female testicles are inside her body they are endowed by nature, however, with little semen; and it is much colder than in men. The male semen is not sufficient by itself for generation unless it is mingled and poured together with the female semen, that is, for composing the offspring. Indeed, neither semen could avail in any way for generation unless the menstrual blood was provided as a sort of material for the creation of the young. For the semen possesses

[34] Cornelius Celsus, *De Medicina* VII. 18.

[35] Paul of Aegina, ed. I. L. Heiberg, *Corpus Medicorum Graecorum* IX. (Leipzig, Teubner, 1921), p. 24.

[36] Probably Xenocrates of Chalcedon, a pupil of Plato, head of the Academy from 339 to 314 B.C., although Xenocrates of Aphrodisias, a physician of Nero's age and that of the Flavians (54–96 A.D.), may be meant. He is mentioned by Galen in *De Simplicium Medicamentorum Temperamentis* etc. (XI K. 793, XII K. 248–250, 258–261) and elsewhere.

[37] Hippocrates, *Aphorisms* VI. 28 (Loeb Classical Library IV, 187).

[38] Galen, *Hippocratis Aphorismi et Galeni in Eos Commentarii*, VI, 28 (XVIII. K. 40); Oribasius, *Oeuvres*, III (ed. Bussemaker et Daremberg, Paris, 1858), pp. 44–46, where Oribasius discusses the effects of castration but does not mention eunuchs as such in the same terms as Galen does.

chiefly the analogy of form.[39] If a woman is not pregnant and is still young she normally expels each month an abundant supply of blood through some very large veins which nature has opened into the matrix. If she has conceived, all that blood, otherwise to be cleansed again at the appointed time, is consumed in the nourishment of the foetus. Hence it is that the menses are forbidden by nature to women who are pregnant. If they flowed this could result only in great harm to the foetus. Nor should you be amazed if that blood, benign in its nature since it is defective not in quality but only in quantity, is expelled from the uterus, impure and smelling quite strongly, if indeed every humor, however pure it may be when it leaves its own vessel, becomes corrupt and thin. At all events, in regard to the female pudenda, I cannot easily repress a smile when I see Galen stoutly contending that they are formed from the male genital parts, something which not even nature, the creatrix of all things, could ever accomplish. Since there are just as many separate parts in the female sexual organs as there are in those of the male, he says the female pudenda are the complete inverse of the male parts and that they have fallen back in the inner region of the female. Both sets of genital parts are endowed by nature with a sexual itching and a sort of libidinous tickling by which we are provoked to desire. Whether or not this itching is caused by some serous humidity, as Galen thought, I should not dare to affirm clearly; I would rather believe that it is due to a flatulent exhalation of the semen. For after the testicles become swollen with very warm semen and are much distended with it they wish to be rid of so great a burden. Therefore they first exhale a warm and steamy vapor to the penis, which is quite full of arteries and nerves. Since this vapor is flatulent by nature it distends the penis and irritates it with such a tickling sensation that the semen is propelled quite easily shortly thereafter. This process reveals also how from day to day women are roused to desire by itching or tickling. We have the support of Plato's[40] opinion for this apparent state of affairs. He says: "The animate [marrow] creates a vital desire for exit in that part where it finds an outlet and thus the desire for procreation has been placed therein." For nature knows how much the human race is disposed toward sloth and indolence; if there were not in sexual relations a certain pleasure as a kind of a reward no one would take the trouble to perform the act, which is something half-divine. Hence in order to prevent the complete destruction of the human race through folly and neglect that supreme cause which governs these lower beings with its nod has endowed men with a libidinous pleasure in sex as a reward and remuneration for their labors in this direction. The genital parts are aroused by this pleasure just as animals are. Plato, well aware of this fact, writes as follows in the *Timaeus*:[41] "For this cause a natural force is present in men, unruly and imperious, and like an animal which will not obey reason the violence of its furious lust seeks to subdue all to its will. The vulva and matrix in women, in the same manner as an animal eager for copulation, if it abstains unseasonably long from sexual relations, withstand the privation with difficulty and are greatly angered." These are Plato's words. The veins in the testicles which I said make the semen in their intertwinings after a long revolution insert themselves at the root of the penis, doubtless near that region which receives the urinary meatus. The top part of the penis is called the acorn, owing to its similarity. The portion which covers it is deservedly called the prepuce. The sinus of women is called uterus in Greek; the caruncula which lie on either side of it are known as nymphae to the Athenians, but the Romans are accustomed to call them little hills (*colliculi*). The entire space hollowed out between the carunculae and the matrix has received the special name of neck. The sinuses of the matrix, into which open, as I said, some very large veins that carry blood, all the physicians have called the wings or horns of the matrix. Among all these parts, however, the carunculae just mentioned possess the keenest sensation just as does the glans of the penis because of the quite dense implantation of nerves into these parts. Especially for this reason they are endowed by nature with every kind of itching and tickling. Having now completed my discussion of the genital parts and examined them in careful detail, we should note that the portion of the vena cava which descends to the lower region supplies those parts with food and nourishment and as it were the material for their operations. It now remains to discuss in similar fashion the other branch of the vena cava which stretches upward since no one doubts that the muscles which constitute the buttocks and the muscles of the legs receive their nourishment from the lower branch of the vena cava. I shall devote a special comment to all of these items as also to the bones which are covered by these muscles. Let me now turn to this purpose. Before I come to the subject, however, I must say something about the diaphragm or transverse septum, which separates the natural force of the life-spirit (*animae*) from the vital and animal forms of it, since the occasion has now arisen. For nature did not consider it sufficient to restrict the most unimportant part of the life-spirit to the lowest region of the central body cavity without placing this septum between it and the

[39] See Aristotle's discussion of the function of semen in *De Generatione Animalium* 4. 738b20–35, upon which de Laguna draws.

[40] Plato, *Timaeus* 91B; cf. 73c, 86c, where the marrow is discussed as the source of human generation.

[41] *Ibid.*

upper region in order to keep it from interfering with the loftier operations of the life-spirit. What I call the transverse septum is a single muscle; since it divides a man crosswise it is named correctly. From its function it is called also perizoma or girdle by some people; the more common term for it is diaphragm, the name given to anything which divides something else, from diaphrattein, which also signifies "to separate." Galen, however, very often calls this member *phren*, as does Plato, signifying "mind," since when this part is injured men are quickly seized with delirium and paraphrosyne [Greek for madness]. This situation led many of the Greeks into a stupid confusion, so that they concluded the mind and the reasoning part of the soul were located in the diaphragm. Many have called this instrument the praecordia [that which is in front of the heart] because it was quite close to the heart and a servant to the latter. It makes no difference what word you use for this part of the body as long as you understand that it refers to that organ which separates the stomach and liver from the heart. This transverse septum performs its greatest function for respiration in whose sole service it is in constant motion. It also does a great deal toward the evacuation of the bowels, for without its aid the expulsive force of the hypogastrium and intestines would be quite weak and would propel the excrements as they overflow with no more power downward than upward, as I have said. To use a familiar comparison, the transverse septum is like the fingers of a hand which presses all the excrements downward so thoroughly that it does not allow even the slightest breath from their contents to float upward. Nor are the regions of the liver and the heart so rigidly separated and delimited by the diaphragm that the powers of these members cannot pass hither and beyond like traveling merchants. Since, as I said, the liver is by far the most abounding in impure sediment of the three viscera I wish you to regard it as that grand Turk, who holds the most dangerous dominion of them all, and the heart as the great Pope, who, since he holds the middle place among all Christians and is so splendid an authority, bestows his favor upon all equally with a rare eagerness, just as the heart bestows vital spirit to all parts of the body. The third of the viscera, the brain, in which the chief power of the soul is established, you may deservedly compare to our most distinguished emperor. Just as these three lands Greece, Italy, and Spain, are not limited by their boundaries so that their inhabitants can pass to and fro among them (for they travel for the sake of commerce, a good many Turks to Italy just as a great number of Italians travel to Spain and likewise Spaniards to Italy) the liver and the heart are not so far apart that a certain very broad vein cannot pass, like a very rich traveling merchant, from the region of the liver to that of the heart. The liver sends forth blood, in which it especially abounds, so that through an exchange not at all regrettable it may receive by arteries from the heart the spirits from the lack of which it suffers. In a similar manner, of course, the heart sends vital spirit to the brain through arteries so that later it may draw from the brain by way of nerves the animal spirit which is by far more valuable. If only the king of Portugal knew how, comfortably and pleasantly, to send leaden swords and clay pots to the Ethiopians so that his ships might return to Portugal, loaded down with the purest gold! This analogy corresponds to the brain and the heart. A very prominent branch of the hollow vein ascends straight through the middle of the diaphragm to the empire of the heart, slanting a little to the right-hand side of the thorax in such a manner that it winds its way into the right auricle of the heart. Furthermore, since every individual is constituted according to a principle of balance the heart occupies the middle region of the thorax although it may seem to verge more and more to the left side on account of that frequent palpitation of the left-hand cavity which seems to strike the left side more than the right side. The heart has only two ventricles, a right and a left. I do not know what is the meaning of the riddle proposed by the people who add a third ventricle to the heart unless perhaps they intend by it those pores which are found in the septum. At any rate, the heart, since it has no blood from itself, receives it in exchange from the hollow vein through the auricle of the right ventricle. Thus transferred hence into the left cavity of the heart, the vital spirits are made [from the blood] and finally conveyed through arteries into the entire constitution of the body. These spirits warm the cold parts and cool the warm parts with their fanning effect. Since the heart is the principal organ of the body and because it is the first to live and the last to die, we must pardon Aristotle to some extent for believing that the chief power of the soul was located nowhere else but in the heart. If indeed from the heart alone rise anger or passion, fear, terror, and sadness; if from it alone spring shame, delight, and joy, why should I say more? In fact, almost all those perturbations which the Greeks call pathemata have their origin in the heart. Anger fills it with a copious supply of blood and spirits boiling in the ventricles, ready to rush forth with a great impulse as if to wound and revenge itself upon the object of wrath. Fear and trepidation, which have to do with generation [and the preservation of the species], exhibit themselves in a contrary manner. For just as in wrathful and furious men the blood, grown extremely fervent, spreads out through the extent of the body, so in those who are stricken with fear the very warm humors are gathered together in rivulets in the body as though in flight from external dangers and are borne to the royal palace of the heart, leaving the exterior parts of the body frigid and pale. In a similar fashion at the siege of Rome [in 1527] while the city was shattered to its foundations, the bishops and cardinals fled in a closely packed huddle to the powerful citadel of Sant' Angelo. Hence it is that often many people who are seized with great fear cease to live since the innate heat of the heart is choked off by an abundant flowing-together of the humors in one place. Those who are afflicted with grief suffer much the same say as those who are frightened to death because sorrow is like a fear of what is going to happen or sadness for what has occurred, which has left some calamity in its wake. Furthermore, joy and excessive laughter are accompanied by an effusive and exultant mind while the blood and the spirits are drawn to the exterior of the body, just as in an access of anger, although for a different reason. In fact, those who are angry, irritated, and exasperated are affected with such a violent disturbance that they behave as if carried beyond themselves with delight and joy. Sometimes they cannot restrain their pleasure and finally die most agreeably and foolishly, as happened to Chilon of Sparta.[42] For while he was embracing his son who had been crowned for his Olympic victory, Chilon died of immoderate joy. Pliny VII, 53[43] reports the same thing about that marvelous poet Sophocles who died of a great and uncontrollable delight when he received news of his triumph in the contest of play-presentation. Philippides, the comic poet, suddenly passed away when he won an unexpected victory in a contest among poets, as Gellius reports. In all these instances, of course, the entire heat of the heart was cut off. I could present numerous other examples were it not for the fact that I might appear to be writing history instead of medical theory. Hence in order to complete my demonstration that all such affections have their origin in the heart I should like to add, with your permission, a few words about another disturbance of the mind which is called shame or modesty. We blush or are affected with modesty when we are praised and commended too scrupulously, although very rarely, or when we commit some shameful or indecent act for which we are marked with ignominy. The cheeks of those afflicted with shame are immediately suffused with an uncontrollable blush, as though the heart, like a most severe judge, were propelling those warm humors which invite the mind to the commission of evil deeds and listen not at all to reason as far as possible from itself and sending them into exile as punishment for their sins. The heart does not admit these humors into its sinuses before they grow cool a bit in a place of correction and become subservient to reason. Aristotle[44] writes that shame is revealed in the eyes alone because those who are shamefast cannot look at anyone. Although it is something extraneous and may seem foreign to my purpose I cannot pass over in silence the witty and most charming jest of a certain Roman painter. When he had painted the images of Saints Peter and Paul so red in the face that they seemed almost to be sweating blood he set them up in his studio for sale. Two very rich cardinals walked down his street and saw the images the painter had thus distorted. "Why didn't you paint them with a normal color on their faces," they asked the painter, "instead of making them so very red? Did you slap on the paint with a bucket instead of a brush?" They continued to ridicule him with other insults. The painter replied: "Don't be astonished if I have sketched Peter and Paul in this manner; the saints above are suffused with a blush much more red than this." "How did you arrive at such a conclusion, O excellent geometer of the vineyards?" the cardinals questioned in turn. "O wearers of the purple," returned the painter, fixing them with a stern gaze, "I'll tell you. These saints are blushing because of their immense shame, for they see their Roman Church, once ruled by the Apostles, now governed by good-for-nothing dissolute rascals like yourselves, and that's why their color has changed to this deep red." This is how that most impudent painter made up a story about the blushes of the Apostles. The substance of the heart is fibrous and quite fleshy. It has all kinds of fibers required for its operation. When it draws the blood very briskly through that vein which is inserted into the right ventricle, the heart uses straight filaments. It holds in the blood by means of the oblique filaments (or fibers) and by means of those valves which are attached to each auricle so that the blood cannot run back to the vena cava. But if any sediment appears in the blood itself the heart disposes of it at once by means of its transverse filaments. That thick blood is not elaborated in the right ventricle but after it has acquired a certain moderate and preliminary condition, a part of it is carried off through the arterial vein to nourish the lungs and that which is left percolates into the left ventricle through that intervening little septum (or membrane) for the purpose of preparing the vital spirits, for the membrane is porous. When these spirits are prepared the nature of the heart sends forth life and ventilation by means of them through very strong arteries to the entire body and through a very slender artery to the lungs. Notice here the artifice and handiwork of nature. She extended only two vessels from the heart to the lungs. One is an arterial vein and the other is a venous artery, that is, made with a single tunic. She knew that the substance of the lungs is very rare and thin and therefore it required a very gentle and slender nourishment. Unless the vein had been made with one tunic the blood would flow still quite thick and fatty to the lungs. Because of the thinness of the membrane

[42] Pliny, *Natural History*, VII. 32. 119.

[43] Pliny, *Natural History*, VII. 53. 180; Aulus Gellius, *Noctes Atticae*, III. 15. 2.

[44] *Problems* 31. 3: "(for shame resides in the eyes) so that they cannot face one" (translated by W. S. Hett, Loeb Classical Library).

nature added another very strong one so that the blood would not flow freely but as though through a very narrow alley and rendered very thin and rare into the lungs. For the purpose followed in making the venous artery I must supply another explanation. In order to avoid any risk or peril in regard to the thick blood in it, so that it might more easily minister to the movement of the heart and lungs and pour back more copious spirits to the lungs this venous artery was made of one tunic only, and that one very thin. Through this artery the heart draws in the spirit elaborated in the lungs for the nourishment of the vital spirit and is not unmindful of that great benefit for by way of that passage whence it draws cold air it sends back as remuneration the warm spirits. An unusual controversy has arisen on this matter. Since only two vessels are extended from the heart to the lungs, one the arterial vein through which thin blood is carried to nourish them, the other being the venous artery by which in the systole of the heart (as Galen himself admits) the vital spirits are dispatched, the question arises as to when or through what place the smoky residue is expelled from the left ventricle of the heart to the lungs themselves. Not through the arterial vein, for the blood alone is sent forth in that vein. Perhaps you say through the venous artery: but it is easy to prove that this is not the channel either, neither in diastole nor in systole. Not in diastole, for at that time the heart dilates and draws in benign air and does not send it forth. Not in systole, when the heart contracts, for then it sends vital spirits to the lungs: hence the heart never [expels the smoky residue]. One must consider that the cold air is very carefully elaborated first in the lungs before it comes to the heart. Since it is very pure after its elaboration it straightway penetrates into the vital spirits so that little or no residue of it is left. From the lungs, however, after this cold air is elaborated or altered by them, this smoky exhalation is expelled; this is the residue of that very benign spirit which was sent to the heart. If you do not approve of this explanation you will no doubt say one of two things: either that more than two vessels are extended from the heart to the lungs or that the heart sends the vital spirits to the lungs by way of this venal artery and that the residue is removed by the same route with those spirits, an explanation which is by no means in accordance with the order of nature. One must likewise note that the heart is not nourished with the vaporous blood elaborated in the right ventricle (for so subtle an aliment, so steamy, would not be suitable for such a thick substance as that of the heart) but by a much thicker blood which distils in greater quantity from the little branch of the vena cava which is attached to the top of the conical point or crown of the right ventricle. The heart elaborates that blood in the right ventricle both for the nourishment of the lungs so that the latter may more freely furnish cool air to the heart as well as for the generation of the vital spirit which originates in the heart. Furthermore, the lungs have no genuine motion of themselves any more than a set of bellows; the latter are dilated by the hands, the former are inflated by the intercostal muscles. Nor are they even inflated or dilated, because the air rushes into them but because they are dilated therefore the spirit or breath is largely precipitated into them also. When the air is at length elaborated in the lungs, it drives back and cools the most ardent heat of the heart. It should not surprise you to learn that sometimes it is warm air that cools the heat of the heart for often that which is less warm cools that which is warmer. The heart is covered in addition by a certain sinewy tunic which the Greeks call pericardion which surrounds the heart like a very strong wall. This tunic or membrane is always found to be full of a tepid moisture lest the heart wearied by its powerful, enormous, and continual motion should dwindle away wretchedly in a state of dryness. The heart is refreshed therefore and kept wet by that very clear moisture which it has near itself. Pliny[45] does not blush to assert in XI, 37, that the mind of man is located in the heart because it is the single human organ which does not waste away and which dies at once when it is wounded, as is well known, according to him. Similarly Chrysippus also declares that the natural and reasoning soul is established in the heart. Galen,[46] in the early books of his account of the opinions of the philosophers, with a certain rare dexterity demolished this frivolous view of the matter. For neither the natural nor animal faculty has its location in the heart since far more suitable members have been provided by nature for the accommodation of these powers. Furthermore, the heart has a double motion, one of its own always in it *per se* and kata tēn physin [in Greek], the other foreign to it and (as they say) joined to it by some accident. I call a motion of its own that which does not require the aid of the brain for its creation. This is precisely the motion of the arteries, which seems properly to be caused by the ebb and flow of the vital spirit through the arteries, when all the spirits regurgitate to the heart. When, however, the heart is contracted there is diastole or dilation of the arteries because at that time the spirits return again to the arteries. The brain appears to furnish no service at all to this motion of the arteries, which is peculiar to the heart. If there were some such assistance from the brain it would be especially by sending animal spirits down through the nerves to cause systole and diastole; but it is easy to show that this assistance does not take place. For if you should tie off or bind all the nerves and arteries which are common to the brain and to the heart, nevertheless systole and diastole would continue in the heart. There is also (as

[45] Pliny, *Natural History* XI. 37. 69. 182: ibi mens habitat.

[46] Galen, *De Placitis Hippocratis et Platonis* (V K. 181–805), especially 308–312.

I said) another motion of the intercostal muscles by which the heart is moved to some extent at the same time. This motion originates without doubt from the brain since it is certainly made by means of the muscles and nerves. Nor should I wish anyone to be deceived when I said that there are two movements in the heart, one its own and one accidental, for this statement is consistent with reason and not unlike many which are proposed as axioms in natural history since there is no astronomer who is ignorant, according to my opinion at any rate, of the fact that the seven planets which are called wandering are set in revolution by one and likewise by another motion: one is cognate or coeval with the planets, that is, motion from west to east, and the second, rapid and violent, is the impetus of the prime mover. Surely by a similar argument the same situation is true of the heart's motion. For the thorax and diaphragm are set in motion not otherwise than the orbs of the seven planets just mentioned, by the swift impetus of the prime mover. This motion is seated within and while it encircles the vital spirits it now repels, now recalls and brings them back. If indeed the motion of the thorax was designed for respiration alone, but the lungs could not be moved by such a motion, why should anyone wonder that the heart also is moved by some means (for these organs are joined by a great and necessary connection) if the heart appears to be moved by contrary motions, although one of the two motions is alien to it and rather follows necessity than use or utility? Therefore, since the pulse of the entire body is the motion of the heart and of the arteries certainly created by the most subtle spirits, but such spirits as are elaborated in the left ventricle of the heart, deservedly then all the arteries draw their origin from that place so that they may carry the spiritual blood, like the great branches of a tree scattered throughout the entire body. If this blood grows warmer than is proper and is rendered more easily movable the motion of the arteries will at once become much more rapid, but if, as happens, the blood becomes greatly chilled its motion will become very slow. You may deduce other differences from the firmness and power of the heart. For if its power is strong the pulse also is strong, appearing to beat very strongly as one touches it with the hand, but if it is depressed and weak the pulse will certainly be sufficiently languid so that one cannot notice that even the arteries are dilated. Similarly, you may discover that the heart's motion which is a mean between these extremes is the normal motion. There are very many other kinds of pulses, such as equal and unequal, frequent and seldom, great and small, soft and hard, which it is not of any advantage to list at the present time; it is no doubt sufficient for my discussion at this point to say that no fever can originate unless the spirits of the left ventricle of the heart first grow warm to some extent so that from them warmth can be scattered throughout the entire body. In order to add the remaining portion of my discourse I should say that the substance of the heart is quite fleshy and not deprived of nerve, artery, vein, and even of bone. It has nerves in order that so noble an organ should not remain insensitive. The bone, however, must strengthen very skillfully the firmness of the ligaments. I have already spoken of the function of the arteries and, further on, the veins in the heart. It seems to me unnecessary to point out now the fact that the heart is carefully placed in the region of the thorax since it is evident to the eyes that it is surrounded as it were by strong walls in a circumvallation and in a level position by the chest, the spinal vertebrae, by the diaphragm and finally by the neck and the shoulder blades. It derives comfort from the lungs which are close to it and serve as a very soft mattress or couch. An especially strong evidence that the heart occupies the middle region of the thorax is the small membrane called the mediastinum. For since it is certain that this membrane divides the entire thorax into two equal parts it also divides the heart equally. This could not be done if the heart did not occupy the exact middle of the thorax. I have said enough on this subject since I think you will be convinced particularly by the evidence of your own eyes and your reason. For I know that there are two instruments by which we always search out the truth: one is reason, the other is experience. If anyone rebels against these especially when their results coincide, he should be banished, in my opinion, as a most bitter enemy of learning. As far as the structure of the lungs is concerned there is no need for me to proceed further since their use is sufficiently known to all. They minister to the heart and furnish it with cool air. Their substance is quite thin and spongy so that they may more easily draw the spirit into themselves, for by all means a light bodily organ was required to withstand such a continuous and persistent motion as that of the lungs. For this reason the very thinnest blood and spirit always distill to the lungs in order to preserve their constitution. Furthermore, on account of many causes the lungs are divided into five lobes. The first of these is that in the case of any calamity to one part another part of the lungs can render assistance. Second, the lungs can move more readily and swiftly than if they were not divided into lobes, similar to the hand which is divided into five fingers and can thus be moved much more completely by every kind of motion than if it had been constructed like a flat writing tablet without divisions. In addition, the division of the rough artery (trachea) seems to follow completely the division of the lungs so that a large portion of the trachea passes into the separate lobes of the lungs. Hence the lungs draw air through the bronchi of the trachea and do not readily collapse as they return that air, carefully elaborated, to the heart in return for the latter's beneficent service. The residues which the lungs are not strong enough to consume or dispose of they expel suddenly through the larynx. But nature did not allow even these sooty residues to remain unused, for after they seem to become valueless to the heart she has designed them very cleverly as material quite suitable for the formation of the voice. The lungs, to return once more to them, cannot be moved by themselves at all, so weak and helpless has nature made them. They are moved, however, by the intercostal muscles which are the instruments of both inspiration and expiration. Hence if these muscles or the small membrane which surrounds them are affected by an inflammation, as happens in the case of pleurisy, these motions are immediately interrupted. Indeed I have often wondered why we cannot by any means cease to breathe in and out without the greatest danger to our lives when every motion of the soul is free and voluntary and when inspiration and expiration are motions of the soul, since they are made by means of the animal spirit. I should easily believe that the necessity of cool air is linked to the necessity for this motion of the lungs which brings that air to us, since a more liberal supply of animal spirits is borne to the intercostal muscles than to the other parts, that motion being analogous to the service of millers and butchers who are permitted no holiday because what they do is always required for the life of mankind. The motion of respiration at any rate seems to be almost natural (involuntary) because without it nature cannot easily subsist. Yet it is not so involuntary that it is not also voluntary in part since we sometimes repress the action of these muscles and check their motion especially when we wish to follow some sight or sound with a more accurate sense-perception. In fact, Galen[47] tells how a certain slave held his breath until he died; hence he called respiration a voluntary motion. But since this motion cannot be stopped without the greatest danger, as I said, it comes to be called partly involuntary and partly voluntary by the physicians. Thus you may clearly conclude how different from each other are the motions of the arteries and of the intercostal muscles. The former is quite involuntary since it is allowed not the slightest intermission even if we wish it. The latter, however, seems to obey a certain choice of the will. The former motion is performed by the heart alone, the latter owes its origin to the brain. Furthermore, the lungs and heart alone fill the region of the thorax, fenced around by the twelve ribs on both sides. Of these, the seven true ribs are fixed, but the five remaining are the false ribs, called also bastard and spurious by physicians according to their common name by analogy, for people call bastard, spurious, or illegitimate a child which is born of a strange or alien woman. Therefore, since the seven upper ribs are attached not only to the dorsal spine but also to the breast bone as if to either parent they are justly called legitimate while the five lower ribs are spurious since they do not recognize one of these parents, that is, the breast bone. Again, because the lung is the weakest of all the parts of the body it is attacked by the most numerous and the most dangerous ills. Since its substance is quite spongy it easily receives any kind of the most impure humors into its sinuses. It has a most suitable location, if anywhere in the body, for fluxions and secretions (rheumata). I pass over the fact that it is in perpetual and constant motion, which contributes considerably to the attraction of humors which cause swelling; lastly, the lung is attacked by more ills than any other part of the body. Since the discussion of these must be reserved for another place, let me turn now to remaining matters. Just as the stomach draws drink and food to itself by means of the esophagus or gullet, so the lungs frequently draw air to themselves through that pipe which the Greeks call rough artery (trachea) and the Latins *arteria aspera*, and, when they have warmed it by an orderly interchange, they breath it out again. This artery is hard and denticulated in its harshness, a feature which nature has designed not in vain. For if it were soft and gentle, for example, made of flesh alone, one would have to fear lest it should finally collapse and become the greatest hindrance to even the slightest amount of air and prevent it from being sent down to the lungs when the muscular fistulae are closed. Nature has by no means failed to foresee this eventuality and has very carefully provided against the great danger that some portion of food or drink should slip down through the trachea to the lungs and disturb or prevent their action while a person is swallowing. For she has fitted a certain epiglottis or small tongue to the end of the rough artery somewhat lower down in the throat which furnishes easy passage to the breath alone but shuts off everything else that it recognizes as harmful to the lungs, holding and thrusting back such a substance like a sort of a lid or covering. It was not possible, however, to construct this safeguard with such precision that nothing of the more liquid substances should not slip down with the air. This lack of precision is of great use for moistening the bronchi of the trachea. If by chance some humidity which is excessive either in quantity or quality passes down the trachea so that it troubles instead of soothing that organ that humidity certainly creates hoarseness, for the breath is sent down unequally by an unequal or uneven path, which constitutes hoarseness. This often happens also if anyone imprudently laughs or roars with mirth while he is swallowing. He may die of sudden strangulation if a crumb of bread or a fishbone or any fragment of food or drink completely stops up the passageway of the vital spirit. This affliction is called suffocation or strangulation. I have not been overly zealous in describing the composition of the epiglottis since Galen par-

[47] Galen, *De Motu Musculorum*, II. 6 (Venice, Giunta, 1625, Vol. I. 315 D; 8 v. in 5, with the excellent index of Antonius Musa Brasavola, more complete than Kühn's.)

ticularly in *De Usu Partium* VII, as well as in many other passages, has discussed it with most abundant detail. In this one matter especially art has imitated nature most perfectly. For just as nature has fitted that little tongue to the upper part of the rough artery, that is, the larynx so that the air is not left free or dispersed but forced or squeezed as through an alley or narrow passageway in order to form the voice so also the makers of flutes attach a sort of slip or reed to the other edge or extremity of a pipe and leave only a single very small hole through which the air may pass very narrowly. The epiglottis seems to furnish the same service in the rough artery as the reed does in musical pipes. And just as the harmony of flutes is not damaged by rubbing them with some kind of liquid, as is usually done, so you should not be surprised if the artery called rough also requires some soothing moisture. Furthermore, as the excrements which overflow from the stomach are for the most part expelled through the esophagus so those which trouble the region of the thorax are very frequently excreted through the trachea when they have reached maturation. Rarely, however, but sometimes, they are thrust out also through the bowels because that passageway is long and intricate with many tortuous bends and turns.

I see that the second portion of my discussion, which was quite troublesome for me, has now been completed in every respect: that portion which dealt with the vital force of the soul or breath of life. There remains the explanation of the third part, concerning the chief faculty of the soul. I have set down as an axiom of procedure the statement that the natural force of the soul resides in the liver and the vital force in the heart on account of its dignity. Later, whatever blood the heart requires from that vena cava both for nourishing the lungs as well as for creating the vital spirits it receives through the auricle of the right ventricle. That portion of the blood which the heart finds superfluous to its own purposes, having taken account of the superior parts, it allows to pass to the brain through that large vein which extends to the throat, in such a way, however, that all the organs which are in the vicinity of this vein receive food and nourishment from it by means of certain rivulets or smaller veins which arise from this very broad vein. Since I have mentioned these adjacent organs I should like you to know that I mean the muscles of the superior ribs and in addition all those parts which are attached to the collar bones, that is, the armpits, shoulders, shoulder blades, and the arms themselves; throughout all of these the blood at its reddest and most copious flows by means of veinlets. Furthermore, since almost never is a vein discovered without an artery for its companion, here of all places is this true, for a large artery full of the most fervent spirit is clearly extended from the left cavity of the heart and presents itself as a companion to that vein. But as soon as this artery and this vein together reach the region of the brain as if it were a marketplace then like very busy merchants each of these goes off in different directions to tend to its business and thus to a certain extent part company with each other. The vein departs straightway to all the parts of the brain in order to revitalize them, but the artery winds in various coils in the shape of a net or plexus to the base of the brain's ventricles around certain small adenous bodies. Galen is accustomed to call this plexus the marvelous net; it is very hard to find. Within this plexus the animal spirits are prepared from the vital spirits, not that the latter are made warmer than they were before (since nothing can be found that is warmer than vital spirit) but that they are accurately tempered by the coolness of the brain. For it is consistent with reason to suppose that the faculty from which motion and sense perception proceed would not excel in heat but would be as temperate as possible, since qualities which are excessive do not effect or promote sense perception but destroy it to the fullest extent. The nerves, furthermore, the vessels of the animal spirits as they come into being, are very cold by nature and could not endure a very warm spirit if that were how the animal spirit turned out to be. In brief, there is no possible way in which the brain, which is very cold and formed according to Aristotle[48] for the refrigeration of the heart (I say this with his own good leave), could render the vital spirits warmer. Otherwise, it would be equivalent to saying that the brain is hotter than the heart if it could make warmer the spirits which have been thoroughly heated already in the heart. Therefore, the spirits are not heated in the brain but rather are weakened in temperature and made cooler. When they have attained such a degree of temperature they are scattered from that net-shaped plexus into the benign substance and ventricles of the brain so that they may be distributed thence as though through rivulets to the separate parts of the body by means of nerves. Thus it is not difficult to conclude that the animal spirit is the chief instrument of the soul (or the life force: *anima*); but the soul which makes use of this instrument I have already said resides in the brain. For if the natural power which is the most imperfect of all has received therefore the lowest position, it is proper in my opinion that the chief force of the soul should have the noblest and most precious position of all because it is the noblest among the body's powers. The head, according to Plato,[49] is beyond controversy the most divine among all the members. The entire body is subject to it and the immortal gods have bidden the body to obey and be subservient to it. I am silent as to the fact that all the nerves originate from the brain; they are the vessels of the animal spirit and as it were the instruments of the brain's motion and sensation;

---

[48] Aristotle, *De Partibus Animalium*, II. 7. 652a25.
[49] Plato, *Timaeus* 39D.

they demonstrate with sufficient brilliance that the soul dwells therein. Do not the poets also represent Pallas as the product of Jove's brain? Why not from his liver or his heart? It was consistent with reason that the goddess who was far the most wise of them all should originate from the most precious site of wisdom. It is a most unworthy error of Aristotle[50] to state that the heart is the origin of nutrition, motion, sensation, and understanding, overturning and disturbing that which both experience and reason have proven to be the truth. I say experience indeed, for who in anatomical investigations would deny that every offshoot of the nerves is derived from the brain, and reason also, for if all the nerves common to both brain and heart were to be tied off with some kind of bond all the parts which pertain to the heart are at once deprived of motion and sensation while those parts above the heart remain uninjured. This, of course, could not take place if the faculty of sense perception proceeded and emanated from the heart and not from the brain. But what else, by the Lord, do fits of apoplexy reveal to us? Or epilepsy? Or that alienation of the mind which the Greeks call phrenitis? What, I say, do all these mean except that the principal force of the soul (or life force) resides in the brain? This is quite clear to judge from the great mass of ignorant people who are, in fact, called insensate, empty-brained, timid or stricken in spirit, finally stupid, showing, of course, that the courageous force of the mind is located in the heart and the chief power of ratiocination is located in the brain. He therefore who is so impudent and so confidently arrogant that he insists that the most perfect power of the life force is situated in the brain should read Galen's[51] commentaries on the opinions of the philosophers, where, unless he was delirious on purpose or was so enormously stubborn, he would be convinced by reasons and authorities. The brain, furthermore, is divided into three sinuses, two of which are located in the anterior region and the third, in which the power of memory is placed, has received its position in the cerebellum, called parencephalis by the Greeks.[52] Almost the entire body of physicians reports that the imagination is located in the anterior sinuses of the brain; the faculty of memory, however, is placed in the posterior sinus as though in the most hidden room of the mind. Hence it is very easy to recognize that these faculties of the soul are distinguished not only by reason but also by their situation since it is clear that when one of them is injured [reading *laesa* for *illaesa*] all the others are affected. The common sense is located toward the anterior parts of the brain corresponding to them all as the center corresponds to the circumference. All the exterior senses render very grateful service to the common sense, for they apprehend sense objects and, without judging them, bring them to the common sense for examination, like henchmen who thrust wrongdoers into jail. For they do not make the slightest inquiry into the life and morals of the offenders but, seizing them violently, immediately drag them before the judge for questioning, and finally betake themselves again to lying in wait for other offenders. In this analogy, of course, the common sense corresponds to the judge. After this common sense has received the forms of sense perception brought to it by the attendants, that is, by the exterior senses (for it receives various images of things) it separates them and assigns various differentiae to them, also dividing contraries from contraries. But if it finds something unusual and worthy of notice in the course of this process the common sense places it, so that it may not immediately slip away, into the ventricle of the brain which is nearby as if into a treasure chest or storehouse. That which is trivial and of no importance the common sense permits to pass away at once. People in whom this posterior part of the brain is at once both rather moist and soft are talented for they are skillful and quick to conceive or imagine, but their memories are poor; on account of the great humidity and softness of their brains the images of things readily slip away from them. But people in whom this part of the brain is rather harder than what is reasonable (as happens principally among the ignorant) follow sense perceptions with their mind somewhat laboriously and thus are quite tenacious and preserve knowledge once they have acquired it. One may compare the situation to the function of a seal: since its form is impressed more quickly upon soft wax than upon a mass of iron it tends to disappear just as quickly. This is very probably the reason why those people who have a soft cerebellum easily retain recent events in their memory but forget those which are much farther in the past; those people with a hard cerebellum behave in an opposite manner in regard to memory. Reason alone governs among all the senses I have mentioned, to which all things appear to render homage as to the most noble emperor and highest monarch of the entire body. Reason, however, has selected no special position for herself in the brain but wanders and runs about throughout its entirety. The reason, furthermore, expunges, obliterates, and cancels whatever it wishes and at its will establishes other things anew. To sum up the matter in a few words, by means of this single gift of the mind man stands forth most completely among brute animals and those deranged and resembles the gods above. If, however, he behaves at times in a shameful manner, that part of the soul which I said was soft and womanish, the reason itself, gives orders to him very severely like a master to his servant. But lest I appear to have entered some deep and rugged ocean from which I may not be able to extricate myself, let me return to

[50] Aristotle, *De Partibus Animalium*, III. 4. 666a10; II. 10. 656a.

[51] Galen, *De Placitis Hippocratis et Platonis* (V K. 181–805).

[52] Aristotle, *De Historia Animalium*, 494b32; Galen, *De Usu Partium* VIII. 11 (III K. 647ff.)

the point whence I digressed. Thus the vision (since for the sake of proper instructional procedure I descend to particular details) first apprehends whiteness and blackness and does not progress further. Secondly, however, the common sense, insofar as it responds to direct vision, judges with a somewhat unskilled intelligence that white is something different from black and that these colors can never blend into one. Next, the intellect with a loftier and greater contemplation examines things worthy of its attention, defines, illuminates, polishes, and most completely investigates each detail, declaring or assigning that to be white which by reason of its color dissipates and divides the vision and that to be black which strengthens or unites the vision. Thus the brain is the origin of all these faculties, although it may occasionally cause some people to be amazed that sensation and ratiocination arise from a member that is insensitive (for the brain itself is completely without sensation). However, when I say that all these faculties arise from the brain, we must understand that they arise from the animal spirits which are elaborated in the brain, for I also said that although the brain is insensitive in its action it does not, nevertheless, lack the power of sensation. For if its substance were hard like the substance of nerves it would possess the most keen sensation, or ability to feel. The substance of the brain and of the nerves is completely similar, with this sole difference, that the substance of the brain is soft and the substance of the nerves is hard and stiff. Furthermore, in this internal organ also all the moods of sleep and drowsiness are generated; hence, if any opinion of Pliny[53] can bolster our belief, that which lacks a brain cannot go to sleep in any way. I must make a complete statement as to how sleep, which is a certain deprivation of wakefulness, takes place in the brain, for it does not come about through a repletion of many vapors in the brain, as Aristotle[54] declared, but rather with a slight distinction to his statement. For some sleep is natural, another kind is connatural or quasi-natural, while a third type is beyond nature or unnatural. A sleep beyond nature takes place when the substance of the brain is permeated by very cold, thick vapors and the soul appears drowsy and marvelously weak. This illness is called lethargy by the Greeks because those afflicted by it are extremely forgetful; the Latins call it, however, an uncontrollable desire to sleep. The quasi-natural form of sleep is created when the cold or hot vapor of food or drink floats about in the substance of the brain, and since men are apt to be affllicted with this kind of sleep for entire single days it seemed best to call it quasi-natural since it departs from the normal custom of man's sleep. The third variety of sleep, which is true and natural, comes about not when the brain is filled with exhalations or vapors but when the animal spirits which are the leaders for the senses are exhaled. When these are exhausted and consumed the instruments or organs by means of which sensations are created necessarily collapse and languish since without those animal spirits these organs have no power to fulfill their functions. I should like to present an example of this process: the machines which men have invented for making flour out of barley or wheat. Some of these require running water to set them in motion, others a strong blast of air. But if one or the other means is lacking these machines will not move. We shall not therefore deprive the organs of sense of their proper motivation as long as the animal spirit which controls them is not dissipated. But if it is dissolved, as happens in violent exercise, assiduous studies, and finally in a long period of wakefulness the organs of sense will without doubt take a holiday until that animal spirit is restored. This is done chiefly in the night time, for then the animal faculty has a complete rest and the natural and vital faculties, if ever, cooperate most actively for the animal spirits, although they are the instrument of the soul, are nevertheless made not by the soul but by nature. All the nerves take their origin from the brain, as everyone agrees, directly or indirectly. There are seven pairs of those nerves which spring directly from the brain. Physicians are not accustomed to make any distinction among those which arise indirectly or by some mediation since they understand that this is unique. These nerves indeed do not arise from the brain but from the spinal medulla which originates from the parencephalis or cerebellum. Since the spinal medulla is soft in nature and differs in almost no way from the substance of the brain, it is not classified as a nerve. But because the nerves which arise from the spinal medulla are innumerable and almost infinite in number I believe I shall make a useful contribution if I turn the entire discussion to those seven pairs which derive directly from the brain. Let us begin with the types into which they are divided. Some nerves are hard, but others are soft. The hard nerves are best adapted for motion but are least fitted for sense perception although these too are sensitive to a certain extent, as are all nerves. On the other hand, the soft nerves are languid and powerless for motion but most suitable for sense perception. Thus nature with the most brilliant foresight in all things used soft nerves for sense perception and hard nerves for motion in putting together the bodily parts, as I said had been done in the construction of the tongue. For it happens very often that sense perception is completely destroyed but motion is fully preserved, and sometimes the direct opposite occurs since these functions of sense perception and motion are of course performed by means of various nerves. If it were necessary for motion and sense perception

---

[53] Pliny, *Natural History*, XI. 37. 49. 135: quae cerebrum non habent non dormiunt.

[54] Aristotle, *De Partibus Animalium* II. 652b10; *De Somno et Vigilia* 455b28ff.

to be performed through the same nerve, then motion would require a greater force of spirit than sense perception. Therefore let me return to my purpose as stated above. The first pair of nerves is of those which send the very soft optic nerves to the eyes. They are sent forth from the anterior part of the brain, which is very humid and soft; the closer those nerves approach to this part of the brain the softer they appear and the farther they proceed from it the harder they become. These are called the optic nerves because they carry to the crystalline humor that spirit which causes vision. But more on these matters later. The second pair of nerves moves the eyes themselves; these are harder than the optic nerves in proportion to the greater hardness of the place from which they originate. The third pair of nerves is extended to the cheeks, temples, and the roots of the teeth; I believe some part of them extends to the skin of the tongue in order to create taste. The fourth pair is inserted into the entire palate. From the fifth pair nerves extend to the auditory canal. The sixth pair, which comes from the extreme posterior part of the brain, extends most abundantly to the intestines, the mesentery, to the stomach (especially to the upper part of its mouth) and finally to all the viscera. According to the extension of these nerves the brain appears to be connected with other organs as can be seen in serious illnesses of the stomach, heart, and liver. Since all of those organs have no need of those nerves which cause free motion, they require no hard nerves but are content simply with those nerves by which they can feel sorrow and pleasure, especially on account of various dispositions or conditions of the body. The seventh and last pair of nerves is the one which from that region where the spine is joined to the brain extends to the larynx and pharynx as well as to the root and muscles of the tongue. Now turning back again to the first pair of nerves which I said are the optic nerves, let us examine their nature and use. The optic nerves originate from the anterior ventricles of the brain, as I said, and extend as far as the roots of the eyes, nor were these twin ventricles formed for the sake of any other purpose than for the origination of these nerves. Those ventricles seem to be pools of the clearest water and the optic nerves to be rivulets or canals which draw water out of those pools. Hence it is that these nerves are not hard, solid, or rough but concave or porous and very soft so that they furnish an easy pathway for the animal spirits to pass through them. But the passage from the brain to the region of the eyes would not be quite safe unless, as is the practice in building canals, we place bricks or something of the sort within them in order to resist external forces. Thus nature places some protection around these nerves, since she did not wish to have neglected their safety. As soon as these nerves issue from the region of the brain the very hard bones of the eyebrows are fitted around them on all sides like a wall. Furthermore, when they have reached the area of the eyes each nerve is disseminated into very many quite thin filaments of various kinds. These later combine mutually among themselves and form a certain internal tunic (if that is the proper word) which Galen[55] calls amphiblestroides or like a net (the retina). This tunic goes around the crystalline and vitreous humors from behind and binds the latter as though in a prison so that this humor cannot leap forward to the pupil, that is, the region in front of the crystalline humor, and thus become a very great hindrance to vision. For the retina (or amphiblestroides tunic) forms in the crystalline humor something like a circle or exact ring which corresponds by analogy to a horizon or diameter which cuts a sphere into equal parts. The examination of the anatomy of the eyes is the most difficult in all the body both on account of the variety of their tunics as well as the number and diversity of their humors which are surrounded by these tunics. I shall, however, try to describe the true composition of the eyes as clearly as it can be done. For in the eyes (to put their number into a single word) there are seven tunics. Three occupy the anterior region and the remaining four the posterior region of the eyes. Since the anterior tunics are much more bright and polished, they reveal more evidently the artifice of nature and are taken into much greater account by physicians for to them alone have been given special names; a brief and merely passing mention is usually made of those tunics which adhere to the rear of the crystalline and vitreous humors. Now let me turn to this task and sum up in a few words for the sake of instructing young people the careful but extensive discussion of the subject contained in Galen, *De Usu Partium* X. The brain itself is covered by two tunics, one quite thick called the dura mater by physicians, the other very soft and thin, which has received the name of pia mater. These tunics are also called choroeides by Galen. But where is this description tending? I shall now tell you. The tunic which first meets our gaze as we examine the eyes, set up like a wall to protect the crystalline humor, draws its origins from the dura mater, which is a sort of protection to the brain. Since it is made up of thin bands or laminae which resemble horn this tunic is appropriately called keratoeides or the cornea. Do not assume that this tunic is single or simple in its form, for if you take a thin scalpel or a lancet with which veins are opened and pluck it apart little by little you will find six or seven tunics; these do not differ, however, from each other in color, substance, or consistency. Thus they are all included under the same term, cornea. Next to this tunic clings another which extends from a thin membrane. It is called choroeides because that soft covering of the brain from which it originates is called chorion. It is also

[55] Galen, *De Usu Partium* VIII. 6, X. 2 (III K. 759ff.)

called rhagoeides, that is, similar to the berry of a grape which is round. Next to the rhagoeides there is the very thin arachnoid tunic, almost imperceptible to the senses, which surrounds, protects, and embraces the anterior region of the crystalline humor toward the pupil like a very light cloak. Lest the very soft crystalline humor should in some way be struck by the very hard tunic of the cornea (for there is nothing between that humor and this tunic except the arachnoid, which presents no resistance, and the rhagoeides, which is perforated) there is a certain watery humidity which nature has very cleverly filled with a highly gleaming spirit. This humidity keeps the ceratoid tunic (cornea) completely distended so that it does not collapse and thus hinder or destroy the instrument of vision. Furthermore, these three tunics which I have called anterior are joined with the four posterior tunics around or near a certain perfect circle like the rainbow which the most learned of the Greeks have called the iris. That whitish tunic which seems to be sprinkled with many little veins and has not yet received a special name seems to perform the function and service of a process or offshoot since in addition to the fact that it covers the hidden parts which it moistens with blood it also serves as a very strong ligament binding those parts to the upper bones. There is furthermore another tunic [conjunctiva] conjoined to this one which, according to Galen, degenerates into a sort of aponeurosis or sinewy filaments attached to the muscles which move the eyes. I have already spoken of the six tunics of the eyes but since it is not easy for anyone to determine what the seventh tunic may be (for the remaining parts of the eye resemble flesh more than a tunic) it seems much more satisfactory and indeed more honest to pass over it now in silence than to impose upon tender youth with some kind of a frivolous fiction. Thus seven tunics of the eyes contain three widely different humors beneath and surround them, that is, the vitreous, which is called also hyaloeides; the aqueous, not unlike the white of an egg; and finally the krystalloeides or crystalline. The last of these three humors, that is, the crystalline, is by far the most precious; hence it occupies the middle position between the other two, for it rests upon the vitreous humor from the rear like a cushion or a pillow and from in front it has the other humor which resembles the white of an egg as a source of protection and cooling. The crystalline humor is in fact nourished by the vitreous humor and cherished and moistened by the other humor they call aqueous, although the latter also provides another service, that is, the distention of the cornea to keep it from pressing upon the crystalline humor through the pupil. This fact enhances the foresight of Prometheus in our behalf, a quite marvelous harmony in regard to the human body. You will also wonder, I know, if you notice how varicolored is the inner circumference of the tunica rhagoides (the uvea). For that very great monarch of all the world concluded that our eyes would be greatly injured by the inner whiteness since it is more remote from external light and that they would be most sharply struck (for whiteness clearly divides the vision); hence he most artfully contrived green, blue, and black colors for the concave part of the rhagoides tunic in order to strengthen the sight of the eyes. For these colors in proportion to their absence of whiteness strengthen and concentrate the sense of sight. Vision itself is a kind of affection or experience of the senses. But since every affection must be created through contact it is necessary for contact to be involved in the experience of sight. Vision depends not only on reception, as Aristotle[56] declares, but on the act of emission, as is consistent both with Galen's[57] statement on the matter and with reason. A certain shining spirit flows down from the brain through the nerves I call optic; this spirit after it has received a certain elaboration by the crystalline humor slips through the pupil of the eye and mingles with the external light as some very diligent messenger or attendant of the brain. Three elements therefore are required for the act of vision. First there is that shining spirit which flows down from the brain through the optic nerves; hence if those nerves are stopped up in any way and obstructed we are immediately deprived of vision. Second, external light itself is required, from which our animal spirit is derived and refreshed. Hence it is that we cannot see in the shadows, deprived of the light of the air, the most familiar guide of the animal spirits; for while the animal spirit is most shining by nature it suddenly fades and scatters when it flits about in air dissimilar to itself. The third element required for sight is the shapes or forms of the things to be seen; whether these reach the brain or the crystalline humor alone I should hesitate to say with any certainty. It would, however, be absurd to deny that the confluence of these forms is necessary for vision. The more there are of them and the more abundant the more firm and clear it is certain the vision will be. It is no slight proof of this fact that a high mountain or many other such objects from which many forms of sight leap forth are much more easily and completely perceived by us than some small object from which few or no forms of sight slip off or escape. When I speak of forms (*species*) I mean visible qualities by which we ourselves are changed. We are changed by them not in reference to our entire bodies but in reference only to the single organ or sensorium, that is, the eye itself, which is the organ of the soul. Nor should you conclude that it is the eye which sees, for it is the

[56] Aristotle, *De Generatione Animalium*, V. 1. 781a5.

[57] Galen, *Methodus Medendi*, I. 6 (Venice, Giunta, 1625, VII. p. 6C) or VII. 1 (Giunta, VII. 43 D); see also *De Hippocratis et Platonis Decretis*, VII. 5 (Giunta, I. 271 B) and the pseudo-Galen, *De Oculis*, 6 (Giunta VII. 188 D–H).

soul[58] alone which sees through the eye as if through some window. This is the reason why in the keenest contemplations of the soul we appear to be deprived of our eyes while the soul is engaged in loftier meditations. Men are for many causes deprived of their vision. Sometimes this happens because of some suffusion which the Greeks call hypochyma when some sooty vapor is caught between the arachnoid tunic, which surrounds the crystalline humor, and the cornea. But I shall speak at greater length elsewhere on suffusions (or cataracts). Sometimes vision is lost because of the great weakness of the brain, especially when the optic nerves are pierced by some weapon in the region of the temples. It also happens not infrequently that the vision or the power of sight is completely lost on account of some sudden transmutation as if we should transfer ourselves from a place of shadows into the brilliant light. Galen[59] writes that this happened to those who were detained in prison by Dionysius, the tyrant of Sicily. It is told that Dionysius constructed a most splendid house above the prison. This house was painted and incrusted with the whitest of lime or whitewash and into it he led the wretched calamitous prisoners after long confinement in chains in order to blind them miserably. While they stood looking with great pleasure at the splendor they had desired for a long time they imprudently blinded themselves. I shall not pass over in silence something that happened to me when I was a small boy. Since I did not have enough money for my childish games and had no source from which I could obtain it I followed my father who was visiting the bedside of a nobleman ill with fever and climbed up with him to the patient's bedroom. The light there was sufficiently bright but it seemed quite dark to me since I had just come in from a brighter place. After I had rested for a while I saw by chance a purse half lying upon the sick man's bed, and because I judged, plausibly enough, that the eyes of the sick man and of those around him were dimmed as mine were (falsely, however, since they had been longer in the room and had accustomed their sight to the shadows) I came up closer and began to handle the purse. But he to whom it belonged (for illness had not deprived him of speech) said: "What are you doing with my purse? Isn't it enough that the druggists have left it thin, without you to empty it completely into your hands?" I blushed and was struck dumb, and began to philosophize very energetically about light and shadows. This incident really happened to me when I was twelve years old. You may learn from this example how useful and necessary is philosophy, the mother of all things, not only to physicians but to thieves and murderers. For I would not have been caught so shamefully in the act if I had known then as I know now the causes of things as well as of their time. Thus the eyes (to return to my discussion) like the keenest watchman of the soul have been allotted the highest position. They do not see themselves, although they are quite bright and shining and seem to sparkle, but the soul sees through them as if through windows, as I said, with three factors contributing harmoniously to the process of vision: the forms of things to be seen, an inner fire, and an external light or glow, as with a rare felicity and beyond controversy the most worthy Plato[60] seems to indicate in the *Timaeus*; these are his indisputable words: "But after the cognate fire has passed out into the night, the ray of light which causes vision vanishes for it bursts out into an air dissimilar to itself and is extinguished and transformed, having no longer any communion with nature when it is deprived of that glow in the air. Therefore it ceases to see and brings on sleep." Those parts of the eyes by which they are closed and made to wink are called eyelids. With the greatest art they protect the eyes with a kind of wall of hair and keep off flies and other such little beasties. The eyes are clothed about entirely by these eyelids; they finally collapse when the faculty of the eyes, that is, their sight-creating spirit, has been completely dispersed or blown away. This easily proves that the motion of the eyelids and vision itself are controlled by one and the same spirit since as long as vision exists so long do the eyelids move but when vision ceases, as in sleep, the eyelids also take a rest, with animal spirit flowing of course to effect each action or to discontinue it, as the case may be. To each of the eyes above the region of the eyelids a single projecting eyebrow adheres to keep dandruff and sweat flowing down from the head from getting into the eyes. Furthermore, the cheeks gently rising below the eyes protect them. Nature in her great wisdom could not have placed the eyes in a better location than where they are, covered over with many very shining membranes, beyond any danger and risk from anything that might strike them. If one eye, either right or left, is destroyed or rooted out the one that remains performs its function much more happily than before since the animal spirit which was distributed earlier to both eyes now flows into only one eye. Therefore, having discussed the eyes abundantly it is reasonable to turn my discourse to the nostrils. Beyond the fact that they contribute to the beauty and comeliness of the human body they seem to be constructed also for use in respiration, by which the animal spirits are nourished and refreshed, as well as for perceiving odors. Just as certain forms from visible things fly up to the eye itself so also certain exhalations from those things which are perceived by the sense of smell flow through and

[58] See Plato, *Timaeus* 45D, 68D and my note 148 to the translation of Benedetti's *Anatomice*.

[59] Galen, *De Usu Partium*, X. 3 (Venice, Giunta, I. 178 H).

[60] Plato, *Timaeus* 45D.

ascend to the borders of the anterior ventricles of the brain, where they create the sense of smell. These exhalations arise through a force of heat, with which all things that give off odor (according to Galen and also Aristotle,[61] twelfth section, fourth question) seem to some extent endowed. Later these exhalations are impressed upon the anterior parts of the brain and thus these obtain the faculty of smell. The heart also draws the familiar spirit to itself through the nostrils not less happily than through the mouth itself, which we can see in those who are sleeping. The ears alone, as Pliny[62] says, are immovable in man so that they may yield more openly to the impulse of the air. Each ear has a certain sinus which resembles a tortuous labyrinth made intricate with many turns and twists contrived by nature to keep the cold and untamed air from making a sudden impact upon the benign substance of the brain before it has been elaborated by means of a long journey. Certain slender nerves extend, as I said, into the passageway of the ears so that they may feel the repercussion of the air. Hearing, in fact, is nothing other than a certain impulse created by the collision of solid bodies striking as sharply as possible upon the auditory nerves or that same very sharp collision of the impulse itself with the nerves mentioned, by reason of which the brain is affected at the same time. Men are affected through this sense with very great pleasure and pain: with pleasure when they perceive some sweet harmony and tuneful modulation of sound, for the mind, consisting as it does in harmony and agreement, is delighted by what is similar to itself. But if anyone strikes our ears beyond method and artifice we bear it quite indignantly. Furthermore, since the brain is of a frigid temperament it gathers many excrements to itself to such a degree that from the purest and most benign blood it renders the phlegm thick, salty, and viscous. Deservedly therefore nature has placed very many broad passages around it so that through them as canals the matter which is harmful can be at once removed. In addition to the ears and nostrils which sometimes expel the superfluities of the brain (although this is not their proper task) there are two fair-sized foramina, later joined into one, which open from the brain above through the palate into the mouth. The substance of the brain is immediately covered by a certain very thin membrane which is sometimes called chorion by Galen. It is quite soft and benign since this membrane is interwoven with the thinnest arteries and veins and hence received the name pia mater from the common people. Another membrane much harder and stronger comes next to the pia mater on its outside; there is the same distance or difference between the two membranes as there is between the pia mater and the brain. It was excellent that such a tunic as this was made for it had to be placed close to the very hard bone of the cranium. This bone was assigned to be a sort of very strong wall on all sides of the entire brain. It is marked by many sutures and commissures for driving out the smoky excrement by means of imperceptible transpiration. Since there are sutures in the cranium it is certain that it is made up of more bones than one, and hence the entire head is in less danger because when one part of it is injured it cannot easily transmit that injury to the neighboring part. Five sutures at most appear in the skull, three of which truly deserve the name, joining its bones with their saw-tooth denticulations. The other two near the temples are called false sutures but are actually the petrous bones set close to or bordering the skull. In this manner they appear to be sutures. A certain straight suture stretches from the sinciput to the occiput; above its extremities descend two other almost parallel lines, one of which slants off to the anterior region from the inner side and makes two right angles. But the other line divides the occiput from the vertex and creates two obtuse angles from the inner side and a single acute angle from the outer side. You may easily recognize the three true sutures in any cadaver. These are the sutures commonly found, although in some skulls more and in other skulls fewer are very often found. In brief, the skull does not have a fixed number of sutures. In fact, as I write these lines with my right hand I grasp in my left the very tall skull of a cadaver which is marked with no more than five lines around it. Of these, three are true sutures but the remaining two are (as I said) merely on the border of the skull. When the number and description of the sutures are known it is not too difficult to discover how many bones the head has, for there are seven in all worthy of note with the exception of the lower jaw bone. One is the bone of the sinciput; another ends at the opposite side of the head at the cerebellum. There are in addition the two bones of the temples which are distinguished by the suture which I said extends lengthwise in the skull. Two other bones called petrous because they are very rough are joined with these bones of the temples toward the lower borders of the skull. Furthermore, the seventh remaining bone combines many bones with itself, bound by very firm ligaments; such bones are those of the eyes, the nostrils, and finally the bones of the palate. Nevertheless all these bones which are placed under the skull are called by the collective term basilar bone. The skull itself is covered by a twofold tunic. The one closest to the skull is sinewy and quite hard and arises of course from the thick membrane of the brain which has

[61] The reference to Aristotle, *Problems* XII, 4, is reversed by de Laguna. Galen, *De Symptomatum Causis* I (VII K. 122) and *De Simplicium Medicamentorum Temperamentis ac Facultatibus* IV (XI K. 698) discusses the relation of heat and smell.

[62] Pliny, *Natural History* XI. 37. 50. 136: Aures homini tantum immobiles.

fallen forward through the sutures as if into certain filaments. The other tunic is fleshy so that with its blood and kindly warmth it may check the hardness and dryness of the skull by moistening it. But so closely joined are these two tunics to each other that you can scarcely ever separate them. Thus Galen everywhere calls both tunics by the names perikranion and periostion, that is, the tunic which surrounds the skull. Lest they should remain naked at any time and thus be injured because of the inclemency of the air these tunics are fortified on all sides by a very abundant growth of hairs. All physicians consider it a fact that these hairs are generated by certain sooty residues which pass daily by expiration through the sutures and pores of the head. Because of the variety of material which is used for generating the hairs their difference is very great, which you may learn especially from the fact that almost all the people of Flanders are blond or white-haired because they abound in colder humors while the Ethiopians and the Spaniards have much darker and rougher hair because they produce a very great deal of sooty material owing to the constant heat of the sun. But the French, who live in a region midway between these peoples enjoy yellow or tawny hair which are colors between white and dark. Hence the hairs seem to be of much importance in determining each and every temperament. But now, since I cannot on account of the size of this book wander about any further in my discussion (for it is already larger than a handbook should be) it is my fixed plan to postpone the arrangement of the legs and of the muscles as well as their structure of nerves and bones and finally all the remaining description of the human body to another treatise, putting a most happy end to this one which has grown longer than is reasonable.

To the most learned and erudite Don Rodrigo de Reinoso, most watchful doctor of the art of medicine, Andrés de Laguna of Segovia sends greetings.

When, day before yesterday after many and varied nightly labors, as is provided for the relief of human weakness when night grows deeper, I had retired to my bed in order to obtain rest for my weary mind and broken body, Reinoso most worthy beyond dispute, some good Genius brought to me my weeping book, thus exclaiming as I lay sleeping with both ears (as they say): "What? Am I thus to be betrayed? Am I thus to flit amid so many hands, so many faces, so many judgments of men? Since this is your intention (some day you shall regret it) where shall I flee at last, featherless, naked, helpless, to be torn to pieces by so many noses, so many pens, finally by so many darts? Shall I be allowed to hide away even for one small hour among the learned Parisians? By no means. For what hope of salvation survives for me if I should come into the hands of the most erudite Jean Tagault, dean of the faculty of medicine due to his admirable foresight and knowledge, or into those of Jean Guinter, most watchful and faithful interpreter of the Greeks? There is certainly none. Therefore I must leave Paris. What if I should seek exile in Flanders? Not even there can I live in safety, for the most noble Luis Vives, whom you know very well, the most redoubtable defender of literature, the ornament and most splendid authority of all the Spaniards, governs those regions in his splendor. Since he sets no moderation or limit to the extermination of false authors and by the skill of his talent renders the most learned of philosophers far more cleansed of error (so great are his gifts) and revises their work if he finds anything in them which is false, trivial, and counterfeit (let the vile passion of calumniators depart hence!), if I chance to reach his ears he will commit me completely to the flames as spurious, illegitimate, false, and hence most worthy of such a destruction. I see no shelter there for me, and hence I must sail elsewhere. Shall I go to that very prosperous market town of the Spaniards, Salamanca? But if you do find there a few people who are distinguished in the more polite studies and in medicine, all the rest are lovers of revilement [philoloidoroi], babblers, and sophists. How could I abolish their jeers and impostures with my small skill in argumentative refutation? From all these last resorts there seemed to remain only one refuge: that I, abortive, mutilated, and badly put together should straightway be burned in the flames since it is better to perish once for all in fire than to be rendered ulcerous by so many sentences and censures. Otherwise, it would come to pass that after so many shipwrecks, so many stormy dangers, after so many turbulent wanderings I should return featherless to you as to the author of my calamities, like that little crow of Horace;[63] or if I may not return to you I may certainly do the following: like young chicks who, afire with a great longing for flight before they have gained any strength or skill for it, fall down into a deep pit, so I, immature, than whom nothing is more unskilled, nothing more lowly, shall try to fly so high that, struck like another son of Apollo by some divine lightning, I shall be thrust down into the depths." While my book was thus bawling out many lamentations of this sort which would have struck great fear even into people who were well used to such behavior I began to tremble and to hesitate, and (as they say), holding a wolf by the ears, I remained in the greatest torment, wondering whether I should acquiesce in the decision my book had presented or, as I had decided earlier, publish it. My book seemed

[63] Horace, *Epistles* 1. 3. 19; the little crow who dressed up in other bird's feathers in the fable Horace uses to reproach a certain Celsus (in his versified letter to his friend Julius Florus) for borrowing literary material from Augustus's library in the temple of Apollo at Rome; the rest of the flock will one day strip him of his borrowed finery.

to ask what was just, under the circumstances, but I was sorely grieved in my turn to have wasted the labor of so many days in writing it. I hung for a long time on this cross; I had almost given up in despair when you, Reinoso, much the most learned of all men, occurred to me. You, the memory of whose name alone will never perish, at once revived my spirit faint unto death with its incredible surge and swell of cares. You, I say, came at once to my mind, you who (to say nothing of your other virtues) far excel the physicians of our time in comeliness of speech, knowledge of Greek literature, great skill in affairs and (as is usually the most outstanding trait in learned men) in your great humanity and benevolence. For in reference to medical practice, indeed, you are so adept in the art that what was customary to be said about that most prudent old man Hippocrates:[64] "Having established many trophies in victory against diseases by means of the instruments of healing, he gained great renown, not by chance but by his skill," since you are no less than Hippocrates, I shall not blush to say about yourself. For I have not forgotten with how much inclination of mind, sincerity of heart, diligence in all things, you urged me to study literature, particularly at that time when, having left Italy, you came in earlier years to Paris. Dependent upon and clinging to these words of advice from you I resumed my spirit and strength although completely cast down in soul, and as though it had been snatched from the flames I immediately, although against its will, saw to it that my book was printed. Therefore it is yours, most noble doctor, if perchance my offspring, poorly nourished as it is, has reached you, so that you may cherish and strengthen it as though it were your own child. For if it should be deprived of you, who alone are most illustrious among physicians, to whom it has fled as to a father, my book can hope for nothing more elsewhere. But if you should decide to read these trifles of mine through for the relaxation of your mind, please read them by carefully scrutinizing everything like a kindly judge with an indulgent pen in your hand, and pardon in your courteous affability the weakness of my intellect and whatever I have written with less than proper style, if indeed you realize the truth of that proverb that "one man does not see all." Farewell, most eminent doctor, and, as you began, continue to deserve well of literature. At Paris on the eighth day before the Calends of October, 1535.

[64] The Greek text with a translation by W. R. Paton can be found in *The Greek Anthology* (Loeb Classical Library, II, 1917, pp. 78–79).

Johannes Dryander

1500–1560

LIFE AND WORKS

JOHANNES Dryander[1] (Eichmann in German) was born June 6, 1500, at Wetter in Oberhessen, Germany, and died on December 20, 1560, at Marburg/Lahn, where he taught medicine for twenty-four years. He was twice married and had five children. Of the Lutheran faith, he studied at Erfurt in 1518, became the assistant of Euricius Cordus, and probably received his master's degree in that city. Later he studied at Bourges and at Paris between 1528 and 1533 while both Vesalius and Andrés de Laguna were also there. In 1533 he took his doctor's degree at Mainz and received the post of physician to Archbishop Johannes von Mettenhausen in Coblenz and Trier. In 1535 he was called to the position of professor of medicine and mathematics at Marburg, becoming rector of the university in 1548. He was also physician at the hospitals of Haina and Merxhausen. He held four public anatomies, among the earliest in Germany, in 1534, 1536, 1539, and 1558. He was also one of the few pre-Vesalian anatomists to use illustrations in his books; Georg Thomas of Basel made the woodcuts in Dryander's *Anatomia capitis humani* (Marburg, 1536), which I present here in English translation, and in his *Anatomia hoc est corporis humani dissectionis pars prior* (Marburg, 1537), perhaps the best pictures to be printed up to that time before those of Vesalius's *Fabrica* with the exception of those in the works of Berengario da Carpi. Dryander plagiarized the illustrations of Vesalius's *Tabulae Sex* (1538) in his own edition of Mundinus (Marburg, 1541).[2]

Dryander was a prolific writer both in medicine and anatomy as well as astronomy and chiromancy. A check list of his published work follows:

1. Ein new Artzney unnd Practicyr Büchlein von allerley Kranckheiten, wie man die erkent und geheylet werden söllen, auss den berumptesten und erfarnesten zu unsernn Zeyten lebenden Medicis . . . in eyn kurtze Summa zuzamen gezogen. Colonię. Apud Eucharium 1527, 1537; Frankfurt am Main, 1563, 1589.
2. De balneis Emsenibus; Marburg, 1535; German translation, Marburg and Mainz, 1535; Strassburg, 1541.
3. Anatomia capitis humani, in Marpurgensi Academia superiori anno publice exhibita . . . Marpurgi, Ex officina Eucharii Cervicorni Agrippinatis, 1536; 28 pp., 14 illustrations. These woodcuts with additional illustrations were published in his *Anatomiae corporis humani* (see below).
4. Novi annuli astronomici nuper anno 29 excogitati . . . Marburg, 1536 (German translation, Marburg, 1546).
5. Zubereitung unnd warer verstanndt eines Quadranten; Frankfurt am Main, 1536 (Latin translation, Marburg, 1542, 1550).
6. Sonnawern allerhandt künstlich zu machen; Frankfurt am Main, 1536; Marburg, 1543.
7. Anatomiae, hoc est, corporis humani dissectionis pars prior, in qua singula quae ad caput spectant recensentur membra, atque singulae partes, singulis suis ad vivum commodissime expressis figuris, deliniantur. Omnia recens nata . . . Item

FIG. 14. Johannes Dryander. Courtesy of National Library of Medicine.

[1] The most recent biographical article on Dryander is that by Robert Herrlinger in *Neue Deutsche Biographie* **4** (Berlin, Duncker and Humblot, 1959): pp. 142–143, who lists the following items: F. W. Strieder, *Grundlage zu einer Hessischen Gelehrten und Schriftsteller Geschichte* (Göttingen and Kassel, 1781 to 1812), 3: pp. 237–242; E. Fuhrmeister, *J. D. Wetteranus*; unpublished medical dissertation, Halle, 1920; Harvey Cushing, *A Bio-bibliography of Andreas Vesalius* (New York, Henry Schuman, 1943), pp. 28–32; E. Zinner, *Dt. u. niederländ. astronom. Instrumente d.* 11–18 Jh., 1956; E. J. Gurlt, *Geschichte der Chirurgie und ihrer Ausübung, etc.* (Berlin, Hirschwald, 1898); A. Hirsch, *Biographisches Lexicon der hervorragenden Aerzte aller Zeiten und Völker etc.* (Vienna, Urban, 1884–1888, 6 v.). A copper engraved portrait by T. de Bry of Dryander is reproduced in J. J. Boissard, *Icones virorum illustrium* (Frankfurt am Main, 1597) and another is to be found in the Hauptbibliothek of the Franckeschen Stiftungen, Halle/s.

[2] Cushing, *op. cit.*, pp. 28–32; C. D. O'Malley, *Andreas Vesalius of Brussels 1514–1564* (Berkeley and Los Angeles, University of California Press, 1964), pp. 89, 221, 455–456 tells the story of Dryander's plagiarisms in detail.

Anatomia porci, ex traditione Cophonis [et Anatomia] infantis, ex Gabriele de Zerbis. Marpurgi, Apud Eucharium Cervicornum, 1537. 72 pages, 23 illustrations, folding table. Beginning: "In praelectionem medicam oratio. . . . " Ending: "Quale sit officium anatomici ex Carpi commentariis supra Mundini expositionem": pp. 63–64.

8. Abano, Pietro d', (1250–1315?) De venenis, atque eorundem commodis remediis. Liber plane aureus, per Joannem Dryandrum medicum, pristino suo nitori restitutus. Marpurgi. Ex officina Eucharii Cervicorni, 1537; Venice, 1537.

9. Annulorum trium diversi generis instrumentorum astronomicorum, componendi ratio atque usus; Marburg, 1537.

10. Johannes Vochs Opusculum . . . de omni pestilentia . . . ; (edition), Cologne, 1537.

11. Cordus, Euricius. (1486–1535) Ein nützlich Buchly, darinn allerley gewüsse unnd bewärte Stuck und Artzny für die grusamme Plag desz Steinwees begriffen: durch . . . Euricium Cordum beschriben, mit einer Vorred Joannis Dryandri medici, [n.p.] 1538; 1542.

12. Anatomiam Mundini ad vetustissimorum aliquot manuscriptorum codicum fidem collatam, justoque ordine restitutam. (with woodcuts from Berengario da Carpi's edition). Marburg, 1541.

13. Antiochus Tibertus, De cheiromantia libri III (edition). Mainz, 1541.

14. Der gantzen Artzenei gemeyner Inhalt, wes einem Artzt, bede in der Theoric und Practic zusteht. Mit Anzeyge bewerter Artzneienn . . . Hiebei beneben des Menschen Cörpers Anatomei, warhafft contrafeyt und beschriben . . . Newlich in Truck verordnet . . . Frankfurt am Main, Christian Egenolph, 1542, 1547, 1557.

15. Von rechtem christlichem Brauch des Artzes und der heylsamen Artzeney . . . Frankfurt, Cyriacus Jacob zum Bartt (preface dated 1543).

16. Libellus de peste; Marburg, 1553: German translation, 1554.

I have translated the 1536, first edition, of Dryander's work on the human head (no. 3 in the checklist), exclusive of dedicatory material. The second, 1537, edition (no. 7 in the checklist) is essentially the same except for added description, some of it culled from Celsus, at the bottom of each plate and a longer introduction dealing with the nature and history of anatomy and its practitioners. The illustrations too are identical through the first eight. In the second edition plates 9, 10, 11, 12 offer somewhat more detail plus an unnumbered plate presenting the sutures of the skull and combining detail from plates 9, 10, and 11 of the first edition. The figure of the complete head ("Universalis Figura Capitis Humani") is the same in both editions except that the 1537 edition adds captions which I have incorporated into my translation.

THE ANATOMY OF THE HUMAN HEAD

Demonstrated Publicly Last Year
in Marburg University
by Johannes Dryander, Physician

Inevitabile Fatum

Like a lily among thorns (printer's motto)

Everything recently published, from the press at Marburg of Eucharius Cervicornus Agrippina, in the month of September, 1536

1 [fig. 15]. It is my intention to complete almost all the anatomy of the human head in twelve figures, and for each figure I have placed letters beside it for the sake of demonstration.

A. thus indicates the cord by which the head is bound round in a circle when it has been shaved of hair on all sides. The cord also indicates the limit of the anatomical operation to be undertaken, which by taking off the skin thus casts into relief the location of the sawing in the cranium.
B. indicates the line of incision from the cord above the nose, creeping through the vertex into the occiput.
C. signifies the transverse line of incision around the ears, the end of it touching the cord on both sides.

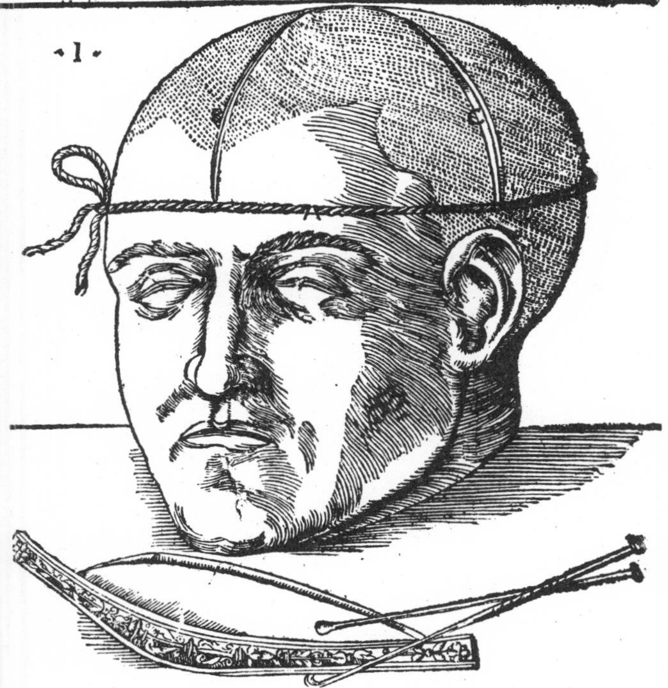

FIGS. 15–26. Illustrations from Dryander, *The Anatomy of the Human Head*. Courtesy of National Library of Medicine.

You see represented in the rest of the picture the instruments we use in anatomical work, such as the razor, the circular saw, curets, (sounds), pincers, forceps and that sort of thing.

2 [fig. 16]. In this second figure you see the two exterior skins of the cranium, lifted up and laid back as far as the circle of the cord.

A. designates on each side of the head the thicker skin which produces the harvest of the hair. This is of a thicker nature because the hairs spring from it with a heavier root. Under this skin there lies a certain rather mucous viscosity. From this innumerable ills are recognized.
B. designates the hard inner membrane of the head bone which the Greeks call pericranium. As this membrane girdles the entire head bone so it also clearly provides sensation.
C. designates the cranium with its sutures and the joinings of the bones. You will note where the

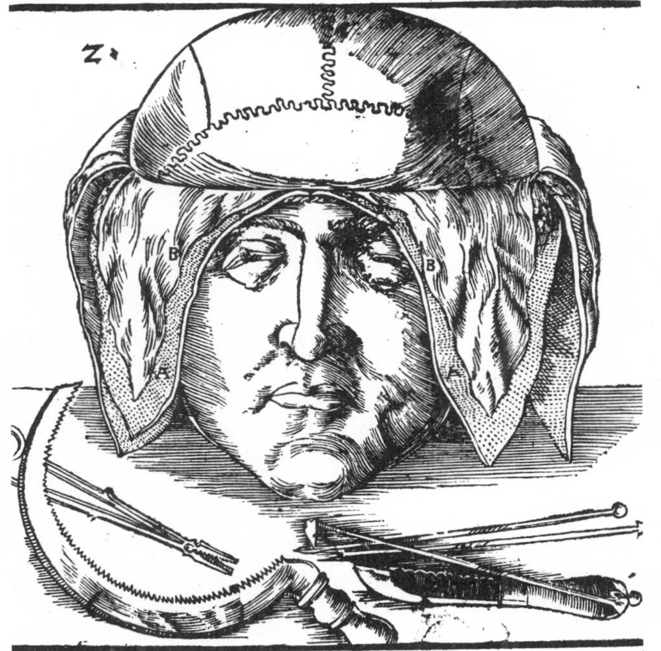

FIG. 16.

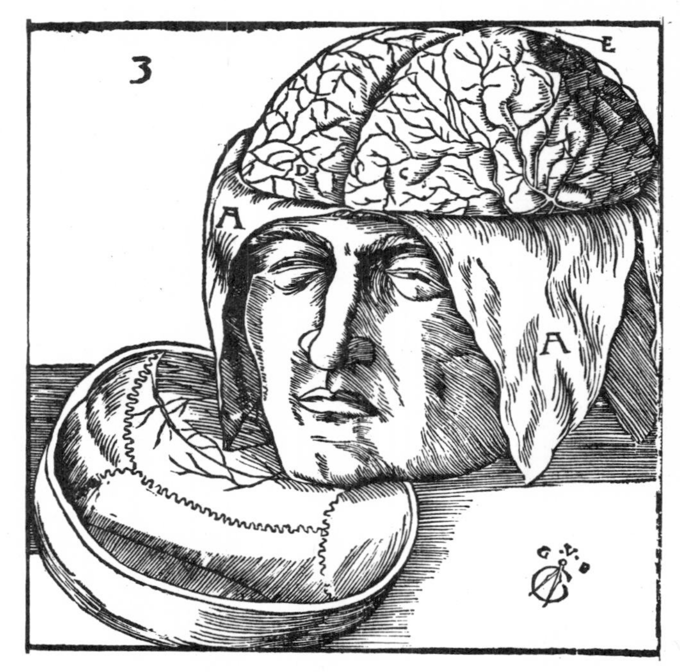

Fig. 17.

suture creeps from the bone of the forehead to the nose quite frequently in the crania of women but differently in the heads of men.

3 [fig. 17]. We thus cut into the head with a circular saw around and above the mark of the cord in such a way that none of the inner skins nor the dura nor the pia mater is torn away, and gradually we remove the upper hemicranium from the rest of the head's trunk.

Here you see that upper part of the cranium, cut with a circular saw and torn away, placed in the lower part of the figure.

A. designates the harder skin within the cranium which they call dura mater. You see the brain is distinctly surrounded by a thin little membrane which they call pia mater and by various and diverse branchlets of veins and arteries. In the substance of the brain you see the right part corresponds to the left part in shape and form.
C. designates the left, D. the right, and E. the middle crosspath of the brain which divides its parts.

4 [fig. 18]. In this figure it is possible to see each inner skin which envelops the brain together with the brain's gray matter itself.

A. thus indicates the dura mater (so-called).
B. designates the pia mater with its branchlets of veins, and each of its skins here torn away from the brain so that a more suitable view of it may be afforded.
C. the gray matter of the left lobe of the brain.
D. the right part or lobe.
E. is the common boundary and division of the brain.

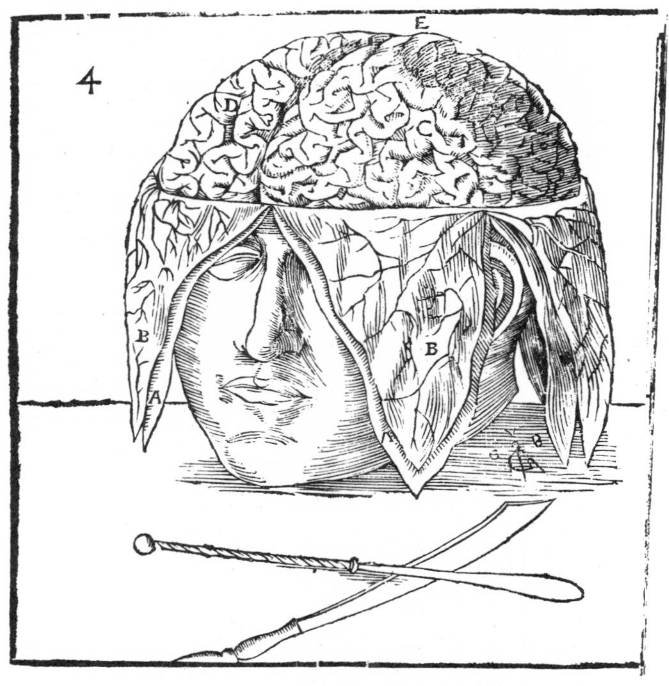

Fig. 18.

5 [fig. 19]. In this figure you see the right-hand substance of the brain, still uninjured. The left-hand substance, however, up to the intermediate ventricles has been almost cut away with a razor.

The place of the sawing, among the panniculi, appears together with the remaining part of the head on all sides.

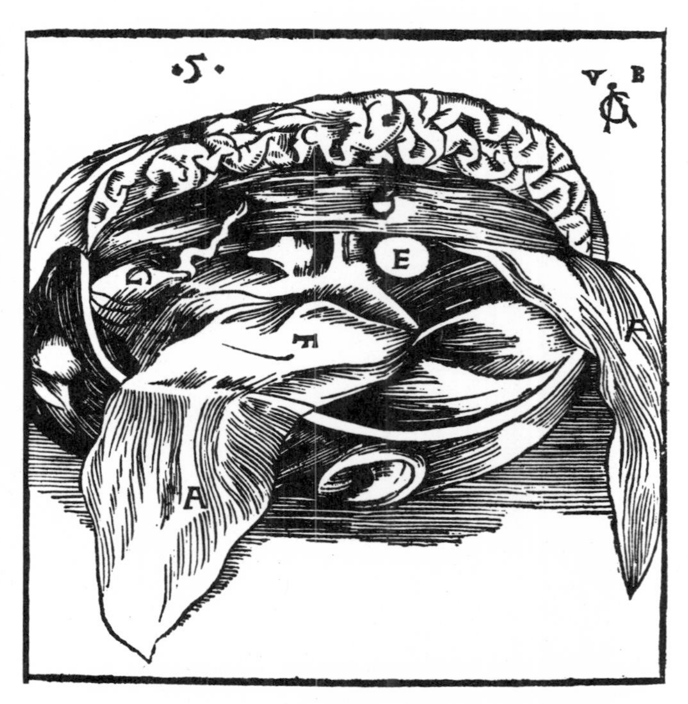

Fig. 19.

C. is the right-hand part of the brain's substance, still uninjured. A.A. indicates the small membrane of the part of the brain which has been removed, the covering which I said is called the pia mater.

D. the little skin which wraps this right-hand part of the brain.

E. indicates the narrow entrance of the ventricles. Above the nose in the cranium itself is a certain notable vacuity as full of air as is possible. They think this air contributes to the sense of smell.

6 [fig. 20]. You see here in the present figure, when you penetrate the brain to a distance of almost three fingers, that there appears a ventricle on each side, curved in the manner of a new moon.

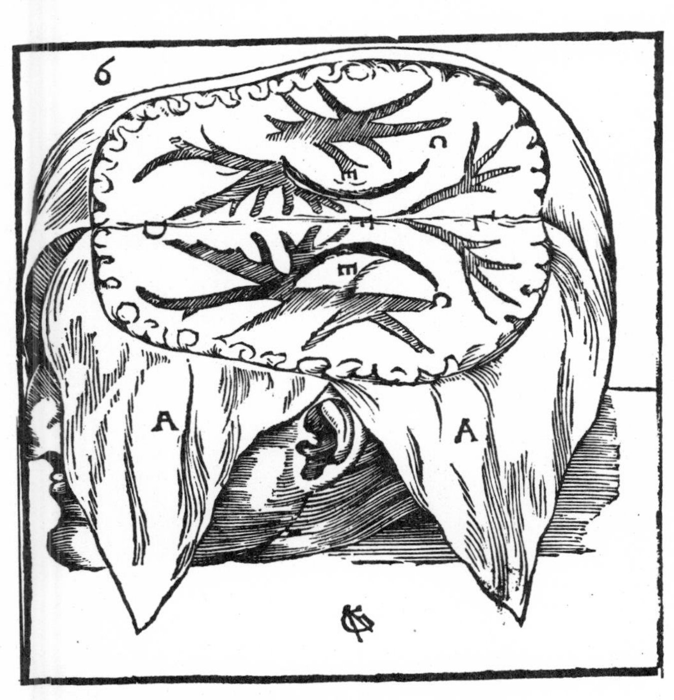

FIG. 20.

EC, EC are the indications of the ventricles on each side.

D.F.L. on both sides around the base (of the brain) is a reddish pellicular substance called the worm, made up of veins and arteries, which extends from one end to the other of each ventricle. This worm according to some has motion and voluntarily opens and closes the ventricles.

Below these worms at their sides is a certain eminent part of the brain which many compare to the human buttocks in its form. This part both in elongation and closing of the ventricles touches its two portions together and separates them in the shortening and dilation of the ventricles. This (is what) Carpi (says) in Isagog. [Carpi, *Isagogae Breves*, for this passage, p. 143, literally plagiarized by Dryander here.]

A.A. is the skin of the pia mater.

You see the ear, the nose, the lips and the chin projecting beyond the skin.

7 [fig. 21]. Just as the dura mater with its accompanying meninge separates the right-hand part of the brain from the left, so also protracted along the side it divides the posterior brain from the anterior. But this second duplication does not seem to be so joined together with veinlets as the first one. It contains within itself, however, a certain vacuity in which terminate many branches issuing from the jugular veins.

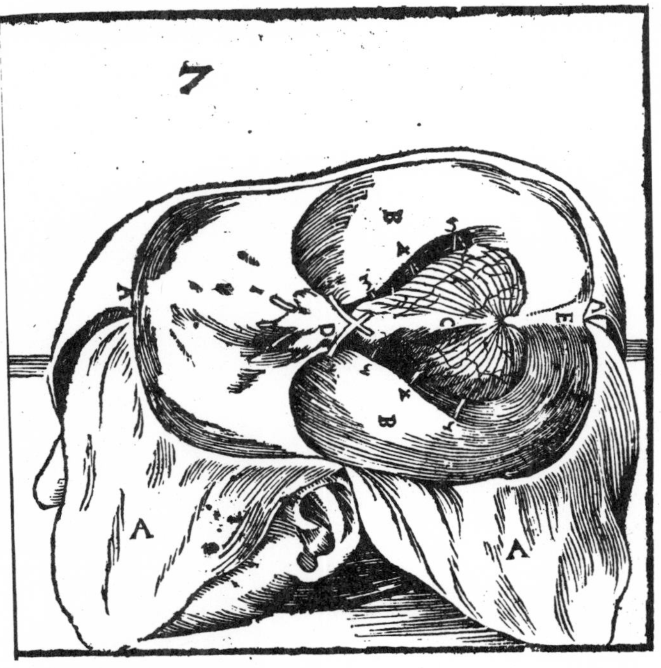

FIG. 21.

Toward the occiput in this duplication is a certain vacuity called the lacuna (lake) in which part of the blood is pressed out. This place is called the torcular (press) of Avicenna.

Around letter D. you see that the mammillary carunculae are indicated, the instruments of smell as some call them. They send forth nerves as you see 1.1. 2.2. called optic and 3.3. 4.4. 5.5. called nerves of vision. You see their numbers set beside each of these nerves of sensation.

AA. as always thus far indicates the little skin (pellicle) of the brain.

BB. the lower brain on both sides.

E. indicates the posterior (of the brain).

8 [fig. 22]. In the posterior part of the middle ventricle is a small foramen which extends toward a certain vacuity. This vacuity as it descends is turned

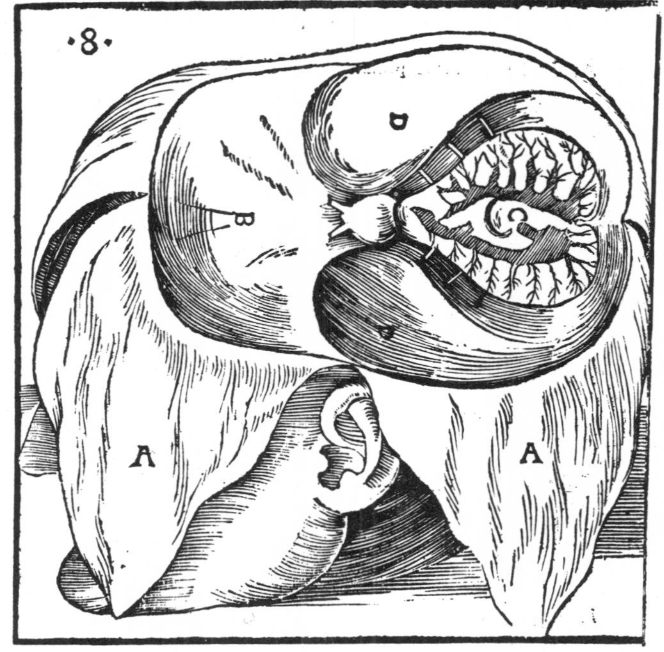

Fig. 22.

toward the place where the nucha [spinal cord] begins. The vacuity is not in the substance of the posterior brain, as many think, nor is it surrounded on all sides by the medullar substance of the brain, but is located between the posterior and anterior brain, notably surrounded toward the posterior brain by the pia mater which covers it (the brain).

C. the cavity of the posterior brain.
DD. the lower substance of each side of the brain.
B. the lowest substance of the anterior part of the brain.
AA. is the pia mater.

9 [fig. 23]. This figure indicates the lower part of the cranium where it is implanted in the neck.

A. expresses the row of teeth.
B. the upper cavity of the mouth.
CC.
DD.
EE. are the bones of the calamoidus (styloid process).
FF. are the two parts of the head where the first vertebra of the neck and the upper vertebra which Galen calls odontoidea receive the entire head.
G. is the foramen through which the nucha [spinal medulla] is distributed to the spine of the back from the brain.
HH. are bones as hard as rocks (mastoid process).
I. is part of the occiput of the cranium.

10 [fig. 24]. The location and shape of the bones and sutures of the true cranium are shown in the following figures.

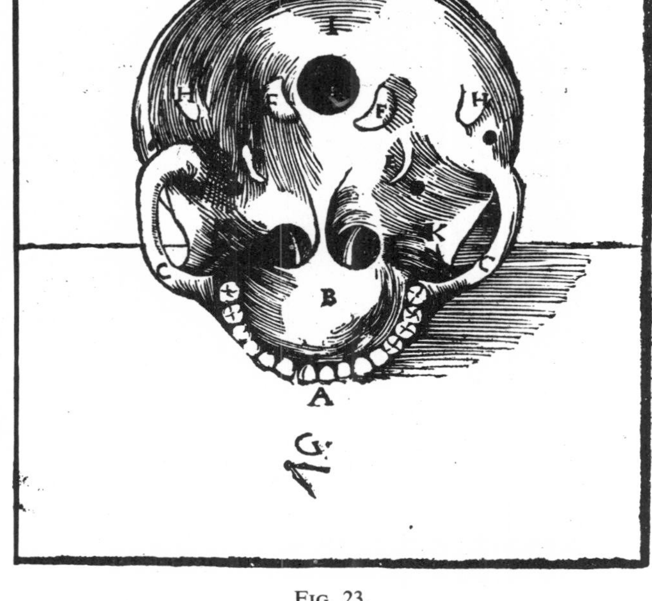

Fig. 23.

The first picture shows the suture of the stephanoid (coronal suture), the frontal bone, the porous bone, and the bone of the vertex.

The second picture shows the eye sockets with the semicircle of the coronal suture.

The third figure indicates the lower row of teeth impacted in their bone.

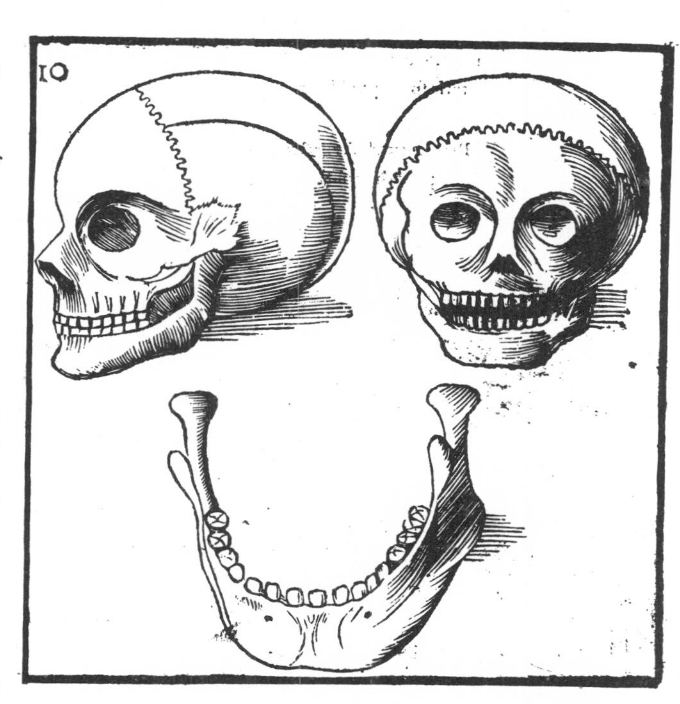

Fig. 24.

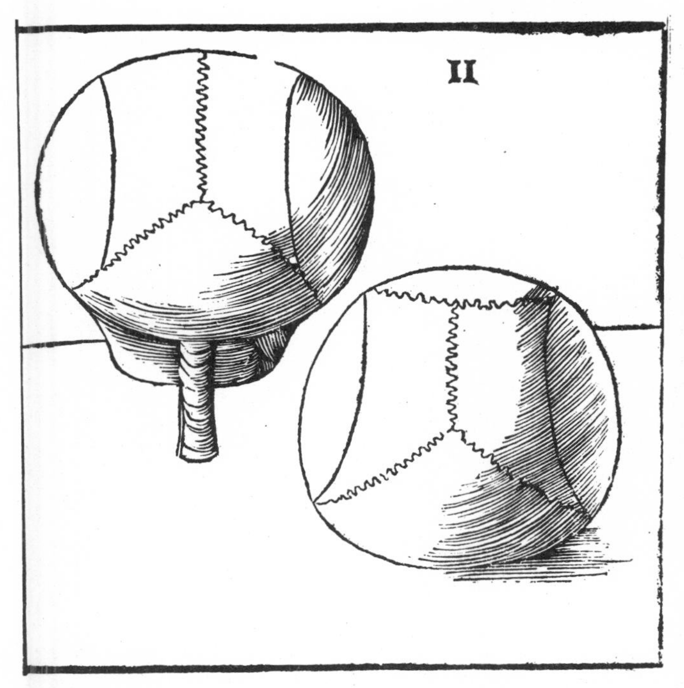

Fig. 25.

Fig. 26.

11 [fig. 25]. In this figure you will see the remaining bones of the cranium which face toward the posterior part of the head, together with their sutures.

The rest is to be sought from our tables (plates).

The complete picture of the human head [fig. 26], whose explanation you will find at the end of the plates which show the head.

A. indicates the vertex of the head in the uppermost part of the cranium.
AB. AB. the first skin of the head hanging down on both sides, in which the hair is fixed.
C. shows the inner side of the first skin.
D. is the outer skin immediately touching the bone of the cranium, that is, the pericranium.
E. designates the bones of the cranium.
F. the hard meninx within the cranium.
G. is the pia mater, the thin softer skin which immediately covers the brain.
H. indicates the substance of the brain.
I. Ventricle of the brain. 1. 2. 3.

Above L you see the representation of the marvelous net.

L. lacuna (lake).
M. pairs of nerves.
N. the mamillary carunculae.
OO. the organ of vision.
P. the organ of hearing.
Q. the sense of smell in the nostrils.
R. the faculty of taste in the tongue.
S. the coronal suture.
T. the sagittal (suture).
V. indicates the lambdoid suture in the occiput.

Feeling is diffused throughout the entire body by means of nerves and muscles.

The end of the figures of the anatomy of the human head. You shall expect in a short time the single parts of the entire human body, completed by anatomical labor, and made as clear as life [no doubt a reference to his next book, *Anatomiae, hoc est, corporis humani dissectionis pars prior* etc., Marburg, 1537].

Giovanni Battista Canano
1515–1579

LIFE AND WORKS

ALMOST all the facts concerning Canano's life and work are set down in the introduction to the facsimile reproduction of his *Musculorum Corporis Pictura Dissectio* (Ferrara, 1541?) by Harvey Cushing and Edward C. Streeter (Florence, R. Lier, 1925: Monumenta Medica, edited by Henry E. Sigerist, Vol. IV). He was born, according to Streeter, at Ferrara in 1515, shortly after the birth of Andreas Vesalius: his ancestors seem to have been Greeks who settled at Ferrara before 1438. His grandfather had been court physician to Matthew Corvinus, King of Hungary. Ludovico Canano (1485–1556), the anatomist's father, educated his son at Ferrara, whose Studium had produced Leoniceno (1428–1524) and other famous men. The boy's uncle Ippolito was a member of the medical faculty, although Antonio Musa Brasavola was Canano's most important teacher. Antonio Maria Canano, another relative, taught anatomy to Giovanni Battista; Antonio Maria, in turn had been taught by Marcantonio dalla Torre in Padua.

Berengario da Carpi, last of the anatomists to whom the study of the muscles was a difficult business because dissections did not ordinarily give much attention to myology, had died at Ferrara in 1530. The Canani, Giovanni Battista and Antonio Maria, set out in 1539 to study the muscles in dissection in the former's home; Vesalius's brother Francis was among those who attended the sessions here. The mentor of this gathering was Bartolomeo Nigrisoli, whose father had taught medicine at Ferrara in the 1470's. Amatus Lusitanus recommended Ferrara at the time for the study of both botany and medicine; he lived six years there (1542–1548).

G. B. Canano succeeded A. M. Canano in the post of professor of anatomy in 1541, having received his doctorate shortly before and taking over the position only because his kinsman wished thus to further the younger man's career. Sometime before 1542 G. B. Canano published his famous work, without date, name of publisher, or place of publication. M. Roth, *Andreas Vesalius Bruxellensis* (1892), p. 127, believes that Vesalius visited his brother at Ferrara in the latter half of 1542 and there showed Canano the proofs of the woodcut illustrations of the *Fabrica*, to be published in the following year. In 1543 Canano was visited by Dr. John Caius of Norwich, who was lecturing on Aristotle at Padua. Caius considered Canano's collection one of the six best libraries in Italy.

Canano probably continued to teach at Ferrara until 1545. He demonstrated the palmaris brevis muscle to Falloppio and discovered the valves in the veins (Vesalius, *Fall. Exam.* 83). Falloppio took over the chair of anatomy in 1548. In 1552 Canano was appointed physician (archiater) to Pope Julius III at Rome, in succession to his teacher Brasavola. There Eustachius experimented with Canano's suggestion that copper plates be used instead of woodblocks for anatomical illustration. Canano stayed in Rome three years, where he was made bishop of Ficarolo, and returned to Ferrara in 1555 on the death of the Pope. In that year both Sylvius in his *Isagoge* and Vesalius in the second edition of the *Fabrica* described the valves in the veins (III, 4). Canano's priority in their discovery was, however, assured for him by Amatus Lusitanus, who had described the function of the valve at the orifice of the azygos vein four years before and attributed its revelation to Canano. Other anatomists, Vesalius, Valesius, Eustachius, insisted that such a valve could not be found and blamed Amatus for his claims. Yet the venous valves were rediscovered by Fabricius in 1574, before the death of Canano in 1579. He had spent the last twenty years of his life in the service of public health as protomedicus of the house of Este, promoting the construction of drainage canals in northern Italy. His last years were full of work owing to pestilence and epidemic pleuro-pneumonia. He died on January 29, 1579, at the age of sixty-four.

The most recent facsimile edition of the *Musculorum Humani Corporis Picturata Dissectio* (the

FIG. 27. Giambattista Canano. Courtesy of National Library of Medicine.

first word is misspelled on the cover) is by Giulio Muratori, director of the Istituto Anatomico of the University of Ferrara (Florence, Sansoni, 1962) and is made from the copy in the Ariostea or Municipal Library of Ferrara. It has a useful bibliography and some notes which add little or nothing to our knowledge of Canano or the date of his book. The English translation of both introduction and notes is frequently quite awkward, as is that of the work itself, which is in both Italian and English.

It should be noted as a final remark on the mystery of the publication of Canano's famous little book that the British Museum *Catalogue of Italian Books published between 1465 and 1600* (London, 1958), pp. 143, 928, lists it as published in Ferrara by Francesco Rossi.

AN ILLUSTRATED DISSECTION OF THE MUSCLES OF THE HUMAN BODY

By Giovanni Battista Canano, Physician of Ferrara, prepared at the request of Bartholomew Nigrisoli, Nobleman of Ferrara, now first published

Giovanni Battista Canano, to Bartholomew Nigrisoli, nobleman of Ferrara, greetings:

Galen of Pergamon, easily the chief of all physicians since Hippocrates, among other commentaries filled with rare and unusual erudition which he composed for the pleasure and use of his friends and of physicians also wrote *On Anatomical Procedures*, at first in two volumes, to please Flavius Boethus, a Roman consul and a man especially interested in anatomical science. Later, however, after the death of Boethus he wrote other books (some of which are extant) at the urgent request of his friends. In them he fully described the method and process of dissecting the members of the ape, an animal very similar to man. Just so, you, Bartholomew Nigrisoli, equally noble and learned, burn no less than that Roman consul Flavius Boethus with a desire for anatomical knowledge. You have often been present at my dissections of the human body and have discovered not only from the testimony of Galen but from that of your own eyes how great is the need and the usefulness of skill in anatomy in the practice of medicine, especially for the surgeon. You also see that most modern surgeons are unskilled in dissection and are therefore responsible more frequently for the destruction rather than the salvation of their patients. For this reason you have often urged me to publish my illustrated anatomy of the human body. In this way you believed that those who practice medicine but who cannot recognize the parts of the body by personal inspection during dissection might gain some knowledge of them at least through pictures and thus become more reliable in their medical consultations for the health of those in their care. I could no longer refuse or resist your frequent and most salutary exhortations; so great is your kindness toward me and the extent of your merits that I thought I should not deny you. Although there are various parts of the human body there are none which suffer more from injury by external objects than the muscles, for they are the first to receive wounds, piercing cuts, or blows. Therefore, to please you and to serve the safety of men I decided to publish the muscles in the form of pictures drawn by Girolamo of Carpi, a painter of our time who is both distinguished and accurate in his work. Thus the muscles of the human members now issue from the press under your auspices, most accomplished of men, the muscles of the human arms and legs which I dissected with as much skill as I could muster while Antonio Maria Canano, physician of Ferrara and a man joined to me by the closest relationship, demonstrated them. I also supervised the printing of the book and I now dedicate the muscles to the immortality of your name. I desire that my readers should not condemn this work of mine before they compare its contents with actual facts collected by frequent personal observation. For in this way I think they will derive some good from my efforts rather than gnash their teeth at them without reason.

Farewell

Candid Reader:

Since they were able to dissect both living and dead bodies, some of the ancients, according to Celsus, denounced each method of doing so, in the first place as cruel and unbefitting a physician, a man who presides over the health of human beings; for they thought it characteristic rather of a brigand than of a physician who wishes to learn the nature of men's inner parts to cut into the body cavity and vital organs of living men, especially when some of these parts sought out with such great violence cannot be recognized at all while others can be recognized even without such wicked behavior. For the qualities of color, softness, hardness, lightness, roughness are not the same in a body which has been dissected as they are when it is intact; if unviolated bodies change with fear, sorrow, hunger, surfeit, weariness, and a thousand other harmful conditions, so much the more do the inner parts, which are softer and upon which the light of day is brought to bear, change under the very severe wounds inflicted by dissection itself. It seems foolish to suppose that any part of a living man remains the same in a dying man or in one who is already dead. If there were any such part, it could be subjected to inspection while the man were still breathing, a situation which often presents itself to physicians. For it sometimes happens that a man is wounded in such a way that some part of his inner region is laid bare, and another part in another man, whence the prudent physician can recognize their location, arrangement, shape, size, and other similar facts while he strives to preserve the health of his patient through his mercy, facts which others recognize by slaughter accompanied with dreadful cruelty. For this reason the ancients used to say that not even the second method of dissection, that is, the laceration of dead bodies, was necessary; if not cruel, it was nevertheless repugnant, since many parts are different in dead bodies and those parts which can be recognized in the living are abundantly demonstrated by the medical care itself which is applied to them. In addition, those who recognize that these facts do not actually clash with the usefulness and necessity of the anatomist's function will readily confute him. For since there is a twofold dissection of bodies, that is, one of those which are alive and one of those which are dead, from the first (if properly carried out) we recognize that the handling of some, although not all, of the internal parts is necessary for a knowledge of their cure, not by dissecting living men (which is cruel and characteristic of a butcher rather than of a physician) but by dissecting brute animals. These in many of their parts and the functions of those parts are very similar to man, as in their kidneys, ureter, bladder, uterus, intestines, mesentery, omentum, spleen, stomach, liver, transverse septum, esophagus, lungs, heart, and other similar parts. But from the second kind of dissection, that of the dead, one may learn the shape of the parts, their number, location, connection, and operation and sometimes (as with muscles drawn to their place of origin) information extremely necessary for a knowledge of their healing. This is not repugnant; indeed, it is excellent, and quite convenient if we consider the cure and recognition of the parts that suffer pain for which chance is by no means a sufficient source of information. For rare are those wounds through which the inner parts of the body are so revealed that you can adequately recognize their shape, position, and relationship. If by chance this occurs with some parts of the body it certainly does not occur with all of them. However, in the dissection of dead bodies we are able to recognize all the parts of man in a short time which by dependence on mere chance any man, however laborious or prudent, could never offer to our view even if he lived as long as Nestor. Since therefore I engaged quite diligently in the dissection both of living animals and of dead human bodies as much as I could during my youth and read many books handed down by the ancients, I found that some of my work was not without value to the community of physicians. Thus I deemed it worthy to impart it to those who practice medicine. Hence it has come to pass that I have published this First Book of the Dissection of the Muscles at the frequent urgings of Bartholomew Nigrisoli, nobleman of Ferrara. The remaining books, already under the engraver's press, will soon be published. Accept this book, candid reader, in a kindly spirit, for your kindness will serve as a sort of spur to impel me toward greater and more assiduous efforts. Farewell meanwhile and pray that the gods may be propitious to me in such great labors, as they have been while the beginning of the work I have undertaken comes to birth.

The First Book of the Illustrated Dissection of the Human Muscles by Giovanni Battista Canano, Physician of Ferrara

Following Galen, I place the muscles of the hand in this first book, but by the word hand (in the manner of Hippocrates and Galen) I mean what is commonly called the arm. This is generally divided into three primary parts, that is, the end of the hand, the hand (the part called the hand proper), and elbow; the arm was called brachion by the Greeks. The end of the hand has three parts, the fingers, the wrist, which the Greeks called karpos, and the part between these two parts called metakarpion by the Greeks and postbrachial by us. The arm, or if you prefer to say it in Greek, brachion, is composed of a single bone, which Celsus called the humerus, but the elbow (which he names the brachium) is composed of two bones. One of these is called kerkis in Greek, radius in Latin; the other is called pekus by the Greeks but cubitus (elbow) by Celsus. At the end of the hand are twenty-seven bones, eight of the brachial or wrist, four of the postbrachial, and fifteen of the fingers, all of which we must learn to recognize for a complete knowledge of the muscles pictured in this first book. With this brief preface I approach the subject proposed.

(1) In this picture [fig. 28] the following muscles are laid bare:

A. The muscle which bends the wrist toward the little finger [flexor carpi ulnaris].
B. The muscle of the palm [palmaris longus].
C. The muscle of the split tendons, under which is a muscle which has tendons that are not split. Both muscles move the fingers [flexor digitorum superficialis].
D. The muscle which bends the wrist near the big finger (or thumb) [flexor carpi radialis].
E. The muscle which bends the hand downward (i.e., closes it) [flexor digitorum profundus].
F. The muscle which bends the hand upward (i. e., opens it) [extensor digitorum profundus].

(2) The muscle of the palm of the hand [fig. 29 which, according to the ancients (as Galen says in *On the Use of the Parts* II and *On Anatomical Procedures* I), bends the five fingers. According to him it is made to strengthen the inner skin of the hand for the sake of a firmer grasp, in order to prevent the growth of the hairs, and to create a more delicate sense of touch. Galen is correct in his contradiction of the ancients in *On Anatomical Procedures* I. He says that, according to them, the tendon of the muscle which moves any joint must be inserted into its bone. But the tendon of the muscle of the palm is not inserted into the bones of the fingers; thus according to their opinion it cannot move the fingers. Galen may be forceful enough in his reasoning against the views of the ancients but it seems nevertheless to be of little value. For a person might say that it is not always necessary for the tendon of a muscle which moves a joint to be inserted into the bone of that joint or joined to it but that it is sufficient if the tendon is joined to something else joined in its turn to the bone, just as the tendon of this muscle is

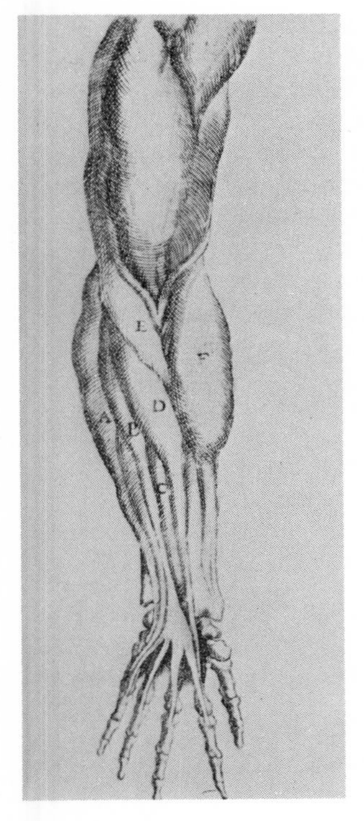

FIG. 28.

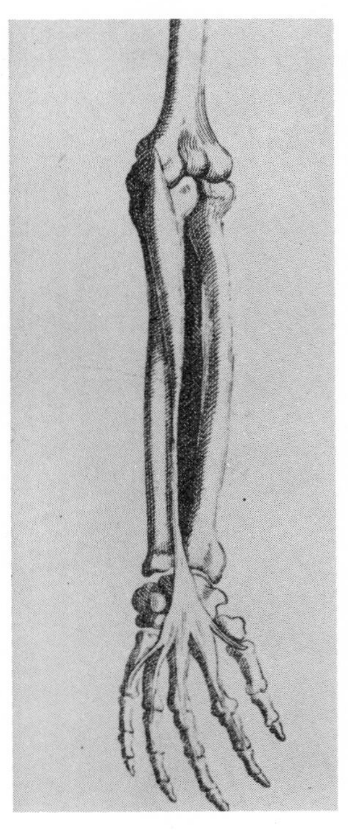

FIG. 29.

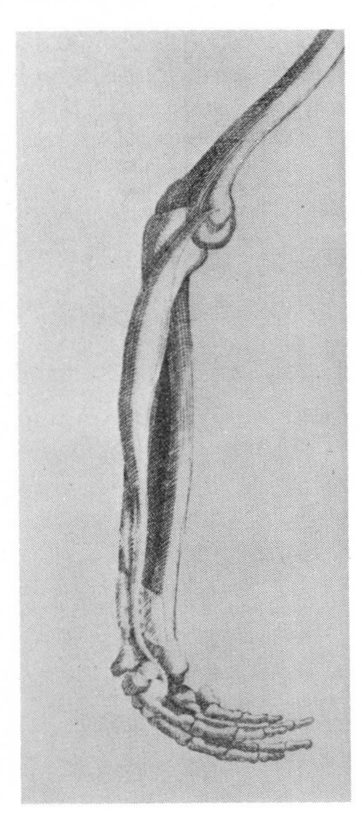

FIG. 30.

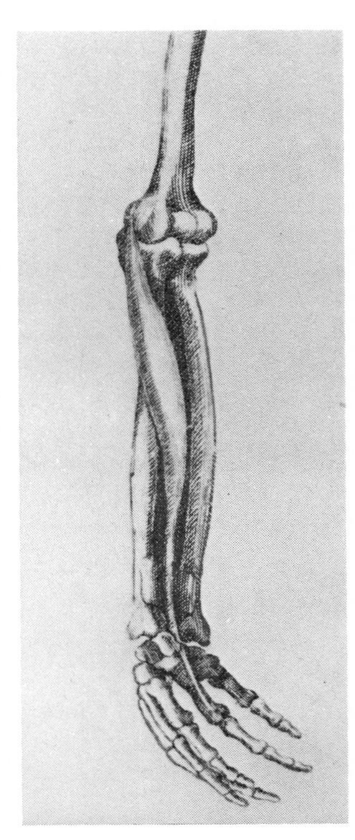

FIG. 31.

FIGS. 28–54. Illustrations from Cannano's *An Illustrated Dissection of the Muscles of the Human Body*. From *Archivio Italiano di Anatomia e di Embriologia* 68 (1962). Courtesy of Professor I. Fazzari.

joined to a ligament which is attached to the bones of the fingers. This muscle originates from the inner node or joint of the arm [epicondylus medialis] and proceeds in the middle between the muscles which bend the wrist. With its tendon it rises above the inner ligament which joins the elbow [ulna] to the radius. Later the tendon of this muscle is dilated in the form of a triangle whose tip is near the aforesaid ligament but whose base at the beginning of the postbrachial is inserted by means of five tendonlike formations [palmar aponeurosis] into the ligament which embraces the tendons of the other muscles which bend the fingers. One may cut this twofold tendon with the skin, of course, and without the skin.

(3) The muscle [fig. 30] which bends the wrist arises from the inner node or joint of the arm, adheres to the elbow, and is inserted in the straight bone with a membrane which arises from the region of the little finger following the process of the elbow which the Greeks called sometimes graphoeidē and sometimes styloeidē.

(4) The muscle [fig. 31] which bends the wrist near the big finger (thumb) arises from the inner joint or node of the arm and proceeds obliquely for a short distance above the elbow and is stretched out to the radius. It is inserted by a tendon into the postbrachial bone in front of the index finger, although at first sight it seems to be inserted into the bone of the wrist near the thumb.

(5) The muscle [fig. 32] which ends in split tendons bends the second joint of the fingers [with the exception of the joint of the thumb, which does not receive a tendon from this muscle] and arises from the inner joint of the arm and touches the elbow [ulna] for a short distance. It is inserted with its tendons into the second internode [phalanx] of the four fingers and is contained under the ligament which contracts the radius and the elbow [retinaculum flexoris] in the inner part.

(6) The muscle [fig. 33] which, according to Galen, bends the first and third joint of the four fingers but only the second and third internodes of the thumb arises from the region of the elbow [ulna] near the protuberance of the latter. Throughout the deeper parts it embraces all the middle region of the radius and of the elbow [ulna] and adheres to the bone of each of these. It is sketched out into three parts, as it were; one of these faces toward the little finger, the second toward the index finger, and the third toward the middle fingers. The tendons of this muscle are contained under a ligament which contracts the radius and the elbow inwardly; and by means of the medial divisions of the tendons which bend the second internode of the four fingers these tendons are carried

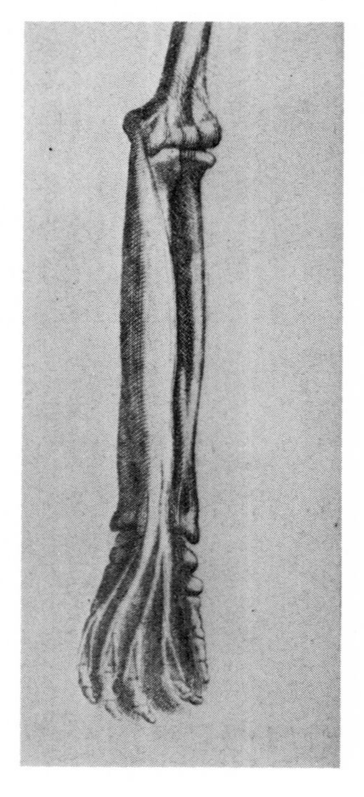

FIG. 32.

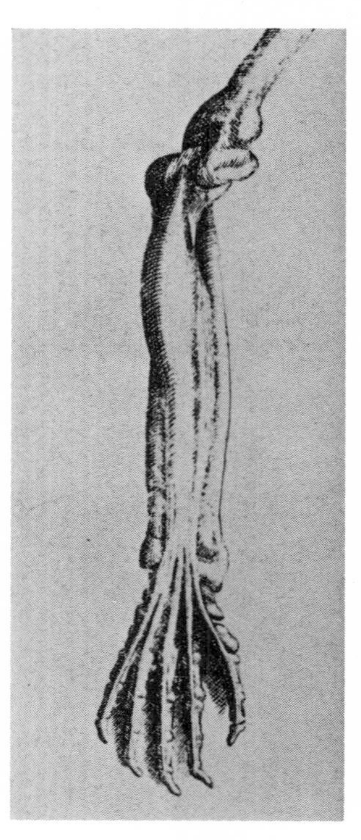

FIG. 33.

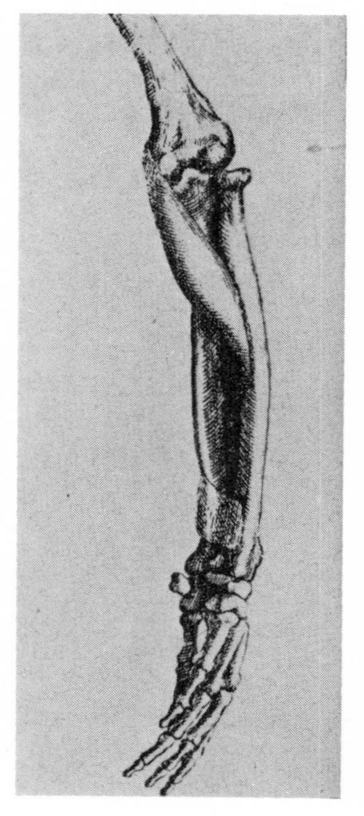

FIG. 34.

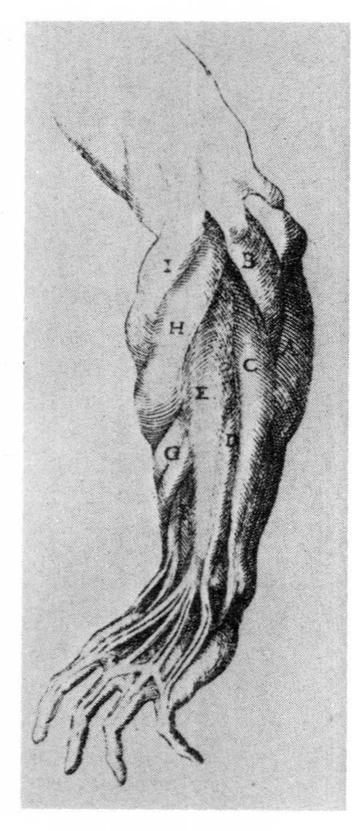

FIG. 35.

to their extremity. Together with the tendons of the preceding muscle these latter tendons are extended under the same ligament from the beginning of the fingers to their ends. Galen insists that by means of this ligament this muscle bends the first internode of the four fingers; but he does not seem to have any more reason for this statement than he had for the previous one, since the tendons of this muscle are included under the same ligament.

(7). Of the muscles [fig. 34] which bend the hand downward one originates from the higher parts of the inner node or joint of the arm and is carried obliquely above the elbow [ulna]. It is inserted in the radius and is extended inwardly to the middle parts of that bone [pronator teres]. The remaining muscle of those which bend the hand downward originates from the lower part of the elbow [ulna] near the wrist and is inserted obliquely into the radius bone [pronator quadratus]. Both muscles have a fleshy terminus but move the bones nevertheless, although Galen, *On the Motion of the Muscles* I, writes that all muscles which move the bones whether large or small terminate in tendons. But on this matter he asserts the opposite opinion in *On the Use of the Parts* II.

(8) In this picture [fig. 35] there appear the following muscles:

A. A muscle from the inner regions which extends five tendons that are not split.
B. A muscle which bends the hand upward.
C. A muscle which bends the wrist toward the little finger.
D. A muscle which extends the little finger and turns it to one side.
E. A muscle which extends the three (middle) fingers.
F. A muscle which turns the index finger to the side.
G. A muscle split in two which extends the thumb by bending it away from the other fingers.
H. A muscle with a double tendon which turns the carpus outward.
I. A muscle which bends the hand upward.

(9) The muscle [fig. 36] which according to Galen has four tendons stretching to the four fingers except the thumb in fact seems to have only three tendons stretched out to three fingers between the thumb and the little finger [extensor digitorum]. Of these tendons two appear to be double, that is, the tendon of the index finger and the tendon of the ring finger, which has two offshoots. One of these offshoots is joined to the tendon which stretches to the middle finger; the other proceeds to that tendon which is near the little finger and originates from its own muscle [extensor digiti minimi]. This muscle is contained under a ligament which stretches the radius and elbow outward [extensor retinaculum]. It originates from the exterior node of the arm [epicondylus lateralis] and stretches out above the radius.

(10) The muscle [fig. 37] which originates from the exterior node of the arm and adheres to the elbow and the radius along their extent is contracted under

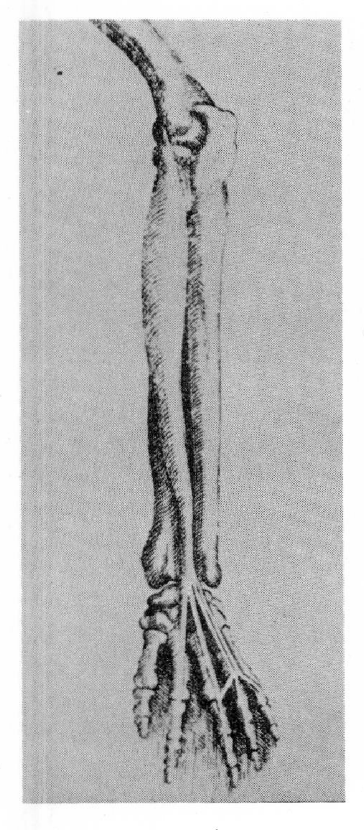

FIG. 36.

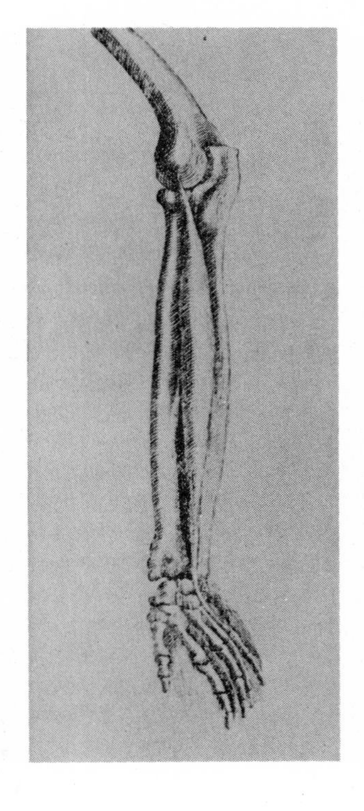

FIG. 37.

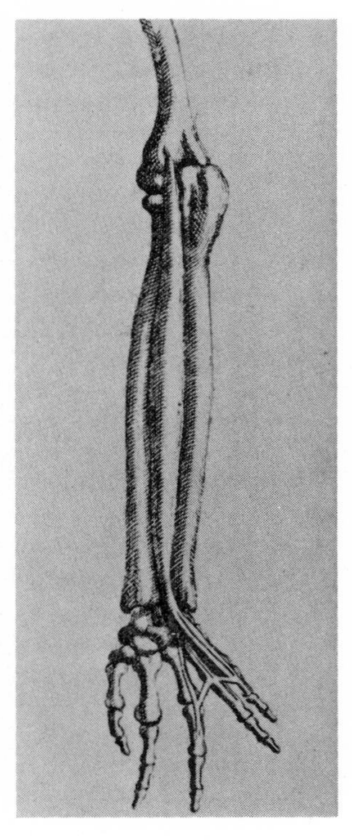

FIG. 38.

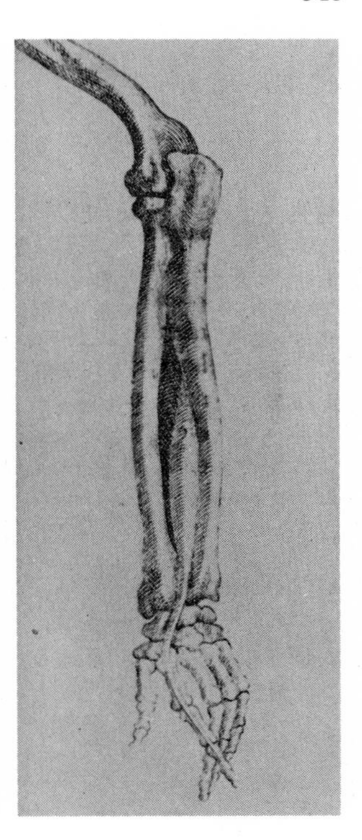

FIG. 39.

its own ligament at the end of the elbow. It is inserted with its tendon in the postbrachial bone in front of the little finger and extends the wrist [extensor carpi ulnaris] toward the little finger.

(11) The muscle [fig. 38] originates from the exterior node of the arm, touches the radius and elbow [ulna] along their extent, and is contained at the end of the elbow [ulna] by its own ligament. It is split into two tendons which stretch toward the little finger and draws the little finger to the side or toward the lower region by bending it away from the other fingers.

(12) The muscle [fig. 39] originates from the elbow [ulna] and adheres to it under a common ligament which contracts [connects] the radius and the elbow. When the muscle [extensor indicis] is stretched it extends a single tendon to the index finger, not two tendons to the index and middle fingers, as Galen asserts. It bends the index finger to the side.

(13) The muscle [fig. 40; extensor pollicis longus] originates from the elbow [ulna] above its middle part, touches the radius, and at the end of the radius is contracted [contained or secured] under its own ligament. It extends a single tendon which stretches throughout the entire thumb and bends the latter to the side, according to Galen.

(14) The muscle [fig. 41] originates from the membrane which divides the interior muscles from the exterior [membranea interossea] ones as well as from the elbow. It ascends the radius and is contained under its own ligament at the end of the radius (where the latter is carved inward). It extends two tendons, one of which is inserted into the extremity of the first bone of the thumb. This tendon does not extend to the ultimate point of the thumb, as Galen insisted. The second tendon is inserted into the bone of the wrist in front of the thumb. This muscle can be regarded as two muscles [extensor pollicis brevis et abductor pollicis longus] says Galen, for by means of one tendon it extends the wrist toward the thumb; by means of the other tendon it bends the thumb away from the other fingers by extending it for a certain distance.

(15) The muscle [fig. 42; extensor carpi radialis] originates from the exterior node of the arm and is stretched out above the radius. It is contracted [contained] by its own ligament at the end of the radius and, inserted by a split tendon into the bones of the postbrachial in front of the index and middle fingers, it moves the entire wrist outward.

(16) The muscle [fig. 43; supinator] originates from the membranous ligaments of the joint, touches the elbow [ulna] for a certain distance, and is inserted into the radius where it sticks out more boldly at its extremity. It bends the hand upward.

(17) The muscle [fig. 44; brachioradialis] which bends the hand upward originates from the arm near its

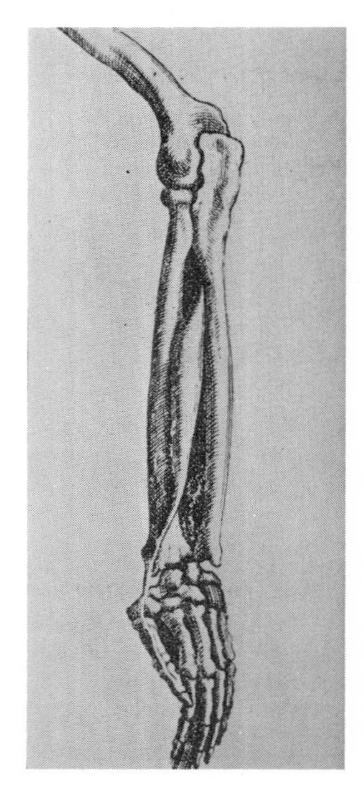

Fig. 40.

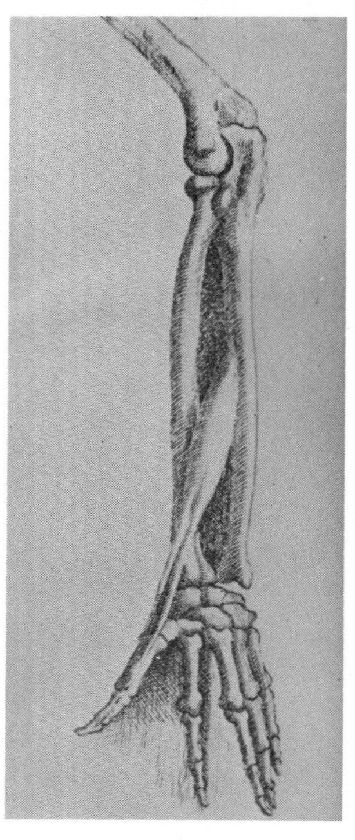

Fig. 41.

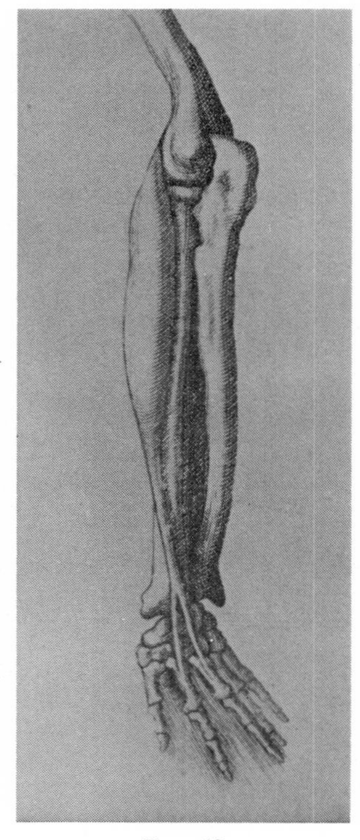

Fig. 42.

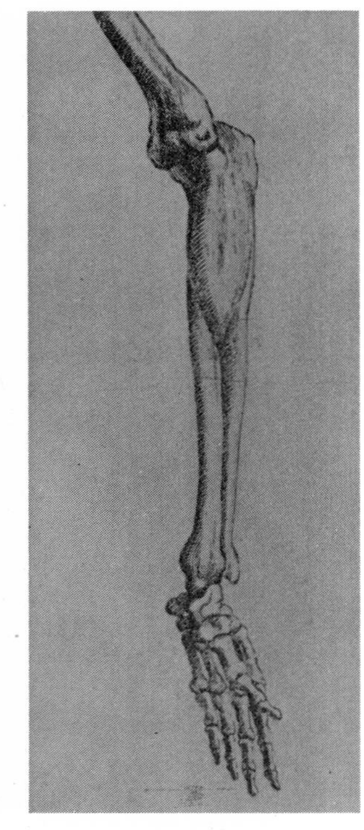

Fig. 43.

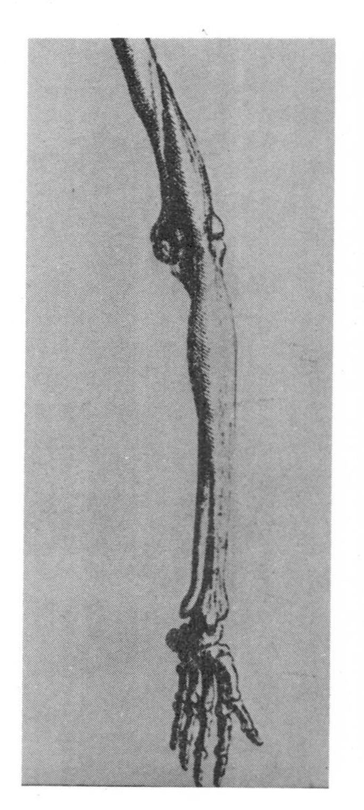

Fig. 44.

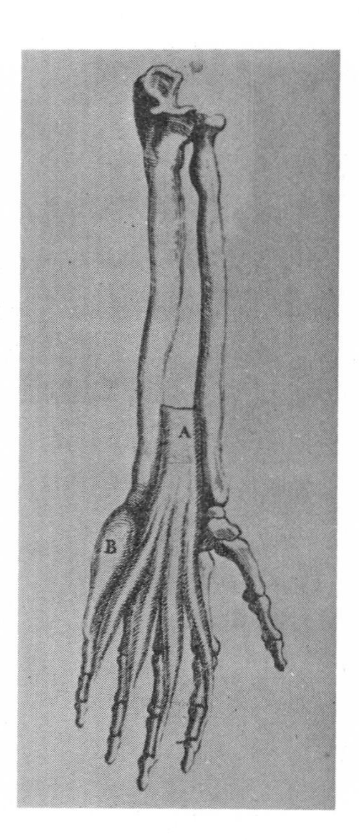

Fig. 45.

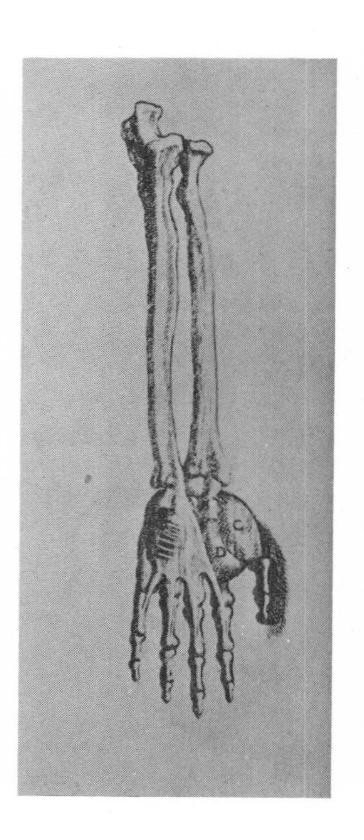

Fig. 46.

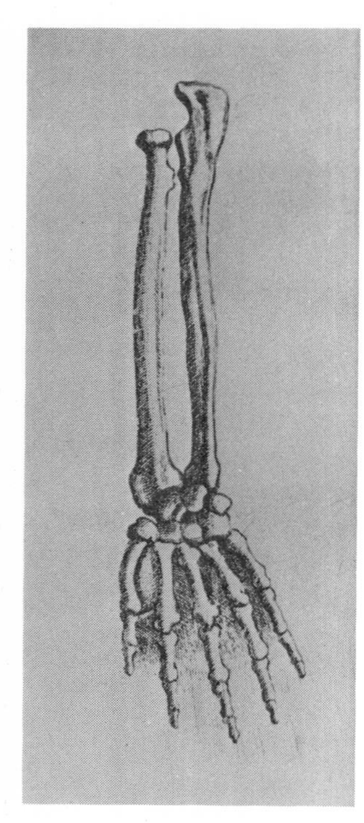

Fig. 47.

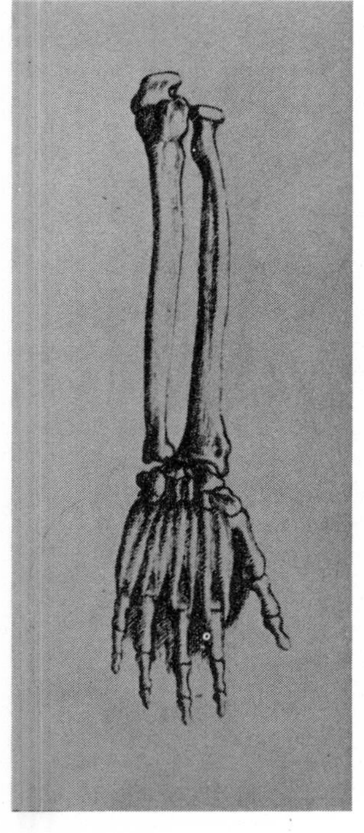

FIG. 48.

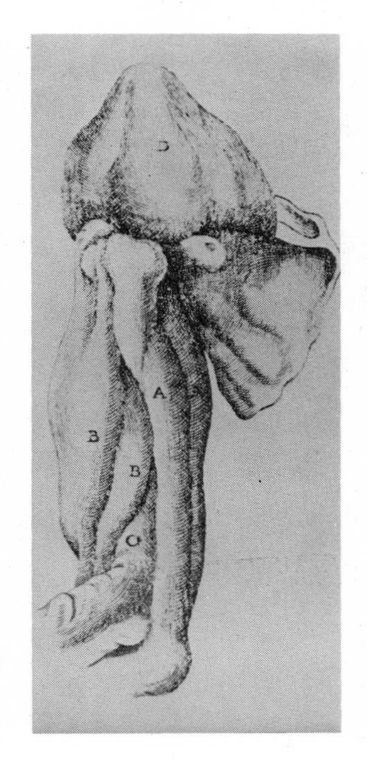

FIG. 49.

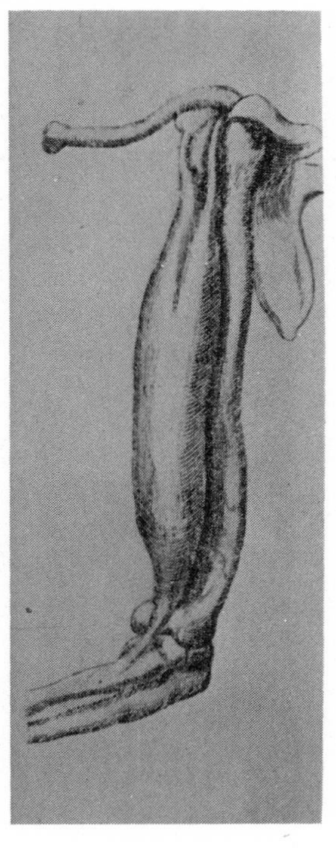

FIG. 50.

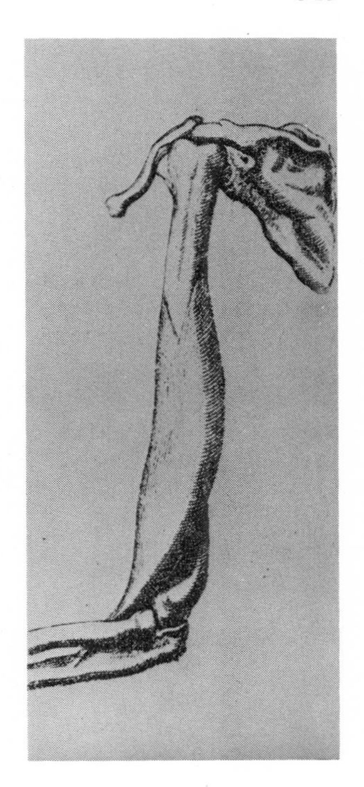

FIG. 51.

middle part. It is extended above the radius and is inserted into that bone at its end as far as the wrist.

(18) A. Four slender muscles [fig. 45; interossei palmares] originate from the membranes which surround the tendons that are not split. These muscles come up under the sides of the fingers, those on the right (when the hand is turned upwards) going toward the left and those on the left going toward the right. Their function is to move the fingers toward the side; those on the right move the fingers to the left and those on the left move the fingers to the right.

B. A muscle [abductor digiti minimi] which originates from the bone to which is attached a tendon of the muscle that bends the wrist to the little finger is inserted into the little finger in the lower part and bends that finger away from the other fingers.

(19) C. A muscle [fig. 46; abductor pollicis brevis et opponens pollicis] which originates from the first bone of the wrist near the thumb to which it is inserted has as its function according to Galen the movement of the thumb away from the other fingers. But it seems rather to be made for drawing the thumb toward the little finger.

D. A muscle [adductor pollicis brevis] fixed with its head in the postbrachial bone in front of the middle finger is inserted with oblique fibers into the first bone of the thumb. This muscle bends the thumb toward the index finger by inclining it inward.

E. A muscle [palmaris brevis] or muscles which are situated above the muscle which bends the little finger away from the others in the inner part of the end of the hand adhere with oblique fibers to the skin. With their tendons these muscles are joined to the tendons of the palm. Galen does not mention these muscles, which appear to be created for the dilation of the tendon of the palm.

(20) A muscle [fig. 47; abductor pollicis longus?] which Galen does not mention is fixed in the exterior part of the hand behind the wrist in front of the index finger. It is inserted with oblique fibers into the first bone of the thumb and bends the latter toward the index finger, inclining it outward and bending it straight at the same time with the second of the preceding muscles.

(21) Eight slender muscles [48; interossei?] which form a fleshy body originate from the coarticulation of the brachial with the postbrachial bones and are inserted by twos into the first abarticulation of the fingers from the inner region, receiving something of the lateral parts. At the same time they bend the first joint straight and turn it apart, inclining it a little to the side.

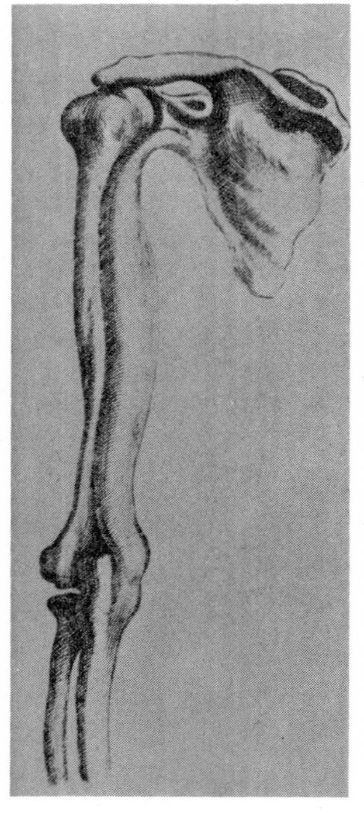

FIG. 52.

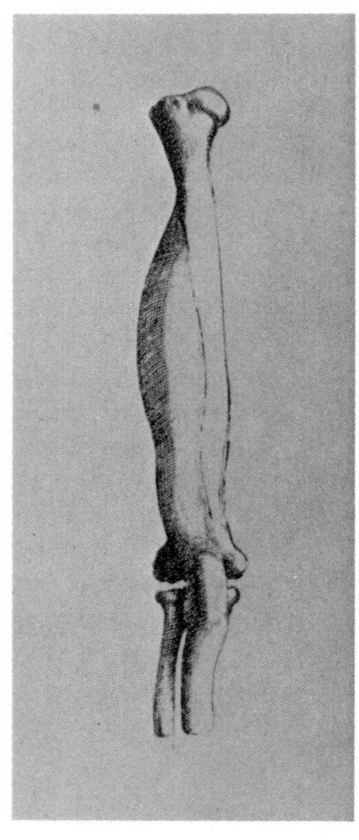

FIG. 53.

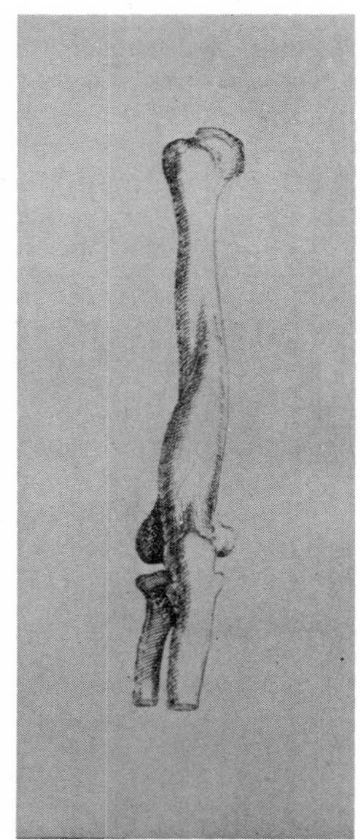

FIG. 54.

(22) In this picture [fig. 49] the following muscles are laid bare:

A. The muscles which extend the elbow.
B. The muscles which bend the elbow.
C. The head of the muscle which bends the hand upward [supinator].
D. The muscle of the neck which some have called deltoeidē since it imitates the form of the Greek letter delta.

(23) The conspicuous muscle [fig. 50] near the humeral vein [cephalic] on this side of the cleavage has two heads [biceps brachii]: one for the open fissure of the arm which proceeds to the high parts of the cervical scapulae, the other for the process called anchor-shaped (anchuroeidē or korakoeidē) [coracoidea]. These heads coming together generate this muscle, which makes a sinewy thinness and procreates a strong tendon. By this tendon it is inserted into the radius. Receiving something of the membranous ligament of the joint it lifts the entire joint, inclining it gradually to the inner region. With the muscle which is laid under it this conspicuous muscle makes a straight flexion.

(24) The muscle [fig. 51; brachialis] is surrounded by two fleshy sources of origin, a higher one in the posterior part, a lower one in the anterior part. These sources coming together generate this muscle, which ends in a sinewy thinness and is inserted into the bone of the elbow [ulna]. It bends the joint by inclining it gradually toward the outer region. With the muscle laid under it this muscle makes a straight flexion.

(25) The muscle [fig. 52; triceps brachii, caput longum] originates from the lower side of the scapulae more or less out of half of their upper part. It is joined to another muscle placed under the head of the arm. Both muscles are inserted by means of a broad tendon into the protuberance of the elbow. This double tendon appears to follow the fibers and receives its external part from this latter muscle but its inner part from the former muscle. If you draw one away from the other you extend the entire hand, as Galen said, with this difference, that the first muscle extends a short distance into the outer region while the second muscle extends into the inner region by deflection downward.

(26) The muscle [fig. 53; anconeus brachii, caput breve] originates under the head of the arm. It is joined to the preceding muscle and to another muscle laid under itself, to such an extent that the latter is considered a part of the former muscle by anatomists. These muscles can be separated, however, along the straight line of their fibers. This muscle has the function, as Galen says, of extending the entire hand by deflecting it obliquely inward.

(27) The muscle [fig. 54] which is considered by anatomists a part of the preceding muscle is continuously fleshy and is inserted into the posterior parts of the protuberance of the elbow and makes a straight flexion of the joint.

End of the First Book

Appendix

I. THE LAST WILL AND TESTAMENT OF ALESSANDRO BENEDETTI[1]

Venezia. Archivio di Stato. Sezione notarile. Testamenti. Notaio Gerolamo de Bosis. B.50, testamento, n. 178.

In nomine Dei eterni amen. Anno ab incarnatione Domini nostri Jhesu Cristi millesimo quingentesimo duodecimo, indictione prima, die vero decimo octavo mensis septembris Rivoalti. Fragilitatis humane cursum dilligentissime considerans et ante oculos habens, ego Magister Alexander de Benedictis de Verona physicus, filius quondam Domini Laurentij de confinio Sancti Pantaleonis Venetiarum, sanus Dei gratia mente et intellectu licet corpore infirmus, volens dum tempus datur rebus meis providere et modo debito ordine disponere, ad me vocare feci Hieronymum de Bossis Venetiarum notarium, ipsumque rogavi ut hoc meum scriberet testamentum, ipsumque post meum obitum compleret et roboraret iuxta formam, tenorem et continentiam presentis cedule bombicine a presente manu mea scripte et eidem notario per me presentate. Interrogatus et informatus ab ipso notario de interrogandis et informandis, respondeo quod volo quod una cum commissariis meis in ipsa cedula annotatis esse debeat commissarius meus Dominus Antonius Moretus sororius meus et Dominus Vincentius Saracenus ducalis secretarius et quod misse Sancti Gregorij celebrentur per sacristam presentem Sancti Pantaleonis. Item declaro quod Lucia uxor mea sit commissaria mea in omni statu et quod ea que sibi legavi, intelligantur legata in omni statu. Item volo quod misse Beate nostre Virginis Marie celebrentur per Reverendum Dominum Magistrum Franciscum Colona fratrem in Sancto Joanne Paulo. Item quia legavi in dicta cedula ducatos sexcentos Julie filie mee volo quod ultra omnes vestes exstimandas et ultra tres petias terre quas sibi ligavi pro computo dicte pecuniarum summe que sunt le nogarete, proris et progolecte volo quod commissarii mei supleant dicte Julie dictam summam ducatorum sexcentorum de terris meis, quas habeo in villa Sancti Bonifacij agri taurisini (*in rasura*) veronensis. Item dimitto prefato Domino Vincentio Saraceno ducatos decem, quos sibi dare debeo ex causa mutue. Tenor autem dicte cedule bombicine talis est infra sequitur de verbo ad verbum videlicet: Testamentum mei Alexandri Benedicti veronensis physici et cetera MDXI die ut in eo. In Christi nomine amen. Individue Trinitati comendo spiritum meum et Marie Virgini. In nomine Dei eterni amen. Anno Domini MDXI die primo julij. Primum testamentum quod alias Iadre condidi obliteratum volo ac irritum. Deinde fateor quod Julia habeat de bonis quae nunc possideo, quae fuere Domine Dionire eius matris primum ducatos ducentos quinquaginta qui sunt pro venditione facta in eximium legum doctorem Dominum Bartolomeum Nordium (?) civem civitatis Austrie pro possessione in contrata de Briliono ut patet instrumento Domini Joannis Marie de Obsteirantiis ut in eo. Item quod ducatis ducentis septuaginta quod pretium fuit pro domo habitata per me in presentiarum reliquum vero quod exbursavi sit pro computo meo ut apparet ad officium commissariorum mercatorum, et cyrographo manu Ser Joannis Francisci Maynenti cum hoc tamen pacto quod si fuerit contenta de summa supradicta hoc est in duabus partitis ducatos quingentos et viginti accipere tot posessiones et affictus hoc est in castro Cruce totum quod ibi possideo pro ducatis centum trigintis, hoc est tres petias terre hoc est nogarete, proris et le progolette cum quella pocha pension che paga ala giesia de Illasi et altri ficti et campi che ascenda ala ditta summa de ducati de ducati [*sic*] cinque cento et vinti computando i soi vestimenti de dosso estimadi per persone idonee sia per conto de sua dota et de tuto le lasso tanto de i mei mobeli che ascenda ala summa de ducati 600 d'oro.

Et non volendo star gita volio et ordeno che la casa comprada a nome de sua madre ma per me compidamente pagada ut supra cum li melioramenti fati per mi sia messo a conto de la sua dote zoe de sua madre cum sit che in libro meo computum sia nota tutto quello ho speso in dicta casa.

Et morendo senze fioli legitimi volio chel dicto superhabundante sia de li altri fradelli over sorelle. Et perchè Marin per me novamente legitimado cum intention che volesse clericar, et non volendo li lasso di mei beni mobeli tanto per summa de ducati cento de esser stimadi per. . . . Item per l'anima mia lasso de i beni mobeli sia fatto uno incanto et se venda per ducati 50 zoe cinquanta da esser dati ala schola di San Marco ducati diexe. Ala scola del Spirito Sancto ducati diexe. Ala scola de Sancta Maria Formosa ducati cinque. Li altri ducati vinticinque per la mia sepultura et spexe del obito da esser partiti come parera al mio infrascritto Comissario. Item el resto che sonno ducati 300 doro posti in tanti beni stabeli in San Bonifacio sonno per conto de Lucia mia consorte. Et lei habia a tuor dove li piace segondo el precio che sonno compradi. Et li soi fitti compradi per essa dispona a sua a sua [*sic*] voglia. De beni mobeli per so conto portadi per essa per ducati 300 li romagna per suo conto ad honesta stima ut supra. El

[1] I am grateful to Dr. Giulia Mirabello, former member of the staff at the Archivio di Stato, Venice, for her transcription of this document, which is difficult to read and where even her expert eye has not been able to distinguish two words which I have conjectured to be "agri" (taurisini) and "Nordium" or "Nardium," after "Dominum Bartolomeum." Dorothy M. Schullian, *op. cit.*, pp. 143–145, has published this will, but her version leaves at least eleven lacunae which my printing of it has filled out, besides correcting a few errors; furthermore, she has not extracted from the will the many details of biography which it yields, merely transcribing it "albeit imperfectly and without interpretation or commentary, in the hope that it may be more readily accessible to historians" (*ibid.*, p. 43).

resto veramente siano de Cornelia et Lucia equalmente, et se una more, vada in laltra. Et se tutte doe vada la mita in dicta Julia et laltra in dicta Lucia mia consorte. Et se tutte tre morisseno senza heredi per la parte che io li don del mio, volio sia ducati 200, zoe duxento de Marin ultra li dicti ducati 100. El resto de dicta Lucia. Item volio che se faci inventario de le mie robe quale se habi a vender al publico incanto per el mio commissario infrascritto. El qual volio habi del mio non per merito, ma per segno de dilection ducati X. Et volio sia anchora commissaria in omnibus cum esso ditta Lucia mia donna. Et sia anco commissario Messer Benedetto Calbo quondam magnifici domini Petri, el qual habi quam cicius poterit adempido ogni cosa, al qual recomando l'anima mia che faci dir de bonis meis le messe de San Gregol et de la Madonna. Item perchè intendo esser morta Faustina sua sorella in ongaria et se quella havesse fato altro testamento per el qual essa lasasse la soa parte ad altri che a la ditta Julia, in questo caxo intendo lhabia tanto de la parte soa quanto ditta Faustina havesse ordinato, azoche non sequite qualche . . . error ne la quantita de essa dote da esser disignada ala ditta Julia perchè se convigneria lassar dove essa Faustina havesse ordenado. Et questo volio sia fermo et rato. Et questa la mia ferma et constante volunta ultima in omnibus et per omnia da esser observada.

Ego Johannes Alexander Benedicti physicus veronensis artium et medicine doctor, corpore et mente sanus manu propria hoc testamentum annotavi die et anno ut supra, quod meis sigillis roboravi et signavi. Clausumque dedi Lucie uxori mee servandum preterea plenissimam virtutem, potestatem et auctoritatem do et confero suprascriptis commissariis meis post mei obitum commissariam meam intromittendi, regendi, gubernandi et administrandi pecunias, res et bona petendi, exigendi et recipiendi et de exactis, habitis et receptis quietandi, liberandi et absolvendi. Et pro ea in omni judicio comparendi, agendi, defendendi, petendi, respondendi, lites contestandi, jura producendi, probandi, in animam meam jurandi, sententias et acta quaelibet fieri faciendi, appellandi et appellata prosequendi usque in finem, ac omnia et singula alia faciendi et exercendi que egomet facere possem si viverem.

Si quis igitur contra hanc mei testamenti cartam ire temptaverit componat cum suis heredibus huic mee commissarie auri libras quinque et nihilominus haec mei testamenti carta in sua permaneat firmitate.

Signum manus suprascripti domini magistri Alexandri Benedicti.

Io Antonio Agnello spicier quondam Ser Agnello da Lonigo testis subscripsi.

(S.T.) Ego Hieronymus de Bossis quondam domini Bartolomei Venetiarum notarius complevi et roboravi.

II. THE LETTERS OF BENEDETTO RIZZONI TO GABRIELE ZERBI

(1)

(fol. 134 r) Magistro Gabrieli de Zerbis physico Padue publice legenti Benedictus Rizonius

Postquam a nobis, vir celeberrime, dudum exiuisti infitiari non possum hoc ipso tempore absentie tue sepissime meminisse, adeo ut nunc utrum esse fatear quod communi sermone refertur et apud Comitium his uerbis scriptum est. Nos homines nostra intelligimus bona dumque in potestate habuimus ea amisimus. Quapropter ne uidear ob huismodi absentiam tuam tantum boni penitus amisisse pro antiqua amicitia et obseruantia erga te mea uisum est mihi post longam intercapedinem hisce meis te lineis reuisere. Quandoquidem magis amici esse existemem memorem se lineis aliquando praestare quam intra pectus amorem diutius silendo retinere que mea ratio impulit officio meo fidentius erga te uti debere qua simul ratione ad hoc scribendi officium fidentius animum applicui. Tam in testimonium mee de te, ut dixi, diuturne memorie quam ut etiam cognoscere queas tue me praeclari doctrine ac medice artis mihi olim ab extremis salutem reddidisti queque meo non solum corpori sed universe curie ut ab omnibus fere praedicatur fuit salutaris hac ipsa tempestate desiderio teneri. Unde mihi quasi in animo est si tempora permitterent tam tui uidendi quam mei recreandi causa istuc proficisci maxime cum ad hec periculosum sit hic in urbe permanere cuius modo et praeteriti temporis periculum. Si tibi ordine recenseri uellem essem fortasse prolixior quam tu publico munere et aliis rebus tuis impeditus a me impresentia haud stare desideras praecipue cum non dubitem hec mala nostra istic quo nos quandoque ea uitandi causa uenturos esse sperabamus lineis et nuntiis palam facere (fol. 134 v) que magis bona erga te mente quam eleganti concinnitate proficiscuntur tue praeclare et exquisite eruditioni non iniocunda fuerint. Vale. Rome tertia Julii 1495.

(2)

Magistro Gabrieli de Zerbis B. Rizonius

(fol. 141r) Tui ad nos, vir praestantissime, optati reditus expectatio quem fore prope diem mihi in his lineis significaueras quas cum consilio tuo dignissimo de meo uiuendi regimine iam dudum accepi hoc unum effecit quod officio meo erga te et gratiarum actione hactenus uti non potui. Modo perspicienti mihi huiusmodi expectationem meam inanem esse ne crimen aliquod ex diuturniori silentio obiiciatur quod coram ut optabam totis affectibus agere non potui visum est denique hisce lineis efficere. Maxime oblata etiam mihi opportunitate a viro humanissimo et iam diebus multis contubernali meo, Vincentio Cremonensi, qui una mecum tui studiosus et excellentissimus esse uidetur quique ad literaria studia suis quibusdam negotiis iamdiu intermissa magno cum desiderio istuc ingreditur a quo propterea multa de rebus urbanis percunctari poteris que ipse breuitate temporis praetermittens ad rem ueniam ubi ex toto animi affectu tibi gratias ago de studio et labore suscepto mee causa salutis quia impresentia satis belle fruor quodque de me tibi alias notum reddidi quasi modo abysse sentio tam ex consilio tuo quod certe gratissimum habui quam ab obseruatione earum rerum quas aut obesse aut prodesse mihi consideraui quod si antea neque fecissem neque tam cito diffidere uoluissem res mea longe meliori loco sita foret. Nam prae timore grauioris egritudinis ac febricula quadam qua hac estate laboraui coartus fui non sine (fol. 141 v) consolatione fore non dubito quia tali socio viro ornatissimo praeditus sis. Aliud pro nostra amicitia ad te scribendum non occurrit nisi hoc quod in fine subticere non possum, id quod tue namque a nobis profectione aliqua causa fuit adhuc in pretio permanet artis hostie impedimento que utinam solo equata esset ut curiales ac reliqui omnes tanto dispendio liberati forent. Si quid uero de familia principis mei tui ad unum excellentissima noscere desideras scito nos omnes bene valere eoque liberius quo ipse princeps secessus et recreationis causa in x Rignano suo ante paucis iamdiu commoratur. Vale. Rome xviii mensis octobre 1495.

(3)

B. Rizonius Magistro Gabrieli de Zerbis

Forte nouum ne dicam in urbanum visum fuit hoc tempore probitati tue quod ego qui multis de causis eidem obnoxius sim te pridie ad gymnasia Patauina legendi causa redeunte in tuo huiusmodi a nobis discessu officium meum praetermiserim, pro cuius rei excusatione, vir amantissime, nulla alia impresentia apud te uti velim nisi quod si eo tunc officio meo uisendi te prius uti non potui propter occupationes meas que tibi note esse possunt hisce nunc lineis illud agere et supplere constitui uelut is quem delicti penitet qui profecto declarat (fol. 142 r) se maluisse non peccare quam penitere, his igitur eam fidem poteris adhibere que uerbis meis ante conspectum tuum fuisset adhibenda quoniam scire te uelim litteras meas profectum ad his similes probatissimos uiros ita bona mente proficisci quia adulationis et uanitatis uitium mea ipsa natura uehementer abhorreat; ea maxime ratione qua hoc idem de me facile tibi persuadere potes quia ab ineunte etate patris mei quem tu nosti virum doctissimum preceptis imbutus ita vitam institui ut illos omni studio diligam omnique beniuolentia prosequar quos merito amari dignos putem, et eo magis quo ab illis me amari intelligam. Nam teste Cicerone digni sunt amicitia ii quibus inest causa cur diligantur [Cicero, *Laelius De Amicitia* 21. 79], quod rarum genus est et omnia praeclara rara. Tu qui omni virtute ornatus ex iis esse uideris immo apertissime constat cum tuarum virtutum multa salutaria experimenta fecisti que cum alibi tum apud nos mihi profectum nota et experta sunt. Animus propterea meus in tui beniuolentia cum obseruatione ita adheret, quod siquid est quod a me hoc tempore absentie tue fieri posse existimes licet tibi, ut opinor, non deficiant quibus res tuas tuto committere possis tamen in testimonium mee erga te fidei sic dicere cogor quia habes me ad omnia obsequia et mandata tua omnimodo fide, studio et opere deditissimum ac siquando aliquid de me experiri uolueris conabor pro uiribus tuam semper expectationem uicere ac opinionem de me superare. Vale.

(4)

Domino Gabrieli de Zerbis

Post longam meam scribendi intercapedinem que non a tuo obliuione sed ab occupationibus meis profecta fuit, quibus modo posthabitis omnibus compulsus sum ea pro nostra antiqua amicitia et obseruatione erga te mea ex tue excellentie significare que eius honori ac tuorum statui et imprimis utilitati (fol. 175 r) haberi ne piget tui conducere uiderentur qui post illam inconsultam sui muneris uenditionem ita ex ingenio suo postmodum se gestiat et impresentia se gerere nititur quia admonitione ne dicam obiurgatione tua agere uideatur ut deinceps studeat ad meliorem uitam animum componere. Non recensebo impresentia que sibi post huiusmodi uenditionem partim sua culpa aduersa contigere mihi; existimem tuas dudum ad aures ea ipsa peruenisse solum ut officio amicitie nostre pro qua ex beneficio salutis sepius mihi tua causa (?) restitute me plurimum tibi debere fateor deinde ut prioribus (?) his diebus assiduis uxoris eius afflicte satisfacerem hec pauca scribenda existimaui que ut in bonam partem accipias oro quoniam ab animo tui beniuolo et amanti corde emanarunt. Vale Rome XXV Martii 1498.

(5)

(fol. 200 r) Domino uel potius Magistro Gabrieli de Zerbis physico. Primum quidem ut meo tam longo tempore non certe obliuione sed nulla potius oblata occasione officio scribendi intermisso, deinde uero nostri cuiusdam amici desiderio qui animi tui deliberationem circa rem que ad te pertinet me auctore intelligere cupit modo satisfaciam has probitati tue cui propter multa suus affectissimus scribere cogitaui, sperans pro nostra ueteri amicitia non iniocundum fore quod ipsi bona mente scribendum existimaui quam testimonio linearum cognoscere poteris quibus memoriam eius a quo beneficia accepimus aliquando renouare non tam diligentis quam amantis esse uidetur. Qua praefatiuncula factus desiderium illius de quo supra dixi tibi liberius aperire incipiam si prius de ualetudine me de qua sollicitus aliquando fuisti quamque etiam diligentissime curasti uerba faciam qua impresentia pro etate ita fruor ut tuo ut alterius artis medici professoris consilio uideam non indigere. Quapropter huius boni largitori tibique optimo amatori gratias ago simulque opto idem de te semper audire et in isto gymnasio ubi legendi causa moratoria duci tibi omnia prospere euenire quecumque optas et praeclare tuo menti requirunt. Hanc ad rem pro qua summo studio me animus adiit sermonem comentum eo sane animo ut tibi autem omnia suaderem in quidquid in rem tuam sit facias, neque enim ullo modo petierim ut alicuius commodis tuo postponas commoda, sed si tua refert desiderio animi satisfacias si nobis significaueris an aedes hic in urbe tuas honesto pretio venumdares ad quod si quando descendere uelles id sibi ut opinor offerentur ex quo uenditio uero commodo tuo confici posset. Quam huic nostro amico uiro probatissimo rem gratissimam efficies si hac de natura uoluntatem tuam rescribere curabis pro qua te etiam atque etiam rogo ut operam des ne uidear incassum ad te scripsisse. Vale.

III. THE LAST WILL AND TESTAMENT OF GABRIELE ZERBI

Archivio antico dell' Università di Padova; Vol. 649, Professori, Artisti e Legisti fino al 1509, folios 55 r-58 v (in margine: Exemplum 1504 Octobris 13. Testamento di Gabriel Zerbo Veronese professore di medicina).

(Fol. 55 r) In Christi nomine amen: anno natuitatis eiusdem millesimo quinquagentesimo quarto mensis octobris die tertio decimo indictione vero septima: Cum vita qua fruimur brevis admodum sit fragilis et incerta, et nihil morte certius, et hora mortis incertius, fitque sapientis non tantum rebus anime salutem spectantibus verum etiam temporalibus bonis recte consulere, et mature prouidere, ea propter eximius ac clarissimus artium et medicine doctor Dominus magister Gabriel de' Zerbis veronensis / premissa omnia reuoluens mente sua, et assidue premeditans / ac preuidens in omnen futurum et euenturum casum diuino beneffició corpore mente, memoria, et intellectu / sanus ac recte valens / suum ultimum nuncupatiuum sine scriptis in hunc modum facere procurauit testamentum iuxta formam tenorem / ac continentiam cuiusdam bombacine cedule / eius propria manu latius verbis scripte et notate / et mihi notario infrascripto presentate clause et tribus eius signis signate: Cuiusquidem bombacine cedule testamentarie tenor huius sequitur, et in omnibus ac per omnia est huiusmodi videlicet: In Christi nomine Dei omnipotentis: / ego Gabriel de Zerbis veronensis phisicus / ac medicus habitator Padue: sanus mente et corpore intendens prouidere saluti anime et corporis / ac voluntati proprie satisfacere cum dispositione bonorum meorum, dum tempus mihi occurit sinceritate mentis et intellectus ac valitudinis corporis (fol. 55 v) per presens nuncupatiuum testamentum: In primis animam meam Deo optimo conmitto ac conmitudo / qui non secundum mea demerita, sed secundum illius infinitam misericordiam eam suscipiat adueniente tempore separationis anime a corpore. Corpus autem meum, si contingat me decedere Padue, sepeliri volo in ecclesia Beatissimi Francisci / non autem in cimiterio sed in loco apparenti dicte ecclesie eleuato / in fabricam autem sepulture mee volo quia heredes et quibus bona mea derelinquo teneantur expendere ducatos sexaginta ut sit honorabilis et eleuata, ac apparens cum memoriali / si vero alibi contingat me decedere eo casu Deus misereatur anime mee et corporis, item de male forte acquisitis aut ablatis aut non solutis, volo quod teneantur heredes mei, et quibus dimitto bona mea dari ducatos duodecim aureos Domino magistro Baptiste de Barziziis: Item alios ducatos triginta octo usque ad summam ducatorum quinquaginta volo dari pauperibus virginibus nubilibus dum maritantur / videlicet filiabus sororis mee Domine Catherine uxoris quae est Domini Joannis de Soriis / si vero tradetur aliqua ipsarum se non propinquioribus affinitatis mee magis egenis dari volo secundum voluntatem commissariorum, e tempore quo videbitur prefactis (fol. 56 r) commissariis: et hec omnia volo fieri cum comoditate eorum quibus dimitto bona mea, usque ad tempus quo commissariis videbitur, post obitum meum secundum arbitrium commissariorum hereditate mee inferius nominatorum / Item volo iure legati, relinquo, iubeo, et instituo nobilem Dominam Helenam de Metaselimis bononiensem matrem filiorum usufructuariam omnium et singulorum bonorum hereditatis mee tam mobilium quam stabilium generis cuiuscunque cum libera et generali potestate et facultate exigendi huiusmodi omnes et quorumque fructus, redditus et prouentus quotcunque et qualitercunque spectantes et pertinentes predicte hereditati / ac de eis omnibus usufructibus pro libito voluntatis cuius disponendi donec vixerit tantum sine contradictione filiorum fratrum et aliorum quorumcunque / et hoc proseguente dicta Domina Helena in bono regimine ac vite honestate / Post mortem autem dicte Domine Helene volo quod ipsa filiis suis Paulo: Hieronymo: Marco et Joanni Aluisio dimittat omnia predicta bona mea tam mobilia quam immobilia / aut titulo hereditatis seu donationis, ymo etiam titulo elemosine quando alio modo non possint derelinqui / hac tandem condictione quod Paulo et Hieronymo dissipantibus bona (fol. 56 v) neque bono et laudabili regimine viventibus equali portione / solum volo eos participare de redditibus bonorum meorum, ita quod nolo eis aluid quicquam dari / si vero uiuant non dispensando neque ludendo aut quovis alio meretricio et inhonesto modo conviventibus volo quod equali portione hereditatis etiam ultra introytus pociantur et gaudeant et hoc totum fiat iuditio commissariorum prout infra / et quod dictum est de Paulo et Hieronymo iam adultis, ita de Marco et Joanne alouisio intellexi volo / quo [quod] bona secundum lineam descendentium, a filiis meis legitimorum et naturalium volo dimitti ex predictis descendentibus masculis tantum, quibus deficientibus, eo casu iure institutionis ordino, iubeo, et volo quod bona mea omnia hereditatis deueniant propinquioribus, et magis nobis attinentibus de familia iubeo masculis tantum / Prohibeo tamen et veto heredibus meis aut successoribus ad quos bona mea peruenerint / quod liceat vendere, donare, aut alienare bona stabilia, etiam urgente maxima necessitate et hoc institutum meum volo imperpetuum seruari debere / itaque etiam neque dicta bona possint pignorari neque ex eis aliquid detrahi posse, neque debere / per eos successores (fol. 57 r) aliquam legittimam trebillianicam et falcidiam vel aliquam aliam partem / quia volo ea mea bona ut dixi conseruari deuolui et dimitti perpetuo tempore domui de zerbis / ut supra in hoc testamento cassans et annullans omnem aliam meam ordinationem circa bona mea si qua inuenirentur / commissarios autem meos, et testamenti huius ultime voluntatis exequatores, instituo et eligo, et volo / excellentem artium et medicine doctorem ac mihi fidum ac dilectissimum compatrem Dominum magistrum Petrum de Mantua, ac nobilem Dominam Helenam suprascriptam matrem filiorum meorum cum potestate et auctoritate ac libertate exequendi vendendi de bonis mobilibus minus damnosis que alicuius heredis aut successoris vel alterius contradictione / absoluendo omnino eosdem commissarios et eorum quemlibet ab omni redditione administrationis de aliquibus bonis predicte hereditatis administratis et ad eorum manus prouenientis, itaque pro administratione in aliqua re molestari seu cogi aut inquietari non possint / per heredes et successores vel alios aliqua ratione vel causa de iure vel de facto. Item volo quod quousque mater mea vixerit subveniatur illi aliquando (fol. 57 v) per commissarios meos de ea portione que videbitur illis iuxta introytus et expensis hereditatis: Item Thadee dilectissime filie mee de bonis meis derelinquo id quod commissariis videbitur secundum posse hereditatis. Item sorori Hermodorie religiose in monasterio sancti ludovici de bononia filie mee dilectissime derelinquo tantum bonorum meorum quod annuatim habeat libras duodecim bononiorum, quoad vixerit / et quod aliquando ex elimosina Domina Helena mittat sibi in anno secundum quod ipsa poterit. / Item illud idem de elimosina faciendo dico de sorore clara religiosa in monasterio sancte Clare de verona ad quas elemosinas tamen volo commissarios posse compelli: Item sorori mee dilectissime Domine Angele dimitto in signum amoris ducatos tres aureos dandos illos cum comoditate hereditatis siue heredum: Item honorando fratri meo Domino Joanni de Zerbis canonico bellunensi dimitto ducatos quatuor dandos illi quando commissariis nostris videbitur: Item Benedicto fratri meo illud idem prout supra de Domino Joanne: Item Francisco filio prefati Benedicti ducatos tres quas pecunias volo dari cum comoditate commissariorum, ex tempore quo illis videbitur: ego idem Gabriel de Zerbis die duodecimo (fol. 58 r) octobris millesimo quingentesimo quarto presens testamentum meum manu propria scripsi quod intendo inuiolabiliter obseruari deberet dans et concedens prefactis dominis clarissimus dominus testator prenominatis dominis eius commissariis in presenti eius testamento nominatis et substitutis post obitum suum et interueniente casu mortis sue quandocunque plenissimam virtutem ac potestatem presentem eius commissariam intromittendi, administrandi, furniendi, dandi, soluendi, quietandi in quocunque iuditio comparendi in animam suam iurandi et ad unguem ac inuiolabiliter adimplendi omnia ac singula per eum superius in presenti eius testamento legata, dimissa, et ordinata que necessaria fuerint et opportuna, et ipsemet Dominus testator faceret ac facere posset et viveret, et presens esset et non secus ac si esset persona propria ipsiusmet Domini testatoris statuens eximium firmum, ratum, et inuiolabile quicquid per eos factum fuerit modo ac forma suprascriptis in omnibus ac singulis premissis.

Actum Venetiis in appotheca aromatarie capitis auri presentibus Domino Joanne Venuitio quondam Domini Petri, domino magistro Jacobo de fabianis artium et medicine doctore, ser alexandro de rezolis deromatario quondam ser Girardi, ser petro de longis quondam ser Girardi aromatario Domino Joanne maria de maphelis vincentis artium et medicine doctore (fol. 58 v) aromatario domino vincento Corrutio quondam Domini Joanni de confinio sancti Simoni prophetae et aliis.

Ego aloysius de archangelis quondam D. Pauli imperialis et curie maioris nostre deputatus ad custodiam scripturarum notarius defunctorum suprascriptum testamentum de quo fuit rogatusque Hieronimus notarius imperialis ex prothocollo suo testamentorum vigore legis extraxi et in publicam formam redigi signo meo apposito consueto.

Die xxvii maii 1513 D. Paulus de bugnis venetus canonicus antiberensis et D. Bartholomeus buerluis canonicus massariensis cognouerunt manum et signum introscripti notarii et ratificati sunt illum esse fidelem A. Amonium.

IV. THE LAST WILL AND TESTAMENT OF NICCOLÒ MASSA

Archivio di Stato, Venezia, Busta 196, no. 870, notaio Marcantonio Cavagnis

In nomine dei eterni Amen: anno ab incarnatione domini nostri Jesu Christi 1569 mensis Julii die 28 indictione 12 Rogati.

Essendo io Nicolo Massa, fisicho, fo de maestro Appollonio che fo de maestro Thomaso, sano della mente e intelleto, benche vechio e infermo del corpo, et volendo far el mio testamento ho fato chiamar e venir in casa mia in contra de san Zaninuovo Marcantonio di Cavanei, fo de maestro Zani Maria nado di Venezia, et quello ho pregato lo scriva e dappoi la morte mia lo compissa e robori secondo li ordenj della terra, dicendo in questa forma, prima rogando l'anima mia al suo creator, alla gloriosa vergene Maria, et a tuta la corte celestial, et prego la sua maesta mi perdoni et remeti li mei peccati fati in questo misero mondo. Constituisso mei comissarii Maria, mia fia, fo maier de maestro Zuani Grifalconi, qual io legitimai nel contrato delle sue nozze, et cosi anche per il presente testamento la legitimo, talche la sia mia fiola sia o nol sia mia fiola, et possi succieder nelli mei beni come fiola legitima senza alcuna contradition. Inanzi l'eccellente maestro Appollonio Massa, mio nepote, et il signor maestro Appollonio Massa, mio nepote, et il signor maestro Lorenzo Massa el mio herredo, fiol de madonna Paula, mia sorella, et l'eccellente maestro Zuane Gratasuol, medico, mio carissimo compadre et da me amato come fiol, quali tuti insieme cuor per la mazor parte loro uogio mandino ad eseqution quanto ordenizo ut supra. Et doue sera l'opinione dedítta Maria, mia fia, quella se intenda per la mazor parte, perche non uogio sia fato cosa alcuna senza la sua opinione. Et prima uogio che tuti li mei fondi, campi, terre, case daie che mi pagano li distrituali da Mestre e anche la schuola de santa Maria da Mestre e stabeli de cadauna sorte cosi in questa terra come de fuera in cadaun luogo non possino esser venduti ne obligati ne allienati ne dimeno sopra dessi tolto pagamento di dasse ne pagato debito di sorte alcuna ma siano galduti per li mei herredi come diro. Inanzi ordeno che la sopra ditta Maria, mia fia, habbia a scuoder tuti li pro de cecha che sonno in nome suo e de Appollonio e Lorenzo Massa, mei herredi, da esser fata de quelli la mia volunta et per lei saluati per spender in palazo in lite, auocati e scriture per deffensione delle litte che fossero mosse alla mia comissaria fin tanto serano terminate tute le litte preditte. Inanzi ordeno e uogio che madonna Paula, mia sorella, decetero habbia a scuoder li 16 ducati de pro che se trouano in so nome alla cecha dadese de quali siano sui come sonno et li altri 4 siano per lei dati a Lugrezia, nostra sorella, fino la uiuera, perche quelli sonno mei. Inanzi lasso alla sopraditta Maria, mia fia, li usufruti della mia frutaria e de tuti li altri mei beni che mi trouo et trouero hauer qui in Venezia per rason de heredita, con obligo de tenir in conzo e colmo il tuto et pagar le angarie che de tempo in tempo serano poste et anche bisognando refar qualche cosa sia tenuta refarla delle intrate. Inanzi li lasso tuti li mei beni mobili de casa che si trouerano oltre li sui et quelli del q. suo fiol che sonno in casa mia nelliquali mei mobili uogio sintendano denari, ori, argenti, zoglie, ornamenti et tapezzarie de qualunche sorte si uoglia. Inanzi ordeno et uoio [uogio] che la preditta Maria non possi esser astreta da alcuno chi esser si uoglia et per qual causa esser si uoglia a render conto della administrazione delli beni mei, denari o altra cosa che havesse per il tempo passato manizata, ma accorrendo alcuna decisione se stia alla sua semplice parolla senza iurar sacramento alcuno ne altro atto iudiziario, perche uogio del mio ordenar a mio muodo a lauda de Iddio omnipotente et del saluator nostro maestro Jesu Christo. Inanzi lasso alla preditta mia fia Maria liberamente el pro che mi pagano li distrituali da Mestre per conto de daie o colte et anche quello che mi pagano li intervenienti della schuola de madonna santa Maria da Mestre per rason de heredita. Constituisso mei herriedi Maria, mia fia sopraditta, e maestro Appollonio e maestro Lorenzo Massa, mei carissimi nepoti e da me amati come fioli delli usufruti della mia possession qual aquistai ultimamente in villa de Pesegia dal offizio del sopragastaldo sopra laqual ho fato da nuovo una falincha da lauorador e chel fondo resti condicionato ut supra, dechiarando che ditti herriedi siano obligati estrazer da ditti usufruti ognanno ducati tre per uno e quelli siano dispensati a' miserabile persone per ditta mia fia come li parera et manchando uno a piu de ditti herriedi la parte sua de ditti usufruti vadi in li altri sopraviventi con la conditione delli ducati 9 ognanno ut supra. Inanzi lasso alla ditta Maria, mia fia, el galdimento e usufruto della mia casa che ho in ditto luogo de Pesegia con il bruolo e campi che sonno congionti e di rason dessa casa. Inanzi li lasso liberamente li mobili che si trouerano in ditta casa delliquali la ne possi disponer el suo voler, eccetuadi perho li instrumenti rurali i quali restar debbino al semifio della possession. Inanzi lasso a madonna Lugrezia e a madonna Paula, mie sorelle, tute le raggion et assion che ho et mi potessero aspetar in qualsi uoglia sorte de beni posti a' Maran, dellequal raggioni et assioni possino disponer ogni suo voler sensa alcuna contradition come cose sue libere. Inanzi lasso alle preditte mie sorella el fitto delle mie due casete che ho in contra de San Piero de Castello appresso el monasterio di frati de San Domenego, con conditione de tenir il tuto in conzo e colmo e pagar le angarie che corerano et morendo una vadi la parte sua ne l'altra et morte tute due le preditte mie sorelle ordeno che per anni x continui siano dati ducati sei all' anno alli piu poueri di douesi si trouerano a parenzo. Et in ditte case siano logate de tempo in tempo tre pouere vedoue, per esser le preditte case in tre soleri, lequale vedoue non possano tenir fioli maschii de eta de 12 anni in suso. Et si alcuna volesse contravenire a' questo mio voler sia espulsa de casa con il mezo della iustizia. Inanzi lasso alla sopraditta Maria, mia fia, ogni mia rason de denari che debbo hauer e scuoder dalle moneghe de San Sezuolo, inanzi el credito che ho da hauer da maestro Marcantonio e Vicenzo di Pizzoni e dal mio lauorador Nardo da Pesegia. Inanzi li crediti che ho in banchi de scrita e nel offizio del proprio e altri offitii quomodocumque et qualitercumque insieme con tuti li altri mei crediti cosi in questa terra come fuera, li quali siano scossi per ditta Maria et siano sui, arecordandoli a far delle elimosine et a sovenir li poureti bisognosi, come sempre ha fato. Inanzi ordeno che fati li conti con Nardo preditto e pagati tuti li debiti che ha con sui li siano date e donate lire dusento de pizzoli per ogni pretension chel potesse hauer per hauersi afatichato et teso alla fabricha nouamente fata per mi e per ogn' altra cosa. Inanzi lasso alli sopraditti Maria, mia fia, e maestro Appollonio e maestro Lorenzo, mei nepoti e comissarii, tuti li mei libri, ma prima ditta Maria possi tirar fuera e ellezer qual solte de libri latini o vulgari li parera. Inanzi prego ditti mei nepoti che vedano con diligentia se in essi libri ne fosse qualcheduno prohibito dalli sacri ordeni del concilio ouero contra bonos mores et trouandone quelli tali siano brusati ouero dessi sia fato quanto e ordinato per il sacrosanto concilio. Et in caso ne hauesse qualcheduno che fosse dello commissario del Redaldi, quelli siano tornati al loco suo. Inanzi uogio che Maria, mia fia, al tempo che maestro Appollonio maridera sua fia li dia ducati vinticinque per comprarglie tanto panno de seda per una vesta o vestura come glie parera, se ben so che non l' ha bisogno; ma questo sia in segno d'amor e de consanguinita. Inanzi lasso al sopraditto maestro Lorenzo Massa, mio nepote et commissario, la mia vesta fodra de martori gioe panno e fodra dellaqual el possi far quello li piasera. Inanzi ordeno e uogio che Maria, mia fia, dia a' maestro Zuane Gratasuol stara dai de farina ogn' anno alle feste de Nadal per anni cinque continui per le molte fatiche fate per mi e chel fara per la mia commissaria, et questo delle entrade della possession posta a Pesegia comprata per mi in compagnia con ditta mia fia. Inanzi ordeno che ditti mei commissarii habbino a servir ditto maestro Zuane Gratasuol delli mei libri de praticha de medicina che li facesse bisogno sicome a' loro commissarii parera. Inanzi ordeno che nella diuisione delli libri che farano ditti mei herredi e fiola cad' anno d'essi habbi a' fuor le materie e volumi intieri e nol smembrati, i quali prego che tuti tre insieme se debbano amar e viver in pace e amorevolezza, aiutandossi l' uno con l' altro da fratelli e sorella con la pace de Dio e di maestro Jesu Christo. Inanzi ordeno che la mia sepultura sia fata nel monasterio de San Domenego da Castello segondo l' ordine descrito in uno mio testamento fato perauanti de man propria, intendendossi leuate quelle parolle de pompa e vanita. Et se arecordino delle mie vertigine al tempo che crederano sia morto, lassandomi dei [dui] giorni sopra terra, accio non si facesse qualche error, et non mi metano in giesia auanti sia passato ditto termine de dui giorni. Inanzi ordeno che me siano dite le messe della madonna e di San Gregorio per el reverendo padre frate Agostino de San Zanipolo, mio confessor, che tende alla pase, alqual li sia dato la elimosina consueta e un ducato de piu, accio habbi causa di pregar Dio per l' anima mia, qual ducato sia per una volta tantum. Inanzi lasso sia dato medesimamente alla sagrestia de San Zaninuovo unaltra elimosina consueta per farmi dir delle altre messe della madonna e di San Gregorio et similmente alla sagrestia di San Domenego per dir le messe soprascritte per l' anima mia e delli altri mei defunti. Inanzi sia mandato alle moneghe del santissimo sepulcro unaltra elimosina simile e ducato uno de piu per dir le messe sopraditte. Inanzi uogio che ognanno el zorno della mia morte Maria, mia fia, spenda sei marcelli in carni o possa si come sera tempo et lo mandi ad honor di Dio et per l' anima mia alle sopraditte moneghe. Inanzi ordeno chel giorno della mia sepultura sia fata elimosina alli mei fratelli della schuola di San Marco et ad altre miserabile persone come parera alli mei commissarii, et ognanno el giorno della mia morte sia mandato sei marcelli alli frati de San Domenego per dir messe per l' anima mia e delli mei morti. Inanzi sia mandato a ditti frati ognanno dui depieri de L. due l' uno per far arder la vigilia di morti sopra la mia sepultura mentre si dirano li offitii e soldi quaranta per dir messe per l' anima mia e di mei morti. Et se li frati ditti di San Domenego nol volessero che si facesse la mia sepultura nel suo monasterio siano privi di quanto li lasso et quello sia fato dove parera a' ditti mei commissarii con le condition sopraditte. Inanzi ordeno siano fate delle altre elimosine come parera alli mei commissarii, accio Dio mi conservi in la sua gratia et mi perdoni li mei peccati. Et perche intendo che la mia commissaria sia perpetua et eterna perho ordeno e uogio che dappoi nol si trouera piu alcuno delli desscendenni delli soprascritti Maria, Appollonio, e Lorenzo, mei herriedi, siano elletti per la bancha e aggionti della mia schuola de San Marco con li dui terzi delle balotte tre commissarii homeni signalati che siano stati guardiani grandi o vicarii o vardiani da Matin, liquali tre commissarii insieme con il guardian grande, vicario e vardian da Matin, che si atrouerano de tempo in tempo dessa schuola habbino a gouernar le intrade de tuti li sopraditti mei beni, saluandone un terzo per tenir in conzo e colmo li fondi e pagar le angarie che correreno de tempo in tempo. Un altro terzo sia per loro dispensato a' poueri bisognisi della schuola sopradetta et l' altro terzo ad altri poueri miserabili de tuta la citta ad honor de Dio. Et manchando ditti commissarii ne siano elleti de gli altri al modo ut supra. Et cosi sia seruato imperpetuo tenendo un libro particular de ditta commissaria. Inanzi uogio e ordeno che la soprascritta Maria, mia fia, habbia appresso di se tute le scriture e testamenti fati perauanti de mia mano, et se vi fosse sopra dessi scrito qualche memorial in materia de opere pie la prego a mandarlo ad esequzione. Interogato dal nado delli lauori pii della terra ho resposo che sia chiamati li puti delli hospedali alle mie esequie et altri poueri che parera alli mei commissarii dandoli delle elimosine come li piacera. Et questo uogio sia el mio ultimo testamento e ultima volunta, cassando et annullando tuti li altri testamenti fati per inanzi. li sigilli et signum et Testes excellentes artium et medicinae doctores dominus Jo. Baptista Peranda et magister dominus Angelus Benedictus quondam magistri domini Joannis Baptistae.

Io Angiolo Benedetti, fo di maestro Giovanni Baptista, fui testimonio, convogado et giurado, agnosco et soprascrisso.

Io Gianbattista Peranda dottore physico fui testimonio pregato et giurato ut supra.

Bibliography

BIBLIOGRAPHY

A. BOOKS AND ARTICLES

ABANO, PIETRO D', *Conciliator*, etc. (Venice, Giunta, 1548).

ACHILLINI, GIOVANNI FILOTEO, in *Collettanee Grece, Latine e Vulgari per diversi Auctori Moderni nella Morte de l'ardente Seraphino Aquilano* (Bologna, C. Bazaleri, 1504).

ADY, CECILIA M., *The Bentivoglio of Bologna, a Study in Despotism* (Oxford University Press, 1937).

[AEGIDIUS] AEGIDII CORBOLIENSIS, *Carmina Medica* (GILLES DE CORBEIL), ed., L. Choulant (Leipzig, 1826).

AGOSTINI, F. GIOVANNI DEGLI, *Notizie Istorico-critiche intorno la vita e le opere degli Scrittori Viniziani*, etc. (Venice, Simone Occhi, 1754).

ALBERICI, GIACOMO, *Catalogo breve degl'illustri et famosi scrittori venetiani*, etc. (Bologna, heirs of Giovanni Rossi, 1605).

ALBERTUS MAGNUS, *Opera Omnia . . . cura et labore Augusti Borgnet* (Paris, L. Vives, 1890–1899).

——, *De Secretis Mulierum* (Lyon, B. Vincentius, 1571).

——, *De Animalibus Libri XXVI nach der Cölner Urschrift*, herausgeg. von Hermann Stadler (Beiträge zur Geschichte der Philosophie des Mittelalters-Texte und Untersuchungen XV, 2 v., Münster i. W., 1916).

ALEXANDER (Pseudo-) OF APHRODISIAS, *Problemata Alexandri Aphrodisei Georgio Valla interprete* (Venice, Albertinus Vercellensis, 1501).

ALIDOSI, GIOVANNI NICOLÒ PASQUALI, *I dottori bolognesi di teologia, filosofia, medicina, e d'arti liberali dall' anno 1000 per tutto marzo del 1623* (Bologna, N. Tebaldini, 1623).

[ANONYMUS] *Anonymi Introductio Anatomica . . . cum notis Dan. Wilh. Trilleri et Jo. Steph. Bernard* (Leiden, Philip Bonk, 1744).

AQUINAS, ST. THOMAS, *Compendium Theologiae*, ed. R. P. Petri Mandonnet (Paris, P. Lethielleux, 1927).

ARCULANUS, IOANNES, *Io. Arculani Omnes qui proximis seculis scripserunt medicos longe excellentis opera . . . Commentarii in Razis Arabis Nonum Librum ad Regem Almansorem*, etc. (Basle, Henricus Petrus, 1540).

ARETAEUS, *On the Causes and Symptoms of Acute Diseases*, translated by Francis Adams (London, The Sydenham Society, 1856).

ARISIO, FRANCESCO, *Cremona Literata*, etc. (Parma, Pazzoni and Monti, 1702).

ARISTOTLE, *Works*, translated into English under the editorship of J. A. Smith and W. D. Ross (Oxford, Clarendon Press, 1908–1931).

——, *Categoriae et Topica cum Porphyrii Isagoge ex recensione I. Bekkeri* (Berlin, G. Reimer, 1843).

——, *Secreta Secretorum* (Lyon, A. Blanchard, 1528).

——, *Secreta Secretorum*, etc., in *Opera Hactenus Inedita Rogeri Baconi*, Fasc. V, ed. Robert Steele (Oxford University Press, 1920).

ASTIUS, D. FRIDERICUS, *Lexicon Platonicum sive Vocum Platonicarum Index* (2 v., Leipzig, Weidmann, 1835–1836).

ATKINSON, JAMES, *Medical Bibliography A. and B.* (London, John Churchill, 1834).

AURELIANUS, CAELIUS, *On Chronic Diseases*, translated by I. E. Drabkin (University of Chicago Press, 1950).

AVERROIS, *Colliget* (Venice, O. Scotus, 1542).

——, *Cordvbensis Commentarium Magnum in Aristotelis De Anima Libros recensuit F. Stuart Crawford* (Cambridge, Mass., Mediaeval Academy of America, 1953).

AVICENNA, *Canon*, translated by Gerard of Cremona, with index by Johannes Costaeus (Venice, Giunta, 1595).

BAEHRENS, AEMILIUS, *Poetae Latini Minores V* (Leipzig, Teubner, 1883).

BANDTLOW, FRIEDRICH WILHELM OSKAR, *Die Schrift des Gabriel de Zerbis: de cautelis medicorum* (Leipzig Inaugural Dissertation, Zeulenroda i. Thür., 1925).

BANKS, JOSEPH, *Catalogus Bibliothecae Historico-Naturalis Josephi Banks* (5 v., London, 1796–1800).

BARDENHEWER, O., *Die pseudo-aristotelische Schrift über das reine Gute bekannt unter dem Namen liber de causis* (Freiburg im Breisgau, 1882).

BARTHOLIN, THOMAS, *On Medical Travel*, translated by Charles D. O'Malley (Lawrence, Kansas, University of Kansas Library Series No. 9, 1961).

BELLONI, LUIGI, *Giovanni Battista Morgagni, Gli inventori anatomici italiani del xvi secolo nel carteggio col medico milanese Bartolomeo Corte; XV Convegno Nazionale della Società Italiana di Anatomia* (Milan, Industrie Grafiche Italiane Stuchi, 1953).

BERNARDI, FRANCESCO, *Prospetto storico-critico dell' origine, facoltà, diversi stati, progressi, e vicende del collegio medico-chirurgico, e dell' arte chirurgica in Venezia*, etc. (Venezia, Domenico Costantini, 1797).

BIADEGO, GIUSEPPE, *Catalogo descrittivo dei manoscritti della biblioteca comunale di Verona* (Verona, G. Civelli, 1892).

Biblioteca Bibliographica Aureliana VII-Index Aureliensis-Catalogus librorum sedecimo saeculo impressorum: Prima Pars, Tomus A, Volumen 1 (Aureliae Aquensis, Librairie Heitz G.M.B.H., Baden Baden, 1962).

Bibliotheca Osleriana (Oxford University Press, 1929).

Biographisches Lexikon der hervorragenden Aerzte aller Zeiten und Völker . . . herausgegeben von dr. August Hirsch (6 v., sec. ed., Berlin and Vienna, Urban, 1929–1935).

BLANKAART, STEPHEN, *Lexicon Medicum*, ed. C. G. Kühn (Leipzig, Schwickert, 1832).

BOETHIUS, *De Disciplina Scholarium*; in J. P. Migne, *Patrologia Latina* 64. 1223–1238.

BOISSARD, J. J., *Icones . . . Virorum Illustrium* (Frankfurt am Main, G. Fitzer, 1597–1599).

BRAMBILLA, G. A., *Storia delle scoperte fisico-medico-anatomico-chirurgiche fatte dagli uomini illustri italiani* (Milan, Nell' imperial Monistero di S. Ambrogio maggiore, 1780–1784, 2 v. in 3).

BROCK, ARTHUR J., *Greek Medicine* (New York, E. P. Dutton, 1929).

BRONZINO, GIOVANNI, *Notitia Doctorum-Universitatis Bononiensis Monumenta* (Varese: Soc. Tip. "Multa Paucis," 1964).

CALCATERRA, CARLO, *Alma Mater Studiorum: l'università di Bologna nella storia della cultura e della civiltà* (Bologna, N. Zanichelli, 1948).

CALEPINO, *Ambrosii Calepini Dictionarium* etc. (Lyon, Philip Tinghus Florentinus, 1580).

Cambridge Modern History, Vol. I, The Renaissance (Cambridge University Press, 1934).

CAPPARONI, P., *Profili bio-bibliografici di medici e naturalisti celebri italiani dal sec. XV al sec. XVIII*, Vol. I (Roma, Istituto Nazionale Medico Farmacologico, 1932).

CASSIRER, E., P. O. KRISTELLER, and JOHN HERMAN RANDALL, JR., *The Renaissance Philosophy of Man.* (Chicago, University of Chicago Press, 1948).

CASTIGLIONI, A., *A History of Medicine* (New York, Alfred A. Knopf, 1946).

Catalogo del museo storico dell' Università di Bologna (Bologna, Tipografia Compositori, 1957).

CAVAZZA, FRANCESCO, *Le scuole dell' antico studio bolognese* (Milan, V. Hoepli, 1896).

CELLINI, BENVENUTO, *Autobiography* (London, E.P. Dutton, 1907).

CENCETTI, GIORGIO, *Gli Archivi dello studio bolognese* (Bologna, Pubblicazioni del R. Archivio di Stato in Bologna 3, Nicola Zanichelli, 1938).

CERVETTO, G., *Frammento storico di alcuni illustri anatomici italiani del decimoquinto secolo* (Verona, Antonelli, 1842).

CHAULIAC, GUY DE, *Ars Chirurgica*, etc. (Venice, Giunta, 1546).

CHOULANT, LUDWIG, *History and Bibliography of Anatomic Illustration*, etc. (New York, Hafner Publishing Co., 1962).

CIGOGNA, EMMANUELE ANTONIO, *Delle Inscrizioni Veneziane raccolte ed illustrate da E.A.C., cittadino Veneto* (Venice, Giuseppe Orlandelli, 1824).

CLARKE, EDWIN, "Aristotelian Concepts of the Form and Function of the Brain," *Bulletin of the History of Medicine* **37** (1963): pp. 1–14.

COLE, F. J., *A History of Comparative Anatomy* (London, Macmillan, 1944).

COLLE, FRANCESCO MARIA, *Storia scientifico-letteraria dello studio di Padova* (4 v., Padua, 1824–1825).

Collectio Ophthalmologica Veterum Auctorum Fasc. VII: Constantini Monachi Montiscassini Liber de Oculis et Galieni Littere ad Corisium De Morbis Oculorum, etc. (Paris, Baillière et fils, 1909–1933).

Commentator in Galenum *De Sectis, Omnia Opera Galeni* (Venice, Diomedes Bonardus, 1490).

CORNER, GEORGE W., *Anatomical Texts of the Earlier Middle Ages* (Washington, D.C., Carnegie Institution, 1927).

CORTE, BARTOLOMEO, *Notizie istoriche intorno a' medici scrittori milanesi e a' principali ritrovamenti fatti in medicina dagl' Italiani*, etc. (Milan, G.P. Malatesta, 1718).

COSENZA, MARIO E., *Biographical and Bibliographical Dictionary of the Italian Humanists*, etc., 5 vols. (Boston, G.K. Hall, 1962).

CRANZ, F. EDWARD, *Alexander Aphrodisiensis*, in *Catalogus Translationum et Commentariorum: Medieval and Renaissance Latin Translations and Commentaries-annotated lists and guides*; Vol. I, ed. P. O. Kristeller (Washington, D.C., Catholic University of America Press, 1960).

CROMBIE, A. C., *Medieval and Early Modern Science* I (New York, Anchor Books, sec. ed., 1959).

CRUMMER, L., translation of Zerbi's *Anatomia Infantis*, in *American Journal of Obstetrics and Gynecology* **13** (1927).

CURTIUS, E. R., *European Literature and the Latin Middle Ages* (London, Routledge and Kegan Paul, 1953).

CUSHING, HARVEY, and E. C. STREETER, *Monumenta Medica* IV, ed. by H. E. Sigerist: facsimile of Canano, *Musculorum Humani Corporis Picturata Dissectio* (Florence, R. Lier and Co., 1925).

DALLARI, V., *I rotuli dei lettori, legisti e artisti dello Studio bolognese dal 1384 al 1799* (Bologna, Merlani, 1888–1924, 4 v. in 5).

DAREMBERG, CH., *Histoire des sciences médicales comprenant l'anatomie, la physiologie, la médecine, la chirurgie et les doctrines de pathologie générale* (Paris, J. B. Bailliére et fils, 1870).

DEMPSTER, W. T., "European Anatomy Before Vesalius," *Annals of Medical History* **6** (1934): pp. 307–319, 448–469.

Dictionnaire Encyclopédique des Sciences Médicales (Paris, 1864–1882, ed. J. Raige-Delorme and A. Dechambre).

DIELS, H., "Die Handschriften der antiken Ärtze. I. Hippokrates and Galenos," *Abhandlungen der königl. Preuss. Akademie der Wissenschaften* (Berlin, 1905) Abh. iii, 58–158.

DIRINGER, DAVID, *The Illuminated Book, Its History and Production* (London, Faber and Faber, 1958).

Dizionario biografico degli italiani (Rome, Istituto della Enciclopedia Italiana fondata da Giovanni Treccani, 1960, Vol. I).

DODOENS, REMBERT, . . . *Medicinalium observationum exempla rara, recognita et aucta* . . . (Cologne, apud Maternum Cholinum, 1581).

DUBLER, CÉSAR E., *La 'Materia Medica' de Dioscórides-Transmisión Medieval y Renacentista*, Vol. IV: *Don Andrés de Laguna y su época* (Barcelona, Tipografía Emporium, 1955).

DUNGLISON, ROBLEY, *A Dictionary of Medical Science*, sec. ed. (Philadelphia, Blanchard and Lea, 1860).

DURLING, RICHARD J., "A Chronological Census of Renaissance Editions and Translations of Galen," *Journal of the Warburg and Courtauld Institutes* **24**, Nos. 3–4 (London, 1961).

EDWARDES, DAVID, *Introduction to Anatomy* (1532): *facsimile reproduction with English translation and an introductory essay on anatomical studies in Tudor England by C. D. O'Malley and K. F. Russell* (Palo Alto, Calif., Stanford University Press, 1961).

ERIKSSON, RUBEN, *Andreas Vesalius' First Public Anatomy at Bologna 1540* etc. (Uppsala and Stockholm, Almquist and Wiksell, 1959).

EUBEL, CONRAD, *Hierarchia Catholica Medii Aevi* (Münster, Libraria Regensbergiana, 1901).

EVANS, ELIZABETH C., "Galen the Physician as Physiognomist," *Transactions and Proceedings of the American Philological Association* **76** (1945): pp. 287–298.

FACCIOLATI, G., *Fasti Gymnasii Patavini. 1517–1756* (Padua, Typis Seminarii apud Joannem Manfré, 1757).

FANTUZZI, G., *Notizie sugli scrittori bolognesi* (9 v., Bologna, Stamperia di S. Tommaso d'Aquino, 1781–1794).

FERRARI, LUIGI, *Onomasticon: repertorio biobibliografico degli scrittori italiani dal 1501 al 1850* (Milan, U. Hoepli, 1947).

FISCH, MAX H., *Nicolaus Pol Doctor* (New York, Herbert Reichner, 1947).

FOESIUS, ANUTIUS, *Oeconomia Hippocratis Alphabeti Serie Distincta* (Frankfurt, A. Wechel, 1588).

FOLIGNO, CESARE, *The Story of Padua* (London, J. M. Dent, 1910).

FORLÌ, JACOPO DA, *Expositio Jacobi Forliuiensis in primum Avicenne Canonem* (Venice, Pentio de Leuco, 1508).

FORSTER, E. S., "The Pseudo-Aristotelian Problems: Their Nature and Composition," *The Classical Quarterly* **22** (1928): pp. 162–165.

FRANCESCHETTI, FR., *La famiglia dei conti Fracanzani di Verona, Vicenza ed Este con notizie dei loro antenati*, etc. (Bari, La direzione del Giorn. Araldico, 1896).

FRANKLIN, ALFRED, *Dictionnaire des noms, surnoms et pseudonymes Latins de l'histoire littéraire du moyen âge [1100 à 1530]* (Paris, Firmin Didot, 1875).

FRATI, LODOVICO, "Indice dei codici latini conservati nella R. Biblioteca Universitaria de Bologna," *Studi italiani di filogia clasica* **16–17** (1908–1909): p. 110.

——, *La Vita Privata in Bologna dal secolo XIII al XVII*, sec. ed. (Bologna, N. Zanichelli, 1928).

Fuhrmeister, E., *J. D. Wetteranus* (Halle Medical Diss., 1920; Universitäts und Landesbibliothek Sachsen-Anhalt, German Democratic Republic).

GALEN, *Opera Omnia*, ed. Petrus Antonius Rusticus Placentinus (Pavia, J. de Burgofranco, 1515–1516).

——, *Opera Omnia*, ed. C. G. Kühn, Vols. 1–20 (Leipzig, C. Cnobloch, 1821–1833).

Galen on Anatomical Procedures-The Later Books, translated by W. L. H. Duckworth, M. C. Lyons, and B. Towers (Cambridge University Press, 1962).

Galen's Hygiene, translated by Robert Montraville Green (Springfield, Ill., Charles C. Thomas, 1951).

GALEN (PSEUDO-), *De Anatomia Vivorum* (Venice, Giunta, 1576; Pavia, 1515).

GARRISON, F. H., *An Introduction to the History of Medicine* (Philadelphia, W. B. Saunders, 1929).

GASELEE, SIR STEPHEN, "The Soul in the Kiss," *The Criterion* **2** (London, 1924): pp. 349–359.

GAURICO, LUCA, *Lucae Gaurici Geophonensis Episcopi Civitatensis Tractatus Astrologicus* (Venice, apud Curtium Troianum Nauò, 1552).

Gesamtkatalog der Preussischen Bibliotheken mit Nachweis des identischen Besitzes der bayerischen Staatsbibliothek und der National-bibliothek in Wien **1** (1931): p. 518.

GHERUS (RANUCCIO GHERO), *Delitiae CC. Italorum Poetarum, pars altera* (Frankfort, J. Rosa, 1608).

GIOMO, GIUSEPPE, *L'Archivio antico della università di Padua* (Venice, Fratelli Visentini, 1893).

GIOVIO, PAOLO, *Elogia Doctorum Virorum*; Opera II (Basle, P. Perna, 1578).

GIUNTINI, FRANCESCO, *Commentaria in Quadripartiti Ptolemaei* (in *Speculum Astrologiae*, Lyon, 1583).

GNUDI, MARTHA TEACH, and JEROME PIERCE WEBSTER, *The Life and Times of Gaspare Tagliacozzi, Surgeon of Bologna, 1545–1599* (Milan, Hoepli; New York, H. Reichner, 1950).

GOSS, CHARLES M., "On Anatomy of Veins and Arteries by Galen of Pergamos [*sic*]," *The Anatomical Record* **141** (1961): pp. 355–366; "On the Anatomy of the Uterus," *ibid.* **144** (1962): pp. 77–83; "On the Anatomy of Muscles for Beginners by Galen of Pergamon," *ibid.* **145** (1963): pp. 477–501; "On Anatomy of Nerves by Galen of Pergamon," *American Journal of Anatomy* **118** (1966): pp. 327–335.

GRAESSE, J. G. Th., *Orbis Latinus oder Verzeichnis der wichtigsten lateinischen Orts-und Ländernamen*, 2 Aufl, F. Benedict (Berlin, R. C. Schmidt, 1909).

GRAUBARD, MARK, *Circulation and Respiration: The Evolution of an Idea* (New York, Harcourt, Brace, 1964).

Greek Anthology, translated by W. R. Paton; Loeb Classical Library (London, Heinemann, 1917).

GUAZZO, MARCO, *Cronica* (Venice, F. Bindoni, 1553).

GUERINI, VINCENZO, *A History of Dentistry* (Philadelphia, Lea and Febiger, 1909).

GUGLIELMINI, J. F., *De Claris Bononiae Anatomicis Oratio* etc. (Bologna, Ex Typografia S. Thomae Aquinatis, 1737).

GURLT, E., *Geschichte der Chirurgie und ihrer Ausübung* (Berlin, Hirschwald, 3 v., 1898).

GURLT, E., and A. HIRSCH, *Biographisches Lexikon der hervorragenden Aerzte*, sec. ed., 5 vols. and supplement (Berlin and Vienna, Urban and Schwarzenberg, 1929–1935).

Gymnasium Patavinum Iacobi Philippi Tomasini Episcopi Aemoniensis Libris V comprehensum (Udine, Ex typographia Nicolai Schiratti, 1654).

HAESER, HEINRICH, *Lehrbuch der Geschichte der Medicin und der Epidemischen Krankheiten I* (Jena, G. Fischer, 3rd ed., 1875–1882).

HALL, A. R., *The Scientific Revolution, 1500–1800* (Boston, Beacon Press, 1956).

HALLER, ALBRECHT VON, *Biblioteca Anatomica*, (2 v., Zurich, apud Orell, Gessner, Füssli et socios, 1774–1776).

HALY FILIUS ABBAS, *Liber Totius Medicine*, etc. (Lyon, J. Myt, 1523).

HALY RODAN, *Commentum in Galenum* (commentary on Galen's *Techni*, with the *Expositio* and *Quaestiones* of Jacopo da Forlì; Pavia, Michele and Bernardino de Garaldi, 1501).

HARRIS, D. F., "On a Latin Translation of the Complete Works of Galen by Andrea Laguna, etc.," *Annals of Medical History* **2** (1919): pp. 384–390.

HECKSCHER, WILLIAM S., *Rembrandt's Anatomy of Dr. Nicolaas Tulp* (New York University Press, 1958).

HEIBERG, J. L., *Beiträge zur Geschichte Georg Valla's und seiner Bibliothek;* Beiheft zum Centralblatt für Bibliothekswesen XVI (Leipzig, Otto Harrassowitz, 1896).

HERNANDO Y ORTEGA, TEOFILO, *Vida y labor médica del doctor Andrés Laguna* (Segovia, Instituto Diego de Colmenares, 1960).

HERRLINGER, ROBERT, article on Johannes Dryander in *Neue Deutsche Biographie* **4** (Berlin, Duncker and Humblot, 1959).

HIPPOCRATES, *Opera*, ed. C. G. Kühn, *Medicorum Graecorum Opera Quae Exstant*, **21–22** (Leipzig, C. Cnobloch, 1825–1826).

——, *Oeuvres complètes. Traduction nouvelle avec le texte grec . . . par E. Littré* (10 v., Paris, J. B. Baillière, 1839–1861).

——, *The Medical Works of*, translated by John Chadwick and W. N. Mann (Springfield, Ill., Charles C. Thomas, 1950).

Hippocrates and the Fragments of Heracleitus, translated by W. H. S. Jones and E. T. Withington (4 v., London, Loeb Classical Library, 1923–1931).

HIPPOCRATES, *On Airs, Waters, Places*, translated by John Chadwick and W. N. Mann, The Medical Works of Hippocrates (Oxford, Blackwell, 1950).

HOLMYARD, E. J., and D. C. MANDEVILLE, *Avicennae De congelatione et conglutinatione lapidum* (Paris, Champion, 1927).

HUGO SENENSIS, *Consilia Saluberrima ad Omnes Aegritudines* (Bologna, 1482; Venice, O. Scotus, 1518).

HYRTL, J. *Das Arabisches und Hebräisches in der Anatomie* (Vienna, Wilhelm Braumüller, 1879).

Index zur Geschichte der Medizin, Naturwissenschaft und Technik (Munich and Berlin, Urban and Schwarzenberg, 1935).

ISAAC, F. S., *An Index to the Early Printed Books in the British Museum*, Part II. 1501–1520, Section II, Italy (London, K. Paul, Trench, Trubner, 1938).

ISAAC ISRAELI, *Omnia Opera* (Lyon, Bartholomeus Trot, 1515).

JACOBO, BARBERI, and OVIDIO MONTALBANI, *Illustriss. Gabellae Syndicorum Nomenclatura pro parte Celeberrimi Collegii Medicae Facultatis ab anno Domini 1508 usque ad annum 1641* (Bologna, Niccolò Tebaldini, 1641).

Janus **2** (1853): pp. 375–424: list of doctors and surgeons of the thirteenth, fourteenth, and fifteenth centuries.

KIBRE, PEARL, *The Library of Pico della Mirandola* (New York, Columbia University Press, 1936).

KICKARTZ, HANS DIETER, *Die Anatomie des Zahn-, Mund- und Kieferbereiches in dem Werk "Historia Corporis Humani Sive Anatomice" von Alessandro Benedetti* (Diss., Institut für Geschichte der Medizin der Medizinischen Akademie in Düsseldorf, 1964).

KONING, P. DE, *Trois traités d'anatomie arabes* (Leiden, Brill, 1903).

KRISTELLER, P. O., *Latin Manuscript Books Before 1600—A List of the Printed Catalogues and Unpublished Inventories of Extant Catalogues* (new ed. rev., New York, Fordham University Press, 1960).

——, *Iter Italicum*, etc., (2 v., Leiden, E. J. Brill, 1963-).

LACOMBE, George, *Aristoteles Latinus* (Cambridge University Press, 1955).

LASSEK, A. M., *Human Dissection: Its Drama and Struggle* (Springfield, Ill., Charles C. Thomas, 1958).

LAUTH, THOMAS, *Histoire de l'Anatomie* (Strasbourg, Levrault, 1815).

LAWN, BRIAN, *The Salernitan Questions: An Introduction to the History of Medieval and Renaissance Problem Literature* (Oxford, Clarendon Press, 1963).

LEONICENUS, NICOLAUS, *Nicolai Leoniceni in libros galeni e greca in latinam a se translatos prefatio communis* (Venice, Jacopo Pentio de Leuco, 1508).

Liber magistri Alvredi de Sareshel ad magistrum magnum Alexandrum Nequam de motu cordis, ed. Clemens Baeumker: "Des Alfred von Sareshel (Alfredus Anglicus) Schrift De Motu Cordis in *Beiträge zur Geschichte der Philosophie des Mittelalters* **23** (Münster i.W., 1923).

Library of Congress Classification P-PA, pp. 118–119.

LIND, L. R., *The Epitome of Vesalius translated*, with notes by C. W. Asling (New York, Macmillan Co., 1949).

——, *A Short Introduction to Anatomy by Jacopo Berengario da Carpi (Isagogae Breves)* translated with an introduction and historical notes (Chicago, Ill., University of Chicago Press, 1959).

——, *Aldrovandi on Chickens: the Ornithology of Ulisse Aldrovandi* (1600) Volume II, Book XIV, translated from the Latin with introduction, contents, and notes (Norman, Oklahoma, University of Oklahoma Press, 1963).

——, *Problemata Varia Anatomica: MS 1165 The University of Bologna*, edited; University of Kansas Publications, Humanistic Studies No. 38 (Lawrence, Kansas, 1968)

——, "The Last Will and Testament and a Letter of Francesco Muratori," *Archivio Italiano di Anatomia e di Embriologia* **76** (1971): pp. 187–200.

——, "The Life and Last Will and Testament of Leonello dei Vittori da Faenza, Pioneer Pediatrician," *Archivio Italiano di Anatomia e di Embriologia* **77** (1972): pp. 285–297.

——, "A New Manuscript Fragment of Constantinus Africanus, *Compendium Megatechni Galeni," Classical Studies Presented to Ben Edwin Perry, etc.* (Urbana, University of Illinois Press, 1969), pp. 103–113.

LOCKWOOD, DEAN PUTNAM, *Ugo Benzi-Medieval Philosopher and Physician* 1376–1439 (University of Chicago Press, 1951).

LOPEZ, LUIS ALBERTI, *La anatomía y los anatomistas españoles del Renacimento* (Madrid, Consejo Superior de Investigaciones Cientificas, 1948).

Luisini, Luigi, *Le cecità dell'ecc. medico m. Luigi Luisini da Udine* (Venice, per Giorgio de' Cavalli, 1569).

MacKinney, L. C., "Sex Determination: A Scientific Superstition," *Medicine Illustrated* **3** (1949): pp. 8–10.

Malagola, Carlo, *L'Archivio di Stato di Bologna dalla sua istituzione a tutto il 1882* (Modena, Vincenzi, 1883).

——, *Statuti dell' Università dei Collegi dello Studio Bolognese* (Bologna, N. Zanichelli, 1888).

Mancini, Clodomiro, "Un codice deontologico del secolo XV (Il "De cautelis medicorum" di Gabriel de Zerbi): *Scientia Veterum*-Collana di Studi di Storia della Medicina diretta e curata da G. Del Guerra, No. 44 (Pisa, Casa editrice Giardini, 1963).

Manitius, Max, *Geschichte der lateinischen Literatur des Mittelalters*, III (Munich, C. H. Beck, 1931).

Manoscritti e stampe venete dell' Aristotelismo e Averroismo (secoli X-XVI); Catalogo di mostra presso la Biblioteca Nazionale Marciana, etc. (Venice, Biblioteca Nazionale Marciana, 1958).

Marco Polo, *Travels*, translated and edited by Sir Henry Yule (3rd ed., London, John Murray, 1903).

Margoliouth, D. S., "The Book of the Apple," *Journal of the Royal Asiatic Society*, 1892: pp. 187–252.

Marini, Gaetano, *Degli archiatri pontifici*, 2 vols. (Rome, Pagliarini, 1784).

Massalongo, Roberto, "Alessandro Benedetti e la medicina veneta del Quattrocento," *Atti del Reale Istituto Veneto di Scienze, Lettere ed Arti* **76**, parte II, fasc. 1 (1916–1917): pp. 197–259.

Martinotti, Giovanni, *L'insegnamento dell' anatomia in Bologna. Studi e memorie per la storia dell' Università di Bologna* (Bologna, Azzoguidi, 1911).

Matsen, Herbert S., *Alessandro Achillini and His Doctrine of "Universals" and "Transcendentals"*; Ph.D. Dissertation, Columbia University, 1969.

Mayer, Claudius F., *Bio-bibliography of XVI. Century Medical Authors*; U. S. Army-Library of the Surgeon General, Index-Catalogue, Fourth Series, Third Supplement, VI, G. (Washington, D. C., U. S. Printing Office, 1944).

Mazzatinti, G., *Inventari dei manoscritti delle biblioteche d'Italia*, 23 vols., in progress, Forlì and Florence, 1890–1915).

Mazzetti, S., *Repertorio di tutti i professori antichi e moderni, della famosa università e del celebre Istituto delle scienze di Bologna, con in fine alcune aggiunte e correzioni alle opere dell' Alidosi, del Cavazza, del Sarti, del Fantuzzi, e del Tiraboschi* (Bologna, Tip. S. Tommaso, 1848).

Mazzuchelli, Giammaria, *Gli scrittori d'Italia cioè Notizie storiche e critiche intorno alle vite e agli scritti dei letterati italiani*, I, 1 (Brescia, Giambattista Bossini, 1753).

Medici, Michele, *Compendio storico della Scuola anatomica di Bologna* (Bologna, Volpe e Sassi, 1857).

Mieli, Aldo, *Gli Scienziati Italiani*, I (Rome, Nardecchia, 1923).

Mondini, Francesco, *De quodam codice Anatomiae Mundini*, etc., in *Novi Commentarii Academiae Scientiarum Instituti Bononiensis Tomus Octavus* (Bologna, 1846).

Muhammad Ibn Zakarīyā, Abū Bakr, al-Rāzī, *Liber Rasis ad Almansorem*, etc. (Venice, Jacopo Pentio de Leuco, 1508).

Münster, Ladislao, "Alessandro Achillini, anatomico e filosofo, professore dello studio di Bologna 1463–1512," *Rivista di storia delle scienze mediche e naturali* **24** (1933): pp. 7–22; 54–77.

——, "Studi e ricerche su Gabriele Zerbi: Nota I, Nuovi contributi biografici: la sua figura," *Rivista di storia delle scienze mediche e naturali* **41** (1950): pp. 64–83.

——, "Il primo trattato pratico compiuto sui problemi della vecchiaia: la "Gerontocomia" di Gabriele Zerbi," *Rivista di Gerontologia e Geriatria* **1** (1951): pp. 38–54.

——, "Alcune considerazioni e precisazioni a proposito di un lavoro su Alessandro Benedetti, con riguardo per la rinoplastica," *Rivista di Storia delle Scienze Mediche e Naturali* **47** (1955): pp. 1–6.

——, "Baverio Maghinardo de'Bonetti, medico Imolese del Quattrocento: La vita, i tempi, il pensiero scientifico," *Atti dell' Associazione per Imola Storicoartista* **7** (Imola, 1956).

——, "Il tema di deontologia medica: Il "De cautelis medicorum" di Gabriele Zerbi," *Rivista di storia delle scienze mediche e naturali* **47** (1956): pp. 60–83.

Muratori, Giulio, facsimile edition of Canano, *Musculorum Humani Corporis Picturata Dissectio* (Florence, Sansoni, 1962).

Nardi, Bruno, "Le opere inedite del Pomponazzi," Estratto dal *Giornale Critico della Filosofia Italiana*, Fasc. III, **8** (1954): pp. 341–355; also see **7** (1953): pp. 45–70, 175–191.

——, *La civiltà veneziana del quattrocento* (Florence, Sansoni, 1957).

——, *Saggi sull'Aristotelismo padovano dal secolo XIV al XVI* (Florence, G. C. Sansoni, 1958).

——, *Studi di filosofia medievale*; Storia e Letteratura—Raccolta di Studi e Testi 78 (Rome, Edizioni di Storia e Letteratura, 1960).

Nardo, L., "Dell' anatomia in Venezia," *L'Ateneo Veneto—Rivista bimestrale di scienze, lettere ed arti,* Anno XX, Vol. 1, Fascicolo 2 (Venice, 1897).

Neander, Johannes, *Antiquissimae et Nobilissimae Medicinae Natalitia . . . auctore Johanne Neandro Bremano* (Bremen, Johannes Wesel. 1623).

Neuburger, Max, "The Latin Poet Maximianus on the Miseries of Old Age," *Bulletin of the History of Medicine* **21** (1947): pp. 113–119.

O'Malley, C. D., *Andreas Vesalius of Brussels 1514–1564* (Berkeley, University of California Press, 1964).

O'Malley, C. D., and Edwin Clarke, "The Discovery of the Auditory Ossicles," *Bulletin of the History of Medicine* **35** (1961): pp. 424–428.

Orlandi da Bologna, Fr. Pellegrino Antonio, *Notizie degli scrittori Bolognesi e dell' opere loro stampate e manoscritte* (Bologna, Costantino Pisarri, 1714).

Panofsky, Erwin, *Meaning in the Visual Arts* (New York, Doubleday, 1955).

Panzani, G., *De Venetae Anatomes Historia et Claris Venetiarum Anatomicis Prolusio Habita in Veneto Anatomico Theatro* (Venice, D. Deregni, 1783).

Papadopoli, Nicolai Comneni, *Historia Gymnasii Patavini* (Venice, Sebastian Colet, 1726).

Paul of Aegina, ed. I. L. Heiberg in *Corpus Medicorum Graecorum* IX (Leipzig, Teubner, 1921).

Pazzini, Adalberto, *Bibliografia di storia della medicina d'Italia* (Rome, Tosi, 1946).

——, *Storia della Medicina*, (2 v., Milan, Soc. Editr. Libr., 1947).

Perry, Ben Edwin, *Aesopica*, etc. (Urbana, Ill., University of Illinois Press, 1952).

Philothei . . . Commentaria in Aphorismos Hippocratis . . . Ludovico Corado Mantuano interprete (Venice, Comino de Tridino, 1549).

Pisani, A., "Gli anatomisti italiani del XVI secolo. Rivendicazioni," *L'Ospedale Maggiore*, No. 12 (Milano, 1930): pp. 553–558.

Ploucquet, Guilielmus Godofredus de, *Literatura Medica Digesta sive Repertorium Medicinae Practicae, Chirurgiae atque Rei Obstetriciae* (Tübingen, J. G. Cotta, 1809).

Podestà, B., "Di alcuni documenti inediti riguardanti Pietro Pomponazzi," *Atti e Memorie della R. Deput. di Storia Patria per le provincie di Romagna* **6** (Bologna, 1868).

Portal, V., *Histoire de l'Anatomie et de la Chirurgie* (6 v., Paris, Didot, 1770–1773).

Premuda, Loris, *Storia dell' Iconografia anatomica* (Milan, Aldo Martello, 1957).

Probst, Otto, Isidors Schrift "de medicina"; *Sudhoffs Archiv für Geschichte der Medizin* **8** (1915): pp. 22–28; see the recent translation of this work of Isidore, with commentary, by William D. Sharpe, *Transactions, American Philosophical Society* **54**, 2, (1964).

PUCCINOTTI, FRANCESCO, *Storia della medicina* (3 v., Livorno, M. Wagner editore, 1850).

PUSCHMANN, TH., M. NEUBURGER, J. PAGEL, *Handbuch der Geschichte der Medizin* (Jena, 1901–1905).

PUTTI, VITTORIO, *Berengario da Carpi: Saggio biografico e bibliografico seguito dalla traduzione del "De fractura calvae sive cranei"* (Bologna, L. Cappelli, 1937).

RANDALL, JOHN HERMAN, JR., *The School of Padua and the Emergence of Modern Science* (Padua, Editrice Antenore, 1961).

RATH, GERNOT, "Pre-Vesalian Anatomy in the Light of Modern Research," *Bulletin of the History of Medicine* **35** (1961): pp. 142–148.

RENZI, SALVATORE DE, *Storia della Medicina in Italia* (3 v., Naples, Filiatre-Sebezio, 1845).

——, *Storia documentata della scuola medica di Salerno* (Naples, Filiatre-Sebezio, sec. ed., 1875).

ROSE, VALENTIN, *Verzeichnis der lateinischen Handschriften der kgl. Bibliothek zu Berlin* (Berlin, 1905).

SANUTO, MARINO, *I diarii* (Venice, a spese degli editori, pubblicato per cura di G. Berchet, 1879).

SARTON, GEORGE, *Introduction to the History of Science* (3 v. in 5, Baltimore, Williams and Wilkins, 1927–1948).

——, *Appreciation of Ancient and Medieval Science During the Renaissance (1450–1600)* (New York, A. S. Barnes, 1961).

SCHENCK, JOHANN GEORG, *Biblia Iatrica sive Bibliotheca Medica Macta, Continuata, Consummata*, etc. (Frankfurt, Johann Spiess, 1609).

SCHULLIAN, DOROTHY, "An Anatomical Demonstration by Giovanni Lorenzo of Sassoferrato, 19 November 1519," in *Miscellanea di Scritti di Bibliografia ed Erudizione in Memoria di Luigi Ferrari* (Florence, L. S. Olschki, 1952), pp. 487–494.

——, *Alessandro Benedetti, Diaria de Bello Carolino*, edited and translated; New York: The Renaissance Society of America (Frederick Ungar Publ. Co., 1967).

SCHWARZ, IGNATZ, *Die medizinischen Handschriften der Kgl. Universitätsbibliothek* in Würzberg, 1907.

SERRA-ZANETTI, ALBERTO, *L'arte della stampa in Bologna nel primo ventennio del Cinquecento* (Bologna, Biblioteca de "L'archiginnasio," Nuova serie, n. 1, 1959).

SIMILI, ALESSANDRO, *I lettori di medicina e chirurgia nello studio di Bologna dal 1460 al 1500* (Bologna, La Grafica Emiliana, 1941).

SINGER, CHARLES, "A Study in Early Renaissance Anatomy, with a New Text: the Anothomia of Hieronymo Manfredi (1490); text transcribed and translated by A. Mildred Westland," *Studies in the History and Method of Science* (Oxford University Press, 1917).

——, *The Fasciculo di Medicina, Venice, 1493 with an Introduction* etc., Part I: . . . Translation of the "Anothomia" of Mondino da Luzzi (Florence, R. Lier and Co., 1925).

——, translation of Antonio Benivieni, *De abditis nonnullis ac mirandis morborum et sanationum causis,* with facsimile reprint of the Latin text and biographical appreciation by Esmond R. Long (Springfield, Ill., C. C. Thomas, 1954).

——, "Brain Dissection Before Vesalius," *Journal of the History of Medicine* **11** (1956): p. 261.

——, *A Short History of Anatomy from the Greeks to Harvey* (New York, Dover Publications, Inc., 1957).

SORANUS, *Gynecology*, translated by Owsei Temkin (Baltimore, Johns Hopkins University Press, 1956).

SORBELLI, ALBANO, and LUIGI SIMEONI, *Storia della Università di Bologna*, I, II (Bologna, N. Zanichelli, 1940, 1947).

SPICER, E. C., *Aristotle's Conception of the Soul* (London, University of London Press, 1959).

SPRENGEL, K., *Versuch einer pragmatischen Geschichte der Arzneykunde* (2 v., 3rd ed., Vienna, C. Gerold, 1837–1840).

STEUDEL, J., "Vesals Reform der anatomischen Nomenklatur," *Zeitschrift der Anatomische Entwicklungsgeschichte* **112** (1943): pp. 675–681.

——, "Der vorvesalische Beitrag zur anatomischen Nomenklatur," *Sudhoffs Archiv für Geschichte der Medizin und der Naturwissenschaften* **36** (1943): pp. 1–42.

STRECKER, KARL, *Introduction to Medieval Latin*, translated by R. B. Palmer (Berlin, Weidmann, 1957; new ed., 1963).

STREETER, EDWARD C., "Francia and Achillini," *The Medical Pickwick:* a monthly literary magazine of wit and wisdom, I (Saranac Lake, New York, 1915).

SUDHOFF, KARL, "Ein unbekannter Druck von Johann Peyligks aus Zeitz "Compendiosa Capitis physici declaratio" auch "Anatomia totius corporis humani" genannt," *Sudhoffs Archiv für Geschichte der Medizin* **9** (1916): pp. 309–314.

SYMONDS, JOHN ADDINGTON, *The Renaissance in Italy* **1** (New York, Modern Library, 1935).

TEMKIN, OWSEI, "Studies on Late Alexandrian Medicine I. Alexandrian Commentaries on Galen's De Sectis ad Introducendos," *Bulletin of the Institute of the History of Medicine* **3** (1935): pp. 405–430.

——, "Geschichte des Hippokratismus im ausgehenden Altertum," *Kyklos: Jahrbuch für Geschichte und Philosophie der Medizin* **4** (1932): pp. 53–71.

——, *The Falling Sickness*, etc. (Baltimore, Md., Johns Hopkins Press, 1945).

Themistii Paraphrases Aristotelis librorum quae supersunt, ed. L. Spengel (Leipzig, Teubner, 1866).

THORNDIKE, LYNN, *A History of Magic and Experimental Science* (6 v., New York, Columbia University Press, 1923–1941).

——, *Science and Thought in the Fifteenth Century* (Columbia University Press, 1929).

——, *The Sphere of Sacrobosco and Its Commentators* (University of Chicago Press, 1949).

THORNDIKE, LYNN, and PEARL KIBRE, *A Catalogue of Incipits of Mediaeval Scientific Writings in Latin* (Cambridge, Mass., Mediaeval Academy of America, 1937; sec. ed., 1963).

TÖPLY, ROBERT RITTER VON, *Studien zur Geschichte der Anatomie im Mittelalter* (Leipzig and Vienna, Franz Deuticke, 1898).

TOSONI, P., *Dell' anatomia degli antichi e della scuola di Padova*; Memoria di Pietro Tosoni pubblicata in occasione della sua laurea in medicina (Padua, Tipografia del Seminario, 1844).

TURISANI (TRUSIANI) *Monaci Cartusiensis plusquam commentum in librum Galieni qui microtechni intitulatur. Cum questione eiusdem de Ypostasi* (Venice, a Philippo Pincio Mantuano, 1512).

VALERIANUS, JOANNES PIERIUS, *De litteratorum infelicitate libri duo* (Venice, Jacob Sarzina, 1620; editio nova curante Domino Egerton Brydges, Geneva, typis Gul. Fick. 1821).

VEDOVA, G., *Biografia degli scrittori padovani*, 2 vols. (Padua, Minerva, 1832–1836).

VITRY, JACQUES DE, *Historia Orientalis et Occidentalis*, ed. Jacques Bongars in *Gesta Dei per Francos* (Douai, Belgium, Beller, 1597).

WAGENINGEN, J. VAN, "De Quattuor Temperamentis," *Mnemosyne*, n.s., **46** (1918): pp. 374–382.

WEINBERGER, BERNHARD WOLF, *An Introduction to the History of Dentistry* (St. Louis, Mo., C. V. Mosby, 1948).

WICKERSHEIMER, ERNEST, *Anatomies de Mondino dei Luzzi et de Guido de Vigevano* (Paris, E. Droz, 1926).

——, *Dictionnaire biographique des médecins en France au moyen âge* **1** (Paris, E. Droz, 1936).

WIGHTMAN, W. P. D., *Science and the Renaissance: an Introduction to the Study of the Emergence of the Sciences in the Sixteenth Century* **1** (New York, Hafner Publ. Co., 1962).

ZACCAGNINI, GUIDO, *Storia dello Studio di Bologna durante il Rinascimento* (Geneva, Olschki, 1930).

ZEMAN, FREDERIC D., "The Gerontocomia of Gabriele Zerbi—A Fifteenth Century Manual of Hygiene for the Aged," *Journal of the Mt. Sinai Hospital* **10** (1943): pp. 710–716.

ZENO, APOSTOLO, *Dissertazioni Vossiane* (2 v., Venice, G. Albrizzi, 1753).

B. MANUSCRIPTS

Bologna

Acta (Libri actorum utriusque collegii, 1481–1604; 4 v., MSS 191–194; Archivio di Stato).
Archiginnasio MSS B. 1401, 1489, 1611.
Archivio di Stato: Indice Masini AR-BE 1500 (rogiti).
Archivio di Stato: Assunteria di Studio, Requisiti dei lettori.
Carrati, Conte Baldassare Antonio Maria de', Genealogie di famiglie nobili bolognesi; Biblioteca del Archiginnasio, MS B. 699.
Formigliari, De illustrioribus Bononiensis Archygymnasii Magistris, Doctoribus et Alumnis qui in philosophie et medicinae floruerunt: MS B. 1287, Bibl. Archiginnasio.
Garzoni, Giovanni, Epistolae; University of Bologna Library MS 1896.
Ghiselli, Antonio Francesco, Memorie antiche manuscritte di Bologna raccolte et accresciute sino a' tempi presenti (1729); University of Bologna Library Cod. 770, vol. XII, 204.
Libri Partitorum; Archivio di Stato.
Libri segreti del collegio delle arti e di medicina 217 (8), 1481–1500 A.D.; 217 (10), 1504–1575 A.D.; Archivio di Stato.
Nomina Doctorum Omnium, incipiendo ab anno 1480 usque ad 1800 laureatorum in medicina et philosophia; Archivio di Stato, Busta 235, (F. 3. 50).
Registri di battesimi in Archivio del Battistero di San Pietro, n. d., Bologna, Italy.
Tioli manuscripts: Notizie della vita e delle miscellanee di Monsignor Pietro Antonio Tioli, etc., University of Bologna Library Cons. ℒ139.
University of Bologna Library MSS 14 (10), 596, 1453, 4099.

Milan

Milano, Biblioteca Ambrosiana, MS A. 236.

Padua

Archivio antico dell' Università di Padova, Vol. 649, fol. 223v.
Archivio antico dell' Università di Padova, Vol. 649, Professori, Artisti e Legisti fino al 1509, ff. 55r–58v, Gabriele Zerbi's will.

Rome-Vatican

Garzoni, Giovanni, Selectae Epistolae; in Joannis Garzonis Bononiensis Selectae Epistolae nunc primum e MS. Codice Bibliothecae S. Dominici erutae [editae] et auctoris Vita illustratae A F. Vincentio Domenico Fassini O. P. anno 1761 [1763]; MS Vat. Lat. 10686.

Venice

Archivio di Stato, Senato terra, Registro 12, f. 109 v; Reg. 13, f. 97r.
Archivio di Stato, Venezia, Busta 50, no. 178 (Alessandro Benedetti's will); Busta 196, no. 870 (Niccolò Massa's will).
Archivio della Curia Vescovile, Vol. 44, f. 247r; Vol. 47, f. 169r.

Verona

Torresani, Antonio, Elogiarum Historicarum Veronae Propagium, MS N. 808, Biblioteca Civica, Verona.

Index of Names and Subjects

INDEX OF NAMES AND SUBJECTS